AF393626

ISNM

INTERNATIONAL SERIES OF NUMERICAL MATHEMATICS
INTERNATIONALE SCHRIFTENREIHE ZUR NUMERISCHEN MATHEMATIK
SÉRIE INTERNATIONALE D'ANALYSE NUMÉRIQUE

Editors:
Ch. Blanc, Lausanne; A. Ghizetti, Roma; A. Ostrowski, Montagnola; J. Todd, Pasadena;
A. van Wijngaarden, Amsterdam

VOL. 20

Linear Operators and Approximation

Proceedings of the Conference
held at the Oberwolfach Mathematical Research Institute, Black Forest,
August 14—22, 1971

Edited by
P. L. Butzer, J.-P. Kahane and **B. Szőkefalvi-Nagy**

Lineare Operatoren und Approximation

Abhandlungen zur Tagung
im Mathematischen Forschungsinstitut Oberwolfach, Schwarzwald,
vom 14. bis 22. August 1971

Herausgegeben von
P. L. Butzer, J.-P. Kahane und **B. Szőkefalvi-Nagy**

1972
**BIRKHÄUSER VERLAG BASEL
UND STUTTGART**

ISBN 978-3-0348-7285-0   ISBN 978-3-0348-7283-6 (eBook)
DOI 10.1007/978-3-0348-7283-6

PREFACE

These proceedings contain the lectures presented at the Conference on *Linear Operators and Approximation* held at the Oberwolfach Mathematical Research Institute, August 14—22, 1971. There were thirty-eight such lectures while four additional papers, subsequently submitted in writing, are also included in this volume. Two of the three lectures presented by Russian mathematicians are rendered in English, the third in Russian. Furthermore, there is a report on new and unsolved problems based upon special problem sessions, with later communications from the participants. In fact, two of the papers included are devoted to solutions of some of the problems posed.

The papers have been classified according to subject matter into five chapters, but it needs little emphasis that such thematic groupings are necessarily somewhat arbitrary. Thus Chapter I on Operator Theory is concerned with linear and non-linear semi-groups, structure of single operators, unitary operators, spectral and ergodic theory. Chapter II on Topics in Functional Analysis includes papers on Riesz spaces, boundedness theorems, generalized limits, and distributions. Chapter III, entitled "Approximation in Abstract Spaces", ranges from characterizations of classes of functions in approximation theory to approximation-theoretical topics connected with extensions to Banach (or more general) spaces. Chapter IV contains papers on harmonic analysis in connection with approximation and, finally, Chapter V is devoted to approximation by splines, algebraic polynomials, rational functions, and to Padé approximation.

A large part of the general editorial work connected with these proceedings was competently handled by Miss F. Fehér, while G. Bragard helped in proof reading. Our particular thanks are due to Mr. C. Einsele of Birkhäuser Verlag, Basel, for his personal interest. Indeed, this is the third time that the proceedings of an Oberwolfach Conference on approximation theory and related topics, conducted from Aachen*), are published by Birkhäuser in their "International Series of Numerical Mathematics". Thanks are again due to the Szeged Printing House for their great care in the production of this book.

October 1971	P. L. BUTZER	J.-P. KAHANE	B. SZ.-NAGY
	Aachen	Paris	Szeged

---

*) The earlier volumes are:

ON APPROXIMATION THEORY, ed. P. L. Butzer and J. Korevaar, Proceedings of the Conference at Oberwolfach, August 4—10, 1963. ISNM, vol. 5, Birkhäuser, Basel 1964.

ABSTRACT SPACES AND APPROXIMATION, ed. P. L. Butzer and B. Sz.-Nagy, Proceedings of the Conference at Oberwolfach, July 18—27, 1968. ISNM, vol. 10, Birkhäuser, Basel 1969.

CONTENTS

I. Operator Theory

II. Topics in Functional Analysis

III. Approximation in Abstract Spaces

IV. Harmonic Analysis and Approximation

V. Spline- and Algebraic Approximation

ZUR TAGUNG

Die Oberwolfacher Tagungen über Approximationstheorie und verwandte Gebiete wie die Funktionalanalysis, die seit 1963 von Aachen aus organisiert werden, sind Dank des ständig wachsenden Interesses zahlreicher Mathematiker in der Zwischenzeit zu einer festen Einrichtung geworden. So war es bei der vierten Tagung dieser Art für die Tagungsleiter eine besondere Freude, neben alten Bekannten wiederum viele neu hinzugekommene Teilnehmer begrüßen zu können. Insgesamt waren 53 Fachkollegen aus 11 Nationen anwesend; für 25 von ihnen war es die erste Tagung in Oberwolfach überhaupt. Durch diesen Teilnehmerkreis wurden bedeutende Schulen der Approximationstheorie aus den USA, der UdSSR, Frankreich, Ungarn und Deutschland vertreten.

Neben der klassischen Approximationstheorie standen die neueren Entwicklungen der abstrakten Theorie im Vordergrund, insbesondere diejenigen Aspekte, die von der Operatortheorie und der Funktionalanalysis wesentliche Impulse erhielten. Die zahlreichen Vorträge verschiedener Prägung gaben den Teilnehmern wertvolle Gelegenheit zu fruchtbaren Diskussionen. Dazu trugen nicht zuletzt auch drei Sitzungen bei, die neuen und ungelösten Problemen gewidmet waren.

Die Tagungsleiter haben es sehr bedauert, daß 45 Teilnahmewünsche, allein 18 davon aus den USA, wegen der begrenzten Kapazität des Oberwolfacher Instituts nicht berücksichtigt werden konnten. Dennoch hoffen sie, bei der nächsten Gelegenheit viele dieser Damen und Herren zu den Gästen zählen zu können.

Unser Dank gilt allen Teilnehmer, die durch ihre Beiträge und ihr Interesse zum Erfolg der Tagung beigetragen haben, insbesondere denen, die zu diesem Zweck eine weite Reise nicht gescheut haben. Besonders auch für die jungen Mathematiker aus Aachen war diese Konferenz eine wertvolle Gelegenheit zur Erweiterung ihrer Kenntnisse. Wenn die angenehme und anregende Atmosphäre der Tagung einerseits allen Teilnehmern zu verdanken ist, so gilt zum anderen unser Dank dem Institutsassistenten Herrn Dr. F. Hartmuth, sowie den Damen und Herren des Oberwolfacher Hauses. Ihre bewährte Gastfreundlichkeit und Hilfsbereitschaft machte es allen leicht, sich rasch zu einer Gemeinschaft zusammenzufinden. Besonders möchten wir auch dem Institutsdirektor, Herrn Professor Dr. M. Barner, für seine stets wohlwollende Unterstützung bei der Vorbereitung der Tagung unseren Dank aussprechen.

Tagungsleiter: P. L. BUTZER   J.-P. KAHANE   B. SZ.-NAGY

TEILNEHMERLISTE

H. Bavinck	Amsterdam
P. Billard	Marseille
P. L. Butzer	Aachen
J. L. B. Cooper	London
F. Deutsch	University Park, Pennsylvania
R. A. DeVore	Edmonton, Canada
J. R. Dorroh	Baton Rouge, Louisiana
F. Ebersoldt	Jülich
F. Fehér	Aachen
G. Freud	Budapest
E. Görlich	Aachen
C. Goulaouic	Paris-Sud
G. Grimeisen	Stuttgart
K. Gustafson	Boulder, Colorado
H. Helson	Berkeley, California
E. Hille	Albuquerque, New Mexico
E. Hölder	Mainz
H. W. Hövel	Aachen
J. Horváth	College Park, Maryland
J. W. Jerome	Evanston, Illinois
J. Johnen	Aachen
M. L. Jonac	Marseille
J. Junggeburth	Aachen
J.-P. Kahane	Paris-Sud
W. Kolbe	Aachen
P. Krée	Nice
L. Leindler	Szeged, Ungarn
G. G. Lorentz	Austin, Texas
C. Micchelli	Yorktown Heights, New York
M. W. Müller	Stuttgart
R. J. Nessel	Aachen
S. M. Nikolskiĭ	Moskau
J. D. Pincus	Stony Brook, New York
Ju. A. Rozanov	Moskau
P. O. Runck	Linz, Österreich

C. Samuel	Marseille
K. Scherer	Aachen
I. J. Schoenberg	Madison, Wisconsin
A. Schönhage	Konstanz
I. Segal	Cambridge, Massachusetts
H. S. Shapiro	Ann Arbor, Michigan
I. Singer	Bukarest
E. L. Stark	Aachen
F. Stummel	Frankfurt
Ju. N. Subbotin	Sverdlovsk, UdSSR
B. Sz.-Nagy	Szeged, Ungarn
W. Trebels	Aachen
R. S. Varga	Kent, Ohio
B. Virot	Orleans
H. J. Wagner	Aachen
H. Wallin	Umeå, Schweden
U. Westphal	Aachen
A. C. Zaanen	Leiden, Niederlande

WISSENSCHAFTLICHES PROGRAMM DER TAGUNG

Sonntag, 15. August

1. Frühsitzung 10^{15}—11 Uhr, Vorsitz: B. Sz.-Nagy
    E. HILLE: Generalizations of Landau's inequality to operators
2. Frühsitzung 11—12^{30} Uhr, Vorsitz: P. L. Butzer
    J. R. DORROH: Semi-groups of nonlinear transformations
    U. WESTPHAL: Der Ergodensatz im Mittel und sein approximationstheoretisches Verhalten
Abendsitzung 19^{45}—20^{30} Uhr, Vorsitz: G. G. Lorentz
    S. M. NIKOLSKIĬ: Nonlinear transformations with the conservation of differential-properties of functions

Montag, 16. August

1. Frühsitzung 9—10^{30} Uhr, Vorsitz: G. G. Lorentz
    G. FREUD: Über gewichtete Approximation durch Polynome
    J. W. JEROME: Singular self-adjoint multipoint boundary value problems: Solutions and approximation schemes
2. Frühsitzung 10^{40}—12^{30} Uhr, Vorsitz: J. L. B. Cooper
    H. BAVINCK: Convolution operators for Fourier—Jacobi expansions
    E. L. STARK: Über Nikolskiĭ-Konstanten positiver singulärer Integrale vom gestörten Fejér-Typ
Nachmittagssitzung 16^{30}—18 Uhr, Vorsitz: I. Segal
    C. GOULAOUIC: Approximation et interpolation de classes de fonctions C^∞
    G. GRIMEISEN: Exchange of unconditional summation with limits in Banach spaces
Abendsitzung 19^{30}—20^{30} Uhr, Vorsitz: E. Hille
    J. SCHOENBERG: On exponential Euler splines

Dienstag, 17. August

1. Frühsitzung 9—10^{30} Uhr, Vorsitz: S. M. Nikolskiĭ
    J. L. B. COOPER: Positive subdefinite functions
    P. KRÉE: Courants et courants projectifs sur les variétés differentielles de dimension infinie

2. Frühsitzung 10^{40}—12^{30} Uhr, Vorsitz: J. Horváth

    H. HELSON: A theorem on boundedness

    J.-P. KAHANE: Projections métriques de $L^1(T)$ sur des sous-espaces invariants par translations

Nachmittagssitzung 17—18^{30} Uhr, Vorsitz: F. Stummel

    M. W. MÜLLER: Sätze vom Bohman—Korowkin-Typ für Banachsche Funktionenräume

    F. DEUTSCH: Some geometric properties of the unit ball and applications to approximation theory

Abendsitzung 19^{30}—20^{30} Uhr, Vorsitz: J.-P. Kahane

    A. C. ZAANEN: Representation theorems for Riesz spaces

Mittwoch, 18. August

1. Frühsitzung 9—10^{30} Uhr, Vorsitz: R. S. Varga

    F. STUMMEL: Discrete approximations of normed spaces and convergence of linear operators

    JU. A. ROZANOV: Some approximation problems in the theory of stationary processes

2. Frühsitzung 10^{40}—12^{30} Uhr, Vorsitz: A. C. Zaanen

    J. HORVÁTH: Endliche Teile von Distributionen

    K. GUSTAFSON: Recent developments on Weyl's Theorems

Donnerstag, 19. August

1. Frühsitzung 9—10^{30} Uhr, Vorsitz: A. Schönhage

    H. JOHNEN: Sätze vom Jackson-Typ auf Darstellungsräumen kompakter zusammenhängender Liegruppen

    R. S. VARGA: Rationale Approximation von ganzen Funktionen in $[0, \infty)$ oder in $(-\infty, \infty)$

2. Frühsitzung 10^{40}—12^{30} Uhr, Vorsitz: I. Singer

    H. WALLIN: Convergence of Padé approximants

    E. GÖRLICH: Logarithmic and exponential versions of Bernstein's inequality and generalized derivatives

Nachmittagssitzung 16^{30}—18^{30} Uhr, Vorsitz: J. Schoenberg

    YU. N. SUBBOTIN: Einige Anwendungen der Theorie der Spline-Approximation

    C. MICCHELLI: The fundamental theorem of algebra for monosplines

    L. LEINDLER: On a certain converse of Hölder's inequality

Abendsitzung 19^{30}—21 Uhr, Vorsitz: J.-P. Kahane

    1. Sitzung: New and Unsolved Problems

Freitag, 20. August

Frühsitzung 9^{30}—12^{30} Uhr, Vorsitz: H. Helson
  I. SEGAL: Singular perturbations of semigroup generators
  B. SZ.-NAGY: Cyclic vectors and commutativity
  J. D. PINCUS: Some applications of operatorvalued functions of two complex
    variables
Nachmittagssitzung 16^{30}—18^{30} Uhr, Vorsitz: J.-P. Kahane
  2. Sitzung: New and Unsolved Problems
  P. O. RUNCK: Eine Bemerkung zum Banachschen Fixpunktsatz
Abendsitzung 19^{30}—20^{30} Uhr, Vorsitz: G. Freud und J.-P. Kahane
  G. G. LORENTZ: Inverse theorems for Bernstein Operators
  3. Sitzung: New and Unsolved Problems

Samstag, 21. August

1. Frühsitzung 9—10^{30} Uhr, Vorsitz: B. Sz.-Nagy
  I. SINGER: On metric projections onto linear subspaces of normed linear spaces
  P. BILLARD: Bases dans H^1 et bases de sous-espaces de dimension finie dans A
2. Frühsitzung 10^{40}—12^{30} Uhr, Vorsitz: P. Krée
  R. A. DEVORE: A pointwise "o" theorem for positive convolution operators
  H. S. SHAPIRO: Fourier multipliers whose multiplier norm is an attained value

I.
Operator Theory

Generalizations of Landau's Inequality to Linear Operators

By

EINAR HILLE

DEPT. OF MATH.
UNIVERSITY OF NEW MEXICO
ALBUQUERQUE

1. Introduction

In 1913 Edmund LANDAU [13] proved that if f is continuous together with its first and second order derivatives in the interval $[0, 1]$, if $\|f\| = 1$, $\|f''\| = 4$, then

$$(1.1) \qquad \|f'\| \leq 4.$$

He showed that 4 is the best constant and the result is not necessarily true for an interval of length < 1. Landau was rounding off earlier results of G. H. HARDY and E. J. LITTLEWOOD (jointly and separately; mostly of the order-of-infinity type but involving also derivatives of higher order). It is customary nowadays to write the inequality in the form

$$(1.2) \qquad \|f'\|^2 \leq 4 \|f\| \|f''\|$$

and the underlying space $\mathfrak{X}$ is a B-space of functions $t \to f(t)$ on an infinite interval. The space $L_\infty(0, \infty)$ was considered by Landau. The inequality is valid for $C[0, \infty]$ and any $L_p(0, \infty)$. The interval $(-\infty, \infty)$ is also admissible. In some spaces the constant "4" may be replaced by a smaller number ≥ 1.

If f has derivatives up to and including the nth order and if $1 \leq k < n$, one can ask for what values of the constants it is true that

$$(1.3) \qquad \|f^{(k)}\|^n \leq C_{n,k}^n \|f\|^{n-k} \|f^{(n)}\|^k.$$

In 1914 J. HADAMARD [3] made a first dent in this problem and there has been many later questioners. For $L_\infty(-\infty, \infty)$ A. N. KOLMOGOROV [10] in 1939 determined the least bounds for all n and k. He also determined the extremal functions. Some of the results are quoted in Section 6. The space $L_\infty(0, \infty)$ had to wait until 1970 when I. SCHOENBERG and A. CAVARETTA [15] determined best constants and extremal functions. For literature and further results, see the works of BECKENBACH—BELLMAN [1], HARDY—LITTLEWOOD—PÓLYA [4] and MITRINOVIĆ [14].

2. Proof and extension of Landau's inequality

Landau based his proof on Taylor's theorem with remainder and this tool also gives most of the extensions.

Let the basic space be $C[0, \infty]$ with the sup norm. If f has first and second order continuous derivatives, we have, for s and t positive,

$$(2.1) \qquad f(t+s) = f(t) + sf'(t) + \int_0^s (s-u)f''(t+u)\,du$$

so that

$$f'(t) = s^{-1}[f(t+s) - f(t)] - s^{-1}\int_0^s (s-u)f''(t+u)\,du.$$

Here the integral is dominated by $\frac{1}{2}s^2\|f''\|$. Hence

$$(2.2) \qquad \|f'\| \leq 2s^{-1}\|f\| + \frac{1}{2}s\|f''\|.$$

If $\|f''\| = 0$, we let $s \to +\infty$ and obtain $\|f'\| = 0$ so that f is a constant and (1.2) obviously holds. If $\|f''\| \neq 0$, it is seen that the right member of (2.2) attains its minimum for

$$(2.3) \qquad s = s_0 = 2\|f\|^{1/2}\|f''\|^{-1/2}.$$

Substituting in (2.2) and squaring one obtains (1.2). The same argument holds for $L_p(0, \infty)$.

In 1967 R. R. KALLMAN and C.-C. ROTA [9] reinterpreted (2.1) in terms of operators and showed how to generalize the argument. The basic observation is that the left member of (2.1) is the result of applying the so-called *shift operator* to f. Here

$$(2.4) \qquad T(s)[f](t) = f(t+s) \qquad (s>0)$$

and the shift operator defines a *continuous semi-group of contractions* on any one of the spaces $C[0, \infty]$ and $L_p(0, \infty)$. For the terminology see below. The infinitesimal generator A of this semi-group is the operator A defined by $A[f](t)=f'(t)$ when f has a derivative belonging to the same space. We can then rewrite (2.1) as follows

$$(2.5) \qquad T(s)[f](t) = f(t) + sA[f](t) + \int_0^s (s-u)T(u)\{A^2[f](t)\}\,du.$$

Now this formula holds for any semi-group of linear bounded operators on a B-space as shown by HILLE and K. YOSIDA. The analogue of (2.2) is now

$$A[f](t) = s^{-1}[T(s)[f](t) - f(t)] - s^{-1}\int_0^s (s-u)T(u)\{A^2[f](t)\}\,du.$$

So far this is general. Suppose now, as in the case of the shift operator, that $T(s)$ is a contraction, i.e.

(2. 6) $$\|T(s)\| \leq 1 \qquad (\forall s \geq 0).$$

We have then

$$\|T(s)[f] - f\| \leq 2\|f\|$$

so that (2. 2) is replaced by

(2. 7) $$\|A[f]\| \leq 2s^{-1}\|f\| + \frac{1}{2}s\|A^2[f]\|.$$

Proceeding as above one finds that $\|A^2[f]\| = 0$ implies $\|A[f]\| = 0$ so that f is left invariant by the operator $T(s)$ for all $s \geq 0$. If instead $\|A^2[f]\| \neq 0$, we minimize the right hand member of (2. 7) and obtain the Landau—Kallman—Rota inequality

(2. 8) $$\|A[f]\|^2 \leq 4\|f\|\,\|A^2[f]\| \qquad (f \in D[A^2]).$$

This is of course trivially satisfied in the special case where $A^2[f] = 0$.

KALLMAN and ROTA devote most of their attention to extensions of (2. 8) in various directions: Luxemburg—Zaanen spaces $\mathfrak{X}$ are considered and elements $f \in \mathfrak{X}$ such that f' and f'' exist as distributions and are elements of $\mathfrak{X}$.

3. Some operator functions

Semi-group theory has been used freely above and it is perhaps appropriate to review some of the notions used, especially since they have also a bearing on cosine and sine operators.

Let there be given a one-parameter family $\{T(s)\}$ of linear bounded operators from the B-space $\mathfrak{X}$ into itself. The operators are defined for $s \geq 0$ so there is a mapping of the non-negative reals into the B-space $\mathfrak{E}(\mathfrak{X})$ of linear bounded operators from $\mathfrak{X}$ into itself. This family is a semi-group if

(3. 1) $$T(s)T(t) = T(t)T(s) = T(s+t) \qquad (0 \leq s, t).$$

$T(0)$ must be an idempotent; we take it to be the unit operator. $T(s)$ is strongly continuous for all $s > 0$, if it is strongly right continuous at $s = 0$, i.e.

(3. 2) $$\lim_{h \downarrow 0} T(h)[f] = f \qquad (\forall f \in \mathfrak{X}).$$

For certain elements of $\mathfrak{X}$

(3. 3) $$\lim_{h \downarrow 0} h^{-1}\{T(h)[f] - f\} \equiv A[f]$$

exists. Actually the domain of A is dense in $\mathfrak{X}$ and the same is true for $\mathfrak{D}[A^n]$. If

$f \in \mathfrak{D}[A]$, so does $T(s)[f]$ for all $s > 0$. Moreover $T(s)[f]$ has a derivative with respect to s, namely

$$(3.4) \qquad A\{T(s)[f]\} = T(s)\{A[f]\}.$$

These functions are continuous in s and hence Riemann—Graves integrable so that

$$(3.5) \qquad T(s)[f] = f + \int_0^s T(u)\{A[f]\}\, du.$$

If $f \in \mathfrak{D}[A^2]$, we can integrate by parts and will obtain (2.5) after simplification. Similarly, if $f \in \mathfrak{D}[A^3]$, we can integrate once more by parts, etc.

$T(s)$ is a contraction operator if (2.6) holds.

Normally A is an unbounded operator. Its spectrum is confined to the closed left half of the complex plane. For λ in the right half-plane, the resolvent of A is the Laplace transform of the semi-group operator so that

$$(3.6 \qquad R(\lambda, A)[f] = (\lambda I - A)^{-1}[f] = \int_0^\infty e^{-\lambda s} T(s)[f]\, ds.$$

This is also a Riemann—Graves integral, absolutely convergent for $\mu > 0$, $\lambda = \mu + iv$. In particular, if $T(s)$ is a contraction operator, so is $\mu R(\mu + iv, A)$, i.e.

$$(3.7) \qquad |\mu|\, \|R(\mu + iv, A)\| \leqq 1.$$

Conversely, if A is a linear operator from $\mathfrak{X}$ to $\mathfrak{X}$ with domain dense in $\mathfrak{X}$ whose resolvent is holomorphic for $R(\lambda) > 0$ where it satisfies (3.7), then A is the infinitesimal generator of a strongly continuous semi-group of contraction operators.

$T(s)$ may be defined for all real s in which case we have a *one-parameter group* of linear bounded operators. Here the spectrum of A is confined to the imaginary axis and the resolvent exists in both half-planes. If $\{T(s)\}$ is a group of contractions, then (3.7) holds in both half-planes. $T(s)$ may also be definable in the complex plane as a holomorphic function. The group case is important for the following.

Let us return to $R(\lambda, A)$ once more. The resolvent possesses an asymptotic series in the sense of Poincaré. If $f \in \mathfrak{D}[A^n]$, repeated use of the identity

$$(3.8) \qquad R(\lambda, A)[f] = \lambda^{-1} f + \lambda^{-1} R(\lambda, A) A[f]$$

gives

$$(3.9) \qquad R(\lambda, A)[f] = \lambda^{-1} f + \lambda^{-2} A[f] + \lambda^{-3} A^2[f] + \cdots + \lambda^{-n} A^{n-1}[f] +$$
$$+ \lambda^{-n} R(\lambda, A) A^n[f].$$

This identity will enable us to prove the existence of analogues of (1.3) for infinitesimal generators of contraction semi-groups.

The semi-group generated by the operator A, say $T(s; A)$, is an interpretation of the symbol exp (sA). We shall also need interpretations of the symbols cos (sA) and sin (sA). Here it is natural to get the interpretation from the formulas of Euler relating the exponential function and the sines and cosines. This means expressing cos (sA) and sin (sA) in terms of the semi-group operators $T(s; iA)$ and $T(-s; iA)$ or, equivalently, in terms of $T(is; A)$ and $T(-is; A)$. It is clear that $T(s; iA)$ must be a group. Define

$$(3.10) \qquad \cos (sA) = \frac{1}{2}[T(s; iA) + T(-s; iA)],$$

$$(3.11) \qquad \sin (sA) = \frac{1}{2i}[T(s; iA) - T(-s; iA)].$$

For these formulas to make sense we need a linear operator A on $\mathfrak{X}$ to $\mathfrak{X}$ with domain dense in $\mathfrak{X}$ whose spectrum is real so that $R(\lambda, A)$ exists for λ in the upper as well as in the lower half-planes. If, in addition,

$$(3.12) \qquad |v| \|R(vi, A)\| \leqq 1 \qquad (v \text{ real} \neq 0),$$

then $T(s; iA)$ and $T(-s; iA)$ are contraction semi-groups and cos (sA) and sin (sA) are contraction operators for s real. If $T(0; iA) = I$ and if $T(s; iA)$ is strongly continuous, then cos $(0A) = I$, sin $(0A) = 0$ and the operators cos (sA) and sin (sA) are strongly continuous.

From our line of approach it is most natural to consider A as the infinitesimal generator of cos (sA) and sin (sA) even if it does not seem to agree with the usage established by S. KUREPA. We shall need the basic properties of these functions.

If $f \in \mathfrak{D}[A]$ or to $\mathfrak{D}[A^2]$ respectively, then

$$(3.13) \quad \lim_{h \to 0} h^{-1} \sin (hA)[f] = A[f], \quad \lim_{h \to 0} h^{-2}\{\cos (hA)[f] - f\} = -\frac{1}{2}A^2[f].$$

The addition formulas are, as could be expected,

$$(3.14) \qquad \cos [(s+t)A] = \cos (sA) \cos (tA) - \sin (sA) \sin (tA),$$

$$(3.15) \qquad \sin [(s+t)A] = \sin (sA) \cos (tA) + \sin (tA) \cos (sA).$$

For $f \in \mathfrak{D}[A]$ we have the expected derivatives

$$(3.16) \qquad \frac{d}{ds}\cos (sA)[f] = -\sin (sA)\{A[f]\} = -A\{\sin (sA)[f]\},$$

$$(3.17) \qquad \frac{d}{ds}\sin (sA)[f] = \cos (sA)\{A[f]\} = A\{\cos (sA)[f]\}$$

and the analogues of (3. 5)

(3. 18)
$$\cos (sA) [f] = f - \int_0^s \sin (uA) \{A[f]\} \, du,$$

(3. 19)
$$\sin (sA) [f] = \int_0^s \cos (uA) \{A[f]\} \, du.$$

In the following $\mathfrak{X}$ will be a space of functions f defined on some interval, usually $(-\infty, \infty)$, and we shall need forms of Taylor's theorem valid for $f \in \mathfrak{D}[A^n]$ with n equal to the order of the highest power of A occurring in the formula. We have

$$(3. 20) \quad \cos (sA)[f](t) = f(t) - \frac{1}{2} s^2 A^2[f](t) + \frac{1}{2} \int_0^s (s-u)^2 \sin (uA)\{A^3[f](t)\} \, du$$

$$(3. 21) \qquad\qquad = f(t) - \frac{1}{2} s^2 A^2[f](t) - \frac{1}{3!} \int_0^s (s-u)^3 \cos (uA)\{A^4[f](t)\} \, du$$

and

$$(3. 22) \qquad \sin (sA)[f](t) = sA[f](t) + \int_0^s (s-u) \sin (uA) \{A^2[f](t)\} \, du$$

$$(3. 23) \qquad\qquad = sA[f](t) - \frac{1}{2} \int_0^s (s-u)^2 \cos (uA) \{A^3[f](t)\} \, du.$$

Having obtained much powerful machinary we can now apply it to the L-K-R problem.

4. The Hadamard—Kolmogorov problem

We shall use the resolvent expansion (3. 9) to make an attack on the problem of generalizing formula (1. 3) replacing differentiation by the infinitesimal generator A of a contraction semi-group.

THEOREM 1. *If A is the infinitesimal generator of a strongly continuous contraction semi-group, if $f \in \mathfrak{D}[A^n]$ and $1 \leq k < n$, then there exist constants $C_{n,k}$ independent of A, so that*

$$(4. 1) \qquad \|A^k[f]\|^n \leq C_{n,k}^n \|f\|^{n-k} \|A^n[f]\|^k.$$

REMARK. We shall give essentially an existence proof. The argument is rather laborious and does not give particularly good bounds for the constants. Thus in the L-K-R case, $n=2$, $k=1$, we get $C_{2,1} = 2\sqrt{2}$ instead of 2. We shall carry through the details of the proof only for $n=3$.

PROOF. We solve (3. 9) for $A^k[f]$ and obtain

$$(4. 2) \qquad A^k[f] = \lambda^k\{\lambda R(\lambda, A)[f] - f\} - \lambda^{k-1} A[f] - \cdots - \lambda A^{k-1}[f] - $$
$$- \lambda^{-1} A^{k+1}[f] - \cdots - \lambda^{k-n+1} A^{n-1}[f] - \lambda^{k-n} \lambda R(\lambda, A) A^n[f].$$

To simplify the notation set

$$(4. 3) \qquad \|f\| = m_0, \quad \|A^j[f]\| = m_j, \qquad j = 1, 2, 3, \ldots .$$

Thus

$$(4. 4) \quad m_k \leqq 2\lambda^k m_0 + \lambda^{k-1} m_1 + \cdots + \lambda m_{k-1} + \lambda^{-1} m_{k+1} + \cdots + \lambda^{k+1-n} m_{n-1} + \lambda^{k-n} m_n ,$$

valid for all $\lambda > 0$ and $k = 1, 2, \ldots, n-1$. We have thus a system of $n-1$ linear inequalities for the $n-1$ unknowns $m_1, m_2, \ldots, m_{n-1}$. We rewrite the system

$$(4. 5) \qquad m_1 - \lambda_1^{-1} m_2 - \lambda_1^{-2} m_3 - \cdots - \lambda_1^{2-n} m_{n-1} \leqq 2\lambda_1 m_0 + \lambda_1^{1-n} m_n ,$$

$$- \lambda_2 m_1 + m_2 - \lambda_2^{-1} m_3 - \cdots - \lambda_2^{3-n} m_{n-1} \leqq 2\lambda_2^2 m_0 + \lambda_2^{2-n} m_n ,$$

$$\cdots\cdots\cdots\cdots\cdots\cdots\cdots\cdots\cdots\cdots\cdots\cdots\cdots\cdots\cdots$$

$$- \lambda_{n-1}^{n-2} m_1 - \lambda_{n-1}^{n-3} m_2 - \lambda_{n-1}^{n-4} m_3 - \cdots + m_{n-1} \leqq 2\lambda_{n-1}^{n-1} m_0 + \lambda_{n-1}^{-1} m_n .$$

We use the last inequality to express m_{n-1} in terms of the other m's. The preceding equation will then yield a similar inequality for m_{n-2} and so on. Finally we have an inequality for m_1 in terms of m_0 and m_n with coefficients which are rational functions of $\lambda_1, \lambda_2, \ldots, \lambda_{n-1}$. Here we set

$$(4. 6) \qquad \lambda_1 = \lambda, \quad \lambda_k = \alpha^{k-1} \lambda \qquad (k = 2, \ldots, n-1),$$

where $0 < \alpha < 1$. The result is then of the form

$$(4. 7) \qquad m_1 \leqq \lambda f_{11}(\alpha) m_0 + \lambda^{1-n} f_{12}(\alpha) m_n$$

where f_{11} and f_{12} are rational functions of α which are positive for $0 < \alpha < 1$. Now λ is still arbitrary and the right hand member has a minimum for

$$(4. 8) \qquad \lambda = \left\{ \frac{(n-1) f_{12}(\alpha)}{f_{11}(\alpha)} \frac{m_n}{m_0} \right\}^{1/n} .$$

Substitution in (4. 7) gives

$$(4. 9) \qquad m_1 \leqq g_{n,1}(\alpha) m_0^{1-1/n} m_n^{1/n}$$

where $g_{n,1}(\alpha)$ is an algebraic function, positive for $0 < \alpha < 1$. If $\gamma_{n,1}$ is its minimum, then

$$(4. 10) \qquad m_1 \leqq \gamma_{n,1} m_0^{1-1/n} m_n^{1/n}.$$

This is (4. 1) for $k = 1$.

Now the inequality preceding (4.7) is of the form

$$(4.11) \qquad m_2 \leqq \lambda m_1 + f_{21}(\alpha)\lambda^2 m_0 + f_{22}(\alpha)\lambda^{2-n} m_n.$$

Again λ is arbitrary, $\lambda > 0$. For ϱ arbitrary, $\varrho > 0$, set

$$(4.12) \qquad \lambda = \varrho m_0^{-1/n} m_n^{1/n}$$

to obtain

$$(4.13) \qquad m_2 \leqq g_{n,2}(\alpha, \varrho) m_0^{1-2/n} m_0^{2/n}.$$

Again (in principle!) we can minimize the multiplier and obtain

$$(4.14) \qquad m_2 \leqq \gamma_{n,2} m_0^{1-2/n} m_n^{2/n}.$$

Continuing in this manner we get (4.1).

To illustrate we consider explicitly the case $n=3$. Here we have the system

$$(4.15) \qquad m_1 - a^{-1} m_2 \leqq 2am_0 + a^{-2} m_3,$$

$$-bm_1 + m_2 \leqq 2b^2 m_0 + b^{-1} m_3.$$

If $b=\alpha x$, $a=x$, then

$$(4.16) \qquad (1-\alpha)m_1 \leqq 2(1+\alpha^2)xm_0 + (1+\alpha^{-1})x^{-2} m_3$$

so that

$$(4.17) \qquad f_{11}(\alpha) = 2\frac{1+\alpha^2}{1-\alpha}, \quad f_{12}(\alpha) = \frac{1+\alpha}{\alpha(1-\alpha)}$$

and

$$(4.18) \qquad g_{3,1}(\alpha) = 3\frac{(1+\alpha)^{1/3}(1+\alpha^2)^{2/3}}{\alpha^{1/3}(1-\alpha)},$$

the minimum $\gamma_{3,1}$ of which is <7.5. Further we have

$$m_2 \leqq bm_1 + 2b^2 m_0 + b^{-1} m_3.$$

The minimum of the right member as a function of b is reached at the root of the equation

$$m_1 + 4bm_0 - b^{-2} m_3 = 0$$

or

$$4m_0 b^3 + \gamma_{3,1} m_0^{1/3} m_3^{2/3} b^2 - m_3 = 0.$$

It follows that

$$b = x_0 \left(\frac{m_3}{m_0}\right)^{1/3},$$

where x_0 is the unique positive root of

$$(4.19) \qquad 4x^3 + \gamma_{3,1} x^2 - 1 = 0.$$

It follows that

$$(4.20) \qquad m_2 \leqq 2x_0^{-1}(1 - 2x_0^3)m_0^{1/3}m_3^{2/3}.$$

COROLLARY. *(4.1) is also valid if A is the infinitesimal generator of a cosine or a sine contraction operator.*

For iA generates a contraction semi-group.

5. Hille—S. Kurepa

KALLMAN and ROTA informed me in May 1969 of their results. It struck me that it would be of some interest to exhibit examples of contraction semi-groups generated by other differential operators. This would give special instances of the L-K-R inequality and, in particular, generalize the Landau inequality. A search of the literature revealed a plethora of such operators. The richest source turned out to be HILLE [7]; since partial differential operators were eliminated in HILLE— PHILLIPS [8] this is less rich. I shall quote only one class of operators: it turns out that Af can be taken as

$$(5.1) \qquad (-1)^{k-1}f^{(2k-1)} \quad \text{or} \quad (-1)^{k-1}f^{(2k)}$$

with the resulting inequalities

$$(5.2) \qquad \|f^{(n)}\|^2 \leqq 4\|f\|\,\|f^{(2n)}\|.$$

If n is even, we are dealing with a group of contractions, if n is odd with a semi-group. In the first case the space should be functions defined on $(-\infty, \infty)$, in the second case on $(0, \infty)$. In (5.2) the constant 4 is almost certainly too large as soon as $n > 1$. See further below. For other admissible infinitesimal generators see [6].

In September 1969 the Seventh International Symposium on Functional Equations met in Canada. I made a report on the L-K-R inequality and on my extension of Landau's inequality to other differential operators [5]. S. KUREPA was there and showed on the spot how (2.8) could be extended from contractions to bounded semi-group operators. If $\|T(s)\| \leqq M$, $\forall s > 0$, then

$$(5.3) \qquad \|Af\|^2 \leqq 2M(M+1)\|f\|\,\|A^2f\|.$$

At a later session of the Symposium [12] he reported on this and also on the possibility of using contraction cosines. In particular he found, using the K-R method, that if $\cos(sA)$ is a contraction, then

$$(5.4) \qquad \|A^2f\|^2 \leqq \frac{4}{3}\|f\|\,\|A^4f\|.$$

If $Af=f'$ which is admissible, then

$$(5.5)\qquad \|f''\|^2 \leq \frac{4}{3}\,\|f\|\,\|f^{(iv)}\|.$$

This is considerably better than (5. 2) for $n=2$. For details, see H. KRALJEVIĆ and S. KUREPA [11].

6. Further use of cosines and sines

We now proceed to a more systematic use of formulas (3. 20)—(3. 23). The method is that of KALLMAN—ROTA. In (3. 20) and (3. 21) we solve for $A^2[f]$ and obtain

$$A^2[f](t) = 2s^{-2}\{f(t)-\cos(sA)[f](t)\} + s^{-2}\int_0^s (s-u)^2 \sin(uA)\{A^3[f](t)\}\,du =$$

$$= 2s^{-2}\{f(t)-\cos(sA)[f](t)\} - \frac{1}{3!}\,2s^{-2}\int_0^s (s-u)^3 \cos(uA)\{A^4[f](t)\}\,du.$$

Here the first equality is valid for $f\in\mathfrak{D}[A^3]$ while the second requires $f\in\mathfrak{D}[A^4]$. We take norms and use the notation of (4. 3). Thus

$$(6.1)\qquad m_2 \leq 4s^{-2}m_0 + \frac{1}{3}\,sm_3,$$

and

$$(6.2)\qquad m_2 \leq 4s^{-2}m_0 + \frac{1}{12}\,s^2m_4,$$

respectively. In the first case $m_3=0$ implies $m_2=0$ while in the second case $m_4=0$ gives $m_2=0$. If these quantities are different from zero, we can minimize the right members and obtain

$$(6.3)\qquad m_2^3 \leq 3m_0 m_3^2,$$

$$(6.4)\qquad m_2^4 \leq \frac{16}{9}\,m_0^2 m_4^2.$$

The second inequality is that of Kurepa (5. 4).

Treating (3. 22) and (3. 23) in a similar manner we obtain

$$(6.5)\qquad m_1^2 \leq 2m_0 m_2,$$

$$(6.6)\qquad m_1^3 \leq \frac{9}{8}\,m_0^2 m_3.$$

Here (6. 3) and (6. 6) are believed to be new while (6. 5) may be known. All three are best possible inequalities by virtue of the results of KOLMOGOROV for the space $L_\infty(-\infty, \infty)$ and the differential operator. In the notation of formula (1. 3) Kolmogorov showed that

$$(6.7) \qquad C_{2,1}^2 = 2,$$

$$(6.8) \qquad C_{3,1}^3 = 9/8, \quad C_{3,2}^3 = 3$$

while

$$(6.9) \qquad C_{4,2}^4 = 25/16.$$

For (6. 7) he gives credit to HADAMARD [3]. It is also true for $L_2(-\infty, \infty)$. See HARDY—LITTLEWOOD—PÓLYA [4] who give the extremal functions. Formula (6. 9) does not throw any light on Kurepa's inequality since $25/16 < 16/9$.

Kolmogorov found the class of extremal functions for $L_\infty(-\infty, \infty)$. For a given n take

$$(6.10) \qquad f_n(t) = \frac{4}{\pi} \sum_{m=0}^{\infty} (2m+1)^{-n-1} \sin\left[(2m+1)t - \frac{1}{2}n\pi\right].$$

Here

$$(6.11) \qquad f_n^{(k)} = f_{n-k}, \quad C_{n,k}^n = \|f_{n-k}\|^n \|f_n\|^{k-n},$$

and the sup norm is given by

$$(6.12) \qquad \|f_n\| = \frac{4}{\pi}\{1 - 3^{-n-1} + 5^{-n-1} - 7^{-n-1} + \cdots\}$$

for n even and

$$(6.13) \qquad \|f_n\| = \frac{4}{\pi}\{1 + 3^{-n-1} + 5^{-n-1} + 7^{-n-1} + \cdots\}$$

for n odd. As is well known these series can be summed in closed form. The sums are rational numbers times a power of π and involve the expansion coefficients of $\pi \operatorname{cosec} \pi z$ (n even) and $\pi \cot \pi z$ (n odd) at the origin. Kolmogorov showed that

$$(6.14) \qquad 1 < C_{n,k} < \frac{1}{2}\pi \qquad (\forall n, k)$$

and also a number of limit relations among which we note

$$(6.15) \qquad \lim_{n\to\infty} C_{2n,n} = (4/\pi)^{\frac{1}{2}}, \quad \lim_{n\to\infty} C_{n,k}^n = (4/\pi)^k$$

for fixed k. These relations may throw some light on (4. 1) for the case where A generates a contraction cosine.

7. The investigations of Everitt

W. N. EVERITT has tackled a problem related to but more general than Landau's original inequality. The following indications are based on a personal communication. The basic idea is to replace the operators f' and f'' by the type of operators that enter in a natural manner in a self-adjoint linear second order differential operator, the underlying space being $L_2(0, \infty)$. Such an operator is of the form

$$(7.1) \qquad M[f] = -(pf')' + qf.$$

Among other conditions, p and q should satisfy the following: (i) $p(t) > 0$, $t \geq 0$, (ii) $p' \in C[0, \infty)$, (iii) $q \in C[0, \infty)$ and (iv) $q(t) \geq -k > -\infty$ for some $k \geq 0$. Let $\mathfrak{D}[M]$ be the submanifold of $L_2(0, \infty)$ where f' is absolutely continuous and $M[f]$ exists and belongs to L_2.

Everitt's problem is now to find if there exists a constant $K = K(p, q)$ such that for all $f \in \mathfrak{D}[M]$

$$(7.2) \qquad \left\{ \int_0^\infty [p(f')^2 + q(f)^2] \, dt \right\}^2 \leq K(p, q) \int_0^\infty f^2 \, dt \int_0^\infty M[f]^2 \, dt.$$

The existence of such a constant depends in the first place on the properties of the solutions of the second order differential equation

$$(7.3) \qquad M[y] \equiv -[py']' + qy = \lambda y$$

on $(0, \infty)$ for $\lambda = \mu + iv$. In addition, the fourth order operator M^2 plays an important role. Extensions are made to $L_2(-\infty, \infty)$. For further details the reader is referred to the forthcoming publication of W. N. EVERITT [2].

REFERENCES

[1] E. F. Beckenbach and R. Bellman, *Inequalities.* Springer, Ergebnisse **30**, 2nd. ed., Berlin—Heidelberg—New York 1965.

[2] W. N. Everitt, *On an extension of an integro-differential inequality in Hardy, Littlewood and Pólya.* Proceedings Royal Society Edinburgh (A) **69**, 23 (1971/72), 295—333.

[3] J. Hadamard, *Sur le module maximum d'une fonction et de ses dérivées.* C. R. des Séances de l'année 1914. Soc. Math. de France (1914), 66—72.

[4] G. H. Hardy, J. E. Littlewood and G. Polya. *Inequalities.* Cambridge University Press 1934.

[5] E. Hille, *Remark on the Landau—Kallman—Rota Inequality.* Reports of Meetings. Seventh International Symposium on Functional Equations. September 1—13, 1969. Aequationes mathematicae **4** (1970), 239—240.

[6] E. Hille, *On the Landau—Kallman—Rota Inequality.* J. Approximation Theory (1972) (in print).

[7] E. Hille, *Functional Analysis and Semi-Groups.* Amer. Math. Soc. Coll. Publ. **31**, New York 1948.

[8] E. Hille and R. S. Phillips, *Functional Analysis and Semi-Groups*. Amer. Math. Soc. Coll. Publ. **31**, Rev. ed., Providence 1957.

[9] R. R. Kallman and G.-C. Rota, *On the inequality* $\|f'\|^2 \leq 4\|f\|\,\|f''\|$. Inequalities II. Ed. by Oved Shisha. Academic Press, New York—London 1970, 187—192.

[10] A. N. Kolmogorov, *On inequalities between the upper bounds of the successive derivatives of an arbitrary function on an infinite interval*. (Russian.) Užen. Zap. Moskov. Gos. Univ. Matematika **30** (1939), 3—13. Amer. Math. Soc. Transl. (1) **2** (1962), 233—243.

[11] H. Kraljević and S. Kurepa, *Semi-Groups on Banach Spaces*. Glasnik Mat. (1970), **5** (25), 109—117.

[12] S. Kurepa, *Remark on the Landau Inequality*. Reports of Meetings. Seventh International Symposium on Functional Equations. September 1—13, 1969. Aequationes mathematicae **4** (1970), 240—241.

[13] E. Landau, *Einige Ungleichungen für zweimal differenzierbare Funktionen*. Proc. London Math. Soc. (2) **13** (1913), 43—49.

[14] D. S. Mitrinović, *Analytic Inequalities*. Grundlehren **165**. Springer, New York—Heidelberg—Berlin 1970.

[15] I. J. Schoenberg and A. Cavaretta, *Solution of Landau's problem concerning higher derivatives on the halfline*. MRC Technical Summary Report. No. 1050. Madison. Wis. 1970.

Semi-Groups of Nonlinear Transformations

By

J. R. DORROH[1])

DEPT. OF MATH.
LOUISIANA STATE UNIVERSITY
BATON ROUGE

Throughout this paper, X denotes a real Banach space, X^* denotes the dual of X, and C denotes a subset of X. By an *operator in X*, we will mean a transformation A with domain $D(A)$ and range $R(A)$ contained in X. By $\mathfrak{J}(C)$, we mean the collection of all transformations from C into C.

A *semi-group in C* means a function T from $[0, \infty)$ into $\mathfrak{J}(C)$ such that $T(0)x = x$ for each x in C, and $T(t)T(s) = T(t+s)$ for each $t, s \geqq 0$. A semi-group T in C is said to be *strongly continuous* if for each x in C, the function $T(\cdot)x$ is a continuous function from $[0, \infty)$ into X. A semi-group T in C is said to be *nonexpansive* if each of the transformations $T(t)$ is nonexpansive; that is,

$$\|T(t)x - T(t)y\| \leqq \|x - y\|$$

for $t \geqq 0$ and x, y in C, where $\| \ \|$ denotes the norm in X. We denote by $Q(C)$ the collection of all strongly continuous nonexpansive semi-groups in C.

If T is a semi-group in C, then the *infinitesimal generator* of T is the operator B in X defined by

$$Bx = \lim_{t \to 0} t^{-1}(T(t)x - x),$$

with domain consisting of all x in C for which this limit exists. If there are no such x, then T has no infinitesimal generator.

The main purpose of the first section of this expository paper is to characterize the infinitesimal generator of a semi-group belonging to $Q(C)$, where C is a closed convex set in Hilbert space. Many of the preliminary theorems, which are interesting in themselves, are proven in greater generality than is necessary for this purpose. We have tried to give the strongest and most general versions possible, and also to be as thorough, as is consistent with a nearly self contained exposition of reasonable length.

---

[1]) The preparation of this paper was partially supported by the National Science Foundation Grant GP-28512.

In the second section, we give a brief discussion of the literature and history of the theory of semi-groups of nonlinear transformations. This theory parallels in many respects the theory of semi-groups of linear transformations; see [12], [26], or [27].

1. The description of infinitesimal generators

We will begin by giving just enough terminology to state the main theorems, Theorems 1. 1—1. 4, and stating enough preliminary propositions to clarify the concepts somewhat. Most of the proofs will be postponed until after the statement of the main theorems.

DEFINITION 1. 1: We denote by F the *duality map* on X; that is, F is the map from X into the subsets of X^* defined by

$$F(x) = \{f \in X^* : \langle x, f \rangle = \|x\|^2 = \|f\|^2\}.$$

PROPOSITION 1. 1 (Kato [28]): *The duality map F is homogeneous; that is, $F(cx) = cF(x)$ for each vector x and real number c. If X^* is strictly convex, then F is single valued; that is, $F(x)$ has only one element for each x in X, so we can consider F as a function from X into X^*. If X^* is uniformly convex, then F is uniformly continuous on bounded sets.*

REMARK: If X is a Hilbert space, then we make the standard identification of X^* with X. It is an easy exercise to show that the duality map is the identity in this case. We will not give a proof of Proposition 1. 1; for a proof, see [28].

DEFINITION 1. 2: If $A \subset X \times X$, then for each x in X, Ax is defined by

$$Ax = \{y : (x, y) \in A\}.$$

The *domain of A*, denoted $D(A)$, is the set of all x in X for which Ax is nonempty. The *range of A*, denoted by $R(A)$, is the set of all y in X such that $y \in Ax$ for some x in $D(A)$. If α is a real number, then

$$\alpha A = \{(x, \alpha y) : (x, y) \in A\}.$$

The *inverse of A* is defined by

$$A^{-1} = \{(y, x) : (x, y) \in A\}.$$

We say that A is *single-valued* if Ax contains only one element for each x in $D(A)$. All the terminology of this definition applies to an operator in X, considering the operator as a subset of $X \times X$; thus A is single-valued if and only if A is an operator in X. If $B \subset X \times X$, then

$$A + B = \{(x, y+z) : (x, y) \in A, (x, z) \in B\},$$

and

$$AB = \{(x, z) : (x, y) \in B, (y, z) \in A \quad \text{for some} \quad y \quad \text{in} \quad X\}.$$

DEFINITION 1.3: A subset A of $X \times X$ is said to be *accretive* if

$$\langle y_1 - y_2, f \rangle \geqq 0$$

for (x_1, y_1), (x_2, y_2) in A, and f in $F(x_1 - x_2)$. A is said to be *hyper-accretive* if A is accretive and $R(I+A) = X$, where I denotes the identity transformation in X. If $D(A) \subset C$, then A is said to be *C-maximal accretive* if A is accretive, and no accretive subset of $C \times X$ properly contains A. A is said to be *maximal-accretive* if A is X-maximal accretive.

DEFINITION 1.4: If $A \subset X \times X$, and Ax has a unique element of minimum norm for each x in $D(A)$, then A^0, the *trace of A*, is the operator defined on $D(A)$ by taking $A^0 x$ to be the element of Ax which has minimum norm. For a general subset E of X, we define

$$|E| = \inf \{\|x\| : x \in E\}.$$

PROPOSITION 1.2: *Let $T \in Q(C)$, and let B be the infinitesimal generator of T. Then $-B$ is accretive.*

PROOF: Let $x_1, x_2 \in D(B)$, and let $f \in F(x_1 - x_2)$. Then

$$\langle [T(t)x_1 - x_1] - [T(t)x_2 - x_2], f \rangle = \langle T(t)x_1 - T(t)x_2, f \rangle - \|x_1 - x_2\|^2 \leqq$$

$$\leqq \|T(t)x_1 - T(t)x_2\| \cdot \|f\| - \|x_1 - x_2\|^2 \leqq 0.$$

REMARK: For the rest of this section, A will denote an accretive subset of $X \times X$, and for each $\lambda > 0$, J_λ and A_λ are defined by

$$J_\lambda = (I + \lambda A)^{-1}, \quad A_\lambda = \lambda^{-1}(I - J_\lambda).$$

PROPOSITION 1.3: *Let A be an accretive subset of $X \times X$. Then:*
  (i) *If $\lambda > 0$, then J_λ is nonexpansive (in particular, it is single-valued).*
  (ii) *If $R(I + \lambda_0 A) = X$ for some $\lambda_0 > 0$, then $R(I + \lambda A) = X$ for each $\lambda > 0$.*
  (iii) *If A is hyper-accretive, then A is maximal accretive.*
  (iv) *If $D(A) \subset C$, and A is C-maximal accretive, then Ax closed and convex for each x in $D(A)$; thus A^0 is defined if X is uniformly convex.*

DEFINITION 1.5: *Let A be an accretive subset of $X \times X$, and let $\lambda_0 > 0$. Then A is said to satisfy*

    (i) *the condition* (C_1, λ_0) *if* $R(I + \lambda A) \supset D(A)$ *for* $0 < \lambda \leqq \lambda_0$,

    (ii) *the condition* (C_2, λ_0) *if* $R(I + \lambda A) \supset \overline{D(A)}$ *for* $0 < \lambda \leqq \lambda_0$, *and*

    (iii) *the condition* (C_3, λ_0) *if* $R(I + \lambda A) \supset C$ *for* $0 < \lambda \leqq \lambda_0$, *where C is the closed convex hull of* $D(A)$.

THEOREM 1.1: *Let X^* be uniformly convex, and let A satisfy (C_1, λ_0). If A is closed, then A satisfies (C_2, λ_0). There is a unique semi-group T belonging to $Q(\overline{D(A)})$ such that*

$$(*) \qquad\qquad T(t)x = \lim \, (I + n^{-1} t A)^{(-n)} x$$

for $t \geqq 0$ and x in $D(A)$; moreover, if $x \in D(A)$, then the convergence is uniform for t in bounded intervals. If A satisfies (C_2, λ_0), then $(*)$ holds for each x in $\overline{D(A)}$ (with the convergence uniform for t in bounded intervals). If $x \in D(A)$, then

$$\|T(t+h)x - T(t)x\| \leqq h \, |Ax|$$

for $t, h \geqq 0$.

REMARK: This theorem is due to OHARU [45] and has been generalized by CRANDALL and LIGGETT [15]; also see [5]. Notice that if A is hyper-accretive, then A satisfies (C_2, λ_0), by (ii) of Proposition 1.3.

PROPOSITION 1.4: *Let $T \in Q(C)$, and let B be the infinitesimal generator of T. If $t \geqq 0$ and $x \in C$, then $T(t)x \in D(B)$ if and only if $T(\cdot)x$ is differentiable from the right at t, in which case*

$$(d^+/dt)(T(t)x) = BT(t)x.$$

If $x \in C$, $0 \leqq s < t$, $T(s)x \in D(B)$, and $T(t)x \in D(B)$, then

$$\|BT(t)x\| \leqq \|BT(s)x\|.$$

PROOF: Let $x \in C$ and $t \geqq 0$. Then $h^{-1}[T(t+h)x - T(t)x] = h^{-1}[T(h)T(t)x - T(t)x]$. This proves the first statement. Now let $x \in C$, $0 \leqq s < t$, and $h > 0$. Then

$$\|T(t+h)x - T(t)x\| = \|T(t-s)T(s+h)x - T(t-s)T(s)x\| \leqq$$

$$\leqq \|T(s+h)x - T(s)x\|.$$

This proves the second statement.

DEFINITION 1.6: *Let $T \in Q(C)$, and let B be the infinitesimal generator of T. An element x of C is said to be regular for T if*

    (i) $T(t)x \in D(B)$ *for all* $t \geqq 0$,

    (ii) $BT(\cdot)x$ *is continuous from the right on* $[0, \infty)$, *and*

    (iii) $BT(\cdot)x$ *is continuous at all but countably many t in* $[0, \infty)$.

REMARK: Of course, if $BT(\cdot)x$ is continuous at $t>0$, then by Proposition 1.4, $T(\cdot)x$ is differentiable at t, and

$$(d/dt)(T(t)x)=BT(t)x.$$

Also by Proposition 1.4, if x is regular for T, then $T(\cdot)x$ is Lipschitz continuous on $[0, \infty)$, with Lipschitz constant $\|Bx\|$.

DEFINITION 1.7: Let B be a subset of $X \times X$, and let $x \in X$. By a *solution of the initial value problem* $P(B, x)$, we mean an absolutely continuous function u from $[0, \infty)$ into X such that

    (i) $u(0)=x$,
    (ii) u is differentiable a.e.,
    (iii) $u(t) \in D(B)$ a.e., and
    (iv) $u'(t) \in Bu(t)$ a.e.

PROPOSITION 1.5: *Let A be an accretive subset of $X \times X$. If $x \in X$, then $P(-A, x)$ has at most one solution; if u is a solution, then $\|u'(\cdot)\|$ is a nonincreasing function. If $x, y \in X$, u is a solution of $P(-A, x)$, and v is a solution of $P(-A, y)$, then*

$$\|u(t)-v(t)\| \leq \|x-y\|$$

for $t \geq 0$. If $P(-A, x)$ has a solution for each x in $D(A)$, then there is a unique semi-group T belonging to $Q(\overline{D(A)})$ such that $T(\cdot)x$ is a solution of $P(-A, x)$ for each x in $D(A)$.

THEOREM 1.2: *Let X^* be uniformly convex, let A and T be as in Theorem 1.1, but also let A be $\overline{D(A)}$-maximal accretive, and let B denote the infinitesimal generator of T. Then A satisfies (C_2, λ_0), $T(t)D(A) \subset D(A)$ for each $t \geq 0$, and $Bx \in -Ax$ for each x in $D(A) \cap D(B)$. If $x \in D(A)$, then $T(\cdot)x$ is the unique solution of $P(-A, x)$. If X is uniformly convex, then $D(A) \subset D(B)$, and if $x \in D(A)$, then $Bx = -A^0 x$, and x is regular for T.*

REMARK: This is very close to a theorem of OHARU in [45], although our proof is a little different; also see [5]. Similar theorems were known earlier; see for instance [17], [21], [28].

PROPOSITION 1.6: *Let C be closed and convex, let G be a nonexpansive transformation from C into C, let $\alpha>0$, and let $B = \alpha[G-I]$. Then B is the infinitesimal generator of a unique semi-group belonging to $Q(C)$.*

PROPOSITION 1.7: *Let X be uniformly convex, and let A satisfy (C_3, λ_0). Then $\overline{D(A)}$ is convex.*

REMARK: This result appears in [5]. It is related to a result of ROCKAFELLAR [48].

THEOREM 1.3: *Let X^* be uniformly convex, and let A, T, and B be as in Theorem 1.2, but also let A satisfy (C_3, λ_0). Let C denote the closed convex hull of $D(A)$, and for* *each positive integer $n \geq \lambda_0^{-1}$, let T_n denote the semi-group belonging to $Q(C)$ whose infinitesimal generator is the restriction of $-A_{1/n}$ to C. If $x \in \overline{D(A)}$, then*

$$(**) \qquad\qquad T(t)x = \lim_{n \to \infty} T_n(t)x$$

for $t \geq 0$, and the convergence is uniform for t in bounded intervals. Furthermore, $B \subset -A$, and if X is uniformly convex, then $\overline{D(A)}$ is convex, and $B = -A^0$.

REMARK: This theorem sems to be a little stronger than what is presently in the literature; see for instance [17], [21], [45]. Remember that if A is hyper-accretive, then A satisfies (C_3, λ_0). Also, by Theorem 1.2, each x in $D(B)$ is regular for T if X is uniformly convex.

THEOREM 1. 4: *Let X be a Hilbert space, let C be closed and convex, let $T \in Q(C)$, and let B be the infinitesimal generator of T. Then $D(B)$ is dense in C, and there is a hyper-accretive subset A of $X \times X$ such that $B = -A^0$.*

REMARK: Notice that by Theorem 1.3, the conditions on B are also sufficient that B be the infinitesimal generator of a $T \in Q(C)$, and that each x in $D(B)$ is regular for T. This is well known; see for instance [6], [17], [18]. By [18], the hyper-accretive set A is unique. Also, KOMURA in [32] has proven that if C is any subset of the Hilbert space X, and $T \in Q(C)$, then there is a closed convex set $C' \supset C$ and a $U \in Q(C')$ such that $U(t) \supset T(t)$ for each $t \geq 0$. Thus, the loss of generality in assuming C to be closed and convex is not as great as it might first appear.

PROOF OF PROPOSITION 1. 3: To prove (i), let (x_1, y_1), $(x_2, y_2) \in A$, and $f \in F(x_1 - x_2)$. Then

$$\|(x_1 + \lambda y_1) - (x_2 + \lambda y_2)\| \cdot \|x_1 - x_2\| \geq \langle x_1 - x_2 + \lambda(y_1 - y_2), f \rangle =$$

$$= \|x_1 - x_2\|^2 + \lambda \langle y_1 - y_2, f \rangle \geq \|x_1 - x_2\|^2.$$

(ii) Let $\lambda > \lambda_0/2$. Then $\lambda^{-1}|\lambda_0 - \lambda| < 1$, and J_{λ_0} is nonexpansive, so $I + \lambda^{-1}(\lambda_0 - \lambda)J_{\lambda_0}$ is a homeomorphism of X onto X. Let $z \in X$, and let (x, y) be an element of A such that

$$x + \lambda_0 y = [I + \lambda^{-1}(\lambda_0 - \lambda)J_{\lambda_0}]^{(-1)}(\lambda_0/\lambda)z.$$

Then

$$(\lambda_0/\lambda)z = x + \lambda_0 y + \lambda^{-1}(\lambda_0 - \lambda)J_{\lambda_0}(x + \lambda_0 y) = x + \lambda_0 y + \lambda^{-1}(\lambda_0 - \lambda)x =$$

$$= \lambda_0 y + (\lambda_0/\lambda)x,$$

$$z = x + \lambda y.$$

Thus $R(I + \alpha A) = X$ if $\alpha > \lambda_0/2$. It follows that $R(I + \alpha A) = X$ for all $\alpha > 0$.

(iii) Suppose $A \subset B \subset X \times X$, and B is accretive. If $(x, y) \in B$, $u = (I + A)^{(-1)}(x + y)$, and $v = x + y - u$, then $(u, v) \in A \subset B$, and $u + v = x + y$. Therefore, by (i), $u = x$, so that $(x, y) = (u, v) \in A$.

(iv) Define the subset B of $C \times X$ by taking Bx to be the closed convex hull of Ax for each x in $D(A)$. Then B is accretive, so $B = A$.

LEMMA 1. 1 (Kato [28]). *Let S be an open real number interval, u a function from S into X, $t \in S$, and $f \in F(u(t))$. Suppose u is weakly differentiable at t with weak derivative $u'(t)$ and that $\|u(\cdot)\|$ is differentiable at t. Then*

$$\|u(t)\| (d/dt)\|u(t)\| = \langle u'(t), f \rangle.$$

PROOF: If $h \neq 0$, and $t + h \in S$, then

$$\|u(t+h)\| \cdot \|u(t)\| \geqq \langle u(t+h), f \rangle,$$
$$\|u(t)\|^2 = \langle u(t), f \rangle,$$
$$\|u(t)\| (\|u(t+h)\| - \|u(t)\|) \geqq \langle u(t+h) - u(t), f \rangle.$$

If we let h be positive (negative), divide both sides by h, and let h decrease (increase) to zero, then we get

$$\|u(t)\| (d/dt) \|u(t)\| \geqq (\leqq) \langle u'(t), f \rangle.$$

PROOF OF PROPOSITION 1. 5: Let $x \in X$, and let u be a solution of $P(-A, x)$. Let $h > 0$, and define

$$\varphi(t) = u(t+h) - u(t)$$

for $t \geqq 0$. Then φ is absolutely continuous and differentiable a.e.; thus, so is $\|\varphi(\cdot)\|$. If u, $u(\cdot + h)$, and $\|\varphi(\cdot)\|$ are differentiable at $t > 0$, and $g \in F(\varphi(t))$, then

$$(d/dt)\|\varphi(t)\|^2 = 2\langle u'(t+h) - u'(t), g \rangle \leqq 0.$$

Therefore $\|\varphi(\cdot)\|$, and thus $\|u'(\cdot)\|$, are nonincreasing.

Now let $x, y \in X$, let u and v be solutions of $P(-A, x)$ and $P(-A, y)$, respectively, and let $\psi = u - v$. Then ψ is absolutely continuous a.e.; thus, so is $\|\psi(\cdot)\|$. If u, v, and $\|\psi(\cdot)\|$ are differentiable at $t > 0$, and $f \in F(\psi(t))$, then

$$(d/dt)\|\psi(t)\|^2 = 2\langle u'(t) - v'(t), f \rangle \leqq 0.$$

This proves that

$$\|u(t) - v(t)\| \leqq \|x - y\|$$

for $t \geqq 0$, and that if $z \in X$, then $P(-A, z)$ has at most one solution. The statement about the semi-group follows from these two facts and the observation that if α is a solution of $P(-A, z)$, $t > 0$, and $\beta(s) = \alpha(t+s)$ for $s \geqq 0$, then β is a solution of $P(-A; \alpha(t))$.

PROOF OF PROPOSITION 1. 6: Since $-B$ is accretive, we have only to show that $P(B, x)$ has a solution for each $x \in C$. We can suppose without loss of generality that C contains the origin.

Let p denote the support functional of C. Then p is a subadditive, positive homogeneous, lower semi-continuous function from X into the extended nonnegative reals, and $C = p^{-1}([0, 1])$.

Let $x \in C$. If $\varphi : [0, \infty) \to X$, $\varphi(0) = x$, and $e^{-\alpha t} \varphi(t) \in C$ for each $t \geq 0$, then define φ^* on $[0, \infty)$ by

$$\varphi^*(t) = x + \int_0^t \alpha e^{\alpha s} G(e^{-\alpha s} \varphi(s)) \, ds.$$

Now $p(\varphi(t)) \leq e^{\alpha t}$ for $t \geq 0$, so using the properties of p, we find that $p(\varphi^*(t)) \leq e^{\alpha t}$ for $t \geq 0$, so that $e^{-\alpha t} \varphi^*(t) \in C$ for $t \geq 0$, and the method of repeated integration applies to yield a function ψ on $[0, \infty)$ such that $e^{-\alpha t} \psi(t) \in C$ for $t \geq 0$, and

$$\psi(t) = x + \int_0^t \alpha e^{\alpha s} G(e^{-\alpha s} \psi(s)) \, ds$$

for $t \geq 0$. The function $u(t) = e^{-\alpha t} \psi(t)$ is a solution of $P(B, x)$.

LEMMA 1.2: *Let X be uniformly convex, let C be closed and convex, and let P_n be a nonexpansive transformation from C into X for each $n = 1, 2, 3, \ldots$. Let*

$$E = \{x \in C : \lim_{n \to \infty} P_n x = x\}.$$

Then E is closed and convex.

PROOF: Clearly, E is closed. Let $x, y \in E$, $x \neq y$, and $z = \alpha x + \beta y$, with $\alpha, \beta \geq 0$ and $\alpha + \beta = 1$. Then

$$\|P_n z - x\| \leq \|P_n z - P_n x\| + \|P_n x - x\| \leq \beta \|x - y\| + \|P_n x - x\|,$$

and

$$\|P_n z - y\| \leq \alpha \|x - y\| + \|P_n y - y\|.$$

Let $\{n_k\}$ be an increasing sequence of positive integers such that $\{P_{n_k} z\}$ is weakly convergent, say to w. Then

$$\|w - x\| \leq \beta \|x - y\|, \quad \|w - y\| \leq \alpha \|x - y\|,$$

$$\|x - y\| \leq \|x - w\| + \|w - y\| \leq \|x - y\|.$$

Thus, since X is uniformly convex, and therefore strictly convex, it follows that $w = \alpha' x + \beta' y$ with $\alpha', \beta' \geq 0$ and $\alpha' + \beta' = 1$. Also,

$$\|w - x\| = \beta' \|x - y\| \leq \beta \|x - y\|,$$

and

$$\|w - y\| = \alpha' \|x - y\| \leq \alpha \|x - y\|,$$

so that $\alpha = \alpha'$, $\beta = \beta'$, and $w = z$. Therefore, $\{P_n z\}$ is weakly convergent to z, and $\{P_n z - x\}$ is weakly convergent to $z - x$. But then

$$\|z - x\| = \beta \|x - y\| \leq \liminf \|P_n z - x\| \leq \limsup \|P_n z - x\| \leq \beta \|x - y\| = \|z - x\|,$$

so that

$$\lim_{n \to \infty} \|P_n z - x\| = \|z - x\|,$$

and thus

$$\lim_{n \to \infty} (P_n z - x) = z - x.$$

LEMMA 1.3: *Let A be an accretive subset of $X \times X$. Then:*

(i) $A_\lambda \subset A J_\lambda$.

(ii) A_λ *is accretive.*

(iii) $\{(J_\lambda x, A_\lambda x): x \in D(J_\lambda)\}$ *is accretive.*

(iv) A_λ *is Lipschitz continuous.*

(v) *If A satisfies (C_1, λ_0), then $\|A_\lambda J_\lambda^n x\| \leq |Ax|$ for x in $D(A)$, $0 < \lambda \leq \lambda_0$, and $n = 0, 1, 2, \ldots$.*

(vi) *If A satisfies (C_1, λ_0), then $\|J_\lambda^{(m+n)} x - J_\lambda^m x\| \leq n\lambda |Ax|$ for x in $D(A)$, $0 < \lambda \leq \lambda_0$, and $m, n = 0, 1, 2, \ldots$.*

(vii) *If A satisfies (C_1, λ_0), then $\|J_\lambda^n x - J_\mu^n x\| \leq n |\lambda - \mu| \cdot |Ax|$ for x in $D(A)$, $0 < \lambda, \mu \leq \lambda_0$, and $n = 1, 2, 3, \ldots$.*

(viii) *If A satisfies $(C_1, \lambda_0) ((C_2, \lambda_0))$, then $\lim_{\lambda \to 0} J_\lambda x = x$ for each x in $D(A) (\overline{D(A)})$.*

PROOF:

(i) $J_\lambda = \{(x + \lambda y, x): (x, y) \in A\}$, $I - J_\lambda = \{(x + \lambda y, \lambda y): (x, y) \in A\}$,

$$A J_\lambda = \{(x + \lambda y, z): (x, y)(x, z) \in A\}.$$

(ii) Clear, since J_λ is nonexpansive.

(iii) Clear from (i).

(iv) Clear, since J_λ is nonexpansive.

(v) Let $(x, y) \in A$ and $0 < \lambda \leq \lambda_0$. Then

$$x - J_\lambda x = J_\lambda(x + \lambda y) - J_\lambda x, \quad \|A_\lambda x\| \leq \|y\|, \quad \|A_\lambda x\| \leq |Ax|.$$

$$A_\lambda J_\lambda^n x = \lambda^{-1}(J_\lambda^n x - J_\lambda^{(n+1)} x), \quad \|A_\lambda J_\lambda^n x\| \leq \lambda^{-1} \|x - J_\lambda x\| = \|A_\lambda x\| \leq |Ax|.$$

(vi) $\qquad \|J_\lambda^{(m+n)} x - J_\lambda^m x\| \leq \|J_\lambda^n x - x\| \leq \sum_{j=1}^{n} \|J_\lambda^j x - J_\lambda^{j-1} x\| \leq n\lambda |Ax|.$

(vii) First, let $u \in D(A)$. Then

$$J_\lambda u - J_\mu u = J_\lambda u - J_\lambda(J_\mu u + \lambda A_\mu u),$$

$$\|J_\lambda u - J_\mu u\| \leq \|u - J_\mu u - \lambda A_\mu u\| = |\mu - \lambda| \cdot \|A_\mu u\|,$$

where we have used $A_\mu u \in A J_\mu u$, from (i). Also

$$\|J_\lambda^n x - J_\mu^n x\| = \|\sum_{j=1}^{n} J_\lambda^j J_\mu^{n-j} x - J_\lambda^{j-1} J_\mu^{n-j+1} x\| \leq \sum_{j=1}^{n} \|(J_\lambda - J_\mu) J_\mu^{n-j} x\| \leq$$

$$\leq \sum_{j=1}^{n} |\lambda - \mu| \cdot \|A_\mu J_\mu^{n-j} x\| \leq n |\lambda - \mu| \cdot |Ax|.$$

(viii) Clear from (v) or (vi) and the fact that each J_λ is nonexpansive.

PROOF OF PROPOSITION 1.7: Let C denote the closed convex hull of $D(A)$, and each $n=1, 2, 3, \dots$, let P_n denote the restriction to C of $J_{\lambda_0/n}$. Let

$$E = \{x \in C : \lim_{n \to \infty} P_n x = x\}.$$

By Lemma 1.3, $E = \overline{D(A)}$, and by Lemma 1.2, E is convex.

LEMMA 1.4: *If A satisfies (C_1, λ_0), and A is closed, then A satisfies (C_2, λ_0).*

PROOF: Let $0 < \lambda \leq \lambda_0$, let z be a limit point of $R(I + \lambda A)$, and let $\{(x_n, y_n)\} \subset A$ be such that $z_n \to z$, where $z_n = x_n + \lambda y_n$ for each n. Since $x_n = (I + \lambda A)^{-1} z_n$, then $\{x_n\}$ is convergent, say to x. Thus $\{y_n\}$ is convergent to an element y of X with $(x, y) \in A$ and $x + \lambda y = z$.

LEMMA 1.5: *If $a, b \in X$, $f \in F(a)$, and $g \in F(b)$, then*

$$2\langle a-b, g \rangle \leq \|a\|^2 - \|b\|^2 \leq 2\langle a-b, f \rangle.$$

PROOF:

$$2\langle a-b, g \rangle = 2\langle a, g \rangle - 2\|b\|^2 \leq 2\|a\| \cdot \|b\| - 2\|b\|^2 \leq \|a\|^2 - \|b\|^2.$$

$$2\langle a-b, f \rangle = 2\|a\|^2 - 2\langle b, f \rangle \geq 2\|a\|^2 - 2\|a\| \cdot \|b\| \geq \|a\|^2 - \|b\|^2.$$

PROOF OF THEOREM 1.1: Let $x \in D(A)$ and $b > 0$. We want to show that $\{J_{(t/n)}^n x\}_{n \geq b\lambda_0^{-1}}$ converges uniformly for $0 \leq t \leq b$.

If $\delta > 0$, then let

$$\varphi(\delta) = \sup_{S(\delta)} \|F(u) - F(v)\|,$$

where

$$S(\delta) = \{(u, v) : \|u - v\| \leq \delta \text{ and } \|u\|, \|v\| \leq 2b|Ax|\}.$$

Then $\lim_{\delta \to 0} \varphi(\delta) = 0$ by Proposition 1.1.

Let m and n be positive integers with $m \geq b\lambda_0^{-1}$, let $0 < t \leq b$, $\lambda = t/mn$, and $\mu = t/m$. If $1 \leq j \leq mn$ and $1 \leq k \leq m$, then by Lemma 1.3, we have

$$\|J_\lambda^j x - J_\mu^k x\| \leq \|J_\lambda^j x - J_\lambda^k x\| + \|J_\lambda^k x - J_\mu^k x\| \leq \lambda |j-k| \cdot |Ax| + k(\mu - \lambda)|Ax| \leq 2b|Ax|.$$

If $1 \leq k \leq m$ and $0 \leq p \leq n-1$, then again by Lemma 1.3, we have

$$\|J_\lambda^{nk} x - J_\lambda^{n(k-1)+p+1} x\| \leq (n-p-1)\lambda|Ax| \leq b|Ax|/m,$$

and thus we have

(1) $$\|F(J_\lambda^{nk} x - J_\mu^k x) - F(J_\lambda^{n(k-1)+p+1} x - J_\mu^k x)\| \leq \varphi(b|Ax|/m).$$

Also, by Lemma 1.3, we have

(2) $$\|A_\lambda J_\lambda^{n(k-1)+p} x - A_\mu J_\mu^{k-1} x\| \leq 2|Ax|.$$

Since $A_\lambda \subset AJ_\lambda$ and $A_\mu \subset AJ_\mu$ by Lemma 1.3, and A is accretive, we have

$$(3) \qquad \langle A_\lambda J_\lambda^{n(k-1)+p} x - A_\mu J_\mu^{(k-1)} x, F(J_\lambda^{n(k-1)+p+1} x - J_\mu^k x)\rangle \geqq 0.$$

Now notice that

$$J_\lambda^{nk} x - J_\lambda^{n(k-1)} x = -\lambda \sum_{p=0}^{n=1} A_\lambda J_\lambda^{n(k-1)+p} x,$$

and that

$$J_\mu^k x - J_\mu^{(k-1)} x = -\mu A_\mu J_\mu^{(k-1)} x = -n\lambda A_\mu^{(k-1)} x,$$

so that

$$(J_\lambda^{nk} x - J_\lambda^{n(k-1)} x) - (J_\mu^k x - J_\mu^{(k-1)} x) = -\lambda \sum_{p=0}^{n-1} \{A_\lambda J_\lambda^{n(k-1)+p} x - A_\mu J_\mu^{(k-1)} x\}.$$

Therefore,

$$(4) \qquad \langle (J_\lambda^{nk} x - J_\mu^k x) - (J_\lambda^{n(k-1)} x - J_\mu^{(k-1)} x), F(J_\lambda^{nk} x - J_\mu^k x)\rangle =$$

$$= -\lambda \sum_{p=0}^{n-1} \langle A_\lambda J_\lambda^{n(k-1)+p} x - A_\mu J_\mu^{(k-1)} x, F(J_\lambda^{n(k-1)+p+1} x - J_\mu^k x)\rangle +$$

$$+\lambda \sum_{p=0}^{n-1} \langle A_\lambda J_\lambda^{n(k-1)+p} x - A_\mu J_\mu^{(k-1)} x, F(J_\lambda^{n(k-1)+p+1} x - J_\mu^k x) - F(J_\lambda^{nk} x - J_\mu^k x)\rangle \leqq$$

$$\leqq 0 + 2n\lambda |Ax| \varphi(b|Ax|/m) \leqq (2b/m)|Ax| \varphi(b|Ax|/m),$$

where we have used (3), (2), and (1).

Therefore, using (4) and Lemma 1.5, we have

$$(7) \qquad \|J_\lambda^{mn} x - J_\mu^m x\|^2 = \sum_{k=1}^{m} \{\|J_\lambda^{nk} x - J_\mu^k x\|^2 - \|J_\lambda^{n(k-1)} x - J_\mu^{(k-1)} x\|^2\} \leqq$$

$$\leqq 2 \sum_{k=1}^{m} \langle (J_\lambda^{nk} x - J_\mu^k x) - (J_\lambda^{n(k-1)} x - J_\mu^{(k-1)} x), F(J_\lambda^{nk} x - J_\mu^k x)\rangle \leqq$$

$$\leqq 2b|Ax| \varphi(b|Ax|/m) = \varepsilon_m^2.$$

Thus if m and n are any two positive integers greater than $b\lambda_0^{-1}$, then we have by (7)

$$\|J_{(t/n)}^n x - J_{t/m}^m x\| \leqq \varepsilon_m + \varepsilon_n.$$

Thus we have proven:

(8) *If* $x \in D(A)$, *then the limit*

$$S(t)x = \lim_{n \to \infty} (I + tn^{-1} A)^{(-n)} x$$

exists, and the convergence is uniform for t in bounded intervals.

Notice that (8) defines a family $\{S(t); t \geqq 0\}$ of transformations from $D(A)$ into $\overline{D(A)}$. Since each J_λ is nonexpansive, then each $S(t)$ is nonexpansive, and thus has a unique extension to a continuous transformation from $\overline{D(A)}$ into $\overline{D(A)}$. Let $T(t)$ denote this extension, and notice that $T(t)$ is also nonexpansive.

If $x \in D(A)$, then

$$\|J_{t/n}^n x - J_{s/n}^n x\| \leq n\,|(t/n)-(s/n)| \cdot |Ax|$$

by Lemma 1.3, so

$$\|T(t)x - T(s)x\| \leq |t-s| \cdot |Ax|.$$

Since each $T(t)$ is nonexpansive it follows that $T(\cdot)x$ is a continuous function on $[0, \infty)$ for each x in $\overline{D(A)}$.

If $x \in D(A)$, $t > 0$, and m is a positive integer, then

$$(J_{(t/n)}^n)^m x = J_{(mt/mn)}^{mn} x,$$

so

$$[T(t)]^m x = T(mt)x.$$

By continuity, this holds for $x \in \overline{D(A)}$. If $x \in \overline{D(A)}$, and r, s, p, and q are positive integers, then

$$T(r/s)T(p/q)x = [T(1/sq)]^{rq}[T(1/sq)]^{ps}x = [T(1/sq)]^{rq+ps}x = T((r/s)+(p/q))x.$$

Thus

$$T(s)T(t)x = T(s+t)x$$

for rational $s, t \geq 0$. By continuity, this holds for all $s, t \geq 0$. Therefore $T \in Q(C)$. The uniqueness is clear.

LEMMA 1.6: *Let X^* be uniformly convex, and let A be $\overline{D(A)}$-maximal accretive. Let $(x_n, y_n) \in A$ for each $n = 1, 2, \ldots$, and let $\lim_{n \to \infty} x_n = x$. Then*

(i) *if* w-$\lim_{n \to \infty} y_n = y$, *then* $(x, y) \in A$,

(ii) *if* $\sup \|y_n\| < \infty$, *then* $x \in D(A)$,

and

(iii) *A is closed.*

PROOF: (i) If $(u, v) \in A$, then

$$\langle y - v, F(x-u) \rangle = \lim_{n \to \infty} \langle y_n - v, F(x_n - u) \rangle \geq 0.$$

(ii) There is an increasing sequence $\{n_k\}$ of positive integers such that $\{y_{n_k}\}$ is weakly convergent. But we still have $\lim_{k \to \infty} x_{n_k} = x$, so (ii) follows from (i).

(iii) Clear from (i).

COROLLARY: *Let X and A be as in Lemma 1.6. Let $\{\lambda_n\}$ be a sequence of positive numbers converging to zero, let $x_n \in D(J_{\lambda_n})$ for each n, and let $\lim_{n \to \infty} x_n = x$. Then*

(i) *if* w-$\lim_{n \to \infty} A_{\lambda_n} x_n = y$, *then* $(x, y) \in A$,

and

(ii) *if* $\sup \|A_{\lambda_n} x_n\| < \infty$, *then* $x \in D(A)$.

PROOF: Let $u_n = J_{\lambda_n} x_n$ for each n. Then $(u_n, A_{\lambda_n} x_n) \in A$ for each n, and $\lim_{n \to \infty} u_n = x$.

PROOF OF THEOREM 1. 2: If $x \in D(A)$, $t > 0$, and $n > t\lambda_0^{-1}$, then by Lemma 1. 3,

$$\|A_{(t/n)} J_{(t/n)}^n x\| \leq |Ax|,$$

so $T(t)x \in D(A)$ by the corollary to Lemma 1. 6. Also, A is closed, by Lemma 1. 6, so A satisfies (C_2, λ_0).

Now let $x \in D(A)$, and for each $n > \lambda_0^{-1}$, define f_n on $[0, \infty)$ by $f_n(kn^{-1} + s) = (I - sA_{1/n})J_{1/n}^k x$ for $0 \leq s \leq n^{-1}$. Notice that $f_n(kn^{-1}) = J_{1/n}^k x$, and that the graph of f_n consists of line intervals joining successive points $(kn^{-1}, f_n(kn^{-1}))$ and $((k+1)n^{-1}, f_n((k+1)n^{-1}))$. Thus $\{f_n\}$ converges uniformly to $T(\cdot)x$ on bounded intervals. Also $f_n'(kn^{-1} + s) = -A_{1/n}J_{1/n}^k x$ for $0 < s < n^{-1}$.

Let $(x_0, y_0) \in A$, and let $x_n = x_0 + n^{-1} y_0$ for $n \geq \lambda_0^{-1}$. Then $J_{1/n} x_n = x_0$ and $A_{1/n} x_n = y_0$. Let $b > 0$. Then

$$\sup_{kn^{-1} \leq t \leq b} [\|J_{1/n}^k x - x_n\| + \|f_n(t) - x_n\|] < \infty,$$

and if $k = k(t, n) = [t/n]$, then

$$\lim_{n \to \infty} \left\| (J_{1/n}^k x - x_n) - (f_n(t) - x_n) \right\| = 0$$

uniformly for $0 \leq t \leq b$, so we have

$$(9) \qquad \left\| F(J_{1/n}^k x - x_n) - F(f_n(t) - x_n) \right\| \leq \varphi(n)$$

for $0 \leq t \leq b$, $k = [t/n]$, where

$$\lim_{n \to \infty} \varphi(n) = 0.$$

Now if $0 < t - kn^{-1} < n^{-1}$, then

$$(d/dt)\|f_n(t) - x_n\|^2 = -2\langle A_{1/n} J_{1/n}^k x, F(f_n(t) - x_n)\rangle =$$

$$= -2\langle A_{1/n} J_{1/n}^k x - A_{1/n} x_n, F(f_n(t) - x_n)\rangle + 2\langle A_{1/n} x_n, F(x_n - f_n(t))\rangle =$$

$$= -2\langle A_{1/n} J_{1/n}^k x - A_{1/n} x_n, F(J_{1/n}^k x - x_2)\rangle +$$

$$+ 2\langle A_{1/n} J_{1/n}^k x - A_{1/n} x_n, F(J_{1/n}^k x - x_n) - F(f_n(t) - x_n)\rangle + 2\langle y_0, F(x_n - f_n(t))\rangle \leq$$

$$\leq 0 + 2(|Ax| + \|y_0\|)\varphi(n) + 2\langle y_0, F(x_n - f_n(t))\rangle = 2\langle y_0, F(x_n - f_n(t))\rangle + \theta(n),$$

where we have used the fact that $A_{1/n}$ is accretive, (9), Lemma 1. 3, and $A_{1/n} x_n = y_0$. Thus,

$$\|f_n(t) - x_n\|^2 \leq \|x - x_n\|^2 + 2 \int_0^t \langle y_0, F(x_n - f_n(s))\rangle \, ds + t\theta(n),$$

so that

$$(10) \qquad \|T(t)x-x_0\|^2 \le \|x-x_0\|^2 + 2\int_0^t \langle y_0, F(x_0-T(s)x)\rangle \, ds.$$

From Lemma 1.5, we have

$$\|T(t)x-x_0\|^2 \ge \|x-x_0\|^2 + 2\langle T(t)x-x, F(x-x_0)\rangle.$$

Combining this with (10), we get

$$\langle T(t)x-x, F(x-x_0)\rangle \le \int_0^t \langle y_0, F(x_0-T(s)x)\rangle \, ds.$$

If $x \in D(B)$, then dividing by t and letting t decrease to 0, we get

$$\langle Bx, F(x-x_0)\rangle \le \langle y_0, F(x_0-x)\rangle,$$

$$0 \le \langle y_0+Bx, F(x_0-x)\rangle,$$

so that $\{(x, -Bx)\} \cup A$ is accretive. Therefore $-Bx \in Ax$. This argument was adapted from CRANDALL and LIGGETT [15].

If $x \in D(A)$, then $T(\cdot)x$ is Lipschitz continuous and differentiable a.e., and $T(\cdot)x \in D(A)$ on $[0, \infty)$, so $T(\cdot)x \in D(A) \cap D(B)$, and $(d/dt)T(t)x = BT(t)x \in -AT(t)x$ wherever $T(\cdot)x$ is differentiable. Therefore, $T(\cdot)x$ is the unique solution of $P(-A, x)$.

Now suppose X is uniformly convex, and let $x \in D(A)$, and let $f(t) = T(t)x$ for $t \ge 0$. Then $\|f'(t)\| \le \|A^0 x\|$ wherever f is differentiable, by Theorem 1.1. Let $\{t_n\}$ be a sequence which decreases to 0 such that f is differentiable at each t_n and such that $\{f'(t_n)\}$ is weakly convergent, say to $-y$. Then $(x, y) \in A$ by Lemma 1.6. Thus $y = A^0 x$. Also $\|f'(t_n)\|$ converges to $\|A^0 x\|$, so $\{f'(t_n)\}$ converges to $-A^0 x$. Since every sequence from the domain of f' which decreases to 0 contains such a sequence, we have

$$\lim_{t \to 0} f'(t) = -A^0 x \qquad (x \in D(B), \, Bx = -A^0 x).$$

Thus $f(t) \in D(B)$ for $t \ge 0$, and $Bf(t) = -A^0 f(t)$. By Proposition 1.4, $\|Bf(\cdot)\|$ is nonincreasing, so a similar argument shows that $Bf(\cdot)$ is continuous from the right. Also, $\|Bf(\cdot)\|$ is continuous except for at most a countable set, and we see that $Bf(\cdot)$ is continuous from the left wherever $\|Bf(\cdot)\|$ is continuous because if $t_0 > 0$, $\{t_n\}$ increases to t_0, and $\{Bf(t_n)\}$ converges weakly to z, then $z = Bf(t_0)$, so $\{Bf(t_n)\}$ converges strongly to $Bf(t_0)$. Therefore, x is regular for T.

PROOF OF THEOREM 1.3: Let $x \in D(A)$, let $f(t) = T(t)x$ for $t \ge 0$, and if $n \ge \lambda_0^{-1}$, then let $f_n(t) = T_n(t)x$ for $t \ge 0$. Then

$$f_n'(t) = -A_{1/n} f_n(t) \in -A_{1/n} J_{1/n} f_n(t)$$

for $t \geqq 0$, and $f'(t) \in -Af(t)$ a.e. on $[0, \infty)$, so

$$\langle f_n'(t) - f'(t), F(J_{1/n}f_n(t) - f(t)) \rangle \geqq 0$$

a.e. on $[0, \infty)$. Also

$$\|J_{1/n}f_n(t) - f(t)\| \leqq \|J_{1/n}f_n(t) - J_{1/n}x\| + \|J_{1/n}x - x\| + \|x - f(t)\| \leqq (n^{-1} + 2t)|Ax|,$$

$$\|f_n(t) - f(t)\| \leqq 2t|Ax|,$$

and

$$\|(J_{1/n}f_n(t) - f(t)) - (f_n(t) - f(t))\| \leqq n^{-1}|Ax|$$

for $t \geqq 0$, by Lemma 1.3, Proposition 1.4, and Theorem 1.1. Therefore,

$$\lim_{n \to \infty} \|F(J_{1/n}f_n(t) - f(t)) - F(f_n(t) - f(t))\| = 0$$

uniformly for t in bounded intervals. Also,

$$\|f_n'(t) - f'(t)\| \leqq 2|Ax|$$

a.e. on $[0, \infty)$. Thus

$$(d/dt)\|f_n(t) - f(t)\|^2 = 2\langle f_n'(t) - f'(t), F(f_n(t) - f(t)) \rangle =$$

$$= 2\langle f_n'(t) - f'(t), F(J_{1/n}f_n(t) - f(t)) \rangle +$$

$$+ 2\langle f_n'(t) - f'(t), F(f_n(t) - f(t)) - F(J_{1/n}f_n(t) - f(t)) \rangle \leqq$$

$$\leqq 4|Ax| \cdot \|F(f_n(t) - f(t)) - F(J_{1/n}f(t) - f(t))\|$$

a.e. on $[0, \infty)$, so that

$$\lim_{n \to \infty} \|f_n(t) - f(t)\| = 0$$

uniformly for t in bounded intervals.

Therefore, if $x \in D(A)$, then

$$T(t)x = \lim T_n(t)x,$$

and the convergence is uniform for t in bounded intervals. Since each transformation $T(t)$, $T_n(t)$ is nonexpansive, the same is true for x in $\overline{D(A)}$.

Now let $x \in D(B)$, $g(t) = T(t)x$ for $t \geqq 0$, and $g_n(t) = T_n(t)x$ for $t \geqq 0$ and $n \geqq \lambda_0^{-1}$. Let $(x_0, y_0) \in A$, and $x_n = x_0 + n^{-1}y_0$ for $n \geqq \lambda_0^{-1}$. Then $J_{1/n}x_n = x_0$, $A_{1/n}x_n = y_0$. Also,

$$(d/dt)\|g_n(t) - x_n\|^2 = -2\langle A_{1/n}g_n(t), F(g_n(t) - x_n) \rangle =$$

$$= -2\langle A_{1/n}g_n(t) - A_{1/n}x_n, F(g_n(t) - x_n) \rangle + 2\langle A_{1/n}x_n, F(x_n - g_n(t)) \rangle \leqq$$

$$\leqq 2\langle y_0, F(x_n - g_n(t)) \rangle.$$

Thus, as in the proof of Theorem 1. 2, we get

$$\|T(t)x - x_0\|^2 \leq \|x - x_0\|^2 + 2 \int_0^t \langle y_0, F(x_0 - T(s)x) \rangle \, ds,$$

$$\langle T(t)x - x, F(x - x_0) \rangle \leq \int_0^t \langle y_0, F(x_0 - T(s)x) \rangle \, ds,$$

$$\langle Bx, F(x - x_0) \rangle \leq \langle y_0, F(x_0 - x) \rangle,$$

$$0 \leq \langle y_0 + Bx, F(x_0 - x) \rangle.$$

Therefore $(x, -Bx) \in A$.

This proves that $B \subset -A$. The rest follows from Theorem 1. 2 and Proposition 1. 7.

REMARK: The following theorem is a strengthening of MINTY's well known theorem that if X is a Hilbert space, then every maximal accretive subset of $X \times X$ is hyper-accretive. This theorem was proved by CRANDALL and PAZY in [18].

THEOREM 1. 5: *Let X be a Hilbert space, and let C be closed and convex. Then every C-maximal accretive subset of $X \times X$ is hyper-accretive. Equivalently, if A is an accretive subset of $C \times X$, then there is a hyper-accretive subset of $C \times X$ which contains A.*

Crandall and Pazy base their proof on the following lemma, which they obtain as a special case of some theorems of BREZIS in [1]. We give a proof here which is a slight modification of an argument used by BROWDER in [10]. Browder's result in [10] is in most respects more general, but he does require C to be compact, a condition not needed in Hilbert space.

LEMMA 1. 7: *Let X be a Hilbert space, let C be closed and convex, let A be an accretive subset of $C \times X$, and let $f \in X$. Then there is an element $x_0 \in C$ such that*

$$\langle f - x_0 - y, x_0 - x \rangle \geq 0$$

for each $(x, y) \in A$.

PROOF: For each $(u, v) \in A$, let

$$N(u, v) = \{x \in X : \langle f - x - v, x - u \rangle < 0\}.$$

Then

$$N(u, v) = \{x \in X : \|x - (f + u - v)/2\| > \|f - u - v\|/2\},$$

and thus $N(u, v)$ is weakly open. Now suppose there exists no $x_0 \in C$ which satisfies the conclusion of the lemma. Then the collection of all these sets $N(u, v)$ is an open cover of C. If $(u, v) \in A$, then $C \setminus N(u, v)$ is just the intersection of C with a closed ball in X, so that $C \setminus N(u, v)$ is weakly compact. Thus there is a finite subset $\{(u_i, v_i)\}_{i=1}^n$ of A such that $\{N(u_i, v_i)\}_{i=1}^n$ is an open cover of C. Let K denote the convex hull of $\{u_i\}_{i=1}^n$, and let $N_i = K \cap N(u_i, v_i)$ for $i=1, 2, \ldots, n$. Thus K, equipped with the topology inherited from the weak topology of X, is a compact Hausdorff space, and $\{N_i\}_{i=1}^n$ is an open cover of K. Let $\{\beta_i\}_{i=1}^n$ be a partition of unity for K which is subordinate to this open cover. Define the maps p and q on K by

$$p(x) = \sum_{i=1}^n \beta_i(x) u_i,$$

$$q(x) = \sum_{i=1}^n \beta_i(x) v_i.$$

Since $p(K) \subset K$, then p has a fixed point x_0. Define λ on K by

$$\lambda(x) = \langle q(x) + x - f, \, x - p(x) \rangle.$$

Thus,

$$\lambda(x) = \sum \beta_i(x) \beta_j(x) \langle v_i + x - f, \, x - u_j \rangle = \sum \lambda_{ij}(x).$$

But,

$$\langle v_i + x - f, \, x - u_j \rangle + \langle v_j + x - f, \, x - u_i \rangle =$$

$$= \langle v_i + x - f, \, x - u_i \rangle + \langle v_j + x - f, \, x - u_j \rangle + \langle v_i - v_j, \, u_i - u_j \rangle \geqq$$

$$\geqq \langle v_i + x - f, \, x - u_i \rangle + \langle v_j + x - f, \, x - u_j \rangle.$$

Thus each $\lambda_{ij} + \lambda_{ji}$ is nonnegative (remember that β_i vanishes off N_i), and for each $x \in K$, we have $\lambda_{ii}(x) > 0$ for some i, namely an i such that $\beta_i(x) > 0$. Therefore, λ is strictly positive. However, $\lambda(x_0) = 0$, a contradiction.

PROOF OF THEOREM 1.5: Let A be a C-maximal accretive subset of $X \times X$, and let $f \in X$. Let $x_0 \in C$ be such that

$$\langle f - x_0 - y, \, x_0 - x \rangle \geqq 0$$

for all $(x, y) \in A$. Then $A \cup \{(x_0, f - x_0)\}$ is an accretive subset of $C \times X$. Therefore $(x_0, f - x_0) \in A$, and $f \in R(I + A)$.

REMARK: The following theorem is due to KOMURA [32], and a simpler proof is the sole subject of the paper [30] of KATO. We do not give a proof here; the reader is referred to [30].

THEOREM 1.6: *Let X be a Hilbert space, let C be closed and convex, and let $T \in Q(C)$. Then the domain of the infinitesimal generator of T is dense in C.*

PROOF OF THEOREM 1.4: Let A be a hyper-accretive subset of $C \times X$ such that $A \supset -B$, such an A exists by Theorem 1.5. Let $S \in Q(C)$ be as in Theorem 1.2. Then $S = T$ by Theorem 1.6, Proposition 1.5, and Theorem 1.2. Thus $B = -A^0$ by Theorem 1.3.

2. History and literature

First we want to briefly describe some recent results of CRANDALL and LIGGETT; see [14], [15], [16], which are very general and very powerful. Their definition of accretive is weaker than the definition used in Section 1, so let us give it here. *A sub-set A of $X \times X$ is said to be* accretive *if*

$$\|(x_1 + \lambda y_1) - (x_2 + \lambda y_2)\| \geqq \|x_1 - x_2\|$$

for $(x_1, y_1)(x_2, y_2)$ in A, and $\lambda > 0$. This is equivalent to

$$\langle y_1 - y_2, f \rangle \geqq 0$$

for some $f \in F(x_1 - x_2)$. A is said to be ω-accretive if $A + \omega I$ is accretive. The results of Crandall and Liggett are for ω-accretive sets, but we will describe them for the special case $\omega = 0$ in order to avoid the necessity of introducing more new terminology. Incidentally, the results of Section 1 also generalize to ω-accretive sets.

THEOREM 2. 1: *Let A be an accretive subset of $X \times X$, and let A satisfy (C_2, λ_0). Then*

$$S(t)x = \lim_{n \to \infty} (I + (t/n)A)^{(-n)}x$$

exists for each x in $\overline{D(A)}$ and defines a semi-group S belonging to $Q(\overline{D(A)})$. Also, $S(\cdot)x$ is Lipschitz continuous for each x in $D(A)$.

THEOREM 2. 2: *Let A and S be as in Theorem 2. 1, but suppose in addition that A is closed and that A satisfies (C_3, λ_0). If $x \in D(A)$, then $S(\cdot)x$ is a solution of $P(-A, x)$ if and only if $S(\cdot)x$ is differentiable a.e. Also, if $P(-A, x)$ has a solution, then the solution is $S(\cdot)x$.*

The last statement of this theorem is due to BREZIS and PAZY [5]. Crandall and Liggett have many other related results in [15] and [16], but these are the main ones.

NEUBERGER's paper [41], which gives an exponential formula for certain non-expansive nonlinear semi-groups was apparently the first on the subject. This was followed by OHARU [43], and DORROH [20], which still dealt mainly with formulas for semi-groups. These were soon followed by KATO [28], [29], KOMURA [31], [32], DORROH [21], CRANDALL and PAZY [17], [18] and OHARU [45]; it is in these papers that the material of Section 1 was largely developed. KOMURA [31] was responsible for the major step of realizing that it was necessary to consider sets or multi-valued operators in order to satisfactorily describe generators. It also turns out that a satisfactory perturbation theory can be obtained only by considering sets; see [17] and [8]. MIYADERA and OHARU have proven many theorems on convergence of semi-groups and representation of semi-groups. The papers [49] of RUTLEDGE and [51] of WEBB are also on this subject.

OHARU's paper [49] deals with semi-groups of nonlinear transformations in topological vector spaces. Other papers listed in the bibliography deal with such topics as convergence of semi-groups, formulas for semi-groups, perturbation of semi-group generators, sufficient conditions that an operator be a semi-group generator, and properties of special semi-groups. Some of these papers deal with the more difficult nonautonomous evolution equation

$$u'(t) = A(t)u(t),$$

and with evolution systems $\{U(t, s)\}$ satisfying

$$U(r, t)U(t, s) = U(r, s).$$

BIBLIOGRAPHY

[1] H. Brezis, *Inéquations variationelles associées à des opérateurs d'évolution.* Proceedings of the NATO Institute on Monotone Operators and their applications, Venice 1968.

[2] H. Brezis, *Sémi-groupes non linéaires et applications.* Symposium sur les problèmes d'évolution, Instituto Nazionale di Alta Matematica, Rome 1970.

[3] H. Brezis, *On a problem of T. Kato.* Comm. Pure Appl. Math. 24 (1971), 1—6.

[4] H. Brezis, *Propriétés régularisantes de certains sémi-groupes non linéaires.* Israel J. Math. 9 (1971), 513—534.

[5] H. Brezis and A. Pazy, *Accretive sets and differential equations in Banach spaces.* Israel J. Math. 8 (1970), 367—383.

[6] H. Brezis and A. Pazy, *Semi-groups of nonlinear contractions on convex sets.* J. Functional Analysis 6 (1970), 237—281.

[7] H. Brezis and A. Pazy, *Convergence and approximation of semi-groups of nonlinear operators in Banach spaces* (to appear).

[8] H. Brezis and A. Pazy, and M. Crandall, *Perturbations of nonlinear maximal monotone sets in Banach space.* Comm. Pure Appl. Math 23 (1970), 123—144.

[9] F. E. Browder, *Nonlinear equations of evolution and nonlinear accretive operators in Banach spaces.* Bull. Amer. Math. Soc. 73 (1967), 867—874.

[10] F. E. Browder, *Nonlinear maximal monotone operators in Banach spaces.* Math. Ann. 175 (1968), 89—113.

[11] F. E. Browder, *Nonlinear operators and nonlinear equations of evolution in Banach spaces.* Proc. Symp. in Pure Math. 18 (1969), Part II, Amer. Math. Soc. (to appear).

[12] P. L. Butzer and H. Berens, *Semi-groups of Operators and Approximation.* Grundlehren 145, Springer, Berlin 1968.

[13] M. G. Crandall, *Differential equations on convex sets.* J. Math. Soc. Japan 22 (1970), 443—455.

[14] M. G. Crandall, *Semi-groups of nonlinear transformations.* Proceedings of 1970 MRC Symposium (to appear).

[15] M. G. Crandall and T. M. Liggett, *Generation of semi-groups of nonlinear transformations on general Banach spaces.* Amer. J. Math. 93 (1971), 265—298.

[16] M. G. Crandall and T. M. Liggett, *A theorem and a counterexample in the theory of semi-groups of nonlinear transformations.* Trans. Amer. Math. Soc. (to appear).

[17] M. G. Crandall and A. Pazy, *Semi-groups of nonlinear contractions and dissipative sets.* J. Functional Analysis **3** (1969), 376—418.

[18] M. G. Crandall and A. Pazy, *On accretive sets in Banach spaces.* J. Functional Analysis **5** (1970), 204—217.

[19] J. R. Dorroh, *Semi-groups of nonlinear transformations.* Michigan Math. J. **12** (1965), 317—320.

[20] J. R. Dorroh, *Some classes of semi-groups of nonlinear transformations and their generators.* J. Math. Soc. Japan **20** (1968), 437—455.

[21] J. R. Dorroh, *A nonlinear Hille—Yosida—Phillips theorem.* J. Functional Analysis **3** (1969), 345—353.

[22] J. R. Dorroh, *A class of nonlinear evolution equations in a Banach space.* Trans. Amer. Math. Soc. **147** (1970), 65—74.

[23] J. R. Dorroh, *Semi-groups of nonlinear transformations with decreasing domain.* J. Math. Anal. Appl. **34** (1971), 396—411.

[24] J. R. Dorroh, *Local groups of differentiable transformations.* Math. Ann. **192** (1971), 243—249.

[25] J. R. Dorroh and T. F. Lin, *Markov processes with quasilinear first order forward differential equations.* Submitted to J. Math. Anal. Appl.

[26] N. Dunford and J. Schwartz, *Linear Operators.* Part I, Interscience, New York 1958.

[27] E. Hille and R. S. Phillips, *Functional Analysis and Semi-groups.* Revised ed., Amer. Math. Soc. Colloquium Publications, vol. **39**, 1957.

[28] T. Kato, *Nonlinear semi-groups and evolution equations.* J. Math. Soc. Japan **19** (1967), 508—520.

[29] T. Kato, *Accretive operators and nonlinear evolution equations in Banach spaces.* Proc. Symposium Pure Math. **18** (1969), Part I, 138—161.

[30] T. Kato, *Note on the differentiability of nonlinear semi-groups.* Proc. Symposium Pure Math. **16** (1970), 91—94.

[31] Y. Komura, *Nonlinear semi-groups in Hilbert space.* J. Math. Soc. Japan **19** (1967), 493—507.

[32] Y. Komura, *Differentiability of nonlinear semi-groups.* J. Math. Soc. Japan **21** (1969), 375—402.

[33] R. H. Martin, Jr., *The logarithmic derivative and equations of evolution in a Banach space.* J. Math. Soc. Japan **22** (1970), 411—429.

[34] R. H. Martin, Jr., *A global existence theorem for autonomous differential equations in a Banach space.* Proc. Amer. Math. Soc. **26** (1970), 307—314.

[35] G. J. Minty, *Monotone (nonlinear) operators in Hilbert space.* Duke Math. J. **29** (1962), 431—346.

[36] I. Miyadera, *Note on nonlinear contraction semi-groups.* Proc. Amer. Math. Soc. **21** (1969), 219—225.

[37] I. Miyadera, *On the convergence of nonlinear semi-groups.* Tôhoku Math. J. **21** (1969), 221—236.

[38] I. Miyadera, *On the convergence of nonlinear semi-groups* II. J. Math. Soc. Japan **21** (1969), 403—412.

[39] I. Miyadera, *Some remarks on semi-groups of nonlinear operators.* (to appear).

[40] I. Miyadera and S. Oharu, *Approximation of semi-groups of nonlinear operators.* Tôhoku Math. J. **22** (1970), 24—47.

[41] J. W. Neuberger, *An exponential formula for one-parameter semi-groups of nonlinear transformations.* J. Math. Soc. Japan **18** (1966), 154—157.

[42] J. W. Neuberger, *Product integral formulas for nonlinear expansive semi-groups and non-expansive evolution systems.* J. Math. Mech. **19** (1969), 403—409.

[43] S. Oharu, *Note on the representation of semi-groups of nonlinear operators.* Proc. Japan Acad. **42** (1966), 1149—1154.

[44] S. Oharu, *A note on the generation of nonlinear semi-groups in a locally convex space.* Proc. Japan Acad. **43** (1967), 847—851.

[45] S. Oharu, *On the generation of semi-groups of nonlinear contractions.* J. Math. Soc. Japan **22** (1970), 526—550.

[46] S. Oharu, *Nonlinear semi-groups in Banach spaces.* (to appear).

[47] B. K. Quinn, *Solutions with shocks: an example of an L_1-contractive semi-group.* Comm. Pure Math. **24** (1971), 125—132.

[48] R. T. Rockafellar, *On the virtual convexity of the domain and range of a nonlinear maximal monotone operator.* Math. Annalen **185** (1970), 81—90.

[49] D. Rutledge, *A generator for a semi-group of nonlinear transformations.* Proc. Amer. Math. Soc. **20** (1969), 491—498.

[50] J. Watanabe, *Semi-groups of nonlinear operators on closed convex sets.* Proc. Japan Acad. **45** (1969), 219—223.

[51] G. F. Webb, *Representation of semi-groups of nonlinear non-expansive transformations in Banach spaces.* J. Math. and Mech. **19** (1969), 159—170.

[52] G. F. Webb, *Product integral representation of time dependent nonlinear evolution equations in Banach spaces.* Pacific J. Math. **32** (1970), 269—281.

[53] G. F. Webb, *Nonlinear evolution equations and product integration in Banach spaces.* Trans. Amer. Math. Soc. **148** (1970), 273—282.

[54] G. F. Webb, *Nonlinear evolution equations and product stable operators in Banach spaces.* Trans. Amer. Math. Soc. **155** (1971), 409—426.

[55] B. Calvert, *Nonlinear evolution equations in Banach lattices.* Bull. Amer. Math. Soc. **76** (1970), 845—850.

[56] B. Calvert, *Semigroups in an ordered Banach space.* J. Math. Soc. Japan **23** (1971), 311—319.

[57] B. Calvert and K. Gustafson, *Multiplicative perturbation of nonlinear m-accretive operators.* J. Functional Anal., **10** (1972), 149—158.

Singular Perturbations of Semi-Group Generators

By

IRVING SEGAL

DEPT. OF MATH.
MASSACHUSETTS INSTITUTE OF TECHNOLOGY
CAMBRIDGE

This is a talk about approximation *and* operators; more specifically, about approximation of semi-groups. For several reasons, the exposition will be in terms of self-adjoint operators in Hilbert space, although the central ideas are applicable to fairly general semi-groups. The self-adjoint case is both illustrative, and of prime importance in the theory of quantum fields, which is the source of many novel problems involving singular perturbations; and it has been developed much more than the general case.

If H is a given self-adjoint operator in a Hilbert space $\mathbf{K}$, and if V is another given self-adjoint operator in $\mathbf{K}$, the additive perturbation of H by V, — formally, the sum of H and V, suitably defined, — has been studied extensively over the past several decades. The problem of determining another self-adjoint operator H' which may appropriately be identified with this perturbation is important in many connections, and is naturally an essential preliminary to the study of the operator H', especially its spectral properties. There is a well-developed theory of such matters, largely originated by RELLICH, with important contributions by KATO, BIRMAN, and HEINZ, among other significant workers in the field. The typical assumption in this work is that V is suitably dominated by H; specifically, that

$$(*) \qquad \|Vx\| \leqq a\,\|Hx\| + b\,\|x\|,$$

where a and b are constants with $a < 1$. The essential results of the theory still hold when $a = 1$, but the theory is best possible of its type in the sense that if $a > 1$, examples show that the desired conclusions (e.g. the essential self-adjointness of $H + V$) may be false, in an apparently irremediable fashion.

Actually, from the point of view of applications to quantum field theory, the condition $(*)$ is hopelessly strong. The perturbations V which arise typically do not satisfy $(*)$ for any values of a and b. In the case of a two-dimensional space-time, for 'scalar' or similar fields, they satisfy an inequality of the form

$$(**) \qquad \|Vx\| \leqq a\,\|H^r x\| + b\,\|x\|,$$

where r depends on V, and may be arbitrarily large; such an inequality is however of very little use. In the physical case of a four-dimensional space-time, the typical perturbations V are, even in the simplest non-trivial cases, not even operators, except in a generalized sense, in which the inequality $(**)$ has no meaning. About the only simple domination one has comes from the formulation of V as a sesquilinear form, and asserts that

$$(***) \qquad |\langle Vx, y\rangle| \leq a\,\|(bI+H)^r x\|\,\|(bI+H)^r y\|$$

for suitable constants a, b, and r (it is no loss of generality to take $b=1$, since $H \geqq 0$). From this it seems fairly evident that nothing like classical perturbation theory is likely to have a significant direct impact on such problems of quantum field theory.

One might be tempted to jump to the conclusion that quantum field theory is a highly special subject, and that one can not reasonably expect to treat key technical aspects by means of a general theory. Thus the simple perturbation $\Delta + V$ on R^n has an extensive theory going far beyond what may be inferred from classical abstract perturbation theory; and one might naturally contemplate constructing a similar theory for the case of quantum fields. There are however several initially visible difficulties in doing this.

First, quantum field theory (of so-called Bose—Einstein fields) corresponds roughly to the case $n = \infty$; more exactly, to function theory in a Hilbert space. While real (and even complex) analysis on a Hilbert space is rather more cogent than one might anticipate, it lacks some important features of analysis on R^n; for example, the relevant measure is not fully countably additive, unless one alters the space in an otherwise inconvenient way. Moreover, the perturbations V in question are not semi-bounded, which can be troublesome even in R^n; the Soboleff inequalities appear to evaporate as $n \to \infty$; etc.

Second, analysis in a Hilbert space covers only the case of Bose—Einstein quantum fields (photons, mesons, and the like); Fermi—Dirac quantum fields (electrons, protons, etc.) are of quite a different character. In the physically relatively best-established quantum field theory, that of quantum electrodynamics, the direct product of Bose—Einstein and Fermi—Dirac fields intervenes. It would be inconvenient and seem unnatural to have special theories applicable only to one type of field or the other. Consequently, an extension of the classical local theory of elliptic partial differential equations to Hilbert space is not only sure to be fraught with serious difficulties, but even if quite successful, might well be limited in its physical application, — although it should be of considerable mathematical interest.

Finally, and perhaps most crucially, a highly intricate, special treatment of quantum field perturbations does violence to the origin of the subject, and might well be an irrelevant technical tour-de-force. What has made non-linear quantum field theory so compelling and interesting, despite its 'divergences' and extremely limited empirical applicability as yet, is primarily its extraordinary combination of logical simplicity, intuitive appeal, and economy of formulation. To show that such a theory 'really exists' in an effective and appropriate sense, it is essential to use mathematical methods which are consonant with the raison-d'etre of the theory. In the case of linear quantum field theory, there have been extensive mathematical developments of this nature, and there is no special reason to doubt that the non-linear (presently largely heuristic) theory will ultimately yield to a similar mathematical approach. In any event, an approach which is special, unintuitive, and applicable only to relatively mild singularities, does nothing to gainsay the still tenable (though I believe incorrect) view that non-linear quantum field theory is basically a specious if persistent illusion.

A rational way out of this dilemma is to use, within a general mathematical framework, more of the structure given in connection with typical concrete perturbations, and which may be neglected in the classical framework of given operators in a fixed Banach space. For example, in the case of the perturbation $\Delta + V$, V is generally a simple multiplication operator, or simple variant thereof; and the given operators have effective actions not only in L_2 but in many other function spaces. This observation, together with an examination of the concrete situation in quantum field theory, leads to an effective scheme for treating quite singular perturbations, based on the following main features.

First, a scale of spaces within the Hilbert space is employed. Specifically, a *calibrated Hilbert space* is defined as a Hilbert space $\mathbf{K}$, together with a mapping $p \to \| \cdot \|_p$ from an interval of the form $[2, p_0)$ to non-negative functionals on $\mathbf{K}$, which functionals are norms of possibly infinite values, having the properties: (1) $\| \cdot \|_2$ is the given norm in $\mathbf{K}$; (2) $\|u\|_p$ is monotone increasing as a function of p, for all u in $\mathbf{K}$; (3) setting $\mathbf{K}_p = [u \in \mathbf{K} : \|u\|_p < \infty]$, $\bigcap_p \mathbf{K}_p$ is dense in $\mathbf{K}$; (4) if $u_n \to u$ in $\mathbf{K}$, then $\|u\|_p \leq \sup_n \|u_n\|_p$, for all p.

An obvious example is provided by the L_p spaces over a given probability measure space, and a calibration technically effective in treating Bose—Einstein quantum fields is essentially a special case of this one. A more general example consists of the L_p spaces over a probability Hilbert algebra, which may be defined briefly as the completions of the spaces of all operators A in a given operator ring $\mathbf{A}$ (in the MURRAY—VON NEUMANN sense), on which is given a central state, — a normalized positive linear functional E such that $E(A^*A) = E(AA^*)$ for all $A \in \mathbf{A}$, — relative to the norm: $\|A\|_p = (E(A^*A)^{p/2})^{1/p}$. This example permits an effective subsumption of the Fermi—Dirac case; $\mathbf{A}$ is then a so-called approximately finite factor, and E is the trace. (The classical case of a probability measure space corresponds to the case in which $\mathbf{A}$ is abelian.)

Now, many concrete semi-groups are of 'finite type' relative to a suitable scale of spaces, in a sense which will be made precise as follows. The continuous semi-group $[V(t):t\geqq 0]$ is of type (P, α, a, t_0), relative to the given calibration in $\mathbf{K}$, — where P is a given subinterval of $[2, p_0)$, and α, a, and t_0 are given real numbers, t_0 being positive, — in case

$$\|V(t)\|_{p, pe^{\alpha t}} \leqq e^{at} \qquad (t \in [0, t_0], \ p \in P).$$

It is of type (P, α, a) if there exists $t_0 > 0$ such that it is of type (P, α, a, t_0). The key parameter here is α; as a function of the generator of the semi-group, it is concave (i.e. its negative is convex), typically. Thus the sum of operators (generators of semi-groups) of finite type is again such, typically. In practice, in connection with the perturbation of H by V, the unperturbed operator H is commonly of relatively 'good' type; and while V is frequently of 'bad' type, as long as it is of finite type, its badness is subject to compensation by the goodness of H.

To implement this idea, a relevant topology in the space of semi-groups is needed. In the case of the self-adjoint operators in $\mathbf{K}$, there is a unique simple notion: $A_n \to A$ if and only if $e^{itA_n} \to e^{itA}$ in the strong (or equivalently weak) operator topology for all $t \in R^1$ is clearly the right notion from the viewpoint of quantum mechanics. This notion, treated in a different form by RELLICH, coincides with generalized strong convergence as defined by KATO, and has also been considered recently by KALLMAN and myself. There is a great variety of comparable notions, all of which are, reassuringly, equivalent. In particular, it is the same to require that $f(A_n) \to f(A)$ strongly for all bounded and continuous functions f; or, in the case that the A_n are uniformly bounded from below, along with A, that $e^{-tA_n} \to e^{-tA}$ for all $t > 0$ (or one $t > 0$). Entirely similar results hold for the resolvents, which however are more technical than inevitable in the consideration of temporal evolution; the exponential functions, although transcendental, are physically more natural, and have the technical advantage of making available the use of the DUHAMEL and LIE—TROTTER formulas. These are easily proved in the case of bounded perturbations, — the only case needed in the proofs, which go by approximation, — and may be stated as follows, B being an arbitrary bounded self-adjoint operator.

Duhamel: $e^{it(A+B)} = e^{itA} + \int e^{i(t-s)A} B e^{is(A+B)} ds$ (integration in strong operator topology).

Lie-Trotter: $e^{it(A+B)} = \lim_{n \to \infty} (e^{itA/n} e^{itB/n})^n$.

It is also useful to note that if $A_n \to A$, then $A_n + B \to A + B$; and that for semi-bounded operators A, there are similar equations for the corresponding real semi-groups.

A general and objective sense in which two given self-adjoint operators H and V determine a 'perturbed' operator H', intuitively the self-adjoint operator corresponding to their sum is as follows. Let $\mathbf{S}$ be a given set of sequences $\{V_n\}$

of bounded self-adjoint operators, each sequence being convergent to V; $\mathbf{S}$ is called an *approximating set* for V. The 'moderated perturbation of H by V', — relative to the given set $\mathbf{S}$, — is defined as the limit $H' = \lim\limits_n (H + V_n)$, if this exists and is unique, for all $\{V_n\} \in \mathbf{S}$, and is denoted $H \mathbin{\tilde{+}} V$. This definition extends readily to the case in which V is merely a generalized operator, and will be useful in this connection later.

It is impossible to summarize briefly the theory of perturbations in a calibrated Hilbert space. Some of the central results may be indicated as follows.

THEOREM 1. *Let H be a self-adjoint operator of type (P, α, a) in the calibrated Hilbert space $\mathbf{K}$. Let $\mathbf{C}$ denote the set of all self-adjoint operators W in $\mathbf{K}$ of type (P, β, b), β and b being given constants, and such that $\|W\|_{q_0, 2} < \infty$, q_0 being a given constant in P; let $\mathbf{C}_0$ denote the set of all bounded elements of $\mathbf{C}$. Let $\mathbf{S}$ be a given approximating set for V, consisting of sequences $\{V_n\}$ such that $V_n \in \mathbf{C}_0$, and $\|V_n - V\|_{q_0, 2} \to 0$. Then*

1. *$H \mathbin{\tilde{+}} V$ exists, and is of type $(I, \alpha + \beta, a + b)$ for a suitable interval I;*

2. *If V' and $\mathbf{S}'$ satisfy the same conditions as V and $\mathbf{S}$, with β and b replaced by β' and b', and if $\alpha + \beta + \beta' > 0$, then the Duhamel formula holds for $e^{-t(H \tilde{+} V)} - e^{-t(H \tilde{+} V')}$ $(t > 0)$.*

If in addition, $V + V'$ is essentially self-adjoint and $V_n + V_n' \to (V + V')^*$ for $\{V_n\} \in S$ and $\{V_n'\} \in S'$, then

$$(H \mathbin{\tilde{+}} V) \mathbin{\tilde{+}} V' = H \mathbin{\tilde{+}} (V + V')^*.$$

3. *If $\alpha + \beta > 0$, and if H is of type (P, α, a, ∞), then every entire vector w for $H \mathbin{\tilde{+}} V$ is in $\mathbf{D}(H) \cap \mathbf{D}(V)$, and $(H \mathbin{\tilde{+}} V)w = Hw + Vw$.*

Some further aspects which may be described briefly are as follows. The associativity in 2) generally fails for the operation $(H, V) \to$ closure of $H + V$, defined on pairs (H, V) such that $H + V$ is essentially self-adjoint, even when all terms on both sides are well-defined. It is not known whether $H + V$ necessarily has a self-adjoint extension, under the hypotheses of Theorem 1; by 3), it can fail to do so only in the case $\alpha + \beta = 0$. When this is not the case, $H + V$ is necessarily extended by $H \mathbin{\tilde{+}} V$. A case of particular interest in connection with quantum-field-theoretic applications, in which the hypotheses of the theorem are relatively easily established, is that in which $\mathbf{K}$ is a Hilbert algebra with unit of unit norm, calibrated by the L_p-norm (the example cited earlier). A quite familiar situation to which the theorem is applicable is that of perturbations of the form $-\Delta + V$ in $L_2(R^n)$, V being a given measurable function. When this problem is transformed into a probability space question by the use of the Gaussian probability measure in R^n in place of Lebesgue measure, it follows that the operator in question is essentially self-adjoint under quite general conditions on V, which conditions are essentially independent of n. They are to the effect that W be very mildly dominated near ∞, and be predominantly non-negative in a rather weak sense.

Having established the perturbed self-adjoint operator H', the next question of importance for applications is that of the perturbation of the spectrum. It is clear from known examples that additional structure must be imposed to obtain cogent results on this matter. The PERRON—FROBENIUS—KREIN theory of positive operators, now much extended by GROSS [1, 2], in particular so that non-commutative Hilbert algebras are covered by the theory, is an appropriate tool for treating the specific perturbations described earlier. A classical result in this theory concerns the one-dimensionality of the highest eigenspace of a positivity-preserving operator; this is entirely invariant under perturbations when suitably formulated in relation to the present theory, where it appears as an assertion about a proper lowest vector of the perturbed operator H'. There is a general theory of this, some impression of which may be given by a fuller specification of the cited one-dimensionality result.

A self-adjoint operator H in $L_2(M)$, M being a given probability space, may be called *indecomposable* in case e^{-tH} maps non-negative functions on M into non-negative functions, for all $t>0$, and if it has the property that $\langle e^{-tH}u, u'\rangle=0$ for all $t>0$ and certain non-negative elements u and u' in $L_2(M)$, then either $u=0$ or $u'=0$. It then follows from Theorem 1 that if H is indecomposable, so also is $H+V$. A simple classical argument shows that if an indecomposable operator has a proper lowest vector, the corresponding invariant subspace is one-dimensional, and the vector may be chosen to be positive a.e. An immediate consequence is the existence, and essential unicity and positivity, of a proper lowest vector for $-\Delta+V$ in $L_2(R^n)$, under the quite general conditions on V relevant to the theorem. With additional structure, existence can be shown in a general context applicable to quantum fields. The more refined theory of the perturbation of the spectrum, — e.g. the presence and size of gaps; the asymptotic dependence of the proper lowest vector on parameters on which V depends linearly, — is largely open, and while of much physical interest, will perhaps be lacking in cogency unless further appropriate mathematical structure is imposed. Indeed, even in the special case of perturbations in $L_2(R^1)$, of the special form $-\Delta+p(x)$, p being a given polynomial of even degree with leading coefficient 1, the asymptotic dependence of the proper lowest vector of this operator on the other coefficients of p, as they become arbitrarily large, is presently quite unknown (this is the simplest non-trivial case of a problem arising in quantum field theory; see [5, II]).

While the foregoing theory can effectively treat perturbations which are much more singular than those of classical perturbation theory, it is still too limited to cope with those of four-dimensional quantum field theory, in which case, as noted earlier, V is merely a generalized operator (or equivalently, a sesqui-linear form). Classical perturbation theory is applicable to perturbations by forms which are sufficiently strongly dominated by the unperturbed operator; but again the requirements are far too stringent to be applicable to any non-trivial instances in quantum field theory. Some idea of the problem may perhaps be derived from the consideration of an analogous, although incomparably simpler, and in part fully resoluble situation in R^1.

At first glance, $-i(d/dx)+\delta$ does not appear appropriately interpretable as a self-adjoint operator in $L_2(R^1)$. The obvious approach is to begin with the corresponding form on $C_0^\infty(R^1)$. However, the operator in L_2 derivable from this form has domain excluding all the functions of key concern here, namely those which do not vanish at zero. Moreover, non-negativity (or sectoriality) considerations are inapplicable. Nevertheless, there is a simple, natural, unique interpretation as a self-adjoint operator for this singular perturbation, in accordance with

THEOREM 2. *Let A_0 denote the self-adjoint generator of the one-parameter group $f(x) \to f(x+t)$ on $L_2(R^1)$, and M_V denote the operation of multiplication by the real bounded measurable function V on R^1.*

There then exists a unique mapping A from the real measurable functions W on R^1 to self-adjoint operators $A(W)$ in $L_2(R^1)$, such that:

a. If W is absolutely continuous with bounded derivative W', then $A(W) = A_0 + M_{W'}$;

b. The map $W \to A(W)$ is continuous (with the topology of sequential star-convergence for the measurable functions, and the earlier given one for the self-adjoint operators).

This can be proved by exhibition of the corresponding one-parameter unitary groups in closed form (from which it is visible that the domain of the self-adjoint operator corresponding to $i(d/dx)+\delta$ contains no function which is non-vanishing and continuous at zero). A more suggestive style of proof may however be based on the use of scattering theory ideas along the lines indicated in [6]; in essence, this consists in the use, in an abstract form, of the method of variation of constants. A key observation is that although V is merely a form, $\int_I e^{itH} V e^{-itH}\, dt$ is for any interval I a self-adjoint operator. Thereby the problem of forming the one-parameter group $e^{it(H+V)}$ is reducible to the solution of an abstract differential equation of the form

$$du(t)=dF(t)u(t),$$

where $F(t)$ is a given self-adjoint operator-valued function on R^1.

In the Hilbert algebra special case of Theorem 1, the approximating sequences V_n are naturally taken to be of the form: $V_n=P_n V P_n$, where $P_n=f_n(V)$, $\{f_n\}$ being a uniformly bounded sequence of real functions converging pointwise to 1 on R^1. If V is merely a generalized operator, $f_n(V)$ is undefined, and this approach is inapplicable. However, there is a natural similar way to form approximating sequences in the case of forms satisfying the condition $(\ast\ast\ast)$ given earlier, namely to take $P_n = =f_n(H)$, where each f_n vanishes to sufficiently high order near ∞. This leads to the

DEFINITION. Let H be a given self-adjoint operator in the Hilbert space $\mathbf{K}$, and let V be a given continuous hermitian sesqui-linear form on the space $\mathbf{D}_\infty(H)$ of all infinitely differentiable vectors for H. The *energetically-moderated* perturbation of H by V is said to exist in case $\lim_n (H+f_n(H)Vf_n(H))$ exists as a self-adjoint operator and is unique, for some given class of sequences $\{f_n\}$ of the type just indicated.

The term 'energetically-moderated' derives from the fact that in many applications, H represents the 'energy' operator, and the approximating operators $V_n = =f_n(H)Vf_n(H)$ are sometimes said to have been derived from V by an 'energy' (or 'momentum') 'cutoff'. This type of perturbation is related to the likewise physically motivated notion of 'adiabatic perturbation' described in [7]. I could now treat, via either of these methods, the perturbation of a power of the laplacian on R^n by an arbitrary distribution of compact support (the power being sufficiently high in relation to the order of the distribution), but shall not do so here.

The work described here was supported in part by the NSF.

REFERENCES

[1] L. Gross, *A non-commutative extension of the Perron—Frobenius theorem.* Bull. Amer. Math. Soc. **77** (1971), 343—347.

[2] L. Gross, *Existence and uniqueness of physical ground states.* (To appear in J. Functional Analysis.)

[3] R. R. Kallman, Jr., *Groups of inner automorphisms of von Neumann algebras.* J. Functional Analysis **7** (1971), 43—60.

[4] T. Kato, *Perturbation theory for linear operators.* Springer, Berlin 1966.

[5] I. Segal, *Construction of nonlinear local quantum processes.* I, Ann. of Math. **92** (1970), 462—481; II (to appear in Invent. Math.)

[6] I. Segal, *Local non-commutative analysis,* in Problems of Analysis, Princeton University Press 1970, 111—130.

[7] I. Segal, Adress at Nice International Congress (1970).

Cyclic Vectors and Commutants

By

BÉLA SZ.-NAGY

JÓZSEF ATTILA TUDOMÁNYEGYETEM
BOLYAI INTÉZETE
SZEGED

1. For a (linear, bounded) operator T on a Banach space X denote by $(T)'$ the commutant of T, i.e. the set of operators on X which commute with T.

If X is finite dimensional then from the Jordan normal form of the matrix of T one can deduce a parametric representation for the operators in $(T)'$. From this it follows easily that $(T)'$ is abelian if and only if the Jordan blocks of T all belong to different eigenvalues of T; this condition in turn is equivalent to the condition that T has a cyclic vector. (A vector x is cyclic for T if $x, Tx, T^2x, \ldots$ span X.)

On the other hand, it is obvious that for any operator T on a reflexive (thus in particular on a finite dimensional) space X, $(T)'$ is abelian if and only if $(T^*)'$ is so.

Hence for T on a finite dimensional Banach space the following three conditions are equivalent:

(i) T has a cyclic vector in X;

(i$_*$) T^* has a cyclic vector in X^*;

(ii) $(T)'$ is abelian.

2. This equivalence breaks down in general for operators on infinite dimensional Banach spaces, and even for operators on an infinite dimensional, separable Hilbert space.

To this end consider first the operator $T = S \oplus S$, where S denotes the simple unilateral shift on the space l^2, viz.

$$S\langle \xi_0, \xi_1, \ldots \rangle = \langle 0, \xi_0, \xi_1, \ldots \rangle.$$

D. E. SARASON observed (see [3], Problem 126) that T^* has cyclic vectors whereas T has none. Moreover, $(T)'$ is not abelian. As a matter of fact, no operator of the form $T = A \oplus A$ (with A acting on a space of dimension ≥ 1) has an abelian commutant: the operators $\begin{bmatrix} 0 & I \\ 0 & 0 \end{bmatrix}$ and $\begin{bmatrix} 0 & 0 \\ 0 & I \end{bmatrix}$ belong to $(T)'$, but do not commute. This counter-example shows that in general (i$_*$) does not imply (i) or (ii) (and equivalently, (i) does not imply (i$_*$) or (ii)). However, it is still an open question whether (i) plus (i$_*$) imply (ii).

That (ii) does not imply (i) or (i$_*$) follows from another recent counter-example, due to J. A. DEDDENS [2], which we indicate in a simplified form; this counter-example answers in the negative a question formulated in [9]. Let H^2 denote the Hardy—Hilbert space for the unit circle, and for any complex number a denote by M_a the operator on H^2 defined by $(M_a f)(z) = (z-a)f(z)$ (z being the variable point on the unit circle).

First we prove that if $a_1 \neq a_2$ then the only operator A on H^2 satisfying the equation

$$AM_{a_1} = M_{a_2}A,$$

or equivalently, the equation

(1) $$AM_0 = M_a A \qquad \text{(for } a = a_2 - a_1 \neq 0\text{)},$$

is $A = 0$.[1])

To this end observe that $a \neq 0$ implies that $|e^{it} - a| \geq q > 1$ on some segment $t_0 \leq t \leq t_1$ of positive length. Set $A1 = u$; (1) implies by recurrence $Az^n = (z-a)^n u$ for $n \geq 0$. Hence

$$\|A\|^2 \geq \|Az^n\|^2 = \|(z-a)^n u\|^2 \geq q^{2n} \int_{t_0}^{t_1} |u(e^{it})|^2 \frac{dt}{2\pi} \qquad (n \geq 0).$$

This forces $u(e^{it})$ to equal 0 a.e. on (t_0, t_1); since $u \in H^2$ this implies that $u \equiv 0$. So we have $Az^n = 0$ for $n \geq 0$. As the set $\{z^n\}_0^\infty$ spans H^2 we conclude that $A = 0$ as asserted.

From what we just proved we infer that

$$(M_0 \oplus M_a)' = (M_0)' \oplus (M_a)' \qquad \text{(for } a \neq 0\text{)}.$$

Now $(M_0)'$ is abelian[2]), and obviously $(M_a)' = (M_0)'$, so we conclude that $(M_0 \oplus M_a)'$ is abelian. Choose $|a| < 1$, $a \neq 0$, and set $T = M_0 \oplus M_a$. Then T has no cyclic vector. Indeed, if $\langle f, g \rangle$ is any candidate for a cyclic vector then choose the numbers α, β such that $\langle \alpha, \beta \rangle$ be a non-zero vector orthogonal to $\langle f(0), g(a) \rangle$ in the usual two-dimensional complex inner-product space. Then $\langle \alpha, \beta/(1-\bar{a}z) \rangle$ is orthogonal in $H^2 \oplus H^2$ to $T^n \langle f, g \rangle$ for $n = 0, 1, \ldots$; indeed we have

$$\int_0^{2\pi} \left[e^{int} f(e^{it}) \bar{\alpha} + (e^{it} - a)^n g(e^{it}) \frac{\bar{\beta}}{1 - ae^{-it}} \right] \frac{dt}{2\pi} =$$

$$= [z^n f(z) \bar{\alpha}]_{z=0} + [(z-a)^n g(z) \bar{\beta}]_{z=a} = \begin{cases} f(0)\bar{\alpha} + g(a)\bar{\beta} = 0 & (n = 0) \\ 0 & (n \geq 1). \end{cases}$$

---

[1]) I am indepted for the following simple proof to Dr. J. Szűcs.

[2]) In fact $(M_0)'$ consists of the operators of multiplication by functions $\varphi \in H^\infty$.

Hence the cyclic candidate fails. Thus if $a \neq 0$, $|a| < 1$, then $T = M_0 \oplus M_a$ satisfies (ii) but not (i).

3. In view of these counter-examples it is of interest to exhibit possibly large classes of operators on Hilbert space $\mathfrak{H}$ for which the conditions (i), (i$_*$), and (ii) do imply each other.

For a *normal* operator N, (i) implies (i$_*$) and conversely; indeed N and N^* have the *same* cyclic vectors. This follows from the fact that $A = N^{*m} - \sum_0^n c_k N^k$ $(m, n = 0, 1, \ldots; c_k$ complex numbers) is then normal, and hence $\|Ax\| = \|A^*x\|$ for every $x \in \mathfrak{H}$, i.e.

$$\left\| N^{*m} x - \sum_0^n c_k N^k x \right\| = \left\| N^m x - \sum_0^n \bar{c}_k N^{*k} x \right\|.$$

For normal N, (i) also implies (ii). Let us make use, to this effect, of the direct integral representation of the (separable) Hilbert space $\mathfrak{H}$, associated with the normal operator N, i.e. let

$$\mathfrak{H} = \int_\sigma^\oplus \mathfrak{H}(\lambda)\, d\mu_\lambda \quad \text{with} \quad N^{*k} N^h \{x(\lambda)\} = \{\bar{\lambda}^k \lambda^h x(\lambda)\},$$

μ being a non-negative, finite, regular measure on the spectrum σ of N. If $x = \{x(\lambda)\}$ is cyclic for N, i.e. if $\{\lambda^n x(\lambda)\}_{n \geq 0}$ span $\mathfrak{H}$, then the values $\lambda^n x(\lambda)$ $(n \geq 0)$ span $\mathfrak{H}(\lambda)$ for μ-almost every fixed point λ in σ. Hence we have dim $\mathfrak{H}(\lambda) \leq 1$ μ-a.e. Now consider any two operators, say A_1 and A_2, which commute with N. By Fuglede's theorem they also commute with N^*, and hence are decomposable:

$$A_i = \int_\sigma^\oplus A_i(\lambda)\, d\mu_\lambda \qquad (i = 1, 2).$$

As dim $\mathfrak{H}(\lambda) \leq 1$ μ-a.e., the component operators $A_1(\lambda)$, $A_2(\lambda)$ commute μ-a.e., therefore A_1, A_2 also commute. Thus, for N, (i) implies (ii).

If we assume (ii) instead of (i) for N then we again infer that dim $\mathfrak{H}(\lambda) \leq 1$ μ-a.e. For in the contrary case $\mathfrak{H}(\lambda)$ has constant dimension ≥ 2 on some set σ_1 of positive μ-measure, and hence we may suppose $\mathfrak{H}(\lambda) \equiv \mathfrak{H}_1$ on σ_1; dim $\mathfrak{H}_1 \geq 2$. Choose two non-commuting operators on $\mathfrak{H}_1$, say B_1 and B_2, and set $A_i(\lambda) = B_i$ on σ_1 and $= 0$ elsewhere $(i = 1, 2)$. The operators $A_i = \int_\sigma^\oplus A_i(\lambda)\, d\mu_\lambda$ $(i = 1, 2)$ do not commute with each other, although both commute with N. Thus (ii) forces $\mathfrak{H}(\lambda)$ to have dimension ≤ 1 μ-a.e. Therefore we have to do in fact with a representation of N as multiplication by the function λ on the space $L_\mu^2(\sigma)$. Now this space is spanned by the functions $\bar{\lambda}^k \lambda^h$ $(k, h = 0, 1, \ldots; \lambda$ on σ), and hence — choosing for x_0 the vector represented by the constant function 1 — the vectors $N^{*k}N^h x_0$ $(k, h = 0, 1, \ldots)$ span $\mathfrak{H}$.

In case N is selfadjoint, this vector x_0 is thus cyclic for N. In case N is unitary, $N = U = \int_0^{2\pi} e^{it}\, dE_t$, we only have that the vectors $U^n x_0$ $(n = 0, \pm 1, \pm 2, \ldots)$ span $\mathfrak{H}$. It follows then by application of the Szegő—Kolmogorov—Kreĭn approximation theorem ([4], p. 49) that the vector

$$x = \int_0^{2\pi} e^{-1/t}\, dE_t x_0$$

is cyclic for U, i.e.

$$\bigvee_{n=0}^{\infty} U^n x = \bigvee_{m=-\infty}^{\infty} U^m x_0 (= \mathfrak{H}).$$

A more involved construction (see BRAM [1]) leads, in the general case too, to a vector x such that

$$\bigvee_{n=0}^{\infty} N^n x = \bigvee_{k,\, h=0}^{\infty} N^{*k} N^h x_0 (= \mathfrak{H}).$$

Hence, for any normal operator on a separable Hilbert space, conditions (i), (i$_*$), *and* (ii) *are equivalent.*

4. Another class of operators on Hilbert space for which the equivalence of conditions (i), (i$_*$), and (ii) was completely established is the class C_0 (for definitions, see [5] or [6]). For these operators a kind of "Jordan model theory" was elaborated by Sz.-NAGY and FOIAŞ [7]—[9]. Although the results show striking analogies to some aspects of matrix theory the methods are entirely different.

Recall (e.g. from [5]) that a standard example $S(m)$ of an operator of class C_0, with a given inner function m as minimal function, is defined on the space

$$\mathfrak{H}(m) = H^2 \ominus mH^2$$

by

$$S(m)u = P_{\mathfrak{H}(m)}(zu) \qquad (u = u(z) \in \mathfrak{H}(m)),$$

where $P_{\mathfrak{H}(m)}$ denotes orthogonal projection from H^2 onto its subspace $\mathfrak{H}^2(m)$.

A result concerning our problem is the following:

For an operator T of class C_0, conditions (i), (i$_*$), *and* (ii) *are equivalent. Moreover, each of these conditions is equivalent to the condition:*

(s) *T is quasi-similar to an operator $S(m)$, notably to that one with $m = m_T$.*

Recall that two operators, say T' on $\mathfrak{H}'$ and T'' on $\mathfrak{H}''$, are called *quasi-similar* if there exist operators $X : \mathfrak{H}' \to \mathfrak{H}''$ and $Y : \mathfrak{H}'' \to \mathfrak{H}'$, such that X, X^*, Y, Y^* have zero kernels, and $T''X = XT'$, $T'Y = YT''$.

Define *multiplicity* μ_T of an operator T on $\mathfrak{H}$ as the minimal cardinality of a set $s \subset \mathfrak{H}$ such that $s, Ts, T^2s, \ldots$ span $\mathfrak{H}$. (Thus cyclicity means $\mu_T = 1$.)

The above result can be extended as follows:

For every operator T of class C_0 we have $\mu_T = \mu_{T^*}$. Moreover, if $\mu_T = K < \infty$, then there exists a unique sequence of non-constant inner functions $m_1, m_2, \ldots, m_K$ such that m_i is a divisor of m_{i-1} for $i = 2, \ldots, K$, and T is quasi-similar to

$$S(m_1) \oplus S(m_2) \oplus \cdots \oplus S(m_K).$$

Let us add that for T of class C_0 with $\mu_T < \infty$ the second commutant $(T)''$ is always abelian, [8]—[9]. Whether the restriction $\mu_T < \infty$ is necessary is not yet known.

5. One more class of operators for which our problem was investigated is the class of C_1., i.e. the class of contraction operators on Hilbert space, such that $T^n x \to 0$ $(n \to \infty)$ for $x = 0$ only.

While C_0 generalizes to some extent the class of operators of finite rank, C_1. generalizes the class of isometric (and in particular unitary) operators.

The following theorem was recently obtained in SZ.-NAGY—FOIAŞ [10], [11]:

For T of class C_1. condition (i) implies conditions (i_*) and (ii) in each of the following two cases:

   α) if there is a point inside the unit circle which is not an eigenvalue of T^*, and
   β) if the unitary part of T has absolutely continuous spectrum.

Note that the simple unilateral shift operator is of class C_1., has cyclic vector, and satisfies *β)* but not *α)*.

6. A contraction T on $\mathfrak{H}$ is called a *weak* one if its spectrum does not cover the unit disc and $I - T^*T$ has finite trace. If T is, moreover, completely non-unitary, then there exist two hyperinvariant subspaces for T, $\mathfrak{H}_0$ and $\mathfrak{H}_1$, such that $\mathfrak{H}_0 \cap \mathfrak{H}_1 = \{0\}$, $\mathfrak{H}_0 \vee \mathfrak{H}_1 = \mathfrak{H}$, and $T|\mathfrak{H}_0$ is of class C_0 while $T|\mathfrak{H}_1$ is of class C_1. (indeed of class C_{11}); see [6]. From the result referred to in sections 4 and 5 it follows by this decomposition that *condition* (i) *implies conditions* (i_*) *and* (ii) *for every completely non-unitary weak contraction.*

REFERENCES

[1] J. Bram, *Subnormal operators.* Duke Math. J., **22** (1955), 75—94.

[2] J. A. Deddens, *Intertwining analytic Toeplitz operators.* (preprint).

[3] P. R. Halmos, *A Hilbert space problem book.* Van Nostrand, Princeton 1967.

[4] K. Hoffman, *Banach spaces of analytic functions.* Englewood Cliffs, N. J. 1962.

[5] B. Sz.-Nagy, Hilbertraum-Operatoren der Klasse C_0, *Abstract Spaces and Approximation.* Birkhäuser, Basel—Stuttgart 1968, 72—81.

[6] B. Sz.-Nagy and C. Foiaş, *Harmonic analysis of operators on Hilbert space.* North-Holland Publishing Co., Amsterdam, Akadémiai Kiadó, Budapest 1970.

[7] B. Sz.-Nagy et C. Foiaş, *Opérateurs sans multiplicité.* Acta Sci. Math. (Szeged) **30** (1969), 1—18.

[8] B. Sz.-Nagy et C. Foiaş, *Modèle de Jordan pour une classe d'opérateurs de l'espace de Hilbert.* ibidem **31** (1970), 91—115.

[9] B. Sz.-Nagy et C. Foiaş, *Compléments à l'étude des opérateurs de classe C_0.* ibidem **31** (1970), 287—296; partie II (to appear).

[10] B. Sz.-Nagy and C. Foiaş, *The "Lifting Theorem" for intertwining operators and some new applications.* Indiana Univ. Math. J. **20** (1971), 901—904.

[11] B. Sz.-Nagy et C. Foiaş, *Vecteurs cycliques et commutativité des commutants.* ibidem **32** (1971), 177—183.

Some Applications of Operator Valued Analytic Functions of Two Complex Variables

By

JOEL D. PINCUS*

DEPT. OF MATHEMATICS
STATE UNIVERSITY OF NEW YORK
STONY BROOK

This survey is meant to illustrate the main ideas of the theory which arises from a certain association of analytic operator valued functions of two complex variables with pairs of operators related by a commutator identity.

No attempt is made to give many proofs here, and some results are stated in their simplest form with more restrictions than necessary.

The results at the beginning on singular integral equations on the line are for the most part already published. Some additional explanation and comment is provided here. The results on singular integral operators over the circle are new and the connection with scattering theory is also new. The statements we make here are intended to be preliminary announcements of forthcoming results whose proofs will be set out in detail elsewhere.

Let $\mathscr{U}$ and $\mathscr{V}$ be bounded self adjoint operators defined on a separable Hilbert space H. Suppose that $\mathscr{V}\mathscr{U}-\mathscr{U}\mathscr{V} = (1/\pi i)\,\mathscr{C}$ where $\mathscr{C}$ is trace class and positive. Suppose, further, that H is the smallest invariant subspace for both $\mathscr{U}$ and $\mathscr{V}$ which contains the range of $\mathscr{C}$.

Let h be an auxilliary coefficient Hilbert space of dimension equal to the maximum of the spectral multiplicity of $\mathscr{U}$ and the dimension of the range of $\mathscr{C}$.

Suppose the Schmidt expansion of $\mathscr{C}$ is $\mathscr{C} = \sum_i \lambda_i^2 \varphi_i(\cdot, \varphi_i)$ where the orthonormal sequence $\{\varphi_i\}$ is completed if necessary, and let the sequence $\{\vartheta_i\}$ denote a complete orthonormal set in h. We define two Hilbert Schmidt mappings $K^*: h \to H$ and $K: H \to h$ by setting $K^*\vartheta_n = \lambda_n \varphi_n$ and $K\varphi_n = \lambda_n \vartheta_n$ and then extending K^* and K by linearity.

The "Determining Function" of the pair $\{\mathscr{U}, \mathscr{V}\}$ is defined to be**

$$E(l, z) = 1 + K(\mathscr{V}-l)^{-1}(\mathscr{U}-z)^{-1}K^*$$

where 1 is the identity operator on h, and $l \notin \sigma(\mathscr{V})$, $z \notin \sigma(\mathscr{U})$.

The following results have been known for some time, [5], [6].

---

* This work was supported by a National Science Foundation Grant.

** A characterization of determining functions purely as operator valued analytic functions of two complex variables is given in [6].

THEOREM 1. *There exists a positive integrable function* $G(\vartheta, \mu)$ *with*
$$\iint G(\vartheta, \mu)\, d\vartheta\, d\mu = 2 \text{ Trace } \mathscr{C} \text{ such that } \det E(l, z) = \exp\left(\frac{1}{2\pi i} \iint G(\vartheta, \mu)\, \frac{d\vartheta}{\vartheta - l}\, \frac{d\mu}{\mu - z}\right).$$

THEOREM 2. $\vartheta \in \sigma(\mathscr{U})$ *if and only if* $\int G(\vartheta, \mu)\, d\mu \neq 0$, $\mu \in \sigma(\mathscr{V})$ *if and only if* $\int G(\vartheta, \mu)\, d\vartheta \neq 0$.

THEOREM 3. *If* $\mathscr{T} = \mathscr{U} + i\mathscr{V}$ *and* $\mathscr{T}' = \mathscr{U}' + i\mathscr{V}'$, *with* $\mathscr{U}'$ *and* $\mathscr{V}'$ *satisfying the hypothesis imposed above on* $\mathscr{U}$ *and* $\mathscr{V}$, *then* $\mathscr{T}$ *is unitarily equivalent to* $\mathscr{T}'$ *if and only if* $E(l, z)$ *is unitarily equivalent to* $E'(l, z)$.

We remark that when $\mathscr{C}$ has one dimensional range then only one eigenvalue of the determining function is different from one, and this eigenvalue then necessarily has the form given in theorem one above. Thus in this case the function $G(\vartheta, \mu)$ above constitutes a complete unitary invariant for the seminormal operator $\mathscr{T}$. Since it is furthermore true that the assumption that $\mathscr{C}$ has one dimensional range and the spectral multiplicity of $\mathscr{U}$ is finite implies that $G(\vartheta, \mu)$ above takes on only the values zero and one, the support of $G(\vartheta, \mu)$ in this case may be regarded as the complete unitary invariant. From this point of view the unit disk emerges as the complete unitary invariant of the unilateral shift.

THEOREM 4. *Under the additional hypothesis that* $\mathscr{C}$ *has one dimensional range and the spectral multiplicity of* $\mathscr{U}$ *is finite the spectral multiplicity of* $\zeta \in \sigma(\mathscr{V})$, $m(\zeta)$, *is computed by means of the following algorithm:*
Let $\mathscr{Q}_\zeta = \{\lambda : G(\zeta, \lambda) = 1\}$. *If* $\mathscr{Q}_\zeta$ *is the union of n disjoint intervals then* $m(\zeta) = n$; *otherwise,* $m(\zeta)$ *is infinite.*

We remark that the assumption that $\mathscr{C}$ has one dimensional range, and even the assumption that $\mathscr{C}$ is positive can be dispensed with — in the sense that a more complicated evaluation of the spectral multiplicity of $\mathscr{V}$ is still possible. See [5] and theorem (14) below.*)

THEOREM 5. $\mathscr{V}$ *is unitarily equivalent to the singular integral operator* $\mathscr{L}$ *defined by setting*
$$\mathscr{L}x(\lambda) = A(\lambda)x(\lambda) + \frac{1}{\pi i} \mathbf{P} \int \frac{K^*(\lambda)K(\mu)}{\mu - \lambda}\, x(\mu)\, d\mu$$

where $x(\lambda) = \{x_1(\lambda), x_2(\lambda), \ldots\} \in h$ *for a.a.* λ *and* $A(\lambda)$ *is a bounded self adjoint operator on h for a.a.* λ *which is weakly measurable in* λ *and such that* $\operatorname*{ess\,sup}_{\lambda} \|A(\lambda)\|_h < \infty$.

Furthermore, $K(\lambda)$ *is a Hilbert-Schmidt operator on h for a.a.* λ *and* $\operatorname*{ess\,sup}_{\lambda} \|K(\lambda)\|_h < \infty$.

---

*) This evaluation becomes completely explicit if additional "smoothness" requirements are imposed. Then one has a sequence of $G_\lambda(\xi, \lambda)$ and an analogue of Theorem 4.

An explicit matrix representation for the operators $K(\lambda)$ and $K(\lambda)^*$ is available:

Let $\mathscr{H}$ be a minimal direct sum decomposition of H into invariant subspaces of $\mathscr{U}$, $\mathscr{H}_i$, each subspace being generated by a cyclic vector. Our hypothesis imply that each $\mathscr{H}_i$ is canonically isometric with a subspace of $L_2\big(\sigma(\mathscr{U})\big)$ and if Q denotes the canonical isomorphism then $Qf=\{Q_1(\lambda), Q_2(\lambda), \ldots\}$ while $Q\mathscr{U}f=\{\lambda Q_1(\lambda), \lambda Q_2(\lambda), \ldots\}$.

For each λ we define the square matrix $K^*(\lambda)$ with i,j^{th} element $(K^*(\lambda))_{i,j}=$ $=\lambda_j Q_i\varphi_j$. If n is the total spectral multiplicity of $\mathscr{U}$ and m is the dimension of the range of $\mathscr{C}$ then i and j range from 1 to $p=\max(n, m)$ and the matrix $K^*(\lambda)$ is filled out with zeros where necessary, i.e. we take $\lambda_{m+1}, \ldots, \lambda_p$ all to be zero, and if the spectral multiplicity of $\lambda \in \sigma(\mathscr{U})$ is less than m we define $Q_i\varphi_j(\lambda)=0$ for $i>m$.

As an elementary example of this theorem consider the case where $\mathscr{U}$ is restricted to have total spectral multiplicity one.

It is immediatly clear from the above that

$$
K^*(\lambda) = \begin{pmatrix} K_{11}^*(\lambda) & K_{12}^*(\lambda) & K_{13}^*(\lambda) & \cdots \\ 0 & 0 & 0 & \cdots \\ 0 & 0 & 0 & \cdots \\ \cdot & \cdot & \cdot & \cdots \end{pmatrix}
$$

hence

$$
\mathbf{P}\int \frac{K^*(\lambda)K(\mu)}{\mu-\lambda}\,x(\mu)\,d\mu = \mathbf{P}\int \begin{pmatrix} \sum\limits_j K_{1j}^*(\lambda)K_{1j}(\mu) & 0 & \cdots \\ 0 & 0 & \cdots \\ 0 & 0 & \cdots \\ \cdot & \cdot & \cdots \end{pmatrix} \begin{pmatrix} x_1(\lambda) \\ x_2(\lambda) \\ x_3(\lambda) \\ \cdot \end{pmatrix} \frac{d\mu}{\mu-\lambda}.
$$

It follows at once that $A(\lambda)$ in theorem (5) above annihilates all vectors of the form $x(\lambda)=(0, x_1(\lambda), x_2(\lambda), \ldots)^*$. Since otherwise there would be a common invariant subspace for $\mathscr{U}$ and $\mathscr{V}$ on which these operators commute, and we have assumed that the smallest invariant subspace of $\mathscr{U}$ and $\mathscr{V}$ which contains the range of $\mathscr{C}$ is the whole space H.

Thus $A(\lambda)x(\lambda)=(a(\lambda)x_1(\lambda), 0, 0, \ldots)^*$ and in theorem (5) above $\mathscr{L}$ is the scalar operator given by

$$
\mathscr{L}_1 q(\lambda) = a(\lambda)q(\lambda) + \frac{1}{\pi i}\int \sum_j \frac{K_{1j}^*(\lambda)K_{ij}(\mu)}{\mu-\lambda}\,q(\mu)\,d\mu,
$$

where $\sum\limits_j |k_{1j}^*(\lambda)|^2$ is essentially bounded.

Exactly similar reasoning can be carried out in the case where $\mathscr{U}$ has higher multiplicity; thus, for example, if $\mathscr{U}$ has multiplicity two then $\mathscr{V}$ is unitarily equivalent to the singular integral operator $\mathscr{L}_2$ given by

$$
\mathscr{L}_2 x(\lambda) = \begin{pmatrix} a_{11}(\lambda) & a_{12}(\lambda) \\ a_{21}(\lambda) & a_{22}(\lambda) \end{pmatrix} \begin{pmatrix} x_1(\lambda) \\ x_2(\lambda) \end{pmatrix} + \frac{1}{\pi i}\,\mathbf{P}\int \begin{pmatrix} \beta_{11}(\lambda,\mu) & \beta_{12}(\lambda,\mu) \\ \beta_{21}(\lambda,\mu) & \beta_{22}(\lambda,\mu) \end{pmatrix} \begin{pmatrix} x_1(\mu) \\ x_2(\mu) \end{pmatrix} \frac{d\mu}{\mu-\lambda}
$$

where

$$\beta_{11}(\lambda, \mu) = \sum_j K^*_{1j}(\lambda) K_{1j}(\mu), \quad \beta_{12}(\lambda, \mu) = \sum_j K^*_{1j}(\lambda) K_{2j}(\mu)$$

$$\beta_{21}(\lambda, \mu) = \sum_j K^*_{2j}(\lambda) K_{1j}(\mu), \quad \beta_{22}(\lambda, \mu) = \sum_j K^*_{2j}(\lambda) K_{2j}(\mu).$$

In these relations we may as well take $a_{12}(\lambda)=a_{21}(\lambda)\equiv 0$ without loss of generality since we can always make an additional λ dependent unitary transformation to achieve this diagonalization.

Theorem 8 below, as we shall see, gives a quite explicit determination of the spectrum of $\mathcal{T}$ in all these cases in terms of the coefficients exhibited here, and theorem 2 above then characterizes the spectrum of singular integral operators of the form $\mathcal{L}_2$, $\mathcal{L}_3$ etc.

We remark that certain generalizations of these results to the case where $\mathcal{U}$ is unbounded are possible.

THEOREM 6. *$E(l, z)$ is the unique (outer) solution of the operator homogeneous Riemann—Hilbert problem*

$$\begin{cases} E(l, \lambda + i0) = E(l, \lambda - i0) \dfrac{1 + K(\lambda)(A(\lambda) - l)^{-1} K^*(\lambda)}{1 - K(\lambda)(A(\lambda) - l)^{-1} K^*(\lambda)} \\ E(l, z) \to 1 \quad \text{as} \quad |\text{Im}\, l\, \text{Im}\, z| \to \infty. \end{cases}$$

Also important in this theory is

THEOREM 7.

$$\det \left(\frac{1 + K(\lambda)(A(\lambda) - l)^{-1} K^*(\lambda)}{1 - K(\lambda)(A(\lambda) - l)^{-1} K^*(\lambda)} \right) = \exp \int G(\vartheta, \lambda) \frac{d\vartheta}{\vartheta - l}$$

and

$$\int G(\vartheta, \lambda)\, d\vartheta = 2\text{Trace}\,[K^*(\lambda) K(\lambda)].$$

We will explain below how the proof of this theorem depends on certain results of M. G. KREIN on perturbation determinants.

To show that theorem 7 actually permits the effective computation of $G(\vartheta, \lambda)$ even when the dimension of the range of $\mathcal{C}$ is not one, consider our illustration above with $\mathcal{U}$ simple.

It is obvious that in this case

$$\det \left(\frac{1 + K(\lambda)(A(\lambda) - l)^{-1} K(\lambda)}{1 - K(\lambda)(A(\lambda) - l)^{-1} K(\lambda)} \right) = \frac{1 + (a(\lambda) - l)^{-1} \sum |K_{j1}(\lambda)|^2}{1 - (a(\lambda) - l)^{-1} \sum |K_{j1}(\lambda)|^2}$$

since $\det(1 + RS) = \det(1 + SR)$ when S and R are completely continuous and RS and SR are trace class (2).

However,

$$\frac{1+\big(a(\lambda)-l\big)^{-1}\sum_{j}|K_{j1}(\lambda)|^2}{1-\big(a(\lambda)-l\big)^{-1}\sum_{j}|K_{j1}(\lambda)|^2} = \exp\int_{A_1}^{A_2}\frac{d\vartheta}{\vartheta-l},$$

where

$$A_1 = a(\lambda)-l-\sum_{j}|K_{j1}(\lambda)|^2, \quad A_2 = a(\lambda)-l+\sum_{j}|K_{j1}(\lambda)|^2.$$

Thus $G(\vartheta, \lambda)$ is the characteristic function of the set of points (ϑ, λ) for which $a(\lambda)-\sum_{j}|K_{j1}(\lambda)|^2 < \vartheta < a(\lambda)+\sum_{j}|K_{j1}(\lambda)|^2$, $\lambda\in\sigma(\mathcal{U})$. Note that the intersection of this set with any vertical line is either an interval or it is empty.

We turn now to results achieved within the past year. In [6] we conjectured and in [8] we proved a result expressing the spectrum of the completely non-normal semi normal operator $\mathcal{U}+i\mathcal{V}$ in terms of the function $G(\vartheta, \lambda)$ of theorems one and seven above.

Let us call the support of any almost everywhere determined version of $G(\vartheta, \lambda)$ a "determining set" for $\mathcal{T}$. Call the essential closure of this set $D(\mathcal{U}, \mathcal{V})$.

THEOREM 8. $D(\mathcal{U}, \mathcal{V}) = \sigma(\mathcal{T})^*$

This result immediately implies that the spectrum of $\mathscr{L}_1$ above is the essential closure of the set of points (ϑ, λ) for which $a(\lambda)-\sum_{j}|K_{j1}(\lambda)|^2 < \vartheta < a(\lambda)+$

$+\sum_{j}|K_{j1}(\lambda)|^2.$

It is of course immediately obvious that the essential closure $D(\mathcal{U}, \mathcal{V})$ may in general have the property that its intersection with vertical lines consists of more than one interval while the "determining set" has the property that its intersection with any vertical is either void or consists of a single interval.

We note the important fact that theorem (3) and theorem (1) above taken together imply that the determining set is the complete unitary invariant of $\mathcal{T}$ when $\mathscr{C}$ has one dimensional range and $\mathcal{U}$ has infinite multiplicity, while the spectrum (which is obtained by adding a set which in general is of positive two dimensional measure) is not a complete invariant.

In the most general situation the complete invariant for $\mathcal{T}$ is the barrier quotient appearing in theorem (6). If the spectral multiplicity of $\mathcal{U}$ is finite and equal to p, then the determinant of this quotient can always be written as the product of p roots that are branches of analytic functions in l with only algebraic singularities (for fixed λ). The number of distinct eigenvalues is a constant independent of l with the exception of some special values of l. Only a finite number of such excep-

---

* The following *conjecture* is of interest in this connection: The essential spectrum of $\mathcal{T}$ is the essential boundary of a determining set for $\mathcal{T}$.

tional points exist in each compact subset of the upper half plane. Each such exceptional point, moreover, is associated with a permutation group on the roots that interchanges branches. Under our assumption that $\mathscr{C}$ is positive it is easy to see that all of these roots lie in the upper half plane, but we note that it is always true that among the roots there will be exactly as many in the upper half plane as there are positive eigenvalues for $K^*(\lambda) \operatorname{Sgn} \mathscr{C} K(\lambda)$ — and as many in the lower half plane as there are negative eigenvalues for this matrix. (Here we refer to the more general form of the barrier quotient given in [6].) Thus we see that the λ-dependent triplet (roots, exceptional points, permutations) is a unitarily invariant structure associated with the operator in the general case.

There is also a dual triplet associated with the determining function.

Given a determining function it is known that there always exists a positive trace class valued measure $dM_\zeta(\cdot)$ on h such that

$$E(\zeta - i0, x) E^*(\zeta - i0, \bar{x}) = 1 + \int \frac{dM_\zeta(\mu)}{\mu - x}$$

and it is known how the spectral multiplicity of $\mathscr{V}$ is determined from this measure.

Here we wish to remark that the operator appearing on the right above can quite easily be shown to arise as a barrier quotient of our previous form, and is therefore associated with a ζ-dependent triplet (roots, exceptional points, permutations) of its own.

In the present paper we wish to do no more than call attention to these triplet structures and the relations that subsist between them.

We wish to indicate now how it is possible to explicitly evaluate $G(\vartheta, \lambda)$ even when the spectral multiplicity of $\mathscr{U}$ is not restricted to be one.

We note first that

$$\frac{1 + K(\lambda)(A(\lambda) - l)^{-1} K^*(\lambda)}{1 - K(\lambda)(A(\lambda) - l)^{-1} K^*(\lambda)} = 1 + \int \frac{dR_\lambda(\vartheta)}{\vartheta - l}$$

where $dR_\lambda(\cdot)$ is for a. a. λ a positive trace class valued measure on h. But it is shown in [6] that

$$(1 - K(\lambda)(A(\lambda) - l)^{-1} K^*(\lambda))^{-1} = 1 + \frac{1}{2} \int \frac{dR_\lambda(\vartheta)}{\vartheta - l}.$$

On the other hand

$$(1 - K(\lambda)(A(\lambda) - l)^{-1} K^*(\lambda))^{-1} = 1 + K(\lambda)(A(\lambda) - K^*(\lambda) K(\lambda) - l)^{-1} K^*(\lambda).$$

Hence,

$$\int \frac{dR_\lambda(\vartheta)}{\vartheta - l} = 2K(\lambda)(A(\lambda) - K^*(\lambda) K(\lambda) - l)^{-1} K^*(\lambda).$$

Now

$$\det\left(1 + 2K(\lambda)(A(\lambda) - K^*(\lambda)K(\lambda) - l)^{-1}K^*(\lambda)\right) =$$

$$= \det\left(1 + 2K^*(\lambda)K(\lambda)(A(\lambda) - K^*(\lambda)K(\lambda) - l)^{-1}\right).$$

Call this last determinant $\Delta_\lambda(l)$.

If we evaluate the residue at infinity we can immediately conclude that

$$\int G(\vartheta, \lambda)\, d\vartheta = 2\mathrm{Trace}\,[K^*(\lambda)K(\lambda)].$$

Perturbation determinant of the form $\Delta_\lambda(l)$ have been studied by many authors, and in particular the representation of the form given in theorem 7 exists by virtue of some results of LIFSHITZ—KREIN. Furthermore, it follows from the results of M. G. KREIN [4] that $G(\vartheta, \lambda) \leq$ rank of $K^*(\lambda)K(\lambda)$. In particular $G(\vartheta, \lambda) \leq$ total spectral multiplicity of $\mathscr{U}$. But $\iint G(\vartheta, \mu)\, d\vartheta\, d\mu = 2\,\mathrm{Trace}\,\mathscr{C}$.

Note that this implies the following remark:*, **

Trace $\mathscr{C} \leq \frac{1}{2}$ (spectral multiplicity of $\mathscr{U}$) $\cdot$ (measure$_2\,\sigma(\mathscr{T})$).

If we choose a single valued harmonic branch of arg $\Delta_\lambda(l)$ in the upper half plane then the "principal function" $G(\vartheta, \lambda)$ can be obtained from the explicit formula

$$G(\vartheta, \lambda) = \lim_{\eta \downarrow 0} \arg \Delta_\lambda(\vartheta + i\eta) + f(\lambda)$$

for some function $f(\lambda)$ which is determined via theorem (7) by requiring that $2\mathrm{Trace}\,[K^*(\lambda)K(\lambda)] = \int G(\vartheta, \lambda)\, d\vartheta$.

For the case of $\mathscr{L}_2$ above, for example, the explit evaluation comes down to the evaluation of determinants of two by two matrices. We will not give the details.

A completely parallel development to the foregoing can be based upon the study of another commutator identity; namely $UV - VU = (1/\pi)DU$ where U is unitary and V is self adjoint, while D is trace class and positive.

This relation can be derived from $\mathscr{V}\mathscr{U} - \mathscr{U}\mathscr{V} = \mathscr{C}/\pi i$ by setting $U = (\mathscr{U} - i)(\mathscr{U} + i)^{-1}$, $D = 2F\mathscr{C}F^*$, $F = (\mathscr{U} - i)^{-1}$ and $\mathscr{V} = V$. However $\mathscr{U}$ bounded corresponds to the restriction: measure $\sigma(U) < 2\pi$. Without this restriction we can still conclude that $H_{ac}(U)$ reduces both U and V and contains the least subspace of H reducing U and containing the range of D. (See C. PUTNAM [10] and T. KATO [3]).

---

* Compare with a result in [3].

** The full inequality for $G(\vartheta, \mu)$ permits us to strengthen this remark in an obvious way We also note that the results of [6] allow us to say something even when $\mathscr{C}$ is no longer positive.

In analogy to theorem 5 above we now have:

THEOREM 9. *V restricted to $H_{ac}(U)$ is unitarily equivalent to the singular integral operator*

$$Lx(t) = \mathfrak{A}(t)x(t) + \frac{1}{\pi i}\int_C \frac{\mathfrak{R}^*(t)\mathfrak{R}(\tau)}{t-\tau}\, x(\tau)\, d\tau$$

where the integration is on the unit circle. $\mathfrak{A}(\lambda)$ is a bounded self-adjoint operator on h weakly measurable in τ with $\operatorname*{ess\,sup}_{\tau}\|\mathfrak{A}(\tau)\|_h < \infty$; and $\mathfrak{R}(\tau)$ is a Hilbert Schmidt operator on h for a.a. τ and $\operatorname*{ess\,sup}_{\tau}\|\mathfrak{R}(\tau)\|_h < \infty$.

THEOREM 10. *Let* $\varphi(l, z) = 1 + \dfrac{1}{\pi}\mathfrak{R}U(U-z)^{-1}(V-l)^{-1}\mathfrak{R}^*$ *where* $\mathfrak{R}^*\mathfrak{R} = D$ *and* $\mathfrak{R}^*: h \to H,\ \mathfrak{R}: H \to h$.

Then

$$\varphi_+(l, \tau) = \frac{1 - \mathfrak{R}(\tau)\left(\mathfrak{A}(\tau) - l\right)^{-1}\mathfrak{R}^*(\tau)}{1 + \mathfrak{R}(\tau)\left(\mathfrak{A}(\tau) - l\right)^{-1}\mathfrak{R}^*(\tau)}\, \varphi_-(l, \tau)$$

where $\varphi_+(l, \tau)$ and $\varphi_-(l, \tau)$ denote the boundary values assumed by $\varphi(l, z)$ as z approaches τ on the unit circle from inside or outside respectively.

THEOREM 11. *There exists a positive summable function $H(\vartheta, \tau)$ such that*

$$\det\left(\frac{1 + \mathfrak{R}(\tau)\left(\mathfrak{A}(\tau) - l\right)^{-1}\mathfrak{R}^*(\tau)}{1 - \mathfrak{R}(\tau)\left(\mathfrak{A}(\tau) - l\right)^{-1}\mathfrak{R}^*(\tau)}\right) = \exp\int H(\vartheta, \tau)\frac{d\vartheta}{\vartheta - l}$$

and

$$\int H(\vartheta, \tau)\, d\vartheta = 2\operatorname{Trace}\mathfrak{R}^*(\tau)\mathfrak{R}(\tau).$$

This result together with theorem 10 immediately implies

THEOREM 12.

$$\det\varphi(l, z) = \exp\frac{1}{2\pi i}\int_C\int_R H(\vartheta, \tau)\frac{d\vartheta}{\vartheta - l}\frac{d\tau}{\tau - z}$$

where the outer integral is over the unit circle.

Let us now define a new operator on h, $R(l)$, by setting $\varphi(l, o)^{-1} = R(l)R^*(l)$ so that $R(l)$ is the unique positive square root of $\varphi(l, o)^{-1}$ when l is real and sufficiently negative. Further, let us define $A(l, z) = \varphi(l, z)R(l)$. Then a simple algebraic computation shows that $A^*(l, 1/\bar{z})A(l, z) = 1$; furthermore, when U is related to $\mathcal{U}$ and V is related to $\mathcal{V}$ as above, then we can show that $R(l) = E(i, l)$, and $A(l, z) = E(\omega, l)$

for $z = (\omega - i)/(\omega + i)$. From the equality of the determinants of $E(\omega, l)$ and $A(l, z)$ follows at once a relation between the "principal functions" $G(\vartheta, \mu)$ and $H(\vartheta, \tau)$,

$$G(\vartheta, \mu) = \left(\vartheta, \frac{\mu - i}{\mu + i}\right).$$

The $A(l, z)$ function introduced above is again a solution (now outer) of the Riemann Hilbert problem appearing in theorem (10), but it is constructed so that certain positivity relations obtain.

THEOREM 13. *There exists a self adjoint essentially bounded operator on h, $\beta(\zeta)$ and a positive family of trace class valued measures on h, $d\Omega_\zeta(\cdot)$ such that*

$$iA^*\left(\zeta - i0, \frac{1}{\bar{y}}\right) A(\zeta - i0, y) = i\beta(\zeta) + \int\limits_0^{2\pi} \frac{e^{i\vartheta} + y}{e^{i\vartheta} - y} \, d\Omega_\zeta(\vartheta).$$

The measure $d\Omega_\zeta(\cdot)$ gives us the spectral multiplicity of V in a way that is exactly analogous to theorem (1. 4) of [5].

The measure $d\Omega_\zeta(\cdot)$ is absolutely continuous with respect to its own trace. If we call the resulting Radon Nikodym derivative $\Omega'_\zeta(\vartheta)$ we can easily see that $0 \leq \Omega'_\zeta(\vartheta) \leq 1$ and if we call the eigenvalues of $\Omega'_\zeta(\vartheta)$, arranged in order of magnitude $\lambda_j(\zeta, \vartheta)$ so that $1 \geq \lambda_1(\zeta, \vartheta) \geq \lambda_2(\zeta, \vartheta) \geq \cdots 0$ we can form a set of positive measures in the following way: for every Borel set on the real axis, let $\mu_\zeta^{(j)}(\Delta) = \int_\Delta \lambda_j(\zeta, \vartheta) \, d\mathrm{Tr}\, \Omega_\zeta(\vartheta)$ and let $H^j(\zeta)$ be the Hilbert space of square integrable functions with respect to this measure.

THEOREM 14. *The spectral multiplicity of $\zeta \in \sigma(V_{ac})$ is equal to $\sum_j \dim H^j(\zeta)$.*

The proof of this result is similar to the proof of the corresponding result in [3] but is sufficiently lengthy so that it can not be given here. The theorem leads immediately to a geometric criterion for multiplicity in the special case where D has one dimensional range and U has finite spectral multiplicity.

THEOREM 15. *If D is of one dimensional range and U has finite spectral multiplicity, then $m(\zeta)$, the spectral multiplicity of ζ in $\sigma(V_{ac})$, is equal to the number of arcs on the unit circle into which the set $\{\tau : H(\zeta, \tau) = 1\}$ decomposes; if this set is not the union of a finite number of disjoint arcs then $m(\zeta)$ is infinite.*

We remark, in addition, that when D has one dimensional range we can also compute the spectral multiplicity of points in the spectrum of U. Let $\Xi_\tau = \{\zeta : H(\zeta, \tau) = 1\}$. If Ξ_τ is the union of p disjoint real intervals then the spectral multiplicity of $\tau \in \sigma(U)$ is p; otherwise, it is infinite.

The preceeding results constitute a generalization of my results in [7]. They furnish a rather complete spectral theory for singular integral operators on the circle, and all the theorems developed above on the line have their analogous forms on the circle. We remark by way of application that we can now treat positive semi-infinite Toeplitz operators with matrix coefficients by means of this theory.

The proofs and further development of these ideas will appear in [9].

*

We wish now to explore certain connections between the theory of determining functions and scattering theory. This is joint work with Rick CAREY.

Let V be a bounded self adjoint operator on H with a purely absolutely continuous spectrum. Let D be trace class and positive. The following strong limits are known to exist:

$$W_\pm = \text{s-lim}_{t \to \pm \infty} e^{it(V+D)} e^{-itV}.$$

These are the wave operators. The S matrix is defined to be $W_+^* W_-$. It is well known that $(V+D)W_+ = W_+ V$, that $(W^\pm)^* W_\pm = 1$ and that $W_\pm (W_\pm)^* = 1 - P$ where P is the projection onto the singular subspace of the perturbed operator, H_s.

Accordingly the wave operators are in general not unitary. We shall show, nevertheless, by means of a certain dilation how the theory developed above can be made to apply to them. Henceforth let us simply call W_-, W.

We will associate with H a new Hilbert space $\mathbf{H}$

$$\mathbf{H} = H \oplus H_s \oplus H_s \oplus H_s \oplus \cdots, \quad f = [f_0, f_1, f_2, f_3, \ldots].$$

Now let $\mathbf{V}$ on $\mathbf{H}$ and $\mathbf{U}$ on $\mathbf{H}$ be defined by setting

$$\mathbf{V}f = \{Vf_0, (V+D)f_1, (V+D)f_2, \ldots\}, \quad \mathbf{D}f = \{Df_0, 0, 0, \ldots\},$$

$$\mathbf{U}f = \{Wf_0 + f_1, f_2, f_3, \ldots\}.$$

Then $\mathbf{U}$ is unitary and we have the commutator identity $\mathbf{UV} - \mathbf{VU} = \mathbf{DU}$, also the spectrum of $\mathbf{U}$ is the entire circle by well known results. Let us define the maps J and J^* by setting

$$Jf_1 = \{f_1, 0, 0, \ldots\}, \quad J^*\{x_1, x_2, x_3, \ldots\} = x_1.$$

Then $JJ^*|_H = 1$ and $J^*J = 1$.

We now define two analytic operator valued functions. Let

$$\varphi(l, z) = 1 + (1/\pi)\mathfrak{K}J^* \mathbf{U}(\mathbf{U}-z)^{-1}(\mathbf{V}-l)^{-1}J\mathfrak{K}^*, \quad \mathfrak{K}^*\mathfrak{K} = \pi D,$$

$$\vartheta(l, z) = 1 + (1/\pi)\mathfrak{K}W(W-z)^{-1}(V-l)^{-1}\mathfrak{K}^*.$$

In the last line $\vartheta(l, z)$ is only defined when $|z| > 1$. Since the spectrum of W is the entire unit disk.

THEOREM 16. $\varphi(l, z) = \vartheta(l, z)$ when $|z| > 1$, and

$$\varphi^{-1}(l, z) = 1 - (1/\pi)\Re(V + D - l)^{-1}(1 - zW^*)^{-1}\Re^* \quad \text{when } |z| < 1.$$

But

$$1 - (1/\pi)\Re(V + D - l)^{-1}\Re^* = (1 + (1/\pi)\Re(V - l)^{-1}\Re^*)^{-1}.$$

Therefore $\varphi(l, 0) = 1 + (1/\pi)\Re(V - l)^{-1}\Re^*$ a symmetrized Aronszajn-Weinstein operator. From the results of M. G. KREIN referred to above it now follows at once that

$$\det \varphi(l, 0) = \det\left(1 + \frac{D}{\pi}(V - l)^{-1}\right) = \exp\int \zeta(\lambda)\frac{d\lambda}{\lambda - l}$$

where $\zeta(\lambda)$ is a positive L_1 function for which $(1/\pi)\,\text{Trace}\,D = \int \zeta(\lambda)\,d\lambda$. This function is called the spectral displacement function (or phase shift) and when V has simple spectral multiplicity the S matrix expressed in the spectral representation of V becomes $S(\lambda) = \exp - \{2\pi i\zeta(\lambda)\}$.

By theorem (12) we see that a relation exists between the spectral displacement function and the "Principal function" associated with the commutator pair, $\{U, V\}$.

THEOREM 17.

$$2\pi\zeta(\lambda) = \int_0^{2\pi} H(\lambda, e^{i\vartheta})\,d\vartheta.$$

Much more than this is true. In the very special case where D has one dimensional range and V has simple spectral multiplicity, we can even show that

$$\frac{\bar{S}(\lambda) - z}{1 - z}\,\varphi(\lambda + i0, z)^{-1} = \varphi(\lambda - i0, z)^{-1} \quad \text{when} \quad |z| > 1.$$

We will examine this simple case when the perturbation produces one eigenvalue outside $\sigma(V) = (a, b)$. $S(\lambda)$ above is identically equal to one outside (a, b) but $\zeta(\lambda)$ will have exactly one interval (r, s) outside (a, b) on which it is identically equal to one. Theorem (17) and the barrier relation above then lead to a complete computation of the principal function $H(\lambda, e^{i\vartheta})$:

$$H(\lambda, e^{i\vartheta}) = \begin{cases} \text{characteristic function of positive the arc from 1 to } \bar{S}(\lambda) \text{ on unit circle,} \\ \lambda \in (a, b) \text{ identically equal to 1 for } \lambda \text{ in } (r, s), \text{ identically zero} \\ \lambda \notin (a, b) \cup (r, s) \end{cases}$$

so that

$$\varphi(l, z) = \begin{cases} \exp\left(\dfrac{1}{2\pi i}\int_a^b \dfrac{d\lambda}{\lambda - l}\int_1^{S(\lambda)} \dfrac{d\tau}{\tau - z}\right) & (|z| > 1) \\[2em] \left(\exp\left(\dfrac{1}{2\pi i}\int_a^b \dfrac{d\lambda}{\lambda - l}\int_1^{S(\lambda)} \dfrac{d\tau}{\tau - z}\right)\right)\left(\dfrac{l - s}{l - r}\right) & (|z| < 1). \end{cases}$$

It is now possible to use the previous results (slightly extended) to compute the spectral multiplicity of the unitary part of the wave operator in terms of the spectral displacement function. We will not show this here but refer the reader instead to a forthcoming paper [1] which deals with a more general situation.

REFERENCES

[1] R. Carey and J. D. Pincus, *The structure of intertwining isometries.* (In Preparation)

[2] I. Gohberg and M. G. Krein, *Introduction to the theory of linear nonselfadjoint operators.* Amer. Math. Soc. Trans. **18** (1969), 162.

[3] T. Kato, *Smooth Operators and Commutators.* Studia Math. **31** (1968), 535—546.

[4] M. G. Kreĭn, *Perturbation Determinants and a Formula for the Traces of Unitary and Self-Adjoint Operators.* Soviet Math. Dokl. 3 **(1962)**, 707—710.

[5] J. D. Pincus, *Commutators and Systems of Singular Integral Equations.* I. Acta Math. **121** (1968), 219—249.

[6] J. D. Pincus, *The Determining Function Method in the Treatment of Commutator Systems.* Colloquia Mathematica Societatis János Bolyai. 5. Hilbert space operators. Tihany (Hungary) (1970), 443—477.

[7] J. D. Pincus, *Singular Integral Operators on the Unit Circle.* Bull. Amer. Math. Soc. **73** (1967), 195—199.

[8] J. D. Pincus, *The Spectrum of Seminormal Operators.* Proc. Nat. Acad. Sci. **68** (1971), 1684—1685.

[9] J. D. Pincus, *Commutators and Systems of Singular Integral Equations.* II. (In Preparation)

[10] C. R. Putnam, *Commutation Properties of Hilbert Space Operators and Related Topics.* Springer, New York 1967.

Weyl's Theorems

By

KARL GUSTAFSON*)

DEPT. OF MATH.
UNIVERSITY OF COLORADO, BOULDER
INST. DE PHYSIQUE THEORIQUE UNIV. DE GENEVE, SUISSE

1. Introduction

The purpose of this paper is to briefly describe some recent results and to give some examples that illustrate and make rather sharp the following three theorems concerning perturbation of spectra of linear operators, each often called Weyl's Theorem in the literature.

THEOREM 1. $\sigma_e(A+B) = \sigma_e(A)$.

THEOREM 2. $\sigma_e(B) = \sigma_e(A) \Rightarrow B = UAU^* - K$.

THEOREM 3. $\sigma_e(A) = \sigma(A) - \pi_{00}(A)$.

Here σ_e denotes the essential spectrum of a closed, densely-defined, linear operator A in a complex Banach space X, σ denotes the spectrum of A, π_{00} denotes the set of isolated eigenvalues of A of finite (geometric) multiplicity, K denotes a compact operator, and U denotes a unitary operator on a separable complex Hilbert space.

That the name Weyl's Theorem is appropriate in each of Theorems 1, 2, and 3 stems from H. WEYL's basic paper [32]. There Theorem 1 was obtained for bounded self-adjoint operators A perturbed by self-adjoint compact operators B; also in [32] a theorem (also usually called Weyl's Theorem), which can be shown to be equivalent to Theorem 2, was obtained for bounded self-adjoint operators A on a separable Hilbert space, namely, that any self-adjoint operator A with continuous spectrum can be converted to one with pure point spectrum by addition of a compact self-adjoint operator B; and in [32] Theorem 3 was obtained for bounded self-adjoint operators A as an alternate characterization of the essential spectrum, which, as defined by Weyl, was the set of all limit points of the spectrum.

Theorem 1, and its extensions to be discussed in section 3, is a basic perturbation theorem useful in the spectral theory of, among others, Schrödinger operators of quantum mechanics, ordinary and partial differential boundary value problems, and singular integral operators. Theorem 2, to be discussed in section 4, is one of several converses of Theorem 1, and is also interesting in other contexts such as,

---

*) Partially supported by NSF GP 15239A1.

for example, the Calkin algebra, quasitriangular operators, and nonabsolutely continuous spectra. Theorem 3, to be discussed in section 5, is a spectral stability theorem characterizing those operators which possess the property that the points in the spectrum which can be removed by arbitrary compact perturbations are exactly the isolated eigenvalues of finite (geometric) multiplicity; this class of operators contains the interesting subclass of formally semi-normal operators which occur, for example, in certain third order ordinary differential equations and as creation operators in quantum field theory.

Before turning to Theorems 1, 2, and 3, we must first make precise in the following section certain facts about the essential spectrum σ_e; the knowledgeable or casual reader may therefore omit section 2, which does, however, contain some definitions to be used later, and also a few new observations. Many further references containing related results are available from the few that we refer to in this paper.

2. The essential spectrum

Several versions of the essential spectrum have occurred in the literature, the choice depending on the particular problems being treated. This has caused some confusion, which we hope to eliminate here by explicitly specifying their different properties.

Following [12], but elaborating a little, the most convenient way to define them seems to be as follows, as complements of sets of λ such that $\lambda - A$ has certain "Fredholm" properties. An operator A is called normally solvable (and let us abbreviate this by writing $A \in F_1$) if A is closed and $R(A)$ is closed; if in addition either $\alpha(A) \equiv \dim N(A)$ is finite or $\beta(A) \equiv \operatorname{codim} R(A)$ is finite, A is called semi-Fredholm, i.e., $A \in F_2$; if in particular $\beta(A) < \infty$, A is called semi-Fredholm β, i.e., $A \in F_\beta$; or if in particular $\alpha(A) < \infty$, A is called semi-Fredholm α, i.e., $A \in F_\alpha$; if in addition both $\alpha(A) < \infty$ and $\beta(A) < \infty$, A is called a Fredholm operator, i.e., $A \in F_3$; if additionally the index $i(A) = \alpha(A) - \beta(A)$ is zero, A is called Fredholm of index zero, i.e., $A \in F_4$; finally, if A is Fredholm of index zero with a deleted neighbourhood of 0 in the resolvent set $\varrho(A)$, we write $A \in F_5$. Let us call a semi-Fredholm operator A with a deleted neighbourhood of 0 in its resolvent set "reducibly nilpotent", since it can be shown that A is reduced by, and nilpotent on, its algebraic nullspace; thus A in F_5 is Fredholm reducibly nilpotent of index zero. Let $\Delta_i(A) = \{\lambda | \lambda - A \in F_i\}$, and let $\sigma_e^i(A)$ be the complement of $\Delta_i(A)$ in the complex plane. Then $\sigma_e^i(A)$ is the essential spectrum of A according to (1) GOLDBERG [9], (2) KATO [19], (β) GUSTAFSON and WEIDMANN [12], (α) WOLF [34, 35] (3) SCHWARTZ [29], (4) SCHECHTER [27], (5) BROWDER [4]. Clearly

$$\sigma_e^1 \subseteq \sigma_e^2 = \sigma_e^\alpha \cap \sigma_e^\beta \subseteq \sigma_e^3 = \sigma_e^\alpha \cup \sigma_e^\beta \subseteq \sigma_e^4 \subseteq \sigma_e^5 \subseteq \sigma.$$

In discussing Weyl's Theorems in this paper, we will use only $\sigma_e(A) \equiv \sigma_e^4(A)$, for several reasons, including simplicity, and also because often the different versions coincide. Also, as SCHECHTER [27] showed, $\sigma_e^4(A) = w(A)$, where $w(A) = \cap \, \sigma(A+B)$, B compact, is the largest subset of the spectrum of A retained after arbitrary compact perturbations. Actually, $\sigma_e^4(A) = \cap \, \sigma(A+F)$, F bounded and of finite rank; this is perhaps easiest to see as follows, using the equivalent formulation $\Delta_4(A) = \cup \varrho(A+F)$. First, any λ in $\varrho(A+F)$ is in $\Delta_4(A+F)$ and therefore is in $\Delta_4(A)$ by the well-known results that both Fredholmness and index are preserved under compact perturbations (see, for example, [19, p. 238] for references). Secondly, for any λ in $\Delta_4(A)$ one can write $X = N(A-\lambda) \oplus M_1 = R(A-\lambda) \oplus M_2$ and take F to be any operator from $N(A-\lambda)$ onto M_2, $F(M_1)=0$, so that $(A-\lambda)+F$ is $1-1$ and onto.

Another advantage we would like to mention here is that $\sigma_e^4(A)$ is large enough to include both the continuous spectrum $\sigma_c(A)$ (which always contains $\overline{\sigma_c(A^*)}$) and the residual spectra $\sigma_r(A)$ and $\overline{\sigma_r(A^*)}$, so that a point λ in $\sigma(A)-\sigma_e^4(A)$ must be an eigenvalue of A and $\bar{\lambda}$ must be an eigenvalue of A^*. To graphically illustrate

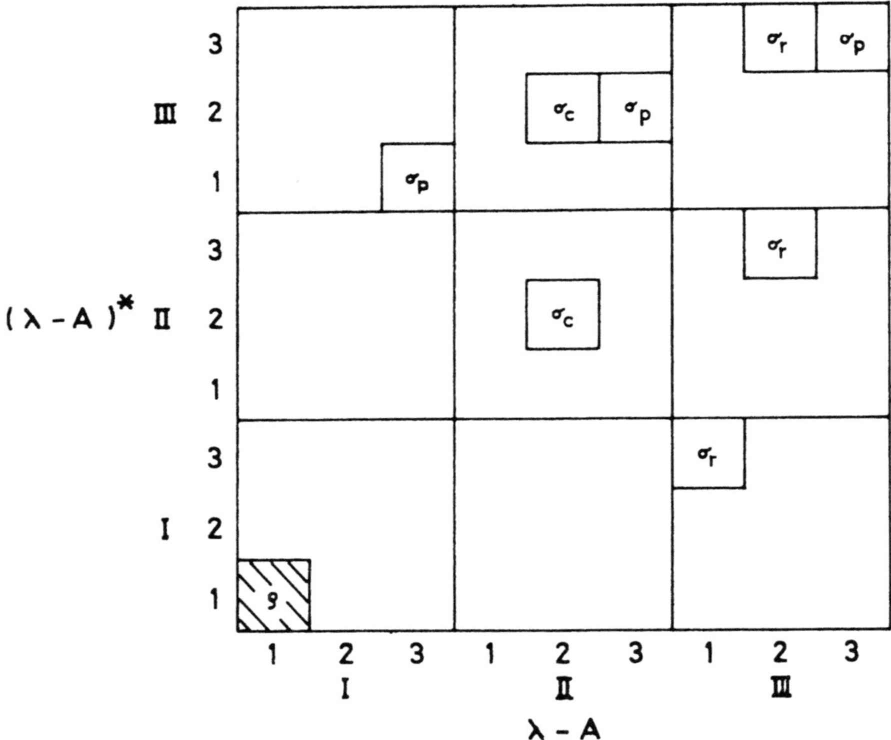

FIGURE 1. State diagram of the spectrum

the last observation, let us resort to the state diagram for $\lambda-A$ versus $(\lambda-A)^*$, where $\lambda-A$ is said to be in "range state" I, II, or III if $R(\lambda-A)$ is all of X, properly dense in X, or not properly dense in X, respectively, and where $\lambda-A$ is said to be in "inverse state" 1, 2, or 3 if $(\lambda-A)^{-1}$ is bounded, exists but is unbounded, or does not exist, respectively. Assigning the same (double, range and inverse) classification scheme to $(\lambda-A)^*$, one has (see, for example, [9] and [11] for further details) only the eight spectral "state" possibilities shown in Fig. 1, for any closed and densely defined linear operator A in a Banach space X, and it is easily verified that Fig. 2 depicts how the same eight states may occur as essential spectrum states; in Fig. 2 we have let $\varDelta_4^s \equiv \varDelta_4^{\text{singular}}$ denote those λ in $\varDelta_4$ which are also eigenvalues. The fact that $\sigma_e \equiv \sigma_e^4$ has $7\frac{1}{2}$ states in Fig. 2 in which the pair $\lambda-A$, $(\lambda-A)^*$ is permitted to reside, contrasted with just 2 possible states for σ_c, is further evidence as to why it is easier to preserve the essential spectrum rather than the continuous spectrum, under perturbation.

In view of Figs. 1, 2 one therefore has the following relationships among parts of

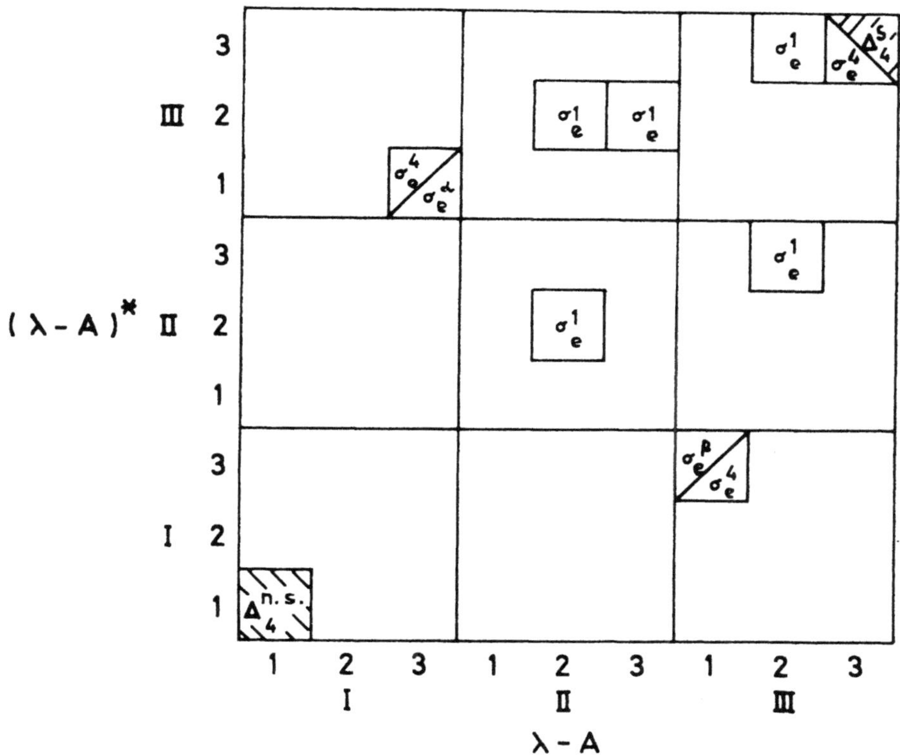

FIGURE 2. State diagram of the essential spectrum

the spectrum and the essential spectrum, letting π denote those λ which are eigenvalues of A such that $\bar\lambda$ is an eigenvalue of A^*.

$$\sigma_c \subset \sigma_e^1 - \pi = \sigma_e^2 - \pi = (\sigma_e^\alpha \cap \sigma_e^\beta) - \pi \subset \sigma_e^3 - \pi = (\sigma_e^\alpha \cup \sigma_e^\beta) - \pi \subset \sigma_e^4 - \pi =$$

$$= \sigma_c(A) \cup \sigma_r(A) \cup \overline{\sigma_r(A^*)} = \sigma_e^5 - \pi.$$

WEYL [32] defined the essential spectrum, for a bounded self-adjoint operator A, to be $\sigma_l(A)$, the set of all "limit points" of the spectrum, including isolated eigenvalues of infinite multiplicity; for general operators A if one regards a point λ in the spectrum $\sigma(A)$ to be a "limit point" unless λ is an isolated point of σ and of finite algebraic (see section 5) multiplicity, one has $\sigma_e^5 = \sigma_l$. Two other "limit point" sets τ and τ' were defined as follows in [2] for bounded operators on a Hilbert space which possess the property that they are reduced by their finite dimensional eigenspaces; let M be the closed linear span of the finite (geometric) dimensional eigenspaces, write $T = T_1 \oplus T_2$, where $T_1 = T_M$, $T_2 = T_{M^\perp}$, let τ' be those λ in $\sigma(T)$ for which either $\lambda \in \sigma(T_2)$ or λ is an accumulation point of eigenvalues of finite multiplicity, and let τ be those λ which are already in τ', or are accumulation points of eigenvalues, or are themselves eigenvalues of infinite multiplicity. As observed in [10], one always has in that case $\sigma_e^5 = \sigma_e^4 \supset \tau \supset \tau'$, and $\sigma_e^5 = \tau'$ if Weyl's Theorem 3 holds. For self-adjoint A one has $\sigma_e^2 = \sigma_e^5$; this situation is easily seen to hold also for (bounded or unbounded) normal A, but not much beyond normal operators, since $\sigma_e^3(A_1) \neq \sigma_e^4(A_1)$ for the unilateral right shift A_1 given by $A_1(e_n) = e_{n+1}$ on l_2. A_1 is quasinormal, the "smallest" class bigger than the normal operators among the classes of normal-like operators; see for example [11] for a tabulation of these operator classes. One can conclude from the discussion in KATO [19, p. 243] that if A has nonempty resolvent and $\sigma_e^2(A)$ is countable, then $\sigma_e^2(A) = \sigma_e^5(A)$, and moreover, Theorem 3 holds. Simply taking A to be an infinite dimensional zero operator illustrates that $\sigma_e^1 \neq \sigma_e^2$, and that $\sigma_e^2 \neq \sigma_e^\beta \neq \sigma_e^\alpha \neq \sigma_e^3$ is shown by A_2 and A_2^*, where $A_2 u = -u''(t)$, $D(A_2) = \{u \in L_2(0, 1),\ u'\ \text{a.c.},\ u'' \in L_2(0, 1),\ u(0) = u(1) = u'(0) = u'(1) = 0\}$; A_2 is a properly closed symmetric operator with closed range, $\beta(A_2^*) = \alpha(A_2) = 0$, and $\alpha(A_2^*) = \beta(A_2) = \infty$ since $R(A_2)$ misses the functions $\sin n\pi x$ for odd n. That $\sigma_e^4 \neq \sigma_e^5$ in general is seen by taking A_3 to be the direct sum of the right shift A_1 and the left shift A_1^*, since then $A_3 \in F_4$ but 0 is a nonisolated point of $\sigma(A_3)$.

The six versions σ_e^2 through σ_e^5 are closed sets; that σ_e^1 is not closed in general can be seen, for example, as follows. Let $A_4 = \bigoplus_{m=1}^{\infty} A^{(m)}$, where $A^{(m)} = \left(\begin{bmatrix} 0 & 1 \\ 0 & m^{-1} + ek^{-1} \end{bmatrix} \right)$; this notation means that $A^{(m)}$ is the infinite direct sum of the kth block shown in brackets. It suffices to show that $R(A_4)$ is closed but that $R(A_4 - 1/n)$ is not closed, $n = 1, 2, 3, \ldots$; for this purpose we recall that a closed operator A has closed range if and only if $\gamma(A) > 0$, where $\gamma(A) = \inf \|Ax\| / d(x, N(A))$, $x \in D(A)$, $x \notin N(A)$, is the minimum modulus of A. It can be easily verified that $N(A^{(m)})=\operatorname{sp}\langle e_{2k-1}\rangle$, $\gamma(A)=1$, $N(A-1/n) = \{0\}$, and that $\gamma(A-1/n) = 0$; to show the last fact, one may use the minimizing sequence $x_k = \bigoplus_{m=1}^{\infty} x^{(m)}$, $x^{(m\neq n)}=0$, $x^{(n)} = \left(\begin{bmatrix} x_{2l-1} \\ x_{2l} \end{bmatrix}\right)$, $x_{2l-1}=x_{2l}=0$ for $l\neq k$, $\begin{bmatrix} x_{2k-1} \\ x_{2k} \end{bmatrix} = \begin{bmatrix} 1 \\ n^{-1} \end{bmatrix}$, for which one obtains $\|(A-1/n)x_k\|/\|x_k\| < e/k$.

3. Theorem 1

That Theorem 1, as shown by WEYL [32], holds for A a bounded self-adjoint operator under perturbation by B a compact self-adjoint operator, may be seen (e.g., see [26]) as follows. Any point λ_0 is shown to be in $\sigma_e(A)$ if and only if it has associated with it a singular sequence $\{x_n\}$, namely, a noncompact normalized sequence such that $x_n \rightharpoonup 0$, $(A-\lambda_0)x_n \to 0$; for more general operators, σ_e^α and σ_e^2 are similarly characterized, see [34] and [19]. Since B is completely continuous, $Bx_n \to 0$, and $\{x_n\}$ is a singular sequence for $A+B$ at λ_0; conversely, a singular sequence $\{x_n\}$ for $A+B$ at λ_0 is clearly a singular sequence for A at λ_0.

The following version of Theorem 1 was subsequently used by WEYL [33], and later by HARTMAN [15], in treating differential operators and, in particular, self-adjoint extensions of closed symmetric operators. If two closed operators A_0 and A are such that there is a λ_0 in $\varrho(A_0)\cap\varrho(A)$ such that $(\lambda_0-A_0)^{-1}-(\lambda_0-A)^{-1}$ is compact, then by Theorem 1 and the spectral mapping theorem one has $\sigma_e(A_0)=\sigma_e(A)$.

WOLF [34] extended Theorem 1, for $\sigma_e=\sigma_e^\alpha$, for the Hilbert space case, to B unbounded but with B relatively compact with respect to the given unperturbed closed operator A; this condition on B, namely, that B is compact on $D(A)$ equipped with the A-norm $\|x\|_A = \|x\|+\|Ax\|$, is also sufficient to guarantee that Theorem 1 holds for σ_e^2, σ_e^3, for a closed operator A in a Banach space, see KATO [19]. BROWDER [4, p. 111] proved that Theorem 1 holds for $\sigma_e=\sigma_e^5$ if A is closed in a Banach space with connected resolvent set $\varrho(A)$ dense in the complex plane, the resolvent set $\varrho(A+B)$ is not empty, and for some λ_0 in the resolvent set of A one has $B(\lambda_0-A)^{-1}$ compact; the last condition is equivalent to the property that B is A-compact. SCHECHTER [27] extended some of these results, and, in particular, showed that Theorem 1 holds for $\sigma_e=\sigma_e^4$; let us state this version here.

THEOREM 1'. *If A is closed in a Banach space and B is A-compact, $\sigma_e(A+B) = \sigma_e(A)$.*

The method of proof of theorems of this type is, roughly speaking, to reduce them to the bounded case by using operator norms, and then to use Fredholm theory to show that $\Delta_i(A+B) = \Delta_i(A)$, as mentioned previously in section 2; in this way one obtains Theorem 1′ for each of the five versions σ_e^2 through σ_e^4. That Theorem 1′ does not hold for σ_e^1 can be seen by letting A be the zero operator, $Be_n = n^{-1}e_n$; to negate Theorem 1′ for σ_e^5, let A be the bilateral shift $Ae_n = e_{n+1}$, $n = 0, \pm 1, \pm 2, \dots$, $Be_0 = -e_1$, $Be_n = 0$ for $n \neq 0$.

Three directions of extension of Theorem 1′ considered in [27] were: to the case where A is not closed but is closeable and (our terminology, for simplicity) B is A-precompact, see also REJTŐ [25]; to B which are only A-pseudo-compact, an extension which turned out to be of a priori value only, as observed in [12]; and to B which are only A^2-compact, as follows.

THEOREM 1″. *If A is closed, B is A^2-compact, and $\varDelta_4(A) \cap \varDelta_4(A+B)$ is non-empty, then $\sigma_e(A+B) = \sigma_e(A)$.*

In GUSTAFSON and WEIDMANN [12] it was shown that for self-adjoint A, one has an extension of Theorem 1″, as follows.

THEOREM 1‴. *If $A = A^*$, $D(B) \supset D(A)$, B is $g(A)$-compact for some locally bounded measurable function g, and $\varDelta_4(A) \cap \varDelta_4(A+B)$ is nonempty, then $\sigma_e(A+B) = \sigma_e(A)$.*

Although stated for σ_e^3 in [12], the proof is the same for $\sigma_e = \sigma_e^4$, and also clearly extends to normal operators A. For applications, the wide choice of g allowed by Theorem 1‴ could be of some value; on the other hand, the proof shows that in fact B is A^p-compact for all $p > 1$.

The following example (see [12]) shows that Theorem 1″ cannot be extended, without additional assumptions, to the case where B is A^3-compact. Let

$$A_5 = \left(\begin{bmatrix} 0 & n \\ 1 & 0 \end{bmatrix}\right), \quad B = \left(\begin{bmatrix} 0 & -n \\ -1 & n \end{bmatrix}\right),$$

$$D(A_5) = D(B) = \left\{ x \in l_2, \; \sum_{k=1}^{\infty} |x_{2k-1}|^2 + \sum_{k=1}^{\infty} |kx_{2k}|^2 < \infty \right\};$$

it follows that

$$A_5^3 = \left(\begin{bmatrix} 0 & n^2 \\ n & 0 \end{bmatrix}\right), \quad A_5 + B = \left(\begin{bmatrix} 0 & 0 \\ 0 & n \end{bmatrix}\right),$$

B is A_5^3-compact, $\sigma_e^4(A_5 + B) = \{0\}$, but $\sigma_e^4(A_5)$ is empty since

$$\alpha(A_5 - \lambda) = \beta(A_5 - \lambda) = 0 \quad \text{for} \quad \lambda \neq \pm \sqrt{n}, \quad n = 1, 2, 3, \dots,$$

and

$$\alpha(A_5 - \lambda) = \beta(A_5 - \lambda) = 1 \quad \text{for} \quad \lambda = \pm \sqrt{n}.$$

Let us add that Kato extended Theorem 1 to perturbation by the class of strictly singular operators, e.g. see GOLDBERG [9]; more generally, for bounded operators on a Banach space, KLEINECKE (Proc. Amer. Math. Soc. 14 (1963)) showed that Theorem 1 holds for perturbation by the class of inessential operators, namely, operators B whose images in the quotient algebra of bounded operators modulo the uniform closure of the finite rank operators belong to the radical.

Theorem 1 is rather important for many applications, other than those mentioned above; for example, we refer the reader to KATO [19] and REJTŐ [25] concerning the essential spectra of Schrödinger operators in quantum mechanics. There are, of course, also many related open questions concerning preservation of continuous and absolutely continuous spectra under perturbation; e.g. see [19] and BALSLEV and COMBES (Commun. math. Phys. **22** (1971)).

4. Theorem 2

WEYL [32] showed that an arbitrary bounded self-adjoint operator A on a separable Hilbert space (which is required throughout this section) could be converted to one with pure point spectra by perturbation by a compact self-adjoint operator B. Later von NEUMANN [22] extended this result to unbounded self-adjoint A and also showed that B could be taken to be Hilbert—Schmidt with arbitrarily small Hilbert—Schmidt norm; let us state that result, usually called the Weyl—von Neumann theorem.

THEOREM 2_0. *Every self-adjoint operator A on a separable Hilbert space can be written as the sum of a diagonal self-adjoint operator B and a compact self-adjoint operator K with K of arbitrarily small Hilbert—Schmidt norm.*

VON NEUMANN [22] then proved that Theorem 2 is true for bounded self-adjoint B and A. By Theorem 1 and Theorem 2_0 above, the proof is immediately reduced to constructing a unitary operator U such that $B - UAU^*$ is compact for the case where A and B both have pure point spectra and identical essential spectra; the latter case is then shown by a permutation argument. On the other hand, for the case of bounded self-adjoint operators, Theorem 2, which we will call here the von Neumann—Weyl Converse, implies Theorem 2_0, as follows; since $\sigma_e(A)$ is compact, one constructs in the usual way (i.e., by letting D have as eigenvalues a dense subset of $\sigma_e(A)$) a diagonal operator D with $\sigma(D)=\sigma_e(D)=\sigma_e(A)$, and then by Theorem 2 one has $A = B+K$, where $B=UDU^*$ is diagonal.

The extension of Theorem 2_0 to normal operators was recently obtained independently by BERG [3] and SIKONIA [31].

THEOREM $2_0'$. *Every bounded normal operator on a separable Hilbert space is the sum of a diagonal normal operator and a compact operator.*

In [3] a special orthonormal basis was constructed via the resolution of the identity of A; in [31] a special Haar basis was constructed in terms of the spectral representation of A; and in HALMOS [14] a simpler proof is given, obtaining the result for the normal case from that for the self-adjoint case. For some additional related investigations, see the discussion in HALMOS [13, Problem 4].

It was observed in GUSTAFSON and WEIDMANN [12], by the example $A_6 = \left(\begin{bmatrix} 0 & 1 \\ 0 & 0 \end{bmatrix} \right)$, that Theorem 2 cannot extend beyond the class of operators for which $\sigma_e(A) = \{0\}$ implies that A is compact. However, BERBERIAN [2] showed that Theorem 2 holds for unitary A and B such that $\sigma_e(A) = \sigma_e(B)$ is a proper subset of the unit circle, and asked whether Theorem 2 was true for normal operators. SIKONIA [31], using Theorem $2'_0$, obtained this extension, which we now state, as did EDGAR, ERNEST, and LEE [8], using Theorem $2'_0$ from [3].

THEOREM 2′. *If A and B are bounded normal operators on a complex separable Hilbert space, with $\sigma_e(A) = \sigma_e(B)$, then there exists a unitary operator U and a compact operator K such that $B = UAU^* - K$.*

The following example shows that Theorem 2 does not even extend to the case where A is a diagonal normal operator and B is a quasinormal operator. Let B be the right shift A_1, and let A be a diagonal operator A_7 with $\sigma(A_7) \equiv \sigma_e(A_7) = = \sigma_e(A_1) =$ the closed unit disc, with $0 \in \sigma_c(A_7)$. It is then impossible that $UA_7U^* = = A_1 + K$ for some compact K and unitary U, because $A_1 + K$, being Fredholm, has closed range, whereas UA_7U^* does not have closed range.

In like manner, let us observe some limitations on the extendability of Theorem 2_0 beyond normal operators. Let B (quasinormal) be the right shift A_1, and suppose $A_1 = D + K$, D diagonal, K compact; then $D = A_1 - K$ is Fredholm with index $i(D) = i(A_1) = -1$, which is impossible. Similarly, let B (triangular) be the left shift $A_8 = A_1^*$; A_8 cannot be of the form $D + K$, because then $D = A_8 - K$ would be Fredholm with index $i(D) = 1$. Somewhat more sharply, consider (quasidiagonal, the class of operators discussed in [13, Problem 4]) $B = A_6$ given above; if $A_6 = D + K$, D diagonal, K compact, then $\sigma_e(D) = \sigma_e(A_6 - K) = \sigma_e(A_6) = \{0\}$, so that D is compact, but A_6 is certainly not compact. A_6 thus serves to illustrate that neither Theorem 2_0 nor Theorem 2 extend to quasidiagonal operators.

Theorem 1 states that if A and B differ by a compact operator, then $\sigma_e(A) = = \sigma_e(B)$. Theorem 2 is a converse statement, stating that if $\sigma_e(A) = \sigma_e(B)$, then A and B essentially differ by a compact operator; Theorem 2_0 is also a converse, stating that among the operators B such that $\sigma_e(A) = \sigma_e(B)$, there is a diagonal operator D such that $A - D$ is compact. Another converse, observed in [12], is as follows.

THEOREM 2_{00}. *If, for a given bounded operator B on a Hilbert space, one has $\sigma_e(A + B) = \sigma_e(A)$ for all bounded operators A, then B is compact.*

We observe here that Theorem 2_{00} is not true in general for Banach spaces, since it can be seen that the class of operators B so determined lies somewhere between the class of strictly singular operators and the class of Riesz operators $(\sigma_e(B) = \{0\})$. Let us add, however, that the "somewhere" can be made precise; that is, Theorem 2_{00} has the following extension to Banach spaces.

THEOREM $2'_{00}$. *For a given bounded operator B on a Banach space, one has $\sigma_e(A+B) = \sigma_e(A)$ for all bounded operators A if and only if B is inessential.*

According to the paper of Kleinecke mentioned at the end of section 3, to prove Theorem $2'_{00}$ it suffices to show that the class $\{B\}$ of bounded operators B satisfying $\sigma_e(A+B) = \sigma_e(A)$ for all bounded operators A is an ideal of operators possessing the property $\sigma_e(B) = \{0\}$. Since for B_1 and B_2 in $\{B\}$ one has $\sigma_e(A + \mu_1 B_1 + \mu_2 B_2) = \sigma_e(A + \mu_1 B_1) = \sigma_e(A)$ for arbitrary bounded A and scalars μ_1 and μ_2, $\{B\}$ is a subspace and $\sigma_e(B) = \{0\}$. $\{B\}$ will also be invariant under left multiplication by bounded operators if $A + CB$ is Fredholm of index zero whenever A is Fredholm of index zero, $B \in \{B\}$, C an arbitrary bounded operator; this follows by applying the Fredholm product formula $i(A_1 A_2) = i(A_1) + i(A_2)$ to the expression $A + CB = (\lambda + C)\ [\lambda I + C)^{-1} (A - \lambda B) + B]$ for $|\lambda| > \|C\|$. The argument for right multiplication is similar.

Let us also mention that Theorem 2_0 was generalized by KURODA [20] to show that, for the self-adjoint case there, one can obtain K arbitrarily small in any cross norm, with the exception of the trace norm or its equivalent. Also, Berg (preprint) has shown that K in Theorem $2'_0$ may be taken from C^p for any $p > 2$, but it remains unknown if K can be taken from the Hilbert—Schmidt class.

5. Theorem 3

Using current terminology, one says that Weyl's Theorem holds for an operator A if the statement in Theorem 3 is true for A; in [32] WEYL showed that Theorem 3 holds for bounded self-adjoint operators on a Hilbert space. Operators satisfying Weyl's Theorem 3 are those, then, whose essential spectra have the same character as in the self-adjoint case, in the sense that the set of points of the spectrum that are removable by arbitrary compact perturbations is exactly the set of isolated eigenvalues of finite (geometric) multiplicity.

That Theorem 3 remains true for normal operators was observed by J. SCHWARTZ [29]; L. A. COBURN [5] extended Theorem 3 to hyponormal operators and Toeplitz operators; V. ISTRATESCU [17] showed (condition (β) of [1]) that if each point of the spectrum $\sigma(A)$ is a bare point of $\sigma(A)$, i.e., if it lies on the circumference of some closed disc that contains $\sigma(T)$, and if (condition α of [1]) the restriction A_M of A to each of its invariant subspaces M has a normaloid resolvent (i.e., $\|(\lambda - A_M)^{-1}\|^{-1} = = d(\lambda, \sigma(A_M)))$, then Theorem 3 holds. S. K. BERBERIAN [1], see also [2], showed that sufficient for Theorem 3 to hold is that (1) (condition (β') of [1]) each eigenvalue of finite multiplicity is a semibare point of $\sigma(A)$, i.e., it lies on the circumference of some closed disc containing no other point of $\sigma(A)$, and (2) (condition (α') of [1]) the restriction A_M of A to each of its reducing subspaces M has a normaloid resolvent. More generally, BERBERIAN [1] showed that Theorem 3 holds for any A a bounded operator on a Hilbert space if (1) A is reduced by each of its finite dimensional eigenspaces, and (2) the restriction A_M of A to each of its reducing subspaces M has the property that every isolated point of $\sigma(A_M)$ is an eigenvalue of A_M (this is condition (α''') of [1]).

The following theorem, see [10], characterizes rather completely, from several viewpoints, when Theorem 3 holds for an arbitrary closed operator A in a Banach space. We recall the following well-known quantities for operators: the minimum modulus function $\gamma(A-\lambda)$ mentioned previously; the ascent of A at λ is the smallest n such that $N((A-\lambda)^n) = N((A-\lambda)^{n+1})$; the descent of A at λ is the smallest n such that $R((A-\lambda)^n) = R((A-\lambda)^{n+1})$; and the algebraic multiplicity of an isolated point λ_0 of $\sigma(A)$ is dim $P(X)$, where $P = (2\pi i)^{-1} \int_{\Gamma} (\lambda - A)^{-1} d\lambda$ for Γ any rectifiable simple closed curve containing λ_0 in its interior and the rest of $\sigma(A)$ in its exterior. Let us also say that A satisfies the "eigenspace gap condition" at an eigenvalue λ_0 if there exists a sequence $\{\lambda_n\}$ such that $\lambda_n \to \lambda_0$ and either $|\lambda_n - \lambda_0| = o(\delta(\lambda_n, \lambda_0))$ or $N(A-\lambda_n) = \{0\}$, where $\delta(\lambda_n, \lambda_0) \equiv \delta(N(A-\lambda_n), N(A-\lambda_0)) \equiv \sup d(x_{\lambda_n}, N(A-\lambda_0))$, $x_{\lambda_n} \in N(A-\lambda_n)$, $\|x_{\lambda_n}\| = 1$, is the gap between the subspaces $N(A-\lambda_n)$ and $N(A-\lambda_0)$.

THEOREM 3'. *Each of the following is a necessary and sufficient condition for* $\sigma_e(A) = \sigma(A) - \pi_{00}(A)$.
  (1) *Partially reducing and closed range.*
       (1i) *For every λ in Δ_4^s, $\{N(A-\lambda)\} - \{R((A-\lambda)^n)\}$ is nontrivial, for some n, which may depend on λ, and*
       (1ii) *for every λ in π_{00}, $R(A-\lambda)$ is closed.*
  (2) *Discontinuity.*
       (2) *$\gamma(A-\lambda)$ is discontinuous at every λ in $\Delta_4^s \cup \pi_{00}$.*
  (3) *Ascent and descent.*
       (3i) *A has finite ascent at every λ in Δ_4^s, and*
       (3ii) *A has finite descent at every λ in π_{00}.*
  (4) *Growth rate.*
       (4i) *Every λ in Δ_4^s satisfies the eigenspace gap condition, and*
       (4ii) *every λ in π_{00} also has finite algebraic multiplicity.*
  (5) *Boundary of spectrum.*
       (5i) *$\Delta_4^s \subset \partial\sigma$, and*
       (5ii) *every λ in π_{00} is a pole of the resolvent operator.*

For related results and corollaries we refer to [10]. Also, in a preprint [28] SCHECHTER has obtained the necessary and sufficient condition (5i), (1ii), and related conditions on invariant and reducing subspaces, more along the lines of [1], [2], [17], but in a Banach space setting.

In [1] and [28] it is shown that Theorem 3 extends to bounded seminormal operators; the following result from [10] extends that fact to unbounded operators.

THEOREM 3″. *For A a formally seminormal operator, $\sigma_e(A) = \sigma(A) - \pi_{00}(A)$.*

Here A is said to be formally seminormal if either A or A^* is formally hyponormal, where A is said to be formally hyponormal if $\|Ax\| \geqq \|A^*x\|$ on $D(A) \subset \subset D(A^*)$. This class of operators includes the formally normal operators studied by CODDINGTON [6] and operators of the type considered in NELSON [21], SEGAL [30], and others; it is also more general than the unbounded seminormal operators recently introduced in ISTRĂŢESCU [18] and DAVID [7]. Let us mention that the spectral substructure possibilities are the same as for the bounded case, since formal hyponormality is translation invariant and the formally hyponormal and formally seminormal operators are easily seen to have the same state diagrams as given in [11] earlier for the bounded case.

By means of the following examples we indicate the sharpness of Theorem 3′ and 3″.

Let A_9 be the weighted left shift operator given by $A_9 e_n = n^{-1} e_{n-1}$, $A_9 e_1 = \{0\}$; then $\sigma(A_9) = \sigma_e(A_9) = \pi_{00}(A_9) = \{0\}$ so that Theorem 3 does not hold for A_9 even though A_9 satisfies both (α''') and (β'), the weakest pair of conditions mentioned in [1, 2]. Let A_{10} be the direct sum of a one dimensional zero operator and a right shift (e.g., A_9^*) weighted as above; then $\sigma(A_{10}) = \sigma_e(A_{10}) = \pi_{00}(A_{10}) = \{0\}$, so that Theorem 3 does not hold even though both of the conditions (1i) and (3i) are satisfied on π_{00}.

The following final counterexample, which is convexoid, linearly normaloid, and in class C_5 of [16], shows that Theorem 3 does not appreciably extend among normal-like operators beyond the seminormal ones; in addition (with $a=1$) it limits extension of the recent result of C. R. PUTNAM [24] that there are no bounded nonnormal seminormal operators with meas $\sigma(A) = 0$. Let A_{11} be the direct sum of a normal "annulus" operator B_1 with $\sigma(B_1) = \sigma_e(B_1) = T_a = \{\lambda | 0 < a \leqq |\lambda| \leqq 1\}$, a one dimensional zero operator B_2, and $B_3 = A_9^*$ as above. Then $\sigma(A_{11}) = \sigma_e(A_{11}) = = T_a \cup \{0\}$, whereas $\overline{W(A_{11})}$ is the closed unit disc; the latter is a spectral set, in the VON NEUMANN sense [23] (also see [26]) for A_{11}, so that A_{11} is in class C_5 of [16]. But Theorem 3 does not hold for A because $\{0\}$ is an isolated eigenvalue of finite (geometric) multiplicity.

The example A_{11} raises the possibility of obtaining for bounded operators on a Hilbert space a characterizing condition (6) in Theorem 3′ in terms of spectral set behaviour, or at least a sufficient condition; this might come about if more were known about the exact relationships between the normal-like operators and the "spectral-set-like" operator classes C_i of [16]. For example, it is an immediate corollary of Theorem 3′ that if A is in class C_6, i.e., $\sigma(A)$ is a spectral set for A, then $\pi_{00}(A) \subset \Delta_4^s(A)$, so that Theorem 3 holds for A if any of the conditions given in Theorem 3 for $\Delta_4^s \subset \pi_{00}$ are satisfied. Also it is known ([11] and [16]) that one has the operator class inclusions $\{\text{normal}\} \subset \{\text{subnormal}\} \subset C_6 \subset C_5 = C_4 \subset C_3$, $\{\text{seminormal}\} \subset C_3$, and by the example A_{11} and Theorem 3″, that $C_5 \not\subset \{\text{semi-normal}\}$.

However, here the example A_3 is a little sharper than A_{11}; since $\Delta_4^s(A_3)$ is the open unit disc, one can observe that Weyl's Theorem 3 does not even extend to the class C_6, and that $C_6 \not\subset \{\text{seminormal}\}$.

REFERENCES

[1] S. K. Berberian, *An extension of Weyl's Theorem to a class of not necessarily normal operators.* Michigan Math. J. **16** (1969), 273—279.

[2] S. K. Berberian, *The Weyl spectrum of an operator.* Indiana Univ. Math. J. **20** (1970), 529—544.

[3] I. D. Berg, *An extension of the Weyl-von Neumann theorem to normal operators.* Trans. Amer. Math. Soc. **160** (1971), 365—371.

[4] F. E. Browder, *On the spectral theory of elliptic differential operators I.* Math. Ann. **142** (1961), 22—130.

[5] L. A. Coburn, *Weyl's theorem for nonnormal operators.* Michigan Math. J. **13** (1966), 285—288.

[6] E. A. Coddington, *Formally normal operators having no normal extensions.* Canad. J. Math. **17** (1965), 1030—1040.

[7] M. David, *Commutators of two operators, one of which is unbounded and seminormal.* Ann. Mat. Pura Appl. (4) **83** (1969), 185—194.

[8] G. Edgar, J. Ernest, and S. G. Lee, *Weighing operator spectra.* Indiana Univ. Math. J. **21** (1971), 61—79.

[9] S. Goldberg, *Unbounded linear operators.* McGraw-Hill, New York 1966.

[10] K. Gustafson, *Necessary and sufficient conditions for Weyl's Theorem.* Michigan Math. J. **19** (1972), 71—81.

[11] K. Gustafson, *State diagrams for Hilbert space operators.* J. Math. Mech. **18** (1968), 33—46.

[12] K. Gustafson and J. Weidmann, *On the essential spectrum.* J. Math. Anal. Appl. **25** (1969), 121—127.

[13] P. Halmos, *Ten Problems in Hilbert space.* Bull. Amer. Math. Soc. **76** (1970), 887—933.

[14] P. Halmos, *Continuous functions of Hermitian Operators.* Proc. Amer. Math. Soc. **31** (1972), 130—132.

[15] P. Hartman, *On the essential spectra of symmetric operators in Hilbert space.* Amer. J. Math. **75** (1953), 229—240.

[16] S. Hildebrandt, *Über den numerischen Wertebereich eines Operators.* Math. Ann. **163** (1966), 230—247.

[17] V. Istrătescu, *Weyl's theorem for a class of operators.* Rev. Roumaine Math. Pures Appl. **13** (1968), 1103—1105.

[18] V. Istrățescu, *On a lemma of O'Raifertaigh and Segal.* J. Indian Math. Soc. **33** (1969), 57—58.

[19] T. Kato, *Perturbation theory for linear operators.* Springer, Berlin 1966.

[20] S. T. Kuroda, *On a theorem of Weyl-von Neumann.* Proc. Japan Acad. **34** (1958), 11—15.

[21] E. Nelson, *Analytic vectors.* Ann. Math. **70** (1959), 572—615.

[22] J. von Neumann, *Charakterisierung des Spektrums eines Integraloperators.* Actualités Sci. Indust. **229** (1935), 1—20.

[23] J. von Neumann, *Eine Spectraltheorie für allgemeine Operatoren eines unitären Raumes.* Math. Nachr. **4** (1951), 258—281.

[24] C. R. Putnam, *An inequality for the area of hyponormal spectra.* Math. Z. **116** (1970), 323—330.

[25] P. Rejtő, *On the essential spectrum of the hydrogen energy and related operators.* Pacific J. Math. **19** (1966), 109—140.

[26] F. Riesz and B. Sz.-Nagy, *Functional analysis.* Ungar, New York 1955.

[27] M. Schechter, *On the essential spectrum of an arbitrary operator I.* J. Math. Anal. Appl. **13** (1966), 205—215.

[28] M. Schechter, *Operators obeying Weyl's Theorem.* (to appear).

[29] J. T. Schwartz, *Some results on the spectra and spectral resolutions of a class of singular interag operators.* Comm. Pure Appl. Math. **15** (1962), 75—90.

[30] I. Segal, *An extension of a theorem of L. O'Raifeartaigh.* J. Functional Analysis **1** (1967), 1—21.

[31] W. Sikonia, *The von Neumann converse of Weyl's Theorem.* Indiana Univ. Math. J. **21** (1971), 121—124.

[32] H. Weyl, *Über beschränkte quadratische Formen, deren Differenz vollstetig ist.* Rend. Circ. Mat. Palermo **27** (1909), 373—392.

[33] H. Weyl, *Über gewöhnliche Differentialgleichungen mit Singularitäten und die zugehörigen Entwicklungen willkürlicher Funktionen.* Math. Ann. **68** (1910), 220—269.

[34] F. Wolf, *On the essential spectrum of partial differential boundary problems.* Comm. Pure Appl. Math. **12** (1959), 211—228.

[35] F. Wolf, *On the invariance of the essential spectrum under a change of boundary conditions of partial differential boundary operators.* Indag. Math. **21** (1959), 142—147.

Some Approximation Problems in the Theory
of Stationary Processes

By

YU. A. ROZANOV

STEKLOV MATHEMATICAL INSTITUTE
MOSCOW

Introduction

We consider a stationary process as the pair $\{A, V_t\}$, where A is a separable subspace of some Hilbert space with a group of unitary operators V_t, $-\infty < t < \infty$. We suppose that «time» runs through $0, \pm 1, \ldots$, though our method gives similar results for continuous parameter t.

For our purposes, we can assume that there exists the so-called *spectral density* f_λ, $-\pi \leq \lambda \leq \pi$. It is convenient to use $f_\lambda^{1/2}$ instead of f_λ; $f_\lambda^{1/2}$, $-\pi \leq \lambda \leq \pi$, can be an arbitrary measurable selfadjoint operator-valued function in the Hilbert space A such that[1])

$$\int_{-\pi}^{\pi} \|f_\lambda^{1/2} a\|^2 \, d\lambda = \|a\|^2 \qquad (a \in A).$$

If $L^2(A)$ is the Hilbert space of all A-valued measurable functions $a(\lambda)$ $\int_{-\pi}^{\pi} \|a(\lambda)\|^2 \, d\lambda < \infty$, with the inner product

$$\int_{-\pi}^{\pi} (a_1(\lambda), a_2(\lambda)) \, d\lambda$$

and $f_\lambda^{1/2} A$ means the subspace in $L^2(A)$ of all functions $a(\lambda) = f_\lambda^{1/2} a$, $a \in A$, then the pair $\{f_\lambda^{1/2} A, e^{i\lambda t}\}$ with the unitary operators of multiplication by the corresponding scalar function $e^{i\lambda t}$ is isometric to the stationary process $\{A, V_t\}$ with the spectral density $f_\lambda = [f_\lambda^{1/2}]^2$:

$$(V_t a_1, a_2) = \int_{-\pi}^{\pi} e^{i\lambda t} (f_\lambda^{1/2} a_1, f_\lambda^{1/2} a_2) \, d\lambda.$$

A similar general model of stationary processes was first suggested in connec-

---

[1]) This condition is not essential; we can only assume that

$$\int_{-\pi}^{\pi} \|f_\lambda^{1/2} a\|^2 \, d\lambda < \infty.$$

tion with the so-called interpolation problem [1, 1957]. We use here the method of that paper by considering the regularity which means that

$$(1) \qquad\qquad \bigcap_t H_t = 0,$$

where

$$H_t = \bigvee_{s<t} e^{i\lambda s} f_\lambda^{1/2} A$$

is the subspace generated by all functions $e^{i\lambda s} f_\lambda^{1/2} a$, $a \in A$, $s < t$.

Recall that this condition appeared in connection with the *extrapolation problem*; namely, the stationary process is called *regular* if for the best approximation $a_t(\lambda)$ of the function $a(\lambda) = f_\lambda^{1/2} a$ by functions $h(\lambda)$ from the subspace H_t satisfying

$$\int_{-\pi}^{\pi} \|a(\lambda) - a_t(\lambda)\|^2 \, d\lambda = \min_{h \in H_t} \int_{-\pi}^{\pi} \|a(\lambda) - h(\lambda)\|^2 \, d\lambda,$$

the following condition holds true

$$\lim_{t \to -\infty} a_t(\lambda) = 0 \quad \text{for all} \quad a \in A.$$

The question on the regularity was posed and completely solved by KOLMOGOROV (1939) for one-dimensional[2]) stationary processes. The new development of multi-dimensional processes started in 1957 and the regularity was considered by many mathematicians (the historical and bibliographic references can be found in [2, 1963]). In particular, it was noticed that condition (1) is equivalent to *the factorization*

$$(2) \qquad\qquad f_\lambda = \varphi_\lambda^* \varphi_\lambda$$

of the spectral density f_λ by an "analytical" operator-valued function φ_λ from A into some Hilbert space B, more exactly, φ_λ is such that the scalar functions $(\varphi_\lambda a, b)$ belong to the so-called analytical class H^2 for all $a \in A$, $b \in B$. The factorization problem being interesting for analysis was considered by many authors (see, for example, [3, 1967]).

With all these researches the question concerning the regularity (factorization) was completely solved in the finite-dimensional case, but one can not say this for the *infinite-dimensional* stationary processes. We give here the general solution which contains all previous results.

1. We need the following definition. Let $\varDelta$ be a subspace in $L^2(A)$ and $S = \{a_k(\lambda),\ k=1, 2, \ldots\}$ a complete system in $\varDelta$. Denote by $\varDelta_S(\lambda)$ the linear manifold of all values $a_k(\lambda) \in A$, $k=1, 2, \ldots$. It is clear that the closure $\overline{\varDelta(\lambda)} = \overline{\varDelta_s(\lambda)}$ does

---

²) The dimension of the stationary process $\{A, V_t\}$ is dim A.

not depend on the system S in the sense that $\overline{A_{S_1}(\lambda)} = \overline{A_{S_2}(\lambda)}$ for almost all λ, where S_1, S_2 are any complete systems in A. We call $\overline{A(\lambda)}$, $-\pi \leq \lambda \leq \pi$, the *space-function* generated by $A \subseteq L^2(A)$.

We assume, for the save of simplicity, that the spectral density f_λ, $-\pi \leq \lambda \leq \pi$, is bounded, though this condition can be omitted after some modification of the space B, which we determine now as the subspace of "analytical" functions $b(\lambda) \in$ $\in L^2(A)$:

$$(3) \qquad \int_{-\pi}^{\pi} e^{i\lambda t}(b(\lambda), a)\, d\lambda = 0 \qquad (t < 0)$$

(i.e. the scalar functions $(b(\lambda), a)$ belong to the class H^2) such that

$$(4) \qquad b(\lambda) \in f_\lambda^{1/2} A \quad \text{for almost all } \lambda$$

and

$$(5) \qquad \int_{-\pi}^{\pi} \| f_\lambda^{-1/2} b(\lambda) \|^2\, d\lambda < \infty,$$

where we take $f^{-1/2}$ as the inverse operator from the subspace $f_\lambda^{1/2} A$ to $f_\lambda^{1/2} A$ that uniquely determines $f_\lambda^{-1/2}$.

Let $\overline{B(\lambda)}$ be the space-function generated by the subspace B.

THEOREM 1. *For the regularity (factorization) it is necessary and sufficient that*

$$(6) \qquad \overline{f_\lambda A} = \overline{B(\lambda)} \quad \text{for almost all } \lambda.$$

Let us discuss this condition before giving a proof of the theorem.

In the case of *one-dimensional* stationary processes with the scalar spectral density $f(\lambda)$ condition (6) means that

$$(7) \qquad \int_{-\pi}^{\pi} \frac{|b(\lambda)|^2}{f(\lambda)}\, d\lambda < \infty$$

for some "analytical" function $b(\lambda) \in H^2$ and, as it follows from (7),

$$\int_{-\pi}^{\pi} \log \frac{|b(\lambda)|^2}{f(\lambda)}\, d\lambda = \int_{-\pi}^{\pi} \log |b(\lambda)|^2\, d\lambda - \int_{-\pi}^{\pi} \log f(\lambda)\, d\lambda < \infty,$$

which is equivalent to the well known condition

$$(8) \qquad \int_{-\pi}^{\pi} \log f(\lambda)\, d\lambda > -\infty.$$

Similarly in the case of n-dimensional stationary processes with a non-degenerated matrix-valued spectral density $f(\lambda)$ the condition (6) means that there is some m-dimensional non-degenerated "analytical" matrix $b(\lambda)=\{b_{k_j}(\lambda)\}$ for which

$$(9) \qquad \int_{-\pi}^{\pi} \operatorname{Sp}\left[b^*(\lambda)f^{-1}(\lambda)b(\lambda)\right] d\lambda < \infty$$

(the vector functions $b_k(\lambda)=\{b_{k1}(\lambda), \ldots, b_{kn}(\lambda)\}$, $k=1, \ldots, n$, are the basis of the corresponding subspace B); as it follows from (9),

$$\int_{-\pi}^{\pi} \log \det\left[b^*(\lambda)f^{-1}(\lambda)b(\lambda)\right] d\lambda =$$

$$= \int_{-\pi}^{\pi} \log |\det b(\lambda)|^2 \, d\lambda - \int_{-\pi}^{\pi} \log\left[\det f(\lambda)\right] d\lambda < \infty$$

and, since the function $\det b(\lambda)$ belongs to the class $H^{2/n}$, we have

$$(10) \qquad \int_{-\pi}^{\pi} \log\left[\det f(\lambda)\right] d\lambda > -\infty.$$

In the general case, the condition (6) gives us the following. For any "analytical" functions $b_1, \ldots, b_n \in B$ the gramian of their values $b_1(\lambda), \ldots, b_n(\lambda)$ belongs to the class $H^{1/n}$ and the subspace $\bigvee_{1 \leq k \leq n} b_k(\lambda)$ has a constant dimension for almost all λ, so

$$\dim B(\lambda)=\dim f_\lambda A =\text{constant}.$$

If we take any "subprocesses" $\{f_\lambda^{1/2} A_0, e^{i\lambda t}\}$, $A_0 \subseteq A$, then

$$(11) \qquad \dim f_\lambda A_0 =\text{constant}$$

for any subspace $A_0 \subseteq A$. Evidently, in the case of *finite-rank* stationary processes:

$$(12) \qquad \dim f_\lambda A =m<\infty,$$

there is some m-dimensional subprocess $\{f_\lambda^{1/2} A_0, e^{i\lambda t}\}$, $A_0 \subseteq A$, such that $\dim f_\lambda A_0 =m$ and hence

$$f_\lambda A_0 =f_\lambda A \quad \text{for almost all } \lambda.$$

If we take some basis $a_1, \ldots, a_m \in A_0$, then the corresponding matrix-valued spectral density $f(\lambda)=\{f_{kj}(\lambda)\}$, where $f_{kj}(\lambda)=(f_\lambda a_k, a_j)$, $k=1, \ldots, m$, satisfies condition (10). Since $f_\lambda a=f_\lambda a(\lambda)$ for the projection $a(\lambda)$ of $a \in A$ on the subspace $f_\lambda A \subseteq A$, we have

$$\inf_{\|a\|=1} \|f_\lambda a\| \leq \inf_{\|a(\lambda)\|=1} \|f_\lambda a(\lambda)\| = \|f_\lambda^{-1}\|^{-1},$$

where f_λ^{-1} is the inverse operator from $f_\lambda A$ to $\overline{f_\lambda A}$; remind that $\delta(\lambda)= \inf_{\|a\|=1} \|f_\lambda a\|,$

$a \in A_0$, is the smallest eigen-value of the matrix $f(\lambda)$ and from (10) we get[3])

$$(13) \qquad \int_{-\pi}^{\pi} \log \|f_\lambda^{-1}\|^{-1} \, d\lambda > -\infty.$$

Suppose that for any stationary process the condition (13) holds true. Then there is a scalar "analytical" function $\theta(\lambda) \in H^2$ such that $|\theta(\lambda)|^2 = \|f_\lambda^{-1}\|^{-1}$; besides, $\|f_\lambda^{-1}\| < \infty$ for almost all λ and the subspace $f_\lambda A$ is closed: $f_\lambda A = \overline{f_\lambda A}$. Suppose that the space-function $f_\lambda A$ is generated by some "analytical" functions $a(\lambda)$, $(a(\lambda), a) \in H^2$ for all $a \in A$. Then $f_\lambda A$ is also generated by the corresponding function $b(\lambda) = \theta(\lambda)a(\lambda)$ and obviously $b(\lambda) \in B$, so the condition (6) of our theorem is satisfied; in particular, for the *non-degenerated* spectral density f_λ we have $f_\lambda A = A$ and we can take $b(\lambda) = \theta(\lambda)a$, $a \in A$, so under the condition (13) the stationary process is regular.

PROOF OF THE THEOREM. Let Δ be some subspace of $L^2(A)$ and $\delta_1(\lambda), \delta_2(\lambda), \ldots$ be some complete system of functions in Δ. Clearly, if the function $a(\lambda)$ belongs to the closed linear manifold $\bigvee\limits_{-\infty < t < \infty} e^{i\lambda t} \Delta$, then for almost all λ it coincides with the limit of some linear form

$$\sum_n \left[\sum_k c_{kn} e^{i\lambda k} \right] \delta_k(\lambda) \in \Delta(\lambda),$$

so that

$$(14) \qquad a(\lambda) \in \overline{\Delta(\lambda)} \quad \text{for almost all } \lambda.$$

LEMMA. *The subspace* $\bigvee\limits_{-\infty < t < \infty} e^{i\lambda t} \Delta$ *consists of all functions* $a(\lambda) \in L^2(A)$ *satisfying condition* (14).

In fact, since the inner product $(a'(\lambda), a''(\lambda))$ is measurable in λ for any $a'(\lambda), a''(\lambda) \in L^2(A)$ the projection $a_n(\lambda)$ of $a(\lambda) \in A$ on $\bigvee\limits_{1 \leq k \leq n} \delta_k(\lambda)$ as a function of λ is measurable. Obviously this function $a_n(\lambda)$ is in the subspace $\bigvee\limits_{-\infty < t < \infty} e^{i\lambda t} \Delta$, and $\lim\limits_{n \to \infty} \|a(\lambda) - a_n(\lambda)\| = 0$ for almost all λ; since

$$\|a(\lambda) - a_n(\lambda)\|^2 \leq \|a(\lambda)\|, \quad \text{where} \quad \int_{-\pi}^{\pi} \|a(\lambda)\|^2 \, d\lambda < \infty,$$

---

[3]) Note that this condition, which was obtained from (6) in the case of a finite-rank, in general is not necessary for regularity. Say, if $f_\lambda a_k = f_k(\lambda)a_k$, $k = 1, 2, \ldots$, where $a_1, a_2, \ldots \in A$ is some basis and $f_1(\lambda) \geq f_2(\lambda) \geq \cdots$ are some "scalar" spectral densities:

$$\int_{-\pi}^{\pi} \log f_k(\lambda) > -\infty, \qquad \int_{-\pi}^{\pi} \lim_{k \to \infty} [\log f_k(\lambda)] \, d\lambda = -\infty,$$

then the stationary process with the non-degenerated spectral density f_λ is regular but the condition (13) is not satisfied.

we have

$$\lim \int_{-\pi}^{\pi} \|a(\lambda)-a_n(\lambda)\|^2 \, d\lambda = 0,$$

i.e.

$$a(\lambda) \in \bigvee_{-\infty < t < \infty} e^{i\lambda t} \Delta.$$

In particular *the space*

$$H = \bigvee_{-\infty < t < \infty} e^{i\lambda t} f_\lambda^{1/2} A$$

consists of all functions $a(\lambda) \in L^2(A)$, *such that*

(15) $$a(\lambda) \in \overline{f_\lambda^{1/2} A} \quad \text{for almost all } \lambda.$$

Applying the method of [1, 1957], consider the subspace

$$\Delta = H \ominus H_0$$

(remember that $H_0 = \bigvee_{s<0} e^{i\lambda s} f_\lambda^{1/2} A$). Evidently the regularity is equivalent to the condition

(16) $$\bigvee_{-\infty < t < \infty} e^{i\lambda t} \Delta = H.$$

Every function $\delta(\lambda) \in \Delta$ is such that

(17) $$\int_{-\pi}^{\pi} e^{i\lambda t}(f_\lambda^{1/2} a, \delta(\lambda)) \, d\lambda = \int_{-\pi}^{\pi} e^{i\lambda t}(a, f_\lambda^{1/2} \delta(\lambda)) \, d\lambda = 0$$

for all $t<0$, so the functions $b(\lambda)=f_\lambda^{1/2} \delta(\lambda)$ satisfy conditions (3)—(5).

Besides, for any function $b(\lambda) \in B$, the corresponding function $\delta(\lambda)=f_\lambda^{-1/2} b(\lambda)$ belongs to the space $H = \bigvee_{-\infty < t < \infty} e^{i\lambda t} f_\lambda^{1/2} A$ and satisfies condition (10), i.e. $\delta(\lambda) \in \Delta$.

The theorem is proved.

The stationary process $\{f_\lambda^{1/2} A, e^{i\lambda t}\}$ is called singular if it can be extrapolated from "the past" $H_0 = \bigvee_{s<0} e^{i\lambda s}[f_\lambda^{1/2} A]$:

$$f_\lambda^{1/2} A \subseteqq H_0.$$

As follows from the proof of theorem 1, the following result holds.

THEOREM 2. *The stationary process is singular if and only if for any "analytical" function* $b(\lambda)$ *satisfying the conditions* (3)—(4),

(18) $$\int_{-\pi}^{\pi} \|f_\lambda^{-1/2} b(\lambda)\|^2 \, d\lambda = \infty.$$

2. We have mentioned that our method is similar to that of the paper [1, 1957], where the interpolation problem was solved. In connection with this problem let us consider the best approximation $a_t(\lambda)$ of the function $a(\lambda)=f_\lambda^{1/2}a$, $a\in A$, by functions from the subspace

$$H_{-t}^t = \bigvee_{|s|>t} e^{i\lambda s}[f_\lambda^{1/2}A] \qquad (t>0).$$

We say that the stationary process is *double-regular* if $\lim_{t\to\infty} a_t(\lambda)=0$, $a\in A$, what is equivalent to the condition

(19)
$$\bigcap_t H_{-t}^t = 0.$$

Let B_0 be the subspace of all polynomials $b(\lambda) = \sum_0^n b_k e^{i\lambda k}$ from B:

$$b(\lambda)\in f_\lambda^{1/2}A \quad \text{for almost all} \quad \lambda$$

and

$$\int_{-\pi}^{\pi} \|f_\lambda^{-1/2}b(\lambda)\|^2\,d\lambda < \infty.$$

Applying the same method as in the proof of theorem 1, we can get the following.

THEOREM 3. *For the double-regularity it is necessary and sufficient that*

(20)
$$\overline{f_\lambda A} = \overline{B_0(\lambda)} \quad \text{for almost all} \quad \lambda.$$

Note that this condition is satisfied if for some scalar polynomial $P(e^{i\lambda})$

(21)
$$\int_{-\pi}^{\pi} \frac{|P(e^{i\lambda})|^2}{\|f_\lambda^{-1}\|^{-1}}\,d\lambda < \infty.$$

It is worth saying about stationary processes which can be interpolated:

(22)
$$f_\lambda^{1/2}A \subseteq H_{-t}^t \qquad (t>0).$$

Similar to Theorem 2, relation (22) is satisfied if and only if

(23)
$$\int_{-\pi}^{\pi} \|f_\lambda^{-1/2}b(\lambda)\|^2\,d\lambda = \infty$$

for any polynomial $b(\lambda)$.

In conclusion we consider a *minimal system* a_j, $j=1, 2, \ldots$, in A (i.e. $a_k \notin \bigvee_{j\neq k} a_j$ for all $k=1, 2, \ldots$).

THEOREM 4. *The complete system of functions* $a_{kj}(\lambda)=e^{i\lambda k}f_\lambda^{1/2}a_j$; $k=0, \pm 1, \ldots$, $j=1, 2, \ldots$, *is minimal in the functional space* $H = \bigvee_{-\infty<t<\infty} e^{i\lambda t}[f_\lambda^{1/2}A]$ *if and only* if $f_\lambda^{1/2} A = A$ for almost all λ and for all elements a_k^*, $k = 1, 2, \ldots$, of the dual system to a_k, $k = 1, 2, \ldots$, in the subspace A

$$(24) \qquad \int_{-\pi}^{\pi} \|f_\lambda^{-1/2} a_k^*\|^2 \, d\lambda < \infty \qquad (k = 1, 2, \ldots).$$

REFERENCES

[1] Yu. A. Rozanov, *On the linear interpolation of stationary processes with discrete time.* Dokl. Akad. Nauk SSSR **116** (1957), 923—926.

[2] Yu. A. Rozanov, *Stationary random processes.* Holden-Day, Inc., San Francisco, Calif. 1967.

[3] B. Sz.-Nagy et C. Foiaş, *Analyse harmonique des opérateurs de l'espace de Hilbert.* Akadémiai Kiadó, Budapest 1967.

Ein Operatorenkalkül für das approximationstheoretische Verhalten des Ergodensatzes im Mittel

Von

P. L. BUTZER und U. WESTPHAL*)

LEHRSTUHL A FÜR MATHEMATIK
TECHNISCHE HOCHSCHULE AACHEN

1. Einleitung

Ist T ein beschränkter linearer Operator, der den Banachraum X in sich abbildet, mit der Eigenschaft $\|T^i\| \leqq M$, $i=0, 1, 2, \ldots$, wobei M eine Konstante ist, so sagt der Ergodensatz im Mittel aus, daß

$$\lim_{n \to \infty} \left\| \frac{1}{n+1} \sum_{i=0}^{n} T^i f - Pf \right\| = 0$$

für jedes $f \in X_0$. Hierbei ist $X_0 = N(I-T) \oplus \overline{R(I-T)}$ und P der beschränkte lineare Projektor (d.h. $P^2 = P$) von X_0 auf $N(I-T)$ entlang $\overline{R(I-T)}$. $N(T)$ bzw. $R(T)$ stehen für den Kern bzw. Wertebereich von T. Dieser Satz ist u. a. deshalb von Bedeutung, weil der Grenzwert $\lim_{n \to \infty} T^n$ in keiner der üblichen Topologien zu existieren braucht (vgl. D. ORNSTEIN [16]).

Der Ergodensatz ist ein Konvergenzsatz; er sagt nichts über die Schnelligkeit der Konvergenz von $(n+1)^{-1} \sum_{i=0}^{n} T^i f$ gegen Pf für $f \in X_0$ aus. Das einzige uns bekannte Ergebnis in dieser Richtung ist ein Saturationssatz in [10, 11].

Das Ziel dieses Vortrages ist, sowohl das optimale (Saturationsfall) als auch nicht-optimale Approximationsverhalten nicht nur der $(C, 1)$-Mittel $(n+1)^{-1} \sum_{i=0}^{n} T^i$ sondern auch der (C, k)-Mittel (k ganzzahlig) gegen P zu untersuchen. Diese Überlegungen sollen für den Banachraum X im Rahmen der Theorie der intermediären Räume durchgeführt werden, genauer gesagt, mittels der diskreten Theorie, wie sie von BUTZER—SCHERER [6, 7, 9] entwickelt wurde. Während diese Theorie das quantitative Verhalten einer beliebigen Folge beschränkter linearer Operatoren T_n auf X gegen den Identitätsoperator I beinhaltet, kommt es im vorliegenden Fall auf die Konvergenz gegen den Projektor P an. Es wird sich jedoch zeigen, daß

---

*) Der Vortrag wurde vom zweitgenannten Verfasser gehalten, der aus Mitteln des Landesamts für Forschung des Landes Nordrhein—Westfalen unterstützt wurde. Die Arbeit stellt einen Beitrag zum Abschlußbericht des Forschungsvorhabens A/3—4452 dar. Für die Erlaubnis zur Veröffentlichung an dieser Stelle sei dem Landesamt freundlichst gedankt.

sich mittels eines geeigneten Operatorenkalküls der Beweis unseres Hauptergebnisses so führen läßt, daß man einen allgemeinen Hauptsatz in [7, 8] anwenden kann. Dies zeigt die Stärke letzteren Satzes, der inzwischen auch auf Probleme aus der Theorie der algebraischen Approximation [17], der fast-periodischen Funktionen [4] und auf Lipschitz Klassen auf kompakten Mannigfaltigkeiten [3, 15] angewandt werden konnte, und zeugt von der Einheitlichkeit der zugrunde liegenden Methode. Mit Hilfe des genannten Satzes ist der Fall der nicht-optimalen Approximation gelöst. Im Saturationsfall werden Methoden benutzt, die sich auf den Begriff der relativen Vervollständigung stützen. Dabei erfordert der Fall $k=1$ besondere Beachtung und ist von gewissem Interesse.

In Abschnitt 2 werden benötigte Begriffe erklärt und die Hauptergebnisse dieses Vortrages formuliert. Im darauffolgenden Abschnitt 3 werden sie bewiesen. Der Vortrag schließt mit einer kurzen Zusammenfassung über das approximations-theoretische Verhalten der Abel-Mittel von $\{T^n\}$.

2. Zusammenstellung der Hauptergebnisse

Die Cesàro-Mittel $\sigma_n^k(T)$ der Ordnung k der Folge $\{T^n\}_{n=0}^\infty$, kurz (C, k)-Mittel genannt, sind gegeben durch

$$(2.1) \quad \sigma_n^0(T) = T^n, \quad \sigma_n^k(T) = \frac{1}{C_n^{k+1}} \sum_{i=0}^{n} C_{n-i}^k T^i \qquad (k=1, 2, \dots,\ n=0, 1, 2, \dots),$$

wobei $C_0^k=1$ und $C_n^k = k(k+1)\cdots(k+n-1)/n!$. Dann heißt die Folge $\{T^n\}$ (C, k)-summierbar, falls

$$(2.2) \qquad (C, k)\text{-}\lim_{n\to\infty} T^n \equiv \text{s-}\lim_{n\to\infty} \sigma_n^k(T)$$

in der starken Operator-Topologie existiert.

Der folgende Ergodensatz im Mittel sagt aus, auf welcher Menge diese Konvergenz stattfindet.

SATZ 1. *Sei T ein beschränkter linearer Operator des Banachraumes X in sich, dessen Potenzen T^i gleichmäßig beschränkt sind. Es gilt*

(a) *Die Folge $\{T^n\}_{n=0}^\infty$ ist für jedes $k=1, 2, \dots$ (C, k)-summierbar gegen den Grenzoperator P genau auf der abgeschlossenen linearen Mannigfaltigkeit $X_0 = = N(I-T) \oplus \overline{R(I-T)}$.*

(b) *X_0 ist äquivalent erklärt als Menge aller $f \in X$, für die $\{\sigma_n^k(T)f, n=0, 1, \dots\}$ für ein festes $k=1, 2, \dots$ schwach folgenkompakt ist.*

(c) *Ist zusätzlich entweder der Operator T schwach kompakt*) oder der Banachraum X reflexiv, so ist $X_0 = X$. Dann gilt für alle $f \in X$.*

$$(2.\,3) \qquad\qquad \operatorname*{s-lim}_{n \to \infty} \sigma_n^k(T)f = Pf.$$

Für die $(C, 1)$-Mittel in reflexiven Banachräumen stammt dieser Ergodensatz wohl von E. Lorch und für die (C, k)-Mittel von L. W. Cohen [12]. Eine Verallgemeinerung auf nicht-reflexive Banachräume (im $(C, 1)$-Fall) geht im wesentlichen auf S. Kakutani und K. Yosida zurück, vgl. auch Dunford—Schwartz [13, p. 728], A. C. Zaanen [18] und die dortigen Anmerkungen. Die Quellen zur zitierten Literatur sind in [11] genau angegeben.

Um die Frage nach dem Grad der optimalen Approximation von P durch $\sigma_n^k(T)$ zu untersuchen, führen wir einen Operator A ein, der dem infinitesimalen Erzeuger in der Theorie der Halbgruppen von Operatoren (vgl. [2, p. 9]) entspricht, und machen uns den Begriff der relativen Vervollständigung (vgl. [1, p. 14], [5, Sec. 10. 4, 13. 4. 1]) zu Nutze.

Zunächst sei erwähnt, daß X_0 als abgeschlossene Mannigfaltigkeit in X ein Banachraum unter derselben Norm wie X ist. Sei T_0 die Restriktion von T auf X_0. Mit Y bezeichnen wir abkürzend (die direkte Summe) $N(I-T) \oplus R(I-T_0)$; sie liegt dicht in X_0. Ist $f \in Y$, so existiert mindestens ein $g \in X_0$, so daß $(I-P)f = (I-T_0)g$. Durch die zusätzliche Forderung $Pg = 0$ ist g eindeutig bestimmt. Dieser Sachverhalt gibt Anlaß zur folgenden

Definition. *Der Operator A mit Definitionsbereich $D(A) = Y$ und Wertebereich in X_0 ist definiert durch $Af = g$, wobei $g \in X_0$ durch $(I-P)f = (I-T_0)g$ und $Pg = 0$ eindeutig gegeben ist.*

Da der lineare Operator A abgeschlossen ist, wie in § 3 gezeigt ist, wird Y ein normalisierter**) Banachunterraum von X_0 unter der Norm $\|f\|_Y = \|f\|_X + \|Af\|_X$. Die Vervollständigung von Y relativ zu X_0***), die mit $\tilde{Y}^{X_0}$ bezeichnet wird, ist die Menge aller Elemente $f \in X_0$, für die eine Folge $\{f_j\}_{j=1}^{\infty} \subset Y$ und eine Konstante $M_0 > 0$ existieren, so daß $\|f_j\|_Y \leqq M_0$ für alle j und $\lim\limits_{j \to \infty} \|f_j - f\|_{X_0} = 0$. Ist Y reflexiv,

---

   *) T heißt schwach kompakt, falls T beschränkte Mengen in schwach folgenkompakte Mengen abbildet.

   **) "normalisiert" heißt, daß für jedes $f \in Y$ gilt $\|f\|_X \leqq \|f\|_Y$.

   ***) Möglicherweise ist Gagliardos Definition der "relativen Vervollständigung" aus dem Begriff der "zwei-Normen-Konvergenz" entwickelt worden, die in Arbeiten von Alexiewicz, Orlicz u. a., auch in Verbindung mit Saks-Räumen, erörtert wurde. Vgl. z. B. Studia Math. 11 (1950), 200—236, 237—272, ibidem 14 (1953/4), 49—56 sowie die dort zitierte Literatur. Auf diesen Zusammenhang machten J. R. Dorroh und A. C. Zaanen während der Tagung aufmerksam.

so ist $\tilde{Y}^{X_0} = Y$. Damit können wir den ersten Hauptsatz dieses Vortrages und zwar folgenden Saturationssatz formulieren.

SATZ 2. *Die Voraussetzungen des Satzes* 1(a) *seien erfüllt.*
(a) *Gibt es ein* $f \in X_0$ *mit der Eigenschaft*

$$\|\sigma_n^k(T)f - Pf\| = o(n^{-1}) \qquad (n \to \infty),$$

dann ist $f \in N(I-T)$.
(b) *Für ein* $f \in X_0$ *sind folgende Behauptungen untereinander äquivalent:*
   (i) $\|\sigma_n^k(T)f - Pf\| = O(n^{-1}) \qquad (n \to \infty)$;
   (ii) $f \in \tilde{Y}^{X_0}$;
   (iii) $f \in \tilde{Y}^X$, *falls* T *zusätzlich schwach kompakt ist;*
   (iv) $f \in Y$, *falls* X *reflexiv ist.*

In anderen Worten, die (C, k)-Mittel $\sigma_n^k(T)$ sind saturiert mit der Ordnung $O(n^{-1})$, und die Saturationsklasse ist gleich $\overline{N(I-T) \oplus R(I-T_0)}^{X_0}$ bzw. gleich $\overline{N(I-T) \oplus R(I-T)}^X$, falls T schwach kompakt ist, und sogar gleich $N(I-T) \oplus \oplus R(I-T)$, falls X reflexiv ist. Interessanterweise ändert sich die Saturationsordnung nicht, wenn man zu einem höheren (C, k)-Mittel übergeht, sie bleibt stets von der Ordnung $O(n^{-1})$. (Wie schon in Satz 1 gesehen, ist der Konvergenzbereich auch der gleiche, nämlich X_0 für alle (C, k)-Verfahren). Diese Beobachtung wurde im Falle der (C, k)-Mittel von Fourierreihen wohl zum ersten Mal von M. ZAMANSKY [19] gemacht.

Um die nicht optimale Approximation betrachten zu können, benötigen wir den Begriff des K-Funktionals; für beliebige Banachräume übernimmt es die Rolle des klassischen Stetigkeitsmoduls und charakterisiert damit die strukturellen Eigenschaften der Elemente $f \in X_0$. Für $0 < t < \infty$ und jedes $f \in X_0$ ist dieses Funktional definiert durch

$$(2.4) \qquad K(t; f; X_0, Y) = \inf_{h \in Y} (\|f-h\|_X + t\|h\|_Y).$$

Es ist eine monoton wachsende, stetige Funktion in t für jedes feste $f \in X_0$ und darüber hinaus für festes t eine Norm auf X_0.

Das K-Funktional ermöglicht es, den folgenden Satz im Rahmen der Theorie der intermediären Räume anzuführen. Mit $(X_0, Y)_{\alpha, q; K}$ bezeichnen wir die Menge aller $f \in X_0$, für die

$$(2.5) \qquad \sup_{0 < t < \infty} (t^{-\alpha} K(t; f; X_0, Y)) < \infty \qquad (0 \leqq \alpha \leqq 1, \ q = \infty)$$

$$(2.6) \qquad \left[\int_0^\infty (t^{-\alpha} K(t; f; X_0, Y))^q \frac{dt}{t} \right]^{1/q} < \infty \qquad (0 < \alpha < 1, \ 1 \leqq q < \infty).$$

$(X_0, Y)_{\alpha, q; K}$ ist unter der Norm (2. 5) bzw. (2. 6) ein Banachraum, der stetig zwischen Y und X_0 eingebettet ist, d.h.

$$Y \subset (X_0, Y)_{\alpha, q, K} \subset X_0,$$

also ein intermediärer Raum von Y und X_0. Es läßt sich zeigen, daß für festes α die $(X_0, Y)_{\alpha, q; K}$ eine Folge monoton wachsender Unterräume von X_0 bilden, wenn q monoton von 1 bis ∞ läuft.

Nach dieser Vorbereitung können wir den Satz über die nichtoptimale Approximation angeben.

SATZ 3. *Unter den Voraussetzungen von Satz 1(a) gilt für jedes* $k = 1, 2, \dots$, $0 < \alpha < 1$ *und* $1 \leqq q \leqq \infty$, *daß folgende Aussagen für ein* $f \in X_0$ *äquivalent sind:*

(i)
$$\begin{cases} \sup_{n=1, 2, \dots} \left(n^\alpha \| \sigma_n^k(T) f - P f \| \right) < \infty & (q = \infty) \\[2ex] \sum_{n=1}^{\infty} \left(n^\alpha \| \sigma_n^k(T) f - P f \| \right)^q \dfrac{1}{n} < \infty & (1 \leqq q < \infty); \end{cases}$$

(ii) $\quad f \in (X_0, Y)_{\alpha, q; K}, \quad$ d.h. $\quad \begin{cases} \sup_{0 < t < \infty} \left(t^{-\alpha} K(t; f; X_0, Y) \right) < \infty & (q = \infty) \\[2ex] \int_0^{\infty} \left(t^{-\alpha} K(t; f; X_0, Y) \right)^q \dfrac{dt}{t} < \infty & (1 \leqq q < \infty); \end{cases}$

(iii)
$$\begin{cases} \sup_{n=0, 1, 2, \dots} \left(2^{n\alpha} \| \sigma_{2^n}^k(T) f - \sigma_{2^{n-1}}^k(T) f \| \right) < \infty & (q = \infty) \\[2ex] \sum_{n=0}^{\infty} \left(2^{n\alpha} \| \sigma_{2^n}^k(T) f - \sigma_{2^{n-1}}^k(T) f \| \right)^q < \infty & (1 \leqq q < \infty). \end{cases}$$

Für $\alpha = 1$, $q = \infty$ gilt: Die Räume $(X_0, Y)_{1, \infty; K}$ und $\tilde{Y}^{X_0}$ sind gleich mit gleichen Normen. Also ist $(X_0, Y)_{1, \infty; K}$ gleich der Saturationsklasse der (C, k)-Mittel.

3. Beweis der Ergodensätze

Um die beiden Sätze zu beweisen, müssen die (C, k)-Mittel der Folge $\{T^n\}_{n=0}^{\infty}$ näher untersucht werden, insbesondere muß ihre Beziehung zu dem in Abschnitt 2 definierten Operator A hergestellt werden. Dazu zeigen wir zunächst eine Identität, die die (C, k)-Mittel mit den $(C, k+1)$-Mitteln verknüpft, und was entscheidender ist, die Menge X mit dem Wertebereich $R(I - T)$ in Zusammenhang bringt.

Erstens sind für jedes $k = 0, 1, \dots$ die (C, k)-Mittel eine gleichmäßig beschränkte Folge von linearen Operatoren von X in sich, da dieselbe Eigenschaft für die Potenzfolge $\{T^n\}_{n=0}^{\infty}$ gilt. In der Tat ergibt sie sich unmittelbar für $k = 0$ wegen $\sigma_n^0(T) = T^n$, für $k = 1, 2, \dots$ mit der Abschätzung

$$(3.\,1) \qquad \| \sigma_n^k(T) f \| \leqq \frac{1}{C_n^{k+1}} \sum_{i=0}^{n} C_{n-i}^k \| T^i f \| \leqq M \| f \| \qquad (f \in X),$$

da $\sum_{i=0}^{n} C_i^k = C_n^{k+1}$ ist. Für den Projektor P als Grenzoperator der $\sigma_n^k(T)$ $(n \to \infty)$ folgt aus (3. 1)

$$(3.2) \qquad \|Pf\| \leqq M\|f\| \qquad (f \in X_0).$$

Weiterhin gehört für $f \in X_0$ auch $\sigma_n^k(T)f$ zu X_0, und da $Pf \in N(I-T)$ ist, gilt

$$(3.3) \qquad P\sigma_n^k(T)f = \sigma_n^k(T)Pf = Pf \qquad (f \in X_0; \; k, n = 0, 1, \ldots).$$

LEMMA 1. *Für* $f \in X$, $k = 0, 1, \ldots$, $n = 1, 2, \ldots$ *gilt die Identität*

$$(3.4) \qquad \frac{n}{k+1} \sigma_{n-1}^{k+1}(T)(I-T)f = f - \sigma_n^k(T)f.$$

BEWEIS. Ihre Gültigkeit beweist man am einfachsten durch Ausmultiplizieren der linken Seite. Für $k = 0$ ergibt sich dabei sofort

$$\sum_{i=0}^{n-1} T^i(I-T)f = f - T^n f.$$

Für $k = 1, 2, \ldots$ folgt

$$\sigma_{n-1}^{k+1}(T)(I-T)f = \frac{1}{C_{n-1}^{k+2}}\left\{\sum_{i=0}^{n-1} C_{n-1-i}^{k+1} T^i f - \sum_{i=0}^{n-1} C_{n-1-i}^{k+1} T^{i+1} f\right\}.$$

Substitutiert man im zweiten Summanden auf der rechten Seite $i+1$ durch i und addiert anschließend die beiden gliedweise, so erhält man

$$\frac{1}{C_{n-1}^{k+2}}\left\{\sum_{i=0}^{n-1} [C_{n-1-i}^{k+1} - C_{n-i}^{k+1}] T^i f + C_n^{k+1} f - C_0^{k+1} T^n f\right\},$$

was gleich

$$(C_{n-1}^{k+2})^{-1}\left\{C_n^{k+1} f - \sum_{i=0}^{n} C_{n-i}^{k} T^i f\right\}$$

ist, da für $k = 0, 1, 2, \ldots$ $C_n^{k+1} - C_{n-1}^{k+1} = C_n^k$ $(n = 1, 2, \ldots)$ und $C_0^{k+1} = C_0^k$ $(n = 0)$. Mit $C_n^{k+1}/C_{n-1}^{k+2} = (k+1)/n$ folgt der Beweis.

Der Zusammenhang der $\sigma_n^k(T)$ mit dem Operator A ist gegeben durch

LEMMA 2. *Für ein* $f \in X_0$ *gilt*

$$(3.5) \qquad \text{s-}\lim_{n \to \infty} \frac{n}{k+1} [\sigma_{n-1}^{k+1}(T)f - Pf]$$

existiert für $k = 1, 2, \ldots$ *dann und nur dann, wenn* $f \in Y (= D(A))$. *In diesem Falle ist der Grenzwert gleich* Af.

BEWEIS. Ist $f \in Y$, so gilt für das Element $g = Af$ nach Definition von A $(I-P)f = (I-T_0)g$ und $Pg = 0$. Wendet man nun die Identität (3. 4) auf g (anstelle von f) an und benutzt (3. 3), so folgt wegen Satz 1, da $g \in X_0$,

$$\text{s-lim}_{n \to \infty} \frac{n}{k+1} [\sigma_{n-1}^{k+1}(T)f - Pf] = \text{s-lim}_{n \to \infty} [g - \sigma_n^k(T)g] = g - Pg = g.$$

Wenn umgekehrt der Grenzwert (3. 5) existiert, den wir mit g bezeichnen, so konvergiert auch die Folge

$$\frac{n}{k+1} \sigma_{n-1}^{k+1}(T)(I-T)(I-P)f$$

im starken Sinne für $n \to \infty$ und zwar gegen $(I-T)g$, da $(I-T)$ ein beschränkter Operator ist. Andererseits sind die Glieder dieser Folge nach (3. 4), angewandt auf $f - Pf$, und nach (3. 3) gleich den Ausdrücken $f - \sigma_n^k(T)f$, die für $n \to \infty$ wegen Satz 1 gegen $f - Pf$ streben. Aus beiden Überlegungen folgt

$$(I-P)f = (I-T)g.$$

Setzt man diese Beziehung in (3. 5) ein und benutzt nochmals (3. 4), diesmal für das Element g, so gilt

$$g = \text{s-lim}_{n \to \infty} \frac{n}{k+1} \sigma_{n-1}^{k+1}(T)(I-T)g = \text{s-lim}_{n \to \infty} g - \sigma_n^k(T)g = g - Pg.$$

Also ist $g \in X_0$ und $Pg = 0$. Damit ist gezeigt, daß $f \in Y$ und $Af = g$.

Es sei betont, daß die Aussage des Lemma 2 über die (C, l)-Mittel $\sigma_n^l(T)$ in ihrem vollen Umfang erst ab $l = 2$ gilt, also für jedes feste ganze $l \geqq 2$. Für $l = 1$ ist nur eine Richtung gültig, nämlich: Existiert $\text{s-lim}_{n \to \infty} n[\sigma_{n-1}^1(T)f - Pf]$, so ist $f \in D(A)$ und der Grenzwert gleich Af.

Im folgenden sind einige einfache Eigenschaften des Operators A, die sich teils unmittelbar aus der Definition, teils aus Lemma 2 ergeben, zusammengestellt.

LEMMA 3.
(a) *Ist $f \in D(A)$, so gelten die folgenden Beziehungen:*
   (i) $(I-T_0)Af = (I-P)f$,
   (ii) $PAf = 0$,
   (iii) $\sigma_n^k(T)f \in D(A)$, $A\sigma_n^k(T)f = \sigma_n^k(T)Af$ $(k, n = 0, 1, \ldots)$.
(b) *Für $f \in X_0$ gilt*
   (iv) $A(I-T_0)f = (I-P)f$,
   (v) $APf = 0$,

$$\text{(vi)} \quad A[f - \sigma_n^k(T)f] = \frac{n}{k+1} \sigma_{n-1}^{k+1}(T)A(I-T_0)f = \frac{n}{k+1} [\sigma_{n-1}^{k+1}(T)f - Pf]$$
$$(k = 0, 1, \ldots, \ n = 1, 2, \ldots).$$

(c) *A ist ein abgeschlossener Operator.*

BEWEIS. (i) und (ii) von Teil (a) folgen unmittelbar aus der Definition von A. Um (iii) zu beweisen, wenden wir die (C, k)-Mittel $\sigma_n^k(T)$ auf die Beziehung (i) an, woraus sich mit Hilfe von (3. 3)

$$\sigma_n^k(T)f = P\sigma_n^k(T)f + (I - T_0)\sigma_n^k(T)Af$$

ergibt. Hieraus sieht man sofort, daß $\sigma_n^k(T)f$ zu $Y = D(A)$ gehört, und da $P\sigma_n^k(T)Af = \sigma_n^k(T)PAf = 0$ wegen (ii) ist, folgt laut Definition von A

$$A\sigma_n^k(T)f = \sigma_n^k(T)Af.$$

Betreffs Teil b) sei $f \in X_0$. Dann ist $(I - T_0)f \in R(I - T_0)$, gehört also sicherlich zu $D(A)$. Aus den simplen Gleichungen

$$(I - P)(I - T_0)f = (I - T_0)(I - P)f$$

$$P(I - P)f = 0$$

folgt nach Definition von A, daß $A(I - T_0)f = (I - P)f$, womit (iv) gezeigt ist. (v) gilt trivialerweise, da $Pf \in N(I - T)$. Die Beziehung (vi) stimmt für $k = 0$ und $n = 1$ mit (iv) überein. Für alle übrigen k und n wird der Beweis wie in (iv) durchgeführt mit Hilfe der Identität (3. 4).

Um zu zeigen, daß A abgeschlossen ist, wählen wir eine beliebige Folge $\{f_j\}$ in $D(A)$, für die $\operatorname{s-lim}_{j \to \infty} f_j = f$ und $\operatorname{s-lim}_{j \to \infty} Af_j = h$ gilt. Dann ist $f \in X_0$, und $\{Pf_j\}$ strebt im starken Sinne gegen Pf; h gehört zu $\overline{R(I - T_0)}$, also ist $Ph = 0$. Wenden wir nun Beziehung (vi) auf f_j an und benutzen die Vertauschbarkeit von A und $\sigma_n^k(T)$ auf $D(A)$ nach (iii), so gilt

$$Af_j - \sigma_n^k(T)Af_j = \frac{n}{k+1}[\sigma_{n-1}^{k+1}(T)f_j - Pf_j],$$

woraus sich für $j \to \infty$

$$h - \sigma_n^k(T)h = \frac{n}{k+1}[\sigma_{n-1}^{k+1}(T)f - Pf]$$

ergibt. Für $n \to \infty$ konvergiert die linke Seite gegen h, also ist auch die rechte konvergent, woraus nach Lemma 2 $f \in D(A)$ und $Af = h$ folgt. Damit ist das Lemma bewiesen.

Für die (C, k)-Mittel $\sigma_n^k(T)$ führen wir nun ein Hilfsverfahren $J_n^k(T)$ ein, auf das die Beweise unserer Approximationssätze reduziert werden können.

Für $k = 0, 1, 2, \ldots$ definieren wir eine Folge $\{J_n^k(T)\}_{n=0}^{\infty}$ beschränkter linearer Operatoren von X_0 in sich durch

$$(3. 6) \qquad J_n^k(T)f = Pf + f - \sigma_n^k(T)f \qquad (f \in X_0).$$

Nach der Identität (3. 4) gilt die äquivalente Darstellung

$$(3.7) \qquad J_n^k(T)f = Pf + \frac{n}{k+1} \sigma_{n-1}^{k+1}(T)(I-T)f \qquad (f \in X_0;\ k=0, 1, \dots,\ n=1, 2, \dots).$$

Die Folge $\{J_n^k(T)\}_{n=0}^{\infty}$ ist bezüglich n gleichmäßig beschränkt, denn wegen (3. 1) und (3. 2) gilt

$$(3.8) \qquad \|J_n^k(T)f\| \leq (2M+1)\|f\| \qquad (f \in X_0;\ k, n=0, 1, \dots).$$

Ferner konvergiert sie nach Satz 1 im starken Sinne auf X_0 gegen den Identitäts-operator, d.h.

$$(3.9) \qquad \operatorname*{s-lim}_{n \to \infty} J_n^k(T)f = f \qquad (f \in X_0,\ k=1, 2, \dots).$$

Da für $f \in X_0$ Pf zu $N(I-T)$ gehört und $f - \sigma_n^k(T)f$ zu $R(I-T_0)$ nach der Identität (3. 4), ersieht man aus der Definitionsgleichung (3. 6) der $J_n^k(T)$ sofort, daß

$$(3.10) \qquad J_n^k(T)f \in Y \qquad (f \in X_0),$$

und nach Lemma 3 gilt

$$(3.11) \qquad A J_n^k(T)f = \frac{n}{k+1}[\sigma_{n-1}^{k+1}(T)f - Pf] = \frac{n}{k+1}[f - J_{n-1}^{k+1}(T)f]$$

$$(f \in X_0;\ k=0, 1, \dots,\ n=1, 2, \dots),$$

bzw.

$$(3.12) \qquad J_n^k(T)f - f = \frac{k}{n+1}[\sigma_{n+1}^{k-1}(T)Af - Af] = -\frac{k}{n+1}J_{n+1}^{k-1}(T)Af$$

$$(f \in Y;\ k=1, 2, \dots,\ n=0, 1, \dots).$$

Damit lautet Lemma 2, auf die Folge $\{J_n^k(T)\}$ übertragen,

Lemma 4. *Für ein $f \in X_0$ existiert*

$$\operatorname*{s-lim}_{n \to \infty} n[J_n^k(T)f - f] \qquad (k=2, 3, \dots)$$

und ist gleich h dann und nur dann, wenn $f \in Y(=D(A))$ und $(-kAf)=h$ ist. Für $k=1$ ist diese Bedingung notwendig, aber nicht hinreichend.

Da nun $\|J_n^k(T)f - f\| = \|\sigma_n^k(T)f - Pf\|$ ist und $\sigma_n^k(T)$ nur in Form dieser Differenz in den Sätzen 2 und 3 auftritt, können ihre Beweise auf das Verfahren $\{J_n^k(T)\}$ abgewälzt werden. Diese Vorbereitungen ermöglichen es uns endlich, Satz 2, den Saturationssatz, und Satz 3, den Satz über das nicht-optimale Approximations-verhalten der $\sigma_n^k(T)$, zu beweisen.

BEWEIS VON SATZ 2. Der Beweis von Teil a) folgt unmittelbar aus Lemma 4. Betreffs Teil b) sei zunächst $f \in \tilde{Y}^{X_0}$. Dann gibt es eine Folge $\{f_j\}_{j=1}^{\infty}$ in Y, so daß $\|f_j\|_Y \leqq M_0$ und $\lim\limits_{j \to \infty} \|f_j - f\|_X = 0$. Nach (3. 12) gilt dann die Abschätzung

$$(n+1)\|J_n^k(T)f_j - f_j\| = k\|J_{n+1}^{k-1}(T)Af_j\| \leqq k(2M+1)\|Af_j\| \leqq$$

$$\leqq k(2M+1)\|f_j\|_Y \leqq k(2M+1)M_0$$

für $k=1, 2, \ldots$ und $n=0, 1, \ldots$, woraus sich die Behauptung (i) für $j \to \infty$ ergibt.

Bezüglich der Umkehrung (i)$\Rightarrow$(ii) gehen wir von (3. 11) aus und setzen darin die Rekursionsformel für die (C, k)-Mittel

$$\sigma_{n-1}^{k+1}(T) = \frac{1}{C_{n-1}^{k+2}} \sum_{i=0}^{n-1} C_i^{k+1} \sigma_i^k(T)$$

ein. Dies führt für $k=0, 1, \ldots$, $n=1, 2, \ldots$ zu

$$AJ_n^k(T)f = \frac{n}{k+1} \frac{1}{C_{n-1}^{k+2}} \sum_{i=0}^{n-1} C_i^{k+1}[f - J_i^k(T)f] \qquad (f \in X_0).$$

Nun ist

$$\frac{C_i^{k+1}}{C_{n-1}^{k+2}} = \frac{(k+1)(i+1)}{n(k+n)} \left[\frac{i+2}{k+i+1} \frac{i+3}{k+i+2} \cdots \frac{n}{k+n-1}\right] \leqq \frac{(k+1)(i+1)}{n(k+n)},$$

womit die Abschätzung

$$\|AJ_n^k(T)f\| \leqq \frac{1}{k+n} \sum_{i=0}^{n-1} (i+1)\|f - J_i^k(T)f\| \qquad (f \in X_0)$$

folgt. Nach der Voraussetzung (i) Teil b) des Satzes existiert zu einem $f \in X_0$ eine natürliche Zahl n_0, so daß für $i \geqq n_0$ gilt $(i+1)\|f - J_i^k(T)f\| \leqq M_1$. Damit ergibt sich für $n \geqq n_0$

$$\|AJ_n^k(T)f\| \leqq \frac{1}{k+n} \sum_{i=0}^{n_0-1} (i+1)\|f - J_i^k(T)f\| + \frac{n-n_0}{k+n} M_1 \leqq M_2.$$

Also ist $\|J_n^k(T)f\|_Y \leqq M_0$, und benutzt man schließlich (3. 9), so folgt $f \in \tilde{Y}^{X_0}$ und damit der Beweis des Satzes.

Für $k \geqq 2$ kann man diesen Satz auch parallel zu einem Saturationssatz in BERENS [1, p. 28] bzw. BUTZER—NESSEL [5, p. 504] beweisen, wenn man die dort benutzten kontinuierlichen Methoden für den hier vorliegenden diskreten Fall aufstellt. Dann nämlich sind (3. 8)—(3. 10) und Lemma 4 genau die entsprechenden Voraussetzungen jenes Satzes für das Verfahren $\{J_n^k(T)\}$. Für den wichtigen Fall $k=1$ gilt Lemma 4 jedoch nur in einer Richtung, so daß man sich zum Beweis von (i)$\Rightarrow$(ii) nicht auf jenen Saturationssatz zurückziehen kann.

BEWEIS VON SATZ 3. Wie in der Einleitung angekündigt, ergibt sich der Beweis als Anwendung eines allgemeinen Satzes über nicht-optimale Approximation von BUTZER–SCHERER [7, p. 180; 9]. In der Tat, erfüllt die Folge $\{J_n^k(T)\}_{n=0}^\infty$ die entsprechenden Voraussetzungen: ihre Mitglieder sind kommutative lineare Operatoren von X_0 in sich, konvergieren gegen den Identitätsoperator und genügen den beiden folgenden Ungleichungen.

LEMMA 5. *Für* $k, n = 1, 2, \ldots$ *gilt die*

(a) *Jackson-Typ-Ungleichung*

$$\|J_n^k(T)f - f\| \leqq D_1 n^{-1} \|f\|_Y \qquad (f \in Y),$$

(b) *Bernstein-Typ-Ungleichung*

$$\|J_n^k(T)f\|_Y \leqq D_2 n \|f\|_{X_0} \qquad (f \in X_0),$$

wobei D_1 *und* D_2 *nur von* k *und* M *abhängige Konstanten sind* $(D_1 = k(M+1)$, $D_2 = 2M/(k+1) + 2M + 1)$.

Diese Ungleichungen, die die Kernstücke des allgemeinen Satzes sind, folgen unmittelbar aus den Identitäten (3. 12) und (3. 11). Dabei ist die Jackson-Ungleichung fundamental für den Beweis der direkten Richtung, also für den Schritt von (ii) nach (i) in Satz 3, die Bernstein-Ungleichung für die Umkehrung.

4. Bemerkungen zum Abel-Ergodensatz

Die vorangegangenen Untersuchungen lassen sich nicht nur für die Cesàro-Mittel, sondern vermutlich für sämtliche Toeplitz-Verfahren, die gewissen zusätzlichen Bedingungen genügen, durchführen. Als Beispiel seien hier die Ergebnisse für die Abel-Summen

$$(4.1) \qquad S_r(T) = (1-r) \sum_{i=0}^\infty r^i T^i \qquad (0 < r < 1)$$

ohne Beweis angegeben. Unter denselben Voraussetzungen an T wie in Satz 1 konvergieren die Abel-Summen, ebenso wie die (C, k)-Mittel, genau auf X_0 gegen den Projektor P, also

$$\operatorname*{s-lim}_{r \to 1-} S_r(T)f = Pf \qquad (f \in X_0).$$

Ausgehend von diesem Konvergenzsatz, läßt sich jeder Schritt des Operatorenkalküls für die (C, k)-Mittel analog für den Abel-Fall vollziehen. Die der fundamentalen Identität (3. 4) entsprechende lautet hier

$$\frac{r}{1-r} S_r(T)(I-T)f = f - S_r(T)f \qquad (f \in X; \ 0 < r < 1).$$

Der in Abschnitt 2 definierte Operator A hängt auch mit dem Abel-Verfahren zusammen und zwar wie folgt: Für ein $f \in X_0$ existiert

$$\operatorname*{s-lim}_{r \to 1-} \frac{r}{1-r} [S_r(T)f - Pf]$$

dann und nur dann, wenn $f \in D(A)$ ist. Der Grenzwert ist gleich Af.

Entsprechend zu der Folge $\{J_n^k(T)\}_{n=0}^\infty$ führt man eine Schar $\{I_r(T); 0 < r < 1\}$ durch

$$I_r(T)f = Pf + f - S_r(T)f \qquad (f \in X_0)$$

ein. Dann läßt sich folgender Satz beweisen:

SATZ 4. *Unter den Voraussetzungen an T wie in Satz* 1a) *gilt*
I. *Der Abel-Summationsprozeß* (4.1) *ist mit der Ordnung* $O((1-r)/r)$ *saturiert, und seine Saturationsklasse ist gleich* $\overline{N(I-T) \oplus R(I-T_0)}^{X_0}$.
II. *Für ein* $f \in X_0$ *sind folgende Aussagen äquivalent:*

(i)
$$\begin{cases} \displaystyle\sup_{0 < r < 1} \left[\left(\frac{r}{1-r} \right)^\alpha \| S_r(T)f - Pf \| \right] < \infty & (q = \infty) \\[2ex] \displaystyle\int_0^1 \left[\left(\frac{r}{1-r} \right)^\alpha \| S_r(T)f - Pf \| \right]^q \frac{dr}{r(1-r)} < \infty & (1 \leq q < \infty); \end{cases}$$

(ii)
$$f \in (X_0, Y)_{\alpha, q; K}.$$

Ein Vergleich der Sätze 2 und 3 einerseits und Satz 4 andererseits zeigt, daß Saturationsordnung und -klasse sowie die Räume zur Charakterisierung der nicht-optimalen Approximation beim Abel-Verfahren dieselben sind wie für die Cesàro-Mittel.

Im Sinne von BUTZER—SCHERER [6] kann man die durch Bedingung II (i) oben definierten Räume als *kontinuierliche Verfahrensräume* und die durch (i) in Satz 3 als *diskrete Verfahrensräume* ansehen.

Zum Schluß sei vermerkt, daß E. HILLE [14] Beziehungen zwischen Cesàro- und Abel-Summierbarkeit bei den entsprechenden Ergodensätzen diskutiert hat.

LITERATUR

[1] H. Berens, *Interpolationsmethoden zur Behandlung von Approximationsprozessen auf Banachräumen.* Lecture Notes in Mathematics No. **64**, Springer, Berlin—Heidelberg—New York 1968.

[2] P. L. Butzer and H. Berens, *Semi-Groups of Operators and Approximation.* Springer, Berlin—Heidelberg—New York 1967.

[3] P. L. Butzer and H. Johnen, *Lipschitz spaces on compact manifolds.* J. Functional Analysis **7** (1971), 242—266.

[4] P. L. Butzer und J. Kemper, *Operatorenkalkül von Approximationsverfahren fastperiodischer Funktionen.* Forschungsberichte des Landes Nordrhein—Westfalen, Nr. **2157**, 1970, 27—53.

[5] P. L. Butzer and R. J. Nessel, *Fourier Analysis and Approximation,* Vol. I: *One-Dimensional Theory.* Birkhäuser, Basel and Academic Press, New York 1971.

[6] P. L. Butzer und K. Scherer, *Approximationsprozesse und Interpolationsmethoden.* Bibliographisches Institut, Mannheim 1968.

[7] P. L. Butzer and K. Scherer, *On the fundamental approximation theorems of D. Jackson and S. N. Bernstein and the theorems of M. Zamansky and S. B. Stečkin.* Aequationes Math. **3** (1969), 170—185.

[8] P. L. Butzer and K. Scherer, *Jackson and Bernstein-type inequalities for families of commutative operators in Banach spaces.* J. Approximation Theory **5** (1971), 308—342.

[9] P. L. Butzer and K. Scherer, *Approximation theorems for sequences of commutative operators in Banach spaces.* In Proceedings of the Conference on Constructive Theory of Functions, Varna, May 1970, 137—145.

[10] P. L. Butzer and U. Westphal, *On the Cayley transform of semigroups of contractions.* In Proceedings of the Conference on Hilbert Space Operators and Operator Algebras, Tihany, September 1970, 89—97.

[11] P. L. Butzer and U. Westphal, *The mean ergodic theorem and saturation.* Indiana Univ. Math. J. **20** (1971), 1163—1174.

[12] L. W. Cohen, *On the mean ergodic theorem.* Ann. of Math. (2) **41** (1940), 505—509.

[13] N. Dunford and J. T. Schwartz, *Linear Operators,* Part I: *General Theory.* New York 1958.

[14] E. Hille, *Remarks on ergodic theorems.* Trans. Amer. Math. Soc. **57** (1945), 246—269.

[15] H. Johnen, *Sätze vom Jackson-Typ auf Darstellungsräumen kompakter zusammenhängender Liegruppen.* Diese Abhandlungen, 254—272.

[16] D. Ornstein, *On the pointwise behaviour of iterates of a self-adjoint operator.* J. Math. Mech. **18** (1968), 473—477.

[17] K. Scherer and H. J. Wagner, *An equivalence theorem on best approximation of continuous functions by algebraic polynomials.* Applicable Anal. **2** (1972) (in print).

[18] A. C. Zaanen, *Linear Analysis.* North-Holland Publishing Co., Amsterdam 1953.

[19] M. Zamansky, *Classes de saturation de certains procédés d'approximation des séries de Fourier des fonctions continues et applications à quelques problèmes d'approximation.* Ann. Sci. École Norm. Sup. (3) **66** (1949), 19—93.

A Note on the Landau—Kallman—Rota—Hille Inequality [1]

By

W. TREBELS and U. WESTPHAL

LEHRSTUHL A FÜR MATHEMATIK
TECHNISCHE HOCHSCHULE AACHEN

In the preceding paper HILLE [6] generalized the Landau inequality for twice differentiable functions on $(0, \infty)$, namely

$$(1) \qquad \left\| \frac{d}{dx} f \right\|^2 \leq 4 \|f\| \left\| \frac{d^2}{dx^2} f \right\|,$$

by replacing the differential operators (d/dx) and $(d/dx)^2$ by semi-group generators A and their powers of arbitrary order in the form

$$(2) \qquad \|A^k f\|^n \leq C^n_{n,k} \|f\|^{n-k} \|A^n f\|^k,$$

where n and k are integers such that $1 \leq k < n$. An intermediate step in this investigation was the Kallman—Rota inequality which is a special case of (2), namely for $n=2$ and $k=1$

$$(3) \qquad \|Af\|^2 \leq 4 \|A^2 f\| \|f\|.$$

For further literature on the Landau inequality and its extensions in various directions the reader is referred to the paper by HILLE [6] where the importance of this inequality in semi-group theory is illustrated in its full generality.

The aim of this note is to give a formula analogous to (2) for fractional (positive) powers of $(-A)$, the proof of which is based upon an operational calculus of fractional derivatives.

The authors are indebted to Professor Hille for discussion on the subject during the Oberwolfach conference, and to Professor Butzer for helpful advice and permission to publish this note in the conference book.

First we cite some definitions and statements on fractional powers. Let $\{T(t); t \geq 0\}$ be a strongly continuous semi-group of uniformly bounded operators (bound M) from a Banach space X into itself with infinitesimal generator A (recall

---

[1] This research was supported by the "Minister für Wissenschaft und Forschung des Landes Nordrhein—Westfalen, Landesamt für Forschung", Grant No. A/3—4452.

[2]). Following [9; I] the fractional power $(-A)$ of order γ, γ a real positive number, is defined by

$$(4) \qquad (-A)^\gamma f = \operatorname*{s-lim}_{\varepsilon \to 0+} K_{\gamma,n}^{-1} \int_\varepsilon^\infty u^{-1-\gamma}[I - T(u)]^n f \, du$$

whenever this limit exists for some $f \in X$, the domain being denoted by $D((-A)^\gamma)$. Here n is any integer $> \gamma$, the limit in (4) being independent of this n, and $K_{\gamma,n}$ is a positive constant given by

$$K_{\gamma,n} = \int_0^\infty u^{-1-\gamma}(1 - e^{-u})^n \, du.$$

For natural numbers $\gamma = m = 1, 2, \ldots$ this definition is equivalent to the usual one given inductively by $A^m = A(A^{m-1})$. Moreover, for arbitrary $\gamma > 0$ it is equivalent to that of BALAKRISHNAN [1]. If $f \in D((-A)^\gamma)$ $(\gamma > 0)$, then $f \in D((-A)^\alpha)$ for every α satisfying $0 < \alpha < \gamma$. In addition to (4) there is a somewhat stronger version, namely

$$(5) \qquad (-A)^\alpha f = K_{\alpha,n}^{-1} \int_0^\infty u^{-1-\alpha}[I - T(u)]^n f \, du \qquad (0 < \alpha < n).$$

As a type of inversion formula we have for $0 < \gamma \leqq n$ (n included)

$$(6) \qquad [I - T(t)]^n f = t^\gamma \int_0^\infty p_{\gamma,n}\left(\frac{u}{t}\right) T(u) \, (-A)^\gamma f \, \frac{du}{t} \qquad (f \in D((-A)^\gamma))$$

where $p_{\gamma,n}$ is a real-valued function in $L(0, \infty)$ given by

$$p_{\gamma,n}(u) = \frac{1}{\Gamma(\gamma)} \begin{cases} \displaystyle\sum_{j=0}^m (-1)^j \binom{n}{j} (u-j)^{\gamma-1} & \left(\begin{matrix} m < u < m+1 \\ m = 0, 1, \ldots, n-1 \end{matrix}\right) \\[2ex] \displaystyle\sum_{j=0}^n (-1)^j \binom{n}{j} (u-j)^{\gamma-1} & (u > n). \end{cases}$$

For proofs of the results above see [9; I].

Now let $f \in D((-A)^\gamma)$ and $\alpha > 0$ such that $\alpha < \gamma$. To obtain an estimate for $\|(-A)^\alpha f\|$ dependent upon $\|f\|$ and $\|(-A)^\gamma f\|$ we start from formula (5) with n such that $n \geqq \gamma$. Having split the integral on the right into two parts: $\int_0^t + \int_t^\infty$ for any $t > 0$, we apply the following

LEMMA. *Let* $0 < \alpha < \gamma \leqq n$. *Then*

$$\|[I - T(t)]^n f\| \leqq \begin{cases} 2^n M \|f\| & (f \in X) \\ t^\gamma M M_{\gamma,n} \|(-A)^\gamma f\| & (f \in D((-A)^\gamma)) \end{cases}$$

where $M_{\gamma,n} = \displaystyle\int_0^\infty |p_{\gamma,n}(u)| \, du.$

PROOF. The first inequality follows by

$$[I - T(t)]^n = \sum_{j=0}^{n} (-1)^j \binom{n}{j} T(jt),$$

the second one by (6) since $\|T(t)\| \leq M$.

From these inequalities we have

(7)
$$\|(-A)^\alpha f\| \leq \frac{M}{K_{\alpha,n}} \left[\frac{2^n}{\alpha} t^{-\alpha} \|f\| + \frac{M_{\gamma,n}}{\gamma - \alpha} t^{\gamma - \alpha} \|(-A)^\gamma f\| \right].$$

If $\|(-A)^\gamma f\| = 0$, then by passing to the limit $t \to \infty$ on the right of (7) we see $(-A)^\alpha f = 0$. If $\|(-A)^\gamma f\| \neq 0$, the right member of (7) attains its minimum for

$$t_0 = (2^n M_{\gamma,n}^{-1})^{1/\gamma} \|f\|^{1/\gamma} \|(-A)^\gamma f\|^{-1/\gamma}.$$

Substituting this in (7) we have proven the following

THEOREM. *Let* $\{T(t); t \geq 0\}$ *be a uniformly bounded strongly continuous semigroup of operators, A its infinitesimal generator. If $f \in D((-A)^\gamma)$ and $0 < \alpha < \gamma$, then*

(8)
$$\|(-A)^\alpha f\|^\gamma \leq C_{\gamma,\alpha}^\gamma \|f\|^{\gamma - \alpha} \|(-A)^\gamma f\|^\alpha,$$

where

$$C_{\gamma,\alpha} = \frac{\gamma}{\alpha(\gamma - \alpha)} \frac{2^{\gamma_0} M}{K_{\alpha,\gamma_0}} \left(\frac{M_{\gamma,\gamma_0}}{2^{\gamma_0}} \right)^{\alpha/\gamma},$$

γ_0 *being the least integer $\geq \gamma$.*

The value of $C_{\gamma,\alpha}$ is not a good bound for the constant as may be seen in case $\alpha = 1$, $\gamma = 2$, $M = 1$ by comparison with (3). Then $K_{1,2} = 2 \log 2$ and $M_{2,2} = 1$, such that (8) reduces to

(9)
$$\|Af\|^2 \leq \left(\frac{2}{\log 2} \right)^2 \|f\| \|A^2 f\|$$

with $C_{2,1} = 2/\log 2 \approx 2{,}885$ instead of 2. But the problem of best constants only seems to be reasonable in concrete spaces; in this field there is a vast literature.

As an example how to improve the constant $C_{\gamma,\alpha}$ in a concrete situation by our methods let us consider the semi-group connected with the singular integral of Cauchy—Poisson given by

$$P(t)f(x) = \Gamma\left(\frac{n+1}{2}\right) \pi^{-\frac{1}{2}(n+1)} t \int_{E_n} (|y|^2 + t^2)^{-(n+1)/2} f(x - y) \, dy \qquad (t > 0),$$

where E_n is the n-dimensional Euclidean space with elements $x = (x_1, \ldots, x_n)$,

$y=(y_1, \ldots, y_n)$, $|y| = \left\{ \sum_{k=1}^{n} y_k^2 \right\}^{1/2}$ and $f \in L^p(E_n)$, $1 < p < \infty$. Its infinitesimal generator has the form (for $n=1$ see [5], for arbitrary n see [3], [8])

$$A_P f(x) = -(\nabla, \mathfrak{R}) f(x) = -\sum_{k=1}^{n} (\partial/\partial x_k) R_k f(x),$$

where the Riesztransformation $\mathfrak{R}=(R_1, \ldots, R_n)$ is defined by

$$R_k f(x) = \lim_{\varepsilon \to 0+} \Gamma\left(\frac{n+1}{2}\right) \pi^{-(n+1)/2} \int_{|y| \geq \varepsilon} y_k |y|^{-n-1} f(x-y)\, dy.$$

Now if $f \in L^p(E_n)$ is twice partially differentiable (in the norm), i.e. $\Delta f \in L^p(E_n)$, where $\Delta = \sum_{k=1}^{n} (\partial/\partial x_k)^2$ is the Laplacian, then (see [8])

$$[(\nabla, \mathfrak{R})]^2 = -\Delta.$$

Thus (9) yields ($\alpha=1$, $\gamma=2$)

(10) $$\|(\nabla, \mathfrak{R}) f\|_p^2 \leq (2/\log 2)^2 \|f\|_p \|\Delta f\|_p.$$

However one can now improve the constant in (10) using our fractional form (8) for $\alpha=1/2$, $\gamma=1$. Observing that the infinitesimal generator of the Weierstrass (contraction) semi-group

$$W(t) f(x) = (4\pi t)^{-n/2} \int_{E_n} \exp\{-|y|^2/4t\} f(x-y)\, dy \qquad (t>0)$$

is given by

$$A_W f(x) = \Delta f(x),$$

it follows that

$$(-\Delta)^{1/2} = (\nabla, \mathfrak{R});$$

thus for $\alpha=1/2$, $\gamma=1$, $M=1$ we obtain $K_{1/2,\,1}=2\sqrt{\pi}$, $M_{1,\,1}=1$ and by an application of (8)

$$\|(\nabla, \mathfrak{R}) f\|_p \leq 4/\sqrt{2\pi}\ \|f\|_p^{1/2} \|\Delta f\|_p^{1/2}.$$

The constant $4/\sqrt{2\pi} \approx 1{,}596$ is a definite improvement upon $C_{2,\,1} \approx 2{,}885$ in (10) and 2 in (3).

As a further application of the above theorem let us deduce Bernstein's inequality for fractional derivatives from the classical one for integers. The latter one reads

(11) $$\left\| \sum_{k=-n}^{n} (ik)^j c_k e^{ikx} \right\| \leq n^j \|t_n(x)\| \qquad (j=1, 2, \ldots),$$

where $t_n(x) = \sum_{k=-n}^{n} c_k e^{ikx}$ is a trigonometric polynomial of degree n and the norm is either $L_{2\pi}^p$, $1 \leq p < \infty$, or $C_{2\pi}$.

Taking the translation semi-group $T(t)f(x) = f(x-t)$ with infinitesimal generator $A = (-d/dx)$, then for fractional α

$$(-A)^\alpha t_n(x) = t_n^{(\alpha)}(x) = \sum_{k=-n}^{n} (ik)^\alpha c_k e^{ikx}$$

which is just the Riemann–Liouville derivative of order α. Thus, by the above theorem and (11) we have for $0 < \alpha < j$

$$\|t_n^{(\alpha)}\|^j \leq C_{j,\alpha}^j \|t_n\|^{j-\alpha} \|t_n^{(j)}\|^\alpha \leq C_{j,\alpha}^j n^{j\alpha} \|t_n\|^j,$$

i. e.

$$\|t_n^{(\alpha)}\| \leq C_{j,\alpha} n^\alpha \|t_n\|.$$

Choosing A and γ appropriately one has by arguing along the same lines

$$\left\| \sum_{k=-n}^{n} |k|^\alpha c_k e^{ikx} \right\| \leq C_\alpha n^\alpha \|t_n\|$$

$$\left\| \sum_{k=-n}^{n} (1 + |k|^2)^{\alpha/2} c_k e^{ikx} \right\| \leq C_\alpha n^\alpha \|t_n\|.$$

For literature in the 2π-periodic case see e.g. [4]. Obviously, replacing $L_{2\pi}^p$ by $L^p(-\infty, \infty)$ and t_n by entire functions of exponential type $\leq n$, one arrives at results of LIZORKIN [7].

Let us finally note that the above arguments may be extended to groups of operators using the analysis of [9; II].

REFERENCES

[1] A. V. Balakrishnan, *Fractional powers of closed operators and the semi-groups generated by them.* Pacific J. Math. **10** (1960), 419—437.

[2] P. L. Butzer and H. Berens, *Semi-Groups of Operators and Approximation.* Grundl. d. math. Wiss., Bd. **145**. Springer, Berlin—Heidelberg—New York 1967.

[3] E. Görlich, *Distributional methods in saturation theory.* J. Appr. Theory **1** (1968), 111—136.

[4] E. Görlich, *Logarithmische und exponentielle Ungleichungen vom Bernstein-Typ und verallgemeinerte Ableitungen.* Habilitationsschrift, TH Aachen 1971.

[5] E. Hille, *On the generation of semi-groups and the theory of conjugate functions.* Proc. R. Physiogr. Soc. Lund. (14) **21** (1951), 130—142.

[6] E. Hille, *Generalizations of Landau's inequality to linear operators.* These Proceedings, 20—32.

[7] R. I. Lizorkin, *Bounds for trigonometrical integrals and an inequality of Bernstein for fractional derivatives* (Russian). Izv. Akad. Nauk SSSR Ser. Mat. **29** (1965), 109—126.

[8] W. Trebels, *Generalized Lipschitz conditions and Riesz derivatives on the space of Bessel potentials L_α^p, II: The case $\alpha > 1$; Riesz derivatives.* Applicable Anal.. **1** (1971), 75—99.

[9] U. Westphal, *Ein Kalkül für gebrochene Potenzen infinitesimaler Erzeuger von Halbgruppen und Gruppen von Operatoren, I: Halbgruppenerzeuger, II: Gruppenerzeuger.* Compositio Math. **22** (1970), 67—103, 104—136.

II.
Topics in Functional Analysis

Representation Theorems for Riesz Spaces

By

A. C. ZAANEN

MATH. INSTITUTE
LEIDEN STATE UNIVERSITY

1. Introduction

First of all, we shall briefly recall several fundamental notions from the theory of Riesz spaces, that is, from the theory of vector lattices. A *Riesz space*, denote it by L, is a real vector space which is at the same time a lattice such that the vector space structure and the lattice structure are compatible (i.e., if $f, g \in L$ satisfy $f \leq g$, then $f+h \leq g+h$ for every $h \in L$ and $af \leq ag$ for every real number $a \geq 0$). The subset $L^+ = (f : f \in L, f \geq 0)$ is called the *positive cone* of L and the elements in the positive cone are called *positive elements*. We shall use the familiar lattice notations $f \vee g$ and $f \wedge g$ for the supremum and infimum of f and g. The notations

$$f^+ = f \vee 0, \quad f^- = (-f) \vee 0, \quad |f| = f \vee (-f)$$

for any $f \in L$ are also familiar, and it is a simple theorem that f^+, f^- and $|f|$ are positive elements in L with $f = f^+ - f^-$ and $|f| = f^+ + f^-$. We say that f is *infinitely small with respect to g* if $f \neq 0$ and $n|f| \leq |g|$ for $n = 1, 2, \ldots$, and f is called an *infinitely small element* if f is infinitely small with respect to some g. A Riesz space containing no infinitely small elements is called *Archimedean*. In the present exposition we shall restrict ourselves to Archimedean Riesz spaces, and we list some examples:

(i) The set $C(X)$ of all real continuous functions on a topological space X, with the usual definitions for the vector space operations and the partial order, is an Archimedean Riesz space.

(ii) We generalize the preceding example by admitting also extended real-valued continuous functions on X (i.e., continuous mappings from X into the topological space R_∞ of all extended real numbers), with the restriction however that if f is a function of this kind, then the open set $(x \mid |f(x)| < \infty)$ should be dense in X. The system of all these functions is denoted by $C^\infty(X)$. In general, $C^\infty(X)$ is no vector space, simply because if f and g in $C^\infty(X)$ are given and we define already $h(x) = f(x) + g(x)$ for all x for which $f(x)$ and $g(x)$ are finite, then it is not always possible to define $h(x)$ in the remaining points x in such a manner that h becomes continuous in the extended sense on X. However, if X is extremally disconnected (i.e., the closure of every open set is open), then $C^\infty(X)$ is a vector space, and so $C^\infty(X)$ is then also a Riesz space. In the other cases, although $C^\infty(X)$ as a whole is not always a vector space, certain subsets of $C^\infty(X)$ may be Riesz spaces.

(iii) If μ is a measure in the point set X and $L_p(X, \mu)$, for some number p satisfying $0 < p < \infty$, is the family of all p-th power μ-integrable real functions on X (or rather equivalence classes of these functions in the well-known manner), then $L_p(X, \mu)$ is a Riesz space. Of course, the partial order has to be defined appropriately, i.e., $f \leq g$ means that $f(x) \leq g(x)$ holds for μ-almost every x.

(iv) If H is a (complex) Hilbert space of dimension at least two and $\mathscr{H}$ is the real vector space of all bounded Hermitian operators on H, partially ordered in the usual manner, then $\mathscr{H}$ is a partially ordered vector space but no Riesz space. Certain appropriate subspaces of $\mathscr{H}$, however, are Riesz spaces, and these subspaces can be used very neatly to give a simple proof of the spectral theorem for Hermitian and normal operators.

In order to conclude the introduction, we state what we shall understand by a representation of a given Archimedean Riesz space L. It will be evident what it means to say that two Riesz spaces are isomorphic. The Riesz space M is now called a *representation* of the given Riesz space L if M is isomorphic to L and, in addition, M is a subspace of $C(X)$ or a subset of $C^\infty(X)$ for some appropriate topological space X. There are several kinds of representation theorems for a given Archimedean space L depending upon which entities are chosen to play the role of the points of the topological space X. In one variant the points of X are prime ideals in the given Riesz space L. It is not always necessary to take all the existing prime ideals; in a rather special case, considered by K. Yosida [5] already in 1941, we restrict ourselves to the maximal ideals, and we obtain a theorem which in some respects is very similar to Gelfand's representation theorem for commutative Banach algebras. A general representation theorem, where arbitrary prime ideals occur as points of the topological space X, was proved by D. G. Johnson and J. E. Kist [1] in 1962 in the United States.

We shall say more about quite a different kind of representation theorem in section 3.

2. Representation theorems of Yosida and Johnson—Kist

We present the definitions of an ideal, a maximal ideal and a prime ideal in a Riesz space (observe the analogy with the corresponding definitions in a commutative ring). The linear subspace I of the Riesz space L is called an *ideal* if it follows from $f \in I$ and $|g| \leq |f|$ that $g \in I$ (equivalently, $f \in I$ holds if and only if $|f| \in I$ holds, and $|f| \wedge |g| \in I$ holds for all $f \in I$ and $g \in L$). The ideal J is *maximal* if J is a proper ideal in L and there exists no other ideal properly between J and L. The ideal P is called a *prime ideal* if it follows from $f \wedge g \in P$ that one at least of f and g is in P. Finally, the prime ideal M is called a *minimal prime ideal* if M does not properly contain any other prime ideal. We denote the set of all proper prime ideals by $\mathscr{P}$, and we list some results holding in $\mathscr{P}$.

For every Riesz space L (not consisting exclusively of the zeroelement) the set $\mathscr{P}$ is non-empty. Actually, $\mathscr{P}$ contains so many elements P that their intersection is the zeroelement, i.e.,

$$\cap (P : P \in \mathscr{P}) = \{0\}.$$

It is convenient to have a special notation for certain subsets of $\mathscr{P}$. For any $u \in L^+$ we define the subset $\{P\}_u$ of $\mathscr{P}$ by

$$\{P\}_u = (P : P \in \mathscr{P}, \ u \notin P).$$

Since $\{P\}_u \cap \{P\}_v = \{P\}_{u \wedge v}$ holds for all $u, v \in L^+$, it follows immediately that every finite intersection of sets of this kind is again of the same kind. Also note that for $u = 0$ the set $\{P\}_u$ is empty. The sets $\{P\}_u$, $u \in L^+$, can be used, therefore, as a base for a topology in $\mathscr{P}$. This topology is called the *hull-kernel topology* in $\mathscr{P}$, because for any subset $\mathscr{D}$ of $\mathscr{P}$ the closure of $\mathscr{D}$ turns out to be the *hull* of the *kernel* of $\mathscr{D}$, where

$$k(\mathscr{D}) = \text{kernel } \mathscr{D} = \cap (P : P \in \mathscr{D})$$

and

$$h\big(k(\mathscr{D})\big) = \text{hull (kernel } \mathscr{D}) = (P : P \supset k(\mathscr{D})).$$

In general, the hull-kernel topology is no Hausdorff topology, but only a T_0-topology; the base sets $\{P\}_u$ are open and compact. It can be proved that the topology is Hausdorff if and only if the topology is T_1, and also if and only if every $P \in \mathscr{P}$ is a maximal ideal, and also if and only if every $P \in \mathscr{P}$ is a minimal prime ideal.

If $\mathscr{R}$ is any non-empty subset of $\mathscr{P}$, we can take the relative topology in $\mathscr{R}$ of the hull-kernel topology in $\mathscr{P}$. This relative topology is then called the hull-kernel topology in $\mathscr{R}$. Thus, it can be proved for example that if we take the hull-kernel topology in the set $\mathscr{M}$ of all minimal prime ideals, we obtain a Hausdorff topology with base sets $\{M\}_u$, $u \in L^+$, that are open and closed.

Any maximal ideal is a prime ideal. However, there exist Archimedean Riesz spaces that possess no maximal ideals at all (e.g. the space $L_1(X, \mu)$ of all μ-integrable functions on X). A sufficient condition for the Archimedean Riesz space L to possess so many maximal ideals J that $\cap J = \{0\}$ holds, is that L has a *strong unit* e (i.e., an element $0 < e \in L^+$ such that every $f \in L^+$ is majorized by a suitable multiple of e, so $f \leq ne$ for some natural number n, where n depends upon f of course). For an Archimedean space L with strong unit e we have the Yosida representation theorem, as follows. We introduce the hull-kernel topology in the set $\mathscr{J}$ of all maximal ideals J. This topology makes $\mathscr{J}$ into a compact Hausdorff space. Given $J \in \mathscr{J}$

and $f \in L$, there exists exactly one real number α satisfying $\alpha e - f \in J$; we denote this number α by $\hat{f}(J)$. For f fixed and J variable in $\mathscr{J}$, we thus obtain a real function $\hat{f}$ on the set $\mathscr{J}$. It can be proved that $\hat{f}$ is continuous on the topological space $\mathscr{J}$. This implies that the mapping $f \to \hat{f}$ is a linear mapping from L into $C(\mathscr{J})$. Since $f \to \hat{f}$ is also a one-one mapping that preserves the lattice structure, the set $\hat{L}$ of all $\hat{f}$ is a Riesz space isomorphic to L, and $\hat{L}$ is a subspace of $C(\mathscr{J})$. In other words, $\hat{L}$ is a representation of L. In general, $\hat{L}$ is a proper subspace of $C(\mathscr{J})$. It can be proved that $\hat{L} = C(\mathscr{J})$ holds if and only if L is *uniformly complete* with respect to the unit e, i.e., every e-uniform Cauchy sequence has a limit, where the sequence $(f_n : n = 1, 2, \ldots)$ in L is called an e-uniform Cauchy sequence if for any given $\varepsilon > 0$ we have $|f_m - f_n| \leqq \varepsilon e$ for m, n large enough. The analogy between the Yosida representation and the Gelfand representation for commutative normed algebras is evident.

The question arises now what we can do if L has no strong unit. First of all, we take an arbitrary element $e > 0$ in L^+ and we investigate to what extent e can play the role of a kind of unit. For this purpose we restrict ourselves for the moment to consider only the prime ideals P that do not contain e, i.e., we consider the set $\{P\}_e$. Given $P \in \{P\}_e$ and $f \in L$, there does not always exist a real number α such that $\alpha e - f \in P$, but as a substitute for this there now exists an extended real number α_0 satisfying

$$\alpha_0 = \sup\,(\alpha : (\alpha e - f)^+ \in P).$$

This number α_0 is denoted by $\hat{f}(P)$. For f fixed and P variable in $\{P\}_e$, we thus obtain an extended real function $\hat{f}$ on the set $\{P\}_e$, and if we equip $\{P\}_e$ with the hull-kernel topology, then $\hat{f}$ turns out to be a member of $C^\infty(\{P\}_e)$. We want to define the function $\hat{f}$, however, on a larger topological space. In order to achieve this, we first choose a *maximal disjoint system* $(e_\alpha : \alpha \in \{\alpha\})$ of positive elements in L, i.e., all e_α satisfy $e_\alpha > 0$, for $\alpha_1 \neq \alpha_2$ we have $e_{\alpha_1} \wedge e_{\alpha_2} = 0$, and furthermore the system is maximal in the sense that there exists no $e_\beta > 0$ satisfying $e_\beta \wedge e_\alpha = 0$ for all α in the index set $\{\alpha\}$. Note that $e_{\alpha_1} \wedge e_{\alpha_2} = 0$ implies

$$\{P\}_{e_{\alpha_1}} \cap \{P\}_{e_{\alpha_2}} = \{P\}_{e_{\alpha_1} \wedge e_{\alpha_2}} = \emptyset.$$

Instead of taking the topological space $\{P\}_e$ for one single e only, we now take the union $\mathscr{R} = \cup_\alpha \{P\}_{e_\alpha}$. Every $P \in \mathscr{R}$ is in exactly one of the $\{P\}_{e_\alpha}$, so $\hat{f}(P)$ can be defined as explained above. The mapping $f \to \hat{f}$ from L into $C^\infty(\mathscr{R})$ is linear and one-one, and it follows easily that the set $\hat{L}$ of all $\hat{f}$ is a representation of L.

In general, the functions of the representation $\hat{L}$ do not *separate the points* of the topological space $\mathscr{R}$, i.e., given two different points P_1 and P_2 of $\mathscr{R}$, there does not always exist an element $f \in L$ such that $\hat{f}(P_1) \neq \hat{f}(P_2)$. If we say, for a moment, that two points P_1 and P_2 of $\mathscr{R}$ are equivalent whenever $\hat{f}(P_1) = \hat{f}(P_2)$

holds for all $f \in L$, then $\mathscr{R}$ is split up into equivalence classes, and if we delete from each equivalence class all points but one, then the functions $\hat{f}$ on the remaining points of $\mathscr{R}$ will still form a representation of L. This "reduced" representation $\hat{L}_{\text{red}}$ will now separate the points of the "reduced" topological space $\mathscr{R}_{\text{red}}$. The problem is how to see whether two points P_1 and P_2 from $\mathscr{R}$ are in the same equivalence class. This can indeed be done, purely in terms of the properties of the points P of $\mathscr{R}$ (i.e., without referring to the functions $\hat{f}$), as follows. The sets $\{P\}_{e_\alpha}$ in $\mathscr{R} = \cup_\alpha \{P\}_{e_\alpha}$ are mutually disjoint; let e be one of the e_α, and consider a point P_0 in that particular $\{P\}_e$. The set $(P : P \in \{P\}_e, P \supset P_0)$ is now linearly ordered by inclusion, and this set has a maximal element $P_{0,\text{max}}$. The element $P_{0,\text{max}}$ is a prime ideal in L, maximal with respect to the property of not containing e (i.e., any ideal properly larger than $P_{0,\text{max}}$ must contain e). It may be said, therefore, that $P_{0,\text{max}}$ is the maximal element in the chain of prime ideals that stands on top of P_0. Different prime ideals P_0 and P_0' in $\{P\}_e$ have different chains on top of them, but from a certain point on the chains may coincide, and so P_0 and P_0' have the same maximal elements on top of their chains in this case, i.e., $P_{0,\text{max}} = P_{0,\text{max}}'$. This property of having the same maximal element on top of their chains obviously defines an equivalence relation in $\{P\}_e$, and it can be proved that this is the same equivalence relation as we had before. Hence, if we desire to have a representation of L that separates the points of the underlying topological space, we must "reduce" the space $\mathscr{R} = \cup_\alpha \{P\}_{e_\alpha}$ by retaining from each equivalence class in $\mathscr{R}$ only one point, then take the hull-kernel topology in the reduced space $\mathscr{R}_{\text{red}}$ and form the functions $\hat{f}$ on $\mathscr{R}_{\text{red}}$ by the method indicated above. The topology then becomes Hausdorff.

3. The representation theorems of Maeda—Ogasawara and Nakano

We briefly indicate how to obtain a representation theorem of a different kind. In this theorem, due to F. MAEDA and T. OGASAWARA ([3], 1942), one considers first of all the set of all bands in the given Archimedean Riesz space L. A *band* (also called an *order closed ideal*) is an ideal B in L with the extra property that if D is a subset of B such that the element sup D exists in L, then this element sup D is contained in B (i.e., B is closed under the operation of taking suprema (and infima), insofar as these exist). The set $\mathscr{B}(L)$ of all bands, partially ordered by ordinary inclusion, is a Boolean algebra. The Boolean algebra $\mathscr{B}(L)$ is order complete, i.e., every subset of $\mathscr{B}(L)$ has a supremum and an infimum. In Boolean algebras there exists a prime ideal theory (with corresponding hull-kernel topology) very similar to the prime ideal theory in Riesz spaces, and even simpler in some respects because in a Boolean algebra the notions of a proper prime ideal and a maximal ideal are identical. Since the Boolean algebra $\mathscr{B}(L)$ is order complete, the topological space Ω of all proper prime ideals ω in $\mathscr{B}(L)$ becomes not only compact and Hausdorff but also extremally disconnected, and this implies that $C^\infty(\Omega)$ is now a Riesz space. Corresponding to any $f \in L$ there is now a function $f^\wedge \in C^\infty(\Omega)$, defined by means of a formula similar to the formula for $\hat{f}$ in the preceding section. The set $L^\wedge$ of all $f^\wedge$ is a representation of L.

The so obtained Maeda—Ogasawara representation has some advantages and some disadvantages when compared with the Yosida or Johnson—Kist representations in the preceding section. We mention one the advantages. First we recall that a partially ordered set S is sometimes called *Dedekind complete* whenever every subset of S which is bounded from above has a supremum, and every subset of S which is bounded from below has an infimum. Most Riesz spaces fail to be Dedekind complete. It is, for example, an easy exercise to show that the space $C(X)$, with $X=[0, 1]$ and the ordinary topology in X, is not Dedekind complete. The set of all rational numbers is of course a standard example of a non-Dedekind complete partially ordered set, and the set of all real numbers is the "Dedekind completion" of the set of rational numbers (proof by means of Dedekind cuts). It can be proved quite similarly by means of "Dedekind cuts" that every Archimedean Riesz space L can be embedded densely in a Dedekind complete space L_{comp}, and L_{comp} is then called the *Dedekind completion* of L. The proof by means of Dedekind cuts is long and laborious. If we observe that we can just as well give a proof for the isomorphic space $L^\wedge$ (the Maeda—Ogasawara representation), then there exists a very short proof, because the Dedekind completion of $L^\wedge$, within the large space $C^\infty(\Omega)$, is immediately visible. It is essential here that $C^\infty(\Omega)$ is a Riesz space. One can have mixed feelings about whether this quick proof is really "better", because the very existence of $L^\wedge$ (and also all the other representations mentioned so far) depends on the use of Zorn's lemma.

One of the disadvantages of the Maeda—Ogasawara representation concerns the separation of points. If we take, for example, the Riesz space $C(X)$ with $X=[0, 1]$ and the ordinary topology in X, then the space $\mathscr{J}$ of all maximal ideals in $C(X)$ is homeomorphic with X, so it follows easily that the Yosida representation separates the points of $\mathscr{J}$. The Maeda-Ogasawara representation, however, does not separate the points of Ω. There is a simple sufficient condition in order that the Maeda—Ogasawara representation will separate the points of Ω. The condition is that the Riesz space L has the *projection property*. We shall explain what this means. The band B in L is called a *projection band* if there exists another band B_1 in L such that $B \cap B_1 = \{0\}$ and $B \oplus B_1 = L$. If there exists a band B_1 with these properties, then this complementary band B_1 is uniquely determined. The Riesz space L is now said to have the projection property if every band in L is a projection band.

A variant of the Maeda—Ogasawara representation is obtained by considering, instead of the Boolean algebra of all bands in L, only the sublattice of all principal bands. For the case that L has the *principal projection property* (i.e., every principal band is a projection band, where by a principal band we mean a band generated by one single element), the functions $f^\wedge$ of $L^\wedge$ are now defined and continuous (as extended realvalued functions) on an appropriate subset of the locally compact space Ω_p, the points of which are the maximal ideals in the sublattice $\mathscr{B}_p(L)$ of all principal bands. This sublattice is a Boolean ring if L has the principal projection property. This representation is essentially due to H. NAKANO ([4], 1941).

Generally, if $L_1^\wedge$ and $L_2^\wedge$ are representations of the same Riesz space L, with $L_1^\wedge$ consisting of functions on the topological space X and $L_2^\wedge$ consisting of functions on the topological space Y, then although $L_1^\wedge$ and $L_2^\wedge$ are isomorphic as Riesz spaces, it is not true in general that X and Y are homeomorphic as topological spaces. It may be asked under which conditions X and Y are homeomorphic such that, for $x \in X$ and $y \in Y$ corresponding points, we have $f^\wedge(x) = f^\wedge(y)$ for every $f \in L$. In this case, of course, we can say that the representations are "identical". We conclude this account by presenting a partial answer. If L has the projection property, then the "reduced" Johnson—Kist representation (mentioned in the preceding section), the Maeda—Ogasawara representation and the Nakano representation are identical in this strong sense. The proof of this result is by no means trivial.

All results mentioned here are presented, with complete proofs, in Volume 1 of a forthcoming book on Riesz spaces by W. A. J. LUXEMBURG and myself [2].

REFERENCES

[1] D. G. Johnson and J. E. Kist, *Prime ideals in vector lattices.* Canad. J. Math. **14** (1962), 517—528.

[2] W. A. J. Luxemburg and A. C. Zaanen, *Riesz Spaces.* Vol. 1, North-Holland Publishing Company, Amsterdam 1971.

[3] F. Maeda and T. Ogasawara, *Representation of vector lattices.* J. Sci. Hiroshima Univ. (A) **12** (1942), 17—35.

[4] H. Nakano, *Eine Spektraltheorie.* Proc. Phys.-Math. Soc. Japan (3) **23** (1941), 485—511.

[5] K. Yosida, *On vector lattice with a unit.* Proc. Imp. Acad. Tokyo **17** (1940—41), 121—124.

Boundedness from Measure Theory

By

HENRY HELSON

DEPT. OF MATH.
UNIVERSITY OF CALIFORNIA
BERKELEY

1. Several recent results in vectorial function theory are obviously of uniform-boundedness type, but their proofs have depended on category arguments rather than on the classical boundedness theorems. This paper is an attempt to find a new boundedness theorem that will simplify and unify these proofs, and perhaps be useful in other contexts.

These results are due to M. SHERMAN [8, 9], SZ.-NAGY and FOIAŞ [2], and D. A. HERRERO [5, 6]. A central theorem [9], conjectured by Sherman and proved by the method of Sz.-Nagy and Foiaş [2], will be reproved here in detail. The application of our method to Herrero's problems will be illustrated rather than carried out. Finally we prove a theorem about functions of several complex variables, similar to a result of KAHANE [7].

Perhaps no one has observed that the Baire category theorem can be replaced, in proving boundedness theorems, by the Steinhaus theorem on the difference set of a set of positive measure. The conclusions are generally similar, but the argument by measure theory gives local information that may sometimes be useful. The exposition below is based on the Steinhaus theorem.

2. THEOREM 1. *Let B be a Banach space and $p(x)$ a Borel function on B with these properties: $p(x) \geqq 0$, $p(tx) = t^\alpha p(x)$, and $p(x-y) \leqq \varphi(p(x), p(y))$ for all x, y in B and $t > 0$, where α is a non-negative number and φ any positive function increasing in each argument. Then p is bounded on each ball of B.*

Suppose on the contrary that $p(x_n)$ tends to ∞ for some bounded sequence $\{x_n\}$ in B. If α is positive we choose the sequence so that $p(x_n) \geqq n^{3\alpha}$. For each sequence $\{r_n\}$ with $0 \leqq r_n < 1$ define

$$(1) \qquad x(r) = \sum_1^\infty r_n n^{-2} x_n,$$

the series being convergent in B. We regard r as an element of the infinite-dimensional torus T^∞ (where coordinates are computed modulo one). $x(r)$ is a Borel mapping of T^∞ into B (the reduction modulo one makes this not quite obvious), and so

$p(x(r))$ is a real Borel function on T^∞. For some positive number K, the set E where $p(x(r))\leqq K$ is a Borel set of positive Haar measure on the torus.

A famous theorem of STEINHAUS [10] asserts in these circumstances that the difference set $E-E$ contains a neighbourhood of 0 in T^∞. The original proof in Euclidean space was not simple, but now one argues as follows. If k is the characteristic function of E, then

$$(2) \qquad \int k(r)k(r+s)\,dr$$

is continuous in s because translation is continuous in $L^2(T^\infty)$, and positive for $s=0$. Therefore the integral is positive on a neighbourhood of 0, from which the assertion follows.

In T^∞ convergence means convergence modulo one in each coordinate. Therefore each neighbourhood of 0 contains all points r satisfying a finite set of inequalities

$$(3) \qquad |\exp 2\pi i r_n - 1| < \varepsilon \qquad (1\leqq n \leqq k),$$

the other coordinates of r being unrestricted. In particular $E-E$ contains all r such that $r_n=0$ $(1\leqq n\leqq k)$, if k is large enough.

Fix $n>k$. Set $r_n=1/2$ and $r_m=0$ for $m\neq n$. Then r is in $E-E$: $r = u - v$ for some u, v in E. This equality in T^∞ is a congruence modulo one in each coordinate. If we express u and v by means of coordinates lying in $[0, 1)$ we shall have the literal equalities $u_m - v_m = 0$ for $m\neq n$, $u_n - v_n = 1/2$ or $-1/2$. Interchange u and v if necessary to obtain $u_n - v_n = 1/2$. Then the algebraic properties of p give

$$(4) \quad (2n^2)^{-\alpha}p(x_n) = p(x(r)) = p(x(u-v)) = p(x(u)-x(v)) \leqq \varphi(p(x(u)), p(x(v))) \leqq$$

$$\leqq \varphi(K, K).$$

The left member tends to infinity with n, whereas the right member is fixed. This contradiction shows that p is bounded on each ball of B.

This theorem is of well-known type [1, p. 463]. The case $\alpha=1$ immediately implies the Banach—Steinhaus theorem, and can be used in a proof of the closed graph theorem. For any positive α the statement of the theorem can easily be reduced to the case $\alpha=1$. But the implications of the result with $\alpha=0$ seem not to have been recognized. We are going to complete, generalize and apply that part of Theorem 1.

3. THEOREM 2. *Suppose in the hypotheses of Theorem 1 that $\alpha=0$ and $\varphi(s, t)=$ $=\max(s, t)$. Then p assumes its least upper bound.*

Let ϱ be the least upper bound of the values of p. If $p(x)<\varrho$ for every x in B, then $q(x) = (\varrho-p(x))^{-1}$ is a new function satisfying the hypotheses of Theorem 1. Hence q is bounded. This contradicts the definition of ϱ, and the theorem is proved.

Actually more is true. *Let $\{x_n\}$ be a bounded sequence from B such that $p(x_n)$ tends to ϱ. Define $x(r)$ by (1). Then $p\big(x(r)\big)=\varrho$ for almost every r in T^∞.* Otherwise there is a positive number γ such that $p\big(x(r)\big) \leqq \varrho-\gamma$ for all r in a set of positive measure in T^∞. The special form of φ means that the same inequality holds on the difference set, which contains a neighbourhood of 0 in T^∞. Hence $p(x_n) \leqq \varrho-\gamma$ for all large n, which is false.

4. Here is an application of Theorems 1 and 2, inspired by HERRERO's work [6]. The first part of this result could have simplified the statement and proof of Theorem 13 and its corollary in [4].

THEOREM 3. *Let $A(z)$ be an operator function in a Banach space B, analytic for $|z|<1$. Suppose for each x in B the vector function $A(z)x$ has an analytic continuation to a neighbourhood (depending on x) of $z=1$. Then this neighbourhood can be chosen independent of x, and $A(z)$ has a continuation there as a bounded operator function. On the other hand if the circle $|z|=1$ is a natural boundary for $A(z)$, the same is true for some vector function $A(z)x$.*

Express $A(z)$ as a Taylor series about the point $z=1/2$:

$$(5) \qquad A(z) = \sum_0^\infty A_n(z-1/2)^n.$$

Applied to a vector x, the series converges in a circle of radius $q(x)>1/2$, where

$$(6) \qquad q(x)^{-1} = \lim \sup \|A_n x\|^{1/n}.$$

Obviously q is homogeneous of degree 0, and it is easy to prove that q is a Borel function. Then $p(x) = \big(q(x)-1/2\big)^{-1}$ has the same properties, and $p(x-y) \leqq$ $\leqq \max\big(p(x), p(y)\big)$. Theorem 1 asserts that p is bounded. That is, all the vector functions $A(z)x$ are continuable to a single neighbourhood of $z=1$.

From (6) it follows that there is a number $\gamma<2$ such that $\|\gamma^{-n}A_n x\|$ is bounded in n for each x. The Banach—Steinhaus theorem implies that $\|A_n\|=O(\gamma^n)$. Hence (5) converges in a circle of radius greater than $1/2$, and the first part of the theorem is proved.

Now suppose that $A(z)$ is not continuable beyond the circle. Let $\{z_k\}$ be a countable dense subset of the open disc. Define $q_k(x)$ to be the radius of analyticity of $A(z)x$ about z_k, or 1 if this quantity exceeds 1. Set $p_k(x) = \big(2-q_k(x)\big)^{-1}$. For each k this function satisfies the hypotheses of Theorem 2.

By the first part of the theorem we can choose a bounded sequence $\{x_n\}$ from B so that the singular points of the functions $A(z)x_n$ are dense on the boundary circle. We define $x(r)$ by (1). By the second proof of Theorem 2, for each k the function $p_k\big(x(r)\big)$ is equal to its upper bound for almost all r. The same statement must hold for all k at once, except in a grand null set of r. Hence $A(z)x(r)$ has the circle for a natural boundary, for almost every r. That concludes the proof.

5. The theorems of this section, generalizing Theorems 1 and 2, are the main results of this paper.

THEOREM 4. *Let (X, S) be a countably generated measure space. For each x in a Banach space B, let μ_x be a finite positive measure on X. Assume that $\mu_x(F)$ is a Borel function on B for each F in S, and also: $\mu_{tx} = \mu_x$, $\mu_{x-y} \leqq \mu_x + \mu_y$ for all x, y in B and $t > 0$. Then there is a finite measure μ on X such that $\mu_x \leqq \mu$ for all x in B.*

Otherwise there is a sequence $\{x_n\}$ from B such that the measures μ_{x_n}, which we shall call μ_n, have no finite majorant. We may assume that $\|x_n\| = 1$ for each n. Define $x(r)$ by (1); fix a positive integer n and a set F in S. Let E be the set of all r in T^∞ such that $\mu_{x(r)}(F) < \frac{1}{2}\mu_n(F)$. E is a Borel set. If the measure of E is positive, $E - E$ contains a neighbourhood of 0 on T^∞, and in particular the sequence having ε in the nth place and zeros elsewhere, for a certain ε satisfying $0 < \varepsilon < 1$. Hence for some u, v in E we have $x(u) - x(v) = tn^{-2}x_n$, where t is ε or $1 - \varepsilon$. Then we have

$$(7) \qquad \mu_n(F) = \mu_{x(u)-x(v)}(F) \leqq \mu_{x(u)}(F) + \mu_{x(v)}(F) < \mu_n(F).$$

This contradiction shows that E was a null set.

Let $\{F_j\}$ be a countable family of sets that generates S. We may assume this family forms a ring, because in any case the smallest ring containing the family is still countable. For each n, j we have $\mu_{x(r)}(F_j) \geqq \frac{1}{2}\mu_n(F_j)$ except in a null set of r. Hence the same inequalities hold at once, except in a grand null set.

For fixed r and n, the sets F such that $\mu_{x(r)}(F) \geqq \frac{1}{2}\mu_n(F)$ form a monotone family. If this family contains the ring of all F_j, then it contains the σ-algebra they generate, namely S [3, p. 27]. That is, $\mu_{x(r)} \geqq \frac{1}{2}\mu_n$.

Thus $2\mu_{x(r)} \geqq \mu_n$ for all n, except in a null set of r. This contradicts the assumption about $\{\mu_n\}$, and the theorem is proved.

THEOREM 5. *Suppose the measures μ_x of Theorem 4 satisfy the stronger condition $\mu_{x-y} \leqq \mu_x \vee \mu_y$ for all x, y in B. Then there is an element y such that $\mu_x \leqq \mu_y$ for all x.*

The theorem follows easily from the following lemma, which has independent interest. SHERMAN proved a special case of the lemma in his first paper [8], and a nearly equivalent result in [9].

LEMMA. *Let x_1, x_2 be any elements of B, $x(r) = r_1 x_1 + r_2 x_2$ for arbitrary real numbers r_1, r_2. Then $\mu_{x(r)} = \mu_{x_1} \vee \mu_{x_2}$ for all r such that $r_1^2 + r_2^2 = 1$ except a countable set.*

We have $\mu_0 = \mu_{x-x} \leqq \mu_x \vee \mu_x = \mu_x$ for all x. Hence $\mu_{-x} = \mu_{0-x} \leqq \mu_0 \vee \mu_x = \mu_x$. We conclude that $\mu_{-x} = \mu_x$ for all x.

Set $v = \mu_{x_1} \vee \mu_{x_2}$. By hypothesis $\mu_{x(r)} \leqq v$ for every r. The Radon—Nikodym theorem asserts that

$$(8) \qquad d\mu_{x(r)} = f_r \, dv, \quad d\mu_{x_1} = g_1 \, dv, \quad d\mu_{x_2} = g_2 \, dv$$

for certain measurable functions f_r, g_1, g_2 taking values between 0 and 1. Furthermore $g_1 \vee g_2 = 1$ almost everywhere (v). We want to show that $f_r = 1$ almost everywhere (v), for all r on the unit circle except a countable set.

If this is not true we can find $\gamma > 0$ such that for infinitely many r with $r_1^2 + r_2^2 = 1$ we have $f_r \leqq 1 - \gamma$ on a set F_r in S such that $v(F_r) \geqq \gamma$. Since v is a finite measure, some pair of distinct F_r must intersect:

$$(9) \qquad r_1^2 + r_2^2 = 1 = s_1^2 + s_2^2, \quad r \neq s, \quad v(F_r \cap F_s) > 0.$$

On $F = F_r \cap F_s$ we have $f_r \leqq 1 - \gamma$, $f_s \leqq 1 - \gamma$.

The set of all r in the plane such that $f_r \leqq 1 - \gamma$ almost everywhere (v) on F is a subgroup of the plane, because $f_{r-s} \leqq f_r \vee f_s$ for all r, s. This subgroup admits multiplication by real numbers, so it is a vector subspace. It contains the independent pair mentioned in (9); therefore it fills the plane. In particular $g_1 \leqq 1 - \gamma$, $g_2 \leqq 1 - \gamma$ almost everywhere (v) on F. This contradiction establishes the lemma.

COROLLARY. *Let* $x(r) = rx_1 + x_2$. *Then* $\mu_{x(r)} = \mu_{x_1} \vee \mu_{x_2}$ *exept for denumerably many* r.

Now we can prove the theorem. Let $\{x_n\}$ be any bounded sequence from B; define $x(r)$ by (1) once more. For fixed $r_2, r_3, \ldots$ the corollary asserts $\mu_{x(r)} \geqq \mu_1$ except for a countable set of r_1. The set of r on which the inequality holds is a Borel set in T^∞ (because S is countably generated); by the Fubini theorem it has to be a null set. Similarly $\mu_{x(r)} \geqq \mu_n$ except in a null set of r, for each n in turn. Finally the inequalities hold all at once except in a grand null set of r.

Let $\varrho = \sup \mu_x(X)$. Choose $\{x_n\}$ so that $\mu_n(X)$ tends to ϱ. Then for almost all r the measures $\mu_{x(r)}$ are identical, have total mass equal to ϱ, and majorize $\{\mu_n\}$.

This measure v is actually a majorant for the family $\{\mu_x\}$. Indeed let x be any element of B. Repeat the argument just given with the new sequence $\{x, x_1, x_2, \ldots\}$. We obtain a new measure v' that is larger than v (because v was the smallest majorant of $\{\mu_n\}$), but has the same total mass ϱ. Hence $v = v' \geqq \mu_x$ as asserted. This completes the proof of the theorem.

6. Theorem 4 generalizes Theorem 1 in the case $\alpha = 0$, $\varphi(r, s) = r + s$. An analogous generalization with $\alpha = 1$ is not true. Let B be $L^1(0, 1)$, $X = (0, 1)$, and S the Borel field of X. For f in B let μ_f be the measure $|f| \, dx$. Then μ_f is homogeneous and subadditive, but the unit ball of B has no majorant.

It is less obvious that Theorem 4 is an *essential* generalization of that part of Theorem 1. The new difficulties of proof, and the applications that follow, provide indirect evidence that it is.

7. Sherman's theorem has to do with vector-valued analytic functions as studied in [4]. $H_{\mathscr{H}}^2$ is the Hardy space of functions on the unit circle with values in separable Hilbert space $\mathscr{H}$. A subspace $\mathfrak{M}$ of $H_{\mathscr{H}}^2$ is called *invariant* if it is closed, and $zF(z)$ belongs to $\mathfrak{M}$ with F.

THEOREM [9]. *If $\mathfrak{M}$ is an invariant subspace with the property that for each x in $\mathscr{H}$ there is a scalar inner function q_x such that $q_x x$ is in $\mathfrak{M}$, then for some single inner q we have qx in $\mathfrak{M}$ for all x.*

Originally SHERMAN conjectured and proved in part a slightly weaker result [8], which recently was established in full by SZ.-NAGY and FOIAŞ [2]. Then Sherman found the theorem above. HERRERO has also given a proof based on his other work.

Let X be the closed unit disc and S the field of Borel subsets of X. For each x in $\mathscr{H}$ we define a measure μ_x on X as follows: μ_x contains the positive singular measure associated with q_x on the boundary of the disc, and a point mass of magnitude $n(1-|z|)$ at each point z of the interior where q_x has a zero of order n. Here q_x is taken to be the weakest inner (or constant) function such that $q_x x$ is in $\mathfrak{M}$. Thus $\mu_x = 0$ just if x belongs to $\mathfrak{M}$.

The measure space is countably generated, and the algebraic hypotheses of Theorem 4 are easy to verify. Assume for the moment that $\mu_x(F)$ is a Borel function of x for each F in S. Then $\{\mu_x\}$ has a finite majorant μ. We may assume that μ, restricted to the boundary of the disc, is singular with respect to Lebesgue measure, because each μ_x is singular. Similarly μ can be taken to consist of weighted integral point masses in the interior of the disc. The finiteness of μ means that μ is the measure associated with an inner function q, which is stronger than each q_x. Hence qx is in $\mathfrak{M}$ for every x, and the theorem is proved.

Now we must prove that $\mu_x(F)$ is a Borel function of x. Our proof depends on the convergence properties of analytic functions, as have other proofs of the theorem.

First we show that $|q_x(z)|$ is a Borel function for each z, $|z| < 1$. Indeed this function is semi-continuous: if x_n tends to x, then $|q_x(z)| \geq \limsup |q_{x_n}(z)|$. If $q_{x_n}(z)$ tends to 0 the relation is obvious. Otherwise, passing to a subsequence if necessary, $q_{x_n}(z)$ converges to a limit $w \neq 0$ and q_{x_n} converges to a function f in H^∞, in the weak topology of L^2. Then $q_{x_n} x_n$ tends to fx in the weak topology of $H_{\mathscr{H}}^2$. Now $\mathfrak{M}$ is closed and therefore weakly closed; hence fx is in $\mathfrak{M}$. Therefore $f = q_x g$ for some g in H^2. On the boundary of the disc $|g| = |f| \leq 1$ almost everywhere. Thus $|w| = |f(z)| \leq |q_x(z)|$, which is the inequality that was to be proved.

Let F be an open set with closure contained in the interior of the disc. Denote by $N(x)$ the number of zeros of q_x in F, counted with multiplicity. We shall show that N is a Borel function. Let K be an infinite compact subset of F. For positive integers r, s let E_r be the set of x in $\mathscr{H}$ where $N(x) \leq r-1$, and $E(r, s)$ the set where $N(x) \leq r-1$ and $\sup_K |q_x(z)| \geq 1/s$. We shall prove that the closure of each $E(r, s)$ is contained in E_r.

Suppose that points x_n in $E(r, s)$ tend to x. Passing to a subsequence if necessary, we may assume that q_{x_n} converges uniformly on interior subsets of the disc to a function f not the null function. By the semi-continuity just proved we have $|f(z)| \leq \leq |q_x(z)|$ for $|z| < 1$. If z is a zero of q_x, f has a zero at z of at least equal multiplicity. By the Hurwitz theorem, q_{x_n} has as many zeros near z as f, if n is large enough. It follows that q_{x_n} has at least as many zeros as q_x in F, for all sufficiently large n. Therefore $N(x) \leq r-1$, and x belongs to E_r.

Obviously E_r is the union over s of $E(r, s)$. We have shown that E_r is the union of the *closures* of all $E(r, s)$. Hence E_r is a Borel set. This fact for each r implies that N is a Borel function.

For $0 < \gamma < 1$ let $N(x, \gamma)$ be the number of zeros of q_x in F with modulus smaller than γ. This function increases in γ for each x and

$$(10) \qquad \mu_x(F) = \int_0^1 (1-\gamma)\, dN(x, \gamma) = \int_0^1 N(x, \gamma)\, d\gamma.$$

Each Riemann sum associated with the integral is a Borel function of x; and a sequence of Riemann sums converges to the value of the integral for all x at once. Therefore the integral is a Borel function.

Now let F be an arc on the boundary of the disc. The singular inner divisor p_x of q_x has the representation

$$(11) \qquad p_x(z) = \lim_{r \uparrow 1} \exp \int_0^{2\pi} \frac{e^{it}+z}{e^{it}-z} \log |q_x(re^{it})|\, dt \qquad (|z| < 1).$$

We have shown that $|q_x(re^{it})|$ is a Borel function of x for fixed r, t, and continuous in t. Unfortunately the logarithm is infinite at the zeros of q_x. Truncate $\log |q_x(re^{it})|$ to obtain a Borel function of x that is still continuous in t. A sequence of Riemann sums approximates the integral (with the truncated function) simultaneously for all x, so the result is a Borel function of x. Let the truncation point tend to $-\infty$, and then let r increase to one. These operations give $p_x(z)$, showing that it is a Borel function for each z.

Therefore

$$(12) \qquad \log |p_x(z)| = -\int_0^{2\pi} P_r(e^{i(u-t)})\, d\mu_x(t) \qquad (z = re^{iu}),$$

where $P_r(e^{iu})$ is the Poisson kernel, is a Borel function. More generally, $\int Q(t)\,d\mu_x(t)$ is a Borel function if $Q(t)$ is a linear combination of $P_r(e^{i(u-t)})$ for different values of r, u. Such linear combinations are dense in the space of real continuous functions on the circle, because a measure orthogonal to all of them is necessarily null. We approximate the characteristic function of F to show that $\mu_x(F)$ is a Borel function.

For two particular types of sets F we know that $\mu_x(F)$ is a Borel function. One can deduce that the same is true for any Borel set F, and this completes the proof of the theorem.

8. Theorem 2 was used in the proof of Theorem 3 to find an analytic function with as many singularities as possible. We shall apply Theorem 5 in a similar way to functions of several complex variables.

Let D be a bounded domain in complex space C^n, and B the set of uniform limits of polynomials on D. B is a Banach space in the nuiform norm, indeed a Banach algebra. The functions in B are holomorphic on D and continuous on its closure $\bar{D}$.

THEOREM 6. *If D is polynomially convex, there is a function f in B whose power series about any point in D converges in no open set containing a boundary point of D.*

We choose and fix a measure μ on C^n with these properties: μ is finite, absolutely continuous with respect to Lebesgue measure, and positive on non-empty open sets. Let $\{z_k\}$ be a sequence of points dense in D.

For each f in B, and each k, let μ_f^k be the measure obtained by restricting μ to the complement of the subset of C^n where the power series for f about z_k converges. For each k this family of measures satisfies the hypotheses of Theorem 5. The only point that is not obvious is that $\mu_f^k(F)$ is a Borel function of f, for each Borel subset F of C^n. This can be proved in a few lines, using the fact that each Taylor coefficient of f is a Borel function of f. Now Theorem 5 asserts that for some element of B, say f_k, the measure $\mu_{f_k}^k$ is a majorant for all the measures μ_f^k.

The functions f_k can be taken to have norm 1. Define $f(r)$ by (1) (with f in place of x). It was shown in the proof of Theorem 5 that $\mu_{f(r)}^k \geqq \mu_{f_j}^k$ for every j, except in a null set of r. This is true for each k, and hence for all k at once except in a grand null set of r. In particular there is an f such that $\mu_f^k \geqq \mu_{f_k}^k$ for all k; and by the property of f_k, we have $\mu_f^k \geqq \mu_g^k$ for all k and for all g in B.

Thus we have proved the following assertion: the region of convergence of the power series for f about z_k is contained in that of g, for every k, and every g in B.

Suppose for some k the power series for f about z_k converges in an open set containing a point of the boundary of D. Then the same is true for every g. This contradicts the polynomial convexity of D. It follows that the power series for f

about an arbitrary point of D cannot converge on a neighbourhood of a boundary point, and the theorem is proved.

A similar theorem has been proved by KAHANE, also using measure theory [7]. His proof seems quite different, but illustrates the use of measure theory instead of category arguments to condense singularities.

9. B has been assumed for convenience to be a Banach space, but the natural domain for the theorems seems to be a complete locally convex topological vector space. This extension would be interesting for applications, for example to the space of all functions holomorphic on a domain of C^n in proving other versions of Theorem 6. But the aim of this paper is only to explain the method and illustrate its applications as simply as possible.

REFERENCES

[1] R. E. Edwards, *Functional Analysis*. Holt, Rinehart and Winston, New York 1965.

[2] C. B. Sz.-Nagy, and Foiaş *Local characterization of operators of class C_0*. J. Functional Analysis **8** (1971), 76—81.

[3] P. R. Halmos, *Measure Theory*. D. van Nostrand, Princeton 1950.

[4] H. Helson, *Lectures on Invariant Subspaces*. Academic Press, New York 1964.

[5] D. A. Herrero, *The exceptional subset of a C_0-contraction*. (to appear).

[6] D. A. Herrero, *Analytic continuation of inner function-operators*. (to appear).

[7] J.-P. Kahane, *Une remarque sur les séries de Rademacher et les domaines d'holomorphie*. Colloq. Math. **19** (1968), 107—110.

[8] M. J. Sherman, *Invariant subspaces containing all analytic directions*. J. Functional Analysis **3** (1969), 164—172.

[9] M. J. Sherman, *Invariant subspaces containing all constant directions*. J. Functional Analysis **8** (1971), 82—85.

[10] H. Steinhaus, *Sur les distances des points des ensembles de mesure positive*. Fund. Math. **1** (1920), 93—104.

Exchange of Infinite Summation with Limits
in Banach Spaces

By

GERHARD GRIMEISEN

MATHEMATISCHES INSTITUT A
UNIVERSITÄT STUTTGART

This is a report on a paper [4] being in preparation.

1. We agree in the following terminology: As usual (see, e.g., MONK [6]), we distinguish between *classes* and *sets*. A class A is a *set* if and only if A is a member of some class B. *Relations* are classes of ordered pairs of sets. *Mappings* are considered as special relations. For each mapping φ and each class X, $\varphi(X)$ is defined to be the class of all sets y such that $(x, y) \in \varphi$, i.e. $y = \varphi(x)$, holds for some $x \in X$. If $(M_i)_{i \in I}$ is a family of sets M_i (I a class) [set-theoretically, such a family is just a mapping], then $\underset{i \in I}{\mathsf{P}} M_i$ denotes the *direct* (or: *cartesian*) product of the sets M_i (which is, by definition, the class of all mappings φ on I into $\bigcup_{i \in I} M_i$ such that $\varphi(i) \in M_i$ for all $i \in I$); in the case when all M_i equal one set M it is usual to write M^I instead of $\underset{i \in I}{\mathsf{P}} M_i$. If $(\mathfrak{b}_i)_{i \in I}$ is a family of filters $\mathfrak{b}_i$ on sets K_i and I is a non-empty set, then we call (as in [2], p. 184) the filter on $\underset{i \in I}{\mathsf{P}} K_i$ generated by the filterbase $\{\underset{i \in I}{\mathsf{P}} \varphi(i) | \varphi \in \underset{i \in I}{\mathsf{P}} \mathfrak{b}_i\}$ to be the *cardinal product of the filters* $\mathfrak{b}_i$ (which coincides with BOURBAKI's "produit de filtres" (see [1], p. 68—69) for finite I). If M is a set, then each ordered triple $(f, I, \mathfrak{a})$ consisting of a non-empty mapping f being a set into M, of its domain I and of a filter $\mathfrak{a}$ on I is called a *filtered family in* M; we designate by ΦM the class of all filtered families in M. If M is a set and I is a class, then each mapping the domain of which is a subclass of M^I into M is said to be an *I-ary partial operation in* M; each I-ary partial operation in M defined on (all of) M^I is said to be an *I-ary operation in* M. There will occur, in this paper, one abuse of the word "operation": If (M, τ) is a Hausdorff space with the topology τ, the *limit operation* $\lim_\tau$ (or just denoted by $\lim$) induced by τ is defined to be the relation consisting of all ordered pairs $(x, (f, I, \mathfrak{a})) \in M \times (\Phi M)$ such that, for each neighbourhood U of x (with respect to τ), there is an $A \in \mathfrak{a}$ with $f(i) \in U$ for all $i \in A$. (Of course, for arbitrary classes X and Y, the *direct product* $X \times Y$ of X and Y is defined to be the class $\{(x, y) | x \in X \ and \ y \in Y\}$.)

2. Let $(E, \| \ \|)$ be a Banach space and $\lim$ be the limit operation induced by the topology being induced by the norm $\| \ \|$. Furthermore, let $\mathfrak{E}$ denote the set of all subsets X of E with card $X \leqq 1$, i.e. of all $X \subseteq E$ such that X has at most one element. Define the *absolute summation in* $(E, \| \; \|)$ to be the relation ${}^{\mathrm{a}}\sum$ such that

$$((a_n)_{n \in N}, a) \in {}^{\mathrm{a}}\sum \quad \textit{if and only if} \quad a = \lim_{m \to \infty} \sum_{n=1}^{m} a_n \quad \textit{and} \quad \sum_{n \in N} \|a_n\| < \infty$$

for all $(a_n)_{n \in N} \in E^N$ and all $a \in E$ $(N = \{1, 2, \dots\})$. In a natural way, the N-ary partial operation ${}^{\mathrm{a}}\sum$ in E (defined on a *subset* of E^N) and the limit operation lim (defined on a *subclass* of the class ΦE) can be "extended" to a certain N-ary operation ${}^{\mathrm{a}}\widehat{\sum}$ in the set $\mathfrak{E}$ (defined on the *whole set* $\mathfrak{E}^N$) [1] and to a certain mapping $\widehat{\lim}$ defined on the *whole class* $\Phi\mathfrak{E}$: Define ${}^{\mathrm{a}}\widehat{\sum}$ by

$$ {}^{\mathrm{a}}\widehat{\sum} (A_n)_{n \in N} = {}^{\mathrm{a}}\sum (\underset{n \in N}{\mathrm{P}} A_n) \quad \textit{for all} \quad (A_n)_{n \in N} \in \mathfrak{E}^N$$

and $\widehat{\lim}$ by

$$\widehat{\lim} (g, K, \mathfrak{b}) = \{x \,|\, x \in E \textit{ and, for some } B \in \mathfrak{b} \textit{ and for some } f \in \underset{k \in B}{\mathrm{P}} g(k), x = \lim (f, B, \mathfrak{b}_B)\}$$

$$\textit{for all} \quad (g, K, \mathfrak{b}) \in \Phi\mathfrak{E},$$

where $\mathfrak{b}_B$ denotes the *trace* $\{B \cap C \,|\, C \in \mathfrak{b}\}$ of the filter $\mathfrak{b}$ in the set B. — If $g \in \mathfrak{E}^N$, which mapping g might be given in the form $(g(n))_{n \in N}$, then, in the following,

$$ {}^{\mathrm{a}}\widehat{\sum}_{n \in N} g(n) \quad \text{stands for} \quad {}^{\mathrm{a}}\widehat{\sum} g.$$

If K is a set, $\mathfrak{b}$ is a filter on K (thus $K \neq \emptyset$) and g is a mapping on K into $\mathfrak{E}$, then, in the following,

$$ {}^{\mathfrak{b}}\widehat{\lim}_{k \in K} g(k) \quad \text{stands for} \quad \widehat{\lim} (g, K, \mathfrak{b}).$$

Since the domain of the mapping ${}^{\mathrm{a}}\widehat{\sum}$ is the whole set $\mathfrak{E}^N$ and the domain of $\widehat{\lim}$ is the whole class $\Phi\mathfrak{E}$, under these agreements, expressions like

$$ {}^{\mathrm{a}}\widehat{\sum}_{n \in N} {}^{\mathrm{a}}\widehat{\sum}_{k \in N} h(n, k) \quad \text{or} \quad {}^{\mathrm{a}}\widehat{\sum}_{n \in N} {}^{\mathrm{a}}\widehat{\lim}_{i \in I} j(n, i) \quad \text{or} \quad {}^{\mathrm{a}}\widehat{\lim}_{i \in I} {}^{\mathrm{a}}\widehat{\sum}_{n \in N} j(n, i)$$

with a mapping h on $N \times N$ into $\mathfrak{E}$ or a mapping j on $N \times I$ (where I is a set) into $\mathfrak{E}$ and a filter $\mathfrak{a}$ on I are defined automatically. The filtered family occuring, implicitly, in the last expression is, of course, $(f, I, \mathfrak{a})$ with $f(i) = {}^{\mathrm{a}}\widehat{\sum}_{n \in N} j(n, i)$ for all $i \in I$.

---

[1] For typographical reasons, we choose the symbol ${}^{\mathrm{a}}\widehat{\Sigma}$ instead of the symbol ${}^{\mathrm{a}}\widehat{\Sigma}$, which would express better that ${}^{\mathrm{a}}\widehat{\Sigma}$ is determined by ${}^{\mathrm{a}}\Sigma$ (see the terminology in [2], p. 185, lines 21—22).

The mapping $^a\sum$ turns out to be *operation-continuous (op-continuous)* [for the terminology ("operationsstetig"), see [2], p. 184], thus one has (by "Satz 2" in [2], p. 186):

THEOREM 1. *If* $\left(g(n, .), K_n, \mathfrak{b}_n\right)_{n \in N}$, *where* $g(n, .)$ *denotes the mapping* $k \to g(n, k)$ $(k \in K_n)$, *is a sequence of elements of* $\Phi\mathfrak{E}$, *then*

$$^a\widehat{\sum_{n \in N}} \,^{\mathfrak{b}_n}\widehat{\lim_{k \in K_n}} \, g(n, k) \subseteqq \,^{\mathfrak{b}}\widehat{\lim_{\varphi \in K}} \,^a\widehat{\sum_{n \in N}} \, g(n, \varphi(n)),$$

where $K = \underset{n \in N}{\mathrm{P}} \, K_n$ *and* $\mathfrak{b}$ *denotes the cardinal product of the filters* $\mathfrak{b}_n$ *with* $n \in N$.

Additionally, one obtains:

THEOREM 2. *In Theorem 1, the equality sign* $=$ *holds instead of* $\subseteqq$, *if* $^{\mathfrak{b}_n}\widehat{\lim_{k \in K_n}} \, g(n, k)$ *is non-empty for all* $n \in N$.

All preceding considerations (in section 2) remain valid, if the absolute summation $^a\sum$ is replaced, consistently, either by the *conditional summation* $^c\sum$ or by the *unconditional summation* $^u\sum$, the relations $^c\sum$ and $^u\sum$ being defined as follows: For all $(a_n)_{n \in N} \in E^N$ and all $a \in E$, let

$$((a_n)_{n \in N}, a) \in \,^c\sum \quad \text{if and only if} \quad a = \lim_{m \to \infty} \sum_{n=1}^{m} a_n$$

and

$$((a_n)_{n \in N}, a) \in \,^u\sum \quad \text{if and only if} \quad a = \lim_{m \to \infty} \sum_{n=1}^{m} a_{\varphi(n)}$$

holds for each permutation φ *of* N.

As an immediate consequence of the definitions, one has

$$^a\widehat{\sum_{n \in N}} \, A_n \subseteqq \,^u\widehat{\sum_{n \in N}} \, A_n \subseteqq \,^c\widehat{\sum_{n \in N}} \, A_n$$

for all $(A_n)_{n \in N} \in \mathfrak{E}^N$. Furthermore (see [5], p. 27, Theorem 6), $^a\sum = \,^u\sum$ holds if and only if the Banach space $(E, \| \, \|)$ is finite dimensional.

For illustration, we are going to reformulate some classical elementary statements about $^a\sum$ in terms of $^a\widehat{\sum}$. Let $(A_n)_{n \in N} \in \mathfrak{E}^N$; then

(1) $\qquad ^a\widehat{\sum_{n \in N}} \, A_n = \,^a\widehat{\sum_{n \in N}} \, A_{\varphi(n)}$ *holds for each permutation* φ *of* N.

For each infinite countable set M and each $(A_n)_{n \in M} \in \mathfrak{E}^M$, one defines $^a\widehat{\sum_{n \in M}} \, A_n$ by

$$^a\widehat{\sum_{n \in M}} \, A_n = \,^a\widehat{\sum_{n \in N}} \, A_{\varphi(n)}, \quad \text{where } \varphi \text{ is some one-to-one mapping on } N \text{ onto } M;$$

by (1), the mapping $(A_n)_{n \in M} \to \,^a\widehat{\sum_{n \in M}} \, A_n \, ((A_n)_{n \in M} \in \mathfrak{E}^M)$ does not depend on the choice of φ. If one defines, for finite non-empty M, $^{\mathrm{a}}\widehat{\sum}_{n \in M} A_n$ to be the set $\{ \sum_{n \in M} \psi(n) | \psi \in \underset{n \in M}{\mathrm{P}} A_n \}$, then (2) stated next holds for each countable non-empty M.

(2) $\quad ^{\mathrm{a}}\widehat{\sum}_{n \in M} A_n \subseteq {}^{\mathrm{a}}\widehat{\sum}_{i \in I} {}^{\mathrm{a}}\widehat{\sum}_{k \in K_i} A_k$ *for each partition* $(K_i)_{i \in I}$ *of the set* M.

(Associative law for $^{\mathrm{a}}\widehat{\sum}$*)*

If I and K are countable non-empty sets and $(A_{ik})_{(i,k) \in I \times K} \in \mathfrak{C}^{I \times K}$, then, as a consequence of (2), one has:

(3) $\quad$ *If* $^{\mathrm{a}}\widehat{\sum}_{(i,k) \in I \times K} A_{ik}$ *is non-empty, then* $^{\mathrm{a}}\widehat{\sum}_{(i,k) \in I \times K} A_{ik} = {}^{\mathrm{a}}\widehat{\sum}_{i \in I} {}^{\mathrm{a}}\widehat{\sum}_{k \in K} A_{ik} = {}^{\mathrm{a}}\widehat{\sum}_{k \in K} {}^{\mathrm{a}}\widehat{\sum}_{i \in I} A_{ik}.$

(Fubini's law for $^{\mathrm{a}}\widehat{\sum}$*)*

Applications especially of Theorem 1, expressed for unconditional summation in the extended real line (thus: not yet for Banach spaces), have been made by the author in the paper [3]. We expect that analogous applications can be made for integrals of the type of a Bochner integral.

REFERENCES

[1] N. Bourbaki, *Topologie générale*. Chap. I—II. 2nd ed. Actual. Sci. Industr. **1142** (1951).

[2] G. Grimeisen, *Darstellung des iterierten Maßintegrals als Limes von iterierten Zwischensummen*. I. J. reine u. angew. Math. **237** (1969), 181—193.

[3] G. Grimeisen, *Darstellung des iterierten Maßintegrals als Limes von iterierten Zwischensummen*. II. J. reine u. angew. Math. **243** (1970), 70—94.

[4] G. Grimeisen, *An exchange theorem for infinite sums and limits in Banach spaces*. (to appear).

[5] J. T. Marti, *Introduction to the Theory of Bases*. Springer, Berlin 1969.

[6] J. D. Monk, *Introduction to set theory*. McGraw-Hill, New York 1969.

Finite Parts of Distributions

By

J. HORVÁTH

DEPT. OF MATH.
UNIVERSITY OF MARYLAND
COLLEGE PARK

0. Introduction

It is well known [16, Example 4. 9. 1, p. 394] that for $n \geqq 3$ an elementary solution of the Laplace operator

$$\Delta = \frac{\partial^2}{\partial x_1^2} + \cdots + \frac{\partial^2}{\partial x_n^2}$$

is the distribution E associated with the function

$$(0. 1) \qquad E(x) = - \frac{\Gamma\left(\dfrac{n-2}{2}\right)}{4\pi^{n/2}} \frac{1}{|x|^{n-2}} \qquad (x \neq 0),$$

where $x = (x_1, \ldots, x_n) \in \mathbf{R}^n$ and $|x| = (x_1^2 + \cdots + x_n^2)^{\frac{1}{2}}$. Indeed, an easy calculation using Green's formula (cf. loc. cit.) shows that for every infinitely differentiable function with compact support φ we have

$$(0. 2) \qquad \int_{\mathbf{R}^n} E(x)\, \Delta\varphi(x)\, dx = \varphi(0),$$

i.e., $\Delta E = \delta$.

For $n = 2$ definition (0. 1) is meaningless but, as was observed by Marcel RIESZ [28, p. 22], if we introduce the function

$$E_\lambda(x) = - \frac{\Gamma\left(\dfrac{2-\lambda}{2}\right)}{2^\lambda \pi \Gamma\left(\dfrac{\lambda}{2}\right)} \frac{1}{|x|^{2-\lambda}} \qquad (x \neq 0),$$

with $0 < |\lambda - 2| < 2$, we can instead of (0. 2) consider

$$(0. 3) \qquad \lim_{\lambda \to 2} \int_{\mathbf{R}^2} \left(E_\lambda(x) - \frac{1}{2\pi(\lambda - 2)} \right) \Delta\varphi(x)\, dx = - \frac{1}{2\pi} \int_{\mathbf{R}^2} \Delta\varphi \left(\log \frac{2}{|x|} - \gamma \right) dx,$$

where γ is Euler's constant. It is again well known that the last integral is equal to $\varphi(0)$, i.e., that the distribution associated with the function $x \mapsto (1/2\pi) \log |x|$ is an elementary solution of the Laplace operator. The procedure amounts to considering the integral

$$(0.4) \qquad\qquad F(\lambda) = \int\limits_{\mathbf{R}^2} E_\lambda(x)\,\varphi(x)\,dx$$

as a function of the complex variable λ. It is meromorphic in $0 < |\lambda-2| < 2$ and at $\lambda=2$ has a simple pole with residue

$$a_{-1} = \frac{1}{2\pi} \int\limits_{\mathbf{R}^2} \varphi(x)\,dx.$$

The regular part of F at $\lambda=2$ is therefore

$$\lim_{\lambda\to 2}\left(F(\lambda) - \frac{a_{-1}}{\lambda-2}\right),$$

this is called the *finite part* of the integral (0.4) at $\lambda=2$ [9, 17.9.1, p. 261] and is denoted by

$$\mathop{\mathrm{Pf}}_{\lambda=2}\int\limits_{\mathbf{R}^2} E_\lambda(x)\,\varphi(x)\,dx.$$

Thus (0.3) can be rewritten in the form

$$\mathop{\mathrm{Pf}}_{\lambda=2}\int\limits_{\mathbf{R}^2} E_\lambda(x)\,\Delta\varphi(x)\,dx = -\frac{1}{2\pi}\int\limits_{\mathbf{R}^2}\Delta\varphi(x)\left(\log\frac{2}{|x|}-\gamma\right)dx.$$

We encounter a similar situation if we consider the elementary solution of the wave operator

$$\Box = \frac{\partial^2}{\partial x_1^2} - \frac{\partial^2}{\partial x_2^2} - \cdots - \frac{\partial^2}{\partial x_n^2}.$$

For $x=(x_1, x_2, \ldots, x_n)\in\mathbf{R}^n$ write $x'=(x_2, \ldots, x_n)\in\mathbf{R}^{n-1}$ and $s(x) = (x_1^2-|x'|^2)^{\frac{1}{2}} = (x_1^2-x_2^2-\cdots-x_n^2)^{\frac{1}{2}}$, and set

$$\gamma_n = 2\pi^{\frac{1}{2}(n-2)}\Gamma\left(\frac{4-n}{2}\right).$$

It is easy to check that for $n=2$ and $n=3$ the distribution associated with the function

$$E(x) = \begin{cases} \dfrac{1}{\gamma_n}\dfrac{1}{s(x)^{n-2}}, & \text{if}\quad x_1<0,\ |x_1|>|x'|, \\[2mm] 0, & \text{otherwise,} \end{cases}$$

is an elementary solution of $\Box$. However, for $n\geq 4$ we run into difficulties: in the first place, if n is even, then γ_n is not defined. Furthermore the function E is not locally integrable; indeed, if we set $|x_1|=t$, $|x'|=\varrho$, denote by C the cone $\{x \in \mathbf{R}^n \mid x_1 < 0,\ |x_1| > |x'|\}$ and by $\mathbf{S}_{n-2}$ the unit sphere in $\mathbf{R}^{n-1}$, then

$$\int_C \frac{dx}{s(x)^{n-2}} = |\mathbf{S}_{n-2}| \int_0^\infty dt \int_0^t \frac{\varrho^{n-2}\, d\varrho}{[(t-\varrho)(t+\varrho)]^{\frac{1}{2}(n-2)}},$$

and $\varrho \mapsto (t-\varrho)^{\frac{1}{2}(2-n)}$ is not integrable in $0 < \varrho < t$.

It is to avoid these difficulties that HADAMARD introduced the notion of finite part of an integral [13]. In the case of n odd, $n \geqq 4$, one proves that there exists a unique function $P(\varepsilon)$, which is a polynomial in ε and in $\log \varepsilon$ such that

$$\lim_{\varepsilon \to 0} \left\{ \frac{1}{\gamma_n} \int_{\substack{x_1 < 0 \\ s(x) \geqq \varepsilon}} \frac{\varphi(x)}{s(x)^{n-2}}\, dx - P(\varepsilon) \right\}$$

is finite; this limit is called the finite part of the integral

$$\frac{1}{\gamma_n} \int_C \frac{\varphi(x)}{s(x)^{n-2}}\, dx$$

and is denoted by

$$\mathrm{Pf}\, \frac{1}{\gamma_n} \int_C \frac{\varphi(x)}{s(x)^{n-2}}\, dx.$$

In particular a solution of the inhomogeneous wave equation $\Box u = f$ is given by

$$(0.\,5) \qquad\qquad u(x) = \mathrm{Pf}\, \frac{1}{\gamma_n} \int_C \frac{f(x-y)}{s(y)^{n-2}}\, dy.$$

For even n a different kind of finite part has to be defined due to a phenomenon called Huygens' principle, according to which the support of the elementary solution of $\Box$ is contained in the nappe $s(x)=0$ of the "light cone".

In his research on the Cauchy problem, Marcel RIESZ [28] found a different representation for the solution of the inhomogeneous wave equation, which is the same for even and odd n. For $\lambda \in \mathbf{C}$ he defines

$$\gamma_n(\lambda) = 2^{\lambda-1} \pi^{\frac{1}{2}(n-2)} \Gamma\left(\frac{\lambda}{2}\right) \Gamma\left(\frac{\lambda+2-n}{2}\right)$$

and considers the integral

$$(0.6) \qquad (I^\lambda f)(x) = \frac{1}{\gamma_n(\lambda)} \int_C \frac{f(y)}{s(x-y)^{n-\lambda}}\, dy.$$

This is a holomorphic function of λ for $\mathrm{Re}\,\lambda > n-2$, and Marcel Riesz proves that it has analytic continuation in the region $\mathrm{Re}\,\lambda \leq n-2$. He proves furthermore that $\Box I^{\lambda+2} = I^\lambda$ and $(I^0 f)(x) = f(x)$, so that $u(x) = (I^2 f)(x)$ is a solution of $\Box u = f$.

In this lecture I want to consider, instead of the analytic continuation of integrals depending on a complex parameter λ, the analytic continuation of distributions $T(\lambda)$, which are holomorphic functions of a complex λ. Of course this has already been done by GELFAND and ŠILOV [11, chap. I, § 3 and appendix 2]. If the function $T: \lambda \mapsto T(\lambda)$ has a pole at $\lambda = \lambda_0$, then we define the *finite part* $\mathop{\mathrm{Pf}}\limits_{\lambda=\lambda_0} T(\lambda)$ or simply $\mathrm{Pf}\,T(\lambda_0)$ of T at $\lambda = \lambda_0$ as the value of the regular part of T at $\lambda = \lambda_0$. This definition of a finite part, at least for distributions defined by functions, seems to have been stated explicitly for the first time in DIEUDONNÉ's recent book [9, chap. 17, sec. 9]. At points where the function T has a pole Gelfand and Šilov use the concept of "canonical regularization", which has the same ad hoc character as Hadamard's finite part of an integral, but in every concrete case the distribution obtained by canonical regularization coincides with the finite part.

Both Hadamard's finite parts of integrals and Riesz's analytic continuation turn out to be aspects of the same technique. Thus if for $\mathrm{Re}\,\lambda > n-2$ we define the distribution $T(\lambda)$ by

$$\langle T(\lambda), \varphi \rangle = \frac{1}{\gamma_n} \int_C \frac{\varphi(x)}{s(x)^{n-\lambda}}\, dx \qquad (\varphi \in \mathscr{D}),$$

then $\lambda \mapsto T(\lambda)$ has a simple pole at $\lambda = 2$, and $\big(\mathop{\mathrm{Pf}}\limits_{\lambda=2} T(\lambda)\big) * f$ is the function u defined in (0.5). On the other hand, if for $\mathrm{Re}\,\lambda > n-2$ we define the distribution $S(\lambda)$ by

$$\langle S(\lambda), \varphi \rangle = \frac{1}{\gamma_n(\lambda)} \int_C \frac{\varphi(x)}{s(x)^{n-\lambda}}\, dx \qquad (\varphi \in \mathscr{D}),$$

then $\lambda \mapsto S(\lambda)$ is regular at $\lambda = 2$, we have $S(2) = \mathop{\mathrm{Pf}}\limits_{\lambda=2} T(\lambda)$, and $S(2) * f$ is the function $I^2 f$ obtained considering the value of the analytic continuation of (0.6) at $\lambda = 2$.

The main feature of Riesz's method consists in dividing by an appropriate meromorphic numerical function, as $\lambda \mapsto \gamma_n(\lambda)$ in (0.6), which makes the singularity at the point $\lambda = \lambda_0$ disappear. Thus he never considers finite parts, only values of the function obtained by analytic continuation. But to obtain simple composition formulas which are valid for all $\lambda \in \mathbf{C}$, we also multiply by certain meromorphic numerical functions of λ and so other poles appear which makes it imperative to consider finite parts.

I shall illustrate the method with the elliptic potentials of Marcel RIESZ [27; 28, p. 16] given by

$$(I^\lambda f)(x) = \frac{\Gamma\left(\dfrac{n-\lambda}{2}\right)}{2^\lambda \pi^{\frac{1}{2}n} \Gamma\left(\dfrac{\lambda}{2}\right)} \int_{\mathbf{R}^n} \frac{f(y)}{|x-y|^{n-\lambda}} \, dy$$

for Re $\lambda > 0$, $\lambda \neq n + 2k$ ($k \in \mathbf{N}$), and which for $\lambda = 2$ becomes the Newtonian potential of f. Analytic continuation gives $I^{-2k}f = (-\Delta)^k f$ for $k \in \mathbf{N}$. For $\lambda = n + 2k$ we have to consider the finite part (as we have already done in the case $n = 2$, $k = 0$) and obtain the misterious formula (VII, 7; 14) which is given in SCHWARTZ's book [30] without explanation. As a second example I shall consider the gradient of the elliptic Riesz distribution, introduced briefly in an earlier note [15], which contains as a particular case the kernel of the Hilbert—Riesz transformation [14], and which was used recently by TREBELS in saturation theory [31]. In a separate note I shall deal with the continuity properties of the convolution operators defined by these distributions. I intend to return in other publications to hyperbolic Riesz distributions, considered also by METHÉE [20, 21], and more generally to distributions associated with general elliptic and hyperbolic polynomials. To conclude this introduction, let me mention that a recent result of ATIYAH [1] and I. N. BERNŠTEĬN and S. I. GELFAND [2] shows that the same method would probably yield results for any real analytic function.

1. Vector-valued holomorphic functions

Let Λ be an open subset of the complex plane $\mathbf{C}$ and let E be a locally convex, quasi-complete [16, Definition 2. 9. 2, p. 128] Hausdorff topological vector space over $\mathbf{C}$. We say that a function $f: \Lambda \to E$ is holomorphic in Λ if at every point $\lambda_0 \in \Lambda$ the derivative

$$\frac{df}{d\lambda}(\lambda_0) = \lim_{\lambda \to \lambda_0} \frac{f(\lambda) - f(\lambda_0)}{\lambda - \lambda_0}$$

exists. If F is a second locally convex, quasi-complete Hausdorff space, $u: E \to F$ a continuous linear map and $f: \Lambda \to E$ a holomorphic function, then $u \circ f: \Lambda \to F$ is holomorphic. Thus in particular for every x' belonging to the dual E' of E, the scalar-valued function $\langle f, x' \rangle : \lambda \mapsto \langle f(\lambda), x' \rangle$ is holomorphic. Conversely if $f: \Lambda \to E$ is such that for every $x' \in E'$ the function $\langle f, x' \rangle$ is holomorphic, then f is holomorphic [12, p. 37; 17, p. 162].

The proofs of the following properties of holomorphic functions can be found in [12]. A holomorphic function $f: \Lambda \to E$ has derivatives of every order at each point $\lambda \in \Lambda$. A function $f: \Lambda \to E$ is holomorphic if and only if it is continuous for the weak topology $\sigma(E, E')$ and for every simple, closed, rectifiable path c in Λ whose interior is contained in Λ we have

$$\int_c f(\lambda)\, d\lambda = 0.$$

Let $f: \Lambda \to E$ be holomorphic and for $\lambda_0 \in \Lambda$ set

$$a_n = \frac{1}{n!}\frac{d^n f}{d\lambda^n}(\lambda_0);$$

if the disk $\{\lambda \mid |\lambda - \lambda_0| < \varrho\}$ lies in Λ and $0 < \sigma < \varrho$, then the family $(a_n(\lambda - \lambda_0)^n)_{n \in \mathbf{N}}$ is summable to $f(\lambda)$, uniformly in $\{\lambda \mid |\lambda - \lambda_0| < \sigma\}$. Conversely, if $f: \Lambda \to E$ is such that for each $\lambda_0 \in \Lambda$ there exists a series $\sum_{n=0}^{\infty} a_n(\lambda - \lambda_0)^n$ which converges weakly to $f(\lambda)$ in some neighbourhood of λ_0, then f is holomorphic.

The function $f: \Lambda \to E$ is holomorphic if and only if for every $\lambda_1 \in \Lambda$ there exists an open neighbourhood N of λ_1 and a bounded, balanced, convex, closed subset B of E such that the function $f: N \to E_B$ is holomorphic, where E_B denotes the Banach space obtained by considering the linear subspace of E generated by B and taking the gauge of B for norm [16, chap. 3, § 5, p. 207]. From here we obtain the following consequence [12, p. 40]:

THEOREM 1. 1. *Let F be a quasi-complete, barrelled, locally convex Hausdorff space and $E = F'$ its dual. If $f: \Lambda \to E$ is such that for every $y \in F$ the numerical function $\langle f, y \rangle : \lambda \mapsto \langle f(\lambda), y \rangle$ is holomorphic, then f is holomorphic if we equip E with the strong topology $\beta(E, F)$.*

Indeed, since F is barrelled, E is quasi-complete for the weak* topology $\sigma(E, F)$ [16, Theorem 3. 6. 1, p. 218], hence f is holomorphic if E is equipped with this topology. But since F is quasi-complete, the bounded sets of E are the same for $\sigma(E, F)$ and $\beta(E, F)$ [16, Theorem 3. 5. 4, p. 210], so the assertion follows from the preceding remark.

The possibility of defining a distribution by analytic continuation is based on the following result, contained essentially in the book of GELFAND and ŠILOV [11, chap. I, app. 2, n° 3, p. 171]:

THEOREM 1. 2. *Let F be a quasi-complete, barrelled, locally convex Hausdorff space and $E = F'$ its dual equipped with the strong topology $\beta(E, F)$. Let Λ and Λ_1 be two domains in $\mathbf{C}$ such that $\Lambda \subset \Lambda_1$, and $f: \Lambda \to E$ a holomorphic function. Assume that for every $y \in F$ the holomorphic numerical function $\lambda \mapsto g(\lambda; y) = \langle f(\lambda), y \rangle$ has an* *analytic continuation into Λ_1. Then there exists a holomorphic function $f_1: \Lambda_1 \to E$ such that $f_1(\lambda) = f(\lambda)$ for $\lambda \in \Lambda$ and*

(1. 1) $$\langle f_1(\lambda), y \rangle = g(\lambda; y)$$

for all $\lambda \in \Lambda_1$ and $y \in F$.

PROOF. By the familiar „Kreiskettenverfahren" it is sufficient to assume that every $\lambda \in \Lambda_1$ belongs to some open disk $\Delta \subset \Lambda_1$ whose center λ_0 lies in Λ. For any $y \in F$ we have

$$g(\lambda; y) = \sum_{n=0}^{\infty} \frac{1}{n!} \frac{d^n}{d\lambda^n} g(\lambda_0; y) (\lambda - \lambda_0)^n = \sum_{n=0}^{\infty} \left\langle \frac{1}{n!} \frac{d^n f}{d\lambda^n} (\lambda_0) (\lambda - \lambda_0)^n, y \right\rangle$$

Thus if we define $s_m(\lambda) \in F'$ by

$$s_m(\lambda) \doteq \sum_{n=0}^{m} \frac{1}{n!} \frac{d^n f}{d\lambda^n} (\lambda_0) (\lambda - \lambda_0)^n \qquad (m \in \mathbf{N}),$$

then $\langle s_m(\lambda), y \rangle$ converges to $g(\lambda; y)$ for every $y \in F$ as $m \to \infty$. By the Banach—Steinhaus theorem [16, Corollary to Proposition 3. 6. 5, p. 215] there exists a linear form $f_1(\lambda) \in F' = E$ such that (1. 1) is satisfied for every $y \in F$. By our assumption the function $\lambda \mapsto \langle f_1(\lambda), y \rangle$ is holomorphic in Δ for every $y \in F$, hence by Theorem 1. 1 the function $\lambda \mapsto f_1(\lambda)$ is holomorphic.

The conditions concerning E in Theorem 1. 2 are satisfied in particular if E is a *reflexive* locally convex Hausdorff space since then E is the dual of $E' = F$ equipped with the topology $\beta(F, E)$, the space F is barrelled [16, Proposition 3. 8. 4, p. 228], quasi-complete [16, Exercise 3. 8. 2, p. 230], and the topology of E is $\beta(E, F)$. This particular case of Theorem 1. 2 is contained also in the following deeper result proved by BOGDANOWICZ [3]:

THEOREM 1. 3. *Let E be a quasi-complete, locally convex Hausdorff space. Let Λ and Λ_1 be two domains in $\mathbf{C}$ such that $\Lambda \subset \Lambda_1$, and $f: \Lambda \to E$ a holomorphic function. Assume that for every $x' \in E'$ the holomorphic numerical function $\lambda \mapsto g(\lambda; x') = \langle f(\lambda), x' \rangle$ has an analytic continuation into Λ_1. Then there exists a holomorphic function $f_1: \Lambda_1 \to E$ such that $f_1(\lambda) = f(\lambda)$ for $\lambda \in \Lambda$ and*

$$\langle f_1(\lambda), x' \rangle = g(\lambda; x')$$

for $\lambda \in \Lambda_1$ and $x' \in E'$.

Actually BOGDANOWICZ proves a more general result which contains both theorems 1. 2 and 1. 3 as special cases.

If f is a function with values in E which is holomorphic in $\Lambda - \{\lambda_0\}$, where $\lambda_0 \in \Lambda$, and if there exists an integer $m \geq 0$ such that $\lim_{\lambda \to \lambda_0} (\lambda - \lambda_0)^{m+1} f(\lambda) = 0$, then there exist $b_j \in E$ $(1 \leq j \leq m)$ such that

$$f(\lambda) = \frac{b_m}{(\lambda - \lambda_0)^m} + \frac{b_{m-1}}{(\lambda - \lambda_0)^{m-1}} + \cdots + \frac{b_1}{\lambda - \lambda_0} + h(\lambda)$$

in some disk $\Delta = \{\lambda \mid |\lambda - \lambda_0| < \varrho\}$, where h is holomorphic in Δ. If $b_j \neq 0$ for some $j \geq 1$, we say that f has a *pole* at λ_0, and the largest integer j such that $b_j \neq 0$ is the order of the pole. A pole of order 1 is also called a *simple pole*. The expression

$$(1. 2) \qquad \sum_{j=1}^{m} \frac{b_j}{(\lambda - \lambda_0)^j}$$

and the holomorphic function h are uniquely determined by f; (1. 2) is the *singular part*, h the *regular part* and b_1 the *residue* of f at λ_0. If f is holomorphic at λ_0, we shall say that f has a pole of order 0 at λ_0; then f coincides with its regular part at λ_0 and its singular part is zero. If f is holomorphic in Λ except for a discrete set of points at which it has poles of order ≥ 1, then f is *meromorphic* in Λ.

2. Distribution-valued holomorphic functions

Let Λ be a domain in $\mathbf{C}$, Ω an open subset of $\mathbf{R}^n$ and $\lambda \mapsto T(\lambda)$ a holomorphic function defined in Λ and taking its value in the space $\mathscr{D}'^m(\Omega)$ of distributions of order $\leq m$ $(0 \leq m \leq \infty)$ in Ω. If Λ_1 is a domain in $\mathbf{C}$ containing Λ and for every $\varphi \in \mathscr{D}^m(\Omega)$ the numerical function $\lambda \mapsto \langle T(\lambda), \varphi \rangle$ has an analytic continuation to Λ_1, then it follows from Theorem 1. 2 that $\lambda \mapsto T(\lambda)$ can be extended to a holomorphic function defined in Λ_1 with values in $\mathscr{D}'^m(\Omega)$. The same applies if instead of $\mathscr{D}'^m(\Omega)$ we consider the space $\mathscr{E}'^m(\Omega)$ of distributions with compact support of order $\leq m$ in Ω, or the space $\mathscr{S}'$ of tempered distributions on $\mathbf{R}^n$, or the space $\mathcal{O}'_c$ of rapidly decreasing distributions on $\mathbf{R}^n$.

Using Theorems 1. 1 and 1. 2 it is easy to verify that analytic continuation preserves the fundamental operations between distributions, and one obtains for instance the following properties:

(2. 1) If A is a closed subset of Ω such that $\mathrm{Supp}\, T(\lambda) \subset A$ for $\lambda \in \Lambda$, then $\mathrm{Supp}\, T(\lambda) \subset A$ for $\lambda \in \Lambda_1$.

(2. 2) If $S(\lambda) = \partial^p T(\lambda)$ for $\lambda \in \Lambda$, then $\lambda \mapsto S(\lambda)$ can also be extended to a holomorphic function in Λ_1 and $S(\lambda) = \partial^p T(\lambda)$ for $\lambda \in \Lambda_1$.

(2. 3) If $S(\lambda)$ is the Fourier transform of $T(\lambda) \in \mathscr{S}'$ for $\lambda \in \Lambda$, then $\lambda \mapsto S(\lambda)$ has an analytic continuation into Λ_1 and $S(\lambda) = \mathscr{F} T(\lambda)$ for $\lambda \in \Lambda_1$.

(2. 4) If $\lambda \mapsto T(\lambda)$ is holomorphic in Λ with values in $\mathscr{D}'^m(\Omega)$ [resp. in $\mathscr{S}'$] and if $\lambda \mapsto \alpha_\lambda$ is holomorphic in Λ with values in $\mathscr{E}^m(\Omega)$ [resp. in $\mathcal{O}_M$], then $\lambda \mapsto S(\lambda) = \alpha_\lambda T(\lambda)$ is holomorphic in Λ with values in $\mathscr{D}'^m(\Omega)$ [resp. $\mathscr{S}'$]. Furthermore if

both $\lambda \mapsto T(\lambda)$ and $\lambda \mapsto \alpha_\lambda$ have analytic continuations into Λ_1, then $\lambda \mapsto S(\lambda)$ has an analytic continuation into Λ_1 and $S(\lambda) = \alpha_\lambda T(\lambda)$ for $\lambda \in \Lambda_1$.

(2. 5) Let Ξ be an open subset of $\mathbf{R}^k$ and H an open subset of $\mathbf{R}^l$. If $\lambda \mapsto S(\lambda)$ is holomorphic in Λ with values in $\mathscr{D}'^s(\Xi)$ and $\lambda \mapsto T(\lambda)$ is holomorphic in Λ with values in $\mathscr{D}'^t(H)$, then $\lambda \mapsto R(\lambda) = S(\lambda) \otimes T(\lambda)$ is holomorphic in Λ with values in $\mathscr{D}'^{s+t}(\Xi \times H)$. Furthermore if both $\lambda \mapsto S(\lambda)$ and $\lambda \mapsto T(\lambda)$ have analytic continuations into Λ_1, then $\lambda \mapsto R(\lambda)$ has an analytic continuation into Λ_1 and $R(\lambda) = S(\lambda) \otimes T(\lambda)$ for $\lambda \in \Lambda_1$.

(2. 6) Let $\lambda \mapsto S(\lambda)$ be holomorphic in Λ with values in $\mathscr{D}'^s(\mathbf{R}^n)$ and $\lambda \mapsto T(\lambda)$ holomorphic in Λ with values in $\mathscr{D}'^t(\mathbf{R}^n)$. Assume that there exist two closed subsets A and B of $\mathbf{R}^n$ such that $\mathrm{Supp}\, S(\lambda) \subset A$ and $\mathrm{Supp}\, T(\lambda) \subset B$ for $\lambda \in \Lambda$ and that A and B verify:

(Σ) for every compact subset K in $\mathbf{R}^n$, the set $A \cap (K - B)$ is compact in $\mathbf{R}^n$ [16, p. 383].

Then $\lambda \mapsto R(\lambda) = S(\lambda) * T(\lambda)$ is holomorphic in Λ with values in $\mathscr{D}'^{s+t}(\mathbf{R}^n)$. Furthermore if both $\lambda \mapsto S(\lambda)$ and $\lambda \mapsto T(\lambda)$ have analytic continuations into Λ_1, then $\lambda \mapsto R(\lambda)$ has an analytic continuation into Λ_1, and $R(\lambda) = S(\lambda) * T(\lambda)$ for $\lambda \in \Lambda_1$.

Let now $T : \lambda \mapsto T(\lambda)$ be a meromorphic function in the domain $\Lambda \subset \mathbf{C}$ with values in $\mathscr{D}'(\Omega)$. For any $\lambda_0 \in \Lambda$ we define the *finite part* of T at λ_0 as the value at λ_0 of the regular part of T at λ_0, and we denote it by $\underset{\lambda = \lambda_0}{\mathrm{Pf}}\ T(\lambda)$ or by $\mathrm{Pf}\, T(\lambda_0)$. Thus if T is holomorphic at $\lambda_0 \in \Lambda$, then $\mathrm{Pf}\, T(\lambda_0) = T(\lambda_0)$. If T has a pole of order m at λ_0, then in some punctured disk $\{\lambda \,\vert\, 0 < |\lambda - \lambda_0| < \varrho\}$ we have

$$(2.\ 7) \qquad T(\lambda) = \frac{S_m}{(\lambda - \lambda_0)^m} + \cdots + \frac{S_1}{\lambda - \lambda_0} + S(\lambda),$$

where $S_j \in \mathscr{D}'(\Omega)$ $(1 \le j \le m)$ and $\lambda \mapsto S(\lambda)$ is holomorphic in $\{\lambda \,\vert\, |\lambda - \lambda_0| < \varrho\}$, and we have by definition

$$\mathrm{Pf}\, T(\lambda_0) = S(\lambda_0) = \lim_{\lambda \to \lambda_0} \left(T(\lambda) - \sum_{j=1}^{m} \frac{S_j}{(\lambda - \lambda_0)^j} \right).$$

The situation which occurs most often in practice is that there exists a subdomain Λ_0 of Λ such that for every $\lambda \in \Lambda_0$ the distribution $T(\lambda)$ is associated with a locally integrable function f_λ, i.e.,

$$(2.\ 8) \qquad \langle T(\lambda), \varphi \rangle = \int_\Omega f_\lambda(x) \varphi(x)\, dx$$

for $\varphi \in \mathscr{D}(\Omega)$ and $\lambda \in \Lambda_0$. Following Laurent SCHWARTZ [29, chap. II, § 2, exemple 2, p. 38; 18, chap. I, § II, pp. 15—21], we then say that T is a *pseudofunction* and very often we shall write f_λ instead of $T(\lambda)$ at points λ in which T is holomorphic but which do not belong to Λ_0. At points where T has a pole, we shall occasionally use the notation $\Pf_{\lambda=\lambda_0} f_\lambda$ or $\Pf f_{\lambda_0}$ for $\Pf T(\lambda_0)$.

The classical concept of the *finite part of a divergent integral* now emerges with clarity: the finite part of the integral which figures in (2. 8) at a point λ which does not belong to Λ_0 is $\langle \Pf T(\lambda), \varphi \rangle$. Of course, if f_λ and $T(\lambda)$ satisfy certain regularity and growth conditions, then more general test functions φ can figure in (2. 8) [4, 5, 25, 26].

Let us list some properties of the finite part of a distribution which are easy to prove. Assume that $T: \lambda \mapsto T(\lambda)$ is holomorphic in $\Lambda - \{\lambda_0\}$ and has a pole at λ_0.

(2. 9) If A is a closed subset of Ω such that $\operatorname{Supp} T(\lambda) \subset A$ for $\lambda \in \Lambda - \{\lambda_0\}$, then $\operatorname{Supp} \Pf T(\lambda_0) \subset A$.

(2. 10) For each $p \in \mathbf{N}^n$ the function $\lambda \mapsto \partial^p T(\lambda)$ has a pole at λ_0 and $\Pf \partial^p T(\lambda_0) = \partial^p \Pf T(\lambda_0)$ [18, (I. II. 10), p. 21].

(2. 11) If T takes its values in $\mathscr{S}'$, then $\lambda \mapsto \mathscr{F} T(\lambda)$ has a pole at λ_0 and $\Pf \mathscr{F} T(\lambda_0) = \mathscr{F} \Pf T(\lambda_0)$.

REMARK. SCHWARTZ [29, p. 42] seems to contradict (2. 10), but he considers $-l \Pf_{\lambda=-l} (x^{\lambda-1})_{x>0}$, whereas the derivative of $\Pf (x^\lambda)_{x>0}$ is $\Pf (\lambda x^{\lambda-1})_{\lambda>0}$. That the two distributions are different appears from

$$\Pf_{\lambda=-l} (\lambda x^{\lambda-1})_{x>0} = -l \Pf_{\lambda=-l} (x^{\lambda-1})_{x>0} + \frac{(-1)^l}{l!} \delta^{(l)},$$

which is formula (II, 2; 28) of [29] if interpreted correctly.

Let me conclude this section with a simple observation which, however, is the basic reason why the Riesz distribution R_{-2k} has its support at the origin (no. 4), and also of Huygens' principle mentioned in the introduction. Let again $T: \lambda \mapsto T(\lambda)$ be holomorphic in $\Lambda - \{\lambda_0\}$ and have a pole at λ_0. Assume that there exists a closed subset A of Ω such that for every closed neighbourhood B of A we can decompose T into the sum of two holomorphic functions T_1 and T_2 defined in $\Lambda - \{\lambda_0\}$, such that $\operatorname{Supp} T_1(\lambda) \subset B$ for $\lambda \in \Lambda - \{\lambda_0\}$ and T_2 is holomorphic at λ_0. Then it is easy to see that the distributions S_j ($1 \leq j \leq m$) which figure in (2. 7) have their supports contained in A.

Suppose now that the preceding conditions are satisfied and that in addition T has a simple pole at λ_0. Let γ be a numerical meromorphic function in Λ, which has a simple pole at λ_0 with residue a. Then clearly

$$\Pf_{\lambda=\lambda_0} \frac{T(\lambda)}{\gamma(\lambda)} = \frac{1}{a} S_1,$$

and the support of this distribution is contained in A.

3. Euclidean pseudofunctions

Let k be a function defined on the unit sphere $\mathbf{S}_{n-1}$ of $\mathbf{R}^n$ and integrable with respect to the surface element $d\sigma$ of $\mathbf{S}_{n-1}$. We extend the definition of k to $\complement\{0\}$ by setting

$$k(x) = k\left(\frac{x}{|x|}\right),$$

and so k becomes a locally integrable function, positively homogeneous of degree 0. For $\alpha = (\alpha_1, \ldots, \alpha_n) \in \mathbf{N}^n$ we define the spherical moment M_α of k by

$$(3.1) \qquad M_\alpha = \int_{\mathbf{S}_{n-1}} \sigma^\alpha k(\sigma)\, d\sigma,$$

where, as usual, $\sigma^\alpha = \sigma_1^{\alpha_1} \cdots \sigma_n^{\alpha_n}$.

For $\operatorname{Re}\lambda > -n$ we define the distribution K_λ by

$$(3.2) \qquad \langle K_\lambda, \varphi \rangle = \int_{\mathbf{R}^n} |x|^\lambda k(x)\varphi(x)\, dx \qquad (\varphi \in \mathscr{D}(\mathbf{R}^n)).$$

Since $x \mapsto |x|^\lambda k(x)$ is a locally integrable function on $\mathbf{R}^n$ for $\operatorname{Re}\lambda > -n$, the distribution K_λ is well-defined and belongs to the space $\mathscr{M}^1(\mathbf{R}^n) = \mathscr{S}_0'^0$ of integrable measures [16, pp. 345 and 410].

We have the following result (cf. also [11], chap. IV, § 1, n° 6, pp. 331—334 or German edition, chap. III, § 3, n° 5, pp. 300—302):

THEOREM 3.1. *Let m be a positive integer or $+\infty$. The distribution K_λ, which is defined by (3.2) for $\operatorname{Re}\lambda > -n$ and $\varphi \in \mathscr{S}^m$, is a holomorphic function of λ in the half-plane $\operatorname{Re}\lambda > -n$ with values in $\mathscr{S}'^m$. It can be extended to a holomorphic function with values in $\mathscr{S}'^m$ in the half-plane $\operatorname{Re}\lambda > -n-m$, with the exception of the points $\lambda = -n-j$ (j integer, $0 \leqq j < m$), which are simple poles with residue*

$$(3.3) \qquad (-1)^j \sum_{|\alpha|=j} \frac{1}{\alpha!} M_\alpha \partial^\alpha \delta.$$

Let us give more explicit expressions for $\operatorname{Pf} K_\lambda$, where $\lambda \in \mathbf{C}$. If λ is not one of the singular values $\lambda = -n-j$ ($j \in \mathbf{N}$), we shall simply write K_λ for the analytic continuation of the pseudofunction defined by (3.2).

First let $\lambda \in \mathbf{C}$ be such that $-n-m < \operatorname{Re}\lambda \leqq -n-m+1$ but $\lambda \neq -n-m+1$, for some $m = 1, 2, 3, \ldots$. Then we have

$$(3.4) \quad \langle K_\lambda, \varphi \rangle = \lim_{\varepsilon \to 0} \left\{ \int_{|x| \geqq \varepsilon} |x|^\lambda k(x)\varphi(x)\, dx + \sum_{j=0}^{m-1} \sum_{|\alpha|=j} \frac{1}{\alpha!} M_\alpha \partial^\alpha \varphi(0) \frac{\varepsilon^{\lambda+n+j}}{\lambda+n+j} \right\}.$$

Next at a pole $\lambda = -n-j$ of K_λ we have

$$\text{(3.5)} \qquad \langle \operatorname{Pf} K_{-n-j}, \varphi \rangle =$$

$$= \lim_{\varepsilon \to 0} \left\{ \int_{|x| \geqq \varepsilon} |x|^{-n-j} k(x)\varphi(x)\,dx + \sum_{v=0}^{j-1} \sum_{|\alpha|=v} \frac{1}{\alpha!} M_\alpha \partial^\alpha \varphi(0) \frac{\varepsilon^{-j+v}}{-j+v} + \right.$$

$$\left. + \sum_{|\alpha|=j} \frac{1}{\alpha!} M_\alpha \partial^\alpha \varphi(0) \log \varepsilon \right\}.$$

At the point $\lambda = -n$ the last formula becomes simply

$$\langle \operatorname{Pf} K_{-n}, \varphi \rangle = \lim_{\varepsilon \to 0} \left\{ \int_{|x| \geqq \varepsilon} |x|^{-n} k(x)\varphi(x)\,dx + M_0\,\varphi(0) \log \varepsilon \right\}.$$

In particular, if

$$M_0 = \int_{S_{n-1}} k(\sigma)\,d\sigma = 0,$$

then $\operatorname{Pf} K_{-n}$ is the Cauchy principal value

$$\langle \operatorname{Pf} K_{-n}, \varphi \rangle = \lim_{\varepsilon \to 0} \int_{|x| \geqq \varepsilon} |x|^{-n} k(x)\varphi(x)\,dx$$

studied by TRICOMI, GIRAUD, MIHLIN and others (for references see [7, 22, 23, 24, 32]).

4. The elliptic pseudofunctions of Marcel Riesz

Now we consider the special case of the distribution K_λ defined by (3.2) when k is the constant function equal to 1, and denote it simply by $|x|^\lambda$. Introducing spherical coordinates we see that the spherical moments M_α defined by (3.1) are zero whenever at least one of the integers $\alpha_1, \ldots, \alpha_n$ is odd, and that

$$M_\alpha = 2 \frac{\Gamma\left(\dfrac{\alpha_1+1}{2}\right) \Gamma\left(\dfrac{\alpha_2+1}{2}\right) \cdots \Gamma\left(\dfrac{\alpha_n+1}{2}\right)}{\Gamma\left(\dfrac{|\alpha|+n}{2}\right)},$$

if all the α_j $(1 \leqq j \leqq n)$ are even, where $|\alpha| = \alpha_1 + \cdots + \alpha_n$. It then follows from (3.3) and the multinomial theorem that the distribution-valued function $\lambda \mapsto |x|^\lambda$ is holomorphic at every point $\lambda \neq -n-2k$, $k \in \mathbf{N}$ of the complex plane, and that at $\lambda = -n-2k$ it has a simple pole with residue

$$\frac{2\pi^{\frac{1}{2}n}}{2^{2k} k! \,\Gamma(\frac{1}{2}n+k)} \Delta^k \delta.$$

In [11, chap. I, § 3, n° 9, pp. 110—114] this residue is computed with the help of Pizetti's formula.

Specializing formulas (3. 4) and (3. 5) we obtain that in the strip $-n-2m < \text{Re } \lambda \leq -n-2m+2$, $\lambda \neq -n-2m+2$ the analytic continuation of $|x|^\lambda$ is given by

$$\langle |x|^\lambda, \varphi \rangle = \lim_{\varepsilon \to 0} \left\{ \int_{|x| \geq \varepsilon} |x|^\lambda \varphi(x)\, dx + \sum_{k=0}^{m-1} \frac{2\pi^{\frac{1}{2}n}}{2^{2k} k! \, \Gamma(\frac{1}{2}n+k)} \Delta^k \varphi(0) \frac{\varepsilon^{\lambda+n+2k}}{\lambda+n+2k} \right\}$$

and that the finite part of $|x|^\lambda$ at the pole $\lambda = -n-2k$ is given by

$$\langle \text{Pf}\, |x|^{-n-2k}, \varphi \rangle =$$

$$= \lim_{\varepsilon \to 0} \left\{ \int_{|x| \geq \varepsilon} |x|^{-n-2k} \varphi(x)\, dx + \sum_{\nu=0}^{k-1} \frac{2\pi^{\frac{1}{2}n}}{2^{2\nu} \nu! \, \Gamma(\frac{1}{2}n+\nu)} \Delta^\nu \varphi(0) \frac{\varepsilon^{-2k+2}}{-2k+2} + \right.$$

$$\left. + \frac{2\pi^{\frac{1}{2}n}}{2^{2k} k! \, \Gamma(\frac{1}{2}n+k)} \Delta^k \varphi(0) \log \varepsilon \right\}.$$

These are the formulas (II, 3; 5) and (II, 3; 6) by which $\text{Pf}\, |x|^\lambda$ is *defined* in [29].

Since the function $\lambda \mapsto \Gamma(\frac{1}{2}\lambda)$ has simple poles with residues $2(-1)^k/k!$ at the points $\lambda = -2k$, $k \in \mathbf{N}$, it follows that the distribution

$$\frac{1}{\Gamma(\frac{1}{2}\lambda)} |x|^{\lambda-n},$$

defined by

$$\left\langle \frac{1}{\Gamma(\frac{1}{2}\lambda)} |x|^{\lambda-n}, \varphi \right\rangle = \frac{1}{\Gamma(\frac{1}{2}\lambda)} \int_{\mathbf{R}^n} |x|^{\lambda-n} \varphi(x)\, dx$$

for $\text{Re } \lambda > 0$ and $\varphi \in \mathscr{S}$, is an entire function of $\lambda \in \mathbf{C}$. At the "critical" points $\lambda = -2k$ we have

(4. 1)
$$\Pf_{\lambda=-2k} \frac{1}{\Gamma(\frac{1}{2}\lambda)} |x|^{\lambda-n} = \frac{2^{-2k} \pi^{\frac{1}{2}n}}{\Gamma\left(\dfrac{n+2k}{2}\right)} (-\Delta)^k \delta$$

(cf. the remark made at the end of section 2).

The last formula suggest that we consider the distribution which for $\text{Re } \lambda > 0$ is associated with the locally integrable function which at $x \in \mathbf{R}^n$ assumes the value

$$\frac{\Gamma\left(\dfrac{n-\lambda}{2}\right)}{2^\lambda \pi^{\frac{1}{2}n} \Gamma(\frac{1}{2}\lambda)} |x|^{\lambda-n}.$$

This distribution is a meromorphic function of $\lambda \in \mathbf{C}$ with simple poles at the poles of $\lambda \mapsto \Gamma(\frac{1}{2}(n-\lambda))$, i.e., at the points $\lambda = n+2k$, $k \in \mathbf{N}$. Its finite part is the *elliptic distribution of Marcel Riesz* [27, 28] and will be denoted by R_λ. It follows from (4. 1) that

$$R_{-2k} = (-\Delta)^k \delta \qquad (k \in \mathbf{N}),$$

and for $n \geqq 3$ the distribution $-R_2$ is the elementary solution (0. 1) of Δ. Using the first two terms of the Laurent series of the gamma-function [10, vol. I, 1. 17, (11), p. 46 and 1. 7, (9), p. 16] we find

$$(4. 2) \quad R_{n+2k} = \frac{2(-1)^k}{\pi^{\frac{1}{2}n} 2^{n+2k} \Gamma\left(\dfrac{n+2k}{2}\right) k!} |x|^{2k} \left\{ \log \frac{2}{|x|} + \frac{1}{2} \left(\sum_{v=1}^{k} \frac{1}{v} - \gamma + \frac{\Gamma'\left(\dfrac{n+2k}{2}\right)}{\Gamma\left(\dfrac{n+2k}{2}\right)} \right) \right\},$$

where γ is Euler's constant. This is the distribution which figures in formula (VII, 7; 14) of [30], and for $n=2$, $k=0$ we obtain the distribution which we considered in connection with formula (0. 3).

A method due to DENY [8, p. 151; 30, p. 113] combined with (2. 11) shows that the Fourier transform of R_λ is the distribution

$$\mathscr{F}(R_\lambda) = \mathrm{Pf}\,(2\pi |\xi|)^{-\lambda},$$

where $\xi=(\xi_1, \ldots, \xi_n) \in \mathbf{R}^n$ and $|\xi|^2 = \xi_1^2 + \cdots + \xi_n^2$. Observe that the "critical" values $\lambda = n+2k$ of R_λ match nicely with the "critical" values $-\lambda = -n-2k$ of $\mathrm{Pf}\,(2\pi |\xi|)^{-\lambda}$. Also the Fourier transform of $R_{-2k}=(-\Delta)^k$ is $(4\pi^2 |\xi|^2)^k$, which is, of course, what it should be [16, example 4. 11. 2, p. 411; 30, formula (VII. 7; 14), p. 114].

5. The gradient of elliptic Riesz distributions

For any $\lambda \in \mathbf{C}$ we consider the distribution

$$N_\lambda = -\operatorname{grad} R_{\lambda+1}$$

introduced in [15]. For $\operatorname{Re} \lambda > 0$ and $\lambda \neq n+2k+1$, $k \in \mathbf{N}$, the distribution N_λ is associated with the locally integrable function which at $x \in \mathbf{R}^n$ assumes the value

$$(5. 1) \qquad \frac{\Gamma\left(\dfrac{n-\lambda+1}{2}\right)}{2^\lambda \pi^{\frac{1}{2}n} \Gamma\left(\dfrac{\lambda+1}{2}\right)} \frac{x}{|x|} |x|^{\lambda-n}.$$

This holds in particular for $\lambda = n-1$, which is a "critical" value of $R_{\lambda+1}$, as we can see from the expression (4. 2) of R_n taking into account that

$$\operatorname{grad} \log \frac{1}{|x|} = -\frac{x}{|x|^2},$$

and will also follow from (2. 2) and Theorem 5. 1 below. For $\lambda=0$ the function (5. 1) is the kernel of the Hilbert—Riesz transform [14] and for $\lambda=1$ it is the kernel of Newtonian force. For $\lambda = -2k-1$, $k \in \mathbf{N}$ we have obviously

$$N_{-2k-1} = (-1)^{k+1} \Delta^k (\operatorname{grad} \delta).$$

Let us consider the distribution which for $\operatorname{Re} \lambda > 0$ is associated with the function

$$(5.\ 2) \qquad\qquad x \mapsto \frac{x}{|x|}\, |x|^{\lambda - n}.$$

In the notation of section 3, we are considering the distribution $K_{\lambda - n}$ corresponding to the vector-valued function $k(\sigma) = \sigma$. With a computation similar to the one indicated in section 4 we obtain from theorem 3. 1 the following

THEOREM 5. 1. *The distribution $K_{\lambda - n}$, which for $\operatorname{Re} \lambda > 0$ is associated with the function (5. 2) is a meromorphic function of $\lambda \in \mathbf{C}$. It has simple poles at the points $\lambda = -2k - 1$ $(k \in \mathbf{N})$ with residues*

$$- \frac{\pi^{\frac{1}{2}n}}{2^{2k} k!\, \Gamma\left(\dfrac{n+2k+2}{2}\right)} \Delta^k (\operatorname{grad} \delta).$$

It follows that the distribution

$$\frac{1}{\Gamma\left(\dfrac{\lambda+1}{2}\right)} K_{\lambda - n}$$

is an entire function of $\lambda \in \mathbf{C}$ and that

$$\operatorname*{Pf}_{\lambda = -2k-1} \frac{1}{\Gamma\left(\dfrac{\lambda+1}{2}\right)} K_{\lambda-n} = - \frac{\pi^{\frac{1}{2}n}}{2^{2k+1} \Gamma\left(\dfrac{n+2k+2}{2}\right)} (-\Delta)^k (\operatorname{grad} \delta).$$

This then leads naturally to considering the distribution N_λ, and since $\Gamma\left(\dfrac{n-\lambda+1}{2}\right)$ has no singularity at $\lambda = n-1$, N_λ at $\lambda = n-1$ is indeed associated with the function (5. 1).

BIBLIOGRAPHY

[1] M. F. Atiyah, *Resolution of singularities and division of distributions.* Comm. Pure Appl. Math. **23** (1970), 145—150.

[2] I. N. Bernšteĭn and S. I. Gel'fand, *Meromorphy of the function P^λ* (in Russian). Funkcional. Anal. i Priložen. **3** (1969), no. 1, 84—85.

[3] W. M. Bogdanowicz, *Analytic continuation of holomorphic functions with values in a locally convex space.* Proc. Amer. Math. Soc. **22** (1969), 660—666.

[4] F. J. Bureau, *Les solutions élémentaires et le problème de Cauchy*. Proc. Internat. Congr. Math.. Amsterdam 1954, vol. III, pp. 58—70, Noordhoff N. V., Groningen and North-Holland Publishing Co., Amsterdam 1956.

[5] F. J. Bureau, *Divergent integrals and partial differential equations*. Comm. Pure Appl. Math. **8** (1955), 143—202.

[6] P. L. Butzer, *Singular integral equations of Volterra type and the finite part of divergent integrals*. Arch. Rational Mech. Anal. **3** (1959), 194—205.

[7] M. Cotlar, *Condiciones de continuidad de operadores potenciales y de Hilbert*. Cursos y seminarios de matemática, fasc. **2**, Universidad Nacional de Buenos Aires 1959.

[8] J. Deny, *Les potentiels d'énérgie finie*. Acta Math. **82** (1950), 107—183.

[9] J. Dieudonné, *Eléments d'analyse*. tome III. Cahiers Scientifiques, fasc. XXXIII, Gauthier-Villars, Paris 1970.

[10] A. Erdélyi, W. Magnus, F. Oberhettinger and F. G. Tricomi, *Higher Transcendental Functions*. McGraw-Hill, New York 1953.

[11] I. M. Gel'fand and G. E. Šilov, *Generalized functions, I. Generalized functions and operations on them*. Gosudarstvennoe izdatel'stvo fiziko-matematičeskoĭ literatury, Moscow 1958; German translation: VEB Deutscher Verlag der Wissenschaften, Berlin 1960.

[12] A. Grothendieck, *Sur certains espaces de fonctions holomorphes. I.* J. Reine Angew. Math. **192** (1953), 35—64.

[13] J. Hadamard, *Le problème de Cauchy et les équations aux dérivées partielles linéaires hyperboliques*. Herman, Paris 1932.

[14] J. Horváth, *Sur les fonctions conjuguées à plusieurs variables*. Indag. Math. **15** (1953), 17—29.

[15] J. Horváth, *On some composition formulas*. Proc. Amer. Math. Soc. **10** (1959), 433—437.

[16] J. Horváth, *Topological vector spaces and distributions. I.* Addison-Wesley, Reading, Mass. 1966.

[17] J. L. Kelley, I. Namioka and co-authors, *Linear topological spaces*. Van Nostrand, New York 1963.

[18] J. Lavoine, *Calcul symbolique: distributions et pseudofonctions*. Monographies du Centre d'Etudes Mathématiques en vue des Applications: B.-Méthodes de Calcul, II, Centre National de la Recherche Scientifique, Paris 1959.

[19] J. Lavoine, *Transformation de Fourier des pseudo-fonctions*. Monographies du Centre d'Etudes Mathématiques en vue des Applications: B.-Méthodes de Calcul, Centre National de la Recherche Scientifique, Paris 1963.

[20] P. D. Méthée, *Sur les distributions invariantes dans le groupe des rotations de Lorentz*. Comment. Math. Helv. **28** (1954), 225—269.

[21] P. D. Méthée, *Transformées de Fourier de distributions invariantes liées à la résolution de l'équation des ondes*. La théorie des équations aux dérivées partielles, Nancy, 9—15 avril 1956, Colloques Internationaux du Centre National de la Recherche Scientifique, LXXI, Paris 1956.

[22] S. G. Mihlin, *Multidimensional singular integrals and integral equations*. Gosudarstvennoe izdatel'stvo fiziko-matematičeskoĭ literatury, Moscow 1962; English translation: Pergamon Press, Oxford 1965.

[23] U. Neri, *Singular integrals: an introduction*. University of Maryland, Department of Mathematics, Lecture Notes No. 3, 1967; Springer Lecture Notes in Mathematics, vol. 200.

[24] U. Neri, *Singular integral operators and distributions*. University of Maryland, Department of Mathematics, Lecture Notes No. 4, 1968; Springer Lecture Notes in Mathematics, vol. 200.

[25] J. O'Keeffe, *Singularities of Hadamard's finite part of improper integrals in the distributions of Schwartz*. Rend. Circ. Mat. Palermo, II. Ser., 6 (1957), 65—82.

[26] J. O'Keeffe, *Distribution theory of the operational calculus.* Rend. Circ. Mat. Palermo, II. Ser., **6** (1957), 157—170.

[27] M. Riesz, *Intégrales de Riemann—Liouville et potentiels.* Acta Sci. Math. (Szeged) **9** (1938), 1—42.

[28] M. Riesz, *L'intégrale de Riemann—Liouville et le problème de Cauchy.* Acta Math. **81** (1949), 1—223.

[29] L. Schwartz, *Théorie des distributions,* I, *2nd ed.* Actualités Scientifiques et Industrielles **1091**, Hermann, Paris 1957.

[30] L. Schwartz, *Théorie des distributions,* II. Actualités Scientifiques et Industrielles **1122**, Hermann, Paris 1951.

[31] W. Trebels, *Generalized Lipschitz conditions and Riesz derivatives on the space of Bessel potentials L_α^p, I. Conditions for elements of L_α^p and their Riesz transforms for* $0 < \alpha \leqq 2$. Applicable Analysis **1** (1971), 75—99.

[32] Singular Integrals. *Proc. Symposia of Pure Math.* vol. **X**, Amer. Math. Soc., Providence, Rhode Island 1967.

Courants et courants cylindriques sur des variétés de dimension infinie

Par

PAUL KRÉE

DEPT. D'INFORMATIQUE
FACULTÉ DES SCIENCES, NICE

Etude de l'espace des fonctions numériques k fois continuement dérivables définies sur une variété banachique paracompacte V. Définition et étude des courants, des courants cylindriques et des courants quotients sur V. Application aux équations aux dérivées partielles.

Soit k un nombre entier $\geqq 0$, et soit V une variété réelle paracompacte de classe k, de dimension quelconque. La motivation de ce travail est la recherche de méthodes permettant l'étude de problèmes d'analyse relatifs à V:

— étude de solutions faibles pour des équations aux dérivées partielles vérifiées par des courants définis sur V (des solutions usuelles pour de telles équations sont étudiées dans L. GROSS [1]);
— étude d'équations intégrales;
— problème de représentation de relation de commutation: voir GELFAND—ŠILOV [1].

On cherche donc à définir en dimension infinie des classes de fonctions généralisées, analogues à certaines classes définies par L. SCHWARTZ en dimension finie. On cherche aussi à définir des prolongements aux notions de mesures et de mesures cylindriques (introduite dans Gelfand—Šilov [1] dans le cas où V est un espace vectoriel topologique).

A cet effet on introduit l'espace des formes linéaires continues sur l'espace $\mathscr{C}^k(V)$, (muni d'une topologie convenable) des fonctions $f:V \to R$ dont les dérivées $f, Df, \ldots, D^k f$ d'ordre $\leqq k$ sont bornées et continues. Ceci oblige (pour $k \geqq 1$) à normer une fois pour toutes, tous les espaces tangents T_x ($x \in V$); car pour tout x, $(D^j f)(x)$ appartient à l'e.v. S^j des formes multilinéaires continues sur T_x. On appelle ces formes linéaires des courants d'ordre au plus k. Ils correspondent en dimension finie à des 0.courants, les p.courants (p entier fini) pouvant être définis de façon analogue en remplaçant $\mathscr{C}^k(V)$ par un espace de p.formes différentielles. Mais l'on ne peut pas définir par ces méthodes, des distributions sur une variété de dimension infinie, car sur une variété de dimension infinie, il n'existe pas de formes extérieures de degré maximum.

On introduit également, si V est muni d'une famille convenable $\mathcal{R}_V$ de relations d'équivalence $(R_i)_{i \in I}$, les courants cylindriques ou systèmes projectifs de courants définis sur les espaces quotients V/R_i, et vérifiant certaines relations de compatibilité. Dans le cas où V est un e.v.t., où $\mathcal{R}_V$ est le système des relations d'équivalence canoniquement associées aux sous espaces fermés de codimension finie de V, et où $k=0$, on retrouve la notion de mesure cylindrique. On définit aussi la classe plus générale des courants quotients, notion pour laquelle on perd toute propriété de projectivité. On indique quelques applications aux solutions élémentaires au paragraphe 5, et quelques problèmes au paragraphe 6.

I — Fonctions dérivables

On utilise les notations de N. BOURBAKI [1] et [3]. Soit k un nombre entier $\geqq 0$. Dans le texte qui suit, si $k \geqq 1$, V désigne une variété réelle paracompacte de classe $\mathscr{C}^k$ et de type E_V, E_V étant un espace de Banach dépendant de V.

(1. 0) Si $k=0$, V désigne un espace topologique séparé.

(1. 1) LEMME. a) *V est métrisable.* b) *Si $k \geqq 1$, il existe une application continue a du fibré tangent $\mathscr{T}(V)$ dans $\mathbf{R}^+$, dont la restriction à chaque espace tangent est une norme. (On enonce cette propriété de la façon suivante: on peut normer tous les espaces tangents à V.)*

PREUVE. a) Soit $(O_i, e_i, E)_{i \in I}$ un certain atlas définissant la structure de variété de V. On peut supposer le recouvrement ouvert $(O_i)_{i \in I}$ localement fini. Soit (φ_i) une partition de l'unité de classe $\mathscr{C}^0$ subordonnée à ce recouvrement. On peut alors normer tous les espaces tangents à V en posant pour tout vecteur $t_x \in \mathscr{T}_x(V)$:

$$\|t_x\| = \sum_{i \in I(x)} \varphi_i(x) \, \|p(_i t_x)\|$$

avec $I(x) = \{i \in I; x \in O_i\}$ où $p_i(t_x)$ est le vecteur de E associé à la carte $(O_i \times E, d_i, E \times E)$ du fibré tangent, qui est canoniquement associée à la carte (O_i, e_i, E) de V.

b) Soit γ un arc de courbe de V, défini par un paramétrage $f: I = [0, 1] \to V$, qui est $\mathscr{C}^0$, et $\mathscr{C}^1$ par morceaux. On peut définir la longueur de cet arc: $\int_0^1 \|f'(t)\| \, dt$; d'où une métrique géodésique sur V.

(1. 2) REMARQUES. a) En général, dès qu'une variété paracompacte est donnée, nous la supposerons munie d'une métrique, les espaces tangents étant normés comme ci-dessus (bien que ces constructions ne soient pas canoniques).

b) Nous ne distinguerons pas deux applications a et b de $\mathscr{T}(V)$ dans $\mathbf{R}^+$, normant chaque espace tangent à V et qui sont équivalentes au sens suivant:

$$\exists C, \quad C' > 0, \quad \forall x, \quad \forall t_x \in T_x, \quad C\|t_x\|_a \leqq \|t_x\|_b \leqq C'\|t_x\|_a.$$

(1. 3) *Espace vectoriel $\mathscr{C}_b^k(V)$.* Nous ne considérons que des fonctions f définies sur V à valeurs réelles. On note S^j l'ensemble des formes multilinéaires symétriques continues sur l'espace de Banach E_V. On note (pour $k \geqq 1$), $\mathscr{S}^j(V)$ le fibré vectoriel sur V, dont les fibres sont isomorphes à S^j, et construit canoniquement à partir du fibré tangent. Vu (1. 1), on peut normer toutes les fibres S_x^j du fibré $\mathscr{S}^j(V)$. Si f est l fois continuement dérivable, $(l \leqq k)$, la dérivée Df définit une section de $\mathscr{S}^j(V)$. On définit $\mathscr{C}^k(V)$ (ou resp $\mathscr{C}_b^k(V)$) comme étant l'e.v. des fonctions $f\colon V \to \mathbf{R}$ admettant des dérivées continues (ou resp des dérivées bornées continues) jusqu'à l'ordre k.

Tout élément f de $\mathscr{C}_b^k(V)$ est caractérisé par la famille:

$$(1.4) \qquad (f, Df, ..., D^k f) \in \Pi^k = \prod_{j=0}^{k} \mathscr{C}_b^0(V, \mathscr{S}^j(V))$$

et ainsi $\mathscr{C}_b^k(V)$ s'identifie à un sous e.v. G de Π^k.

(1. 5) *Topologies $\mathscr{T}_K$, $\mathscr{T}_\infty$ et $\mathscr{T}$ sur $\mathscr{C}_b^k(V)$.* Vu (1. 4), pour mettre une topologie sur $\mathscr{C}_b^k(V)$, il suffit de mettre une topologie sur l'espace $F_j = \mathscr{C}_b^0(V, \mathscr{S}^j(V))$ des sections continues bornées d'un fibré du type $\mathscr{S}^j$ (pour $j=0$, $\mathscr{C}^0(V, \mathscr{S}^0(V)) = \mathscr{C}_b^0(V)$); puisqu'alors on pourra munir $\mathscr{C}_b^k(V)$ de la topologie induite par la topologie produit sur Π^k.

Nous définissons à présent trois topologies sur $\mathscr{C}^0(V, \mathscr{S}^j(V))$:

a) La topologie $\mathscr{T}_k^j$ de la convergence uniforme sur les compacts de V. Pour tout $\eta > 0$ et pour tout compact K de V, on peut définir la semi-boule:

$$B_K^j(\eta) = \{g \in F_j;\ \sup_{x \in K} \|g(x)\| \leqq \eta\}.$$

b) La topologie $\mathscr{T}_\infty^j$ de la convergence uniforme. Pour tout $\eta > 0$, on peut considérer la boule:

$$B_\infty^j(\eta) = \{g \in F_j;\ \sup_{x \in V} \|g(x)\| \leqq \eta\}.$$

c) La topologie $\mathscr{T}^j$ localement convexe la plus fine qui coïncide sur $B_\infty^j(1)$ avec $\mathscr{T}_k^j$. La topologie $\mathscr{T}^j$ existe car c'est la borne supérieure de la famille (non vide) des topologies localement convexes qui coïncident avec $\mathscr{T}_K^j$ sur $B_\infty^j(1)$.

d) Notons que $\mathscr{T}_K^j \subseteq \mathscr{T}^j \subseteq \mathscr{T}_\infty^j$.

e) Dans ces notations, on supprimera parfois l'indice j si aucune confusion n'est possible. On définit alors la topologie $\mathscr{T}_K$ (et resp $\mathscr{T}$, et resp $\mathscr{T}_\infty$) sur $\mathscr{C}_b^k(V)$ comme étant la trace sur G de la topologie sur Π^k produit des topologies $\mathscr{T}_K^j$ (et resp $\mathscr{T}^j$, et resp $\mathscr{T}_\infty^j$). On définit de même dans $\mathscr{C}_b^k(V)$ les boules et semi-boules:

$$B_\infty(\eta) = \{f \in \mathscr{C}_b^k(V);\ \|D^l f\|_\infty = \sup_{x \in V} \|(D^l f)(x)\| \leqq \eta \text{ pour } l = 0, ..., k\},$$

$$B_K(\eta) = \{f \in \mathscr{C}_b^k(V);\ \|D^l f\|_K = \sup_{x \in K} \|(D^l f)(x)\| \leqq \eta \text{ pour } l = 0, ..., k\}.$$

f) D'une façon générale, si $\mathscr{U}$ est une topologie localement convexe sur l'e.v. M, on notera $(M, \mathscr{U})$ l'e.v. localement convexe associé, et $(M, \mathscr{U})'$ le dual topologique de cet espace.

Comme on a des injections à image dense:

$$(1.6) \qquad (\mathscr{C}_b^0(V, \mathscr{S}^j(V)), \mathscr{T}_\infty) \overset{\text{dense}}{\to} (\mathscr{C}_b^0(V, \mathscr{S}^j(V)), \mathscr{T}) \overset{\text{dense}}{\to} (\mathscr{C}_b^0(V), \mathscr{S}^j(V), \mathscr{T}_K)$$

d'où par transposition:

$$(1.7) \qquad (\mathscr{C}_b^0(V), \mathscr{S}^j(V), \mathscr{T}_K)' \overset{\text{dense}}{\to} (\mathscr{C}_b^0(V, \mathscr{S}^j(V)), \mathscr{T})' \overset{\text{dense}}{\to} (\mathscr{C}_b^0(V, \mathscr{S}^j(V)), \mathscr{T}_\infty)'.$$

Pour tout élément l de ce dernier espace E_j', on désigne par $\|l\|$ sa norme dans le dual de l'espace normé $Ej = (\mathscr{C}_b^0(V, \varphi^j(V)), \mathscr{T}_\infty)$. Le théorème qui suit est très utile pour reconnaître dans E_j' les formes linéaires continues sur $F_j = (\mathscr{C}_b^0(V, \mathscr{S}^j(V)), \mathscr{T})$ et les parties équicontinues du dual de cet espace.

(1.8) THÉORÈME. *Soit G un e.v.l.c. et soit L un ensemble d'applications linéaires de $F_j = \mathscr{C}_b^0(V, \mathscr{S}^j(V))$ dans G. Alors L est une partie équicontinue de $\mathscr{L}((F_j, \mathscr{T}); G)$ si et seulement si L est une partie équicontinue de $\mathscr{L}((F_j, \mathscr{T}_\infty); G)$ et si L vérifie la condition suivante notée (ε, K):*

(ε, K) *Quel que soit $\varepsilon > 0$, quelle que soit la semi-norme continue p sur G, il existe un compact K de V tel que pour toute f de la boule unité de F_j, nulle sur K on ait pour tout $l \in L$, $p(l(f)) < \varepsilon$.*

PREUVE. Désignant par $\mathscr{C}$ le système des voisinages de 0 dans $(F_j, \mathscr{T})$, il suffit de montrer l'équivalence des trois propriétés suivantes:

$$(1.9) \qquad \forall \varepsilon > 0, \quad \forall p, \quad \exists C \in \mathscr{C}, \quad \forall l \in L, \quad |p(l(C))| < \varepsilon$$

$$(1.10) \quad \text{i)} \qquad \forall p, \quad \sup_{l,\ \|f\| \leq 1} p(l(f)) < \infty$$

$$\text{ii)} \qquad \forall \varepsilon, \quad \forall p, \quad \exists B_K(\eta), \quad \forall l \in L, \quad p[l(B_K(\eta))] < \varepsilon$$

$$(1.11) \quad \text{i)} \qquad \forall p, \quad \sup_{l,\ \|f\|_\infty < 1} p(l(f)) < \infty$$

$$\text{ii)} \qquad L \text{ vérifie } (\varepsilon, K).$$

Il est clair que (1.9) entraîne (1.10). Montrons la réciproque. Supposons qu'on ait (1.10). Soit $\mathscr{U}$ la topologie de G. Alors $\bigcap_{l \in L} l^{-1}(\mathscr{U})$ est une topologie localement convexe sur F_i, qui coïncide avec $\mathscr{T}$ sur $B_\infty(1)$ d'après (1.10). Vue la définition de $\mathscr{T}$, on a $\mathscr{T} \supseteq \bigcap l^{-1}(\mathscr{U})$ soit (1.9). Comme (1.10) entraîne (1.11), il suffit de montrer la réciproque; supposons qu'on ait (1.11). Etant donné ε et p, il s'agit de trouver $\eta > 0$ tel que:

$$(*) \qquad \forall f \in B_K(\eta) \cap B_\infty(1), \quad \forall l \in L, \quad p(l(f)) < 2\varepsilon.$$

On introduit la fonction numérique continue $g:V \to [0, 1]$ telle que:

$$(1.12) \qquad g(x) = \begin{cases} 1 & \text{si} \quad \|f(x)\| \leqq \eta \\ \dfrac{4\eta^2 - \|f(x)\|^2}{3\eta^2} & \text{si} \quad \eta \leqq \|f(x)\| \leqq 2\eta \\ 0 & \text{si} \quad \|f(x)\| \geqq 2\eta. \end{cases}$$

On écrit $p\big(l(f)\big) \leqq p[l(fg)] + p[l(f(1-g))]$. Comme $\|f\| < 2\eta$ si $g \neq 0$, on a $p[l(fg)] \leqq \leqq 2\alpha\eta$ et comme K est contenu dans le fermé où $g = 1$, la fonction $f(1-g)$ est nulle sur K; et comme L vérifie la condition (ε, K) on a $p[l(f(1-g))] < \varepsilon$. Finalement $p\big(l(f)\big) \leqq \varepsilon + 2\alpha\eta$ et l'on a $(*)$ si $\eta \leqq \varepsilon/2\alpha$.

(1.13) LEMME. *Soit L une famille de formes linéaires sur $(\mathscr{C}_b^k(V), \mathscr{T})$. Alors pour que L soit équicontinue, il faut et il suffit qu'il existe pour $j = 0, 1, \ldots, k$, des parties équicontinues L_j de $F_j' = (\mathscr{C}_b^0(V, \varphi^j(V)), \mathscr{T})'$ telles que:*

$$(1.14) \qquad \forall l \in L, \quad \exists (l_j) \in \prod_{j=0}^{k} L_j, \quad \forall f \in \mathscr{C}_b^k(V), \quad l(f) = \sum_{j=0}^{k} l_j(D^j f).$$

PREUVE. La condition est évidemment suffisante. Reciproquement, supposons que L est une famille équicontinue de formes linéaires sur $(\mathscr{C}_b^k(V), \mathscr{T})$. La famille L définit une famille équicontinue de formes linéaires sur le sous e.v.t. G de Π_k. Vu le théorème de Hahn-Banach, cette famille est la famille des restrictions à G d'un ensemble équicontinu de formes linéaires sur Π_k. Comme $\Pi_k = \prod_{j=0}^{k} F_j$ cet ensemble est défini par des familles équicontinues $L_j \subseteq F_j'$.

(1.15) *La catégorie $\mathscr{V}$ des variétés V.* a) On considère la catégorie $\mathscr{V}$ dont les objets sont pour $k \geqq 1$ les variétés banachiques réelles paracompactes de dimension quelconque, de classe k. Les flèches (ou morphismes) de la catégorie $\mathscr{V}$ sont les applications k-fois continuement dérivables dont les dérivées d'ordre $1, 2, \ldots, k$ sont uniformément bornées. Pour une flèche $f:V \to W$, on a donc:

$$\|D^l f\|_\infty = \sup_{x \in V} \|D^l f(x)\| < \infty \qquad (l = 1, \ldots, k).$$

On voit que pour toute $\varphi \in \mathscr{C}_b^k(W)$, $\varphi \circ f \in \mathscr{C}_b^k(V)$.

b) On dit qu'un système de flèches est compatible si tous les diagrammes formés par ces flèches sont commutatifs. On dit qu'un système de flèches f_i est uniformément borné si:

$$\forall h \leqq k \quad \sup_i \|D^h f_i\| < \infty.$$

c) Lorsqu'on parlera d'un système projectif $(V_i f_{ij})_{i, j \in I}$ dans la catégorie $\mathscr{V}$, on supposera toujours que les flèches f_{ij} sont uniformément bornées.

11*

II — Courants d'ordre fini

(2. 1) *La topologie $\mathcal{T}'$, définition des courants.* a) *Une famille équicontinue de courants d'ordre au plus k sur la variété V est une famille L de formes linéaires sur $\mathscr{C}_b^k(V)$ qui vérifient les conditions suivantes:*

i) $\displaystyle\sup_{l \in L} \|l\|_k < \infty$.

ii) *Pour tout $\varepsilon > 0$, il existe un compact K tel que pour toute $f \in B_\infty(1)$, nulle à l'ordre k sur K (i.e. $f = Df = \cdots = D^k f = 0$ sur K), on a $|l(f)| < \varepsilon$ pour tout $l \in L$.*

b) *La topologie $\mathcal{T}'$ sur $\mathscr{C}_b^k(V)$ est la topologie localement convexe sur $\mathscr{C}_b^k(V)$ admettant pour familles équicontinues de formes linéaires, les familles équicontinues de courants.*

c) *L'ensemble des courants d'ordre $\leqq k$ est noté $\mathscr{C}_b'^k(V)$.*

(2. 2) REMARQUES. a) Les polaires des familles équicontinues de courants d'ordre au plus k, constituent un système fondamental de voisinages de zéro.

b) Vus (1. 5. d), (1. 8) et (1. 13), on a $\mathcal{T}_k \subseteqq \mathcal{T} \subseteqq \mathcal{T}' \subseteqq \mathcal{T}_\infty$.

c) Les inclusions à image dense:

$$(2.3) \qquad (\mathscr{C}_b^k(V), \mathcal{T}_\infty) \overset{\text{dense}}{\subseteqq} (\mathscr{C}_b^k(V), \mathcal{T}') \overset{\text{dense}}{\to} (\mathscr{C}_b^k(V), \mathcal{T}_k)$$

montrent l'existence d'injections à image dense:

$$(2.4) \qquad (\mathscr{C}_b^k(V), \mathcal{T}_K)' \overset{\text{dense}}{\to} \mathscr{C}_b'^k(V) \to (\mathscr{C}_b^k(V), \mathcal{T}_\infty)'.$$

Toute $l \in (\mathscr{C}_b^k(V), \mathcal{T}_K)'$ est telle qu'il existe un compact K de V tel que $l(f) = 0$ pour toute f qui s'annule à l'ordre k sur K: $(\mathscr{C}_b^k(V), \mathcal{T}_K)'$ est l'ensemble des courants à support compact, d'ordre au plus k: voir L. NACHBIN [1].

(2. 5) *Exemples de courants.* a) Si $k = 0$, d'après un résultat de L. LE CAM (voir N. BOURBAKI [2]), $\mathscr{C}_b'^0(V)$ est l'espace des mesures de Radon bornées sur V.

b) Supposons que V est une variété de dimension finie dénombrable à l'infini. Les topologies $\mathcal{T}_\infty$ et $\mathcal{T}_K$ ont même restriction $\mathcal{T}''$ à $\mathscr{C}_0^k(V) = \mathscr{D}^k(V)$. Toute forme l linéaire continue sur $(\mathscr{C}_0^k(V), \mathcal{T}'')$ se prolonge d'une infinité de manières (par Hahn Banach) en une forme linéaire continue sur $(\mathscr{C}_b^k(V), \mathcal{T}')$. Cependant, on peut définir un prolongement *canonique* l' par la formule $l'(f) = \lim_i (f\zeta_i)$ où ζ_i est une fonction de $\mathscr{C}_0^k(V)$ valant 1 sur un compact, et où la limite est prise sur le filtre défini par l'ordre: $i < j$ si ζ_j vaut 1 dans un voisinage de ζ_i. (Pour montrer que la limite existe, il suffit de noter que pour tout $\varepsilon > 0$, il existe φ_0 dans la boule unité $B_\infty(1)$ de $\mathscr{C}_0^k(V)$ avec $l(\varphi_0) \geqq \|l\| - \varepsilon$, et par conséquent pour toute f de $B_\infty(1)$, à support disjoint de celui (noté $\underline{\varphi}_0$) de φ_0, on a $|l(f)| < \varepsilon$.) Par le même type de raisonnement, et en utilisant la condition (ε, K), on voit que toute forme linéaire continue sur $(\mathscr{C}_b^k(V), \mathcal{T}')$ est le prolongement canonique de sa restriction à $\mathscr{C}_0^k(V)$. En conclusion si V est de dimension finie, on peut *définir* l'ensemble des courants d'ordre au plus k, comme l'ensemble des formes linéaires continues sur $\mathscr{C}_0^k(V)$, muni de la topologie de la convergence uniforme (des fonctions et de leurs dérivées d'ordre au plus k). D'ailleurs, on peut remplacer dans cette définition $\mathscr{C}_0^k(V)$ par son complété, c'est-à-dire par l'ensemble des fonctions f de classe $\mathscr{C}^k$, telles que $D_j f$ tende vers 0 à l'infini.

c) La variété V est quelconque, et l'on fait $k=2$. Pour tout point x de V et tout vecteur x_0 tangent en x, on note $D_{x_0}(\delta_x)$ le courant $f: f \mapsto D_{x_0} f(x)$. Si l'on a n vecteurs x_i tangents en x, et si l'on a une forme quadratique $\sum\limits_{i,j=1}^{n} a_{ij} u_i u_j$ alors on peut noter $\sum a_{ij} D_{x_i x_j} \delta_x$ le courant:

$$f \mapsto \sum a_{ij} D_{x_i x_j} f(x) = \sum_{i,j=1}^{n} a_{ij} D^2 f(x; x_i, x_j).$$

On étudie à présent quelques opérations qui prolongent naturellement les opérations sur les mesures de Radon (en dimension infinie), et sur les distributions bornées (en dimension finie).

(2. 6) *Image par un courant.* Soit a une flèche $V \to W$ dans la catégorie $\mathscr{V}$ et soit l un courant d'ordre au plus k sur V. Pour toute $g \in \mathscr{C}_b^k(W)$ on pose $\tilde{l}(g) = l(g \circ a)$. On voit que l'application $l \mapsto \tilde{l}$ transforme toute partie équicontinue de $\mathscr{C}_b'^k(V)$ en une partie équicontinue de $\mathscr{C}_b'^k(W)$; par conséquent l'application

$$\mathscr{C}_b^k(V) \xleftarrow{a'} \mathscr{C}_b^k(W)$$

(2. 7)

$$g \circ a \mapsfrom g$$

est continue. L'image du courant l sur V par l'application a est le transformé de l par la transposée de a'. Pour simplifier les notations (et puisque $(a')'$ prolonge a), cette transposée est notée a).

(2. 8) *Extension à une variété plus grande.* Si en particulier $i:V \to W$ est une immersion et si $l \in \mathscr{C}_b'^k(V)$, $i(l)$ est appelée l'extension de l à W.

(2. 9) *Produit par une fonction $f \in \mathscr{C}_b^k(V)$.* Soit $l \in \mathscr{C}_b'^k(V)$. On définit de même lf comme étant l'image de l par la transposée de:

$$\mathscr{C}_b^k(V) \to \mathscr{C}_b^k(V)$$

$$\varphi \mapsto f\varphi.$$

(2. 10) *Dérivation selon un champ de vecteurs.* Au champ de vecteurs X, de classe $\mathscr{C}^k$, sur la variété V est associée:

$$\mathscr{C}_b^k(V) \xrightarrow{\alpha} \mathscr{C}_b^{k-1}(V), \quad \varphi \mapsto -X\varphi.$$

La transposée de cette application transforme toute partie équicontinue de $\mathscr{C}_b'^{k-1}(V)$

en une partie équicontinue, et en particulier le courant l d'ordre au plus $k-1$, en un courant noté X, d'ordre au plus k.

(2. 11) *Systèmes projectifs de courants.* Soit $(V_i, f_{ij})_{i \in I}$ un système projectif dans la catégorie V. Un système projectif (l_i, f_{ij}) de courants (d'ordre au plus k) sur les V_i, est par définition une famille de courants l_i sur les V_i, qui sont compatibles, (soit $i > j$ entraîne $l_j = f_{ij}(l_i)$) et uniformément bornés (soit $\sup \|l_i\| < \infty$). En général, un tel système projectif n'admet pas de limite projective. On a tout de même le:

(2. 12) THÉORÈME. *Soit $V \in \mathcal{V}$ et soit $f_i : V \to V_i$ $(i \in I)$ un système de flèches uniformément bornées qui sont compatibles avec les f_{ij} (soit $i > j$ entraîne $f_j = f_{ij} \circ f_i$):*

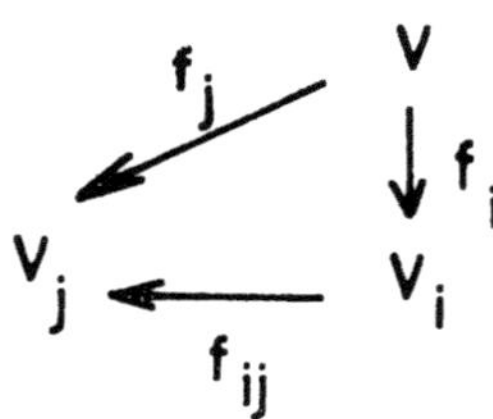

On se donne un système projectif (l_i, f_{ij}) de courants d'ordre au plus k sur V_i. S'il existe un courant sur V tel que $f_i(l) = l_i$ pour tout i; alors nécessairement:

(ε, K) $\forall \varepsilon, \ \exists K \subseteq V, \ \forall i \in I, \ \forall \Phi_i \in \mathscr{C}_b^k(V_i)$

avec $\|\Phi\|_k \leq 1$, Φ s'annulant à l'ordre k sur $f_i(K)$, on a $|l_i(\Phi)| < \varepsilon$.

En fait (2. 12) est une conséquence directe de la définition des courants: la condition indiquée est nécessaire car si l existe, la forme linéaire correspondante sur $(\mathscr{C}_b^k(V), \mathscr{T}')$ vérifie la condition (ε, K). Par conséquent pour tout ε, pour tout K, pour toute Φ_i avec $\|\Phi_i\| \leq 1$, Φ_i s'annulant à l'ordre k sur K, on a $|l_i(\Phi_i)| = = |l(\Phi_i \circ f_i)| < \varepsilon$.

III — Quelques variantes de la définition des courants

Parmi toutes les variantes possibles pour les définitions du paragraphe 2 (on peut remplacer $\mathscr{T}'$ par une autre topologie ...) nous en indiquons deux qui nous semblent importantes.

3 A. *Courants à trace*

L'exemple (2. 5 c) montre que les définitions précédentes ne permettent pas de définir le laplacien d'une fonction $f : V \to R$, si dim $V = +\infty$. Le cadre de GROSS [1], conduit alors aux définitions suivantes:

S_N^j *désigne l'ensemble des formes multilinéaires symétriques sur E_V qui*
(3. 1) *proviennent d'un élément du produit tensoriel complété*

$$E_V' \hat{\underset{\pi}{\otimes}} E_V' \ldots \hat{\underset{\pi}{\otimes}} E_V' \quad (N \text{ facteurs}).$$

S_N^j est muni de la norme correspondante (inférieure ou égale à la norme de S^j).
$\mathscr{S}^j(V)$ désigne le fibré vectoriel correspondant sur la variété V. $\mathscr{C}_{b,N}^k(V)$ designe
le sous e.v. de $\mathscr{C}_b^k(V)$ formé par les fonctions f telles que $D^j f \in \mathscr{C}^\circ(V, \mathscr{S}_N^j)$ pour
$j=0, 1, \ldots, k$ (noter que $S_N^\circ = S^\circ$ et $S_N^1 = S^1$). On peut alors définir une boule
unité $B_{\infty, N}(1)$, ce qui permet (en introduisant l'indice N dans (2. 1)) de définir
l'ensemble $\mathscr{C}_{b,N}'^k(V)$ des courants nucléaires d'ordre au plus k sur V.

(3. 2) EXEMPLE. *Si V est un espace de Hilbert réel, on définit $\Delta\delta_0$ comme étant le
courant nucléaire d'ordre 2 qui à $f \in \mathscr{C}_{b,N}^2(H)$ associe* $\operatorname{tr}(D^2 f(0))$.

3 B. *Courants pondérés*

(3. 3) *Définition des poids p_i et multipoids $p=(p_0, \ldots, p_k)$.* a) un poids (non
trivial) sur V est une fonction continue $p_i: V$ qui vérifie la condition (ε, K) suivante:

$$(3. 4) \qquad \forall \varepsilon, \quad \exists K \subseteq V, \quad \forall x \in K, \quad \frac{1}{p_i(x)} \leqq \varepsilon$$

b) un poids q_i sur V est dit plus fort que le poids p_i sur V si

$$(3. 5) \qquad \forall \varepsilon, \quad \exists K \subseteq V, \quad \forall x \in K, \quad \frac{p_i(x)}{q_i(x)} \leqq \varepsilon$$

(un poids est donc plus fort que le poids trivial, égal identiquement à 1).

c) un multipoids sur V est une famille $p=(p_0, \ldots, p_k)$ de $(k+1)$ poids. Le
multipoids $q=(q_0, \ldots, q_k)$ est dit plus fort que le multipoids p, si pour tout i, q_i
est plus fort que p_i. Chacun de ces poids est destiné à controler la croissance d'une
dérivée de $f \in \mathscr{C}^k(V)$:

(3. 6) *Définition de $\mathscr{C}_p^k(V)$ et de $\mathscr{C}_p'^k(V)$.* $\mathscr{C}_p^k(V)$ est l'ensemble des fonctions
$f: V \to R$ de classe $\mathscr{C}^k$, telles que

$$\forall i = 0, \ldots, k, \quad \sup_{x \in V} \frac{\|D^i f(x)\|}{p_i(x)} = \|D^i f\|_{p_i} < \infty.$$

On pose $B_{\infty, p}(1) = \{f \in \mathscr{C}_p^k(V), \forall i, \|D^i f\|_{p_i} \leqq 1\}$. On peut alors en introduisant naturel-
lement les poids p_i dans la définition (2. 1) définir l'ensemble $\mathscr{C}_p'^k(V)$ des $\binom{k}{p}$ courants
sur V. Si q est un multipoids plus fort que p_i, on a des injections:

$$\mathscr{C}_b'^k(V) \to \mathscr{C}_p'^k(V) \to \mathscr{C}_q'^k(V).$$

(3. 7) *Une propriété de compacité.*

L'ensemble M des courants l d'ordre au plus k tel que $\alpha = \sup\limits_{l \in M} \|l\|_p < \infty$ est relativement compact dans $\mathscr{C}_b'^k(V)$ muni de la topologie faible.

En effet pour ε donné, il existe K tel que l'on ait (3. 4) pour $i=0, \ldots, k$. Pour tout φ dans $B_\infty(1)$ nulle à l'ordre k sur K, on a:

$$\left| \int \varphi \, dm \right| \leq \alpha \|\varphi\|_p = \alpha \sup_{x \notin K} \frac{|\varphi(x)|}{p(x)} \leq \alpha\varepsilon.$$

Vu (2. 1) M est équicontinu. Vu le théorème d'Alaoglu, M' est relativement compact.

IV — Courants cylindriques et courants quotients

On donne d'abord la définition générale des courants cylindriques sur une variété réelle V de dimension infinie. Etant donnée une certaine famille de relations d'équivalences (R_i) sur V, la donnée d'un tel courant équivaut à la donnée d'un système projectif de courants sur les ensembles E/R_i, que l'on suppose munis de structures de variétés.

On définit également la notion plus générale de courant quotient, qui correspond aux distributions de POULSEN [1].

(4. 1) *Le système $\mathscr{R}_V$ des relations d'équivalence sur V.* (Tout système de relations d'équivalence sur un ensemble X est muni de l'ordre $R_i \leq R_j$ si R_j est plus fine que R_i). Soit $\mathscr{R}_V = (R_i)_{i \in I}$ une famille de relations d'équivalence sur la variété V, ayant les propriétés suivantes.

a) La famille est filtrante à droite.

b) Pour tout R_i, toute chaîne de relations se terminant en R_i ($R_{i_1} < R_{i_2}, \ldots$
$\ldots, R_{i_n} = R_i$) n'a qu'un nombre fini d'éléments.

c) Pour tout i, $V/R_i \in \mathscr{V}$, la structure de V/R_i étant canonique.

d) Pour tout i, la surjection canonique f_i de V sur V/R_i est un morphisme de $\mathscr{V}$, la famille (f_i) étant uniformément bornée.

e) Si R_j est plus fine que R_i, la surjection $f_{ij}: V/R_j \to V/R_i$ est un morphisme et la famille (f_{ij}) est uniformément bornée.

f) Pour tout i, tout compact de V/R_i est l'image par f_i d'un compact de V.

(4. 2) *Les fonctions cylindriques sur V.* On note $\mathscr{C}_{cyl}^k(V, R_i)$ l'image de

$$\mathscr{C}_b^k(V) \xleftarrow{\alpha} \mathscr{C}^k(V/R_i)$$

$$\varphi \circ f_i \longmapsfrom \varphi.$$

On dit que $\varphi \circ f_i$ est une fonction cylindrique de directrice R_i. L'axiome (4. 1. b) entraîne que toute fonction cylindrique a une seule directrice maximale. L'axiome

(4. 1. f) entraîne que l'application α est un morphisme de l'e.v.t. $\mathscr{C}_b^k(V/R_i)$ sur Im α: d'où une topologie $\mathscr{T}_i$ sur $\mathscr{C}_b^k(V/R_i)$. L'axiome (4. 1. a) entraîne que l'e.v. des fonctions cylindriques est une algèbre.

(4. 3) *Courant quotient d'ordre au plus k.* On note $\mathscr{C}_{cyl,m}^k(V; R_i)$ le sous espace de $\mathscr{C}_{cyl}^k(V; R_i)$ de directrice maximale R_i; on note $\mathscr{T}_i'$ la topologie induite par $\mathscr{T}_i$ sur ce sous espace. On munit $\prod_{i \in I} \mathscr{C}_{cyl,m}^k(V; R_i)$ de la topologie somme des topologies $\mathscr{T}_i'$.

On appelle courant quotient toute forme linéaire continue sur cet espace. Un courant quotient sur V est donc représentable par une famille $(l_i)_{i \in I}$ de courants sur les espaces V/R_i. Ces courants l_i ne vérifient aucune condition de compatibilité. Notons qu'on peut introduire des poids.

(4. 4) *Courants cylindriques d'ordre au plus k.* Soit $(R_i)_{i \in I}$ un système de relations d'équivalence sur V, système ayant les propriétés (4. 1). Un courant cylindrique d'ordre au plus k est la donnée d'un système projectif (l_i, f_{ij}) de courants d'ordre au plus k sur le système projectif $(V/R_i, f_{ij})$ de $\mathscr{V}$. (On suppose $\sup_i \|l_i\| < \infty$.)

(4. 5) *Courants cylindriques et formes linéaires.* Il est naturel de représenter l'e.v. $\mathscr{C}_{cyl}^k(V)$ des fonctions cylindriques sur V, non pas comme la somme des ensembles $\mathscr{C}_{cyl,m}^k(V, R_i)$, mais comme la limite inductive du système inductif d'e.v. $(\mathscr{C}_{cyl}^k(V, R_i), f_{ij}')$, f_{ij}' étant l'injection naturelle de $\mathscr{C}_{cyl}^k(V, R_i)$ dans $\mathscr{C}_{cyl}^k(V, R_j)$ définie si $i > j$. On voit alors qu'un courant cylindrique d'ordre au plus k sur V définit naturellement une forme linéaire sur l'algèbre des fonctions cylindriques sur V.

(4. 6) *Projectivité et croissance à l'infini.* En vue d'accroître les possibilités de croissance à l'infini des courants l_i, on peut se demander si l'on peut, dans la définition des courants cylindriques, remplacer les espaces $\mathscr{C}_b^k(V/R_i)$ ou $\mathscr{C}_{cyl}^k(V, R_i)$ par des espaces de fonctions plus décroissantes à l'infini (en utilisant des poids). En fait, un tel remplacement est impossible en général, si l'on veut conserver la propriété de projectivité du système (l_i, f_{ij}). Par exemple, V étant un espace de Banach et $(R_i)_{i \in I}$ le système d'équivalences associées aux sous espaces fermés V_i de codimension finie, chaque e.v. $E_i = V/V_i$ est muni de la norme quotient. Soit $k = 0$. Un courant d'ordre 0 est défini par (μ_i, f_{ij}), μ_i étant une forme linéaire continue sur $\mathscr{C}_b^0(E_i)$. On ne peut remplacer cet espace par:

$$\mathscr{C}_{\varepsilon}^0(E_i) = \left\{ f \in \mathscr{C}^0(E_i) \text{ avec } \sup \frac{|f(x)|}{(\|x\|^2 + 1)^{\varepsilon}} < \infty \right\}$$

car par exemple si $E_i = R_x^2$, la mesure $\mu_i = \sum_{n=1}^{\infty} n^{-1} \delta_0(x_1, x_2 - n)$ définit une forme linéaire continue sur $\mathscr{C}_{\varepsilon}^0(R^2)$ et cependant l'image de μ_i par la projection canonique:

$(x_1, x_2) \rightarrow x_1$, n'est pas définie. En conclusion pour un système (l_i, f_{ij}) la propriété de projectivité (soit $l_j = f_{ij}(l_i)$ si $i < j$) et la propriété de croissance à l'infini des l_i, sont antagonistes.

(4. 7) *Exemples de courants cylindriques.* (On rappelle (1. 0)).

a) *Mesures cylindriques et probabilités cylindriques.* Soit X un e.v.l.c.s. La famille $(X_i)_{i \in I}$ des sous e.v. fermés de codimension finie de X vérifie (4. 1). Un système cohérent μ_i de lois de probabilités sur les e.v. X/X_i est un système projectif de lois de probabilité. Une "mesure cylindrique" sur X (voir N. BOURBAKI [2], L. SCHWARTZ [1]...) est donc un courant d'ordre zéro. Nous préférons employer dans ce cas le terme de "probabilité cylindrique", afin de réserver le terme "mesure cylindrique" pour désigner un courant d'ordre zéro défini par un système cohérent μ_i de mesures sur les e.v. X/X_i (avec $\sup_i \|\mu_i\| < \infty$).

b) *Courants cylindriques de classe k sur les e.v.n.* Un courant cylindrique l sur l'e.v.n. X est la donnée d'un système projectif (l_i, f_{ij}) de courants de classe k (ou distributions bornées d'ordre au plus k) sur le système projectif $(X/X_i, f_{ij})$, où X_i décrit l'ensemble des sous espaces fermés de codimension finie de X, f_{ij} étant toujours la surjection canonique: $X/X_i \rightarrow X/X_j$, définie pour $X_i \subseteq X_j$. Cette définition se prolonge au cas où X est un e.v.l.c.s.; et pour $k = 0$, on retrouve les mesures cylindriques. Ces courants cylindriques sur les e.v.t. ont été introduits indépendamment de nous par C. SUNYACH (communication personnelle, Mai 1971).

c) *Probabilités X-cylindriques sur $X \times Y$.* Soit X un espace topologique séparé et soit Y un e.v.t.l.c.s. Soit (Y_i) le système des sous e.v. fermés de codimension finie de Y et soit φ_{ij} la surjection canonique de Y/Y_i sur Y/Y_j si $Y_i \subseteq Y_j$. Soit Id l'application identique de X. Alors on a un système projectif d'espaces topologiques $(X \times Y/Y_i, f_{ij})$ avec $f_{ij} = Id \times \varphi_{ij}$. Une probabilité X-cylindrique sur $X \times Y$ est un système cohérent de lois de probabilités sur les espaces $X \times Y/Y_i$ (et c'est aussi un courant d'ordre 0 sur $X \times Y$). La notion de probabilité X-cylindrique introduite par L. SCHWARTZ [2], joue un rôle important dans la théorie des équations linéaires (par exemple aux dérivées partielles) à coefficients aléatoires: voir KRÉE [1] et [2].

d) *Courants projectifs sur des sphères.* Soit S la sphère unité d'un espace de Hilbert H et soient s et n deux points diamétralement opposés de S. La projection stéréographique (inversion de pole n, de puissance 2) applique $S' = S \setminus \{n\}$ sur l'hyperplan K tangent en s à S. Au feuilletage de H par des sous espaces fermés parallèles de codimension finie, il correspond par cette projection un feuilletage de S'. A partir de la notion de courant cylindrique sur un espace de Hilbert K, on peut donc construire par transport de structure une notion de courant projectif sur S'.

(4. 8) *Réalisation d'un courant cylindrique sur un e.v. à l'aide d'un processus linéaire.* Ce qui suit prolonge la définition d'une probabilité cylindrique à l'aide

d'un processus stochastique linéaire (voir A. BADRIKIAN [1], L. SCHWARTZ [1]). Soient X et Y deux e.v. en dualité séparante. Soit b une application linéaire de Y dans $\mathscr{C}_b^k(V)$, avec V quelconque dans $\mathscr{V}$. La restriction b_i de b à un sous e.v. Y_i de dimension finie N de Y est definie par des systèmes équivalents:

$$\sum_{i=1}^{N} x_i \otimes \varphi_i \quad x_i \in Y_i' \simeq X/Y_i^{\perp}, \quad \varphi_i \in \mathscr{C}_b^k(V)$$

c'est-à-dire par une application $\underline{b}_i$ de V dans $X/Y_i^{\perp}$. Soit T_0 fixé dans $\mathscr{C}_b'^k(V)$. Le système projectif $(\underline{b}_i(T_0), f_{ij})$ est un courant cylindrique d'ordre au plus k sur l'e.v. X, muni de la topologie $\sigma(X, Y)$.

(4. 9) *Topologie k-cylindrique.* Soit (l^j) une famille de courants cylindriques d'ordre au plus k sur $(V, \mathscr{R}_V)$, chaque courant l^j étant définie par un système projectif (l_i^j, f_{ij}). On dit que $(l^j)_j$ converge vers le courant l défini par (l_i, f_{ij}) si pour tout i, l_i^j converge k-étroitement vers l_i.

(4. 10) *Propriétés des courants cylindriques.* Nous n'étudions pas en détail ces propriétés qui prolongent naturellement les propriétés des probabilités cylindriques (voir L. SCHWARTZ [1], A. BADRIKIAN [1], N. BOURBAKI [2], P. KRÉE [2].

a) *Produit par une fonction cylindrique.* Pour tout courant cylindrique l d'ordre au plus k sur V et pour toute $f \in \mathscr{C}_{cyl}^k(V)$, on pose

$$(lf, \varphi) = (l, f\varphi) \quad \text{pour toute} \quad \varphi \in \mathscr{C}_{cyl}^k(V).$$

b) *Image par un morphisme* $f: V \to W$ *de* $\mathscr{V}$, ce morphisme étant tel que pour tout $R_j \in \mathscr{R}_W$, $f^{-1}(R_j) = R_i \in \mathscr{R}_V$. Si g désigne le morphisme $V/R_i \to W/R_j$ associé à f, on pose $\mu(l_i) = m_i$, (l_i) étant le système projectif définissant le courant l sur V. Alors le système (m_i) définit un courant projectif sur W, appelé image de l par f.

c) *Produit direct.* On se donne deux courants cylindriques d'ordre au plus k:

$$l = (l_i) \quad sur \quad (V, \mathscr{R}_V); \quad m = (m_j) \quad sur \quad (W, \mathscr{R}_W).$$

On suppose que les espaces quotients V/R et W/R_j sont localement compacts. Alors le système des $l_i \times m_j = l_i \otimes m_j$ définit sur $(V \times W, \mathscr{R}_V \times \mathscr{R}_W)$ un courant projectif d'ordre au plus k, noté $l \times m$ ou $l \otimes m$.

d) *Transformation de Fourier.* Soit $l = (l_i)$ un courant cylindrique d'ordre au plus k sur l'e.v. X, supposé en dualité séparante avec l'e.v. Y. On appelle transformée de Fourier de l, la fonction $\hat{l}: Y \to \mathbf{C}$ telle que

$$\forall y \in Y \quad \hat{l}(y) = \int_{-\infty}^{+\infty} e^{-it}(l(y))(t)\, dt.$$

172 P. KRÉE

(4. 11) *Pour qu'une fonction m, définie sur Y soit la transformée de Fourier d'un courant cylindrique d'ordre au plus k sur X, il faut et il suffit que pour tout sous e.v. Y_i de dimension finie de Y, $m|Y_i$ soit la transformée de Fourier d'une distribution bornée sur $X/Y_i^\perp$, d'ordre au plus k.*

(4. 12) *Si l et m sont deux courants d'ordre au plus k sur X, le courant l*m (défini comme l'image de l$\times$m par l'application somme) a pour transformée de Fourier $\hat{l}\cdot\hat{m}$.*

V — Application aux équations aux dérivées partielles

(5. 1) *Opérateurs différentiels opérant entre sections de fibrés d'une variété riemannienne V de dimension infinie.* Evoquons cette définition sur l'exemple d'un opérateur différentiel scalaire P homogène d'ordre 2, défini sur un espace de Hilbert réel H. Un tel opérateur P est défini par exemple par une fonction $\underline{P} \in \mathscr{C}_b^2(H, (H \overset{\hat{}}{\underset{\pi}{\otimes}} H)')$. L'opérateur P fait correspondre à $f \in \mathscr{C}_b^2$, tr (H), la fonction $x \mapsto (\underline{P}(x), D^2f(x))$. Le symbole de P est la fonction:

$$H \times H' \to C$$

$$(x, y) \mapsto Q(x; y, y)$$

où $Q(x, \cdot, \cdot)$ est la forme bilinéaire associée à $K \in (H \overset{\hat{}}{\underset{\pi}{\otimes}} H)' \cong \mathscr{L}(H)$. Par exemple l'opérateur $\varDelta : f \mapsto$ tr D^2f, a pour symbole $\|y\|^2$. C'est aussi l'opérateur de convolution par $\varDelta\delta_0$(vu (4. 11)) et l'opérateur se prolonge aux courants cylindriques.

(5. 2) *Solution élémentaire ou parametrix d'un opérateur différentiel scalaire défini sur un espace de Hilbert H.* Soit $P(D)$ un opérateur différentiel à coefficients constants sur H d'ordre au plus k et soit $P(D)\delta_0$ le courant associé. On dit que le courant E (resp P) sur H est une solution élémentaire (resp un paramétrix) de $P(D)$ si:

$$E * P(D)\delta_0 = P(D)\delta_0 * E$$

est la distribution de Dirac δ_0 (resp est la somme de δ_0 et d'un courant qui a une transformée de Fourier à décroissance rapide).

(5. 3) EXEMPLES. a) La fonction $y \mapsto -\|y\|^{-2} = -(\widehat{\varDelta\delta_0})^{-1}$ ne vérifie pas la condition (4. 10) donc on ne peut obtenir par ce procédé une solution élémentaire (mais l'on peut obtenir un paramétrix; ainsi que pour l'opérateur de la chaleur $(\partial_t\delta_0 - \varDelta\delta_0) *$ défini sur $R_t \oplus H$).

b) La fonction $y \mapsto -(\|y\|^2 + 1)^{-2}$ vérifie (4. 10) donc l'opérateur non homogène $\varDelta - 1$ admet une solution élémentaire E. Notons que E n'est pas un courant et que l'équation $\Delta E - E = \delta_0$ n'a de sens que dans le cadre de la théorie des courants cylindriques.

c) L'opérateur $\Delta - \left(\dfrac{\partial}{\partial t}\right)^2$, défini sur $R_t \oplus H$, a une solution élémentaire.

Quelques problèmes

a) *Interpolation et applications p-sommantes.* Soient (A_0, A_1) et (B_0, B_1) deux couples d'interpolation d'espaces de Banach et soit T un opérateur linéaire continu de A_i dans B_i ($i=0$ et 1). Soient p_0 et p_1 deux réels >0 et soit p_ϑ tel que $p_\vartheta^{-1} = (1-\vartheta)^{-1} p_0 + \vartheta^{-1} p_1$ avec $0 < \vartheta < 1$. On suppose que $T\colon A_i \to B_i$ est p_i-sommante (voir L. Schwartz [1] exposé 7). Est ce que T restreinte à $[A_0, A_1]_\vartheta$ (voir A. P. Calderon [1]) ou à $(A_0, A_1)_{\vartheta, p}$ (voir Lions—Peetre [1]) est p_ϑ-sommante à valeurs dans $[B_0, B_1]_\vartheta$ ou $(A_0, A_1)_{\vartheta, p}$? De même, quelle relation existe-t-il entre les produits tensoriels completés $\hat{\bigotimes}_{g_{p_\vartheta}}$ (voir Saphar [1] et S. Chevet [1]) des espaces interpolés $[A_0, A_1]_\vartheta$ et $[B_0, B_1]_\vartheta$, et les interpolés de $A_0 \hat{\bigotimes}_{g_p} B_0$ et $A_1 \hat{\bigotimes}_{g_p} B_1$?

b) *Problème d'approximation.* L. Nachbin [2] a donné une extension du théorème de Stone Weierstrass aux fonctions de classe $\mathscr{C}^k$ définies sur une variété différentiable de dimension finie: peut-on étendre ces théorèmes en dimension infinie?

c) Les théorèmes de Whitney (Amer. J. Math. 1948) permettant de définir et prolonger des fonctions de classe $\mathscr{C}^k$ sur un compact K d'une variété de dimension finie. Existe-t-il des propriétés analogues en dimension infinie?

d) D'une façon générale l'extension en dimension infinie, des propriétés des fonctions numériques de classe $\mathscr{C}^k$ sur une variété de dimension finie, faciliterait l'étude des courants sur une variété banachique.

BIBLIOGRAPHIE

A. Badrikian

[1] *Séminaire sur les fonctions aléatoires linéaires et les mesures cylindriques.* Lecture notes **139**, Springer, Berlin 1970.

N. Bourbaki

[1] *Variétés différentielles et analytiques* — paragraphes 1 à 7. Hermann, Paris 1971.

[2] *Intégration sur les espaces topologiques séparés* — livre **VI** — chapitre **9** Hermann, Paris 1969.

[3] *Variétés différentielles et analytiques* — § 8 à 15. Hermann, Paris 1971.

A. P. Calderon

[1] *Intermediate spaces and interpolation.* The complex method — Stud. Math. Vol. **24** — fasc. 2 (1964), 113—190.

S. Chevet

[1] *Sur certains produits tensoriels topologiques d'espaces de Banach.* Z. Wahrscheinlichkeits-theorie und Verw. Gebiete **11** (1969), 120—138.

D. H. Fremlin, D. J. H. Garling et R. G. Haydon

[1] *Regular Borel measures on topological spaces.* Actes (à paraître dans les mémoires de la S.M.F.) des conférences internationales de Bordeaux (avril 1971).

I. M. Gelfand et N. J. Vilenkin

[1] *Les distributions* — Vol. IV (traduit du russe). Dunod, Paris 1968.

L. Gross

[1] *Abstract Wiener measure and infinite dimensional theory.* Lecture notes in Mathematics **140**, Springer, Berlin 1970, 88—116.

P. Krée

[1] *Equations linéaires à coefficients aléatoires.* Instituto Nazionale di Alta Matematica, Symposia Mathematica. Vol. VII (1971), 515—546.

[2] *Image de probabilités cylindriques par certaines applications non linéaires.* Séminaire L. Schwartz, École Polytechnique de Paris (janvier 1972).

J. L. Lions et J. Peetre

[1] *Sur une classe d'espaces d'interpolation.* Publications bleues de l'I.H.E.S. (Paris) n° **19** (1964).

L. Nachbin

[1] *Topology on spaces of holomorphic mappings.* Springer, Berlin (1969).

[2] *Sur les algèbres denses de fonctions différentiables sur une variété.* Note C.R. Acad. Sci. Paris **228** (1949), 1549—1551.

E. Thue Poulsen

[1] *Distributions sur les e.v.t.* Proceedings of an International Summer Institute held in Lisbon September 1964 (Fondation Gulbenkian).

P. Saphar

[1] *Produits tensoriels d'espaces de Banach et classes d'applications linéaires.* Studia math. **38** (1970), 71—100.

L. Schwartz

[1] *Les applications radonifiantes.* Séminaire à l'Ecole Polytechnique de Paris (1969—1970).

[2] *Probabilités X-cylindriques.* (communication personnelle — août 71).

Subdefinite Functions

By

J. L. B. COOPER

CHELSEA COLLEGE, UNIVERSITY OF LONDON

A celebrated result of BOCHNER states that if $f(x)$ is a continuous function on the real line such that for any choice of real $x_1, \ldots, x_m$ and complex $\zeta_1, \ldots, \zeta_m$,

$$(1) \qquad \sum \zeta_r \bar{\zeta}_s f(x_r - x_s) \geqq 0$$

then f, which is called a positive definite function, has a representation of the form

$$(2) \qquad f(x) = \int e^{iux} \varrho(du)$$

where ϱ is a positive bounded measure on the real line. It is easily shown that if f is continuous then (1) is equivalent to the requirement that

$$(3) \qquad F(\varphi) = \iint f(x-y)\varphi(x)\overline{\varphi(y)}\,dx\,dy \geqq 0$$

for all $\varphi \in C_c$ the class of continuous functions with compact support. We shall discuss conditions under which analogous results hold when f is not continuous, or when (3) holds for classes of functions other than C_c.

For any class Φ of functions defined on a locally compact abelian group G, we shall write Φ_c for the set of members of Φ that have compact support, and Φ_{loc} for the set of functions on G whose restrictions to compact sets are in Φ.

We say $f \in P(\Phi)$, or that f is positive definite for Φ, if (3) holds for every $\varphi \in \Phi$. More generally, if

$$(4) \qquad F(\varphi) = \langle f, \varphi * \varphi^* \rangle = \iint f(x-y)\varphi(x)\,\overline{\varphi(y)}\,dx\,dy$$

exists for all $\varphi \in \Phi$, we say $f \in K(\Phi)$; here $\varphi^*(x) = \overline{\varphi(-x)}$. If $f \in P(\Phi)$, this form is hermitian; and if the class Φ separates points then $\overline{f(x)} = f(-x)$ almost everywhere. Now suppose that f is hermitian and in $K(\Phi)$; for any choice of $\varphi_1, \ldots, \varphi_m$ the form $Q(\zeta) = F(\varphi)$ with $\varphi = \sum \zeta_r \varphi_r$ defines a hermitian quadratic form in ζ.

If there is a choice of $\varphi_1, \ldots, \varphi_m$ such that $Q(\zeta)$ is a sum of squares of which just k have negative coefficients, and no choice for which it is a sum of squares more than k of which have negative coefficients, then we say that $f \in P_k(\Phi)$. Our purpose is to discuss the manner in which these classes vary for Φ. The classical case is that for which $\Phi = C_c$; in the theory of distributions the case of infinitely differentiable classes Φ is discussed, but without the requirement that f be a function.

To begin with, we investigate the properties of $K(\Phi)$.

THEOREM. *Let Φ be such that for any compact C there is a nonnegative $\varphi \in \Phi$ such that φ has compact support, is summable and strictly positive on C. Then $K(\Phi) \subset L_{\mathrm{loc}}$.*

For, if C is compact, let φ have the stated properties for $(C + C) \cup C$; let $\varphi(x) > 1$ on this set. Then $\int \varphi(s + t)\overline{\varphi(t)}\, dt \geq m(C)$ if $s \in C$. Now $f(s) \int \varphi(s + t)\overline{\varphi(t)}\, dt \in L(G)$ by hypothesis, hence $f \in L(C)$ and so since C is arbitrary $f \in L_{\mathrm{loc}}$.

It is easy to see that if $\Phi_1 \subset \Phi_2$, $K(\Phi_1) \supset K(\Phi_2)$, $P(\Phi_1) \supset P(\Phi_2)$, $P_k(\Phi_2) \subset \subset \bigcup_{r \leq k} P_r(\Phi_1)$.

If Φ_1 is dense in Φ_2, then $K(\Phi_2) \cap P_k(\Phi_1) = P_k(\Phi_2)$. For let $\varphi_1, \ldots, \varphi_m \in \Phi_2$, and let (φ_j^n) be a sequence in Φ_1 tending to φ_j. $Q(\zeta) = F(\sum \zeta_r \varphi_r)$ is the limit of $F(\sum \zeta_r \varphi_r^n)$ and so must have at least as many negative squares as this last form for large enough n, so that $P_k(\Phi_2) \subset K(\Phi_2) \cap P_k(\Phi_1)$. The opposite inclusion follows from the formulae given above.

If $p \geq 2$, $K(L_c^p) = K(C_c) = L_{\mathrm{loc}}$; if $G = R^n$ and D is the Schwartz test space for distributions, $K(D) = L_{\mathrm{loc}}$. This follows because convolutions of functions in these spaces are all bounded and with compact supports. On the other hand $K(L_c^p)$ does vary with p for $p < 2$. If we write $F(\varphi, \psi)$ for the polar form derived from $F(\varphi)$, then if $f \in K(\Phi)$ this form is bilinear on Φ; if $f \in K(L_c)$ then for any compact sets C_1, C_2 $F(\varphi, \psi)$ is bilinear on $L(C_1) \times L(C_2)$ and so by two applications of the uniform boundedness theorem there is an $M(C_1, C_2) < \infty$ such that $|F(\varphi, \psi)| \leq \leq M(C_1, C_2) \|\varphi\|_1 \|\psi\|_1$ if supp $\varphi \subset C_1$, supp $\psi \subset C_2$. If $|f(x)| > M$ on a set E of positive measure with compact E then E can be covered by a finite number of translates of C, one of which must have nonzero measure, say E'. Let $\varphi(x) = \chi_{E'}(x)$ sgn $\overline{f(x)}/m(E')$; then $|\int f(x)\varphi(x)\, dx| > M$ and so $|\int f(x - y)\varphi(y)\, dy| > M$ on some compact neighbourhood V of 0. We can choose ψ with support in $C_2 \cap V$ so that $F(\varphi, \psi) > M$ and so $M \leq M(C_1, C_2)$. It follows that $f(x)$ is locally essentially bounded.

If $f \in P(L_{\mathrm{loc}})$, it can be shown that f is essentially bounded and moreover that it is almost everywhere equal to a continuous function; in this case the theory reduces to that of functions that are almost everywhere equal to Bochner positive definite functions. In other cases the functions in $P(\Phi)$ need not be locally bounded.

If $1 \leq p_1 < p_2 \leq 2$, then the inclusion $K(L_c^{p_1}) \subset K(L_c^{p_2})$ is strict. If $\varphi \in L_c^p$ then according to a wellknown inequality of Young $\varphi * \varphi^* \in L_c^{q'}$ where we write $\dfrac{1}{q} + \dfrac{1}{q'} = 1$, as usual, and where $q = p/(2p - 1)$; thus $L_{\mathrm{loc}}^q \subset K(L_c^p)$.

Functions in $K(L_c^p)$ for $1<p<2$ have properties close to those of L_{loc}^q; for instance it is proved in [1] that if $f \in K(L_c^p)$, $f(x)|x-a|^{-\alpha} \in L_{\text{loc}}$ for any $a \in R^n$ if $\alpha q'<1$ but Dr. J. STEWART has proved that $K(L_c^p)$ contains functions not in L_{loc}^q.

Expressions of the form (2) are valid for functions positive definite in this wider sense. It is shown in [1] that if f is a locally summable positive definite function — that is, $f \in P(C_c) = P(L_c^2)$ — then an expression (2) holds, and is valid in the sense of $(C, 1)$ summability; moreover $\varrho(0, u) = o(u)$ as $|u| \to \infty$. Dr. J. STEWART [2] has extended this result to general locally compact abelian G; his condition is that, for each compact C, $\varrho(u+C) \to 0$ as $u \to \infty$. For the real line it is proved in [1] that if $f \in P(L_c^p)$, $1<p<2$, then $\varrho(0, u) = o(u^{1/q+\varepsilon})$ for any $\varepsilon>0$. These estimates on the ϱ are necessary but no sufficient estimates for the ϱ are known.

Results concerning functions of type $P_k(\Phi)$ are incomplete save for the case in which $G=R^n$ and either f is continuous [3] or Φ is the Schwartz space $D=C_c^\infty$ [4] in which case f can be taken to be a distribution. In this case, f is conditionally positive definite, that is, there is a polynomial in the operator of differentiation $i\dfrac{d}{dx}$, $Q\left(i\dfrac{d}{dx}\right)$, such that $\overline{Q\left(i\dfrac{d}{dx}\right)}Q\left(i\dfrac{d}{dx}\right)$ is positive definite in the usual sense. In the case $G=R$, f has an expression of the form

$$h\varrho(x) + \int [e^{i\lambda x} - S_\varrho(x, \lambda)] \frac{\sigma(d\lambda)}{[Q_0(\lambda)]^2}$$

where σ is a tempered measure, $Q(z) = Q_0(z)Q_1(z)$ where $Q_0(z)$ is the product of all the real factors in $Q(z)$, $Q\left(i\dfrac{d}{dx}\right)h(x)=0$ and $S_\varrho(x, \lambda)$ is a correction term. Dr. J. STEWART has shown that if f is a function then $\int_0^a \varrho(du)/Q_0(u)^2 = o(a)$ as $a \to \infty$.

REFERENCES

[1] J. L. B. Cooper, *Positive definite functions of a real variable*. Proc. London Math. Soc. (3) **10** (1960), 53—66.

[2] J. Stewart, *Unbounded positive definite functions*. Canad. J. Math. **21** (1969), 1309—1318.

[3] M. G. Krein, *The integral representation of a continuous Hermitian-indefinite function with a finite number of negative squares*. Dokl. Akad. Nauk SSSR **125** (1959), 31—34.

[4] Shah Tao-Shing, *On conditionally positive definite generalized functions*. Sci. Sinica **11** (1962), 1147—1168.

Eine Bemerkung zum Banachschen Fixpunktsatz

Von

PAUL OTTO RUNCK

MATHEMATISCHES INSTITUT
HOCHSCHULE LINZ (ÖSTERREICH)

1. Zunächst geben wir ein Fixpunktproblem an, auf das der Banachsche Fixpunktsatz in seiner bekannten Form nicht angewendet werden kann, weil die Kontraktionseigenschaft nicht erfüllbar ist. Sodann betrachten wir einen speziellen Fixpunktsatz, bei dem wir die Kontraktionsbedingung in der Weise abschwächen, daß der zugrunde liegende Raum Y als Banachunterraum eines Banachraumes X angesehen wird und in X eine Kontraktionsbedingung, jedoch in Y nur eine Beschränktheitsbedingung gefordert wird. Dieser spezielle Fixpunktsatz läßt sich auf das angegebene Problem anwenden.

2. Das Problem der konformen Abbildung des Inneren des Einheitskreises auf ein Gebiet G, das durch eine bezüglich 0 sternige Jordankurve Γ berandet wird, führt man auf die Ränderzuordnungsfunktion zurück. Bedeuten $(1, \Phi)$ bzw. $(\varrho(\theta), \theta)$ die Polarkoordinaten des Einheitskreises bzw. von Γ, so führt die Ränderzuordnungsfunktion, gegeben durch $\Phi \mapsto \theta(\Phi)$, $\Phi \in [0, 2\pi]$, auf die Integralgleichung von THEODORSEN

$$(1) \qquad \theta(\Phi) = \Phi + \frac{1}{2\pi} \int\limits_{0 (H)}^{2\pi} \log \varrho(\theta(\delta)) \operatorname{ctg} \frac{\Phi - \delta}{2} \, d\delta =: (T\theta)(\Phi) \quad (\Phi \in [0, 2\pi]).$$

Für die Lösung dieser Integralgleichung hat WARSCHAWSKI folgendes Ergebnis angegeben. (Vgl. hierzu D. GAIER [5], S. 69 und die in [5] angegebene Literatur).

SATZ (WARSCHAWSKI): *(1) läßt sich lösen durch das Iterationsverfahren*

$$(2) \qquad \theta_{n+1}(\Phi) = (T\theta_n)(\Phi) = \Phi + \frac{1}{2\pi} \int\limits_{0 (H)}^{2\pi} \log \varrho(\theta_n(\delta)) \operatorname{ctg} \frac{\Phi - \delta}{2} \, d\delta$$

$$(n \in N, \ \Phi \in [0, 2\pi]),$$

$\theta_0(\Phi)$ ($\Phi \in [0, 2\pi]$) *absolut stetig mit* $\theta_0' \in L_2(0, 2\pi)$, *wenn für die sternige Jordankurve* Γ *die folgenden Einschränkungen gelten:*

i) $\varrho(\theta)$ ($\theta \in [0, 2\pi]$) *absolut stetig und* $\left| \dfrac{\varrho'(\theta)}{\varrho(\theta)} \right| \leqq \varepsilon$ *f. ü. in* $[0, 2\pi]$ *mit* $\varepsilon \in {]}0, 1[$.

ii) $\dfrac{a}{1+\varepsilon} \leqq \varrho(\theta) \leqq a(1+\varepsilon)$ *für ein* $a \in R_+$.

Für $\theta_0(\Phi) = \Phi$ ($\Phi \in [0, 2\pi]$) *gilt die Fehlerabschätzung*

$$|\theta_n(\Phi) - \theta(\Phi)| \leq 2 \left(\frac{\pi}{\sqrt{1 - \varepsilon^2}} \right)^{1/2} \varepsilon^{n/2 + 1} \qquad (n \in N).$$

Warschawski verwendet für seinen Beweis die folgende Ungleichung: Ist $\theta(\Phi)$ ($\Phi \in [0, 2\pi]$, $\theta(0) = \theta(2\pi)$) absolut stetig mit $\theta' \in L_2(0, 2\pi)$, so gilt

$$|\theta^2(\Phi_1) - \theta^2(\Phi_2)| \leq 2\pi \|\theta\|_{L_2} \cdot \|\theta'\|_{L_2} \qquad (\Phi_1, \Phi_2 \in [0, 2\pi]).$$

Will man den Banachschen Fixpunktsatz zum Beweis der Konvergenz heranziehen, so stellt man fest, daß die Kontraktion von T in L_2 mit der Kontraktionskonstanten ε (< 1) leicht verifizierbar ist, aber nicht im Raum der absolutstetigen Funktionen, deren Ableitungen L_2 angehören. Bezüglich θ'_n gilt lediglich

$$(3) \qquad \|\theta'_n - 1\|_{L_2}^2 \leq \frac{\varepsilon^2}{1 - \varepsilon^2} \|\theta'_0\|_{L_2} \qquad \text{(gleichm. in } n\text{)}.$$

Somit läßt sich der Fixpunktsatz in der bekannten Form nicht anwenden, um θ absolut stetig mit $\theta' \in L_2(0, 2\pi)$ nachzuweisen.

Wir betrachten nun einen speziellen Fixpunktsatz, der sich mit Erfolg auf dieses Problem anwenden läßt.

3. Zunächst bringen wir einige Vorbereitungen (man vgl. hierzu ARONSZAJN—GAGLIARDO [1], BERENS [2], BUTZER—BERENS [3], GAGLIARDO [4]).

DEFINITION: Ein Banachraum Y heißt *Banachunterraum* des Banachraumes X, wenn i) der lineare Raum Y ein Unterraum des linearen Raumes X ist und wenn ii) Y stetig in X eingebettet ist (d.h. wenn die Identitätsabbildung $I: Y \to X$ stetig ist).

Für die Normen folgt hierfür die Beziehung: Es existiert ein $M \in R_+$, sodaß für ein beliebiges Element $f \in Y$ gilt $\|f\|_X \leq M \|f\|_Y$.

Man kann sich o. E. d. A. auf den Fall $M = 1$ beschränken. Man definiert deshalb

DEFINITION: Der Banachunterraum Y von X heißt *normalisiert*, wenn für ein beliebiges Element $f \in Y$ gilt

$$\|f\|_X \leq \|f\|_Y.$$

BEZEICHNUNGSWEISEN (vgl. [1], [2]): Sind W und V Hausdorffräume, so bedeutet $W \underset{c}{\subset} V :\Leftrightarrow W$ ist stetig eingebettet in V. Sind W und V Banachräume, so setzen wir

$$W = V :\Leftrightarrow [W \underset{c}{\subset} V \text{ und } V \underset{c}{\subset} W],$$

$$W \cong V :\Leftrightarrow [W \underset{c}{\subset} V, \ V \underset{c}{\subset} W \text{ und } \|f\|_V = \|f\|_W].$$

Nun bringen wir die folgende auf E. GAGLIARDO [4] zurückgehende

DEFINITION: Es sei X ein Banachraum und Y ein Banachunterraum von X. Die *Vervollständigung* von Y *relativ* zu X, bezeichnet durch $\tilde{Y}^X$, wird definiert als die Menge aller $f \in X$, die in der Abschließung bezüglich X einer beschränkten, aber beliebigen Kugel von Y liegen. Somit

$$\tilde{Y}^X := \bigcup_{R \in \mathbf{R}_+} \overline{S_Y(R)}^X \quad \text{mit} \quad S_Y(R) := \{f \in Y \mid \|f\|_Y \leq R\}.$$

Aus dieser Definition ergibt sich die Äquivalenz $f \in \tilde{Y}^X \leftrightarrow$ i) $f \in X$, ii) es existiert eine Folge $(f_n)_{n \in N} \subset Y$ mit $\|f_n\|_Y \leq R$ $(R \in \mathbf{R}_+)$ und $\|f_n - f\|_X \xrightarrow[n \to \infty]{} 0$. Mit der Norm

$$\|f\|_{\tilde{Y}^X} := \inf\{R \mid f \in \overline{S_Y(R)}^X\} =$$
$$= \inf\{\sup_{n \in N} \|f_n\|_Y \mid (f_n)_{n \in N} \subset Y \quad \text{mit} \quad \|f_n - f\|_X \to 0\}$$

ist $\tilde{Y}^X$ ein Banachunterraum von X und Y ein normalisierter Banachunterraum von $\tilde{Y}^X$.

Für die Vervollständigung von Y relativ zu X sind mehrere Eigenschaften bekannt (vgl. z. B. [1], [2]). Wir geben einige für uns besonders wichtige Beziehungen an:

 i) $\tilde{Y}^X \cong \left(\overline{\tilde{Y}^X}\right)^X$ (Vervollständigung von $\tilde{Y}^X$ relativ zu X).

 ii) Ist Y reflexiv, so folgt $\tilde{Y}^X \cong Y$.

 iii) $\tilde{Y}^X = X$ genau dann, wenn $Y = X$.

4. Spezieller Fixpunktsatz

Es sei X ein Banachraum und Y ein Banachunterraum von X, für den $Y \cong \tilde{Y}^X$ gelte. $T: X \to X$ sei weiter eine in X kontrahierende Abbildung. Dann folgt:

i) *Es existiert genau ein Fixpunkt f^* in X. Für eine beliebige Iterationsfolge $(f_\nu)_{\nu \in N_0}$ in X mit $f_{\nu+1} = Tf_\nu$ $(\nu \in N)$ gilt hierfür*

$$(*) \qquad\qquad\qquad \|f_\nu - f^*\|_X \xrightarrow[\nu \to \infty]{} 0.$$

ii) *Erfüllt die Iterationsfolge die zusätzliche Voraussetzung*
$(\overset{*}{*})$ *$f_\nu \in Y$ $(\nu \in N_0)$ und $\|f_\nu\|_Y \leq M$ gleichmäßig in ν $(M \in \mathbf{R}_+)$, dann gilt*

$$f^* \in Y \quad \text{mit} \quad \|f^*\|_Y \leq M.$$

BEWEIS: i) ist die Aussage des Banachschen Fixpunktsatzes für den Banachraum X.

ii) ergibt sich wie folgt: (f_ν) ist wegen $(*)$ und $(\overset{*}{*})$ eine Folge, die einen Grenzwert in $\tilde{Y}^X$, der Vervollständigung von Y relativ zu X, besitzt. Wegen der Voraussetzung $Y \cong \tilde{Y}^X$ gilt dann $f^* \in Y$ und $\|f^*\|_Y = \|f^*\|_{\tilde{Y}^X} \leq M$.

5. Anwendung

a) Für die reflexiven B-Räume $X = L_p(0, 2\pi)$, $1 < p < \infty$, und $Y = (L_p(0, 2\pi))^{(1)}$ (B-Raum, bestehend aus den Funktionen f, die f. ü. eine Ableitung besitzen, mit der Norm $\|f\|_Y = \|f\|_{L_p} + \|f'\|_{L_p}$) gilt $\tilde{Y}^X \cong Y$.

b) Für die Räume $X = L_p(0, 2\pi) \cap C(0, 2\pi)$, $1 < p < \infty$, und $Y = (L_p(0, 2\pi))^{(1)} \cap C(0, 2\pi)$, wobei $f(0) = f(2\pi)$ gelte, folgt ebenfalls $\tilde{Y}^X \cong Y$. (Hierbei beachte man: Gilt für zwei B-Räume V, W: $V \underset{c}{\subset} \mathfrak{X}$, $W \underset{c}{\subset} \mathfrak{X}$, wobei $\mathfrak{X}$ ein linearer Hausdorffraum sei, so ist auch $V \cap W$ ein B-Raum bei der Norm $\|f\|_{V \cap W} = \max(\|f\|_V, \|f\|_W)$. Hier wähle man $\mathfrak{X} = L_p(0, 2\pi)$).

V ist der Raum der absolut stetigen Funktionen f mit $f' \in L_p(0, 2\pi)$, $1 < p < \infty$. (Man vgl. hierzu z. B. RIESZ—SZ.-NAGY [6]).

Man beachte: Im Fall $p = 1$ gilt $\tilde{Y}^X \neq Y$. Hier gilt $\tilde{Y}^X = BV(0, 2\pi) \cap C(0, 2\pi)$ (mit $f(0) = f(2\pi)$).

BEMERKUNG: Um die Konvergenz einer Folge aus dem Raum X (aus b)) zu zeigen, genügt es, die Konvergenz der Folge in $L_p(0, 2\pi)$ sowie die Stetigkeit der Grenzfunktion nachzuweisen.

c) Die Konvergenz des im Satz von Warschawski angegebenen Iterationsverfahrens (2) gegen eine absolut stetige Funktion θ mit $\theta' \in L_2(0, 2\pi)$, die somit eine eindeutig bestimmte Lösung der Integralgleichung (1) ist, folgt nach dem speziellen Fixpunktsatz und b) aus der Kontraktion von T in $L_2(0, 2\pi)$, der Stetigkeit von θ, $(\theta_n)_{n \in N_0} \subset Y$ und der Beziehung (3). Weiter läßt sich (3) mit θ anstelle von θ'_n folgern.

LITERATUR

[1] N. Aronszajn and E. Gagliardo, *Interpolation spaces and interpolation methods*. Ann. Mat. Pura Appl. (4) **68** (1965), 51—117.

[2] H. Berens, *Interpolationsmethoden zur Behandlung von Approximationsprozessen auf Banachräumen*. Lecture Notes in Math. **64**, Springer, Berlin 1968, 6—18.

[3] P. L. Butzer and H. Berens, *Semi-Groups of Operators and Approximation*. Grundlehren Bd. **145** Springer, Berlin 1967, 223—224.

[4] E. Gagliardo. *A unified structure in various families of function spaces, compactness and closure theorems*. Proc. Internat. Symp. on Linear Spaces, Jerusalem (1960), 237—241, New York 1961.

[5] D. Gaier, *Konstruktive Methoden der konformen Abbildung*. Springer Tracts in Natural Philosophy, Vol. 3 Springer, Berlin 1964, 61—73.

[6] F. Riesz—B. Sz.-Nagy, *Vorlesungen über Funktionalanalysis*. Deutscher Verlag der Wissenschaften, Berlin 1968, 66—71.

On a Certain Converse of Hölder's Inequality

By

L. LEINDLER

JÓZSEF ATTILA TUDOMÁNYEGYETEM
BOLYAI INTÉZETE
SZEGED

Recently A. PRÉKOPA [1] has proved the following interesting integral inequality

$$(1) \qquad \frac{1}{2} \int\limits_{-\infty}^{\infty} \sup_{x+y=t} f(x)g(y)\,dt \geqq \Big[\int\limits_{-\infty}^{\infty} f^2(x)\,dx \Big]^{1/2} \Big[\int\limits_{-\infty}^{\infty} g^2(y)\,dy \Big]^{1/2},$$

where $f(x)$ and $g(y)$ are arbitrary measurable non-negative functions. The proof of (1) is quite difficult and long.

The purpose of the present paper is to prove an inequality of similar type and, using our method of proof, to give a certain reversibility of a theorem of W. H. YOUNG.

THEOREM 1. *Suppose that* $1 \leqq r,\ s \leqq \infty$ *and* $\dfrac{1}{r}+\dfrac{1}{s}=1+\dfrac{1}{\gamma}$, *where* $1 \leqq \gamma \leqq \infty$. *Then for every pair of non-negative measurable functions* $f(x)$ *and* $g(x)$ *we have*).*

$$(2) \qquad \int\limits_{-\infty}^{\infty} \Big[\int\limits_{-\infty}^{\infty} (f(x)g(t-x))^\gamma \, dx \Big]^{1/\gamma} dt \geqq \Big[\int\limits_{-\infty}^{\infty} f^r(x)\,dx \Big]^{1/r} \Big[\int\limits_{-\infty}^{\infty} g^s(x)\,dx \Big]^{1/s}.$$

We remark that the special case $r=s=2$ and $\gamma=\infty$ of Theorem 1 does not include inequality (1), namely (1) states more with the factor 2.

We guess that the following inequality is also valid:**)

$$(2^*) \qquad \int\limits_{-\infty}^{\infty} \sup_{x+y=t} f(x)g(y)\,dt \geqq p^{1/p} q^{1/q} \Big[\int\limits_{-\infty}^{\infty} f^p(x)\,dx \Big]^{1/p} \Big[\int\limits_{-\infty}^{\infty} g^q(x)\,dx \Big]^{1/q},$$

where $p, q \geqq 1$ and $\dfrac{1}{p}+\dfrac{1}{q}=1$. This would already include (1) as a special case.

---

*) If $\gamma=\infty$ then $\Big[\int\limits_{-\infty}^{\infty} k^\gamma(x)\,dx \Big]^{1/\gamma}$ with $k(x) \geqq 0$ means the essential upper bound of $k(x)$.

**) During the printing of this paper we proved inequality (2^*). It is easy to prove that $p^{1/p} q^{1/q}$ is the best possible constant.

W. H. YOUNG [2] proved that if $f(x)$ and $g(x)$ are 2π-periodic functions and $f(x) \in L^r$ and $g(x) \in L^s$, where $r, s \geq 1$, $\dfrac{1}{r} + \dfrac{1}{s} = 1 + \dfrac{1}{\gamma}$, $\gamma \geq 1$, then the function

$$h(x) = \frac{1}{2\pi} \int_0^{2\pi} f(x-t)g(t)\,dt$$

belongs to L^γ, and

$$(3) \qquad \left\{ \frac{1}{2\pi} \int_0^{2\pi} |h(x)|^\gamma \, dx \right\}^{1/\gamma} \leq \left\{ \frac{1}{2\pi} \int_0^{2\pi} |f(x)|^r \, dx \right\}^{1/r} \left\{ \frac{1}{2\pi} \int_0^{2\pi} |g(x)|^s \, dx \right\}^{1/s}.$$

Using the same method as in the proof of Theorem 1, taking the suitable integrals on the interval $[0, 2\pi]$ instead of $(-\infty, \infty)$, we obtain "a certain converse of (3)". This can be formulated as follows:

THEOREM 2. *Suppose that* $1 \leq r$, $s \leq \infty$ *and* $\dfrac{1}{r} + \dfrac{1}{s} = 1 + \dfrac{1}{\gamma}$, *where* $1 \leq \gamma \leq \infty$.
Then for every non-negative 2π-periodic measurable functions $f(x)$ *and* $g(x)$ *we have*

$$\frac{1}{2\pi} \int_0^{2\pi} \left\{ \frac{1}{2\pi} \int_0^{2\pi} (f(x)g(y-x))^\gamma \, dx \right\}^{1/\gamma} dy \geq \left\{ \frac{1}{2\pi} \int_0^{2\pi} f^r(x) \, dx \right\}^{1/r} \left\{ \frac{1}{2\pi} \int_0^{2\pi} g^s(x) \, dx \right\}^{1/s}.$$

PROOF OF THEOREM 1. We may assume that the integral on the left-hand side of (2) has finite value and that the functions $f(x)$ and $g(x)$ do not vanish almost everywhere. Let us define $f_n(x)$ by

$$f_n(x) = \begin{cases} \min(n, f(x)) & \text{if} \quad |x| \leq n, \\ 0 & \text{if} \quad |x| > n; \end{cases}$$

and $g_m(x)$ similarly.

First we assume that r and s are finite. Choose n and m such that

$$A_n = \left[\int_{-\infty}^{\infty} f_n^r(x) \, dx \right]^{1/r} > 0 \quad \text{and} \quad B_m = \left[\int_{-\infty}^{\infty} g_m^s(x) \, dx \right]^{1/s} > 0.$$

Now we prove that for any y

$$(4) \qquad \int_{-\infty}^{\infty} f_n^r(x) g_m^s(y-x) \, dx \leq A_n^{r-1} B_m^{s-1} \left[\int_{-\infty}^{\infty} (f_n(x) g_m(y-x))^\gamma \, dx \right]^{1/\gamma}.$$

If $\gamma = 1$, then $r = s = 1$, and therefore (4) is trivial. If $1 < \gamma \leq \infty$, then we put $\gamma' = \gamma/(\gamma-1)$ $(1 \leq \gamma' < \infty)$ and using Hölder's inequality we get

$$(5) \qquad \int_{-\infty}^{\infty} f_n^r(x) g_m^s(y-x) \, dx = \int_{-\infty}^{\infty} f_n(x) g_m(y-x) f_n^{r-1}(x) g_m^{s-1}(y-x) \, dx \leq$$

$$\leq \left[\int_{-\infty}^{\infty} f_n^\gamma(x) g_m^\gamma(y-x) \, dx \right]^{1/\gamma} \left[\int_{-\infty}^{\infty} f_n^{\gamma'(r-1)}(x) g_m^{\gamma'(s-1)}(y-x) \, dx \right]^{1/\gamma'}.$$

If $r=1$, then $s=\gamma$; and if $s=1$, then $r=\gamma$; therefore (5) gives, in these cases, inequality (4). If $r, s>1$, then using Hölder's inequality with $p = \dfrac{r}{(r-1)\gamma'}$ and $q = \dfrac{s}{(s-1)\gamma'}$, we obtain that

$$\int_{-\infty}^{\infty} f_n^{\gamma'(r-1)}(x) g_m^{\gamma'(s-1)}(y-x)\, dx \le$$

$$\le \left[\int_{-\infty}^{\infty} f_n^r(x)\, dx\right]^{1/p} \left[\int_{-\infty}^{\infty} g_m^s(y-x)\, dx\right]^{1/q} \le A_n^{(r-1)\gamma'} B_m^{(s-1)\gamma'}.$$

Hence and from (5), inequality (4) follows.

By (4) we have

$$(6) \qquad C_{n,m} \equiv \int_{-\infty}^{\infty}\left[\int_{-\infty}^{\infty} f_n^r(x) g_m^s(y-x)\, dx\right] dy \le$$

$$\le A_n^{r-1} B_m^{s-1} \int_{-\infty}^{\infty}\left[\int_{-\infty}^{\infty} (f_n(x) g_m(y-x))^\gamma\, dx\right]^{1/\gamma} dy.$$

On the other hand a straightforward computation gives that $C_{n,m}=A_n^r B_m^s$; thus, by (6) and because $f_n(x)\le f(x)$ and $g_m(x)\le g(x)$, the inequality

$$\left[\int_{-\infty}^{\infty} f_n^r(x)\, dx\right]^{1/r}\left[\int_{-\infty}^{\infty} g_m^s(x)\, dx\right]^{1/s} \le \int_{-\infty}^{\infty}\left[\int_{-\infty}^{\infty} (f(x)g(y-x))^\gamma\, dx\right]^{1/\gamma} dy$$

follows. This implies (2) by letting $n\to\infty$ and $m\to\infty$.

If $r=\infty$ (or $s=\infty$), then $\gamma=\infty$ and $s=1$ ($r=1$). Then inequality (2) can be deduced from the finite case by taking the limit $r\to\infty$ and $s\to 1$ ($s\to\infty$ and $r\to 1$) with $\gamma=\infty$. Thus we have completed our proof.

REFERENCES

[1] A. Prékopa, *Logarithmic Concave Measures with Application to Stochastic Programming.* Acta Sci. Math. **32** (1971), 301—316.

[2] W. H. Young, *Sur la généralisation du théorème de Parseval.* C. R. Acad. Sci. Paris Sér. A—B **155** (1912), 30—33.

III.
Approximation in Abstract Spaces

On Some Geometric Properties of the Unit Sphere

By

FRANK DEUTSCH[1]

DEPT. OF MATH.
PENNSYLVANIA STATE UNIVERSITY
PENNSYLVANIA

1. Introduction

The purpose of this note is to provide an elementary exposition of two geometric properties of normed linear spaces: the QP-property which was introduced in [1] and the more general Q-property introduced in [4]. A QP-space is "polyhedral" in the sense of KLEE [6], and it represents a generalization of the unit ball being a "convex polytope" in the sense of MASERICK [8]. A finite-dimensional space is a QP-space if and only if its unit ball is a polyhedron. We will observe that these properties are also related to property (P) of BROWN [3] and hence to the continuity of set-valued metric projections onto finite-dimensional subspaces. For a more comprehensive study of these and other properties see [1], [4], [3], and [2].

Throughout this paper, X will denote a real normed linear space, X^* its dual space, $S(X) = \{x \in X : \|x\| = 1\}$, $B(X) = \{x \in X : \|x\| \leq 1\}$, and for each $x \in X$ and $\varepsilon > 0$, $B(x, \varepsilon) = \{y \in X : \|x - y\| < \varepsilon\}$. Given $x \in X$, we denote the *peak set* of x by $P(x) = \{x^* \in S(X^*) : x^*(x) = \|x\|\}$. The set of extreme points of a set K will be denoted by ext K.

2. The QP-property

DEFINITION [1]. A point $x \in S(X)$ is called a *QP-point* if there exists $\varepsilon > 0$ such that $\|y\| < 1$ whenever $\|x - y\| < \varepsilon$ and $x^*(y) < 1$ for every $x^* \in P(x)$. The space X is called a *QP-space* if each $x \in S(X)$ is a QP-point. (QP is an abbreviation for "quasi-polyhedral.")

Given $x \in S(X)$, define the open cone

$$K(x) = \{y \in X : x^*(y) < 1 \quad \text{for every} \quad x^* \in P(x)\}.$$

(In the notation of [1], $K(x)$ was denoted by $K(x, 0)$.)

[1]) Supported in part by a grant from the National Science Foundation.

2.1 LEMMA. *Let $x \in S(X)$. The following statements are equivalent.*

1) *x is a QP-point.*

2) *There exists $\varepsilon > 0$ such that $K(x) \cap B(x, \varepsilon) \subset B(0, 1)$.*

3) *There exists $\varepsilon > 0$ such that for every two-dimensional subspace Y which contains x, x is the intersection of two line segments in $S(Y)$(which may be collinear) each having length at least ε.*

In particular, the QP-property is hereditary.

Recall that a finite dimensional space X is called *polyhedral* if its unit ball $B(X)$ is the intersection of a finite number of half-spaces. More generally, KLEE [6] calls a normed linear space *polyhedral* if each of its finite-dimensional subspaces is polyhedral.

2.2 THEOREM. *Every QP-space is polyhedral.*

PROOF. By a well-known result of KLEE [5], it suffices to show that every two-dimensional subspace Y is polyhedral. By the equivalence of 1) and 3) of Lemma 2.1 and the compactness of $S(Y)$, one sees that $S(Y)$ consists of a finite number of line segments and hence Y is polyhedral. $\square$

Before showing that the converse is also true if X is finite-dimensional, it is convenient to observe certain equivalent formulations of a QP-point.

2.3 LEMMA. *Let $x \in S(X)$. The following statements are equivalent.*

1) *x is a QP-point*

2) *There exists $\varepsilon > 0$ such that $B(x, \varepsilon) \cap K(x) = B(x, \varepsilon) \cap B(0, 1)$*

3) *There exists $\varepsilon > 0$ such that $B(x, \varepsilon) \cap \overline{K(x)} = B(x, \varepsilon) \cap \overline{B(0, 1)}$*

4) *There exists $\varepsilon > 0$ such that $B(x, \varepsilon) \cap \mathrm{bd}\, K(x) = B(x, \varepsilon) \cap S(X)$* (where $\mathrm{bd}\, K(x)$ *denotes the boundary of* $K(x)$).

PROOF. The equivalence of 1) and 2) is clear.

2)$\Rightarrow$3). Let $B(x, \varepsilon) \cap K(x) = B(x, \varepsilon) \cap B(0, 1)$. Since $K(x) \supset B(0, 1)$, $B(x, \varepsilon) \cap \overline{K(x)} \supset B(x, \varepsilon) \cap \overline{B(0, 1)}$. Now let $y \in B(x, \varepsilon) \cap \overline{K(x)}$. Then there exist $y_n \in B(x, \varepsilon) \cap K(x)$ such that $y_n \to y$. Thus $y \in \overline{B(x, \varepsilon) \cap K(x)} = \overline{B(x, \varepsilon) \cap B(0, 1)} \subset \overline{B(0, 1)}$. Thus $y \in B(x, \varepsilon) \cap \overline{B(0, 1)}$ and $B(x, \varepsilon) \cap \overline{K(x)} \subset B(x, \varepsilon) \cap \overline{B(0, 1)}$.

3)$\Rightarrow$2). If the result were false, there would exist $y \in B(x, \varepsilon) \cap K(x)$ such that $\|y\| \geqq 1$. Then since $B(x, \varepsilon) \cap K(x)$ is open, there exists $\delta > 0$ such that $B(y, \delta) \subset B(x, \varepsilon) \cap K(x)$. Hence $(1 + \delta/2)y \in B(x, \varepsilon) \cap K(x)$ but $\|(1 + \delta/2)y\| > 1$ which contradicts 3).

4) follows by subtracting 2) from 3).

4)$\Rightarrow$1). If 4) holds, then $\varnothing = B(x, \varepsilon) \cap \mathrm{bd}\, K(x) \cap K(x) = B(x, \varepsilon) \cap S(X) \cap K(x)$ so $B(x, \varepsilon) \cap K(x) \subset B(0, 1)$, i.e. x is a QP-point. $\square$

2.4 THEOREM [1]. *Let X be finite-dimensional. Then X is* QP *if and only if X is polyhedral.*

PROOF. Let $B(X) = \bigcap_1^n E_i$, where $E_i = \{y \in X : x_i^*(y) \leqq 1\}$ and $x_i^* \in S(X^*)$. Let $H_i = x_i^{*-1}(1)$. Given any $x \in S(X)$, let $I_0 = \{i : x \in H_i\}$ and let $\varepsilon = \text{dist}(x, \bigcup_{i \notin I_0} H_i)$. Then $\varepsilon > 0$ and $\varepsilon = \min_{i \notin I_0} [1 - x_i^*(x)]$ since $\text{dist}(x, H_i) = 1 - x_i^*(x)$. Now

$$(*) \qquad B(x, \varepsilon) \cap \left(\bigcap_1^n E_i \right) = B(x, \varepsilon) \cap \left(\bigcap_{i \in I_0} E_i \right)$$

for otherwise there would exist $y \in B(x, \varepsilon) \cap (\bigcap_{i \in I_0} E_i) \sim \bigcap_1^n E_i$. Thus $y \notin E_i$ for some $i \notin I_0$ implies $x_i^*(y) > 1$ and

$$\varepsilon \leqq 1 - x_i^*(x) < x_i^*(y) - x_i^*(x) \leqq \|y - x\|$$

which is a contradiction. But $(*)$ is equivalent to

$$B(x, \varepsilon) \cap \overline{B(0, 1)} = B(x, \varepsilon) \cap \overline{K(x)}.$$

Hence x is a QP-point. This shows that X is a QP-space. The converse follows from Theorem 2.2. $\square$

MASERICK [8] has defined a "convex polytope" P as an intersection of a family of half-spaces: $P = \bigcap_{i \in I} E_i$ (corresponding to the hyperplanes $\{H_i : i \in I\}$), such that for every $x \in X$, there is a finite subcollection $I_0 \subset I$ with $x \in \bigcap_{i \notin I_0} E_i$.

2.5 THEOREM [1]. *Let $B(X)$ be a Maserick convex polytope. Then X is a QP-space.*

PROOF. Properties 2.3, 2.4, and 2.5 in [8] assert that if $x \in S(X)$, then $I_0 = \{i \in I : x \in H_i\}$ is a nonempty finite family and $\bigcup_{i \notin I_0} H_i$ is a closed set. Setting $\varepsilon = \text{dist}(x, \bigcup_{i \notin I_0} H_i)$, one observes that exactly the same proof as in Theorem 2.4 shows that x is a QP-point. $\square$

There are QP-spaces X such that $B(X)$ is not a Maserick convex polytope. In fact, if X is the l_1-product of the real numbers with c_0, $X = (R \times c_0)_{l_1}$, then X is a QP-space [1], and the vertex $x = (1; 0, 0, \ldots)$ is an extreme point of $B(X)$. But Maserick convex polytopes in infinite-dimensional spaces have no extreme points [8].

3. The Q-property

DEFINITION. A point $x \in S(X)$ is called a *Q-point* if for each $y \in S(X)$ with $x^*(y) = 1$ for every $x^* \in P(x)$ there exists $\lambda > 0$ such that $\|x - \lambda y\| = 1 - \lambda$. X is called a *Q-space* [4] if each $x \in S(X)$ is a Q-point.

It is well-known and easy to verify (cf. [4]) that the minimal extremal subset of $S(X)$ which contains x is given by

$$E(x) = \{y \in S(X) : x = \lambda y + (1 - \lambda)v \quad \text{where} \quad v \in S(X) \quad \text{and} \quad 0 < \lambda < 1\} =$$

$$= \{y \in S(X) : \|x - \lambda y\| = 1 - \lambda \quad \text{for some} \quad 0 < \lambda < 1\}.$$

For a given $x \in S(X)$, let $Q(x)$ denote the intersection of all exposed sets in $S(X)$ which contain x, i.e.

$$Q(x) = \bigcap_{x^* \in P(x)} \{y \in S(X) : x^*(y) = 1\}.$$

3. 1 LEMMA [4]. *Let $x \in S(X)$. The following statements are equivalent.*
1) *x is a Q-point.*
2) *$E(x) = Q(x)$.*
3) *For each two-dimensional subspace Y which contains x, either x is an exposed point of $S(Y)$ or $x \notin \operatorname{ext} S(Y)$.*

From this result, we see that the Q-property is hereditary. Further, there follows from Lemma 2. 1 (although a simple direct proof could have also been given) the

3. 2 THEOREM. *Every QP-point of $S(X)$ is a Q-point. In particular, [4] every QP-space is a Q-space (but not conversely).*

Every strictly convex space is a Q-space. It is easy to construct examples of two-dimensional Q-spaces which are neither strictly convex nor polyhedral.

It is an interesting fact [4] that in *a Banach space with the Q-property, every (convex) extremal subset E of $S(X)$ is of the form $E = Q(x)$ for some $x \in E$. Also, in a separable Q-space, each extreme point of $S(X)$ is an exposed point.*

4. Property (P)

DEFINITION. A point $x \in X$ is called a *(P)-point* if for each $z \in X$ with $\|x + z\| \leq \|x\|$ There exist positive constants ε, δ such that $\|y + \delta z\| \leq \|y\|$ whenever $\|x - y\| < \varepsilon$. the space X is called a *(P)-space*, or has *property* (P) (cf. BROWN [3]), if each $x \in X$ is a (P)-point.

It is clear that property (P) is hereditary. We now give a useful "normalized" reformulation of property (P).

4. 1 LEMMA. *Let $x \in S(X)$. The following statements are equivalent.*

1) *x is a (P)-point*

2) *For each $z \in S(X)$, there exist positive constants ε, δ such that $\|y+\delta(z-x)\| \leqq$ $\leqq \|y\|$ whenever $\|x-y\| < \varepsilon$*

3) *For each $z \in S(X)$ with $z \neq x$ and $\|\frac{1}{2}(x+z)\| = 1$, there exist positive constants ε, δ such that $\|y+\delta(z-x)\| \leqq \|y\|$ whenever $\|x-y\| < \varepsilon$.*

Moreover, X is a (P)-space if and only if each $x \in S(X)$ is a (P)-point.

PROOF. 1)$\Rightarrow$2). Let $z \in S(X)$. Setting $z' = z-x$, we have that $\|x+z'\| = 1 = \|x\|$ so there exist positive numbers ε, δ such that $\|y+\delta z'\| \leqq \|y\|$ (i. e. $\|y+\delta(z-x)\| \leqq$ $\leqq \|y\|$) whenever $\|x-y\| < \varepsilon$.

2)$\Rightarrow$3). This is obvious.

3)$\Rightarrow$1). Let $z \in X$ with $\|x+z\| \leqq \|x\| = 1$. If $z=0$ any ε, $\delta > 0$ work in the definition of (P)-point. Thus assume $z \neq 0$. Choose $\varrho > 0$ such that $\|x+\varrho z\| = 1$ and set $z' = x+\varrho z$. Then $z' \in S(X)$ and $z' \neq x$.

Case 1: $\|\frac{1}{2}(x+z')\| = 1$.

Then there exist positive numbers ε', δ' such that $\|y+\delta'(z'-x)\| \leqq \|y\|$ (i.e. $\|y+\delta'\varrho z\| \leqq \|y\|$) whenever $\|x-y\| < \varepsilon'$. Take $\delta=\delta'\varrho$ and $\varepsilon=\varepsilon'$.

Case 2: $\|\frac{1}{2}(x+z')\| < 1$.

Let $\varepsilon' = \frac{1}{2}[1-\|\frac{1}{2}(x+z')\|]$. If $\|x-y\| < \varepsilon'$, then

$$\|y+\frac{1}{2}(z'-x)\| = \|y-x+\frac{1}{2}(z'+x)\| \leqq \|y-x\| + \|\frac{1}{2}(z'+x)\| \leqq$$

$$\leqq \|y-x\| + 1 - 2\|x-y\| = 1 - \|x-y\| \leqq \|y\|,$$

i.e. $\|y+\frac{1}{2}\varrho z\| \leqq \|y\|$ whenever $\|x-y\| < \varepsilon'$. Take $\delta=\frac{1}{2}\varrho$ and $\varepsilon=\varepsilon'$.

This proves 1).

Finally, suppose each $x \in S(X)$ is a (P)-point and let $x \in X$. If $x=0$, it is obviously a (P)-point. Thus assume $x \neq 0$. Let $z \in X$ be such that $\|x+z\| \leqq \|x\|$. Then $\left\|\dfrac{x}{\|x\|}+\dfrac{z}{\|x\|}\right\| \leqq 1$. Choose ε', $\delta' > 0$ such that $\left\|\dfrac{y}{\|x\|}+\delta'\dfrac{z}{\|x\|}\right\| \leqq \left\|\dfrac{y}{\|x\|}\right\|$ whenever $\left\|\dfrac{x}{\|x\|}-\dfrac{y}{\|x\|}\right\| < \varepsilon'$. Then $\|y+\delta'z\| < \|y\|$ whenever $\|x-y\| < \varepsilon'\|x\|$. Taking $\delta=\delta'$ and $\varepsilon=\varepsilon'\|x\|$ we get that x is a (P)-point. $\square$

REMARK. It is clear that in the definition of (P)-point and in Lemma 4. 1, one may replace the two constants ε, δ by a single one. (In fact, $\lambda=\min \{\varepsilon, \delta\}$ works.)

For each $x \in X$, let

$$m(x) = \sup \{x^*(x) : X^* \in \text{ext } S(X^*) \sim P(x)\} = \sup \{x^*(x) : x^* \in \text{ext } S(X^*) \sim \text{ext } P(x)\}.$$

Note that $m(x) \leqq \|x\|$.

4.2 THEOREM. *Let $x \in S(X)$. If $m\left(\frac{1}{2}(x+z)\right) < 1$ for each $z \in S(X)$ with $z \neq x$ and $\|\frac{1}{2}(x+z)\| = 1$, then x is a (P)-point.*

PROOF. Let $z \in S(X)$. If $z=x$, any $\varepsilon, \delta > 0$ work in the definition of (P)-point. Thus assume $z \neq x$ and set $\mu = m\left(\frac{1}{2}(x+z)\right)$. Then $\mu < 1$. Set $\varepsilon = \frac{1}{2}(1-\mu)$ and suppose $\|x-y\| < \varepsilon$. If $x^* \in P\left(\frac{1}{2}(x+z)\right)$, then $x^* \in P(x) \cap P(z)$ so

$$x^*[y + \tfrac{1}{2}(z-x)] = x^*(y) \leqq \|y\|.$$

If $x^* \in S(X^*) \sim P\left(\frac{1}{2}(x+z)\right)$, then

$$x^*[y + \tfrac{1}{2}(z-x)] = x^*[y - x + \tfrac{1}{2}(z+x)] \leqq \|y-x\| + \mu < \varepsilon + \mu = 1 - \varepsilon < \|y\|.$$

Thus $\|y + \frac{1}{2}(z-x)\| \leqq \|y\|$. $\square$

4.3 COROLLARY. *Let $x \in S(X)$ be such that $\|\frac{1}{2}(x+z)\| < 1$ for every $z \in S(X)$ with $z \neq x$. Then x is a (P)-point. In particular, [3] every strictly convex space is a (P)-space.*

4.4 COROLLARY. *If $m(x) < 1$ for every $x \in S(X)$, then X is a (P)-space.*

4.5 COROLLARY [3]. *Every finite-dimensional polyhedral space is a (P)-space.*

PROOF. Since X is polyhedral, so is X^* [5] and thus $S(X^*)$ has a finite number of extreme points. Hence $m(x) < 1$ for every $x \in S(X)$ and Corollary 4.4 yields the result. $\square$

Let M be a finite-dimensional subspace of a normed linear space X and for each $x \in X$ let

$$T_M(x) = \{y \in M : \|x-y\| = \text{dist}\,(x, M)\}.$$

The set-valued mapping $T_M : X \to 2^M$ is called the *metric projection*, or best approximation operator, onto M. We say that $T = T_M$ is *continuous* (in the Hausdorff metric) at x if $x_n \to x$ implies

$$\max\left\{ \sup_{y \in T(x)} \text{dist}\,(y, T(x_n)), \quad \sup_{y \in T(x_n)} \text{dist}\,(y, T(x))\right\} \to 0.$$

A *selection* of the metric projection T_M is any function $s : X \to M$ such that $s(x) \in \in T_M(x)$ for every x.

4.6 THEOREM. *The following statements are equivalent.*
1) *X is a (P)-space*
2) *The metric projection onto each finite-dimensional subspace of X is continuous*
3) *The metric projection onto each one-dimensional subspace of X is continuous.*

The equivalence of 1) and 2) was stated by BROWN [3], while the equivalence of 1) and 3) was asserted by BLATTER, MORRIS, and WULBERT [2]. Brown also showed [3] that *in (P)-spaces, the metric projection onto a finite-dimensional subspace has a continuous selection.*

5. The spaces $C_0(T)$ and $L_1(T, \Sigma, \mu)$

Let T be a locally compact Hausdorff space and let $C_0(T)$ denote the space of all real valued continuous functions on T which vanish at infinity and endowed with the supremum norm. For any $x \in C_0(T)$, we denote

$$\operatorname{crit} x = \{t \in T : |x(t)| = \|x\|\}.$$

5.1 LEMMA. *Let $x \in S(C_0(T))$. The following statements are equivalent:*
1) x *is a QP-point*
2) x *is a Q-point*
3) $\operatorname{crit} x$ *is open.*

PROOF. The implication 1)$\Rightarrow$2) follows from Theorem 3.2.
2)$\Rightarrow$3). We have

$$Q(x) = E(x) = \overset{\infty}{\underset{n=1}{\bigcup}} E_{1/n}(x),$$

where

$$E_\lambda(x) = \{y \in S(X) : \|x - \lambda y\| \leq 1 - \lambda\}.$$

By the Baire category theorem, there exists an integer N, an $\varepsilon > 0$, and $z \in E_{1/N}(x)$ such that

$$B(z, \varepsilon) \cap Q(x) \subset E_{1/N}(x).$$

We will first show that $Q(x) \subset E_\lambda(x)$, where $\lambda = \varepsilon/(2+\varepsilon)N$. Let $y \in Q(x)$ and $w = (\varepsilon/(2+\varepsilon))y + (\varepsilon/(2+\varepsilon))z \in Q(x)$. Then $\|w - z\| = (\varepsilon/(2+\varepsilon))\|y - z\| < \varepsilon$ and so $w \in E_{1/N}(x)$. Hence

$$\|x - \lambda y\| \leq \left\|x - \frac{1}{N}w\right\| + \frac{2}{(2+\varepsilon)N} \leq 1 - \frac{1}{N} + \frac{2}{(2+\varepsilon)N} = 1 - \lambda$$

so $y \in E_\lambda(x)$. This shows $Q(x) \subset E_\lambda(x)$.

Now suppose crit x is not open. Then there exists $t_0 \notin \operatorname{crit} x$ such that $|x(t_0)| > 1 - \lambda$. Choose a neighbourhood U of crit x which misses t_0. By Urysohn's lemma, there is $f \in C_0(T)$ such that $0 \leq f \leq 1, f = 1$ on crit x, and $f = 0$ off U. Setting $y = xf$, one sees that $y \in S(X)$ and $y = x$ on crit x, i.e. $y \in Q(x)$. But

$$|x(t_0) - \lambda y(t_0)| = |x(t_0)| > 1 - \lambda$$

implies $y \notin E_\lambda(x)$, which is a contradiction.
3)$\Rightarrow$1). Since crit x is open, there is $\varepsilon > 0$ such that

$$\sup \{|x(t)| : t \notin \operatorname{crit} x\} = 1 - 2\varepsilon.$$

If $x^*(y)<1$ for every $x^*\in P(x)$ and $\|x-y\| < \varepsilon$, then $|y(t)|<1$ for all $t\in$ crit x. Also, if $t\notin$ crit x, then

$$|y(t)| \leqq |y(t)-x(t)|+|x(t)| < \varepsilon+1-2\varepsilon < 1.$$

Hence $\|y\|<1$. We've shown that

$$B(x, \varepsilon)\cap K(x) \subset B(0, 1),$$

i.e. x is a QP-point. $\square$

The equivalence of statements 1), 2), and 4) of the following theorem is now readily deduced from Lemma 5. 1.

5. 2 THEOREM. *The following statements are equivalent.*
1) $C_0(T)$ *is a QP-space.*
2) $C_0(T)$ *is a Q-space.*
3) $C_0(T)$ *is a (P)-space.*
4) T *is discrete.*

The equivalence of 1) and 4) (resp. 2) and 4), 3) and 4)) was first established in [1] (resp. [4], [2]).

For the final results, let (T, Σ, μ) be a σ-finite measure space and $L_1=L_1(T, \Sigma, \mu)$ be the space of (equivalence classes of) real valued integrable functions with the norm

$$\|x\| = \int_T |x|\, d\mu.$$

For any $x\in L_1$, we define the *support* of x, modulo a set of measure zero, by

$$\operatorname{supp} x=\{t\in T : x(t)\neq 0\}.$$

Recall that an *atom* is a set $A\in\Sigma$ such that $0<\mu A<\infty$ and if $B\subset A$, either $\mu B=0$ or $\mu B=\mu A$.

5. 3 LEMMA. *Let* $x\in S(L_1)$. *The following statements are equivalent:*
1) x *is a QP-point.*
2) x *is a Q-point.*
3) $\operatorname{supp} x$ *is a union of finitely many atoms.*

PROOF. The implication 1)$\Rightarrow$2) is a consequence of Theorem 3. 2.

2)$\Rightarrow$3). It is a routine exercise to deduce from the definitions that $y\in Q(x)$ if and only if $\operatorname{sgn} y=\operatorname{sgn} x$ a.e. on $\operatorname{supp} y$; and $y\in E(x)$ if and only if $y\in Q(x)$ and there exists $\lambda>0$ such that $\lambda|y|\leqq|x|$ a.e. If $\operatorname{supp} x$ were not a finite union of atoms, there would exist a disjoint sequence of sets (A_n) in $\operatorname{supp} x$ each having finite positive measure. Since

$$\sum_1^\infty \int_{A_n} |x|\, d\mu \leqq \int_T |x|\, d\mu = 1,$$

194   F. DEUTSCH

it follows that $\int_{A_n} |x|\, d\mu \to 0$. By considering only a subsequence of (A_n) if necessary, we may assume that

$$\int_{A_n} |x|\, d\mu < \frac{1}{n^3} \quad \text{for every} \quad n.$$

Define $y_1 = \sum_1^\infty n x \chi_{A_n}$. Then

$$\int_T |y_1|\, d\mu = \sum_1^\infty n \int_{A_n} |x|\, d\mu < \sum_1^\infty \frac{1}{n^2} < \infty$$

so $y_1 \in L_1$. Set $y = (y_1/\|y_1\|) \in S(L_1)$. Then $\operatorname{sgn} y = \operatorname{sgn} x$ on $\operatorname{supp} y$ so $y \in Q(x)$. Given any $\lambda > 0$, choose n so that $n > \frac{1}{\lambda} \|y_1\|$. Then, on A_n, we have

$$\lambda |y| = \lambda \frac{|y_1|}{\|y_1\|} = \frac{\lambda n}{\|y_1\|} |x| > |x|$$

so $y \notin E(x) = Q(x)$, which is a contradiction.

$3) \Rightarrow 1)$. Let $\operatorname{supp} x = \{A_1, \dots, A_n\}$, where the A_i are disjoint atoms. Let $x(A_i)$ denote the constant value which x assumes a.e. on A_i. Set $\varepsilon = \min\{|x(A_i)|\mu A_i : 1 \leq i \leq n\} > 0$. If $y \in L_1$ and $\|x - y\| < \varepsilon$, then $\operatorname{sgn} y = \operatorname{sgn} x$ a.e. on $\operatorname{supp} x$. If also $x^*(y) < 1$ for every $x^* \in P(x)$, then

$$\|y\| = \int_{\operatorname{supp} x} y \operatorname{sgn} x\, d\mu + \int_{T \sim \operatorname{supp} x} |y|\, d\mu < 1.$$

This shows that x is a QP-point. $\square$

The equivalence of statements 1), 2), and 4) in the next theorem are an immediate consequence of this lemma.

5. 4 Theorem. *The following statements are equivalent.*
1) L_1 *is a QP-space.*
2) L_1 *is a Q-space.*
3) L_1 *is a (P)-space.*
4) L_1 *is finite-dimensional.*

The equivalence of 1) and 4) (resp. 2) and 4), 3) and 4)) was first established in [1] (resp. [4], [2]).

We should mention that various *products* of spaces with the QP, Q, or (P) property also have this property (see [1], [4], and [2]).

Recently Wegmann [9] has studied the peak-point mapping $x \to P(x)$. Among other things, he characterized the QP and P-properties in terms of this map and proved that every QP-space is a P-space.

An open problem. Is the converse to Theorem 2. 2 true? That is, is the concept of a polyhedral space identical with that of a QP-space? Related to this, LINDEN- STRAUSS has constructed an interesting example of a polyhedral space whose unit ball has infinitely many extreme points [7; p. 242]. It is easy to verify that his example is also a QP-space. Hence there exists a QP-space whose unit ball has infinitely many extreme points.

REFERENCES

[1] D. Amir and F. Deutsch, *Suns, moons, and quasi-polyhedra* (to appear in J. Approximation Theory).

[2] J. Blatter, P. Morris, and D. Wulbert, *Continuity of the set-valued metric projection.* Math. Ann. **178** (1968), 12—24.

[3] A. L. Brown, *Best n-dimensional approximation to sets of functions.* Proc. London Math. Soc. (3) **14** (1964), 577—594.

[4] F. Deutsch and R. Lindahl, *Minimal extremal subsets of the unit sphere* (to appear in Math. Ann.)

[5] V. Klee, *Some characterizations of convex polyhedra.* Acta Math. **102** (1959), 79—107.

[6] V. Klee, *Polyhedral sections of convex bodies.* Acta Math. **103** (1960), 243—267.

[7] J. Lindenstrauss, *Notes on Klee's paper "Polyhedral sections of convex bodies".* Israel J. Math. **4** (1966), 235—242.

[8] P. H. Maserick, *Convex polytopes in linear spaces.* Illinois J. Math. **9** (1965), 623—635.

[9] R. Wegmann, *Some properties of the peak-set-mapping* (to appear in J. Approximation Theory).

Diskrete Konvergenz linearer Operatoren III

Von

FRIEDRICH STUMMEL

MATHEMATISCHES SEMINAR DER UNIVERSITÄT
FRANKFURT

Einleitung

Mit Teil III dieser Arbeit soll jetzt die Theorie der diskreten Konvergenz linearer Operatoren auf eine Reihe bekannter Probleme angewandt werden. In § 1 behandeln wir Folgen linearer Abbildungen eines normierten Raumes E in einen normierten Raum F und die zugehörige Störungstheorie. Ausgangspunkt ist hierzu Abschnitt I-3. 3 über die diskrete Konvergenz linearer Operatoren, die in diesem Rahmen auf einen verallgemeinerten Begriff der starken Operatorkonvergenz führt. Damit können wir zunächst die Ergebnisse von ATKINSON [4] und ANSELONE-PALMER [3] über kollektiv kompakte Folgen von Operatoren aus unserer Theorie erhalten. Weiter zeigt sich in Abschnitt 1. 2, daß unsere Begriffsbildungen in natürlicher Weise die Theorie der verallgemeinerten starken Konvergenz in KATO [16], VIII-§ 1 ergeben. Darüber hinaus erhalten wir über [16] hinausgehende Resultate für Folgen abgeschlossener Operatoren mit vollstetiger Resolvente, die der Kompaktheitsbedingung 1. 2. (20) genügen.

Die Projektionsmethode enthält eine Reihe klassischer Näherungsmethoden der Analysis wie z. B. die Verfahren von RITZ und GALERKIN und läßt sich sehr einfach in die Theorie der diskreten Approximation normierter Räume einordnen. Abschnitt 2. 1 spezialisiert Sätze aus Teil I, II dieser Arbeit für diesen Fall. Die Anwendung von Satz II-3. 2. (8) ergibt hier z. B. die Konvergenz von Eigenwerten und zugehörigen Hauptvektoren der algebraischen Vielfachheit nach für Approximationen des allgemeinen, nicht notwendig symmetrischen Eigenwertproblems $Au = \lambda Bu$ in normierten Räumen. Abschnitt 2. 2 ordnet einige spezielle Anwendungen in diesen Rahmen ein. Unter anderem erhalten wir in Beispiel 3 die Ergebnisse von GRIGORIEFF [12] für das GALERKIN-Verfahren. Ebenso folgen die entsprechenden Ergebnisse von PETRYSHYN [20, 22] aus unserer Theorie, was wir hier allerdings nicht näher ausführen.

Die Quadraturformelmethode für Integraloperatoren in § 3 verallgemeinert Ergebnisse von ANSELONE [2], BRAKHAGE [7], [8] u. a. Hier können wir die in Abschnitt I-5. 2.1 eingeführten diskreten Approximationen von Räumen stetiger Funktionen verwenden. Gleichzeitig erhält man damit einfache konkrete Beispiele für Satz I-3. 1. (4) bzw. Satz II-2. 2. (1) und die Sätze aus Abschnitt II-3. 2.

1. Verallgemeinerte starke Konvergenz linearer Operatoren

Als erstes wollen wir die Theorie der diskreten Konvergenz auf die Störungstheorie linearer Abbildungen eines normierten Raumes E in einen normierten Raum F anwenden. Ausgangspunkt ist Abschnitt I-3. 3 über die diskrete Konvergenz unbeschränkter Operatoren, der durch die Bedingung (d) den verallgemeinerten Konvergenzbegriff $T_\iota \xrightarrow{d} T$ ($\iota \in I$) für Folgen linearer Abbildungen T, T_ι, $\iota \in I$, ergibt. Unabhängig davon läßt sich dieser neue Konvergenzbegriff in Satz (3) durch die Graphen der Operatoren T, T_ι charakterisieren. Satz (14) wird dann die Äquivalenz der verallgemeinerten starken Konvergenz $T_\iota \xrightarrow{s} T$ im Sinne von KATO mit den beiden Bedingungen $T_\iota \xrightarrow{d} T$ und $\Delta_b \cap P(T) \neq \emptyset$ zeigen.

1. 1. *Definition und Eigenschaften.* Seien E, F normierte Räume über demselben Körper $\mathbf{K}=\mathbf{R}$ bzw. $\mathbf{K}=\mathbf{C}$, sei $I=I_0$ eine abzählbar unendliche Indexfolge und seien T, T_ι lineare Abbildungen von E in F mit Definitionsbereich $D(T)$, $D(T_\iota)$ aus E und Bildbereich $R(T)$, $R(T_\iota)$ aus F für $\iota \in I$. Damit definieren wir den Begriff der *verallgemeinerten starken Konvergenz* dieser Operatoren folgendermaßen.

(1) *Die Operatoren T_ι, $\iota \in I$, konvergieren im verallgemeinerten Sinne stark gegen T, abgekürzt $T_\iota \xrightarrow{d} T$ ($\iota \in I$), wenn für jedes $u \in D(T)$ eine Folge von Elementen $u_\iota \in D(T_\iota)$ existiert mit der Eigenschaft*

$$(2) \qquad \|u_\iota - u\|_E + \|T_\iota u_\iota - Tu\|_F \to 0 \qquad (\iota \in I).$$

Die verallgemeinerte starke Konvergenz einer Folge linearer Abbildungen von E in F läßt sich in einfacher Weise mit Hilfe der Graphen dieser Operatoren charakterisieren. Seien also $G(T)$, $G(T_\iota)$ in $E \times F$ die Graphen der Operatoren T, T_ι für $\iota \in I$. Diese Graphen sind lineare Teilräume des Produktraumes $E \times F$, den wir in der üblichen Weise als normierten Raum auffassen. Dann hat man mit dem Begriff der diskreten Approximation durch Teilräume aus Abschnitt I-5. 1. 2 die folgende Charakterisierung.

(3) *Genau dann konvergiert $T_\iota \xrightarrow{d} T$ ($\iota \in I$), wenn die Folge der Graphen $G(T_\iota)$, $\iota \in I$, eine diskrete Approximation des Graphen $G(T)$ in $E \times F$ ist.*

BEWEIS. Die Teilräume $G(T_\iota) \subseteq E \times F$, $\iota \in I$, bilden genau dann eine diskrete Approximation von $G(T)$ in $E \times F$, wenn zu jedem Element $\{u, Tu\} \in G(T)$ eine Folge von Elementen $\{u_\iota, T_\iota u_\iota\} \in G(T_\iota)$, $\iota \in I$, existiert mit der Eigenschaft

$$\|u_\iota - u\|_E + \|T_\iota u_\iota - Tu\|_F \to 0 \qquad (\iota \in I).$$

Dies ist aber gerade die Bedingung (2) der verallgemeinerten starken Konvergenz $T_\iota \xrightarrow{d} T$ ($\iota \in I$).

Im Anschluß an Abschnitt I-3. 3 bezeichnen wir wieder mit D den durch die Graphennorm

$$\|u\|_D = \|u\|_E + \|Tu\|_F \qquad (u \in D(T))$$

normierten Definitionsbereich $D(T) \subseteq E$. Entsprechend seien die normierten Räume D_ι mit den Definitionsbereichen $D(T_\iota) \subseteq E$ und den Operatoren $T_\iota : E \to F$ für $\iota \in I$ erklärt. Die obige Definition der verallgemeinerten starken Konvergenz gestattet dann weiter den folgenden Satz.

(4) *Die Bedingung $T_\iota \xrightarrow{d} T$ ($\iota \in I$) ist notwendig und hinreichend dafür, daß die normierten Definitionsbereiche D, D_ι eine diskrete Approximation $\mathscr{A}(D, \Pi D_\iota, R^D)$ bilden mit*

$$(5) \qquad R^D(u) = \{(u_\iota)_{\iota \in I} \,|\, u_\iota \in D(T_\iota),\ \|u_\iota - u\|_E + \|T_\iota u_\iota - Tu\|_F \to 0\ (\iota \in I)\}.$$

BEWEIS. Unter der Voraussetzung $T_\iota \xrightarrow{d} T$ ($\iota \in I$) wird offenbar durch die Vorschrift (5) jedem Element $u \in D = D(T)$ eine nichtleere Klasse von Folgen aus ΠD_ι zugeordnet. Neben der Bedingung (A1) sind dabei auch die Bedingungen (A2), (A3) erfüllt, so daß die diskrete Approximation $\mathscr{A}(D, \Pi D_\iota, R^D)$ im Sinne von Abschnitt I-1. 1 erklärt ist. Wenn umgekehrt die normierten Definitionsbereiche D, D_ι, $\iota \in I$, eine diskrete Approximation mit der Abbildung R^D aus (5) bilden, dann existiert auf Grund der Bedingung (A1) notwendig zu jedem $u \in D = D(T)$ eine Folge von Elementen mit der in (2) und (5) genannten Eigenschaft.

Die diskrete Konvergenz einer Folge von Elementen in der diskreten Approximation $\mathscr{A}(D, \Pi D_\iota, R^D)$ is gleichbedeutend mit der Konvergenzaussage

$$(6) \qquad u_\iota \to u \text{ in } (D_\iota \to D) \Leftrightarrow u_\iota \to u \text{ in } E,\quad T_\iota u_\iota \to Tu \text{ in } F \qquad (\iota \in I).$$

In Abschnitt I-3. 3 wurde die Bedingung (d) eingeführt. Hiermit hat man weiter die Äquivalenz

$$(7) \qquad\qquad\qquad (d) \Leftrightarrow T_\iota \xrightarrow{d} T \qquad (\iota \in I).$$

Somit gelten im Rahmen der verallgemeinerten starken Konvergenz die Sätze (13), (16) aus Abschnitt I-3. 3 mit der Voraussetzung $T_\iota \xrightarrow{d} T$ ($\iota \in I$) anstelle von (d). Dabei ist in Satz (13) die diskrete Konvergenz in $(E_\iota \to E)$ bzw. in $(F_\iota \to F)$ gleich der starken Konvergenz in E bzw. in F. Das folgende hinreichende Kriterium verallgemeinert den entsprechenden Satz 1. 5 in KATO [16], VIII-§ 1. 1 auf den hier betrachteten Fall der verallgemeinerten starken Konvergenz.

(8) *Es konvergiert $T_\iota \xrightarrow{d} T$ ($\iota \in I$), wenn ein dichter Teilraum $\Phi \subseteq D$ existiert, so daß es zu jedem Element $\varphi \in \Phi$ ein Endstück $I_1 = I_1(\varphi)$ gibt mit der Eigenschaft*

$$(9) \qquad\qquad\qquad \varphi \in D(T_\iota),\quad T_\iota \varphi \to T\varphi \qquad (\iota \in I_1).$$

BEWEIS. Unter diesen Bedingungen kann man durch die Vorschrift

$$(10) \qquad\qquad r_\iota^E \varphi = \varphi,\quad \iota \in I_1(\varphi),\quad r_\iota^E \varphi = 0,\quad \iota \in I - I_1(\varphi),$$

eine Folge von Restriktionsoperatoren $r_\iota^E \colon \Phi \to E_\iota$ erklären. Für diese Abbildungen r_ι erhält man unmittelbar die Eigenschaften I-4. 1. (1), (2). Weiter gilt für jedes $\varphi \in \Phi$ die Beziehung

$$r_\iota^E \varphi \to \varphi \ \text{ in } E, \quad T_\iota r_\iota^E \varphi \to T\varphi \ \text{ in } F \qquad (\iota \in I).$$

Damit sind die Voraussetzungen von Satz I-3. 3. (11) erfüllt, welcher die Bedingung (d) ergibt, was nach (7) äquivalent zur Konvergenz $T_\iota \xrightarrow{\ d\ } T$ ($\iota \in I$) ist.

Anschließend betrachten wir noch zwei Beispiele, die eine Reihe von bekannten Resultaten in unsere Theorie einordnen.

BEISPIEL 1. Mit diesem Beispiel erhält man aus unserer Theorie die Ergebnisse von ANSELONE [2], ATKINSON [4] u. a. über die Konvergenz der Quadraturformelmethode. Sei $E = F$ ein normierter Raum, der nicht vollständig zu sein braucht, und $T, T_\iota \in \mathscr{B}(E)$ für $\iota \in \mathbf{N} = I$ eine Folge von Operatoren. Wenn die Folge $(T_\iota)_{\iota \in \mathbf{N}}$ beschränkt ist, dann wird offenbar die verallgemeinerte starke Konvergenz (1) äquivalent zur üblichen Konvergenz

$$T_\iota \xrightarrow{\ d\ } T \Leftrightarrow T_\iota u \to Tu, \quad u \in E \qquad (\iota \to \infty).$$

In den obengenannten Arbeiten ist die Folge $(T_\iota)_{\iota \in \mathbf{N}}$ diskret kompakt, und die Operatoren T, T_ι, $\iota \in \mathbf{N}$, sind vollstetig. Damit können wir unsere Theorie anwenden mit $E = F = E_\iota = F_\iota$, $\iota \in \mathbf{N}$, und der starken Konvergenz in E als diskreter Konvergenz (s. I-1. 1. (3)). Weiter setzen wir $A = A_\iota = 1$ und $B = T$, $B_\iota = T_\iota$, $\iota \in \mathbf{N}$. In diesem Falle existieren für $z = 0$ trivialerweise die Operatoren $A(z)^{-1} = 1$, $A_\iota(z)^{-1} = 1$, $\iota \in \mathbf{N}$, so daß $z = 0 \in \varDelta_b \cap \mathsf{P}(A, B)$ ist. Weiter hat man die Konvergenz $A_\iota \to A$, $B_\iota \to B$ ($\iota \to \infty$), also die Stabilität und Konsistenz dieser Operatorfolgen. Hiermit sind die Voraussetzungen von Satz II-2. 2. (1) erfüllt, der die Bedingungen ergibt

$$\varDelta_s = \varDelta_b = \mathsf{P}(1, T), \quad (1 - zT_\iota)^{-1} \to (1 - zT)^{-1} \qquad (\iota \to \infty), \ z \in \mathsf{P}(1, T).$$

Entsprechend sind für $\mathbf{K} = \mathbf{C}$ die Voraussetzungen von Satz II-3. 2. (1) erfüllt, so daß die Sätze aus Abschnitt II-3. 2, II-3. 3 über die Konvergenz von Eigenwerten und zugehörigen Eigenräumen anwendbar sind. Wenn es eine Abbildung J von E in E' gibt mit der Eigenschaft

$$\langle Tu, Jv \rangle = \overline{\langle Tv, Ju \rangle}, \quad \langle u, Jv \rangle = \overline{\langle v, Ju \rangle} \qquad (u, v \in E),$$

und $\langle u, Ju \rangle > 0$ für $0 \neq u \in E$ ist, dann wird das Paar 1, T trivialerweise J-definit für $\alpha = 1$, $\beta = 0$ (s. II-1. 3. (18)). Übrigens wird in diesem Falle durch $\langle u, Jv \rangle$ ein Skalarprodukt für E erklärt.

BEISPIEL 2. Mit etwas anderen Bedingungen können wir die Arbeit von ANSELONE-PALMER [3] in unsere Theorie einordnen. Dazu sei jetzt E ein Banachscher Raum und $T, T_\iota \in \mathscr{B}(E)$, $\iota \in \mathbf{N}$, eine Folge von Operatoren mit den Eigenschaften: $T_\iota u \to Tu$

($\iota \to \infty$) für $u \in E$, die Folge $(T_\iota - T)_{\iota \in \mathbf{N}}$ sei diskret kompakt und die Operatoren $T_\iota - T$, $\iota \in \mathbf{N}$, seien vollstetig. Damit setzen wir $A = T$, $B = 1$ und $A_\iota = T_\iota$, $B_\iota = 1$, $\iota \in \mathbf{N}$. Also konvergiert $A_\iota \to A$, $B_\iota \to B$ ($\iota \to \infty$), so daß Stabilität und Konsistenz der Folgen gegeben sind. Weiter hat man für jede Zahl $z \in \mathsf{P}(A, B) = \mathsf{P}(T)$ die Darstellung

$$T_\iota - z1 = T - z1 - (T - T_\iota) = (1 - K_\iota)(T - z1), \quad K_\iota = (T - T_\iota)(T - z1)^{-1} \qquad (\iota \in \mathbf{N}).$$

Nach Voraussetzung sind die Operatoren K_ι, $\iota \in \mathbf{N}$, vollstetig und die Folge $(K_\iota)_{\iota \in \mathbf{N}}$ ist diskret kompakt. Weiter konvergiert $K_\iota \to 0$ ($\iota \to \infty$). Mit Hilfe von Satz I-3. 1. (4) für $K = 0$, $\lambda = 1$ folgt dann die Existenz eines Endstücks $\mathbf{N}_1$ von $\mathbf{N}$ sowie die Existenz und Konvergenz der Inversen

$$(T_\iota - z1)^{-1} = (T - z1)^{-1}(1 - K_\iota)^{-1} \to (T - z1)^{-1} \qquad (\iota \to \infty), \ z \in \mathsf{P}(T).$$

Folglich ist $\mathsf{P}(T) \subseteq \Delta_b$ und $\mathsf{P}(T) = \Delta_b \cap \mathsf{P}(T)$. Damit sind die Voraussetzungen von Satz II-2. 1. (3) und Satz II-2. 3. (4) erfüllt. Für jeden isolierten Teil des Spektrums $\Sigma(A, B) = \Sigma(T)$ und ein zugehöriges Cauchygebiet Δ mit $\Gamma = \partial \Delta \subseteq \mathsf{P}(T)$ ist hier $\Gamma = \partial \Delta \subseteq \Delta_b$, also $\Gamma \subseteq \mathsf{P}(T_\iota)$ und $\Sigma(T_\iota) \cap \Delta$ ein isolierter Teil des Spektrums von T_ι für fast alle $\iota \in \mathbf{N}$. Also konvergieren die zugehörigen Projektionsoperatoren II-2. 3. (10) und reduzierten Operatoren II-2. 3. (11),

$$P_\iota \to P, \quad T_\iota P_\iota \to TP \qquad (\iota \to \infty).$$

Schließlich kann man noch zeigen, daß die Operatoren $P_\iota - P$, $T_\iota P_\iota - TP$ für $\iota \in \mathbf{N}_1$ vollstetig und die zugehörigen Folgen diskret kompakt sind.

1. 2. *Verallgemeinerte starke Konvergenz im Sinne von* KATO. In Anlehnung an KATO [16], Kapitel VIII, beschränken wir uns jetzt auf den Fall eines normierten Raumes F über dem Körper $\mathbf{K} = \mathbf{C}$ der komplexen Zahlen. Über die Operatoren T, T_ι in F mit Definitionsbereichen $D(T), D(T_\iota) \subseteq F$ für $\iota \in I$ setzen wir voraus: $T, T_\iota, \iota \in I$, abgeschlossen und F vollständig oder die Operatoren $T, T_\iota, \iota \in I$, besitzen eine vollstetige Resolvente. Die Resolventenmenge $\mathsf{P}(T)$ besteht hier aus denjenigen Zahlen $z \in \mathbf{C}$, für die $T(z)^{-1} = (T - z1)^{-1}$ in $\mathscr{B}(F)$ existiert. Entsprechend sind die Resolventenmengen $\mathsf{P}(T_\iota)$, $\iota \in I$, erklärt. Das Beschränktheitsgebiet Δ_b besteht aus den Zahlen $z \in \mathbf{C}$, für die ein Endstück I_b der Folge I und eine Zahl $\gamma > 0$ existieren mit

$$(11) \qquad \Delta_b: \quad z \in \mathsf{P}(T_\iota), \quad \|T_\iota(z)^{-1}\| \leqq \frac{1}{\gamma} \qquad (\iota \in I_b),$$

und entsprechend das Konvergenzgebiet Δ_s aus den Zahlen $z \in \mathbf{C}$, für die ein Endstück I_s von I und ein Operator L existieren mit

$$(12) \qquad \Delta_s: \quad z \in \mathsf{P}(T_\iota), \quad T_\iota(z)^{-1} \to L \qquad (\iota \in I_s).$$

Hier bedeutet $T_\iota(z)^{-1} \to L$, daß die Folge $\left(T_\iota(z)^{-1}\right)_{\iota \in I_s}$ beschränkt ist und $T_\iota(z)^{-1} w \to$ $\to Lw$ in F konvergiert für jedes $w \in F$. Mit diesen Bezeichnungen wird in [16], VIII-§ 1. 1 die folgende Definition der verallgemeinerten starken Konvergenz $T_\iota \underset{s}{\to} T$ $(\iota \in I)$ gegeben.

(13) *Es konvergiert* $T_\iota \underset{s}{\to} T$ $(\iota \in I)$, *wenn die Menge* $\Delta_b \cap \mathsf{P}(T)$ *nicht leer ist und für ein* $z \in \Delta_b \cap \mathsf{P}(T)$ *gilt*

$$T_\iota(z)^{-1} \to T(z)^{-1} \qquad (\iota \in I_s).$$

Wir zeigen jetzt, daß diese verallgemeinerte starke Konvergenz $T_\iota \to T$ $(\iota \in I)$ ein spezieller Fall der von uns in Abschnitt 1. 1 eingeführten verallgemeinerten starken Konvergenz $T_\iota \underset{d}{\to} T$ $(\iota \in I)$ ist. Genauer gilt der folgende Satz.

(14) *Die beiden anschließenden Bedingungen sind äquivalent*

$$T_\iota \underset{s}{\to} T \Leftrightarrow \Delta_b \cap \mathsf{P}(T) \neq \emptyset \quad und \quad T_\iota \underset{d}{\to} T \qquad (\iota \in I).$$

BEWEIS. (i) Mit $T_\iota \underset{s}{\to} T$ $(\iota \in I)$ hat man eine Zahl $z \in \Delta_b \cap \mathsf{P}(T)$, so daß die inversen Operatoren $T_\iota(z)^{-1} \to T(z)^{-1}$ $(\iota \in I)$ konvergieren. Damit kann man zu jedem $u \in D(T)$ die Folge $u_\iota = T_\iota(z)^{-1} T(z) u \in D(T_\iota)$, $\iota \in I_s$, erklären. Für diese Folge konvergiert dann $u_\iota \to u$ sowie

$$T_\iota u_\iota = T(z)u + z u_\iota \to T(z)u + zu = Tu \qquad (\iota \in I_s).$$

Folglich strebt $T_\iota \underset{d}{\to} T$ $(\iota \in I)$ gemäß Definition (1).

(ii) Mit $T_\iota \underset{d}{\to} T$ $(\iota \in I)$ gilt für jede Zahl $z \in \mathbf{C}$ auch $T_\iota(z) \underset{d}{\to} T(z)$ $(\iota \in I)$. Weiter ist nach (7) damit die Bedingung (d) aus Abschnitt I-3. 3 erfüllt. Für eine Zahl $z \in \Delta_b \cap$ $\cap \mathsf{P}(T)$ wird der Operator $T(z)$ surjektiv und $\left(T_\iota(z)\right)_{\iota \in I}$ eine Folge stetig invertierbarer bijektiver Operatoren, die der Stabilitätsbedingung $\gamma \|u\| \leq \|T_\iota(z)u\|$ für $u \in D(T_\iota)$, $\iota \in I$, genügt. Mit Satz I-3. 3. (16) für $E = F = E_\iota = F_\iota$, $\iota \in I$, folgt dann noch die Konvergenz von $T_\iota(z)^{-1} w \to T(z)^{-1} w$ in F für jedes $w \in F$, also die starke Konvergenz im Sinne von (13).

Die Sätze dieser Arbeit lassen sich nun auf stark konvergente Folgen $T_\iota \underset{d}{\to} T$ $(\iota \in I)$ in der folgenden Weise anwenden. Man setzt $F = F_\iota$, $\iota \in I$, und für die diskrete Konvergenz in $\mathscr{A}(F, \Pi F_\iota, R^F)$ die starke Konvergenz in F. Weiter seien D, D_ι die mit der Graphennorm versehenen Definitionsbereiche $D(T)$, $D(T_\iota)$ für $\iota \in I$. Nach Satz (4) ist so eine diskrete Approximation $\mathscr{A}(D, \Pi D_\iota, R^D)$ erklärt mit der diskreten Konvergenz (5) bzw. (6) für $E = F$. Wie in Abschnitt I-3. 3 ausgeführt wurde, definieren die Operatoren T, T_ι dann beschränkte lineare Abbildungen $A \in \mathscr{B}(D, F)$, $A_\iota \in \mathscr{B}(D_\iota, F)$, $\iota \in I$. Schließlich setzen wir $B = J$, $B_\iota = J_\iota$ mit den Operatoren J, J_ι der natürlichen Einbettung von D, D_ι in F für $\iota \in I$. Auf Grund von (7) gilt dann mit der Bedingung I-3. 3. (d) die diskrete Konvergenz $A_\iota \to A$, $J_\iota \to J$ $(\iota \in I)$. Diese Operatoren genügen daher immer der Stabilitäts- und Konsistenzbedingung. Für die zugehörigen Resolventenmengen hat man offenbar $\mathsf{P}(A, J) = \mathsf{P}(T)$, $\mathsf{P}(A_\iota, J_\iota) = \mathsf{P}(T_\iota)$,

$\iota \in I$, und entsprechende Gleichungen gelten für die Spektren, das Beschränktheitsgebiet und das Konvergenzgebiet.

Wenn die Operatoren T, T_ι abgeschlossen sind und F vollständig ist, dann sind D, D_ι und $F = F_\iota$, $\iota \in I$, Banachsche Räume. Andernfalls sollen die Operatoren T, T_ι, $\iota \in I$, eine vollstetige Resolvente besitzen. In diesem Falle sind auch die Operatoren J, J_ι der natürlichen Einbettung von D in F, D_ι in F, $\iota \in I$, vollstetig, wie man leicht zeigt.

Nach diesen Vorbereitungen können wir die Sätze dieser Arbeit auf den hier betrachteten Fall linearer Operatoren übertragen. Zum Beispiel ergibt Satz II-2. 1. (5) mit Hilfe von (13), (14) jetzt das folgende Ergebnis.

(15) *Wenn die Menge $\Delta_b \cap \mathsf{P}(T)$ nicht leer ist, dann wird die Konvergenz $T_\iota \underset{d}{\rightarrow} T$ ($\iota \in I$) äquivalent zu den Bedingungen*

$$(16) \qquad \Delta_s = \Delta_b \cap \mathsf{P}(T), \quad T_\iota(z)^{-1} \rightarrow T(z)^{-1} \quad (\iota \in I_s),\ z \in \Delta_s.$$

Weiter lassen sich die Sätze aus Abschnitt II-2. 3 über die Existenz von Resolventenintegralen und Projektionsoperatoren zu isolierten Teilen der Spektren anwenden. Sei also Δ ein Cauchygebiet mit der Eigenschaft $\Gamma = \partial \Delta \subseteq \mathsf{P}(T) \cap \Delta_b$. Zu den isolierten Teilen $\Sigma(T) \cap \Delta$, $\Sigma(T_\iota) \cap \Delta$ der Spektren von T, T_ι haben dann die zugehörigen Projektionsoperatoren Q, $Q_\iota \in \mathscr{B}(F)$ die Darstellung

$$(17) \qquad Q = -\frac{1}{2\pi i} \int_\Gamma T(z)^{-1}\, dz, \quad Q_\iota = -\frac{1}{2\pi i} \int_\Gamma T_\iota(z)^{-1}\, dz \quad (\iota \in I_1).$$

Die Operatoren C, C_ι in II-2. 3. (8) sind hier von der gleichen Gestalt wie Q, Q_ι. Der Unterschied besteht nur darin, daß im ersten Fall diese Operatoren aufgefaßt werden als Abbildungen von F nach $D = D(T)$, $D_\iota = D(T_\iota)$ und im zweiten Fall als Abbildungen in F. Damit haben wir hier den Satz

(18) *Es konvergiere $T_\iota \underset{d}{\rightarrow} T$ ($\iota \in I$) und es sei Δ ein Cauchygebiet mit der Eigenschaft $\Gamma = \partial \Delta \subseteq \mathsf{P}(T) \cap \Delta_b$. Dann sind die zugehörigen Folgen der Operatoren Q_ι, $T_\iota Q_\iota$, $\iota \in I$, beschränkt und konvergieren in F mit*

$$(19) \qquad Q_\iota \rightarrow Q, \quad T_\iota Q_\iota \rightarrow TQ \quad (\iota \in I_1).$$

Für die Anwendung der Sätze in Abschnitt II-2. 3, II-3. 2 benötigen wir noch die diskrete Kompaktheit der zur Folge T_ι, $\iota \in I$, gehörenden Folgen von Einbettungsoperatoren $J_\iota \in \mathscr{B}(D_\iota, F)$, $\iota \in I$. Dieses bedeutet hier, daß für jede Teilfolge $I' \subseteq I$ und jede in den Graphennormen beschränkte Folge $u_\iota \in D(T_\iota)$, $\iota \in I'$, eine Teilfolge $I'' \subseteq I'$ und ein Element $w \in F$ existieren mit der Eigenschaft

$$(20) \qquad \|u_\iota\| + \|T_\iota u_\iota\| \leqq \beta, \quad \iota \in I' \quad \Rightarrow u_\iota \rightarrow w \text{ in } F \quad (\iota \in I'').$$

Damit kann man Satz II-2. 2. (1) hier die folgende Gestalt geben.

(21) *Seien T, T_ι, $\iota \in I$, Operatoren mit vollstetiger Resolvente, die Folge $(T_\iota)_{\iota \in I}$ genüge der Kompaktheitsbedingung* (20), *und das Beschränktheitsgebiet Δ_b sei nicht leer. Dann ist die Konvergenz $T_\iota \underset{d}{\to} T$ $(\iota \in I)$ notwendig und hinreichend für die Beziehung*

$$(22) \qquad \Delta_s = \Delta_b = \mathsf{P}(T), \quad T_\iota(z)^{-1} \to T(z)^{-1} \qquad (\iota \in I_s), \ z \in \mathsf{P}(T).$$

Operatoren mit vollstetiger Resolvente haben ein diskretes Spektrum aus Eigenwerten endlicher Vielfachheit, die keinen Häufungspunkt im Endlichen besitzen. Damit sind hier die Voraussetzungen von Abschnitt II-3. 2 und insbesondere Satz II-3. 2. (8) erfüllt, der dementsprechend folgende Aussage über die Konvergenz der Eigenwerte aus $\Sigma(T_\iota)$, $\iota \in I$, der algebraischen Vielfachheit nach gestattet.

(23) *Die Voraussetzungen von Satz* (21) *seien erfüllt und λ ein Eigenwert von T mit der algebraischen Vielfachheit m. Dann gibt es für jede beschränkte abgeschlossene Umgebung U von λ mit $\Sigma(T) \cap U = \{\lambda\}$ ein Endstück I_1 von I und für jedes $\iota \in I_1$ genau m Eigenwerte $\lambda_1^\iota, \ldots, \lambda_m^\iota$ von T_ι mit der Eigenschaft*

$$(24) \qquad \Sigma(T_\iota) \cap U = \{\lambda_1^\iota, \ldots, \lambda_m^\iota\}, \quad \lambda_k^\iota \to \lambda, \qquad k = 1, \ldots, m, \ (\iota \in I_1).$$

Unter den Voraussetzungen von Satz (21) ist mit $\mathsf{P}(T) = \Delta_b$ auch noch Satz (18) anwendbar, der die Konvergenz der Projektionsoperatoren $Q_\iota \to Q$ $(\iota \in I_1)$ der zugehörigen algebraischen Eigenräume ergibt.

2. Projektionsmethode

Eine Reihe von bekannten Näherungsverfahren der Analysis läßt sich mit der Projektionsmethode behandeln (s. KANTOROWITSCH—AKILOW [15], Kapitel XIV, MIKHLIN—SMOLITSKIY [18]). Die Projektionsmethoden enthalten bekanntlich als Spezialfall das Näherungsverfahren von RITZ und allgemeiner das Verfahren von GALERKIN zur Lösung von inhomogenen Gleichungen und zur Bestimmung von Eigenwerten und Eigenvektoren, was wir im folgenden Abschnitt 2. 2 an Beispielen erläutern. Auch gewisse Kollokationsverfahren lassen sich mit der Projektionsmethode behandeln. Wir betrachten anschließend die Lösung inhomogener Gleichungen und das allgemeine Eigenwertproblem mit einem Operator im Eigenwertterm. Damit gilt unsere Theorie zum Beispiel für das GALERKIN-Verfahren bei den K.-p. d. und non K.-p. d. Operatoren von PETRYSHYN [20], [22]. In einfacher Weise erhält man auch die Resultate von GRIGORIEFF [12] aus unserer Theorie, was wir in Beispiel 3 von Abschnitt 2. 2 näher ausführen.

2. 1. *Allgemeine Sätze.* Seien E, F normierte Räume über demselben Körper $\mathbf{K} = \mathbf{R}$ bzw. $\mathbf{K} = \mathbf{C}$. Wie wir in Abschnitt I-5. 1. 3 bereits ausgeführt haben, betrachtet man für die Projektionsmethode lineare stetige Projektionsoperatoren P_0, P_ι in F auf Teilräume $F_0 = P_0(F)$, $F_\iota = P_\iota(F)$, $\iota \in \mathbf{N}$, die der folgenden Stabilitäts- und Konvergenzbedingung genügen

$$(1) \qquad \|P_\iota\| \leqq \beta, \quad \|P_\iota v - P_0 v\|_F \to 0 \qquad (\iota \to \infty), \quad v \in F.$$

Damit bilden die Teilräume F_ι, $\iota \in \mathbf{N}$, eine diskrete Approximation $\mathscr{A}(F_0, \Pi F_\iota, R^F)$ des Teilraumes F_0 mit der Konvergenz in F als diskreter Konvergenz. Weiter sei eine Folge von Teilräumen E_0, $E_\iota \subseteq E$, $\iota \in \mathbf{N}$, gegeben, die eine diskrete Approximation $\mathscr{A}(E_0, \Pi E_\iota, R^F)$ im Sinne von Abschnitt I-5. 1. 2 bilden. Die diskrete Konvergenz dazu ist gleich der Konvergenz in E.

Die Projektionsmethode betrachtet zu einem Paar beschränkter linearer Abbildungen $A, B \in \mathscr{B}(E, F)$ die Operatoren $A_\iota, B_\iota \in \mathscr{B}(E_\iota, F_\iota)$, erklärt durch die Gleichungen

$$(2) \qquad A_\iota u = P_\iota Au, \quad B_\iota u = P_\iota Bu \qquad (u \in E_\iota, \; \iota = 0, 1, 2, \ldots).$$

In Abschnitt I-5. 1. 3 haben wir bereits die diskrete Konvergenz dieser Operatoren gezeigt

$$(3) \qquad A_\iota \to A_0, \quad B_\iota \to B_0 \qquad (\iota \to \infty).$$

Die Operatorfolgen $(A_\iota)_{\iota \in \mathbf{N}}$, $(B_\iota)_{\iota \in \mathbf{N}}$ sind also stabil, und die Operatoren A_0, A_ι, $\iota \in \mathbf{N}$, sowie B_0, B_ι, $\iota \in \mathbf{N}$, sind konsistent. Im folgenden schreiben wir wieder

$$A(z) = A - zB, \quad A_\iota(z) = A_\iota - zB_\iota \qquad (\iota = 0, 1, 2, \ldots, \; z \in \mathbf{K}).$$

Die Beziehung (3) ist dann äquivalent zur diskreten Konvergenz $A_\iota(z) \to A_0(z)$ $(\iota \to \infty)$, $z \in \mathbf{K}$.

Mit $A|E_\iota$, $B|E_\iota$ bezeichnen wir die Restriktion der Abbildungen auf E_ι, $\iota = 0, 1, 2, \ldots$. Diese sind beschränkte lineare Abbildungen aus $\mathscr{B}(E_\iota, F)$. Dementsprechend kann man die Operatoren $P_\iota A|E_\iota$, $P_\iota B|E_\iota \in \mathscr{B}(E_\iota, F_\iota)$ in (2) als *Restriktion* der Abbildungen A, B in den Raum $\mathscr{B}(E_\iota, F_\iota)$ auffassen und die Bedingung (3) in der Form schreiben

$$(4) \qquad P_\iota A | E_\iota \to P_0 A | E_0, \quad P_\iota B | E_\iota \to P_0 B | E_0 \qquad (\iota \to \infty).$$

Die Projektionsmethode läßt sich ohne Schwierigkeiten auch noch auf Folgen von Operatoren $\hat{A}_\iota, \hat{B}_\iota \in \mathscr{B}(E_\iota, F_\iota)$ anwenden, die in der Norm der Operatoren aus $\mathscr{B}(E_\iota, F_\iota)$ der Bedingung genügen

$$(5) \qquad \|\hat{A}_\iota - P_\iota A | E_\iota\| \to 0, \quad \|\hat{B}_\iota - P_\iota B | E_\iota\| \to 0 \qquad (\iota \to \infty).$$

Der Einfachheit halber beschränken wir uns im folgenden jedoch auf den durch (2) definierten Fall.

Im Rahmen der Projektionsmethode nehmen viele Sätze aus den früheren Abschnitten eine besonders einfache Gestalt an, da die Konsistenzbedingung immer erfüllt ist und die Folgen $(A_\iota)_{\iota \in \mathbf{N}}$, $(B_\iota)_{\iota \in \mathbf{N}}$ unter den obigen Bedingungen stabil sind. Wenn der Operator B vollstetig ist, dann brauchen wir im folgenden nicht die Vollständigkeit der normierten Räume E, F vorauszusetzen. Andernfalls machen wir die *Voraussetzung*, daß die Räume E, F, E_0, F_0 und E_ι, F_ι, $\iota \in \mathbf{N}$, vollständig sind. Die Anwendung von Satz II-2. 1. (5) ergibt hier den folgenden Satz.

(6) *Wenn die Menge $\Delta_b \cap \mathrm{P}(A_0, B_0)$ nicht leer ist, dann hat man die Beziehung $\Delta_s =$ $= \Delta_b \cap \mathrm{P}(A_0, B_0)$ und die Konvergenzaussage*

$$A_\iota(z)^{-1} \to A_0(z)^{-1} (\iota \to \infty), \quad z \in \Delta_s.$$

Im Falle vom RITZ—GALERKIN-Verfahren (s. [15], XIV-§ 1. 8) werden die Teilräume E_ι, F_ι für $\iota \in \mathbf{N}$ endlichdimensional, so daß man das Beschränktheitsgebiet Δ_b der Resolventenfolgen $(A_\iota(z)^{-1})_{\iota \in \mathbf{N}}$ noch einfacher als in II-2. 1. (1) charakterisieren kann.

(7) *Für endlichdimensionale Räume E_ι, F_ι, $\iota \in \mathbf{N}$, ist genau dann die Zahl $z \in \Delta_b$, wenn für fast alle $\iota \in \mathbf{N}$ die Gleichung gilt*

(8) $$\dim E_\iota = \dim F_\iota$$

und mit einer positiven Zahl γ die Ungleichung besteht

(9) $$\gamma \|u\| \leqq \|A_\iota(z)u\| \quad (u \in E_\iota).$$

BEWEIS. Für $z \in \Delta_b$ existiert die Folge der inversen Operatoren $A_\iota(z)^{-1}$ in $\mathscr{B}(F_\iota, E_\iota)$ für fast alle $\iota \in \mathbf{N}$ und ist stabil. Dies ergibt die Isomorphie der Räume E_ι, F_ι und die Ungleichung (9) für fast alle $\iota \in \mathbf{N}$. Umgekehrt wird bekanntlich im Falle endlichdimensionaler Räume E_ι, F_ι gleicher Dimension mit (9) der Operator $A_\iota(z)$ nicht nur injektiv sondern auch gleichzeitig surjektiv, so daß $A_\iota(z)^{-1}$ in $\mathscr{B}(F_\iota, E_\iota)$ existiert und auf Grund von (9) durch $1/\gamma$ beschränkt ist. Damit ist dann $z \in \Delta_b$.

Jetzt betrachten wir etwas spezieller in normierten Räumen E, F ein Paar von Operatoren A, $B \in \mathscr{B}(E, F)$ mit einem vollstetigen Operator B. Hiermit werden auch die Operatoren $B_\iota \in \mathscr{B}(E_\iota, F_\iota)$ für $\iota = 0$, $\iota \in \mathbf{N}$, vollstetig. Weiter wird in diesem Falle die Folge $(B_\iota)_{\iota \in \mathbf{N}}$ diskret kompakt (s. I-5. 1. 3). Mit Satz II-2. 2. (1) hat man so für die Projektionsmethode den Satz

(10) *Der Operator $B \in \mathscr{B}(E, F)$ sei vollstetig, die Resolventenmenge $\mathrm{P}(A_0, B_0)$ und das Beschränktheitsgebiet Δ_b seien nicht leer. Dann gilt die Gleichung $\Delta_s = \Delta_b =$ $= \mathrm{P}(A_0, B_0)$, so daß für jedes $z \in \mathrm{P}(A_0, B_0)$ bzw. $z \in \Delta_b$ die inversen Operatoren $A_\iota(z)^{-1}$ für fast alle $\iota \in \mathbf{N}$ existieren und gegen $A_0(z)^{-1}$ konvergieren.*

Für die Lösung inhomogener Gleichungen ergibt Satz (10) mit Hilfe von Satz II-1. 1. (8) die anschließende Aussage: Immer wenn die Zahl $z \in \mathbf{K}$ kein Eigenwert von A_0, B_0 ist, wird die inhomogene Gleichung

$$(11) \qquad A_0 u - z B_0 u = v_0$$

für jede rechte Seite $v_0 \in F_0$ eindeutig und stetig lösbar. Dazu gibt es weiter ein Endstück $\mathbf{N}_1$ von $\mathbf{N}$, so daß für jedes $\iota \in \mathbf{N}_1$ und jedes $v_\iota \in F_\iota$ die approximierende Gleichung

$$(12) \qquad A_\iota u_\iota - z B_\iota u_\iota = v_\iota$$

eine eindeutig bestimmte Lösung u_ι in E_ι besitzt. Wenn die rechten Seiten konvergieren, dann konvergieren auch die Lösungen u_ι von (12) gegen die Lösung u von (11), also

$$(13) \qquad v_\iota \to v \text{ in } F \Rightarrow u_\iota \to u \text{ in } E \qquad (\iota \to \infty).$$

Unter den Bedingungen von Satz (10) besteht das Spektrum $\Sigma(A_0, B_0)$ aus einer höchstens abzählbar unendlichen Folge von Eigenwerten endlicher algebraischer Vielfachheit, die keinen Häufungspunkt im Endlichen besitzt. Die gleiche Eigenschaft haben die Spektren $\Sigma(A_\iota, B_\iota)$ der approximierenden Operatoren für jedes $\iota \in \mathbf{N}_b \subseteq \mathbf{N}$ mit dem Endstück $\mathbf{N}_b$ des Beschränktheitsgebietes für ein $z_0 \in \Delta_b$. Unter den Voraussetzungen von Satz (10) mit $\mathbf{K} = \mathbf{C}$, also für komplexe normierte Räume E, F, sind dann die Voraussetzungen von Satz II-3. 2. (1) erfüllt. Damit gelten für die Projektionsmethode in diesem Fall und mit A_0, B_0, $\mathbf{N}$ anstelle von A, B, I wörtlich der Satz II-3. 2. (1) über die Konvergenz der Resolventenmengen, der Hauptsatz II-3. 2. (8) über die Konvergenz der Eigenwerte und zugehöriger Systeme von Hauptvektoren sowie für J-definite Paare A_0, B_0 der Satz II-3. 2. (13) über die Konvergenz der geordneten Spektren. Zum Beispiel lautet der Hauptsatz für die Projektionsmethode folgendermaßen.

(14) *Die Voraussetzungen von Satz* (10) *seien erfüllt und es sei* λ *ein Eigenwert von* A_0, B_0 *mit der algebraischen Vielfachheit* m *sowie* $w_1, \ldots, w_m$ *eine Basis des zugehörigen algebraischen Eigenraums. Weiter sei* U *eine beliebige beschränkte abgeschlossene Umgebung von* λ *mit* $\Sigma(A_0, B_0) \cap U = \{\lambda\}$. *Dann gibt es ein Endstück* $\mathbf{N}_1$ *von* $\mathbf{N}$ *und für jedes* $\iota \in \mathbf{N}_1$ *genau* m *Eigenwerte* $\lambda_1^\iota, \ldots, \lambda_m^\iota$, *wenn diese ihrer algebraischen Vielfachheit entsprechend oft wiederholt werden, sowie linear unabhängige Vektoren* $w_1^\iota, \ldots, w_m^\iota$ *aus der Summe der algebraischen Eigenräume von* $\lambda_1^\iota, \ldots, \lambda_m^\iota$ *mit der Eigenschaft*

$$\Sigma(A_\iota, B_\iota) \cap U = \{\lambda_1^\iota, \ldots, \lambda_m^\iota\} \qquad (\iota \in \mathbf{N}_1)$$

und

$$\lambda_k^\iota \to \lambda, \quad w_k^\iota \to w_k \qquad (\iota \in \mathbf{N}_1, \; k = 1, \ldots, m).$$

Im folgenden Abschnitt 2. 2, Beispiel 3, geben wir auch ein Beispiel für J-definite Paare von Operatoren A_0, B_0. Bei der Anwendung weiterer Sätze auf die Projektionsmethode ist zu beachten, daß die hier eingeführten Projektionsoperatoren P_ι natürlich nichts mit den ebenso bezeichneten Projektionsoperatoren zu isolierten Teilen von Spektren oder algebraischen Eigenräumen zu tun haben.

2. 2. *Spezielle Anwendungen.* Anschließend wollen wir einfache Beispiele für die Anwendung der Projektionsmethode zur Lösung spezieller inhomogener Gleichungen und Eigenwertaufgaben angeben.

BEISPIEL 1. Die Projektionsmethode gestattet die Lösung der inhomogenen Gleichung $u - zK_0 u = w$ sowie der zugehörigen Eigenwertaufgabe mit einem vollstetigen Operator K in einem normierten Raum $E = F$ und $K_0 = P_0 K | E_0$ in $E_0 = F_0$. Wenn die zugehörige homogene Gleichung $v - zK_0 v = 0$ eine nichttriviale Lösung hat, so ist hier z eine charakteristische Zahl des Operators K_0, z ist notwendig ungleich Null und $1/z$ ein Eigenwert von K_0. Für die Anwendung der Sätze in Abschnitt 2. 1 setzt man weiter $E_\iota = F_\iota$ für $\iota \in I$, $A = 1$ mit der Identität $1 \in \mathscr{B}(E)$ sowie $B = K \in \mathscr{B}(E)$. Damit existieren trivialerweise für $z = 0$ die zugehörigen Operatoren $A_0(z)^{-1} = 1$ in $\mathscr{B}(E_0)$ und $A_\iota(z)^{-1} = 1_\iota$ in $\mathscr{B}(E_\iota)$ für $\iota \in I$. Folglich ist die Zahl $z = 0$ aus der Resolventenmenge $\mathsf{P}(1, K_0)$ und aus dem Beschränktheitsgebiet Δ_b der Folge $(A_\iota(z)^{-1})_{\iota \in I}$. Hiermit sind die Voraussetzungen von Satz (10) erfüllt und dementsprechend die Folgerungen daraus, die wir oben in Abschnitt 2. 1 formuliert haben.

BEISPIEL 2. In KANTOROWITSCH—AKILOW [15], XIV-§ 2. 3 wird die Projektionsmethode auf die Lösung der Gleichung (11) angewandt unter den folgenden Bedingungen:

(i) Es sei $E_0 = E$, $F_0 = F$ und $P_0 = 1$, so daß mit (1) dann $P_\iota \to 1$ $(\iota \to \infty)$ konvergiert.

(ii) Der Operator $B_0 = B \in \mathscr{B}(E, F)$ sei vollstetig und der Operator $A_0^{-1} = A^{-1}$ existiert in $\mathscr{B}(F, E)$.

(iii) Die Teilräume E_ι, F_ι sind endlichdimensional und der Operator A bildet den Teilraum E_ι bijektiv ab auf F_ι für jedes $\iota \in \mathbf{N}$.

Mit Bedingung (iii) wird E_ι isomorph F_ι und folglich dim $E_\iota = $ dim F_ι, $\iota \in \mathbf{N}$, so daß die Bedingung (8) erfüllt ist. Weiter existiert auf Grund von (ii) eine positive Zahl γ mit der Eigenschaft $\gamma \|u\| \le \|Au\|$, $u \in E$. Für $z = 0$ hat man dann mit Hilfe von (iii) die Darstellung

$$A_\iota(z)u = P_\iota Au = Au \qquad (u \in E_\iota, \quad \iota \in \mathbf{N}, \quad z = 0),$$

womit auch die Ungleichung (9) für $z = 0$ besteht. Nach Satz (7) ist dann $z = 0 \in \Delta_b$. Die Existenz von $A_0^{-1} = A^{-1}$ ergibt auch $0 \in \mathsf{P}(A_0, B_0) = \mathsf{P}(A, B)$. Damit sind wieder die Voraussetzungen aus Satz (10) erfüllt und die entsprechenden Folgerungen in Abschnitt 2. 1 gültig.

BEISPIEL 3. Jetzt wird die Projektionsmethode angewandt auf die Approximation eines reellen bzw. komplexen Hilbertschen Raumes $E=E_0$, mit $\mathbf{K}=\mathbf{R}$ bzw. $\mathbf{K}=\mathbf{C}$ und Skalarprodukt $(\cdot,\ \cdot)$ durch abgeschlossene Teilräume $E_\iota \subseteq E, \iota \in \mathbf{N}$. Hierzu existieren Operatoren P_ι der orthogonalen Projektion von E auf E_ι für $\iota \in \mathbf{N}$. Damit wird

$$\|P_\iota\| \leqq 1, \quad |u, E_\iota| = \|u - P_\iota u\|, \qquad u \in E, \quad \iota \in \mathbf{N},$$

und die Folge von Teilräumen $E_\iota, \iota \in \mathbf{N}$, bildet genau dann eine diskrete Approximation von $E=E_0$ im Sinne von Abschnitt I-5. 1. 2, wenn mit $P_0=1$ die Bedingung (1) erfüllt ist, also $P_\iota \to 1$ $(\iota \to \infty)$ konvergiert. Das GALERKIN-Verfahren betrachtet speziell eine aufsteigende Folge endlichdimensionaler Teilräume $E_\iota, \iota \in \mathbf{N}$, deren Vereinigung Φ dicht in E ist (s. I-5. 1. 2. (6)). Für die Aufgaben (15), (16) unter der Bedingung (17) wird dieser Fall mit weiteren Beispielen in GRIGORIEFF [12] behandelt.

Die Projektionsmethode aus Abschnitt 2. 1 wird nun angewandt auf die Lösung der inhomogenen Gleichung

$$(15) \qquad\qquad a(\varphi, u) - \bar{z}b(\varphi, u) = l(\varphi) \qquad (\varphi \in E)$$

und der zugehörigen Eigenwertaufgabe

$$(16) \qquad\qquad a(\varphi, w) = \lambda b(\varphi, w) \qquad (\varphi \in E)$$

mit einer beschränkten Sesquilinearform a auf $E \times E$ und einer vollstetigen Sesquilinearform b auf $E \times E$. Die zugehörigen quadratischen Formen sollen mit einer positiven Zahl γ und einer Zahl $z_0 \in \mathbf{K}$ der a-priori-Ungleichung genügen

$$(17) \qquad\qquad \gamma \|u\|^2 \leqq |a(u) - \bar{z}_0 b(u)| \qquad (u \in E).$$

Das Paar a, b ist damit regulär und koerzitiv im Sinne von STUMMEL [27], § 3. 1. Die Sesquilinearformen lassen sich bekanntlich darstellen durch beschränkte lineare Operatoren $A, B \in \mathscr{B}(E)$. Für eine vollstetige Form b wird der Operator B vollstetig in E, und aus (17) erhält man die Ungleichung

$$(18) \qquad\qquad \gamma \|u\|^2 \leqq |(u, A(z_0)u)| \qquad (u \in E).$$

Folglich ist die Zahl $z_0 \in \mathrm{P}(A, B)$. Mit der Beziehung

$$(u, A(z)v) = (u, P_\iota A(z)v) = (u, A_\iota(z)v) \qquad (u, v \in E_\iota, \iota \in I, z \in \mathbf{K})$$

folgt aus der letzten Ungleichung die entsprechende Ungleichung für die Operatoren $A_\iota(z_0)$ in E_ι, so daß $A_\iota(z_0)^{-1}$ in $\mathscr{B}(E_\iota)$ existiert und $\|A_\iota(z_0)^{-1}\| \leqq 1/\gamma$ wird für $\iota \in \mathbf{N}$. Also ist die Zahl $z_0 \in \Delta_b$.

Damit sind für $A=A_0$, $B=B_0$ und $E_\iota=F_\iota$, $\iota \in \mathbf{N}$, die Voraussetzungen von Satz (10) erfüllt und dementsprechend die zugehörigen Folgerungen gültig. Die inhomogenen Gleichungen (12) lassen sich hier in der Gestalt schreiben

$$a(\varphi, u_\iota) - \bar{z}b(\varphi, u_\iota) = l_\iota(\varphi) \qquad (\varphi \in E_\iota, \ \iota \in \mathbf{N})$$

und die zugehörigen approximierenden Eigenwertaufgaben in der Form

$$a(\varphi, w_\iota) = \lambda b(\varphi, w_\iota) \qquad (\varphi \in E_\iota, \ \iota \in \mathbf{N}).$$

Für einen komplexen Hilbertschen Raum E und hermitesche Sesquilinearformen a, b mit der Eigenschaft (17) werden A, B J-definite Abbildungen. Definiert man nämlich die Abbildung J durch die Gleichung

$$\langle u, Jv \rangle = (u, v) \qquad (u, v \in E),$$

so erhält man die Darstellung

$$a(u, v) = (Au, v) = \langle Au, Jv \rangle \qquad (u, v \in E)$$

und eine entsprechende Darstellung für b. Aus a, b hermitesch folgt die Bedingung II-1. 3. (17). Die Ungleichung (17) ergibt weiter die Beziehungen

$$|a(u)| + |b(u)| > 0 \leftrightarrow |a(u) + ib(u)| > 0 \qquad (0 \neq u \in E).$$

Nach dem Satz von HAUSDORFF über die Konvexität des Wertebereichs quadratischer Formen existiert eine reelle Zahl ψ mit der Eigenschaft

$$\mathrm{Re}\left\{ e^{-i\psi}(a(u) + ib(u)) \right\} > 0 \leftrightarrow \cos \psi \, a(u) + \sin \psi \, b(u) > 0 \qquad (0 \neq u \in E).$$

Also besteht die Ungleichung II-1. 3. (18) mit $\alpha = \cos \psi$, $\beta = \sin \psi$.

3. Quadraturformelmethode für Integraloperatoren

Ein schönes Beispiel für die diskrete Störungstheorie erhält man mit der Quadraturformelmethode bei Integralgleichungen. Diese ist in einfacher Form bereits von FREDHOLM und VOLTERRA betrachtet worden (s. [23], S. 171). Für neuere Arbeiten hierzu verweisen wir auf ANSELONE [2] und BRAKHAGE [7], [8]. In der Arbeit [2] wird der Begriff kollektiv kompakter Operatorfolgen eingeführt, um die Konvergenz der Näherungslösungen inhomogener Gleichungen zu zeigen. Die Konvergenz von Eigenwerten und zugehörigen algebraischen Eigenräumen von kollektiv kompakten Operatorfolgen wurde dann in ATKINSON [4] bewiesen, worauf wir in Abschnitt 1. 1, Beispiel 1, bereits eingegangen sind. Das anschließende Beispiel soll nun zeigen, daß die Theorie diskreter Approximationen mit Leichtigkeit die Konvergenz einer sehr viel allgemeineren Klasse von Approximationen zu beweisen gestattet, als sie anscheinend bisher betrachtet wurden. Dies gilt in noch stärkerem

Maße für diskrete Approximationen von Integraloperatoren in den Räumen $L^p(X)$ durch Operatoren in den Räumen $l^p(X_\iota)$, die in Beispiel I-5. 2. 2. (ii) angegeben wurden. Hierauf können wir jedoch im Rahmen dieser Beispiele leider nicht eingehen.

3. 1. *Konsistenz und diskrete Kompaktheit.* Sei Ω eine beschränkte offene Teilmenge des n-dimensionalen Zahlenraumes $\mathbf{R}^n$ und $X = \bar{\Omega}$ die Abschließung von Ω. Mit $E = C(X)$ bezeichnen wir dann den Banachschen Raum der reell- bzw. komplexwertigen stetigen Funktionen auf X mit der Maximumnorm. Weiter sei M ein beschränktes abgeschlossenes Intervall des Zahlenraumes $\mathbf{R}^n$ mit $X \subseteq M$, und für eine geeignete Indexfolge $I = I_0$ sei X_ι, $\iota \in I$, eine Folge diskreter, endlicher Punktmengen aus M, die eine zugehörige Folge konvergenter Quadraturformeln Σ_ι gestatten mit der Eigenschaft

$$(1) \qquad \Sigma_\iota(f) = \sum_{x \in X_\iota} f(x)\mu_\iota(x) \to \int_X f(x)\,dx \qquad (\iota \in I)$$

für jedes $f \in C(M)$. Damit wird bekanntlich eine schwach konvergente Folge linearer stetiger Funktionale Σ_ι auf $C(M)$ definiert. Diese Folge ist notwendig beschränkt, so daß mit einer positiven Zahl μ die Ungleichung besteht

$$(2) \qquad \|\Sigma_\iota\| = \sum_{x \in X_\iota} |\mu_\iota(x)| \leqq \mu \qquad (\iota \in I).$$

Für jedes $\iota \in I$ bezeichnen wir entsprechend mit $E_\iota = C(X_\iota)$ den Banachschen Raum der reell- bzw. komplexwertigen Funktionen auf der Menge X_ι mit der Maximumnorm.

Mit den Bezeichnungen von Abschnitt I-5. 2. 1. kann man dann den für das folgende grundlegenden Satz aussprechen.

(3) *Die Räume $C(X_\iota)$, $\iota \in I$, bilden eine diskrete Approximation $\mathscr{A}(C(X), \Pi C(X_\iota), R)$ im Sinne von Satz I-5. 2. (4), wenn die Punktmengen X_ι, $\iota \in I$, noch der Voraussetzung genügen*

$$(4) \qquad d_0(X_\iota, X) = \sup_{x \in X_\iota} |x, X| \to 0 \qquad (\iota \in I).$$

BEWEIS. (i) Aus der Eigenschaft (1) der Folge von Quadraturformeln folgt, daß zu jedem Punkt $x \in X$ der Abstand $|x, X_\iota| \to 0$ strebt für $\iota \in I$. Sonst gäbe es nämlich einen Punkt $z \in X$, eine Teilfolge $I' \subseteq I$ und eine positive Zahl ε_0 mit $|z, X_\iota| \geqq \varepsilon_0$, $\iota \in I'$. Dazu existiert dann nach dem Satz von TIETZE—URYSOHN eine stetige Funktion $f \in C(M)$ mit $f(z) = 1$, $f(x) = 0$, $|x - z| \geqq \varepsilon_0$, und $0 \leqq f(x) \leqq 1$, $x \in M$. Die Beziehung (1) führt damit auf den Widerspruch

$$0 = \Sigma_\iota(f) \to \int_X f(x)\,dx > 0 \qquad (\iota \in I').$$

(ii) Weiter hat man die Konvergenz von

$$(5) \qquad d_0(X, X_\iota) = \sup_{x \in X} |x, X_\iota| \to 0 \qquad (\iota \in I).$$

Da X kompakt ist, existierten sonst nämlich eine Teilfolge $I' \subseteq I$, eine Zahl $\varepsilon_0 > 0$ und ein Punkt $x \in X$ mit der Eigenschaft $|x, X_\iota| \geqq \varepsilon_0 > 0$, $\iota \in I'$, was nach Teil (i) ausgeschlossen ist. Mit den beiden Bedingungen (4), (5) konvergiert dann der HAUSDORFF-Abstand $d(X_\iota, X) \to 0$ $(\iota \in I)$, womit die Voraussetzungen von Satz I-5. 2. (4) erfüllt sind.

Im Banachschen Raum $C(X)$ betrachten wir den Integraloperator K, erklärt durch die Gleichung

$$(6) \qquad (Ku)(x) = \int\limits_X K(x, y) u(y)\, dy \qquad (x \in X)$$

für $u \in C(X)$ mit einem stetigen Integralkern $K(\cdot, \cdot)$ auf $X \times X$. Nach der *Quadraturformelmethode* erhält man dann diskrete Approximationen des obigen Integraloperators in der Gestalt

$$(7) \qquad (K_\iota u)(x) = \sum_{y \in X_\iota} K_\iota(x, y) u(y) \mu_\iota(y) \qquad (x \in X_\iota)$$

für $u \in C(X_\iota)$ und mit Matrizen $K_\iota(\cdot, \cdot)$ mit reell- bzw. komplexwertigen Elementen $K_\iota(x, y)$, $x, y \in X_\iota$, für $\iota \in I$. Der Operator K ist bekanntlich eine vollstetige lineare Abbildung im Raum $C(X)$. Entsprechend definiert die Gleichung (7) vollstetige lineare Abbildungen im Raum $C(X_\iota)$ für jedes $\iota \in I$. Nach dem Erweiterungssatz von TIETZE—URYSOHN gibt es für den Integralkern $K \in C(X \times X)$ stetige Fortsetzungen $\hat{K} \in C(M \times M)$ mit $\hat{K}(x, y) = K(x, y)$, $x, y \in X$. Damit setzen wir nun die folgende *Konsistenzbedingung* voraus.

(8) *Die Matrizen $K_\iota(\cdot, \cdot)$ sollen mit einer Fortsetzung $\hat{K} \in C(M \times M)$ des Integralkerns K der Bedingung genügen*

$$\max_{x \in X_\iota, \, y \in X_\iota} \sum |K_\iota(x, y) - \hat{K}(x, y)| \, |\mu_\iota(y)| \to 0 \qquad (\iota \in I).$$

Man kann übrigens leicht zeigen, daß diese Bedingung nicht von der speziell gewählten Fortsetzung $\hat{K}$ abhängt. Die Bedingung (8) ist trivialerweise erfüllt, wenn man für K_ι die Restriktion von $\hat{K}$ auf $X_\iota \times X_\iota$ wählt,

$$(9) \qquad K_\iota(x, y) = \hat{K}(x, y) \qquad (x, y \in X_\iota, \, \iota \in I).$$

Falls die Teilmengen X_ι in X enthalten sind, dann erübrigt sich die Einführung einer Fortsetzung $\hat{K}$ des Integralkerns K. Insbesondere hat man nun unter den obigen Voraussetzungen für die diskrete Approximation $\mathscr{A}(C(X), \Pi C(X_\iota), R)$ den anschließenden Satz.

14*

(10) *Mit der Bedingung* (8) *werden die Operatoren* K, K_ι, $\iota \in I$, *konsistent und die Folge* $(K_\iota)_{\iota \in I}$ *diskret kompakt.*

BEWEIS. (i) Wir zeigen zunächst, daß die Folge $(K_\iota)_{\iota \in I}$ diskret kompakt ist. Sei also $I' \subseteq I$ eine beliebige Teilfolge und $(u_\iota)_{\iota \in I'}$ eine beschränkte Folge von Funktionen $u_\iota \in C(X_\iota)$ mit $\|u_\iota\| \leq \gamma$, $\iota \in I'$. Durch die Vorschrift

$$(11) \qquad \hat{g}_\iota(x) = \sum_{y \in X_\iota} \hat{K}(x, y) u_\iota(y) \mu_\iota(y) \qquad (x \in M)$$

wird dann eine Folge stetiger Funktionen $\hat{g}_\iota \in C(M)$, $\iota \in I$, definiert. Mit Hilfe von Ungleichung (2) ist diese Folge beschränkt durch

$$|\hat{g}_\iota(x)| \leq \mu\gamma \max_{x, y \in M} |\hat{K}(x, y)| \qquad (x \in M, \ \iota \in I).$$

Die Funktion $\hat{K}$ ist gleichmäßig stetig auf $M \times M$, so daß zu jedem $\varepsilon > 0$ ein $\delta > 0$ existiert mit

$$|\hat{K}(x, y) - \hat{K}(x', y')| < \varepsilon, \quad |x - x'| + |y - y'| < \delta \qquad (x, x', y, y' \in M).$$

Hiermit wird dann

$$|\hat{g}_\iota(x) - \hat{g}_\iota(x')| \leq \mu\gamma\varepsilon \qquad (|x - x'| < \delta, \qquad x, x' \in M, \ \iota \in I).$$

Also ist die Funktionenfolge $(\hat{g}_\iota)_{\iota \in I'}$ gleichmäßig gleichgradig stetig auf M. Nach dem Satz von ARZELÀ—ASCOLI existiert dann eine Teilfolge $I'' \subseteq I'$ und eine stetige Funktion $\hat{w} \in C(M)$ mit der Eigenschaft

$$\max_{x \in X_\iota} |\hat{g}_\iota(x) - \hat{w}(x)| \leq \max_{x \in M} |\hat{g}_\iota(x) - \hat{w}(x)| \to 0 \qquad (\iota \in I'').$$

Setzt man nun $g_\iota = K_\iota u_\iota$, so folgt mit der Bedingung (8) die Beziehung

$$(12) \qquad \max_{x \in X_\iota} |g_\iota(x) - \hat{g}_\iota(x)| \leq \gamma \max_{x \in X_\iota} \sum_{y \in X_\iota} |K_\iota(x, y) - \hat{K}(x, y)| |\mu_\iota(y)| \to 0$$

für $\iota \in I$ und folglich mit der Funktion $w = \hat{w}|X \in C(X)$ die diskrete Konvergenz $K_\iota u_\iota \to w$ im Sinne von I-5. 2. (5). Hiermit ist die Eigenschaft I-3. 1. (k) der diskreten Kompaktheit der Folge $(K_\iota)_{\iota \in I}$ gezeigt.

(ii) Sei jetzt $u \in C(X)$ eine beliebige stetige Funktion, sei $\hat{u} \in C(M)$ eine Fortsetzung von u und $u_\iota(x) = \hat{u}(x)$, $x \in X_\iota$, so daß $u_\iota \to u$ strebt im Sinne der diskreten Konvergenz für $\iota \in I$. Hiermit setzen wir wie in (i) dann $g_\iota = K_\iota u_\iota$ und $\hat{g}_\iota$ wie in (11). Für jedes $x \in M$ konvergieren auf Grund von (1) die Näherungssummen

$$\hat{g}_\iota(x) = \sum_{y \in X_\iota} \hat{K}(x, y) \hat{u}(y) \mu_\iota(y) \to \int_X \hat{K}(x, y) u(y)\, dy = \hat{g}(x)$$

für $\iota \in I$ mit der stetigen Funktion $\hat{g} = \widehat{Ku} \in C(M)$. Weiter ist nach Abschnitt (i) die Folge $(\hat{g}_\iota)_{\iota \in I}$ beschränkt und gleichmäßig gleichgradig stetig auf M, so daß nach einem bekannten Satz die gleichmäßige Konvergenz von $\hat{g}_\iota(x) \to \hat{g}(x)$ $(\iota \in I)$ folgt für $x \in M$ und die Konvergenz von

$$\max_{x \in X_\iota} |\hat{g}_\iota(x) - \hat{g}(x)| \leq \max_{x \in M} |\hat{g}_\iota(x) - \hat{g}(x)| \to 0 \qquad (\iota \in I).$$

Zusammen mit der Beziehung (12) erhält man so die diskrete Konvergenz von $K_\iota u_\iota =$ $= g_\iota \to g = Ku$ $(\iota \in I)$, so daß die Konsistenzbedingung I-1. 2. (c) für die Operatoren K, K_ι, $i \in I$, erfüllt ist.

3. 2. *Konvergenz*. Nach diesen Vorbereitungen können wir jetzt unsere Sätze anwenden auf die Frage nach der Konvergenz der Quadraturformelmethode zur Lösung der Fredholmschen Integralgleichung zweiter Art

$$(13) \qquad u(x) - z \int_X K(x, y)u(y)\, dy = f(x) \qquad (x \in X)$$

mit einer gegebenen Funktion $f \in C(X)$ und einer Zahl $z \in \mathbf{K}$. Die zu dieser Gleichung gehörige homogene Gleichung erhält man für $f = 0$. Wenn diese eine nichttriviale Lösung besitzt, dann ist $z \neq 0$ und z ein Eigenwert von A, $B = 1$, K bzw. z eine charakteristische Zahl und $\lambda = 1/z$ ein Eigenwert des Integraloperators K. Es gibt eine höchstens abzählbar unendliche Folge von charakteristischen Zahlen, welche keinen Häufungspunkt im Endlichen besitzt (Satz II-1. 1. (8)).

Die Quadraturformelmethode approximiert die Gleichung (13) durch eine Folge von Gleichungen der Gestalt

$$(14) \qquad u_\iota(x) - z \sum_{y \in X_\iota} K_\iota(x, y)u_\iota(y)\mu_\iota(y) = f_\iota(x) \qquad (x \in X_\iota)$$

mit $u_i, f_i \in C(X)$ für $i \in I = I_0$. Unter den oben genannten Voraussetzungen, insbesondere den Bedingungen (1), (4) und der Konsistenzbedingung (8) ist dann Satz I-3. 1. (4) anwendbar mit dem folgenden Ergebnis.

(15) *Wenn die zu (13) gehörige homogene Gleichung nur die triviale Lösung besitzt, dann sind für fast alle $\iota \in I$ die Gleichungen (14) für jede rechte Seite f_ι eindeutig lösbar. Wenn die Folge der rechten Seiten f_ι, $\iota \in I$, diskret gegen f konvergiert, dann konvergieren auch die Lösungen $u_\iota \to u$, also gegen die Lösung u von (13) mit*

$$\max_{x \in X_\iota} |u_\iota(x) - \hat{u}(x)| \to 0 \qquad (\iota \in I).$$

Dieses Ergebnis kann man übrigens auch aus Satz II-2. 2. (1) erhalten. Dazu setzt man $A = 1$, $B = K$ und $A_\iota = 1_\iota$, $B_\iota = K_\iota$ für $\iota \in I$. Für $z = 0$ ist $A_\iota(z) = 1_\iota$, $\iota \in I$, so daß der Punkt $z = 0$ trivialerweise im Beschränktheitsgebiet Δ_b der Folge $A_\iota(z)^{-1} =$ $= (1 - zK_\iota)^{-1}$, $\iota \in I$, liegt. Hiermit sind die Voraussetzungen von Satz II-2. 2. (1) erfüllt, der die diskrete Konvergenz der Lösungen $u_\iota \to u (\iota \in I_s(z))$ für jede diskret konvergente rechte Seite $f_\iota \to f$ $(\iota \in I)$ ergibt, wenn $z \in \mathsf{P}(A, B)$, also z keine charakteristische Zahl des Operators K ist.

Weiter sind unter den hier betrachteten Bedingungen für $\mathbf{K} = \mathbf{C}$ die Voraussetzungen des Hauptsatzes II-3. 2. (8) erfüllt. Sei dann $\lambda = z$ eine charakteristische Zahl von K, also die Gleichung

$$(16) \qquad w(x) - \lambda \int_X K(x, y) w(y) \, dy = 0 \qquad (x \in X)$$

nichttrivial lösbar. Der zugehörige algebraische Eigenraum läßt sich hier in äquivalenter Weise charakterisieren als Nullraum von $(\mathbf{1} - \lambda K)^s$ für eine geeignete Zahl s und dementsprechend die algebraische Vielfachheit m von λ durch $m = \dim N((\mathbf{1} - \lambda K)^s)$. Dann erhält man für die Näherungsgleichungen

$$(17) \qquad w_\iota(x) - \lambda^\iota \sum_{y \in X_\iota} K_\iota(x, y) w_\iota(y) \mu_\iota(y) = 0 \qquad (x \in X_\iota, \ \iota \in I)$$

im Sinne von Satz II-3. 2. (8) die folgende Aussage über die Konvergenz der Eigenwerte der algebraischen Vielfachheit nach.

(18) *Für jede beschränkte abgeschlossene Umgebung U der charakteristischen Zahl λ mit $\Sigma(\mathbf{1}, K) \cap U = \{\lambda\}$ und der algebraischen Vielfachheit m von λ gibt es ein Endstück I_1 von I und für jedes $\iota \in I_1$ genau m charakteristische Zahlen $\lambda_1^\iota, \ldots, \lambda_m^\iota$ von (17) mit der Eigenschaft*

$$(19) \qquad \Sigma(\mathbf{1}_\iota, K_\iota) \cap U = \{\lambda_1^\iota, \ldots, \lambda_m^\iota\}, \quad \lambda_k^\iota \to \lambda \qquad (k = 1, \ldots, m; \ \iota \in I_1).$$

Weiter folgt aus Satz II-3. 2. (8) die diskrete Konvergenz $w_k^\iota \to w_k$ eines zugehörigen Systems von m Hauptvektoren, also die Konvergenzaussage

$$(20) \qquad \max_{x \in X_\iota} |w_k^\iota(x) - \hat{w}_k(x)| \to 0 \qquad (\iota \in I_1; \ k = 1, \ldots, m).$$

Wenn schließlich der Integralkern $K(\cdot, \cdot)$ hermitesch ist, also der Bedingung genügt

$$(21) \qquad K(x, y) = \overline{K(y, x)} \qquad (x, y \in X),$$

dann erhält man mit der Abbildung J von $C(X)$ in $C(X)'$, erklärt durch

$$(22) \qquad \langle u, Jv \rangle = \int_X u(x) \overline{v(x)} \, dx \qquad (u, v \in C(X))$$

die J-Definitheit des Paares $A = 1$, $B = K$ mit $\alpha = 1$, $\beta = 0$ in II-1. 3. (18). In diesem Falle gilt auch noch Satz II-3. 2. (13) über die Konvergenz der geordneten Spektren.

LITERATUR

[1] S. Agmon, *Lectures on elliptic boundary value problems.* Van Nostrand, Princeton 1965.

[2] P. M. Anselone, *Convergence and error bounds for approximate solutions of integral and operator equations.* Proc. Symp. Wisconsin, Madison 1965, 231—252.

[3] P. M. Anselone and T. W. Palmer, *Spectral analysis of collectively compact, strongly convergent operator sequences.* Pacific. J. Math. **25** (1968), 423—431.

[4] K. E. Atkinson, *The numerical solution of the eigenvalue problem for compact integral operators* Trans. Amer. Math. Soc. **129** (1967), 458—465.

[5] J. P. Aubin, *Approximations des espaces de distributions et des opérateurs différentiels.* Bull. Soc. Math. France **12** (1967), 1—139.

[6] I. Babuška and R. Výborný, *Continuous dependence of eigenvalues on the domain.* Czechoslovak Math. J. **15** (1965), 169—178.

[7] H. Brakhage, *Über die numerische Behandlung von Integralgleichungen nach der Quadraturformelmethode.* Numer. Math. **2** (1960), 183—196.

[8] H. Brakhage, *Zur Fehlerabschätzung für die numerische Eigenwertbestimmung bei Integralgleichungen.* Numer. Math. **3** (1961), 174—179.

[9] J. Céa, *Approximation variationnelle des problèmes aux limites.* Ann. Inst. Fourier (Grenoble) **14** (1964), 345—444.

[10] S. Goldberg, *Unbounded linear operators, theory and applications.* McGraw-Hill, New York 1966.

[11] R. D. Grigorieff, *Approximation von Eigenwertproblemen und Gleichungen zweiter Art in Hilbertschen Räumen.* Math. Ann. **183** (1969), 45—77.

[12] R. D. Grigorieff, *Über die Lösung regulärer koerzitiver Rand- und Eigenwertaufgaben mit dem Galerkinverfahren.* Manuscripta Math. **1** (1969), 385—411.

[13] R. D. Grigorieff, *Die Konvergenz des Rand- und Eigenwertproblems linearer gewöhnlicher Differenzengleichungen.* Num. Math. **15** (1970), 15—48.

[14] R. D. Grigorieff, *Über die Koerzitivität gewöhnlicher Differenzenoperatoren und die Konvergenz von Mehrschrittverfahren.* Num. Math. **15** (1970), 196—218.

[15] L. W. Kantorowitsch und G. P. Akilow, *Funktionalanalysis in normierten Räumen.* Akademie Verlag, Berlin 1964.

[16] T. Kato, *Perturbation theory for linear operators.* Springer, Berlin—Heidelberg—New York 1966.

[17] G. Köthe, *Topologische lineare Räume.* I. Zweite Auflage. Springer, Berlin—Heidelberg—New York 1966.

[18] S. G. Mikhlin and K. L. Smolitskiy, *Approximate methods for solution of differential and integral equations.* American Elsevier, New York 1967.

[19] J. Nečas, *Les méthodes directes en théorie des équations elliptiques.* Masson, Paris 1967.

[20] W. V. Petryshyn, *On a class of K-p. d. and non-K-p. d. operators and operator equations.* J. Math. Anal. Appl. **10** (1965), 1—24.

[21] W. V. Petryshyn, *Projection methods in nonlinear numerical functional analysis.* J. Math. Mech. **17** (1967), 353—372.

[22] W. V. Petryshyn, *On the eigenvalue problem $Tu - \lambda Su = o$ with unbounded and nonsymmetric operators T and S.* Philos. Trans. Roy. Soc. London **262** (1968), 413—458.

[23] F. Riesz et B. Sz.-Nagy, *Leçons d'analyse fonctionnelle.* Akadémiai Kiadó, Budapest 1952.

[24] F. Stummel, *Über die Differenzenapproximation des Dirichletproblems für eine lineare elliptische Differentialgleichung zweiter Ordnung.* Math. Ann. **163** (1966), 321—339.

[25] F. Stummel, *Elliptische Differenzenoperatoren unter Dirichletrandbedingungen.* Math. Z. **97** (1967), 169—211.

[26] F. Stummel, *Lineare elliptische Differenzenoperatoren.* Frankfurt: Vorlesung WS 1966/67.

[27] F. Stummel, *Rand- und Eigenwertaufgaben in Sobolewschen Räumen.* Lecture Notes in Mathematics **102.** Springer, Berlin—Heidelberg—New York 1969.

[28] F. Stummel, *Diskrete Konvergenz linearer Operatoren* I, II. Math. Ann. **190** (1970), 45—92, Math. Z. **120** (1971), 231—264.

[29] A. E. Taylor, *Functional analysis.* Wiley, New York 1961.

On Set-Valued Metric Projections

By

IVAN SINGER

INSTITUTUL DE MATEMATICĂ
BUCUREŞTI

Introduction

Let E be a normed linear space and G a *proximinal* linear subspace of E, i.e. such that

$$(1) \qquad \mathscr{P}_G(x) = \{g_0 \in G \mid \|x - g_0\| = \inf_{g \in G} \|x - g\|\} \neq \emptyset \qquad (x \in E).$$

The mapping $\mathscr{P}_G : x \to \mathscr{P}_G(x)$ of E into the collection 2^G of all non-void closed subsets of G is called *the set-valued metric projection* of E onto G.

We recall (see e.g. [9]) that if E, G are two metric spaces, a mapping $\mathscr{U} : E \to 2^G$ is called *upper semi-continuous* (u.s.c.), respectively *lower semi-continuous* (l.s.c.) if the set $\{x \in E \mid \mathscr{U}(x) \subset M\}$ is open for each open subset M of G, respectively closed for each closed subset M of G, or, equivalently, if the set $\{x \in E \mid \mathscr{U}(x) \cap N \neq \emptyset\}$ is closed for each closed subset N of G, respectively open for each open subset N of G. Let us also recall that a continuous mapping $u : E \to G$ is said to be a *continuous selection* for a set-valued mapping $\mathscr{U} : E \to 2^G$ if $u(x) \in \mathscr{U}(x)$ for all $x \in E$. For $G \subset E$ proximinal and the set-valued metric projection $\mathscr{U} = \mathscr{P}_G$ these properties have been relatively little studied, the starting basic papers being essentially [12], [3], [8] in 1964 and then [11], [14] and [2] in 1968 (see also [1]).

In the present paper we shall further our study of semi-continuity properties of $\mathscr{P}_G$ [12] and of continuous selections for $\mathscr{P}_G$ [14]. Among other results, we shall give characterizations of the proximinal linear subspaces G of a normed linear space E, for which $\mathscr{P}_G$ is upper or lower semi-continuous or admits a continuous selection. As a consequence, we shall obtain relations between properties of set-valued metric projections and of set-valued Hahn—Banach extension maps.

1. Upper semi-continuity of $\mathscr{P}_G$

THEOREM 1. *Let G be a proximinal linear subspace of a normed linear space E, such that $\mathscr{P}_G$ is u.s.c. Then for each $x \in E$ the set $\mathscr{P}_G(x)$ is compact.*

PROOF. Assume that there exists an element $x_0 \in E$ for which $\mathscr{P}_G(x_0)$ is not compact and thus there is a sequence $\{g_n'\} \subset \mathscr{P}_G(x_0)$ which contains no convergent subsequence. We may also assume, without loss of generality (making a translation, if necessary), that $0 \notin \mathscr{P}_G(x_0)$.

Observe now that for any sequence $\{\alpha_n\}$ such that $1 \leqq \alpha_n < \infty$ $(n=1, 2, \ldots)$, the sequence $\{\alpha_n g'_n\}$ contains no convergent subsequence; indeed, if $\{\alpha_{n_k} g'_{n_k}\}$ converges, then, since $0 < \dfrac{1}{\alpha_{n_k}} \leqq 1$ $(k=1, 2, \ldots)$, some subsequence $\left\{\dfrac{1}{\alpha_{n_{k_m}}}\right\}$ of $\left\{\dfrac{1}{\alpha_{n_k}}\right\}$ converges, whence $\{g'_{n_{k_m}}\} = \left\{\dfrac{1}{\alpha_{n_{k_m}}}(\alpha_{n_{k_m}} g'_{n_{k_m}})\right\}$ converges, in contradiction with our assumption on $\{g'_n\}$.

Now put

$$\text{(2)} \qquad \beta_n = \sup_{\beta g'_n \in \mathscr{P}_G(x_0)} \beta \qquad (n=1, 2, \ldots),$$

$$\text{(3)} \qquad g_n = \frac{n+1}{n} \beta_n g'_n \qquad (n=1, 2, \ldots).$$

Then, since $\{g'_n\} \subset \mathscr{P}_G(x_0)$ and $0 \notin \mathscr{P}_G(x_0)$, and since $\mathscr{P}_G(x_0)$ is bounded, we have $1 \leqq \beta_n < \infty$ $(n=1, 2, \ldots)$, whence, by the above observation $\left(\text{for } \alpha_n = \dfrac{n+1}{n} \beta_n > 1\right)$, $\{g_n\}$ contains no convergent subsequence. Furthermore, since $\dfrac{n+1}{n} \beta_n > \beta_n$ $(n=1, 2, \ldots)$, by (2) we have $\{g_n\} \subset G \setminus \mathscr{P}_G(x_0)$. Put

$$\text{(4)} \qquad x_n = \frac{n+1}{n} x_0 \qquad (n=1, 2, \ldots),$$

$$\text{(5)} \qquad N = \{g_n\} \qquad (n=1, 2, \ldots).$$

Then $x_n \to x_0$ and, by the above, N is a closed subset of G, such that $\mathscr{P}_G(x_0) \cap \cap N = \emptyset$. However, we shall show that $g_n \in \mathscr{P}_G(x_n)$, whence $\mathscr{P}_G(x_n) \cap N \neq \emptyset$ $(n=1, 2, \ldots)$, and thus $\mathscr{P}_G$ is not u.s.c., which will complete the proof. Indeed, observe first that for each n the sup in (2) is attained, i.e. $\beta_n g'_n \in \mathscr{P}_G(x_0)$. For, if $\beta_n = 1$, the statement is obviously true; if $\beta_n > 1$, then, taking

$$\{\beta_n^{(k)}\}_{k=1}^{\infty} \quad \text{with} \quad 1 \leqq \beta_n^{(k)} < \beta_n \ (k=1, 2, \ldots), \quad \lim_{k \to \infty} \beta_n^{(k)} = \beta_n,$$

we have, by (2) and the convexity of $\mathscr{P}_G(x_0)$, $\beta_n^{(k)} g'_n \in \mathscr{P}_G(x_0)$ $(k=1, 2, \ldots)$, whence, since $\mathscr{P}_G(x_0)$ is closed, $\beta_n g'_n \in \mathscr{P}_G(x_0)$. Consequently, for any $g \in G$ we have

$$\|x_n - g_n\| = \left\|\frac{n+1}{n} x_0 - \frac{n+1}{n} \beta_n g'_n\right\| = \frac{n+1}{n} \|x_0 - \beta_n g'_n\| \leqq$$

$$\leqq \frac{n+1}{n} \left\|x_0 - \frac{n}{n+1} g\right\| = \left\|\frac{n+1}{n} x_0 - g\right\| = \|x_n - g\|,$$

and thus $g_n \in \mathscr{P}_G(x_n)$ $(n=1, 2, \ldots)$, which completes the proof of theorem 1.

We observe that theorem 1 permits to construct easily, in various concrete spaces, examples of proximinal linear subspaces G with $\mathscr{P}_G$ non-u.s.c.; in this connection, see also corollary 3 below.

P. D. MORRIS has proved ([11], theorem 3) that if G is a proximinal linear subspace in a normed linear space E, such that the closed set

$$(6) \qquad \mathscr{P}_G^{-1}(0) = \{x \in E \mid 0 \in \mathscr{P}_G(x)\}$$

is boundedly compact (i.e., every bounded sequence in $\mathscr{P}_G^{-1}(0)$ contains a convergent subsequence), then a) $\mathscr{P}_G$ is u.s.c. and b) for each $x \in E$ the set $\mathscr{P}_G(x)$ is compact and, if codim $G < \infty$, the converse is also true, i.e., a) and b) imply that $\mathscr{P}_G^{-1}(0)$ is boundedly compact (we recall that, by definition, codim $G = \dim E/G$). Theorem 1 above sharpens this result of Morris, by reducing conditions a) and b) to condition a), that is, we obtain the following result:

COROLLARY 1. *If G is a proximinal linear subspace of finite codimension in a normed linear space E, $\mathscr{P}_G$ is u.s.c. if and only if $\mathscr{P}_G^{-1}(0)$ is boundedly compact.*

Now we shall prove that if G is a hyperplane in E (that is, if codim $G = 1$), the converse of theorem 1 is also true, i.e. we have

THEOREM 2. *Let G be a proximinal hyperplane in a normed linear space E, such that for each $x \in E$ the set $\mathscr{P}_G(x)$ is compact. Then $\mathscr{P}_G$ is u.s.c.*

PROOF. By corollary 1 (or by the above mentioned result of MORRIS) it is sufficient to prove that $\mathscr{P}_G^{-1}(0)$ is boundedly compact. Assume the contrary, i.e. that there exists a bounded sequence $\{x_n\} \subset \mathscr{P}_G^{-1}(0)$ which contains no convergent subsequence; we may assume (by excluding 0 from $\{x_n\}$, if necessary) that $x_n \neq 0$ $(n = 1, 2, \ldots)$. Since G is a hyperplane, there exists an $f \in E^*$ such that $G = \{x \in E \mid f(x) = 0\}$. Then, since $x_n \in \mathscr{P}_G^{-1}(0)$ and $\mathscr{P}_G^{-1}(0) \cap G = \{0\}$, we have $f(x_n) \neq 0$ $(n = 1, 2, \ldots)$. Take any $x_0 \in E$ such that $f(x_0) = 1$ and put

$$(7) \qquad g_n = x_0 - \frac{1}{f(x_n)} x_n \qquad (n = 1, 2, \ldots).$$

Then $f(g_n) = f(x_0) - 1 = 0$ and thus $g_n \in G$ $(n = 1, 2, \ldots)$. Furthermore, for any n and any $g \in G$ we have, by $x_n \in \mathscr{P}_G^{-1}(0)$,

$$\|x_0 - g_n\| = \left| \frac{1}{f(x_n)} \right| \|x_n\| \leq \left| \frac{1}{f(x_n)} \right| \|x_n - f(x_n)(g - g_n)\| =$$

$$= \left\| \frac{1}{f(x_n)} x_n + g_n - g \right\| = \|x_0 - g\|,$$

and thus $g_n \in \mathscr{P}_G(x_0)$ $(n=1, 2, \ldots)$. However, we shall show that $\{g_n\}$ contains no convergent subsequence, whence $\mathscr{P}_G(x_0)$ is not compact, which will complete the proof.

Indeed if $\left\{ \dfrac{1}{f(x_{n_k})} x_{n_k} \right\}$ converges, then, since $\sup_k |f(x_{n_k})| \leq \|f\| \sup_k \|x_{n_k}\| < \infty$, some subsequence $\{f(x_{n_{k_m}})\}$ of $\{f(x_{n_k})\}$ converges, whence

$$\{x_{n_{k_m}}\} = \left\{ f(x_{n_{k_m}}) \left(\frac{1}{f(x_{n_{k_m}})} x_{n_{k_m}} \right) \right\}$$

converges, in contradiction with our assumption on $\{x_n\}$. This completes the proof of theorem 2.

REMARK 1. Theorem 2 is no longer true for $2 \leq \operatorname{codim} G < \infty$. Indeed, observe that if G is a Čebyšev subspace of E, i.e. such that $\mathscr{P}_G(x)$ is a singleton for each $x \in E$, say $\mathscr{P}_G(x) = \{\pi_G(x)\}$, then the upper semi-continuity of $\mathscr{P}_G$ reduces to the usual continuity of $\pi_G : E \to G$ and it is known (see e.g. [11]) that there exist normed linear spaces E containing, for each m with $2 \leq m < \infty$, a Čebyšev subspace G with $\operatorname{codim} G = m$, for which π_G is discontinuous.

We recall that a linear subspace G of a normed linear space E is called a pseudo-Čebyšev subspace if G is proximinal and if $\dim \mathscr{P}_G(x) < \infty$ for each $x \in E$. From theorem 2 it follows obviously

COROLLARY 2. *If G is a pseudo-Čebyšev hyperplane in a normed linear space E, then $\mathscr{P}_G$ is u.s.c.*

From theorems 1 and 2 we infer

COROLLARY 3. *For a normed linear space E the following statements are equivalent:*

$1°$. *For every proximinal hyperplane $G \subset E$, $\mathscr{P}_G$ is u.s.c.*

$2°$. *Every face of the unit cell $S_E = \{x \in E \mid \|x\| \leq 1\}$ (i. e., every maximal convex subset of $\operatorname{Fr} S_E = \{x \in E \mid \|x\| = 1\}$) is compact.*

PROOF. Assume that S_E has a non-compact face, say F. By Mazur's theorem, there exists a support hyperplane H of S_E such that $H \supset F$. Then H is a proximinal linear manifold (i.e., translated linear subspace) and $\mathscr{P}_H(0) = H \cap S_E \supset F \cap S_E = F$, whence $\mathscr{P}_H(0)$ is non-compact. Hence, by translation, there exist a proximinal hyperplane G (through the origin) and an element $x_0 \in E$ with $\mathscr{P}_G(x_0)$ non-compact. Consequently, by theorem 1, $\mathscr{P}_G$ is not u.s.c.

Conversely, assume that there exists in E a proximinal hyperplane G for which $\mathscr{P}_G$ is not u.s.c. Then, by theorem 2, there is an element $x_0 \in E$ such that $\mathscr{P}_G(x_0)$ is non-compact. Since by (1) $\mathscr{P}_G(x_0) = G \cap \operatorname{Fr} S(x_0, d)$, where $S(x_0, d)$ is the cell with center x_0 and radius $d = \inf_{g \in G} \|x_0 - g\|$, and since $\mathscr{P}_G(x_0)$ is convex, it follows that any face of the cell $S(x_0, d)$, containing $\mathscr{P}_G(x_0)$, is non-compact. Hence, by translation and homothety, the unit cell S_E also has a non-compact face, which completes the proof.

Let us define now, for any proximinal subspace G of E, a set-valued mapping $\mathscr{V}_G$ of E/G into $2^{\mathscr{P}_G^{-1}(0)}$, by

$$(8) \qquad \mathscr{V}_G(x+G) = x - \mathscr{P}_G(x) = \{x - g_0 \,|\, g_0 \in \mathscr{P}_G(x)\} \qquad (x+G \in E/G).$$

Observe that $\mathscr{V}_G$ is well defined since $\mathscr{P}_G(x+g) = \mathscr{P}_G(x)+g$ $(g \in G)$. It is easy to see that $\mathscr{V}_G(x+G) \in 2^{\mathscr{P}_G^{-1}(0)}$, i.e., is a non-void closed subset of $\mathscr{P}_G^{-1}(0)$. Indeed, $\mathscr{V}_G(x+G) \neq \emptyset$ since G is proximinal. Furthermore, $\mathscr{V}_G(x+G) \subset \mathscr{P}_G^{-1}(0)$, since for every $g_0 \in \mathscr{P}_G(x)$ we have $0 \in \mathscr{P}_G(x-g_0)$, and thus $x-g_0 \in \mathscr{P}_G^{-1}(0)$. Finally, if $x_n - g_n \to \to x \in E$, where $g_n \in \mathscr{P}_G(x_n)$ $(n=1, 2, ...)$, then, since $\mathscr{P}_G^{-1}(0)$ is closed, we have $x \in \mathscr{P}_G^{-1}(0)$, whence $x = x-0 \in \mathscr{V}_G(x+G)$, which proves that $\mathscr{V}_G(x+G)$ is closed.

Observe that $\mathscr{V}_G$ is nothing else than the set-valued mapping $E/G \to 2^{\mathscr{P}_G^{-1}(0)}$ induced by $I - \mathscr{P}_G$, where I is the identical mapping of E onto itself. In the sequel we shall show that there are close connections between the properties of the mappings $\mathscr{P}_G$ and $\mathscr{V}_G$. For this purpose we shall need the following result, similar to theorem 1:

THEOREM 1'. *Let G be a proximinal linear subspace of a normed linear space E, such that $\mathscr{V}_G$ is u.s.c. Then for each $x \in E$ the set $\mathscr{P}_G(x)$ is compact.*

PROOF. Assume that there exists an element $x_0 \in E$ for which $\mathscr{P}_G(x_0)$ is not compact and construct $g_n \in \mathscr{P}_G(x_n) \setminus \mathscr{P}_G(x_0)$ and $x_n \in E$ $(n=1, 2, ...)$ as in the above proof of theorem 1. Put

$$(9) \qquad\qquad\qquad N_0 = \{x_n - g_n\}.$$

Then, since $g_n \in \mathscr{P}_G(x_n)$, we have $x_n - g_n \in \mathscr{P}_G^{-1}(0)$ $(n=1, 2, ...)$ and thus $N_0 \subset \mathscr{P}_G^{-1}(0)$. Furthermore, N_0 is closed, since $\{x_n - g_n\}$ contains no convergent subsequence (indeed, if $\{x_{n_k} - g_{n_k}\}$ converges then, since $x_n \to x_0$, the sequence $\{g_{n_k}\} = \{x_{n_k} - (x_{n_k} - g_{n_k})\}$ also converges, in contradiction with the fact established in the proof of theorem 1 that $\{g_n\}$ contains no convergent subsequence). Observe now that $x_n + G \to x_0 + G$ (since $x_n \to x_0$) and that $\mathscr{V}_G(x_n + G) \cap N_0 \neq \emptyset$, since it contains $x_n - g_n$ $(n=1, 2, ...)$. We shall show that $\mathscr{V}_G(x_0 + G) \cap N_0 = \emptyset$, whence $\mathscr{V}_G$ is not u.s.c., which will complete the proof. In other words, this amounts to show that there is no $g_0 \in \mathscr{P}_G(x_0)$ such that $x_0 - g_0 \in N_0$, but one can show even more, that there is no $g_0 \in G$ such that $x_0 - g_0 \in N_0$. Indeed, if there existed such a $g_0 \in G$, then, by (9) and (4), there would exist an index $n_0 \geq 1$ such that $x_0 - g_0 = x_{n_0} - g_{n_0} = (n_0+1/n_0)x_0 - g_{n_0}$, whence $(1/n_0)x_0 = g_{n_0} - g_0 \in G$, in contradiction with the assumption that $\mathscr{P}_G(x_0)$ is non-compact. This completes the proof of theorem 1'.

Let us give now two lemmas on relations between the mapping $\mathscr{V}_G$ and the canonical mapping $\omega_G : E \to E/G$, which we shall also use in the sequel. Observe first that $\omega_G|_{\mathscr{P}_G^{-1}(0)}$ maps $\mathscr{P}_G^{-1}(0)$ *onto* E/G, since for any $x+G \in E/G$ and any $g_0 \in \mathscr{P}_G(x)$ (recall that we assume G proximinal) we have $x-g_0 \in \mathscr{P}_G^{-1}(0)$ and $\omega_G(x-g_0) = x+G$. Hence $(\omega_G|_{\mathscr{P}_G^{-1}(0)})^{-1}$ is a set-valued mapping defined on the whole quotient space E/G.

LEMMA 1. *Let G be a proximinal linear subspace of a normed linear space E. Then*

$$(10) \qquad \mathscr{V}_G(x+G) = \omega_G^{-1}(x+G) \cap \mathscr{P}_G^{-1}(0) \qquad (x+G \in E/G),$$

where $\omega_G^{-1}(x+G) = \{z \in E | \omega_G(z) = x+G\} = \{x+g | g \in G\}$. *Consequently, we have*

$$(11) \qquad \mathscr{V}_G = (\omega_G|_{\mathscr{P}_G^{-1}(0)})^{-1}.$$

PROOF. The inclusion $\mathscr{V}_G(x+G) \subset \omega_G^{-1}(x+G) \cap \mathscr{P}_G^{-1}(0)$ was observed after formula (8). Conversely, let $z \in \omega_G^{-1}(x+G) \cap \mathscr{P}_G^{-1}(0)$. Then $z = x+g$ for some $g \in G$, whence, since $0 \in \mathscr{P}_G(z)$, we obtain $-g \in \mathscr{P}_G(z-g) = \mathscr{P}_G(x)$ and therefore $z = x - -(-g) \in x - \mathscr{P}_G(x) = \mathscr{V}_G(x+G)$. Thus we have (10). Now, since by definition for any $x+G \in E/G$ we have

$$(\omega_G|_{\mathscr{P}_G^{-1}(0)})^{-1}(x+G) = \{z \in \mathscr{P}_G^{-1}(0) | \omega_G(z) = x+G\} = \omega_G^{-1}(x+G) \cap \mathscr{P}_G^{-1}(0),$$

(11) follows from (10), which completes the proof of lemma 1.

LEMMA 2. *Let G be a proximinal linear subspace of a normed linear space E. Then for any set $A \subset E$ we have*

$$(12) \qquad \omega_G(A) = \{x+G \in E/G | \omega_G^{-1}(x+G) \cap A \neq \emptyset\}.$$

Consequently, for any set $A \subset \mathscr{P}_G^{-1}(0)$ we have

$$(13) \qquad \omega_G|_{\mathscr{P}_G^{-1}(0)}(A) = \{x+G \in E/G | \mathscr{V}_G(x+G) \cap A \neq \emptyset\}.$$

PROOF. Let $x+G \in \omega_G(A)$. Then $x+G = \omega_G(z)$ for some $z \in A$, whence $z \in \omega_G^{-1}(x+G)$ and therefore $\omega_G^{-1}(x+G) \cap A \neq \emptyset$. Conversely, if $z \in \omega_G^{-1}(x+G) \cap A$, then $x+G = \omega_G(z) \in \omega_G(A)$. Thus we have (12). Now, if $A \subset \mathscr{P}_G^{-1}(0)$, then $A = \mathscr{P}_G^{-1}(0) \cap A$, whence, by (12),

$$\omega_G|_{\mathscr{P}_G^{-1}(0)}(A) = \omega_G(A) = \{x+G \in E/G | \omega_G^{-1}(x+G) \cap \mathscr{P}_G^{-1}(0) \cap A \neq \emptyset\}$$

which, together with lemma 1, formula (10), implies (13), completing the proof of lemma 2.

Now we are ready to prove our main characterizations of upper semi-continuity of $\mathscr{P}_G$:

THEOREM 3. *For a proximinal linear subspace G of a normed linear space E the following statements are equivalent:*

1°. $\mathscr{P}_G$ *is u.s.c.*

2°. $\mathscr{V}_G$ *is u.s.c.*

3°. $\omega_G|_{\mathscr{P}_G^{-1}(0)}$ *carries closed sets onto closed sets.*

These conditions are implied by — and if codim $G < \infty$, *they are equivalent to — the following:*

4°. $\mathscr{P}_G^{-1}(0)$ *is boundedly compact.*

The above conditions imply — and if codim $G = 1$, *they are equivalent to — the following:*

5°. $\mathscr{P}_G(x)$ *is compact for every* $x \in E$.

PROOF. The implication $4^\circ \Rightarrow 1^\circ$ is due to P. D. MORRIS [11] and for codim $G < \infty$ the implication $1^\circ \Rightarrow 4^\circ$ is contained in corollary 1 above. Furthermore, the implication $1^\circ \Rightarrow 5^\circ$ and for codim $G = 1$ the implication $5^\circ \Rightarrow 1^\circ$, are nothing else than theorems 1 and 2 above. Observe also that the equivalence $2^\circ \leftrightarrow 3^\circ$ is an immediate consequence of lemma 2, formula (13) and the definition of upper semi-continuity. Thus it remains to prove the equivalence $1^\circ \leftrightarrow 2^\circ$.

Assume that we have 1°, but not 2°. Then there exist a closed set $N_0 \subset \mathscr{P}_G^{-1}(0)$ and a sequence $\{x_n + G\} \subset E/G$ converging to an element $x_0 + G \in E/G$, such that $\mathscr{V}_G(x_n + G) \cap N_0 \neq \emptyset$ $(n = 1, 2, \ldots)$, $\mathscr{V}_G(x_0 + G) \cap N_0 = \emptyset$; thus, by the definition of $\mathscr{V}_G$, there exist elements $g_n \in \mathscr{P}_G(x_n)$ such that $x_n - g_n \in N_0$ $(n = 1, 2, \ldots)$, but there is no $g_0 \in \mathscr{P}_G(x_0)$ such that $x_0 - g_0 \in N_0$. This latter condition means that $g_0 \notin x_0 - N_0$ for all $g_0 \in \mathscr{P}_G(x_0)$, i.e.

$$(14) \qquad \mathscr{P}_G(x_0) \cap (x_0 - N_0) = \emptyset.$$

Now, since $x_n + G \to x_0 + G$, there exist elements $g'_n \in G$ such that $x_n - x_0 - g'_n \to 0$ (e.g. one can take any $g'_n \in \mathscr{P}_G(x_n - x_0)$), hence

$$(15) \qquad x_n - g'_n \to x_0.$$

We claim that there exists a positive integer K such that

$$(16) \qquad \mathscr{P}_G(x_0) \cap (x_n - g'_n - N_0) = \emptyset \qquad (n \geq K).$$

Indeed, otherwise there would exist an infinite sequence $\{n_k\}$ and a sequence $\{y_k\} \subset N_0$ such that $x_{n_k} - g'_{n_k} - y_k \in \mathscr{P}_G(x_0)$ $(k = 1, 2, \ldots)$. Then, since by theorem 1 $\mathscr{P}_G(x_0)$ is compact, this sequence would have a subsequence $\{x_{n_{k_m}} - g'_{n_{k_m}} - y_{k_m}\}$ converging to a $g_0 \in \mathscr{P}_G(x_0)$, whence, by (15), $y_{k_m} = x_{n_{k_m}} - g'_{n_{k_m}} - (x_{n_{k_m}} - g'_{n_{k_m}} - y_{k_m}) \to x_0 - g_0$. Since N_0 is closed and $\{y_{k_m}\} \subset N_0$, it would follow that $x_0 - g_0 \in N_0$, in contradiction with the assumption that there is no such $g_0 \in \mathscr{P}_G(x_0)$. This proves our claim (16).

Put

$$(17) \quad N = \left\{ \bigcup_{n \geq K} (x_n - g'_n - N_0) \cup (x_0 - N_0) \right\} \cap G = \left\{ (\{x_n - g'_n\}_{n \geq K} \cup \{x_0\}) - N_0 \right\} \cap G.$$

Since by (15) the set $\{x_n - g'_n\}_{n \geq K} \cup \{x_0\}$ is compact and since N_0 is closed, it follows (see e.g. [6], p. 414, lemma 3) that the set $\{(\{x_n - g'_n\}_{n \geq K} \cup \{x_0\}) - N_0\}$ is closed, whence N is a closed subset of G in the relative topology of G. Furthermore, by (14) and (16) we have $\mathcal{P}_G(x_0) \cap N = \emptyset$. However, we shall show that $\mathcal{P}_G(x_n - g'_n) \cap N \neq \emptyset$ $(n \geq K)$, whence, by (15), $\mathcal{P}_G$ is not u.s.c., in contradiction with $1°$, which will prove that $1° \Rightarrow 2°$. Indeed, by $x_n - g_n \in N_0$ we have $g_n \in x_n - N_0$, whence $g_n - g'_n \in (x_n - g'_n - N_0) \cap G \subset N$ for all $n \geq K$, and by $g_n \in \mathcal{P}_G(x_n)$ we have $g_n - g'_n \in \mathcal{P}_G(x_n) - g'_n = \mathcal{P}_G(x_n - g'_n)$ for all n, which proves that $\mathcal{P}_G(x_n - g'_n) \cap N \neq \emptyset$ $(n \geq K)$. Thus, $1° \Rightarrow 2°$.

Conversely, assume now that we have $2°$, but not $1°$. Then there exist an open subset M of G (in the relative topology of G) and a sequence $\{x_n\} \subset E$ converging to an element $x_0 \in E$, such that $\mathcal{P}_G(x_n) \not\subset M$ $(n = 1, 2, \ldots)$, $\mathcal{P}_G(x_0) \subset M$; thus, since G is proximinal, there exist elements $g_n \in \mathcal{P}_G(x_n) \setminus M$ $(n = 1, 2, \ldots)$. Let M_0 be any open set in E such that $M = G \cap M_0$. We claim that there exists a positive integer K such that

$$(18) \qquad \mathcal{P}_G(x_0) + (x_n - x_0) \subset M_0 \qquad (n \geq K).$$

Indeed, since $\mathcal{P}_G(x_0) \subset M \subset M_0$ and since M_0 is open in E, for each $g_0 \in \mathcal{P}_G(x_0)$ there exists a cell $S(G_0, r_{g_0})$ with center at g_0 such that $S(g_0, r_{g_0}) \subset M_0$. Now, since by theorem $1'$ $\mathcal{P}_G(x_0)$ is compact, the covering $\bigcup_{g_0 \in \mathcal{P}_G(x_0)} S(g_0, r_{g_0})$ of $\mathcal{P}_G(x_0)$ contains a finite subcovering of $\mathcal{P}_G(x_0)$. If r is the minimal radius of the cells of this finite subcovering, then for all n satisfying $\|x_n - x_0\| \leq r$ we have obviously $\mathcal{P}_G(x_0) + (x_n - x_0) \subset M_0$. Thus, if we take K such that $\|x_n - x_0\| \leq r$ for all $n \geq K$ (which is possible, since $x_n \to x_0$), then we obtain (18).

Put

$$(19) \qquad N_0 = \overline{\bigcup_{n \geq K} \{x_n - g_n\}}.$$

Then, since $x_n - g_n \in \mathcal{P}_G^{-1}(0)$ (by $g_n \in \mathcal{P}_G(x_n)$) and since $\mathcal{P}_G^{-1}(0)$ is closed, N_0 is a closed subset of $\mathcal{P}_G^{-1}(0)$ and clearly $x_n - g_n \in \mathcal{V}_G(x_n + G) \cap N_0 \neq \emptyset$ $(n \geq K)$. Hence, since $x_n + G \to x_0 + G$ (by $x_n \to x_0$) and since we assumed $2°$, it follows that $\mathcal{V}_G(x_0 + G) \cap N_0 \neq \emptyset$, i.e. there exists an element $g_0 \in \mathcal{P}_G(x_0)$ such that $x_0 - g_0 \in N_0$. Now, by the definition (19) of N_0 there are two cases:

α) There exists an index $n \geq K$ such that $x_0 - g_0 = x_n - g_n$. Then by (18), $g_n = g_0 + (x_n - x_0) \in \mathcal{P}_G(x_0) + (x_n - x_0) \subset M_0$, whence $g_n \in M$, in contradiction with the definition of g_n.

β) There exists an infinite subsequence $\{x_{n_k} - g_{n_k}\}$ of $\{x_n - g_n\}$ such that $x_{n_k} - g_{n_k} \to x_0 - g_0$. Then, since $x_{n_k} \to x_0$, it follows that $g_{n_k} \to g_0$, whence, since $g_0 \in \mathcal{P}_G(x_0) \subset M_0$ and since M_0 is open, we obtain that $g_{n_k} \in M_0$ and hence $g_{n_k} \in M$, for all sufficiently large n_k, in contradiction with the definition of g_{n_k}.

Thus, in both cases we have arrived to a contradiction, which proves the implication $2° \Rightarrow 1°$. This completes the proof of theorem 3.

REMARK 2. Since the above proof of the equivalence $1° \leftrightarrow 2°$ is somewhat lengthy, it is of some interest to mention that in the particular cases when codim $G < \infty$ or dim $G < \infty$ one can give the following shorter proofs of this equivalence:

a) *The case when* codim $G < \infty$. Assume that we have $1°$. Let $N_0 \subset \mathscr{P}_G^{-1}(0)$ be a closed set and let $\{x_n + G\} \subset E/G$ be a sequence converging to an $x_0 + G \in E/G$, such that $\mathscr{V}_G(x_n + G) \cap N_0 \neq \emptyset$ $(n = 1, 2, \ldots)$; thus, there exist elements $g_n \in \mathscr{P}_G(x_n)$ with $x_n - g_n \in N_0$ $(n = 1, 2, \ldots)$. Then, since $g_n \in \mathscr{P}_G(x_n)$, and $\{x_n + G\}$ converges, we have $\sup_n \|x_n - g_n\| = \sup_n \|x_n + G\| < \infty$, whence, since by corollary 1 $\mathscr{P}_G^{-1}(0)$ is boundedly compact, $\{x_n - g_n\}$ contains a subsequence $\{x_{n_k} - g_{n_k}\}$ converging to an element $y \in N_0$ (because N_0 is closed). Then

$$\omega_G(y) = \omega_G\left(\lim_{k \to \infty} (x_{n_k} - g_{n_k})\right) = \lim_{k \to \infty} \omega_G(x_{n_k} - g_{n_k}) = \lim_{k \to \infty} (x_{n_k} + G) = x_0 + G,$$

and hence there is an element $g_0 \in G$ such that $x_0 - g_0 = y \in N_0$. Furthermore,

$$\|x_0 - g_0\| = \lim_{k \to \infty} \|x_{n_k} - g_{n_k}\| = \lim_{k \to \infty} \|x_{n_k} + G\| = \|x_0 + G\| = \inf_{g \in G} \|x - g\|,$$

whence $g_0 \in \mathscr{P}_G(x_0)$. Thus, $\mathscr{V}_G(x_0 + G) \cap N_0 \neq \emptyset$ (since it contains $x_0 - g_0$), which proves that $1° \Rightarrow 2°$.

Conversely, assume now that we have $2°$, but not $1°$. Then, by corollary 1, there exists a bounded sequence $\{x_n\} \subset \mathscr{P}_G^{-1}(0)$ containing no convergent subsequence. We may assume that $\{x_n + G\}$ has an infinite subsequence consisting of distinct elements, since otherwise (omitting, if necessary, a finite number of elements of $\{x_n + G\}$) we have $x_1 + G = x_2 + G = \ldots$ and then we can replace $\{x_n\}$ by the sequence $\{n^{-1}(n+1)x_n\} \subset \mathscr{P}_G^{-1}(0)$, which will have all the required properties; indeed, clearly $\{n^{-1}(n+1)x_n\}$ is bounded and contains no convergent subsequence and if for a pair of indices n, m we have $n^{-1}(n+1)x_n + G = m^{-1}(m+1)x_m + G$ then, taking into account that $x_n + G = x_m + G$, we obtain $x_n, x_m \in G \cap \mathscr{P}_G^{-1}(0) = \{0\}$, in contradiction with the assumption that the x_n are distinct. This proves our claim. Therefore, we may assume that the sequence $\{x_n + G\}$ itself consists of distinct elements. Since codim $G < \infty$ and $\sup_n \|x_n + G\| \leq \sup_n \|x_n\| < \infty$, $\{x_n + G\}$ has a subsequence $\{x_{n_k} + G\}$ converging to an element $x_0 + G$; we may assume (omitting, if necessary, one element of the sequence $\{x_{n_k} + G\}$), that $x_0 + G \neq x_{n_k} + G$ $(k = 1, 2, \ldots)$. Now put

$$N_0 = \{x_{n_k}\}. \tag{20}$$

Then $N_0 \subset \mathscr{P}_G^{-1}(0)$ is closed and since $0 \in \mathscr{P}_G(x_{n_k})$, we have $x_{n_k} = x_{n_k} - 0 \in \mathscr{V}_G(x_{n_k} + G) \cap N_0 \neq \emptyset$ $(k = 1, 2, \ldots)$, whence, since $x_{n_k} + G \to x_0 + G$ and since $\mathscr{V}_G$ is assumed u.s.c., we obtain $\mathscr{V}_G(x_0 + G) \cap N_0 \neq \emptyset$, i.e. there exists an element $g_0 \in \mathscr{P}_G(x_0)$ such that $x_0 - g_0 \in N_0$. However, this is not possible even for an element $g_0 \in G$, since if $x_0 - g_0 = x_{n_{k_0}}$, then $x_{n_{k_0}} + G = x_0 + G$, in contradiction with our assumption on $x_0 + G$. Thus, $2° \Rightarrow 1°$, which completes the proof.

b) *The case when* dim $G < \infty$. Since by [12], theorem 1, for any linear subspace $G \subset E$ with dim $G < \infty$, $\mathscr{P}_G$ is u.s.c., it will be sufficient to show that for any such subspace $\mathscr{V}_G$ is u.s.c. Let $N_0 \subset \mathscr{P}_G^{-1}(0)$, $x_n + G \to x_0 + G$ and $g_n \in \mathscr{P}_G(x_n)$ be as in the necessity part of a) above. It will be sufficient to prove that $\{x_n - g_n\}$ contains a convergent subsequence, since then the argument of the necessity part of a) above applies. Since dim $G < \infty$, there exists a bounded linear projection p_G of E onto G. Let us write

$$x_n = p_G(x_n) + (I - p_G)(x_n) = g_n' + y_n \qquad (n = 0, 1, 2, \ldots). \tag{21}$$

Then, since $\sup_n \|x_n - g_n\| = K < \infty$ (see a) above), we have

$$\|g_n' - g_n\| = \|p_G(x_n) - g_n\| = \|p_G(x_n - g_n)\| \leq K \|p_G\| \qquad (n = 1, 2, \ldots),$$

whence, since $\dim G < \infty$, $\{g_n' - g_n\}$ contains a convergent subsequence $\{g_{n_k}' - g_{n_k}\}$.

On the other hand, since $x_n + G \to x_0 + G$ and since $(I - p_G)(E)$ is isomorphic to E/G by the mapping $\omega_G|_{(I-p_G)(E)} : y \to y + G$, we have

$$y_n = (\omega_G|_{(I-p_G)(E)})^{-1}(x_n + G) \to (\omega_G|_{(I-p_G)(E)})^{-1}(x_0 + G).$$

Consequently, $\{x_n - g_n\} = \{g_n' - g_n + y_n\}$ contains a convergent subsequence, which completes the proof.

REMARK 3. In the particular case when G is a Čebyšev subspace of E (see remark 1), the equivalence $1° \leftrightarrow 2°$ of theorem 3 reduces to the theorem of R. B. HOLMES ([7], theorem 6) according to which the metric projection π_G onto a Čebyšev subspace $G \subset E$ is continuous if and only if $(\omega_G|_{\pi_G^{-1}(0)})^{-1}$ is a continuous one-valued mapping of E/G onto $\pi_G^{-1}(0)$, or, equivalently, $\omega_G|_{\pi_G^{-1}(0)}$ is a homeomorphism of $\pi_G^{-1}(0)$ onto E/G. However, the techniques used above for arbitrary proximinal G and set-valued $\mathscr{P}_G$ are completely different from those of HOLMES [7] and more complicated, due to the greater generality. Let us also mention that for a Čebyšev subspace G the implications $4° \Rightarrow 1°$ and, for codim $G < \infty$, $1° \Rightarrow 4°$, of theorem 3, reduce to the theorem of E. W. CHENEY and D. E. WULBERT ([5], proposition 7, 5° and theorem 8) that π_G is continuous if — and, whenever codim $G < \infty$, if and only if — $\pi_G^{-1}(0)$ is boundedly compact. Finally, for a Čebyšev subspace G 5° always holds and thus, whenever codim $G = 1$, we have 1°, that is, π_G is continuous; however, it is known (see e.g. [13], p. 142, theorem 6. 2), that in this case π_G is even linear.

We shall deduce now from theorem 3 a relation between upper semi-continuity of metric projections and upper semi-continuity of Hahn—Banach extension maps. For this purpose let us first give

LEMMA 3. a) *For any closed linear subspace G of a normed linear space E we have*

$$(22) \qquad \mathscr{P}_G^{-1}(0) = \{x \in E \mid \|\omega_G(x)\| = \|x\|\},$$

i.e. $\mathscr{P}_G^{-1}(0)$ coincides with that subset of E, on which the restriction of ω_G (the canonical mapping of E onto E/G) is norm-preserving.

b) *For any proximinal linear subspace $G \subset E$ we have*

$$(23) \qquad \mathscr{V}_G(x + G) = \{z \in E \mid \omega_G(z) = x + G, \quad \|z\| = \|x + G\|\} \qquad (x \in E).$$

c) *For any w^*-closed (hence proximinal) linear subspace Γ of E^* we have*

$$(24) \qquad \mathscr{P}_\Gamma^{-1}(0) = \{f \in E^* \mid \|f\| = \|f|_{\Gamma_\perp}\|\},$$

$$(25) \qquad \mathscr{V}_\Gamma(f + \Gamma) = f - \mathscr{P}_\Gamma(f) = \{h \in E^* \mid h|_{\Gamma_\perp} = f|_{\Gamma_\perp}, \quad \|h\| = \|f|_{\Gamma_\perp}\|\} \qquad (f \in E^*),$$

where $\Gamma_\perp \{x \in E \mid f(x) = 0 \ (f \in \Gamma)\}$.

PROOF. a) We have, by definition,

$$\mathscr{P}_G^{-1}(0) = \{x \in E \mid 0 \in \mathscr{P}_G(x)\} = \{x \in E \mid \|x\| = \inf_{g \in G} \|x - g\|\},$$

and thus it remains only to use that, again by definition, $\|\omega_G(x)\| = \inf\limits_{g \in G} \|x-g\|$ for all $x \in E$.

b) By lemma 1, formula (11) and part a) above we have

$$\mathscr{V}_G(x+G) = (\omega_G|_{\mathscr{P}_G^{-1}(0)})^{-1}(x+G) = \{z \in \mathscr{P}_G^{-1}(0)\,|\,\omega_G(z) = x+G\} =$$

$$= \{z \in E\,|\,\omega_G(z) = x+G,\ \|\omega_G(z)\| = \|z\|\} \qquad (x \in E),$$

whence (23), which proves b). We mention that b) can be also proved directly, by observing that

$$\mathscr{V}_G(x+G) = x - \mathscr{P}_G(x) = \{x-g_0\,|\,g_0 \in \mathscr{P}_G(x)\} =$$

$$= \{z \in E\,|\,z-x \in G,\ \|z\| = \|x+G\|\} \qquad (x \in E).$$

c) Using for $G \subset E$ the notation $G^\perp = \{f \in E^*\,|\,f(x)=0\,(x \in G)\} = \{f \in E^*\,|\,f|_G=0\}$, we have, since Γ is w^*-closed, $\Gamma=(\Gamma_\perp)^\perp$, whence

$$(26) \quad f+\Gamma = f+(\Gamma_\perp)^\perp = \{h \in E^*\,|\,(h-f)|_{\Gamma_\perp} = 0\} = \{h \in E^*\,|\,h|_{\Gamma_\perp} = f|_{\Gamma_\perp}\} \quad (f \in E^*).$$

Consequently,

$$(27) \qquad \|\omega_\Gamma(f)\| = \inf\limits_{\gamma \in \Gamma} \|f+\gamma\| = \inf\limits_{\substack{h \in E^* \\ h|_{\Gamma_\perp}=f|_{\Gamma_\perp}}} \|h\| = \|f|_{\Gamma_\perp}\| \qquad (f \in E^*),$$

and hence, by (22), it follows (24).

Finally, by (23), (26), and (27) we have

$$\mathscr{V}_\Gamma(f+\Gamma) = \{h \in E^*\,|\,h-f \in \Gamma,\ \|h\| = \|f+\Gamma\|\} =$$

$$= \{h \in E^*\,|\,h|_{\Gamma_\perp} = f|_{\Gamma_\perp},\ \|h\| = \|f|_{\Gamma_\perp}\|\} \qquad (f+\Gamma \in E^*/\Gamma),$$

i.e. (25), which completes the proof of lemma 3.

We mention that it is worth while to rewrite formula (25) also in the following equivalent form:

$$(25)' \qquad \mathscr{P}_\Gamma(f) = f - \{h \in E^*\,|\,h|_{\Gamma_\perp} = f|_{\Gamma_\perp},\ \|h\| = \|f|_{\Gamma_\perp}\|\} \qquad (f \in E^*).$$

Now we are ready to give the following corollary of theorem 3:

COROLLARY 4. *For a w^*-closed linear subspace Γ of E^*, $\mathscr{P}_\Gamma$ is u.s.c. if and only if the set-valued Hahn—Banach extension map $\varphi \in (\Gamma_\perp)^* \to \{f \in E^*\,|\,f|_{\Gamma_\perp}=\varphi,\ \|f\|=\|\varphi\|\}$ is u.s.c.*

PROOF. By theorem 3, $\mathscr{P}_\Gamma$ is u.s.c. if and only if $\mathscr{V}_\Gamma : E^*/\Gamma \to 2^{\mathscr{P}_\Gamma^{-1}(0)}$ is u.s.c. On the other hand, since Γ is w^*-closed, there exists a canonical linear isometry $(\Gamma_\perp)^* \equiv E^*/(\Gamma_\perp)^\perp = E^*/\Gamma$, given by

$$\tau : \varphi \to \{h \in E^*\,|\,h|_{\Gamma_\perp} = \varphi\} = f_\varphi + \Gamma \qquad (\varphi \in (\Gamma_\perp)^*),$$

228   I. SINGER

where f_φ is an arbitrary extension of φ to the whole space E, i.e. $f_\varphi|_{\Gamma_\perp} = \varphi$,and therefore, $\mathscr{V}_\Gamma$ is u.s.c. if and only if $\mathscr{V}_\Gamma \tau : (\Gamma_\perp)^* \to 2^{\mathscr{P}_\Gamma^{-1}(0)}$ is u.s.c. However, by lemma 3 c), formula (25), we have

$$\mathscr{V}_\Gamma \tau(\varphi) = \mathscr{V}_\Gamma(f_\varphi + \Gamma) = \{h \in E^* \,|\, h|_{\Gamma_\perp} = f_\varphi|_{\Gamma_\perp} = \varphi, \ \|h\| = \|f_\varphi|_{\Gamma_\perp}\| = \|\varphi\| \}$$

$$(\varphi \in (\Gamma_\perp)^*),$$

i.e. $\mathscr{V}_\Gamma \tau$ is nothing else than the Hahn—Banach extension map $(\Gamma_\perp)^* \to E^*$, which completes the proof of corollary 4.

2. Lower semi-continuity of $\mathscr{P}_G$

We recall (see [12]) that if E, G are two metric spaces, a mapping $\mathscr{U} : E \to 2^G$ s called *upper (K)-semi-continuous* (u. (K)-s.c.), respectively *lower (K)-semi-continuous* (l. (K)-s.c.) if the relations $x_n \to x$, $y_n \in \mathscr{U}(x_n)$ $(n=1, 2, \ldots)$, $y_n \to y$ imply $y \in \mathscr{U}(x)$, respectively, if the relations $x_n \to x$, $y \in \mathscr{U}(x)$ imply the existence of a sequence $\{y_n\}$ with $y_n \in \mathscr{U}(x_n)$ $(n=1, 2, \ldots)$ such that $y_n \to y$. Here (K) stands for "Kuratowski", who has studied these notions of semi-continuity and has proved among other results, that every u.s.c. (l.s.c.) mapping is u. (K)-s.c. (respectively, l. (K)-s.c.) and that, if G is compact, the converse is also true. For non-compact G and upper semi-continuity this latter statement is no longer valid, since in [12] we have proved that *for any proximinal $G \subset E$, $\mathscr{P}_G$ is u. (K)-s.c.* (it is easy to see that $\mathscr{V}_G$ is also u.s.c., since if $x_n + G \to x_0 + G$, $x_n - g_n \in \mathscr{V}_G(x_n + G)$ for $n=1, 2, \ldots$ and $x_n - g_n \to z$, then

$$x_0 + G = \lim_{n \to \infty} (x_n + G) = \lim_{n \to \infty} (x_n - g_n + G) = z + G,$$

whence $x_0 - z = g_0 \in G$ and

$$\|x_0 - g_0\| = \|z\| = \lim_{n \to \infty} \|x_n - g_n\| = \lim_{n \to \infty} \|x_n + G\| = \|x_0 + G\| = \inf_{g \in G} \|x_0 - g\|,$$

i.e. $g_0 \in \mathscr{P}_G(x_0)$ and thus $z = x_0 - g_0 \in \mathscr{V}_G(x_0 + G))$, but we have seen in § 1 above that in general $\mathscr{P}_G$ *(and $\mathscr{V}_G$) need not be u.s.c.* However, we shall now show that for lower semi-continuity the situation is different, namely, we have

LEMMA 4. *If E, G are two metric spaces, a mapping $\mathscr{U} : E \to 2^G$ is l.s.c. if and only if it is l.(K)-s.c.*

PROOF. We have already observed that l.s.c. $\Rightarrow$ l. (K)-s.c., but for the sake of completeness let us give a proof. Assume that $\mathscr{U}$ is l.s.c. and let $x_n \to x_0$, $g_0 \in \mathscr{U}(x_0)$. Then, since $\mathscr{U}$ is l.s.c., for each p the set $A_p = \{x \in E \,|\, \mathscr{U}(x) \cap M_p \neq \emptyset\}$ is open, where M_p denotes the open cell $\left\{ g \in G \,\middle|\, \mathrm{dist}\,(g, g_0) < \dfrac{1}{p} \right\}$ in G. Hence, since $x_0 \in A_p$ (because

$g_0 \in \mathscr{U}(x_0) \cap M_p)$ and since $x_n \to x_0$, it follows that for each p there is an index $K_p >$ $> K_{p-1}$ such that $x_n \in A_p$ $(n > K_p)$; thus, there exist elements $g_n^{(p)} \in \mathscr{U}(x_n) \cap M_p$ $(n > K_p$; $p = 1, 2, \ldots)$. Now let $\{g_n\}_1^{K_1}$ be an arbitrary sequence such that $g_n \in \mathscr{U}(x_n)$ $(n = 1, \ldots, K_1)$ and let $\{g_n\}_{K_1+1}^{\infty}$ be the sequence $g_{K_1+1}^{(1)}, \ldots, g_{K_2}^{(1)}, g_{K_2+1}^{(2)}, \ldots, g_{K_3}^{(2)}, g_{K_3+1}^{(3)}, \ldots$. Then clearly $g_n \in \mathscr{U}(x_n)$ $(n = 1, 2, \ldots)$ and $g_n \to g_0$, which proves that $\mathscr{U}$ is l. (K)-s.c.

Conversely, assume now that $\mathscr{U}$ is l. (K)-s.c., but not l.s.c., that is, there exists a closed set N in G such that $\{x \in E \,|\, \mathscr{U}(x) \subset N\}$ is not closed. Then there exists a sequence $x_n \to x_0$ such that $\mathscr{U}(x_n) \subset N$ $(n = 1, 2, \ldots)$, but $\mathscr{U}(x_0) \not\subset N$; thus, there is an element $g_0 \in \mathscr{U}(x_0)$ such that $g_0 \notin N$. Since $\mathscr{U}$ is l. (K)-s.c., it follows that there exist elements $g_n \in \mathscr{U}(x_n)$ $(n = 1, 2, \ldots)$ with $g_n \to g_0$. Hence, by $\mathscr{U}(x_n) \subset N$, we have $g_n \in N$ $(n = 1, 2, \ldots)$ and $g_n \to g_0 \notin N$, in contradiction with the assumption that N is closed in G. This completes the proof of lemma 4.

For $G \subset E$ and $\mathscr{U} = \mathscr{P}_G$ lemma 4 was observed, essentially, by A. L. BROWN ([3], proposition 1. 1), but we shall also apply it here for $\mathscr{U} = \mathscr{V}_G$. In the sequel we shall freely use, without further references to lemma 4, the identity of lower semi-continuity and lower (K)-semi-continuity.

Corollary 3 above shows that whenever the unit cell S_E has a non-compact face, the space E contains a proximinal hyperplane G such that $\mathscr{P}_G$ is not u.s.c. For lower semi-continuity the situation is different, namely, we have

THEOREM 4. *For every proximinal hyperplane G in a normed linear space E the metric projection $\mathscr{P}_G$ is l.s.c.*

PROOF. Let $x_n \to x_0$ and $g_0 \in \mathscr{P}_G(x_0)$. If $x_0 \in G$, then $g_0 = x_0$ and hence for any $g_n \in \mathscr{P}_G(x_n)$ $(n = 1, 2, \ldots)$ we have (see e.g. [13], p. 140, theorem 6. 1 b))

$$\|x_n - g_n\| = \big|\,\|x_n - g_n\| - \|x_0 - g_0\|\,\big| \leq \|x_n - x_0\| \to 0.$$

Consequently, $\|g_n - g_0\| \leq \|g_n - x_n\| + \|x_n - g_0\| \to 0$ and thus $g_n \to g_0$.

Assume now that $x_0 \in E \setminus G$. Then, since G is a hyperplane, there exists (see e.g. [13], p. 18, theorem 1. 1) an $f \in E^*$ such that $\|f\| = 1$, $G = \{x \in E \,|\, f(x) = 0\}$ and $f(x_0 - g_0) = \|x_0 - g_0\|$. Put

$$(28) \qquad g_n = x_n - \frac{f(x_n)}{f(x_0)}(x_0 - g_0) \qquad (n = 1, 2, \ldots).$$

Then $f(g_n) = f(x_n) - f(x_n) + \dfrac{f(x_n)}{f(x_0)} f(g_0) = 0$ (since $g_0 \in G$) and thus $g_n \in G$ $(n = 1, 2, \ldots)$. Furthermore, since

$$\|x_n - g_n\| = \left|\frac{f(x_n)}{f(x_0)}\right| \|x_0 - g_0\| = |f(x_n)| = |f(x_n - g)| \leq \|x_n - g\| \qquad (g \in G),$$

we have $g_n \in \mathscr{P}_G(x_n)$ $(n=1, 2, \ldots)$. Finally, since $x_n \to x_0$, we have $f(x_n) \to f(x_0)$, whence

$$\|g_n - g_0\| = \left\| x_n - \frac{f(x_n)}{f(x_0)} x_0 - \left(1 - \frac{f(x_n)}{f(x_0)} g_0\right) \right\| \leq$$

$$\leq \|x_n - x_0\| + \left\| \left(1 - \frac{f(x_n)}{f(x_0)}\right)(x_0 - g_0) \right\| \to 0,$$

and therefore $g_n \to g_0$. Thus, in both cases the relations $x_n \to x_0$ and $g_0 \in \mathscr{P}_G(x_0)$ imply the existence of $g_n \in \mathscr{P}_G(x_n)$ $(n=1, 2, \ldots)$ with $g_n \to g_0$, i.e. $\mathscr{P}_G$ is l. (K)-s.c., whence also l.s.c., which completes the proof of theorem 4.

REMARK 4. Theorem 4 is no longer true for $2 \leq \operatorname{codim} G < \infty$. Indeed, if G is a Čebyšev subspace of E, then the lower semi-continuity of $\mathscr{P}_G$ reduces to the usual continuity of $\pi_G : E \to G$ (see remark 1) and in suitable spaces E we can take, for each m with $2 \leq m < \infty$, Čebyšev subspaces G with $\operatorname{codim} G = m$, for which π_G is discontinuous.

Now we shall prove the following characterization of lower semi-continuity of $\mathscr{P}_G$, similar to theorem 3 (but having a simpler proof):

THEOREM 5. *For a proximinal linear subspace G of a normed linear space E the following statements are equivalent:*
1°. $\mathscr{P}_G$ *is l.s.c.*
2°. $\mathscr{V}_G$ *is l.s.c.*
3°. $\omega_G|_{\mathscr{P}_G^{-1}(0)}$ *is open (i.e., carries open sets onto open sets).*

PROOF. The equivalence $2° \leftrightarrow 3°$ is an immediate consequence of lemma 2, formula (13), and the definition of lower semi-continuity and thus it remains to prove the equivalence $1° \leftrightarrow 2°$.

Assume that we have 1° and let $x_n + G \to x_0 + G$, $x_0 - g_0 \in \mathscr{V}_G(x_0 + G)$, with $g_0 \in \mathscr{P}_G(x_0)$. Since the canonical mapping $\omega_G : E \to E/G$ is open, from lemma 2, formula (12) it follows that the set-valued mapping $\mathscr{W}_G : E/G \to 2^E$ defined by

$$(29) \qquad \mathscr{W}_G = (\omega_G)^{-1} : x + G \to \{x + g \mid g \in G\}$$

is l.s.c., whence also l. (K)-s.c. Thus, since $x_n + G \to x_0 + G$ and $x_0 - g_0 \in \mathscr{W}_G(x_0 + G)$, there exist elements $x_n - g_n \in \mathscr{W}_G(x_n + G)$, with $g_n \in G$ $(n=1, 2, \ldots)$, such that $x_n - g_n \to x_0 - g_0$. Put $x_n' = x_n - g_n$ $(n=1, 2, \ldots)$, $x_0' = x_0 - g_0$, $g_0' = 0$. Then $x_n' \to x_0'$ and $g_0' = 0 \in \mathscr{P}_G(x_0 - g_0) = \mathscr{P}_G(x_0')$, whence, since by 1° $\mathscr{P}_G$ is l. (K)-s.c., there exist elements $g_n' \in \mathscr{P}_G(x_n') = \mathscr{P}_G(x_n - g_n) = \mathscr{P}_G(x_n) - g_n$ $(n=1, 2, \ldots)$ with $g_n' \to g_0' = 0$. Consequently, $g_n + g_n' \in \mathscr{P}_G(x_n)$, i.e. $x_n - (g_n + g_n') \in \mathscr{V}_G(x_n + G)$ $(n=1, 2, \ldots)$ and $x_n - (g_n + g_n') = x_n' - g_n' \to x_0' - g_0' = x_0 - g_0$ and thus $\mathscr{V}_G$ is l. (K)-s.c., whence also l.s.c.

Conversely, assume that we have 2° and let $x_n \to x_0$, $g_0 \in \mathscr{P}_G(x_0)$. Then $x_n + G \to x_0 + G$ and $x_0 - g_0 \in \mathscr{V}_G(x_0)$, whence, since by 2° $\mathscr{V}_G$ is l. (K)-s.c., there exist elements $x_n - g_n \in \mathscr{V}_G(x_n + G)$, with $g_n \in \mathscr{P}_G(x_n)$ $(n=1, 2, \ldots)$, such that $x_n - g_n \to x_0 - g_0$.

Since $x_n \to x_0$, it follows that $g_n \to g_0$ and thus $\mathscr{P}_G$ is l. (K)-s.c. whence also l.s.c., which completes the proof of theorem 5.

REMARK 5. In the particular case when G is a Čebyšev subspace of E, the equivalence $1° \leftrightarrow 2°$ of theorem 5 reduces to the theorem of R. B. HOLMES mentioned in remark 3. Again, the techniques used above for arbitrary proximinal G and set-valued $\mathscr{P}_G$ are different from those of Holmes.

From theorem 5 it results the following relation between lower semi-continuity of metric projections and lower semi-continuity of Hahn—Banach extension maps:

COROLLARY 5. *For a w^*-closed linear subspace Γ of E^*, $\mathscr{P}_\Gamma$ is l.s.c. if and only if the set-valued Hahn—Banach extension map $\varphi \in (\Gamma_\perp)^* \to \{f \in E^* | f|_{\Gamma_\perp} = \varphi, \|f\| = \|\varphi\|\}$ is l.s.c.*

The proof is similar to that of corollary 4.

3. Continuous selections for $\mathscr{P}_G$

Since for every $x \in E$ the set $\mathscr{P}_G(x)$ is closed and convex, from a well-known selection theorem of E. A. MICHAEL ([10], theorem 3. 2″) it results the following sufficient condition for the existence of a continuous selection for $\mathscr{P}_G$: *If G is a proximinal linear subspace of a Banach space E, such that $\mathscr{P}_G$ is l.s.c., then $\mathscr{P}_G$ admits a continuous selection* (it is known that the converse is not true, even if dim $G=1$). Hence, by theorem 4, *for every proximinal hyperplane G in a Banach space E, the metric projection $\mathscr{P}_G$ admits a continuous selection*; however, much more is known, namely (see e.g. [13], p. 142, theorem 6. 2), for every proximinal hyperplane G in a normed linear space E, the metric projection $\mathscr{P}_G$ admits even a linear selection.

Let us prove now the following result, corresponding to theorems 3 and 5:

THEOREM 6. *For a proximinal linear subspace G of a normed linear space E the following statements are equivalent:*
1°. *$\mathscr{P}_G$ admits a continuous selection.*
2°. *$\mathscr{V}_G$ admits a continuous selection.*

PROOF. Assume that $\mathscr{P}_G$ admits a continuous selection, say $\pi_G^{(0)}$. Since the mapping $\mathscr{W}_G : E/G \to 2^E$ defined by (29) is l.s.c. (see the proof of theorem 5), it admits, by a selection theorem of BARTLE and GRAVES (which is a particular case of the above mentioned selection theorem of MICHAEL [10]), a continuous selection, say w_G. Then, clearly, the mapping $v_G : E/G \to \mathscr{P}_G^{-1}(0)$ defined by

$$(30) \qquad v_G(x+G) = w_G(x+G) - \pi_G^{(0)} w_G(x+G) \qquad (x+G \in E/G),$$

is a continuous selection for $\mathscr{V}_G$.

Conversely, assume that $\mathscr{V}_G$ admits a continuous selection, say $v_G(x+G) = = x - \pi_G^{(0)}(x)$ $(x+G \in E/G)$, where $\pi_G^{(0)}(x) \in \mathscr{P}_G(x)$ $(x \in E)$. Then $\pi_G^{(0)}$ is a continuous selection for $\mathscr{P}_G$; indeed, if $x_n \to x_0$, then

$$\|\pi_G^{(0)}(x_n) - \pi_G^{(0)}(x_0)\| \leq \|\pi_G^{(0)}(x_n) - \pi_G^{(0)}(x_0) - (x_n - x_0)\| + \|x_n - x_0\| =$$
$$= \|v_G(x_0 + G) - v_G(x_n + G)\| + \|x_n - x_0\| \to 0,$$

which completes the proof of theorem 6.

REMARK 6. Theorem 6 improves the following theorem on continuous selections for set-valued metric projections, which we have given in [14] (see [14], theorem 3, equivalence $1° \leftrightarrow 3°$): Let $G \subset E$ be proximinal. In order that $\mathscr{P}_G$ admit a continuous selection it is sufficient and, if E/G is reflexive, also necessary that there exist a norm-continuous mapping $\psi : (G^\perp)^* \to E$ such that

$$\tag{31} f(\psi(\Phi)) = \Phi(f) \qquad (\Phi \in (G^\perp)^*, \ f \in G^\perp),$$

$$\tag{32} \|\psi(\Phi)\| = \|\Phi\| \qquad (\Phi \in (G^\perp)^*).$$

Indeed, if such a mapping ψ exists, then, since $(G^\perp)^* \equiv (E/G)^{**}$, ψ induces a continuous mapping $(E/G)^{**} \to E$, whence, by restriction, a continuous mapping $v_G : E/G \to E$ which satisfies, by (31) and (32),

$$\tag{33} f(v_G(x+G)) = f(x) \qquad (x \in E, \ f \in G^\perp),$$

$$\tag{34} \|v_G(x+G)\| = \|x+G\|.$$

Then, by (33), $v_G(x+G) - x \in (G^\perp)_\perp = G$, whence $v_G(x+G) = x - g_0(x)$, where $g_0(x) \in G$ $(x \in E)$, and by (34) $\|x - g_0(x)\| = \|v_G(x+G)\| = \|x+G\| = \inf_{g \in G} \|x-g\|$, whence $g_0(x) \in \mathscr{P}_G(x)$ $(x \in E)$. Thus, v_G is a continuous selection for $\mathscr{V}_G$, whence, by theorem 6, $\mathscr{P}_G$ admits a continuous selection. Conversely, if E/G is reflexive and $\mathscr{P}_G$ admits a continuous selection, then, by theorem 6, $\mathscr{V}_G$ admits a continuous selection, say v_G and, since $E/G \equiv (E/G)^{**} \equiv (G^\perp)^*$, v_G induces a continuous mapping $\psi : (G^\perp)^* \to E$ satisfying (31) and (32). Theorem 6 deals with $v_G : E/G \to E$ (rather than $\psi : (G^\perp)^* \to E$), but it does not assume the reflexivity of E/G.

REMARK 7. In the particular case when G is a Čebyšev subspace of E, theorem 6 reduces to the theorem of R. B. Holmes mentioned in remark 3. Also, from theorem 6 it results, as a corollary, the following relation between continuous selections for metric projections and for Hahn—Banach extension maps, due to J. LINDENSTRAUSS [8]: For a w^*-closed linear subspace Γ of E^*, $\mathscr{P}_\Gamma$ admits a continuous selection if and only if the set-valued Hahn—Banach extension map $\varphi \in (\Gamma_\perp)^* \to \to \{f \in E^* \mid f|_{\Gamma_\perp} = \varphi, \ \|f\| = \|\varphi\|\}$ admits a continuous selection.

We conclude this paper with the remark that it seems to be of some interest to consider the following two problems, related to the above results: a) Extend to arbitrary proximinal G and set-valued $\mathscr{P}_G$ the following known result (see HOLMES [7], theorem 14) on the continuity of π_G for Čebyšev subspaces G: For a closed linear subspace G of finite codimension in a reflexive strictly convex Banach space E (hence G is a Čebyšev subspace) π_G is continuous if and only if $t|_{G^\perp}$, the restriction of the "spherical image map" $t : E^* \to E$ to $G_\perp$, is continuous. We recall that for each $f \in E^*$, $t(f)$ is defined as the unique element $x \in E$ satisfying $f(x) = \|f\| \|x\|$, $\|f\| = \|x\|$. In the general case of our problem, we assume only that E is reflexive (not necessarily strictly convex), hence G is only proximinal (not necessarily Čebyšev), and we replace t by the set-valued spherical image map

$$\tag{35} T(f) = \{x \in E \mid f(x) = \|f\| \|x\|, \ \|f\| = \|x\|\} \qquad (f \in E^*).$$

b) Extend to arbitrary proximinal G and set-valued $\mathscr{P}_G$ the known results on other properties of π_G for Čebyšev subspaces G; e.g., defining weakly continuous, Lipschitzian, differentiable or linear selections for $\mathscr{P}_G$ in the natural way, give characterizations of those proximinal linear subspaces G for which $\mathscr{P}_G$ admits such selections. Or, find suitable extensions of these notions concerning π_G (for Čebyšev subspaces G) to set-valued metric projections $\mathscr{P}_G$ (e.g., linearity could be extended to a relation between $\mathscr{P}_G(x+y)$ and $\mathscr{P}_G(x)+\mathscr{P}_G(y)$, for all $x, y \in E$) and then give characterizations of those proximinal linear subspaces G for which $\mathscr{P}_G$ has one of these properties. It is likely that the set-valued mapping $\mathscr{V}_G : E/G \to 2^{\mathscr{P}_G^{-1}(0)}$ defined by (8) will be a useful tool in the solution of these problems.

Added in proof: In the meantime problem a) above has been solved, see F. DEUTSCH, W. POLLUL and I. SINGER, *On set-valued metric projections, Hahn-Banach extension maps and spherical image maps.* (to appear).

REFERENCES

[1] J. Blatter, *Zur Stetigkeit von mengenwertigen metrischen Projektionen.* Schr. Rh.-Westf. Inst. für Instrum. Math. Univ. Bonn, Ser. A, Nr. **16** (1967), 19—38.

[2] J. Blatter, P. D. Morris and D. E. Wulbert, *Continuity of the set-valued metric projection.* Math. Ann. **178** (1968), 12—24.

[3] A. L. Brown, *Best n-dimensional approximation to sets of functions.* Proc. London Math. Soc. **14** (1964), 577—594.

[4] A. L. Brown, *On continuous selections for metric projections in spaces of continuous functions* (to appear).

[5] E. W. Cheney and D. E. Wulbert, *Existence and unicity of best approximations.* Math. Scand. **24** (1969), 113—140.

[6] N. Dunford and J. Schwartz, *Linear operators. Part I: General theory.* Interscience Publ., New York (1958).

[7] R. B. Holmes, *On the continuity of best approximation operators.* Symp. on infinite dimensional topology. Princeton Univ. Press (to appear).

[8] J. Lindenstrauss, *Extension of compact operators.* Mem. Amer. Math. Soc. **48** (1964).

[9] E. A. Michael, *Topologies on spaces of subsets.* Trans. Amer. Math. Soc. **71** (1951), 152—182.

[10] E. A. Michael, *Continuous selections. I.* Ann. of Math. **63** (1956), 361—382.

[11] P. D. Morris, *Metric projections onto subspaces of finite codimension.* Duke Math. J. **35** (1968), 799—808.

[12] I. Singer, *Some remarks on approximative compactness.* Rev. roumaine Math. Pures Appl. **9** (1964), 167—177.

[13] I. Singer, *Best approximation in normed linear spaces by elements of linear subspaces.* Publ. House Acad. Soc. Rep. Romania, Bucharest and Springer, Grundl. Math. Wiss. **171**, Berlin—Heidelberg—New York (1970).

[14] I. Singer, *On metric projections onto linear subspaces of normed linear spaces.* Proc. Confer. on "Projections and related topics" held in Clemson. Aug. 1967. Preliminary Edition (January 1968).

On the Comparison of Approximation Processes in Hilbert Spaces

By

P. L. BUTZER, R. J. NESSEL, and W. TREBELS*)

LEHRSTUHL A FÜR MATHEMATIK
TECHNISCHE HOCHSCHULE AACHEN

Dedicated to Professor F. Reutter on his 60-th birthday

1. Introduction

In two papers Jean FAVARD [16, 17] suggested the study of the comparison of approximation processes in Banach spaces. We pick up this problem in the case of Hilbert spaces. Strictly speaking we will deal with the following problem:

Let H be an arbitrary (real or complex) Hilbert space, [H] the set of all bounded linear operators on H into itself. Let Γ be an index set of real numbers with accumulation point $+\infty$ and $\{T_\gamma; \gamma \in \Gamma\} \subset [H]$ be a strong approximation process, i.e., the operators T_γ are uniformly bounded and

$$(1.1) \qquad \lim_{\gamma \to \infty} \|T_\gamma f - f\| = 0 \qquad (f \in H).$$

Let $\{S_\gamma; \gamma \in \Gamma\} \subset [H]$ be a further strong approximation process. We seek for conditions upon T_γ, S_γ such that relations of type

$$(1.2) \qquad \|T_\gamma f - f\| \leq C \|S_\gamma f - f\| \qquad (\gamma \in \Gamma; \ f \in H)$$

hold, where the constant $C > 0$ should then be independent of γ and f; in other words, we look for some relation which furnishes us with some type of absolute continuity of $\{T_\gamma - I\}$ with respect to $\{S_\gamma - I\}$. To fix matters let us call an approximation process $\{T_\gamma; \gamma \in \Gamma\}$ better than the process $\{S_\gamma; \gamma \in \Gamma\}$ provided that (1.2) holds. If $\{T_\gamma\}$ is better than $\{S_\gamma\}$ and the latter is in turn better than $\{T_\gamma\}$, we call the approximation processes equivalent and write

$$\|T_\gamma f - f\| \sim \|S_\gamma f - f\| \qquad (\gamma \in \Gamma; \ f \in H).$$

Concerning contributions to the general comparison problem posed by FAVARD there should be mentioned:

i) the two papers of FAVARD [16, 17] where some first results in connection with the Banach—Steinhaus theorem are given,

---

*) The research of this author was supported by the "Landesamt für Forschung bei dem Minister für Wissenschaft und Forschung des Landes Nordrhein—Westfalen" Grant No. A/3—4379. Thanks are due to the Landesamt for permission to publish the results in these Proceedings.

ii) work of SHAPIRO [29; Ch. 9] who considers a homogeneous Banach space so that convolution with a measure is defined and comparison theory reduces to a discussion of the Fourier transforms of the measures involved (for precursory material cf. LÖFSTRÖM [23a]),

iii) a paper by the authors [12], in which for a Banach space with a given fundamental, total biorthogonal system $\{f_k, f_k^*\}$ (or, more generally, with a total system $\{P_k\} \subset [H]$ of orthogonal projections) summation processes of the (formal) Fourier series of f

$$f \sim \sum_{k=0}^{\infty} f_k^*(f) f_k$$

are studied, i.e., the operators T_γ (S_γ, respectively) admitted are assumed to possess only a discrete spectrum and the eigenfunctions of T_γ (S_γ, respectively) coincide with the given (unconditional or Schauder or Cesàro) basis $\{f_k\}: T_\gamma f_k = t_\gamma(k) f_k$, thus

(1. 3) $$T_\gamma f \sim \sum_{k=0}^{\infty} t_\gamma(k) f_k^*(f) f_k.$$

In this paper we do not make any assumptions upon the spectra of the operators S_γ, T_γ in advance, but it is assumed that S_γ, T_γ possess a simultaneous spectral representation which may be considered as a substitute for (1. 3). Thus, apart from notations and definitions, Sec. 2 gives a general comparison theorem based upon Def. 2. 1. According to the Hilbert space structure, the multiplier condition consists of a simple boundedness condition of some quotient (see (2. 7)), i.e., it corresponds to the case of an unconditional basis in [12]. The theory to be presented here is based upon typical Hilbert space arguments; their application to the above problem was inspired by some work of RIVKIND [27, 28] on the problem of "rough convergence" of approximation processes — a problem which may, more precisely, be regarded as the saturation problem (compare e.g. [11; p. 434]).

In Sec. 3 a wide range of examples is considered. Thus particular approximation processes are compared in the spaces $L^2(0, 2\pi)$, $L^2(0, \infty)$, $L^2(-\infty, \infty)$, and in the space of almost periodic (continuous) functions. Here expansions in Fourier, Laguerre, and Hermite series as well as representations as Fourier or Hankel integrals yield explicit simultaneous spectral representations for special types of operators.

Let us conclude with a brief review of what is known in concrete spaces. BERMAN [3] compares in $L^p(0, 2\pi)$, $1 \leq p \leq \infty$, polynomial operators $U_n^1(f; x)$, $U_n^2(f; x)$ which coincide provided f is a trigonometric polynomial of degree $\leq n$. As an application of a general theorem he deduces that an approximation process of integral type is equivalent to its discrete analogue. The same question from the point of view of best constants is pursued by GANZBURG [18]; in the algebraic case there is a paper of KAL'NIBOLOCKAJA [21]. BOMAN—SHAPIRO [6] compare strong approximation processes in $\mathsf{L}^p(E_n)$ and $\mathsf{L}^p(T_n)$, $1 \leq p \leq \infty$, by Fourier analytic methods. Let us also mention results concerning a comparison of the (mere) convergence of linear summation methods of orthogonal series (see e.g. SHMANDIN [30]), as well as, for a certain class of operators $\{T_\gamma;\ \gamma \in \Gamma\}$, estimates of $\|T_\gamma f - f\|$ from below and (or) above by a linear combination of the best approximation of f, where the element of best approximation is constructed with the aid of a fixed eigenfunction system of a Sturm—Liouville operator (see e.g. BRAß [7, 8]), or by moduli of continuity and classical best approximation (see e.g. STEČKIN [33]). In this respect, let us remark that for particular approximation processes such as those of Fejér, Abel—Poisson, Weierstrass, Bernstein, and so on, a number of estimates by means of best approximation and moduli of continuity are known, and there exists a vast literature.

The authors are indebted to Professors G. LUMER and B. SZ.-NAGY for fruitful discussions during the occasion of the Aberdeen Conference (July 5—10, 1971) and this Conference, respectively.

2. Comparison theorems

Let $f, g, \dots$ denote elements of an arbitrary Hilbert space H with inner product (f, g) and norm $\|f\| = (f, f)^{1/2}$. Let $\mathbf{Z}$, $\mathbf{P}$, $\mathbf{N}$ be the sets of all, of all non-negative, of all positive integers, respectively. Consider two families $\{S_\gamma;\ \gamma \in \Gamma\}$, $\{T_\gamma;\ \gamma \in \Gamma\}$ of bounded linear operators on H which need not necessarily be strong approximation processes.

In order to arrive at an estimate of type (1. 2), the idea of a simultaneous spectral representation of certain families of operators seems to be useful; theorems concerning this matter may be found in standard books on functional analysis, for example in [14], [26], [35].

Thus let A be a self-adjoint operator with domain $\mathsf{D}(A)$ (dense in H; cf. [35; p. 28]). In view of the spectral theorem for these operators there exists a (unique) family of projections $\{E_\lambda;\ \lambda \in (-\infty, \infty)\}$ on H — called resolution of the identity (corresponding to A) — such that

$$\text{(i)} \qquad E_\lambda E_\mu f = E_\mu E_\lambda f = E_\lambda f \qquad (\lambda \leq \mu,\ f \in \mathsf{H})$$

$$(2.\ 1) \qquad \text{(ii)} \qquad \lim_{\mu \to \lambda+} E_\mu f = E_\lambda f \qquad (f \in \mathsf{H})$$

$$\text{(iii)} \qquad \lim_{\lambda \to -\infty} E_\lambda f = 0, \quad \lim_{\lambda \to \infty} E_\lambda f = f \qquad (f \in \mathsf{H}),$$

and one has $Af = \int_{-\infty}^{\infty} \lambda\, dE_\lambda f$ for all $f \in \mathsf{D}(A)$ as well as $f \in \mathsf{D}(A)$ if and only if $\int_{-\infty}^{\infty} \lambda^2\, d\|E_\lambda f\|^2 < \infty$, in which case the latter integral equals $\|Af\|^2$. If $t(\lambda)$ is measurable and bounded with respect to the spectral measure E_λ (i.e. with respect to $\|E_\lambda f\|^2$ for all $f \in H$), then

$$(2.2) \qquad T = t(A) = \int_{-\infty}^{\infty} t(\lambda)\,dE_\lambda$$

defines a bounded operator on H with norm

$$(2.3) \qquad \|T\| = \text{ess. sup}\,|t(\lambda)|,$$

the supremum being taken relative to the measure E_λ. Thus a number of operators may be constructed from some given resolution of the identity. The converse problem gives rise to

DEFINITION 2. 1. *A given family of operators* $G \subset [H]$ *is said to have a simultaneous spectral representation if there exists a family of projections* $\{E_\lambda\}$ *on* H *satisfying* (2. 1) *such that every* $T \in G$ *admits the representation*

$$(2.4) \qquad T = \int_{-\infty}^{\infty} t(\lambda)\,dE_\lambda$$

with some (complex-valued) function $t(\lambda)$ *measurable and bounded with respect to* E_λ.

In particular it follows with the aid of the operational calculus (see [14; Ch. IX], [26; Sec. 127]) that

$$(2.5) \qquad \|Tf\|^2 = \int_{-\infty}^{\infty} |t(\lambda)|^2\,d\|E_\lambda f\|^2 \qquad (f \in H).$$

Interpreting the two families $\{S_\gamma\}$, $\{T_\gamma\}$ as members of one family G, one arrives at the following comparison theorem.

THEOREM 2. 2. *Let* $G = \{S_\gamma;\ \gamma \in \Gamma\} \cup \{T_\gamma;\ \gamma \in \Gamma\}$ *possess a simultaneous spectral representation. Then*

$$(2.6) \qquad \|T_\gamma f - f\|^2 \le \text{ess. sup}_\lambda \frac{|t_\gamma(\lambda) - 1|^2}{|s_\gamma(\lambda) - 1|^2} \|S_\gamma f - f\|^2 \qquad (f \in H),$$

where the functions $s_\gamma(\lambda)$, $t_\gamma(\lambda)$ *are connected via* (2. 4) *with* S_γ, T_γ, *respectively, and the supremum is taken with respect to the measure* E_λ.

Indeed, on account of the hypothesis and (2. 5) one has

$$\|T_\gamma f - f\|^2 = \int_{-\infty}^{\infty} |t_\gamma(\lambda) - 1|^2\,d\|E_\lambda f\|^2 \le \text{ess. sup}_\lambda \frac{|t_\gamma(\lambda) - 1|^2}{|s_\gamma(\lambda) - 1|^2} \int_{-\infty}^{\infty} |s_\gamma(\lambda) - 1|^2\,d\|E_\lambda f\|^2$$

which yields the assertion by (2. 5).

Let us point out that the hypothesis "G has a simultaneous spectral representation" is not so restrictive as it looks. For if G is a countable, commuting family of normal operators and their adjoints in an arbitrary Hilbert space, then G has a simultaneous spectral representation (see [35; p. 67], [26; Sec. 130]). Moreover, since in this paper S_γ and T_γ are assumed to be bounded operators and therefore may be interpreted as elements of a suitable B*-algebra in [H], there exists a simultaneous spectral representation without countability restrictions (see [14; p. 895]). Since all operators considered later in the applications are self-adjoint, bounded, commutative, they satisfy the hypothesis of Theorem 2. 2.

It is to be noted that if the quotient in (2. 6)

$$(2.\,7) \qquad\qquad \operatorname*{ess.\,sup}_{\lambda} |t_\gamma(\lambda) - 1|^2 \, |s_\gamma(\lambda) - 1|^{-2} = C_\gamma$$

is not finite, then the statement of Theorem 2. 2 is trivial. As indicated by the index γ, the constant C_γ generally depends upon the parameter γ. Relation (1. 2) is guaranteed if the numbers C_γ are uniformly bounded with respect to γ; its verification is usually easy in case of appropriately chosen, explicit simultaneous spectral representations. Indeed, the problem in applications actually is the following: Given two families of operators, one has to construct explicitly a simultaneous spectral representation for these families such that not only does Theorem 2. 2 apply but also that (2. 7) may be calculated conveniently. This is particularly easy and indeed trivial in case the operators possess an expansion of type (1. 3).

Indeed, let $\{f_k\}_{k=0}^{\infty} \subset \mathsf{H}$ be an orthonormal (not necessarily complete) system. Let us consider operators $T \in [\mathsf{H}]$ of type

$$(2.\,8) \qquad\qquad Tf = \sum_{k=0}^{\infty} t(k)\,(f, f_k) f_k \qquad (f \in \mathsf{H}).$$

Now operators S_γ of type (2. 8) (with coefficients $s_\gamma(k)$) form a strong apprroximation process if and only if the orthonormal system $\{f_k\}$ is complete in H, $|s_\gamma(k)| \leq M$ uniformly for $\gamma \in \Gamma$, $k \in \mathsf{P}$, and $\lim_{\gamma \to \infty} s_\gamma(k) = 1$.

But to show that just (2. 8) represents a simultaneous spectral representation for the operators T in question, it is unessential whether $\{f_k\}$ is complete or not. Indeed, following the standard procedure, define $P_k f = (f, f_k) f_k$. Obviously $\{P_k\}$ is a system of orthogonal, symmetric projections. Denoting the closed linear manifold spanned by $\{f_k\}$ by L, and its orthogonal complement with respect to H by $\mathsf{L}^{\perp}$, each $f \in \mathsf{H}$ is uniquely decomposable as $f = f^1 + f^2$ where $f^1 \in \mathsf{L}$ and $f^2 \in \mathsf{L}^{\perp}$. Thus, defining $P_{-1} f = f^2$, it is clear that P_{-1} is a symmetric projection orthogonal to the other P_k. Setting

$$(2.\,9) \qquad\qquad E_\lambda = \sum_{k \leq \lambda} P_k,$$

with E_λ being the zero operator if the sum is empty, $\{E_\lambda\}$ satisfies (2. 1), and (2. 8) may be rewritten as

$$T = \int_{-\infty}^{\infty} t(\lambda)\, dE_\lambda,$$

where $t(\lambda) = 0$ for $\lambda \leq -1$ and $t(\lambda)$ is a polygonal curve connecting $t(k)$ and $t(k+1)$.

For operators of type (2. 8) let us examine the quotient (2. 7).

LEMMA 2. 3. *Let $S_\gamma, T_\gamma \in [H], \gamma \in \Gamma$, be given via (2. 8) with $s_\gamma(k), t_\gamma(k)$, respectively. For fixed γ let 1 be excluded as an accumulation point of the eigenvalues $s_\gamma(k)$.*
   a) *For each $\gamma \in \Gamma$ one has*

$$(2.10) \qquad \|T_\gamma f - f\| \leq C_\gamma \|S_\gamma f - f\| \qquad (f \in H)$$

if and only if

$$(2.11) \qquad N(S_\gamma - I) \subseteq N(T_\gamma - I),$$

$N(T) = \{f \in H; \ Tf = 0\}$ *being the null manifold of $T \in [H]$.*
   b) *C_γ in (2. 10) is uniformly bounded in γ if and only if there exists at least one simultaneous spectral representation of $\{S_\gamma\}$, $\{T_\gamma\}$ with corresponding measure E_λ^* such that*

$$(2.12) \qquad \text{ess. sup} \left| \frac{t_\gamma^*(\lambda) - 1}{s_\gamma^*(\lambda) - 1} \right| \leq C^*,$$

the supremum being taken with respect to E_λ^*.

PROOF. Since (2. 10) obviously implies (2. 11), let (2. 11) hold. If $f \in N(S_\gamma - I)$, then (2. 10) is trivial; if $f \notin N(S_\gamma - I)$ then, on account of (2. 8) and (2. 11),

$$\|T_\gamma f - f\|^2 = \sum |t_\gamma(k) - 1|^2 |(f, f_k)|^2 + \|P_{-1}\|^2 \leq \max\left\{ 1; \sup \left| \frac{t_\gamma(k) - 1}{s_\gamma(k) - 1} \right|^2 \right\} \|S_\gamma f - f\|^2,$$

the sum and the supremum being extended over $\{k \in P; \ k \neq k', \ s_\gamma(k') = 1\}$. Since 1 is not an accumulation point of $s_\gamma(k)$ there exists $\varepsilon_\gamma > 0$ such that $|s_\gamma(k) - 1| \geq \varepsilon_\gamma$ for all $k \neq k'$. Thus the quotient remains bounded since $|t_\gamma(k)| \leq M_\gamma$, and (2. 10) is proved.

b) Obviously (2. 12) implies the inequality $C_\gamma \leq C^*$ by Theorem 2. 2. Conversely, suppose $C_\gamma \leq C$ does not imply the uniform boundedness of the quotient in (2. 12) for any simultaneous spectral representation. Choose the one constructed above. Then we can define the quotient in (2. 12) to be 1 if $s_\gamma(k') = 1$. Denote by k_γ (where now $s_\gamma(k_\gamma) \neq 1$) the point for which the value of the quotient is equal to or close to the value of the supremum. Then

$$\|T_\gamma f_{k_\gamma} - f_{k_\gamma}\| = \left| \frac{t_\gamma(k_\gamma) - 1}{s_\gamma(k_\gamma) - 1} \right| \|S_\gamma f_{k_\gamma} - f_{k_\gamma}\|,$$

and the latter quotient tends to infinity as $\gamma \to \infty$, in contradiction to the hypothesis $C_\gamma \leq C$ for all $f \in H$.

Concerning condition (2. 11) for all $\gamma \in \Gamma$ it is obvious by (2. 10) that it is necessary for a uniform bound of C_γ; but it is not sufficient as will be shown by a counterexample in Sec. 3. 2.

3. Applications

The first applications deal with strong approximation processes which possess discrete spectra only and have an orthonormal system of eigenfunctions.

3. 1 *Laguerre series.* Choose $H = L^2(0, \infty)$, $(f, g) = \int_0^\infty f(x)\overline{g(x)}\,dx$, and consider the Laguerre polynomials of order $\alpha \geq 0$ defined by (see [34; § 5. 1])

$$L_k^{(\alpha)}(x) = \frac{1}{k!}\,e^x\,x^{-\alpha}\left(\frac{d}{dx}\right)^k (e^{-x}\,x^{k+\alpha}).$$

They form an orthogonal system on $(0, \infty)$ with respect to the weight function $x^\alpha e^{-x}$, namely

$$\int_0^\infty L_k^{(\alpha)}(x)\,L_m^{(\alpha)}(x)\,x^\alpha e^{-x}\,dx = \Gamma(\alpha+1)\binom{m+\alpha}{m}\delta_{km}.$$

Equivalently, the system $\{\varphi_k^{(\alpha)}\}_{k=0}^\infty$ given by

$$\varphi_k^{(\alpha)}(x) = \left\{\Gamma(\alpha+1)\binom{k+\alpha}{k}\right\}^{-1/2} L_k^{(\alpha)}(x)\,x^{\alpha/2}\,e^{-x/2}$$

is an orthonormal system on $(0, \infty)$. It is also complete (see e.g. [1]), i.e.

$$\lim_{n\to\infty} \int_0^\infty \left| f(x) - \sum_{k=0}^n (f, \varphi_k^{(\alpha)})\,\varphi_k^{(\alpha)}(x) \right|^2 dx = 0.$$

Therefore we can apply Lemma 2. 3 b) to general approximation processes of type (2. 8), in particular to those of convolution type in the sense of DEBNATH [13]. However, we restrict ourselves to the Cesàro means of $f \in L^2(0, \infty)$, namely

$$(3. 1) \qquad (C, 1)_n(f; x) = \sum_{k=0}^n \left(1 - \frac{k}{n+1}\right)(f, \varphi_k^{(\alpha)})\,\varphi_k^{(\alpha)}(x),$$

and compare them with the Poisson integral (see [34; § 5. 1])

$$(3. 2) \quad P_r(f; x) = \frac{(-r)^{-\alpha/2}}{1-r}\int_0^\infty \exp\left\{-\frac{1+r}{2(1-r)}(x+y)\right\} J_\alpha\left(\frac{2\sqrt{-rxy}}{1-r}\right)f(y)\,dy,$$

where

$$(3. 3) \qquad J_\alpha(z) = \sum_{k=0}^\infty \{k!\,\Gamma(k+\alpha+1)\}^{-1}(-1)^k(z/2)^{2k+\alpha}$$

is the Bessel function of the first kind. Since

$$P_r(f; x) = \sum_{k=0}^\infty r^k (f, \varphi_k^{(\alpha)})\,\varphi_k^{(\alpha)}(x)$$

and setting $r=e^{-1/n}$, according to Lemma 2. 3 b) we have to examine the suprema of

$$q_n(k) = (1-e^{-k/n})^{-1} \begin{cases} \dfrac{k}{n+1}, & 0 \leq k \leq n \\ 1, & k \geq n+1, \end{cases}$$

and $[q_n(k)]^{-1}$ with respect to $k \in \mathbf{P}$ and $n \in \mathbf{N}$. Extending $q_n(k)$ from $k \in \mathbf{P}$ to $\lambda \in (-\infty, \infty)$ and from $n \in \mathbf{N}$ to $\gamma \in (0, \infty)$ by setting

$$(3.4) \qquad q_\gamma(\lambda) = (1-e^{-|\lambda|/\gamma})^{-1} \begin{cases} \dfrac{|\lambda|}{\gamma+1}, & 0 \leq |\lambda| \leq \gamma+1 \\ 1, & |\lambda| \geq \gamma+1, \end{cases}$$

it is easy to show that $q_\gamma(\lambda)$ as well as $[q_\gamma(\lambda)]^{-1}$ are uniformly bounded in $\lambda \in (-\infty, \infty)$, $\gamma \in (0, \infty)$. Hence

COROLLARY 3. 1. *The Cesaro means* (3. 1) *and the Poisson integral* (3. 2) *approximate* $f \in \mathsf{L}^2(0, \infty)$ *strongly with the same order, or*

$$\int_0^\infty |(C,1)_n(f;x) - f(x)|^2 \, dx \sim \int_0^\infty |P_r(f;x) - f(x)|^2 \, dx \qquad (r=e^{-1/n}).$$

On the other hand, we may choose for f measurable on $(0, \infty)$

$$(3.5) \qquad \mathsf{H} = \left\{ f; \int_0^\infty |f(x)|^2 x^\alpha e^{-x} \, dx < \infty \right\}$$

with inner product

$$(f, g) = \int_0^\infty f(x) \overline{g(x)} \, x^\alpha e^{-x} \, dx.$$

In H a complete orthonormal system is given by $\{\psi_k^{(\alpha)}\}_{k=0}^\infty$ with

$$\psi_k^{(\alpha)}(x) = \left\{ \Gamma(\alpha+1) \binom{k+\alpha}{k} \right\}^{-1/2} L_k^{(\alpha)}(x)$$

(see [34; § 5. 7]). Therefore if the Cesàro means are defined analogously to (3. 1) by replacing $\varphi_k^{(\alpha)}$ by $\psi_k^{(\alpha)}$, and the Poisson integral by

$$(3.6) \qquad P_r(f;x) = \frac{1}{1-r} \int_0^\infty (-rxy)^{-\alpha/2} \exp\left\{ \frac{-r}{1-r}(x+y) \right\} J_\alpha\left(\frac{2\sqrt{-rxy}}{1-r} \right) f(y) y^\alpha e^{-y} \, dy$$

with eigenvalues r^k, one has as above

COROLLARY 3. 2. *Let* H *be given by* (3. 5) *and* $P_r(f;x)$ *by* (3. 6). *Then for* $f \in \mathsf{H}$ *and* $r=e^{-1/n}$

$$\int_0^\infty \left| \sum_{k=0}^n \left(1-\frac{k}{n+1}\right)(f, \psi_k^{(\alpha)})\psi_k^{(\alpha)}(x) - f(x) \right|^2 x^\alpha e^{-x} \, dx \sim \int_0^\infty |P_r(f;x)-f(x)|^2 x^\alpha e^{-x} \, dx.$$

Let us mention that the system $\{\varphi_k^{(\alpha)}\}$ is also complete in $L^p(0, \infty)$ for $4/3 < p < 4$ (see [1]), whereas $\{\psi_k^{(\alpha)}\}$ is only complete in H defined by (3. 5) (see [25]) but not in any space of measurable functions with finite norm

$$|||f|||_p = \left\{ \int_0^\infty |f(x)|^p x^\alpha e^{-x} \, dx \right\}^{1/p} \qquad (p \neq 2).$$

3.2 *Trigonometric system.* Let $H = L^2(0, 2\pi)$ with $(f, g) = \int_0^{2\pi} f(x)\overline{g(x)} \, dx$ and

$$(3. 7) \qquad f_{2m}(x) = (2\pi)^{-1/2} e^{imx}, \quad f_{2m+1}(x) = (2\pi)^{-1/2} e^{-i(m+1)x} \qquad (m \in P).$$

$\{f_k\}_{k=0}^\infty$ is a complete orthonormal system so that P_{-1} (of the construction in Sec. 2) is the zero operator. Here we would like to indicate how the familiar approximation processes in connection with this (complex) trigonometric system fit into the general scheme of Sec. 2. The operators in question are integrals of Fourier convolution type, i.e.

$$(3. 8) \qquad Tf(x) = \frac{1}{2\pi} \int_0^{2\pi} f(x - y) d\mu(y)$$

with μ being of 2π-periodic bounded variation (cf. [11; p. 14]). Classical approximation processes such as those of Fejér, Abel—Poisson, Gauss—Weierstrass, etc. are of this type (the kernels μ of these integrals are indeed absolutely continuous). In view of (3. 7), Tf is of type (2. 8), for indeed

$$Tf_k(x) = [d\mu]^\wedge \left((-1)^k \left[\frac{k+1}{2} \right] \right) f_k(x) \qquad (k \in P),$$

with $[\lambda]$ the largest integer less than or equal to λ and Fourier-coefficients

$$[d\mu]^\wedge(m) = (2\pi)^{-1} \int_0^{2\pi} e^{-imx} d\mu(x) \qquad (m \in Z).$$

Thus we only need to discuss these coefficients (on Z) in order to apply Lemma 2. 3 b). For example, to compare the singular integral of Fejér

$$(3. 9) \qquad F_n(f; x) = \frac{1}{2\pi(n+1)} \int_0^{2\pi} f(x - y) \left[\frac{\sin((n+1)y/2)}{\sin(y/2)} \right]^2 dy \qquad (n \in P)$$

with that of Abel—Poisson

$$(3. 10) \qquad P_r(f; x) = \frac{1}{2\pi} \int_0^{2\pi} f(x - y) \frac{1 - r^2}{1 - 2r \cos y + r^2} \, dy \qquad (r = e^{-1/n}),$$

the Fourier coefficients of the kernels being given by (see e.g. [11; pp. 516])

$$\begin{cases} \left(1 - \dfrac{|m|}{n+1}\right), & |m| \leq n \\ 0, & |m| > n \end{cases} , \quad e^{-|m|/n} \qquad (m \in Z),$$

respectively, one has to examine the suprema of $q_n(\lambda)$ given by (3. 4) and that of $[q_n(\lambda)]^{-1}$ for $\lambda \in \mathbf{Z}$. Since both are uniformly bounded as above, we arrive at the well-known result

COROLLARY 3. 3. *The processes F_n and P_r given by (3. 9) and (3. 10) are equivalent on* $\mathsf{L}^2(0, 2\pi)$, *i.e.*

$$\int_0^{2\pi} |F_n(f; x) - f(x)|^2 \, dx \sim \int_0^{2\pi} |P_r(f; x) - f(x)|^2 \, dx \qquad (r = e^{-1/n}).$$

Comparing the singular integral of Abel—Poisson with that of Gauss—Weierstrass which may be expanded in the form

$$W_n(f; x) = \sum_{k=0}^{\infty} \exp\left\{-\left[\frac{k+1}{2}\right]^2 / n^2\right\} (f, f_k) f_k,$$

one is led to the counterexample announced at the end of Sec. 2. For on account of the uniqueness of the Fourier coefficients the null manifolds of $\{P_r - I\}$, $r = e^{-1/n}$, and $\{W_n - I\}$ consist of the constants and thus coincide for all n; but

$$\sup_{m \in \mathbf{Z}} \left| \frac{1 - \exp\{-|m|/n\}}{1 - \exp\{-m^2/n^2\}} \right| \geqq \frac{1 - e^{-1/n}}{1 - e^{-1/n^2}} \geqq \frac{n}{2} \qquad (n \geqq 2)$$

is not uniformly bounded in n.

3. 3 *Hermite series.* Choose $\mathsf{H} = \mathsf{L}^2(-\infty, \infty)$, $(f, g) = \int_{-\infty}^{\infty} f(x)\overline{g(x)} \, dx$, and consider the Hermite polynomials defined by (cf. [34; § 5. 2])

$$H_k(x) = (-1)^k e^{x^2} \left(\frac{d}{dx}\right)^k e^{-x^2}.$$

$\{H_k\}_{k=0}^{\infty}$ is an orthogonal system with respect to the weight function $\exp\{-x^2\}$, thus

$$\int_{-\infty}^{\infty} H_k(x) H_m(x) e^{-x^2} \, dx = 2^m m! \sqrt{\pi} \, \delta_{km},$$

or equivalently, the system $\{\varphi_k\}_{k=0}^{\infty}$,

$$\varphi_k(x) = \left(2^k k! \sqrt{\pi}\right)^{-1/2} e^{-x^2/2} H_k(x)$$

is orthonormal on $(-\infty, \infty)$, and indeed complete in $\mathsf{L}^2(-\infty, \infty)$. We again apply Lemma 2. 3 b) only to the Cesàro means defined analogously to (3. 1) and the Poisson integral [36; p. 79]

$$P_r(f; x) = \{\pi(1 - r^2)\}^{-1/2} \int_{-\infty}^{\infty} \exp\left\{\frac{x^2 - y^2}{2} - \frac{(x - ry)^2}{1 - r^2}\right\} f(y) \, dy$$

with eigenvalues r^k. Hence

COROLLARY 3. 4. *For $f \in L^2(-\infty, \infty)$ and $r = e^{-1/n}$*

$$\int_{-\infty}^{\infty} |(C, 1)_n(f; x) - f(x)|^2 \, dx \sim \int_{-\infty}^{\infty} |P_r(f; x) - f(x)|^2 \, dx.$$

Since $\{\psi_k\}_{k=0}^{\infty}$, where

$$\psi_k(x) = \{2^k k! \sqrt{\pi}\}^{-1/2} H_k(x),$$

is a complete orthonormal system in the Hilbert space

$$(3.11) \qquad \mathsf{H} = \left\{ f; \int_{-\infty}^{\infty} |f(x)|^2 e^{-x^2} \, dx < \infty \right\}$$

(see [34; § 5. 7]) $\left(\text{and not complete in } \left\{ f; \int_{-\infty}^{\infty} |f(x)|^p e^{-x^2} \, dx < \infty, p \neq 2 \right\}, \text{ cf. [25]} \right)$ with inner product

$$(f, g) = \int_{-\infty}^{\infty} f(x) \overline{g(x)} \, e^{-x^2} \, dx,$$

one may again apply Lemma 2. 3 b) for this case. Defining $(C, 1)_n$ analogously to (3. 1) and

$$(3.12) \qquad P_r(f; x) = \int_{-\infty}^{\infty} k_r(x, y) f(y) e^{-y^2} \, dy,$$

where

$$k_r(x, y) = \{\pi(1 - r^2)\}^{-1/2} \exp\{-(r^2 x^2 - 2rxy + r^2 y^2)/(1 - r^2)\},$$

we obtain the equivalence $(r = e^{-1/n})$

$$(3.13) \qquad \int_{-\infty}^{\infty} \left| \sum_{k=0}^{n} \left(1 - \frac{k}{n+1}\right) (f, \psi_k) \psi_k(x) - f(x) \right|^2 e^{-x^2} \, dx \sim \int_{-\infty}^{\infty} |P_r(f; x) - f(x)|^2 e^{-x^2} \, dx$$

on account of the uniform boundedness of $q_n(\lambda)$ (given by (3. 4)) and $[q_n(\lambda)]^{-1}$.

Now let us consider a modified Poisson integral of Muckenhoupt [24]

$$(3.14) \qquad P_\delta^*(f; x) = \int_{-\infty}^{\infty} \left(\int_0^1 (1 - r^2)^{1/2} \frac{\delta \exp\{\delta^2/2 \log r\}}{\sqrt{2} \, r(-\log r)^{3/2}} k_r(x, y) \, dr \right) f(y) e^{-y^2} \, dy.$$

Its expansion into a Hermite series reads

$$P_\delta^*(f; x) = \sum_{k=0}^{\infty} e^{-\sqrt{2k} \, \delta} (f, \psi_k) \psi_k(x).$$

Setting

$$q_{n,\delta}(k) = (1 - e^{-\sqrt{2k} \, \delta})^{-1} \begin{cases} \dfrac{k}{n+1}, & 0 \leq k \leq n \\[2mm] 1, & k > n \end{cases}$$

and choosing $\delta = n^{-1}$ we see that $q_{n,\delta}(k)$ is not uniformly bounded in $k \in \mathsf{P}$ and $n \to \infty$, but $[q_{n,\delta}(k)]^{-1}$

is uniformly bounded. However taking $\delta = n^{-1/2}$, then $q_{n,\delta}(k)$ is uniformly bounded but $[q_{n,\delta}(k)]^{-1}$ is not. Thus, on account of (3. 13), we arrive at

COROLLARY 3. 5. *Let* H *be given by* (3. 11), P_r *by* (3. 12), *and* P_δ^* *by* (3. 14). *Then*

$$(3.\,15) \qquad \int_{-\infty}^{\infty} |P_r(f;x) - f(x)|^2 e^{-x^2}\,dx \leqq C \int_{-\infty}^{\infty} |P_\delta^*(f;x) - f(x)|^2 e^{-x^2}\,dx$$

if $r = e^{-1/n}$, $\delta = n^{-1/2}$, *and conversely if* $r = e^{-1/n}$, $\delta = n^{-1}$.

Thus we are faced with the following problem: given two approximation processes S_γ, T_ϱ with different parameters γ, ϱ, respectively, how does one associate the parameters and thus compare these processes?

In the general case this seems to be rather difficult. However, if we consider the particular instance that the functions $s_\gamma(\lambda)$, connected with S_γ via (2. 4), and $t_\varrho(\lambda)$, associated with T_ϱ, can be transformed to $s_\gamma(\lambda) = s(\lambda/\eta)$ $(\eta = \Phi(\gamma)$, Φ monotonely increasing) and $t_\varrho(\lambda) = t(\lambda/\eta)$ $(\varrho = \Phi^*(\eta)$, Φ^* monotonely increasing), then by the theory of Fourier series and integrals it seems reasonable to link γ and ϱ by $\varrho = \Phi^*(\Phi(\gamma))$ and then to apply the quotient test (2. 7). In this sense, the Poisson integral $P_r(f; x)$ of (3. 12) approximates $f \in$ H as given by (3. 11) better than the modified $P_\delta^*(f; x)$ of (3. 14) in case one sets $n = (1-r)^{-1}$ and $\delta^{-1} = \sqrt{n}$.

3. 4 *Almost periodic functions.* Let F be the set of almost periodic functions with inner product

$$(3.\,16) \qquad (f, g) = M\{f(x)\overline{g(x)}\}, \qquad M\{f(x)\} = \lim_{T \to \infty} \frac{1}{2T} \int_{-T}^{T} f(x)\,dx.$$

F is an uncomplete inner product space which may be completed to a Hilbert space by the standard procedure; but this is not essential here. For every $f \in$ F there exists an at most countable set of Fourier exponents $\lambda_k \in (-\infty, \infty)$ with

$$(3.\,17) \qquad \lim_{n \to \infty} \left\| f(x) - \sum_{k=0}^{n} (f, e^{i\lambda_k x}) e^{i\lambda_k x} \right\| = 0,$$

the norm being induced by the inner product (3. 16). (For proofs of the basic facts here see [26; § 101—102], [9]).

In distinction to the series treated above there is no fixed basis in F. Since the Fourier exponents of one $f \in$ F may have countably many accumulation points, it seems difficult to quote a strong approximation process which is non-trivial for all $f \in$ F. Therefore we restrict ourselves to the subspace Q of those functions $f \in$ F for which the Fourier exponents λ_k only accumulate at infinity on the supposition that $\{\lambda_k\}$ is ordered in the following sense: $|\lambda_k| \leqq |\lambda_{k+1}|$, and if $|\lambda_k| = |\lambda_{k+1}|$ then $\lambda_k < \lambda_{k+1}$. Now let us consider convolution operators on Q

$$(3.\,18) \qquad Tf(x) = \frac{1}{\sqrt{2\pi}} \int_{-\infty}^{\infty} f(x-y)\chi(y)\,dy$$

with $\chi \in L^1(-\infty, \infty)$. Let $\{\lambda_k\}$ be the Fourier exponents of a fixed f. Then Tf is of type (2. 8) since

$$(3.19) \qquad (Te^{i\lambda_k u})(x) = \chi^{\hat{}}(\lambda_k)e^{i\lambda_k x}, \quad \chi^{\hat{}}(v) = \frac{1}{\sqrt{2\pi}} \int_{-\infty}^{\infty} e^{-ivx}\chi(x)\,dx.$$

Now Lemma 2. 3 b) may be applied to operators of type (3. 18). As before, we compare the Fejér integral

$$(3.20) \qquad F_\gamma(f; x) = \frac{2}{\pi\gamma} \int_{-\infty}^{\infty} f(x-y)\left[\frac{\sin \gamma y/2}{y}\right]^2 dy$$

with the Cauchy—Poisson integral

$$(3.21) \qquad P_\delta(f; x) = \frac{\delta}{\pi} \int_{-\infty}^{\infty} f(x-y)(\delta^2+y^2)^{-1}\,dy \qquad (\delta=\gamma^{-1}).$$

In analogy with the Fourier series situation Lemma 2. 3 b) leads to an examination of $q_\gamma(\lambda)$, given by (3. 4), as well as $[q_\gamma(\lambda)]^{-1}$ for all λ with $(f, e^{i\lambda x})\neq 0$ (compare [11; p. 516] for Fourier transform representation of kernels). Both are uniformly bounded in $\lambda \in (-\infty, \infty)$ and $\gamma > 0$, independently of the Fourier exponents of a particular f. Thus

COROLLARY 3. 6. *Let* $f \in Q$, F_γ *be given by* (3. 20), P_δ *by* (3. 21). *Then for* $\delta = \gamma^{-1}$

$$\lim_{T\to\infty} \frac{1}{2T} \int_{-T}^{T} |F_\gamma(f; x) - f(x)|^2\,dx \sim \lim_{T\to\infty} \frac{1}{2T} \int_{-T}^{T} |P_\delta(f; x) - f(x)|^2\,dx.$$

As a final remark to our applications to series expansions note that the point spectrum of a fixed operator T in Sec. 3. 1—3. 3 consisted of at most countably many eigenvalues, whereas in the case of almost periodic functions the set of eigenvalues is not countable.

Now let us turn to operators which also have a continuous spectrum.

3. 5 *Fourier integrals.* Choose $H = L^2(-\infty, \infty)$, $(f, g) = \int_{-\infty}^{\infty} f(x)\overline{g(x)}\,dx$ as in the Hermite series case. Consider operators T of convolution type

$$(3.22) \qquad Tf(x) = \frac{1}{\sqrt{2\pi}} \int_{-\infty}^{\infty} f(x-y)\,d\mu(y),$$

where $\mu \in BV(-\infty, \infty)$. The existence of a simultaneous spectral representation for operators (3. 22) for all $\mu \in BV$ is ensured by the remarks following Theorem 2. 2.

To construct some appropriate one explicitly, set

$$(3.23) \qquad E_\lambda = \mathfrak{F}^{-1}\mathfrak{P}_\lambda\mathfrak{F}$$

with $\mathfrak{F}$ being the Fourier—Plancherel transformation

$$(3.24) \qquad \lim_{N_1 \to -\infty,\, N_2 \to \infty} \left\| (2\pi)^{-1/2} \int_{N_1}^{N_2} f(x) e^{-ivx}\, dx - \mathfrak{F}[f](v) \right\| = 0,$$

$\mathfrak{F}^{-1}$ its inverse, and $\mathfrak{P}_\lambda$ the projection operator

$$(3.25) \qquad \mathfrak{P}_\lambda f(x) = p_\lambda(x) f(x), \quad p_\lambda(x) = \begin{cases} 1, & -\infty < x < \lambda \\ 0, & \lambda \le x < \infty. \end{cases}$$

Since $\mathfrak{F}$, $\mathfrak{P}_\lambda$, $\mathfrak{F}^{-1}$ are linear and bounded, this is also true for E_λ. Furthermore,

$$E_\lambda E_\mu = \mathfrak{F}^{-1}\mathfrak{P}_\lambda \mathfrak{P}_\mu \mathfrak{F} = E_\mu E_\lambda = E_\lambda \qquad (\mu \ge \lambda)$$

and E_λ is a projection operator. E_λ is continuous from the right since $\|\mathfrak{F}^{-1}f\| = \|f\|$ and hence for $\mu > \lambda$

$$\|E_\mu f - E_\lambda f\| = \|\{p_\mu(v) - p_\lambda(v)\}\mathfrak{F}[f](v)\| = o(1) \qquad (\mu \to \lambda+)$$

by Lebesgue's dominated convergence theorem. Moreover, since $\lim_{\lambda \to -\infty} p_\lambda(v) = 0$, $\lim_{\lambda \to \infty} p_\lambda(v) = 1$, it follows that

$$\lim_{\lambda \to -\infty} E_\lambda f = 0, \quad \lim_{\lambda \to \infty} E_\lambda f = f \qquad (f \in L^2(-\infty, \infty)).$$

Therefore, since with the aid of the convolution and inverse theorems (cf. [11; Ch. 5]) (3.22) may be rewritten as

$$Tf(x) = (2\pi)^{-1/2} \int_{-\infty}^{\infty} [d\mu]^{\wedge}(\lambda) \mathfrak{F}[f](\lambda) e^{i\lambda x}\, d\lambda,$$

with $[d\mu]^{\wedge}(\lambda) = (2\pi)^{-1/2} \int_{-\infty}^{\infty} e^{-i\lambda x}\, d\mu(x)$ and, by definition,

$$E_\lambda f(x) = (2\pi)^{-1} \int_{-\infty}^{\lambda} e^{ixv} \left(\int_{-\infty}^{\infty} f(y) e^{-ivy}\, dy \right) dv$$

(the integrals being interpreted in the sense of (3.24)), we come to

$$T = \int_{-\infty}^{\infty} [d\mu]^{\wedge}(\lambda)\, dE_\lambda$$

which is an explicit simultaneous spectral representation for convolution operators (3.22) for all $\mu \in \mathrm{BV}(-\infty, \infty)$.

Now one may apply Theorem 2.2 to the Cauchy—Poisson integral P_δ on the line (see (3.21) with $f \in Q$ replaced by $f \in L^2(-\infty, \infty)$) which solves Dirichlet's problem for the upper half-plane, and a singular integral which arises in the problem of approximation by functions harmonic in the strip $\{(x, \delta);\ -\infty < x < \infty,\ 0 < \delta < \eta\}$, namely

$$(3.26) \qquad I_\delta(f; x) = \frac{\sqrt{2}}{\pi} \int_{-\infty}^{\infty} f(x-y) \frac{\sigma \sin \sigma\delta \cosh \sigma y}{\sinh^2 \sigma y + \sin^2 \sigma\delta}\, dy \qquad (\sigma = \pi/\eta)$$

(cf. [10]). Though $I_\delta(f; x)$ approximates f also for $\delta \to \eta-$, in comparison with P_δ

we are only interested in the behaviour for $\delta \to 0+$. Since P_δ and I_δ are of type (3. 22) and

$$[d\mu(P_\delta)]\hat{\ }(\lambda) = e^{-\delta|\lambda|}, \quad [d\mu(I_\delta)]\hat{\ }(\lambda) = \frac{\sinh(\eta-\delta)\lambda + \sinh \delta\lambda}{\sinh \eta\lambda}$$

it follows by Theorem 2. 2 that we have to examine the function

$$q_\delta(\lambda) = \left| \frac{(1-e^{-\delta|\lambda|})\sinh \eta\lambda}{\sinh \eta\lambda - \sinh(\eta-\delta)\lambda - \sinh \delta\lambda} \right|;$$

$q_\delta(\lambda)$ is divergent in $\lambda=0$. However, $[q_\delta(\lambda)]^{-1}$ is uniformly bounded in $0<\delta\leq\eta/2$, $-\infty<\lambda<\infty$. Thus,

COROLLARY 3. 7. *Let P_δ be given by* (3. 21), *I_δ by* (3. 26). *Then, for all $f\in L^2(-\infty, \infty)$ and $0<\delta\leq\eta/2$*

$$(3.27) \qquad \int_{-\infty}^{\infty} |I_\delta(f; x)-f(x)|^2\, dx \leq C \int_{-\infty}^{\infty} |P_\delta(f; x)-f(x)|^2\, dx.$$

In [10] it was shown that the saturation classes of P_δ and I_δ (defined on $L^p(-\infty, \infty)$, $1\leq p<\infty$) coincide, and, by modifying I_δ, an estimate inverse to (3. 27) was obtained.

REMARK. By the same procedure as above we could treat the generalized Fourier transformation in the sense of KODAIRA [22, 23] since the corresponding expansion theorem contains analogues to Plancherel's theorem and Parseval's formula. We list some of its particular forms on $L^2(0, \infty)$ (see [14; pp. 1388]):
(i) the Fourier—Sine transformation $\mathfrak{S}$

$$\mathfrak{S}[f](v) = \underset{N\to\infty}{\text{l.i.m.}} \sqrt{\frac{2}{\pi}} \int_0^N f(x) \sin vx\, dx,$$

(ii) the Fourier—Cosine transformation $\mathfrak{C}$

$$\mathfrak{C}[f](v) = \underset{N\to\infty}{\text{l.i.m.}} \sqrt{\frac{2}{\pi}} \int_0^N f(x) \cos vx\, dx,$$

(iii) a further generalization $\mathfrak{U}_k$, $0<k<\infty$ fixed,

$$\mathfrak{U}_k[f](v) = \underset{N\to\infty}{\text{l.i.m.}} \sqrt{\frac{2}{\pi}} (1+k^2 v^2)^{-1/2} \int_0^N f(x) \{\sin vx + kv \cos vx\}\, dx,$$

(iv) the Hankel transformation $\mathfrak{H}_\nu$ which we now treat in greater detail.

3. 6 *Hankel integrals*. Choose, for $\nu\geq 0$ fixed, $H^\nu = \left\{f; \int_0^\infty |f(x)|^2\, d\mu_\nu(x)<\infty\right\}$ with

$$\mu_\nu(x) = [2^{\nu+1/2}\Gamma(\nu+3/2)]^{-1} x^{2\nu+1}, \quad {}_\nu(f, g) = \int_0^\infty f(x)\overline{g(x)}\, d\mu_\nu(x),$$

and $_v\|f\|=_v(f,f)^{1/2}$. Consider operators T of convolution type

$$(3.28) \qquad Tf(x) = \int_0^\infty \int_0^\infty f(y)g(z)D(x,y,z)\,d\mu_v(y)\,d\mu_v(z),$$

where $g \in L^1(d\mu_v)$, i.e., g is measurable on $(0, \infty)$ with $\int_0^\infty |g(x)|\,d\mu_v(x) < \infty$, and

$$D(x,y,z) = \frac{2^{(3v-5/2)}\Gamma(v+1/2)^2}{\Gamma(v)\pi^{1/2}}(xyz)^{-2v+1}A(x,y,z)^{2v-2}.$$

Here $A(x, y, z)$ is the area of a triangle whose sides are x, y, z if there is such a triangle, and zero otherwise. The space H^v, the convolution concept as well as the subsequent definition of the Hankel transform are taken over from HIRSCHMAN [20]. T is a bounded linear operator on H^v into H^v since

$$(3.29) \qquad _v\|Tf\| \leq \left[\int_0^\infty |g(x)|\,d\mu_v(x)\right]_v\|f\|.$$

Defining the Hankel transform $\mathfrak{H}_v[f]$ by

$$(3.30) \qquad \lim_{N\to\infty}\ _v\left\|\int_0^N x^{-v}y^v f(y)(xy)^{1/2}J_{v-1/2}(xy)\,dy - \mathfrak{H}_v[f](x)\right\| = 0,$$

this limit exists for all $f \in H^v$, $\mathfrak{H}_v$ is self-reciprocal, i.e., $\mathfrak{H}_v[\mathfrak{H}_v(f)](x)=f(x)$, and there holds Parseval's formula $_v(f, g)=_v(\mathfrak{H}_v[f], \mathfrak{H}_v[g])$ (see [4; p. 226], [22]). Setting

$$(3.31) \qquad E_\lambda = \mathfrak{H}_v\mathfrak{P}_\lambda\mathfrak{H}_v,$$

$\mathfrak{H}_v$ being given by (3. 30) and $\mathfrak{P}_\lambda$ by (3. 25) (where f is extended to the real line by setting $f(x) \equiv 0$ for $x < 0$), then it follows as in Sec. 3. 5 that $\{E_\lambda\}$ satisfies (2. 1). Now, Tf has the Hankel representation

$$(3.32) \qquad \mathfrak{H}_v[Tf](x) = g^\vee(x)\mathfrak{H}_v[f](x),$$

where

$$g^\vee(x) = \int_0^\infty x^{-v}y^v g(y)(xy)^{1/2}J_{v-1/2}(xy)\,dy \qquad (g \in L^1(d\mu_v))$$

exists for all $x \in [0, \infty)$ (see [20]). (3. 32) follows by the standard argument: consider the truncated function $f_N(x)=f(x)$ for $0 \leq x \leq N$ and zero otherwise. Since $f_N \in L^1(d\mu_v)$ one applies the convolution theorem for $L^1(d\mu_v)$ due to HIRSCHMAN, and then uses the continuity of $\mathfrak{H}_v$. Just as in the Fourier integral case this leads to an explicit simultaneous spectral representation

$$(3.33) \qquad T = \int_{-\infty}^\infty g^\vee(\lambda)\,dE_\lambda$$

of operators of type (3. 28).

REMARK. Let E_n be the n-dimensional Euclidean space, $\xi=(\xi_1, \ldots, \xi_n)$, $\zeta= =(\zeta_1, \ldots, \zeta_n)$ its elements, $\xi \cdot \zeta = \sum_{k=1}^{n} \xi_k \zeta_k$, $|\xi| = (\xi \cdot \xi)^{1/2}$. Consider radial functions on E_n, i.e., $f(\xi)=f(|\xi|)$, square integrable on E_n with respect to Lebesgue measure: $f \in L^2(E_n)$. By Plancherel's theorem its Fourier transform is defined by (see e.g. [5; pp. 120], cf. (3. 24))

$$\lim_{N \to \infty} \left\{ \int_{E_n} \left| (2\pi)^{-n/2} \int_{|\xi| \leq N} f(\xi) e^{-i\xi \cdot \zeta} \, d\xi - \mathfrak{F}_{(n)}[f](\zeta) \right|^2 d\zeta \right\}^{1/2} =$$

$$= \lim_{N \to \infty} C_{\frac{n-1}{2}} \left\| \int_0^N x^{-\frac{1}{2}(n-1)} y^{\frac{1}{2}(n-1)} f(y) (xy)^{1/2} J_{\frac{n-2}{2}}(xy) \, dy - \mathfrak{F}_{(n)}[f](x) \right\| = 0$$

with a certain constant C and $|\zeta|=x$, $|\xi|=y$. In view of (3. 30) one has $\mathfrak{F}_{(n)}=\mathfrak{H}_{(n-1)/2}$, which is a well-known formula if the Hankel transform is (formally) defined on $L^2(0, \infty)$ by (see [15; p. 3], and for a modified version [31; p. 48])

$$\mathfrak{H}_v^*[f](x) = \int_0^\infty f(y) (xy)^{1/2} J_{v-1/2}(xy) \, dy;$$

for in this case we have

$$\mathfrak{H}_{\frac{n-1}{2}}^* \left[|\xi|^{\frac{n-1}{2}} f(|\xi|) \right](|\zeta|) = |\zeta|^{\frac{n-1}{2}} \mathfrak{F}_{(n)}[f](|\zeta|),$$

$$\mathfrak{H}_v^*[y^v f(y)](x) = x^v \mathfrak{H}_v[f](x).$$

In particular $\mathfrak{H}_0^*=\mathfrak{H}_0$ is the Fourier cosine transformation.

To give an example for comparison theorems on H^v, let $v = (n-1)/2$, $n \in \mathbf{N}$, and consider radial functions $f \in L^2(E_n)$. Define the Gauss–Weierstrass integral by

(3. 34) $$W_\tau(f; \xi) = (2\sqrt{\pi}\tau)^{-n} \int_{E_n} f(\xi-\zeta) \exp\{-|\zeta|^2/4\tau^2\} \, d\zeta,$$

and

(3. 35) $$B_\tau(f; \xi) = \Gamma((n+1)/2) \pi^{-(n+1)/2} \int_{E_n} f(\xi-\zeta) k_\tau(\zeta) \, d\zeta$$

with

$$k_\tau(\zeta) = \frac{\tau}{(|\zeta|^2+\tau^2)^{(n+1)/2}} + \frac{n\tau^2-|\zeta|^2}{(|\zeta|^2+\tau^2)^{(n+3)/2}}.$$

Let us observe that for $n=2$ the Gauss–Weierstrass integral $W_\tau(f; \xi)$ is the solution of the heat-conduction equation in the plane with initial value f (radial), whereas $B_\tau(f; \xi)$ is related to a problem in the theory of elasticity, namely to finding the distribution of stress in the semi-infinite elastic solid $\tau \geq 0$ when the bounding surface is deformed in a prescribed manner. More precisely, $B_\tau(f; \xi)$ can be interpreted as the solution of one of the stress components with boundary value f (radiality is achieved by the assumption of symmetry) in "Boussinesq's First Problem" (see [32; (88. 29)]).

$W_\tau(f; \xi)$ and $B_\tau(f; \xi)$ are radial functions since on account of the convolution theorem for Fourier transforms (on E_n) and the radiality of the functions occuring

$$\mathfrak{F}_{(n)}[W_\tau(f; \cdot)](\zeta) = e^{-\tau^2 |\zeta|^2} \mathfrak{F}_{(n)}[f](|\zeta|)$$

$$\mathfrak{F}_{(n)}[B_\tau(f; \cdot)](\zeta) = (1 + \tau |\zeta|) e^{-\tau |\zeta|} \mathfrak{F}_{(n)}[f](|\zeta|).$$

Furthermore, as $\mathfrak{F}_{(n)} = \mathfrak{H}_{(n-1)/2}$ and the kernels involved are radial and belong to $L^1(d\mu_{(n-1)/2})$, (3. 34) and (3. 35) are of convolution type (3. 28). Therefore, we may use the particular simultaneous spectral representation (3. 33), and in order to apply Theorem 2. 2 we have to examine the function

$$q_\tau(\lambda) = \left| \frac{1 - e^{-\tau^2 \lambda^2}}{1 - (1 + \tau\lambda) e^{-\tau\lambda}} \right|.$$

$q_\tau(\lambda)$ as well as $[q_\tau(\lambda)]^{-1}$ are uniformly bounded in $\tau \geq 0$, $0 \leq \lambda < \infty$. Hence we have

COROLLARY 3. 8. *Let $f \in L^2(E_n)$ be radial. Then*

$$\int_0^\infty |W_\tau(f; |\xi|) - f(|\xi|)|^2 \, d\mu_{\frac{n-1}{2}}(|\xi|) \sim \int_0^\infty |B_\tau(f; |\xi|) - f(|\xi|)|^2 \, d\mu_{\frac{n-1}{2}}(|\xi|).$$

In other words

$$\int_{E_n} |W_\tau(f; \xi) - f(\xi)|^2 \, d\xi \sim \int_{E_n} |B_\tau(f; \xi) - f(\xi)|^2 \, d\xi.$$

Thus with the aid of one-parameter methods a comparison theorem for the closed subspace of radial functions in $L^2(E_n)$ is derived.

REFERENCES

[1] R. Askey and S. Wainger, *Mean convergence of expansions in Laguerre and Hermite series.* Amer. J. Math. 87 (1965), 695—708.

[2] S. K. Berberian, *Introduction to Hilbert Space.* Oxford Univ. Press, Oxford 1961.

[3] D. L. Berman, *Some remarks on the problem of speed of convergence of polynomial operators* (Russian). Izv. Vysš. Učebn. Zaved. Matematika 1961, no. 5 (24), 3—5.

[4] S. Bochner, *Lectures on Fourier Integrals* (with an author's supplement on: *Monotonic Functions, Stieltjes Integrals, and Harmonic Analysis*). Princeton Univ. Press, Princeton 1959.

[5] S. Bochner and K. Chandrasekharan, *Fourier Transforms.* Princeton Univ. Press, Princeton 1949.

[6] J. Boman and H. S. Shapiro, *Comparison theorems for a generalized modulus of continuity.* Ark. Mat. 9 (1971), 91—116.

[7] H. Braß, *Approximation mittels Linearkombinationen von Projektionsoperatoren.* Abh. Braunschweig. Wiss. Gesell. 18 (1966), 50—69.

[8] H. Braß, *Approximation durch Linearkombinationen von Eigenfunktionen.* Abh. Braunschweig. Wiss. Gesell. **21** (1969), 429—479.

[9] P. L. Butzer und J. Kemper, *Operatorenkalkül von Approximationsverfahren fastperiodischer Funktionen.* In: Forschungsber. des Landes Nordrhein—Westfalen **2157**, pp. 23—53, Westdeutscher Verlag, Köln 1970.

[10] P. L. Butzer, W. Kolbe, and R. J. Nessel, *Approximation by functions harmonic in a strip.* Arch. Rational Mech. Anal. **44** (1972), 329—336.

[11] P. L. Butzer and R. J. Nessel, *Fourier Analysis and Approximation,* Vol. I. Birkhäuser, Basel 1971.

[12] P. L. Butzer, R. J. Nessel, and W. Trebels, *On summation processes of Fourier expansions in Banach spaces, I: Comparison theorems.* Tôhoku Math. J. **24** (1972), 127—140.

[13] L. Debnath, *On Faltung theorem of Laguerre transform.* Studia Univ. Babeş—Bolyai Ser. Math.-Phys. **14** (1969), 41—45.

[14] N. Dunford and J. T. Schwartz, *Linear Operators,* Vol. II: *Spectral Theory.* Interscience Publ., New York 1963.

[15] A. Erdélyi, F. Oberhettinger, W. Magnus, and F. G. Tricomi, *Tables of Integral Transforms,* Vol. II. Mc Graw-Hill, New York 1954.

[16] J. Favard, *Sur la comparaison des procédés de sommation.* In: *On Approximation Theory* (P. L. Butzer and J. Korevaar, Eds.) ISNM **5**, pp. 4—11, Birkhäuser, Basel 1964.

[17] J. Favard, *On the comparison of the processes of summation.* SIAM J. Numer. Anal. Ser. B **1** (1964), 38—52.

[18] I. M. Ganzburg, *On a certain relation in the theory of linear processes of approximation constructed on the basis of Fourier series and their interpolational analogues* (Russian). In: *Studies of Modern Problems of Constructive Theory of Functions* (Russian), pp. 126—129. Fizmatgiz, Moscow 1961.

[19] I. M. Ganzburg, *On the relation between upper bounds of approximations by certain linear processes* (Russian). In: *Studies Contemporary Problems Constructive Theory of Functions* (Proc. Second All-Union Conf., Baku 1962) (Russian), pp. 346—350. Izdat. Akad. Nauk Azerbaïdžan. SSR, Baku 1965.

[20] I. I. Hirschman, Jr., *Variation diminishing Hankel transforms.* J. Analyse Math. **8** (1960/61), 307—336.

[21] L. A. Kal'niboločkaja, *Summation analogue of integral operators* (Russian). In: *Studies of Modern Problems of Constructive Theory of Functions* (Russian), pp. 28—31, Fizmatgiz, Moscow 1961.

[22] K. Kodaira, *The eigenvalue problem for ordinary differential equations of the second order and Heisenberg's theory of S-matrices.* Amer. J. Math. **71** (1949), 921—945.

[23] K. Kodaira, *On ordinary differential equations of any even order and the corresponding eigenfunction expansions.* Amer. J. Math. **72** (1950), 502—544.

[23a] J. Löfström, *Some theorems on interpolation spaces with applications to approximation in* L_p. Math. Ann. **172** (1967), 176—196.

[24] B. Muckenhoupt, *Poisson integrals for Hermite and Laguerre expansions.* Trans. Amer. Math. Soc. **139** (1969), 231—242.

[25] H. Pollard, *The mean convergence of orthogonal series II.* Trans. Amer. Math. Soc. **63** (1948), 355—367.

[26] F. Riesz und B. Sz.-Nagy, *Vorlesungen über Funktionalanalysis.* 2. rev. ed., Deutscher Verlag der Wissenschaften, Berlin 1968.

[27] Ja. I. Rivkind, *On a question in the theory of approximation of functions* (Russian). Uspehi Mat. Nauk **14**, no. 6 (90) (1959), 185—190.

[28] Ja. I. Rivkind, *On a rough convergence of sequences of operators and polynomials* (Russian). In: *Studies of Modern Problems of Constructive Theory of Functions* (Russian), pp. 338—342, Fizmatgiz, Moscow 1961.

[29] H. S. Shapiro, *Topics in Approximation Theory*. Springer, Berlin 1971.

[30] Yu. M. Shmandin, *Summability of orthogonal series by some linear methods*. Math. Notes **5** (1969), 450—456.

[31] I. N. Sneddon, *Fourier Transforms*. McGraw-Hill, New York 1951.

[32] I. N. Sneddon, *Functional analysis*. In: *Handbuch der Physik: Mathematische Methoden* II, pp. 198—348. Springer, Berlin 1955.

[33] S. B. Stečkin, *The approximation of periodic functions by Fejér sums*. Amer. Math. Soc. Transl. (2) **28** (1963), 269—282.

[34] G. Szegő, *Orthogonal Polynomials* (3. ed.). (*Amer. Math. Soc. Colloq. Publ.* **23**) Amer. Math. Soc., Providence, R. I. 1967.

[35] B. Sz.-Nagy, *Spektraldarstellung linearer Transformationen des Hilbertschen Raumes* (rev. ed.). Springer, Berlin 1967.

[36] E. C. Tichmarsh, *Introduction to the Theory of Fourier Integrals* (2. ed.). Oxford Univ. Press, Oxford 1959.

Sätze vom Jackson-Typ auf Darstellungsräumen kompakter, zusammenhängender Liegruppen*)

Von

H. JOHNEN

LEHRSTUHL A FÜR MATHEMATIK
TECHNISCHE HOCHSCHULE AACHEN

1. Einleitung

Ein Banachraum E heißt Darstellungsraum einer topologischen Gruppe G, wenn es einen Homomorphismus T von G in die Algebra $\mathscr{E}$(E) der beschränkten linearen Transformationen auf E gibt, also

$$T(ab) = T(a)T(b) \qquad (a, b \in G).$$

T heißt Darstellung von G und stetig, wenn zusätzlich die Abbildung

$$a \to T(a)f$$

von G in E für alle $f \in$ E stetig ist. Aus der Stetigkeit der Darstellung folgt, daß $\{\|T(a)\|$, $a \in G\}$ für eine kompakte Gruppe G beschränkt ist. Da die Größe der oberen Schranke im folgenden unwesentlich ist, sei stets angenommen

$$\|T(a)\| \leqq 1 \qquad (a \in G).$$

Jede stetige Darstellung T einer Liegruppe G induziert in natürlicher Weise eine Abbildung dT der Liealgebra g von G in die Liealgebra der linearen abgeschlossenen Transformationen auf E durch

$$dT(X)f = \text{s-}\lim_{t \to 0} t^{-1}(T(\exp tX) - I)f \qquad (X \in \mathfrak{g}).$$

Sind $Y_1, \ldots, Y_r$ beliebige Elemente von g, so sei

$$dT(Y_1 \ldots Y_r) = dT(Y_1) \ldots dT(Y_r)$$

und $D(dT(Y_1 \ldots Y_r))$ der Definitionsbereich dieses Operators. Bezeichnet $\mathfrak{g}^r$ das r-fache kartesische Produkt von g, so verstehen wir unter dem Raum der r-mal differenzierbaren Elemente von E den linearen Raum

$$E^{(r)} = \bigcap_{(Y_1, \ldots, Y_r) \in \mathfrak{g}^r} D(dT(Y_1 \ldots Y_r)).$$

---

*) Der Verfasser dankt der Deutschen Forschungsgemeinschaft für die Förderung unter Bu 166/19 und Herrn W. Tebels für wertvolle Hinweise und Ratschläge.

Für eine Basis $X_1, \ldots, X_m$ von $\mathfrak{g}$ gilt (siehe LANGLANDS [16])

$$E^{(r)} = \bigcap_{i_1, \ldots, i_r = 1}^{m} D\big(dT(X_{i_1} \ldots X_{i_r})\big).$$

Die vorliegende Arbeit beschäftigt sich nun mit dem Problem, für Approximationsverfahren, deren Approximationsverhalten auf den Räumen $E^{(r)}$ vorgegeben ist, Aussagen über die Güte der Konvergenz für beliebige Elemente des Darstellungsraumes E zu erhalten. Genauer, ist

$$\|f^{(r)}\| = \sum_{i_1, \ldots, i_r = 1}^{m} \|dT(X_{i_1} \ldots X_{i_r})f\| \qquad (f \in E^{(r)})$$

die definierende Halbnorm von $E^{(r)}$, so handelt es sich um folgendes Problem:

Gegeben sei eine Folge $\{p_n; n \in \mathbf{N}\}$ von gleichmäßig beschränkten Halbnormen auf E (i.e. $p_n(f) \leqq M\|f\|, f \in E$), für die gilt

$$(J) \qquad\qquad p_n(f) \leqq c\varphi(n)^r \|f^{(r)}\| \qquad (f \in E^{(r)}),$$

wobei $\mathbf{N}$ die Menge der natürlichen Zahlen, c eine positive Konstante und $\varphi(n)$ eine durch 1 nach oben beschränkte positive Funktion sei. Gefragt ist nach einer Abschätzung von $p_n(f)$ für beliebige f durch Größen, die die Eigenschaft von E, Darstellungsraum von G zu sein, zum Ausdruck bringen. Eine solche Größe ist der Stetigkeitsmodul eines $f \in E$. Ist B eine invariante, positiv definite Bilinearform auf $\mathfrak{g}$, d.h. $B([X, Y], Z) = B(X, [Y, Z])$ für $X, Y, \in \mathfrak{g}$, so induziert B eine translationsinvariante Riemannsche Metrik ϱ auf G (vgl. HELGASON [12]). Der r-te Stetigkeitsmodul von $f \in E$ bezüglich dieser Metrik ist definiert durch

$$\omega_r(t, f) = \omega_r(t, f, \varrho) = \sup \{\|(T(a) - I)^r f\|, \ a \in G, \ \varrho(e, a) \leqq t\},$$

wobei e das Identitätselement von G ist. Es sei bemerkt, daß die Stetigkeitsmodule für verschiedene Riemannsche Metriken äquivalent sind, da diese Metriken selbst äquivalent sind.

2. Hauptresultate

Wir sind nun in der Lage, die Hauptergebnisse zu formulieren.

SATZ 2.1. *Ist G kommutativ und erfüllt die Folge der gleichmäßig beschränkten Halbnormen $\{p_n\}_{n \in \mathbf{N}}$ die Bedingung (J), dann gibt es eine Konstante C, so daß*

$$(2.1) \qquad\qquad p_n(f) \leqq C\omega_r(\varphi(n), f) \qquad (f \in E).$$

SATZ 2. 2. *Ist G nicht kommutativ und erfüllt die Folge der gleichmäßig beschränkten Halbnormen $\{p_n\}_{n \in \mathbf{N}}$ die Bedingung (J), dann gibt es eine Konstante C, so daß*

$$(2. 2) \qquad p_n(f) \leqq C\left[\omega_r(\varphi(n), f) + \varphi(n)^3 \int\limits_{\varphi(n)}^{\infty} \frac{\omega_r(s, f)}{s^2}\, ds\right] \qquad (f \in E).$$

Ist überdies $j \in \{1, \ldots, r-1\}$ und

$$(2. 3) \qquad \int\limits_0^{\infty} s^{-(j+1)} \omega_r(s, f)\, ds < \infty,$$

dann gibt es eine von f unabhängige Konstante C, so daß

$$(2. 4) \quad p_n(f) \leqq C\left[\omega_r(\varphi(n), f) + \varphi(n)^{3(j+1)} \int\limits_{\varphi(n)}^{\infty} \frac{\omega_r(s, f)}{s^{j+2}}\, ds + \varphi(n)^{3j} \int\limits_0^{\varphi(n)^3} \frac{\omega_r(s, f)}{s^{j+1}}\, ds\right].$$

Definieren wir für eine monoton wachsende, positive Funktion $\omega(t)$, $t > 0$, den linearen Unterraum

$$\mathsf{Lip}\,(\omega, r, E) = \left\{ f \in E,\ \sup_{0 < t < \infty} \frac{\omega_r(t, f)}{\omega(t)} = |f|_{\omega, r} < \infty \right\},$$

so ergibt sich aus Satz 2. 2 sofort

FOLGERUNG 2. 3. *Genügt $\omega(t)$ der Bedingung*

$$(2. 5) \qquad t^3 \int\limits_t^{\infty} s^{-2} \omega(s)\, ds \leqq c\omega(t) \qquad (0 < t \leqq 1)$$

oder aber der Bedingung

$$(2. 6) \qquad \begin{cases} t^{3(j+1)} \int\limits_t^{\infty} s^{-(j+2)} \omega(s)\, ds \leqq c\omega(t) \\[2mm] \qquad\qquad\qquad\qquad\qquad\qquad\qquad (0 < t \leqq 1) \\[2mm] t^{3j} \int\limits_0^{t^3} s^{-(j+1)} \omega(s)\, ds \leqq c\omega(t) \end{cases}$$

mit $j \in \{1, \ldots, r-1\}$, dann folgt aus (J) für $f \in \mathsf{Lip}\,(\omega, r, E)$

$$(2. 7) \qquad\qquad\qquad p_n(f) \leqq c\omega(\varphi(n)) |f|_{\omega, r}.$$

Die Voraussetzung (2. 5) verbietet ein zu schnelles Abfallen der Funktion $\omega(t)$ für $t \to 0$, während in (2. 6) das Abfallen für $t \to 0$ nach oben und unten eingeschränkt wird. Die Voraussetzungen (2. 5) und (2. 6) sind insbesondere dann erfüllt, wenn $\omega(t)$ für $t > 1$ beschränkt ist und Konstanten $c_1 > 0$, $c_2 > 0$, $r > \beta \geqq 0$, $\beta \neq 1, 2, \ldots, r$ existieren, so daß

$$c_1 t^{\beta+2} \leqq \omega(t) \leqq c_2 t^{\beta}, \qquad (0 < t \leqq 1).$$

Im folgenden bezeichne $X_1, \ldots, X_m$ immer eine bezügliche B orthonormierte Basis von $\mathfrak{g}$. Bezüglich dieser Basis definieren wir für $f \in \mathsf{E}^{(j)}$

$$(2.8) \qquad \omega_r(t, f^{(j)}) = \sum_{i_1, \ldots, i_j = 1}^{m} \omega_r(t, dT(X_{i_1} \ldots X_{i_j})f),$$

und erhalten

FOLGERUNG 2.4. *Ist G kommutativ, dann folgt aus* (J) *für alle* $f \in \mathsf{E}^{(j)}$, $j \in \{0, 1 \ldots, r\}$

$$(2.9) \qquad p_n(f) \leqq c\varphi(n)^j \omega_{r-j}(\varphi(n), f^{(j)}).$$

Ist G nicht kommutativ, dann folgt aus (J) *für alle* $f \in \mathsf{E}^{(j)}$, $j \in \{0, 1, \ldots, r\}$

$$(2.10) \qquad p_n(f) \leqq c\varphi(n)^j \omega_{r'}(\varphi(n), f^{(j)}),$$

wobei $r' = \min(3, r-j)$.

Das Problem, von (J) auf (2.1) zu schließen, wurde für den eindimensionalen Torus im Falle $r=2$ im wesentlichen von FREUD [9] gestellt und von SUNOUCHI [22] für beliebige $r \in \mathbf{N}$ gelöst. Bezeichnet $p_n(f)$ die beste Approximation einer stetigen Funktion durch trigonometrische Polynome vom Grade n, so sind diese Fragestellungen schon bei ACHIESER [1, p. 237] und ZYGMUND [24, p. 117] zu finden. Wandelt man die Voraussetzungen in (J) dahin ab, daß die definierende Halbnorm von $\mathsf{E}^{(r)}$ durch eine (die) Banachnorm von $\mathsf{E}^{(r)}$ ersetzt wird, so wurden Fragen des Typs, von Aussagen (J) auf die Art der Konvergenz für beliebige Elemente von E zu schließen, auf allgemeinen Banachräumen mit Hilfe des K-Funktionals von BUTZER— SCHERER [5, 6, 7] und SCHERER [21] behandelt. Ist die Ungleichung (J) in schwächerer Form vorgegeben, d.h. die Halbnorm in $\mathsf{E}^{(r)}$ ist durch die Banachnorm von $\mathsf{E}^{(r)}$ ersetzt, so erhält man natürlich etwas schwächere Versionen der Sätze 2.1—2.2. Solche Untersuchungen wurden vom Verfasser in der Arbeit [14] für Darstellungsräume E einer (nicht notwendig kompakten) Liegruppe G mit kompakter Liealgebra vorgenommen. Auf diese Arbeit sei im folgenden besonders für hier nicht durchgeführte Beweise generell verwiesen.

3. Beweis der Hauptresultate

Wir benötigen zunächst einige Ergebnisse über Stetigkeitsmodule.

LEMMA 3. 1. *Für jedes* $f \in E$ *ist* $\omega_r(t,f)$ *eine monoton wachsende, stetige Funktion mit* $\lim_{t \to 0} \omega_r(t,f) = 0$ *und genügt den Ungleichungen*

$$(3. 1) \qquad \omega_r(\lambda t,f) \leqq (1+\lambda)^r \omega_r(t,f) \qquad (\lambda > 0),$$

$$(3. 2) \qquad \omega_{r+k}(t,f) \leqq 2^k \omega_r(t,f),$$

$$(3. 3) \qquad \omega_r(t,f) \leqq c t^r \int_t^\infty s^{-(r+1)} \omega_{r+k}(s,f)\, ds \qquad (r,k \geqq 1).$$

Berücksichtigen wir, daß $\omega_r(s,f) = \omega_r(d(G),f)$ für $s \geqq d(G)$ gilt, wobei $d(G)$ definiert sei durch

$$d(G) = \sup_{a \in G} \varrho(e,a),$$

dann ergibt sich aus (3. 1) und (3. 3)

$$(3. 4) \qquad t^k \omega_r(t,f) \leqq c \omega_{r+k}(t,f) \qquad (0 < t \leqq 1),$$

$$(3. 5) \qquad \omega_r(t^k,f) \leqq c \omega_{rk}(t,f) \qquad (0 < t \leqq 1).$$

LEMMA 3. 2. *Ist* $f \in E^{(j)}$, $j \in \{0, 1, 2, \ldots, r\}$, *dann gilt*

$$(3. 6) \qquad \omega_r(t,f) \leqq t^j \omega_{r-j}(t,f^{(j)}).$$

Für $f \in E$ *mit* $\int_0^\infty s^{-(j+1)} \omega_r(s,f)\, ds < \infty$, $j \in \{1, \ldots, r-1\}$, *folgt* $f \in E^{(j)}$, *und für beliebige* $Y_1, \ldots, Y_j \in \mathfrak{g}$ *mit* $B(Y_i, Y_i) = 1$, $i = 1, \ldots, j$, *hat man*

$$(3. 7) \qquad \|dT(Y_1 \ldots Y_j)f\| \leqq c \int_0^\infty s^{-(j+1)} \omega_r(s,f)\, ds,$$

$$(3. 8) \qquad t^j \omega_r(t, dT(Y_1 \ldots Y_j)f) \leqq c \left[\omega_r(t,f) + t^j \int_0^t s^{-(j+1)} \omega_r(s,f)\, ds \right].$$

Mit (3. 3) erhalten wir aus (3. 8)

FOLGERUNG 3. 3. *Unter der Voraussetzung* $\int_0^\infty s^{-(j+1)} \omega_r(s,f)\, ds < \infty$ *ergibt sich für* $l \in \{1, \ldots, r-1\}$

$$(3. 9) \qquad \omega_l(t, dT(Y_1 \ldots Y_r)f) \leqq c \left[t^l \int_t^\infty \frac{\omega_r(s,f)}{s^{j+l+1}}\, ds + \int_0^t \frac{\omega_r(s,f)}{s^{j+1}}\, ds \right].$$

Aus den Ungleichungen (3. 1)—(3. 9) folgt mit Hilfe der Kompaktheit von G eine leicht abgewandelte, verschärfte Aussage eines entsprechenden Lemmas in [14].

LEMMA 3.4. *Es seien* $F_i(t)$, $i=1, 2, \ldots, l$, *Abbildungen des Intervalls* $[0, \delta]$, $\delta \leq 1$, *in* $\mathfrak{g}$ *mit* $B(F_i(t), F_i(t)) < c$ *gleichmäßig für* $t \in [0, \delta]$ *und* $i=1, \ldots, l$. *Ist* $\nu \in \{1, \ldots, r\}$, *dann gibt es eine Konstante* c, *so daß für alle* $f \in E$ *mit* $\int_0^\infty s^{-(j+1)} \omega_r(s, f)\, ds < \infty$ *gilt*

$$(3.10) \qquad \left\| \sum_{k=0}^{\nu} (-1)^{\nu-k} \binom{\nu}{k} T(\exp ktF_1(t)\ \exp k^2 t^3 F_2(t) \ldots \exp k^l t^{l+1} F_l(t))f \right\| \leq$$

$$\leq c \left[\omega_\nu(t, f) + t^{3(j+1)} \int_t^\infty \frac{\omega_r(s, f)}{s^{j+2}}\, ds + t^{3j} \int_0^{t^3} \frac{\omega_r(s, f)}{s^{j+1}}\, ds \right] \qquad (0 < t \leq \delta).$$

Sind $X = x^1 X_1 + \cdots + x^m X_m$ und $Y = y^1 X_1 + \cdots + y^m X_m$ aus $\mathfrak{g}$ mit $B(X, X) \leq d$, $B(Y, Y) \leq d$, dann existiert ein $\delta = \delta(k, d, l) > 0$ so (siehe [14]), daß für alle $0 \leq t \leq \delta$

$$\exp(tX + ktY) = \exp tX\ \exp ktF_1(x, y, t)\ \exp k^2 t^3 F_2(x, y, t) \ldots$$

$$\exp k^l t^{l+1} F_l(x, y, t)\ \exp t^{l+2} R_{l+1}(x, y, t, k).$$

Hierbei hängen die Elemente F_i, $i=1, \ldots, l$, und R_{l+1} in analytischer Weise für $0 \leq t \leq \delta$ und $\sum_{i=1}^m (x^i)^2 \leq d$, $\sum_{i=1}^m (y^i)^2 \leq d$ von diesen Variablen ab. Es gilt also insbesondere $B(F_i, F_i) \leq c$, $i=1, \ldots, l$ und $B(R_{l+1}, R_{l+1}) \leq c$ auf diesen Bereichen. So erhalten wir aus (3.3) und (3.10) mit $l = \nu$

LEMMA 3.5. *Sind* $X, Y \in \mathfrak{g}$ *mit* $B(X, X) \leq d$, $B(Y, Y) \leq d$, *dann gibt es ein* $\delta > 0$, $\delta = \delta(d, r) \leq 1$ *so, daß für alle* t *mit* $0 \leq t \leq \delta$

$$(3.11) \quad \left\| \sum_{k=0}^{\nu} (-1)^{\nu-k} \binom{\nu}{k} T(\exp(tX + ktY))f \right\| \leq c \left[\omega_\nu(t, f) + t^3 \int_t^\infty s^{-2} \omega_r(s, f)\, ds \right] \qquad (1 \leq \nu \leq r).$$

Erfüllt $f \in E$ *außerdem die Bedingung* $\int_0^\infty s^{-(j+1)} \omega_r(s, f)\, ds < \infty$ *für* $j \in \{1, \ldots, r-1\}$, *so hat man für* $1 \leq \nu \leq r$, $0 \leq t \leq \delta$

$$(3.12) \qquad \left\| \sum_{k=0}^{\nu} (-1)^{\nu-k} \binom{\nu}{k} T(\exp(tX + ktY))f \right\| \leq$$

$$\leq c \left[\omega_\nu(t, f) + t^{3(j+1)} \int_t^\infty \frac{\omega_r(s, f)}{s^{j+2}}\, ds + t^{3j} \int_0^{t^3} \frac{\omega_r(s, f)}{s^{j+1}}\, ds \right]$$

BEMERKUNG 3.6. *Für* $f \in E$ *und* $\nu = 1, 2, 3$ *folgt aus Lemma 3.1 und* (3.11)

$$(3.11^*) \qquad \left\| \sum_{k=0}^{\nu} (-1)^{\nu-k} \binom{\nu}{k} T(\exp(tX + ktY))f \right\| \leq c\omega_\nu(t, f).$$

Ist $f \in E^{(1)}$ *und* $\nu \geq 4$, *so erhält man*

$$(3.12^*) \quad \left\| \sum_{k=0}^{\nu} (-1)^{\nu-k} \binom{\nu}{k} T(\exp(tX + ktY))f \right\| \leq c[\omega_\nu(t, f) + t^3 \omega_{\nu-2}(t, f^{(1)}) + t^3 \omega_1(t^3, f^{(1)})].$$

17*

Bemerkung 3.7. *Falls G kommutativ ist, gilt unter den Voraussetzungen des Lemmas*

$$(3.13) \qquad \left\| \sum_{k=0}^{v} (-1)^{v-k} \binom{v}{k} T(\exp(tX+ktY)) \right\| \leq c\omega_v(t,f).$$

Mit Hilfe der orthonormierten Basis $X_1, \ldots, X_m$ von $\mathfrak{g}$ ordnen wir nun jedem $f \in E$ eine Abbildung $f^{\sim}$ von $\mathbf{R}^m$ in E zu, definiert durch

$$f^{\sim}(x) = f^{\sim}(x^1, \ldots, x^m) = T(\exp(x^1 X_1 + \cdots + x^m X_m))f.$$

Für $h \in \mathbf{R}^m$, $r \in \mathbf{N}$ setzen wir

$$(3.14) \qquad \Delta_h^r f^{\sim}(x) = \sum_{k=0}^{r} (-1)^k \binom{r}{k} f^{\sim}(x+(r-k)h).$$

Bezeichnen wir mit e_i, $i=1, \ldots, m$, die Einheitsvektoren in Richtung der Koordinatenachsen von $\mathbf{R}^m$, dann sei für $\gamma \in \mathbf{N}^m$ und $s \in \mathbf{R}$

$$(3.15) \qquad \Delta_s^\gamma f^{\sim}(x) = \Delta_{se_1}^{\gamma^1} \ldots \Delta_{se_m}^{\gamma^m} f^{\sim}(x).$$

Ferner erklären wir das Integralmittel der Funktion $f^{\sim}$ bezüglich der i-ten Koordinate durch

$$St_{i,t} f^{\sim}(x) = \int_0^1 f^{\sim}(x+\sigma t e_i) \, d\sigma$$

und setzen für $\gamma \in \mathbf{N}^m$

$$(3.16) \qquad St_t^\gamma f^{\sim}(x) = St_{1,t}^{\gamma^1} \ldots St_{m,t}^{\gamma^m} f^{\sim}(x).$$

Ist ξ eine in einer genügend großen Umgebung des Punktes $x \in \mathbf{R}^m$ definierte stetige Funktion, so sei

$$(3.17) \qquad \Diamond_t^\gamma(\xi) f^{\sim}(x) = \int_0^1 \ldots \int_0^1 \xi\left(x+t \sum_{i=1}^{m} \sum_{j=1}^{\gamma^i} \sigma^{ij} e_i\right) \nabla_t f^{\sim}(x) \prod_{i=1}^{m} \prod_{j=1}^{i} d\sigma^{ij},$$

wobei

$$(3.18) \qquad \nabla_t f^{\sim}(x) = \left(\prod_{i=1}^{m} \prod_{j=1}^{\gamma^i} \Delta_{t(1-\sigma^{ij})e_i} f^{\sim}\right)\left(x + \sum_{k=1}^{m} \sum_{l=1}^{\gamma^k} \sigma^{kl} e_k\right).$$

Mit diesen Bezeichnungen zeigen wir das fundamentale

Lemma 3.8. *Sind $X_{i_1} \ldots X_{i_r}$ Elemente der orthonormierten Basis von $\mathfrak{g}$, dann ist die Abbildung*

$$f \to dT(X_{i_1} \ldots X_{i_r}) St_t^{\{r\}} f^{\sim}(0),$$

mit $\{r\} \in \mathbf{N}^m$ definiert durch $\{r\} = (r, \ldots, r)$, für kleines $t>0$ eine beschränkte lineare Transformation von E in E, die folgender Identität genügt

$$(3.19) \qquad dT(X_{i_1} \ldots X_{i_r}) St_t^{\{r\}} f^{\sim}(0) =$$

$$= \sum_{1 \leq |\gamma| \leq r} St_t^{\{r\}-\gamma} \left(\sum_{0 \leq \beta \leq \gamma} \binom{\gamma}{\beta} t^{-|\gamma-\beta|} \Diamond_t^\beta (\partial^\beta \xi_{i_1 \ldots i_r}^\gamma) \Delta_t^{\gamma-\beta} f^{\sim} \right)(0).$$

Hierbei sind $\xi_{i_1 \ldots i_r}^\gamma$ in einer Umgebung des Nullpunktes beliebig oft differenzierbare Funktionen und
$\partial^\beta = (\partial/\partial x^1)^{\beta^1} \ldots (\partial/\partial x^m)^{\beta^m}$

BEWEIS. Die rechte Seite der Ungleichung (3. 19) stellt einen beschränkten linearen Operator auf E dar. Da $E^{(r)}$ dicht in E ist und $St^{\{r\}}f^\sim(0) \in E^{(r)}$ für beliebiges $f \in E$ gilt, genügt es, die Identität für $f \in E^{(r)}$ zu beweisen. Ist $f \in E^{(r)}$ und $|x|$ genügend klein, so haben wir

$$dT(X_{i_1} \ldots X_{i_r})f^\sim(x) = \sum_{1 \le |\gamma| \le r} \xi^\gamma_{i_1 \ldots i_r}(x)\partial^\gamma f^\sim(x),$$

also

$$(3.\,20) \qquad dT(X_{i_1} \ldots X_{i_r})St^{\{r\}}_t f^\sim(0) = \sum_{1 \le |\gamma| \le r} St^{\{r\}}_t(\xi^\gamma_{i_1 \ldots i_r}\partial^\gamma f^\sim)(0).$$

Durch partielle Integration und Anwenden der Taylorformel erhält man

$$(3.\,21) \qquad \int_0^1 \xi(x+\sigma te_i)\partial^\gamma f^\sim(x+\sigma te_i)\,d\sigma =$$

$$= t^{-1}\xi(x)\Delta_{te_i}\partial^{\gamma-e_i}f^\sim(x) + \int_0^1 \partial^{e_i}\xi(x+\sigma te_i)\Delta_{t(1-\sigma)e_i}\partial^{\gamma-e_i}f^\sim(x+\sigma te_i)\,d\sigma.$$

Nach (3. 17) ist der zweite Ausdruck auf der rechten Seite dieser Identität $\Diamond^{e_i}_t(\partial^{e_i}\xi)\partial^{\gamma-e_i}f^\sim(x)$, und wir erhalten durch wiederholte Anwendung der Formel (3. 21) schließlich aus (3. 20) die gewünschte Identität (3. 19) für $f \in E^{(r)}$.

Wir kommen nun zum eigentlichen Beweis der Sätze 2. 1—2. 2. Dazu genügt es, ein Element $f_t \in E^{(r)}$ so zu konstruieren, daß für t klein genug, sagen wir $0 \le t \le \le \delta \le 1$,

$$(3.\,22) \qquad \|f-f_t\| \le c\omega_r(t,f)$$

gilt und im kommutativen Fall

$$(3.\,23) \qquad t^r\|f_t^{(r)}\| \le c\omega_r(t,f)$$

bzw. im nichtkommutativen Fall

$$(3.\,24) \qquad t^r\|f_t^{(r)}\| \le c\left[\omega_r(t,f)+t^3\int_t^\infty s^{-2}\omega_r(s,f)\,ds\right]$$

oder unter der Voraussetzung (2. 3)

$$(3.\,25) \qquad t^r\|f_t^{(r)}\| \le c\left[\omega_r(t,f)+t^{3(j+1)}\int_t^\infty \frac{\omega_r(s,f)}{s^{j+2}}\,ds+t^{3j}\int_0^{t^3}\frac{\omega_r(s,f)}{s^{j+1}}\,ds\right].$$

Ohne Einschränkung der Allgemeinheit kann dann wegen Ungleichung (3. 1) angenommen werden, daß auch $0 < \varphi(n) \le \delta$, und es gilt mit $t = \varphi(n)$ nach (J)

$$p_n(f) \le p_n(f-f_{\varphi(n)})+p_n(f_{\varphi(n)}) \le M\|f-f_{\varphi(n)}\|+c\varphi(n)^r\|f^{(r)}_{\varphi(n)}\|,$$

d.h. Satz 2. 1 und Satz 2. 2 wären bewiesen. Wir geben nun ein Element f_t an, für das diese Eigenschaften erfüllt sind. Dazu setzen wir

$$(3.\,26) \qquad -f_t = \sum_{k=1}^r (-1)^k \binom{r}{k} St^{\{r\}}_{kt}f^\sim(0).$$

Trivialerweise genügt f_t der Abschätzung (3. 22). Andererseits gilt aber nach Lemma 3. 8

$$(3.27) \qquad -dT(X_{i_1} \dots X_{i_r})f_t =$$

$$= \sum_{k=1}^{r} (-1)^k \binom{r}{k} \sum_{1 \le |\gamma| \le r} St_{kt}^{|r|-\gamma} \left(\sum_{0 \le \beta \le \gamma} \binom{\gamma}{\beta} (kt)^{-|\gamma-\beta|} \mathcal{V}_{kt}^{\beta} (\partial^{\beta} \xi_{i_1 \dots i_r}) \Delta_{kt}^{\gamma-\beta} f^{\sim} \right)(0).$$

Sind $h_1, \dots, h_r$ beliebige Vektoren aus $\mathbf{R}^m$, dann gibt es (siehe BAISHANSKI [3]) eine ganze Zahl $r_0 = r_0(r)$, $2r_0$ Linearformen Y_i, Z_i und reelle Zahlen C_i, $i = 1, \dots, r_0$, so daß

$$\Delta_{h_1} \dots \Delta_{h_r} f^{\sim}(x) = \sum_{i=1}^{r_0} C_i \Delta_{Y_i(h_1, \dots, h_r)}^{r} f^{\sim}(x + X_i(h_1, \dots, h_r)).$$

Demnach erhalten wir mit

$$kty = kt \sum_{k=1}^{m} \sum_{\lambda=1}^{\beta^k} \sigma^{k\lambda} e_k + kt \sum_{\mu=1}^{m} \sum_{\nu=\gamma^\mu+1}^{r} \sigma^{\mu\nu} e_\mu$$

die Identität

$$(3.28) \qquad \left[\prod_{i=1}^{m} \prod_{j=1}^{\beta^i} \Delta_{kt(1-\sigma^{ij})e_i} \Delta_{kt}^{\gamma-\beta} f^{\sim} \right](kty) = \sum_{i=1}^{|\gamma|_0} C_i \Delta_{ktY_i}^{|\gamma|} f^{\sim}(kty + ktZ_i).$$

Ist G kommutativ, so folgt, daß die in (3. 27) auftretenden Funktionen $\xi_{i_1, \dots, i_r}^{\gamma}$ für $|\gamma| < r$ identisch verschwinden, während für $|\gamma| = r$ gilt $\xi_{i_1 \dots i_r}^{\gamma} = 1$. Deshalb ergibt sich aus (3. 27) mit (3. 28) und (3. 13), daß für kommutatives G das in (3. 26) definierte Element f_t die Ungleichung (3. 23) erfüllt. Ist G nicht kommutativ, so wenden wir auf (3. 27) die Umformung (3. 28) an und schätzen dann mit Lemma 3. 5 und Ungleichung (3. 4) nach oben ab. Also sind für das in (3. 26) konstruierte f_t die Ungleichungen (3. 24) bzw. (3. 25) erfüllt, d.h. Satz 2. 1 und Satz 2. 2 sind bewiesen.

Die Folgerung 2. 3 erhielt man unmittelbar aus Satz 2. 1 und Satz 2. 2. Zum Beweis von Folgerung 2. 4 beachten wir, daß (2. 9) sofort aus (2. 1) und (3. 6) folgt. Um schließlich Ungleichung (2. 10) zu beweisen, treffen wir drei Fallunterscheidungen

(i) $j = 0$. Wir beachten, daß

$$(3.29) \qquad t^3 \int_t^{\infty} \frac{\omega_r(s,f)}{s^2} \, ds \le t^3 \int_t^{d(G)} \frac{\omega_r(s,f)}{s^2} \, ds + t^3 d(G)^{-1} \omega_r(d(G), f),$$

wobei $d(G) = \sup_{a \in G} \varrho(e, a)$. Ist $r = 1, 2, 3$, so folgt mit (3. 1) aus (3. 29)

$$t^3 \int_t^{\infty} s^{-2} \omega_r(s, f) \, ds \le c\omega_r(t, f) \qquad (0 < t \le 1).$$

Ist $r>3$, dann wenden wir auf (3. 29) die Ungleichung (3. 2) an und erhalten

$$t^3 \int\limits_t^\infty s^{-2}\omega_r(s,f)\,ds \;\leqq\; 2^{r-3}\left[t^3 \int\limits_t^{d(G)} s^{-2}\omega_3(s,f)\,ds + t^3 d(G)^{-1}\omega_3(d(G),f)\right].$$

Wenden wir nun wieder wie vorher auf die rechte Seite dieser Ungleichung die Ungleichung (3. 1) mit $\lambda=s/t$ bzw. $\lambda=d(G)/t$ an, dann folgt (2. 10).

(ii) $j=1$. In diesem Fall wissen wir nach (3. 11*), daß die rechte Seite der Identität (3. 28) für $|\gamma|\leqq 3$ durch $\omega_{|\gamma|}(t,f)$ abgeschätzt werden kann, für $r\geqq|\gamma|\geqq 4$ durch $\omega_{|\gamma|}(t,f)+t^3\omega_{|\gamma|-2}(t,f^{(1)})+t^3\omega_1(t^3,f^{(1)})$ nach (3. 12*). Nach (3. 4) und (3. 5) gilt dann für $0<t\leqq 1$

$$t^{r-|\gamma|}\omega_{|\gamma|}(t,f)\;\leqq\; c\omega_r(t,f)$$

$$t^{r+3-|\gamma|}\omega_{|\gamma|-2}(t,f^{(1)})\;\leqq\; ct^2\omega_{r-1}(t,f^{(1)})$$

$$t^{r-|\gamma|}t^3\omega(t^3,f^{(1)})\;\leqq\; ct\omega_5(t,f^{(1)}),$$

und es ergibt sich für das in (3. 26) definierte f_t aus (3. 27)

$$t^r\|f_t\|\;\leqq\; ct[\omega_{r-1}(t,f^{(1)})+\omega_5(t,f^{(1)})].$$

Setzen wir $t=\varphi(n)$, so folgt a fortiori (2. 10).

(iii) $j\geqq 2$. Da in diesem Falle nach (3. 6)

$$\int\limits_0^\infty s^{-j}\omega_r(s,f)\,ds\;\leqq\;\infty \qquad (r>j),$$

erhalten wir aus (2. 4)

$$p_n(f)\;\leqq\; c\left[\omega_r(\varphi(n),f)+\varphi(n)^{3j}\int\limits_{\varphi(n)}^\infty \frac{\omega_r(s,f)}{s^{j+1}}\,ds+\varphi(n)^{3(j-1)}\int\limits_0^{\varphi(n)^3}\frac{\omega_r(s,f)}{s^j}\,ds\right].$$

Wegen (3. 6) und der Monotonie folgt ferner

$$\varphi(n)^{3(j-1)}\int\limits_0^{\varphi(n)^3} s^{-j}\omega_r(s,f)\,ds\;\leqq\;\varphi(n)^{3j}\omega_{r-j}(\varphi(n)^3,f^{(j)}).$$

Ebenfalls aus (3. 6) schließt man mit (3. 1) auf

$$\varphi(n)^{3j}\int\limits_{\varphi(n)}^\infty s^{-(j+1)}\omega_r(s,f)\,ds\;\leqq\; c\varphi(n)^j\omega_{r'}(\varphi(n),f^{(j)})$$

mit $r' = \min(2j, r-j)$. Folglich ist (2. 10) für alle $j\in\mathbf{N}$ bewiesen.

4. Anwendungen

In erster Anwendung spezialisieren wir den Banachraum E, den Darstellungsraum von G. Ist H eine abgeschlossene Untergruppe von G, so wählen wir E als
einen der Banachräume $\mathsf{C}(G/H)$ oder $\mathsf{L}_p(G/H)$, $1 \leq p < \infty$, wobei unter $\mathsf{C}(G/H)$ der
Raum der stetigen Funktionen auf dem homogenen Raum G/H verstanden werde
und die Räume $\mathsf{L}_p(G/H)$ bezüglich des invarianten Maßes auf G/H (siehe HELGASON
[12, p. 369]) gebildet seien. Ist f aus einem dieser Räume, so definieren wir die Darstellung T von G durch

$$(4.\,1) \qquad\qquad (T(a)f)(bH) = f(a^{-1}bH) \qquad (bH \in G/H).$$

Diese Darstellung ist stetig und es gilt

$$\|T(a)\| \leq 1 \qquad (a \in G).$$

Für $X \in \mathfrak{g}$ ist $dT(X)$ definiert durch

$$(4.\,2) \qquad\qquad (dT(X)f)(bH) = \operatorname*{s-lim}_{t \to 0} t^{-1}(f(\exp(-tX)bH) - f(bH)).$$

Als Anwendung von Folgerung 2.4 erhalten wir daher

SATZ 4.1. *Es sei* E *einer der Räume* $\mathsf{C}(G/H)$, $\mathsf{L}_p(G/H)$, $1 \leq p < \infty$. *Ist* $\{p_n\}_{n \in \mathbf{N}}$
eine Folge gleichmäßig beschränkter Halbnormen auf E *und gilt*

$$(4.\,3) \qquad\qquad p_n(f) \leq c\varphi(n)^r \| f^{(r)} \| \qquad (f \in \mathsf{E}^{(r)}),$$

dann gibt es eine Konstante C, *so daß für alle* $f \in \mathsf{E}^{(j)}$, $j \in \{0, 1, \ldots, r\}$,

$$(4.\,4) \qquad\qquad p_n(f) \leq C\varphi(n)^j \omega_{r'}(\varphi(n), f^{(j)}),$$

wobei $r' = \min(3, r-j)$.

Wählen wir speziell $G = SO(\nu+1)$, die Gruppe der Drehungen im $\nu+1$-di-
mensionalen euklidischen Raum $\mathbf{R}^{\nu+1}$ und $H = SO(\nu)$ als die Drehungen aus G,
die eine vorgegebene Richtung, z. B. die Richtung der $(\nu+1)$-ten Koordinate in
$\mathbf{R}^{\nu+1}$, invariant lassen, dann läßt sich der Quotienten-Raum G/H mit der ν-dimen-
sionalen Kugel S^ν in $\mathbf{R}^{\nu+1}$ identifizieren. Die Liealgebra $\mathfrak{so}(\nu+1)$ von $SO(\nu+1)$
ist gegeben (vgl. HELGASON [12, p. 341]) durch die Menge aller reellen $(\nu+1)\times(\nu+1)$
Matrizen X, die schiefsymmetrisch sind. Sind $X, Y \in \mathfrak{so}(\nu+1)$, so definieren wir
$B(X, Y) = -1/2 \operatorname{Sp}(XY)$, wobei Sp die Spur einer Matrix bedeutet. Bezeichnet E_{ij}
die Matrix, die am Schnittpunkt der i-ten Zeile und j-ten Spalte eine Eins besitzt
und deren andere Elemente verschwinden, so bilden die Elemente $(E_{ij} - E_{ji}) = X_{ij}$,
$1 \leq i < j \leq \nu+1$, eine orthonormierte Basis von $\mathfrak{so}(\nu+1)$ bezüglich B. Die Drehung
$\exp tX_{ij}$ ist gegeben durch

$$\exp tX_{ij} = \sum_{k=0}^{\infty} \frac{t^k}{k!} (X_{ij})^k,$$

d.h., ist $x=(x^1, \ldots, x^{\nu+1})\in\mathbf{R}^{\nu+1}$, dann bleiben alle Koordinaten von x außer der i-ten und j-ten Koordinate unter dieser Drehung invariant. Die i-te Koordinate geht über in $y^i = x^i \cos t + x^j \sin t$, die j-te Koordinate in $y^j = -x^i \sin t + x^j \cos t$, $\exp tX_{ij}$ ist also eine Drehung um den Winkel t in der (x^i, x^j)-Ebene.

In einem nächsten Schritt konkretisieren wir die Folge der Halbnormen auf den Räumen $\mathsf{C}(S^\nu)$ und $\mathsf{L}_p(S^\nu)$, $1\leqq p<\infty$. Bezeichnen wir mit $\mathscr{P}_n$ die Menge aller Linearkombinationen von Kugelfunktionen der Ordnung $\leqq n$ (Definition vgl. [4]), dann sei diese Folge gegeben auf E ($\mathsf{E}=\mathsf{C}(S^\nu)$, $\mathsf{E}=\mathsf{L}_p(S^\nu)$) durch

$$(4.5) \qquad E_n(f, \mathsf{E}) = \inf\{\|f-P_n\|,\ P_n\in\mathscr{P}_n\} \qquad (f\in\mathsf{E}).$$

SATZ 4.2. *Es gibt eine Konstante* C, *so daß für alle* $f\in\mathsf{E}^{(j)}$, $j\in\mathbf{N}$,

$$(4.6) \qquad E_n(f, \mathsf{E}) \leqq Cn^{-j}\omega_{r'}(n^{-1}, f^{(j)}) \qquad (r'=1, 2, 3).$$

BEWEIS. $E_n(f, \mathsf{E})$ ist eine Folge gleichmäßig beschränkter Halbnormen auf E und sie erfüllt für jedes $r\in\mathbf{N}$ die Ungleichung (J) mit $\varphi(n)=n^{-1}$ (vgl. [4]). Damit ist (4.6) eine Folgerung aus (4.4).

Zur Illustration dieses Ergebnisses betrachten wir den Fall $\nu=2$. Bezeichnen wir X_{12} mit J_3, X_{13} mit J_2 und X_{23} mit J_1, so genügen J_1, J_2, J_3 den aus der Quantentheorie bekannten Relationen (vgl. MILLER [17, p. 31]) $[J_1, J_2]=J_3$, $[J_1, J_3] = -J_2$, $[J_2, J_3]=J_1$ und wir haben

$$\exp tJ_3 = \begin{pmatrix} \cos t & \sin t & 0 \\ -\sin t & \cos t & 0 \\ 0 & 0 & 1 \end{pmatrix}$$

$$\exp tJ_2 = \begin{pmatrix} \cos t & 0 & \sin t \\ 0 & 1 & 0 \\ -\sin t & 0 & \cos t \end{pmatrix}$$

$$\exp tJ_1 = \begin{pmatrix} 1 & 0 & 0 \\ 0 & \cos t & \sin t \\ 0 & -\sin t & \cos t \end{pmatrix}$$

und, falls für $x\in S^2$ gilt

$$x^1=\cos\vartheta\cos\varphi, \qquad x^2=\cos\vartheta\sin\varphi, \qquad x^3=\sin\vartheta$$

mit $0<\vartheta<\pi$, $0<\varphi<2\pi$, dann sind die Differentialoperatoren $dT(J_1)$, $dT(J_2)$, $dT(J_3)$ gegeben durch

$$[dT(J_3)f](\vartheta, \varphi) = +\left[\frac{\partial}{\partial\varphi}f\right](\vartheta, \varphi)$$

$$[dT(J_2)f](\vartheta, \varphi) = +\cos\varphi\,\frac{\partial f}{\partial\vartheta}(\vartheta, \varphi) + \operatorname{tg}\vartheta\sin\varphi\,\frac{\partial f}{\partial\varphi}(\vartheta, \varphi)$$

$$[dT(J_1)f](\vartheta, \varphi) = +\sin\varphi\,\frac{\partial f}{\partial\vartheta}(\vartheta, \varphi) - \operatorname{tg}\vartheta\cos\varphi\,\frac{\partial f}{\partial\varphi}(\vartheta, \varphi).$$

Wir wenden uns nun den speziellen homogenen Räumen mit $H=\{e\}$ zu, d.h.
der Darstellungsraum $\mathsf{C}(G/H)$ wird zum Raum der stetigen Funktionen $\mathsf{C}(G)$ auf
G, die Räume $\mathsf{L}_p(G/H)$ gehen über in die Räume $\mathsf{L}_p(G)$, $1 \leq p < \infty$. Als Darstellung
von G auf diesen wählen wir hier die in (4.1) definierte Darstellung T und
untersuchen Approximation durch trigonometrische Polynome vom Grade n.

Ist D eine treue, endlich dimensionale, unitäre, stetige Darstellung von G, so
ordnen wir dieser zu jedem $n \in \mathbf{N}$ die Darstellung D^n zu, definiert durch

$$(4.7) \qquad D^n(a) = \overset{n}{\bigotimes} (D(a) + \bar{D}(a) + D(e)) \qquad (a \in G),$$

wobei $\overset{n}{\bigotimes}$ das n-te Kroneckerprodukt, $+$ die direkte Summe (siehe MURNAGHAN
[18, p. 68]) und $\bar{D}$ die konjugiert komplexe Darstellung zur Darstellung D bedeute.
Führen wir auf dem Darstellungsraum V von G bezüglich der Darstellung D eine
orthonormierte Basis ein, dann ist der Darstellung D eindeutig eine Matrixfunktion
zugeordnet, die ebenfalls mit D bezeichnet sei. Jede Funktion auf G, die Linear-
kombination von Koeffizienten der Matrix D^n ist, heiße ein trigonometrisches Polynom vom Grade n (siehe RAGOZIN [19]), und die Menge aller trigonometrischen
Polynome von Grade n sei mit $\mathscr{P}_n$ bezeichnet. $\mathscr{P}_n$ ist unabhängig von der Wahl
der Basis in V.

Bezeichnet $\mathbf{T}$ einen maximalen Torus in G, dann zerfällt die Restriktion der
Darstellung D auf $\mathbf{T}$ in eine direkte Summe von eindimensionalen, irreduziblen,
unitären Darstellungen. Deshalb gibt es eine orthonormierte Basis von V, so daß
die Restriktion von D auf $\mathbf{T}$ Diagonalgestalt besitzt. Ist V $\varkappa$-dimensional und sind
$\vartheta_1, \ldots, \vartheta_\varkappa$ die Elemente dieser Diagonalen, so gilt also für $x, y \in \mathbf{T}$, daß $\vartheta_k(xy) =$
$= \vartheta_k(x)\vartheta_k(y)$ mit $|\vartheta_k(x)| = 1$. Die Funktionen ϑ_k sind irreduzible Darstellungen
von $\mathbf{T}$, $k = 1, \ldots, \varkappa$, so daß die aus ihnen gebildete Diagonalmatrix eine treue Darstellung von $\mathbf{T}$ ist. Ist der Torus $\mathbf{T}$ l-dimensional, dann können wir $\mathbf{T}$ mit der Menge
aller Elemente $x = (x^1, \ldots, x^l) \in \mathbf{R}^l$ identifizieren, für die gilt $\frac{1}{2} \leq |x^j| < \frac{1}{2}$, $j = 1, \ldots, l$,
und die Gruppenverknüpfung auf $\mathbf{T}$ ist dort definiert durch die koordinatenweise
Addition modulo 1. Jede irreduzible Darstellung von $\mathbf{T}$ in dieser Gestalt ist gegeben
durch Funktionen $e^{i2\pi(\mu_1 x^1 + \cdots + \mu_l x^l)}$, wobei $\mu_1, \ldots, \mu_l$ ganze Zahlen sind. Es gibt
also ganze Zahlen $\mu_{k1}, \ldots, \mu_{kl}$, $k = 1, \ldots, \varkappa$, so daß (vgl. [8, p. 213])

$$\vartheta_k(x) = e^{i2\pi(\mu_{k1} x^1 + \cdots + \mu_{kl} x^l)}.$$

LEMMA 4.3. *Ist G eine kompakte, zusammenhängende Liegruppe vom Rang l
und gibt es eine orthonormierte Basis für V, so daß die Funktionen ϑ_k die Gestalt
haben*

$$(4.8) \qquad \vartheta_k(x) = e^{i2\pi x^k} \qquad (k = 1, \ldots, l),$$

dann gelten für die Funktionen

$$(4.9) \qquad L_n(a) = \left| \frac{\det(D(e) - D(a^{n+1}))}{\det(D(e) - D(a))} \right|^{2(\mu_0+2)}$$

die asymptotischen Formeln

$$(4.10) \qquad \int_G L_n(a)\, da \approx n^{2((\mu_0+2)\varkappa - \mu_0) - l},$$

$$(4.11) \qquad \int_G L_n(a)\varrho^2(e, a)\, da \approx n^{2((\mu_0+2)\varkappa - \mu_0) - l - 2}.$$

Hierbei bezeichne μ_0 die Anzahl der verschiedenen Wurzeln von G.

BEWEIS. Sind θ_μ, $\mu = 1, \ldots, \mu_0$ die verschiedenen Wurzeln von G und ist f eine Klassenfunktion auf G, dann haben wir (siehe ADAMS [2, p. 142])

$$\int_G f(a)\, da = \int_T f(x)u(x)\, dx,$$

wobei $u(x) = \delta(x)\overline{\delta}(x)/|w|$ mit

$$\delta(x) = \prod_{\mu=1}^{\mu_0} (e^{i\pi\theta_\mu(x)} - e^{-i\pi\theta_\mu(x)})$$

und $|w|$ die Ordnung der Weylgruppe ist. Als Koeffizienten von irreduziblen Darstellungen von T induziert durch die adjungierte Darstellung von G haben die Funktionen $e^{i2\pi\theta_\mu(x)}$ die Darstellung (siehe Adams [2, p. 76, 82])

$$(4.12) \qquad e^{i2\pi\theta_\mu(x)} = e^{i2\pi(v_{\mu 1}x^1 + \cdots + v_{\mu l}x^l)}$$

mit ganzen Zahlen $v_{\mu 1}, \ldots, v_{\mu l}$. Somit erhalten wir mit Hilfe der Eindeutigkeit des Haarintegrals auf T

$$(4.13) \qquad \int_G L_n(a)\, da = c \int_{-1/2}^{1/2} \cdots \int_{-1/2}^{1/2} \prod_{k=1}^{\varkappa} \left(\frac{\sin\left(\pi(n+1)\sum_{v=1}^{l}\mu_{kv}x^v\right)}{\sin\left(\pi\sum_{v=1}^{l}\mu_{kv}x^v\right)} \right)^{2(\mu_0+2)} \times$$

$$\times \prod_{\mu=1}^{\mu_0} \sin^2 \pi\left(\sum_{\lambda=1}^{l} v_{\mu\lambda}x^\lambda\right) dx^1 \ldots dx^l.$$

Da für $l + 1 \leq k \leq \varkappa$

$$\left| \frac{\sin\left(\pi(n+1)\sum_{v=1}^{l}\mu_{kv}x^v\right)}{\sin\left(\pi\sum_{v=1}^{l}\mu_{kv}x^v\right)} \right| \leq n+1$$

und für $-1/2 \leq x^k \leq 1/2$

$$\left| \frac{\sin\pi(n+1)x^k}{\sin\pi x^k} \right| \leq \frac{\pi}{2} \left| \frac{\sin\pi(n+1)x^k}{\pi x^k} \right|,$$

folgt aus (4. 13) nach Substitution die Abschätzung

$$(4.14) \quad \int\limits_G L_n(a)\,da \leq c\,(n+1)^{2\,((\mu+2)\varkappa - \mu_0)\,-\,l} \int\limits_{-(n+1)\pi}^{(n+1)\pi} \cdots \int\limits_{-(n+1)\pi}^{(n+1)\pi} \prod_{k=1}^{l} \left(\frac{\sin x^k}{x^k}\right)^{2\,(\mu_0+2)} \times$$

$$\times \prod_{\mu=1}^{\mu_0} \left(\sum_{\lambda=1}^{l} v_{\mu\lambda} x^\lambda\right)^2 dx \ldots dx^l.$$

Das Integral auf der rechten Seite dieser Ungleichung läßt sich durch eine Konstante abschätzen die unabhängig von n ist, so daß gilt

$$\int\limits_G L_n(a)\,da \leq c\,(n+1)^{2\,((\mu+2)\varkappa - \mu_0)\,-\,l}.$$

Wählen wir andererseits $\varepsilon > 0$ so, daß für $|x^k| \leq \varepsilon$, $k = 1, \ldots, l$ gilt

$$|v_{\mu 1} x^1 + \cdots + v_{\mu l} x^l| \leq 1/2 \qquad (\mu = 1, \ldots, \mu_0),$$

dann folgt aus (4. 13) mit Hilfe der Ungleichungen

$$\left(\sin \pi \sum_{\lambda=1}^{l} v_{\mu\lambda} x^\lambda\right)^2 \geq \left(\frac{2}{\pi}\right)^2 \left(\pi \sum_{\lambda=1}^{l} v_{\mu\lambda} x^\lambda\right)^2,$$

$$\left(\frac{\sin\left(\pi(n+1) \sum\limits_{v=1}^{l} \mu_{kv} x^v\right)}{\sin\left(\pi \sum\limits_{v=1}^{l} \mu_{kv} x^v\right)}\right)^2 \geq \left(\frac{\sin\left(\pi(n+1) \sum\limits_{v=1}^{l} \mu_{kv} x^v\right)}{\pi \sum\limits_{v=1}^{l} \mu_{kv} x^v}\right)^2$$

die Abschätzung

$$(4.15) \quad \int\limits_G L_n(a)\,da \geq c \int\limits_{-\varepsilon/(n+1)}^{\varepsilon/(n+1)} \cdots \int\limits_{-\varepsilon/(n+1)}^{\varepsilon/(n+1)} \prod_{k=1}^{\varkappa} \left(\frac{\sin \pi(n+1) \sum\limits_{v=1}^{l} \mu_{kv} x^v}{\pi \sum\limits_{v=1}^{l} \mu_{kv} x^v}\right)^{2\,(\mu_0+2)} \times$$

$$\times \prod_{\mu=1}^{\mu_0} \left(\sum_{\lambda=1}^{l} v_{\mu\lambda} x^\lambda\right)^2 dx^1 \ldots dx^l.$$

Definieren wir für $\mu_{k1} = \cdots = \mu_{kl} = 0$

$$\frac{\sin \pi(n+1) \sum\limits_{v=1}^{l} \mu_{kv} x^v}{\pi \sum\limits_{v=1}^{l} \mu_{kv} x^v} = n+1,$$

dann ergibt sich aus dieser Ungleichung durch Substitution

$$\int\limits_G L_n(a)\,da \geq c\,(n+1)^{2\,((\mu_0+2)\,k - \mu_0)\,-\,l}$$

und die Formel (4. 10) ist bewiesen. Zum Beweis der Formel (4. 11) bemerken wir, daß auch $\varrho(e, a)$ eine Klassenfunktion ist und daß der Abstand $\varrho(e, x)$ für $x \in T$ äquivalent ist zu dem Abstand $\varrho_0(e, x)$ definiert durch

$$\varrho_0(e, x) = \sqrt{\sum_{k=1}^{l} \sin^2 \pi x^k}.$$

Deshalb erhält man auch in diesem Fall Abschätzungen, die den Ungleichungen (4. 14) und (4. 15) entsprechen, wobei die Integrale durch von n unabhängige Konstanten abgeschätzt werden können.

LEMMA 4. 4. *Ist G eine zusammenhängende, kompakte Liegruppe mit einer treuen, endlich dimensionalen, unitären Darstellung D, dann gibt es eine Folge von gleichmäßig beschränkten linearen Operatoren $\{I_n\}_{n\in N}$, die den Raum E ($E=C(G)$ oder $E=L_p(G)$, $1\leq p<\infty$) in die Menge der trigonometrischen Polynome vom Grade n abbilden, so daß*

$$(4.\ 16) \qquad \|I_n f - f\| \leq c\omega_2(n^{-1}, f) \qquad (f\in E).$$

BEWEIS. Es sei zunächst angenommen, daß D die Voraussetzung von Lemma 4. 3 erfüllt. Zu jedem $f\in E$ definieren wir

$$I'_{n'} f(a) = \int\limits_G J'_{n'}(b) f(b^{-1}a)\, db,$$

wobei $J'_{n'}(b)=\lambda_{n'} L_{n'}(b)$ und $\lambda_{n'} = \int\limits_G L_{n'}(b)db$. Da $L_{n'}(b)=\left|\det\left(\sum\limits_{k=0}^{n'} D(b^k)\right)\right|^{2(\mu_0+2)}$ gilt, gehört $I_n f$ zum Raum $\mathscr{P}_{2(\mu_0+2)\times n'}$ für alle $f\in E$. Setzen wir also n' als die größte ganze Zahl, die kleiner oder gleich $n/2(\mu_0+2)\varkappa$ ist und bezeichnen den so erhaltenen Operator mit I_n, so gilt $I_n f\in\mathscr{P}_n$, $\|I_n\|\leq 1$ und

$$\|I_n f(a)-f(a)\| = \tfrac{1}{2}\left\|\int\limits_G J_n(b)\,(T(b)-I)^2 f(ba)\, db\right\| \leq \tfrac{1}{2}\int\limits_G J_n(b)\,\omega_2\big(\varrho(e,b),f\big)db.$$

Auf Grund der Ungleichung (3. 1) der Formeln (4. 10) und (4. 11) ergibt sich hieraus

$$\|I_n f - f\| \leq \tfrac{1}{2}\int\limits_G J_n(b)\,(1+n\varrho(e,b))^2\, db\,\omega_2(n^{-1}, f) \leq c\omega_2(n^{-1}, f).$$

Erfüllt D selbst nicht die Voraussetzungen von Lemma 4. 3, dann gibt es aber ein n_0, so daß D^{n_0} diese Voraussetzungen erfüllt. Bildet man mit D^{n_0} statt mit D den Operator $I'_{n'}$, so gilt für alle $f\in E$, daß $I'_{n'}f\in\mathscr{P}_{2(\mu_0+2)(3\varkappa)^{n_0}n'}$. Wählen wir hier n' als die größte ganze Zahl, die kleiner als $n/2(\mu_0+2)(3\varkappa)^{n_0}$ ist, so gelten für die so konstruierte Operatorfolge I_n die gleichen Abschätzungen wie oben.

FOLGERUNG 4. 5. *Unter den Voraussetzungen von Lemma 4. 4 gibt es zu jedem $r\in N$ eine Folge gleichmäßig beschränkter linearer Operatoren $\{I_n^{(r)}\}_{n\in N}$ von E in $\mathscr{P}_n$, so daß gilt*

$$(4.\ 17) \qquad \|I_n^{(r)} f - f\| \leq c_r n^{-r}\omega_2(n^{-1}, f^{(r)}) \qquad (f\in E^{(r)}).$$

BEWEIS. Ist I_n der in Lemma 4. 4 definierte Operator, so setzen wir $I_n^{(r)} = I - (I-I_n)^{r+1}$. Da nach (4. 16)

$$\|I_n^{(r)} f - f\| \leq c\omega_2(n^{-1}, (I-I_n)^r f)$$

und nach (3. 6) und (3. 2)

$$\omega_2\big(n^{-1}, (I-I_n)^r f\big) \leq 2n^{-1}\,\|(I-I_n)^r f^{(r)}\|,$$

folgt (4. 17) für $r=1$ aus der Beobachtung, daß

$$\|(I-I_n)^r f^{(1)}\| = \sum_{i_1=1}^{m} \|(I-I_n)^r\, dT(X_{i_1})f\|.$$

Für $r>1$ folgt (4. 17) also durch Induktion, da $\|(I-I_n)^r dT(X_{i_1})f\| = \|I_n^{(r-1)} dT(X_{i_1})f - dT(X_{i_1})f\|$ und $dT(X_{i_1})f \in \mathsf{E}^{(r-1)}$.

Man könnte nun Satz 2. 2 mit seinen Folgerungen oder Satz 4. 1 anwenden, um schärfere Aussagen von Folgerung 4. 5 für die Operatorenfolge $\{I_n^{(r)}\}_{n\in \mathbf{N}}$ abzuleiten. Wir möchten uns aber jetzt dem Problem der besten Approximation durch trigonometrische Polynome aus $\mathscr{P}_n$ zuwenden. Sei für $f\in \mathsf{E}$ ($\mathsf{E}=\mathsf{C}(G)$ oder $\mathsf{E}=\mathsf{L}_p(G)$, $1\leq p<\infty$) die beste Approximation definiert durch

$$E_n(f, \mathsf{E}) = \inf\{\|f-P_n\|;\ P_n\in\mathscr{P}_n\},$$

dann folgt aus (4. 17) die Existenz einer Konstanten c_r, so daß

$$(4.18)\qquad\qquad E_n(f, \mathsf{E}) \leq c_r n^{-r}\|f^{(r)}\| \qquad (f\in \mathsf{E}^{(r)}),$$

d.h. Ungleichung (J) ist für die gleichmäßig beschränkte Folge der Halbnormen $E_n(f, \mathsf{E})$ mit $\varphi(n)=n^{-1}$ erfüllt. Deshalb folgt insbesondere aus Satz 4. 1 eine Verschärfung eines Resultates von PETER—WEYL über die Approximation von Funktionen durch Koeffizienten irreduzibler unitärer Darstellungen von G.

SATZ 4. 6. *Ist G eine zusammenhängende, kompakte Liegruppe, dann gibt es eine Konstante $c_j>0$, so daß für alle $f\in \mathsf{E}^{(j)}$ ($\mathsf{E}=\mathsf{C}(G)$ oder $\mathsf{E}=\mathsf{L}_p(G)$, $1\leq p<\infty$)*

$$(4.19)\qquad\qquad E_n(f, \mathsf{E}) \leq c_j n^{-j}\omega_r(n^{-1}, f^{(j)}) \qquad (r=1, 2, 3).$$

Auch hier könnten weitere Aussagen mit Hilfe von Satz 2. 2 und seinen Folgerungen gewonnen werden. Aussagen vom Typ (4. 19) wurden im Falle $\mathsf{E}=\mathsf{C}(G)$ für $G=SU(2)$ und $r=1, 2$ von RAGOZIN [20], für $G=U(m)$ und $r=1, 2$ vom Verfasser [13] bewiesen. Die für den Beweis benutzte Operatorenfolge I_n geht auf GONG—SHENG [10] zurück, der sie für $G=U(m)$, $r=1$ und $j=0$ zu ähnlichen Abschätzungen verwandte. Ist $G=\mathbf{T}$ ein m-dimensionaler Torus und wählt man für $\mathbf{T}$ die Darstellung

$$D(x) = \begin{pmatrix} e^{i2\pi x^1} & 0 & \cdots & & 0 \\ 0 & e^{i2\pi x^2} & & & 0 \\ \vdots & & \ddots & & \vdots \\ & & & \ddots & 0 \\ 0 & & \cdots & 0 & e^{i2\pi x^m} \end{pmatrix}$$

$x=(x^1, \ldots, x^m) \in \mathbf{T}$, dann sind die mit Hilfe dieser Darstellung gebildeten trigonometrischen Polynome vom Grade n die normalen trigonometrischen Polynome dieses Grades. Wir erhalten deshalb aus (4. 18) mit Satz 2. 1

SATZ 4. 7. *Ist* $\mathbf{T}$ *ein m-dimensionaler Torus, dann gibt es zu jedem* $r \in \mathbf{N}$ *eine Konstante* $c_r > 0$, *so daß für* $\mathsf{E} = \mathsf{C}(\mathbf{T})$ *oder* $\mathsf{E} = \mathsf{L}_p(\mathbf{T})$, $1 \leqq p < \infty$,

$$(4.\ 20) \qquad E_n(f, \mathsf{E}) \leqq c_r \omega_r(n^{-1}, f) \qquad (f \in \mathsf{E}).$$

Für $m=1$ ergibt sich hieraus eine verschärfte Form des Satzes von Jackson über die Approximation von periodischen Funktionen durch trigonometrische Polynome, der den Satz von Weierstraß verbessert.

Während in den bisherigen Anwendungen immer die Funktion $\varphi(n) = n^{-1}$ in Ungleichung (J) einzusetzen war, soll zum Abschluß eine Anwendung gegeben werden, bei der dies nicht der Fall ist. Ist $\gamma = (\gamma^1, \ldots, \gamma^m)$ ein Multiindex und für $x \in \mathbf{T}$

$$e^{i2\pi \langle \gamma, u \rangle} = e^{+i2\pi(\gamma^1 u^1 + \cdots + \gamma^m u^m)},$$

so definieren wir für $n \in \mathbf{N}, f \in \mathsf{L}^1(\mathbf{T})$

$$(4.\ 21) \qquad W_n f(x) = \int_{\mathbf{T}} \sum_{\gamma} e^{-n \langle \gamma, \gamma \rangle} e^{2\pi i \langle \gamma, u \rangle} f(x - u)\, du.$$

SATZ 4. 8. *Ist* E *einer der Räume* $\mathsf{C}(\mathbf{T})$ *oder* $\mathsf{L}_p(\mathbf{T})$, $1 \leqq p < \infty$, *dann gibt es eine Konstante* c, *so daß*

$$\| W_n f - f \| \leqq c \omega_2(n^{-1/2}, f) \qquad (f \in \mathsf{E}).$$

BEWEIS. Für $f \in \mathsf{E}^{(2)}$ gilt (siehe GÖRLICH [11])

$$\| W_n f - f \| \leqq c n^{-1} \| f^{(2)} \|.$$

LITERATURVERZEICHNIS

[1] N. I. Achieser, *Vorlesungen über Approximationstheorie*. 2. verbesserte Auflage. Akademie Verlag, Berlin 1967.

[2] J. F. Adams, *Lectures on Lie Groups*. W. A. Benjamin, Inc., New York 1969.

[3] B. M. Baishanski, *The asymptotic behavior of the n-th order difference*. Enseignement XV (1969), 29—41.

[4] P. L. Butzer and H. Johnen, *Lipschitz spaces on compact manifolds*. J. Functional Anal. 7 (1971), 242—266.

[5] P. L. Butzer und K. Scherer, *Approximationssätze und Interpolationsmethoden*. B—I Hochschulskripten, Mannheim-Zürich 1968.

[6] P. L. Butzer und K. Scherer, *Über die Fundamentalsätze der klassischen Approximationstheorie in abstrakten Räumen*. In *Abstract Spaces and Approximation* (P. L. Butzer, B. Sz.-Nagy Eds.), ISNM **10**, Birkhäuser, Basel 1969.

[7] P. L. Butzer and K. Scherer, *Jackson and Bernstein-type inequalities for families of commutative operators in Banach spaces*. J. Approx. Theory (im Druck).

[8] Cl. Chevalley, *Theory of Lie Groups*. Princeton Univ. Press, Princeton 1946.

[9] G. Freud, An unsolved problem. In: *On Approximation Theory* (P. L. Butzer—J. Korevaar, Eds.), ISNM **5**, Birkhäuser, Basel 1964, S. 182.

[10] Gong Sheng, *Fourier analysis on the unitary group, IV. On the Peter—Weyl theorem*. Chinese Math.-Acta 4 (1963), 351—359.

[11] E. Görlich, *Zur Saturation von Summationsverfahren mehrdimensionaler Fourierreihen*. Oesterreich. Akad. Wiss. Math.-Natur. K. S.-B. II. 77 (1968), 171—202.

[12] S. Helgason, *Differential Geometry and Symmetric Spaces*. Academic Press, New York 1962.

[13] H. Johnen, *Best approximation on the unitary group*. Colloquia Mathematica Societatis János Bolyai, 5. Hilbert Space Operators. Tihany (Hungary) 1970, 295—303.

[14] H. Johnen, *Darstellungen von Liegruppen und Approximationsprozesse auf Banachräumen*. J. Reine Angew. Math. 254 (1972), 160—187.

[15] S. Kobayashi and K. Nomizu, *Foundations of Differential Geometry, II*. Interscience Publishers, New York 1969.

[16] R. P. Langlands, *On Lie semi-groups*. Canad. J. Math. **12** (1960), 686—693.

[17] W. Miller, *Lie Theory and Special Functions*. Academic Press, New York 1968.

[18] F. D. Murnaghan, *The Theory of Group Representations*. John Hopkins Press, Baltimore 1938.

[19] D. L. Ragozin, *Approximation theory on compact manifolds and Lie groups with applications to harmonic analysis*. Ph. D. Thesis Harvard Univ. Cambridge, Mass. 1967.

[20] D. L. Ragozin, *Approximation theory on SU (2)*. J. Approximation Theory 1 (1968), 464—475.

[21] K. Scherer, *Über die Dualen von Banachräumen, die durch lineare Approximationsprozesse erzeugt werden, und Anwendungen für periodische Distributionen*. Acta Math. Szeged (im Druck).

[22] G. Sunouchi, Remark to Freud's problem. Ibidem ISNM **5** (1964), S. 183. (siehe [9]).

[23] H. Weyl, *The Classical Groups*. Princeton Univ. Press, Princeton 1946.

[24] A. Zygmund, *Trigonometric Series*. Reprint with corrections and some additions of the second edition, Cambridge Univ. Press, Cambridge 1968.

Approximation d'espaces de fonctions C^∞ et interpolation

Par

M. S. BAOUENDI et C. GOULAOUIC

UNIVERSITÉ PARIS-SUD
CENTRE D'ORSAY

Summary. We give a characterization of C^∞ functions and analytic functions on suitable compact subsets of R^n by their polynomial approximation. Using these approximation properties and interpolation between spaces of sequences, we construct interpolation spaces between the space of C^∞ functions and the space of analytic functions. When the compact subset of R^n is very regular, we give more precise properties of these interpolation spaces, by using degenerate differential operators.

Il n'est pas possible ici de donner des démonstrations détaillées des théorèmes (qui font l'objet de publications parues ou à paraître [1], [2], [4]); on a choisi de décrire les lignes générales des démonstrations les moins techniques et de renvoyer à la bibliographie pour les autres démonstrations. Une partie des résultats décrits ici a été annoncée au Colloque d'Analyse fonctionnelle de Bordeaux (1971).

Dans une première partie, on donne une caractérisation des fonctions indéfiniment différentiables et des fonctions analytiques sur un compact de R^n, par leur approximation polynomiale (rapidement convergente dans le cas C^∞ et exponentiellement convergente dans le cas analytique).

Dans une deuxième partie, on décrit les espaces d'interpolation entre l'espace des fonctions C^∞ et l'espace des fonctions analytiques: ce sont des espaces de Gevrey dans le cas d'une variété compacte sans bord et des espaces "un peu différents" dans le cas d'une variété à bord. Pour cela on rappelle quelques résultats d'interpolation puis on utilise d'une part les caractérisations obtenues par approximation polynomiale et d'autre part des propriétés des itérés d'opérateurs elliptiques dégénérés.

1. Approximation de fonctions C^∞ et analytiques sur un compact de $\mathbf{R}^n$

On note $\mathscr{P}_k$ l'espace des polynômes en n variables à coefficients complexes et de degré au plus k.

Soient E un ensemble borné intégrable dans $\mathbf{R}^n$, $1 \leqq p \leqq \infty$ et $f \in L^p(E)^*$; on désigne la distance de f à $\mathscr{P}_k$ dans $L^p(E)$ par:

$$d_{p,E}(f, \mathscr{P}_k) = \inf_{P \in \mathscr{P}_k} \|f - P\|_{L^p(E)}.$$

Pour un ouvert Ω de $\mathbf{R}^n$, on désigne par $C^\infty(\Omega)$ (respectivement: $\mathscr{A}(\Omega)$) l'espace des fonctions indéfiniment différentiables (respectivement: analytiques) dans Ω, à valeurs complexes; pour un ensemble quelconque K de $\mathbf{R}^n$, en particulier pour un compact, on désigne par $C^\infty(K)$ (respectivement: $\mathscr{A}(K)$) l'espace des fonctions qui sont restrictions à K de fonctions C^∞ (respectivement: analytiques) dans un voisinage ouvert de K.

On note $\jmath$ l'espace des suites à décroissance rapide (la suite (C_k) est dans $\jmath$ si et seulement si, pour tout $\alpha > 0$, la suite $(k^\alpha C_k)$ est bornée); on note aussi Exp l'espace des suites à décroissance exponentielle (la suite (C_k) est dans Exp si et seulement si on peut trouver $a \in \,]0, 1[$ tel que la suite $(C_k a^{-k})$ soit bornée).

Rappelons d'abord un résultat bien connu, dû essentiellement à BERNSTEIN [3]:

PROPOSITION 1. *Soient* $I = \,]-1, +1[$ *et* $f \in L^p(I)$; *on a:*

1°) *Pour que la fonction f soit dans $C^\infty(\bar{I})$ il faut et il suffit que la suite* $(d_{p,I}(f, \mathscr{P}_k))$ *soit dans $\jmath$.*

2°) *Pour que la fonction f soit dans $\mathscr{A}(\bar{I})$ il faut et il suffit que la suite* $(d_{p,I}(f, \mathscr{P}_k))$ *soit dans Exp.*

On voit aisément que cette proposition 1 est valable également si on remplace l'intervalle I de $\mathbf{R}$ par un cube (ou un parallélépipède) Π de $\mathbf{R}^n$.

La caractérisation des fonctions C^∞ (1ére partie de la proposition 1) a été généralisée par M. ZERNER [7] et est prouvée lorsqu'on remplace I par un ouvert Ω borné de $\mathbf{R}^n$, à frontière lipschitzienne.

Nous avons démontré que la caractérisation des fonctions analytiques (2ème partie de la proposition 1) est valable aussi pour des domaines assez généraux de $\mathbf{R}^n$.

En particulier, il est possible d'énoncer les résultats suivants:

PROPOSITION 2. *Soit K un compact de $\mathbf{R}^n$; on a:*

1°) *Pour une fonction $f \in C^\infty(K)$, la suite $(d_{p,K}(f, \mathscr{P}_k))$ est dans $\jmath$.*

2°) *Pour une fonction $f \in \mathscr{A}(K)$, la suite $(d_{p,K}(f, \mathscr{P}_k))$ est dans Exp.*

---

*) Dans la suite, on aura toujours $1 \leqq p \leqq \infty$; $L^p(E)$ désigne l'espace de Lebesgue usuel et dans le cas $p = \infty$, on peut aussi remplacer $L^\infty(E)$ par l'espace des fonctions continues bornées sur E muni de sa norme usuelle.

PROPOSITION 3. *Soit Ω un ouvert borné de $\mathbf{R}^n$ tel que $\bar{\Omega}$ soit une variété à bord lipschitzien; soit $f \in L^p(\Omega)$; on a:*

1°) *La fonction f est dans $C^\infty(\bar{\Omega})$ si et seulement si la suite $\left(d_{p,\Omega}(f, \mathscr{P}_k)\right)$ est dans $\jmath$.*

2°) *La fonction f est dans $\mathscr{A}(\bar{\Omega})$ si et seulement si la suite $\left(d_{p,\Omega}(f, \mathscr{P}_k)\right)$ est à décroissance exponentielle.*

Nous renvoyons à [2] pour une démonstration de ces résultats; remarquons que les hypothèses de la proposition 3 peuvent être considérablement affaiblies, surtout pour la caractérisation des fonctions analytiques (cf. [2] pour les détails qui ne peuvent être exposés ici); cependant, la réciproque de la proposition 2 n'est pas vraie sans hypothèses sur K, comme le montrent les exemples suivants:

EXEMPLE 1. (Donné dans [7] pour le cas C^∞.) Soit $\Omega = \{(x, y) \in \mathbf{R}^2; \; 0 < x < 1$ et $0 < y < e^{-1/x}\}$. On peut trouver une fonction $f \in L^p(\Omega)$ dont la distance aux polynômes est à décroissance rapide et qui n'est pas dans $C^\infty(\bar{\Omega})$; on peut par exemple, prendre $f(x, y) = y/x$.

Par contre, les fonctions analytiques sur $\bar{\Omega}$ sont caractérisées par leur approximation polynomiale.

EXEMPLE 2. Soient $K_1 = \{(x, y) \in \mathbf{R}^2; \; x^2 + y^2 \leqq 1\}$, $K_2 = \{(x, y) \in \mathbf{R}^2; \; y = 0$ et $0 \leqq x \leqq 2\}$ et $K = K_1 \cup K_2$.

On peut trouver une fonction f continue sur K, dont la distance aux polynômes, dans l'espace des fonctions continues, est à décroissance exponentielle et qui n'est pas analytique sur K; on prend $f(x, y) = yg(x)$ avec g analytique sur un voisinage de K_1 mais pas analytique sur un voisinage de K.

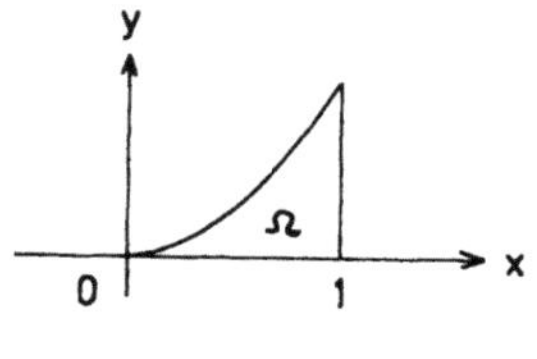

EXEMPLE 1

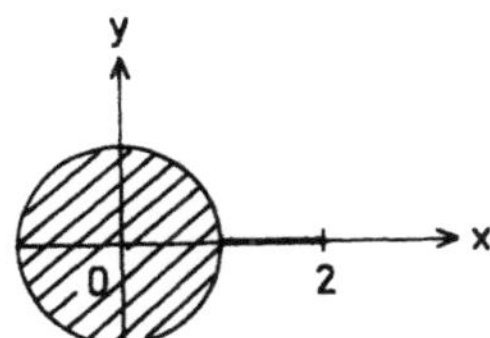

EXEMPLE 2

18*

2. Espaces d'interpolation

1. *Interpolation entre espaces de suites.* On rappelle, sous une forme plus simple, les résultats de [4] que l'on utilisera:

Soit d'abord un couple compatible d'espaces de Banach $(E_0, E_1)*$; on définit, pour $t > 0$ et $a \in E_0 + E_1$

$$K_2(t, a) = \inf_{\substack{a = a_0 + a_1 \\ a_0 \in E_0 \\ a_1 \in E_1}} (\|a_0\|_{E_0}^2 + t^2 \|a_1\|_{E_1}^2)^{1/2};$$

et soit γ une mesure sur $]0, \infty[$, non nulle, positive et vérifiant

$$\int_0^\infty \min(1, t^2)\, d\gamma < +\infty;$$

on note:

$$\Phi_{\gamma, 2}[E_0, E_1] = \left\{ a \in E_0 + E_1; \ \int_0^\infty (K_2(t, a))^2\, d\gamma < \infty \right\}$$

et on sait que cet espace, muni de la norme évidente, est un espace d'interpolation entre E_0 et E_1 et que l'on associe ainsi à la mesure γ un foncteur d'interpolation $\Phi_{\gamma, 2}$ défini sur les couples compatibles d'espaces de Banach.

Considérons maintenant des espaces de suites; pour toute suite $(Q(k))$ de réels strictement positifs, nous notons

$$l^2_{Q(k)} = \left\{ (C_k) \in \mathbf{C}^N; \ \sum_{k=0}^\infty |C_k|^2 Q(k) < \infty \right\}$$

muni de la norme hilbertienne naturelle. Soient $m > 0$ et $a > 0$; on montre que l'on a

$$\begin{cases} \Phi_{\gamma, 2}[l^2_{k^m}, l^2_{a^k}] = l^2_{\Phi(a, m, k)} \\ \text{avec } \Phi(a, m, k) = \int_0^\infty \min(k^m, t^2 a^k)\, d\gamma. \end{cases}$$

Etant donné alors $r > 1$, on peut construire la mesure γ de façon que $\Phi(a, m, k)$ soit "équivalent" à $a^{k^{1/r}}$; on note plus brièvement Φ_r le foncteur d'interpolation correspondant. Ensuite, par prolongements par limite projective et puis par limite inductive, on en déduit un foncteur d'interpolation $\tilde{\Phi}_r$ (défini sur des couples compatibles d'espaces $\mathfrak{LF}$) tel que:

$$\tilde{\Phi}_r[\varprojlim_m l^2_{k^m}, \varinjlim_a l^2_{a^k}] = \varinjlim_a l^2_{a^{k^{1/r}}}$$

---

*) On peut se contenter ici de considérer des couples (E_0, E_1) tels que $E_1 \hookrightarrow E_0$, le symbole $\hookrightarrow$ désignant l'injection continue.

et même, pour tout $\alpha > 0$,

(1)
$$\tilde{\tilde{\Phi}}_r[s, \varinjlim l^2_{ak^\alpha}] = \varinjlim_a l^2_{ak^{\alpha/r}}.$$

Remarquons que l'on pourrait obtenir d'autres espaces d'interpolation entre s et Exp, mais ils nous intéressent moins pour la suite.

2. Espaces $\mathscr{A}_{r,p}(\overline{\Omega})$. Pour plus de clarté, nous nous limitons désormais ici à des ouverts Ω bornés dans R^n tels que $\overline{\Omega}$ soit une variété à bord lipschitzien.

Définition 1. *Soit un nombre réel $r \geq 1$; on note:* $\mathrm{Exp}_r = \{(c_k) \in C^N;\ \exists a \in]0, 1[$ *tel que* $(c_k a^{-k^{1/r}}) \in l^\infty\}$, *et on désigne par* $\mathscr{A}_{r,p}(\overline{\Omega})$ *l'espace des fonctions* $f \in L^p(\Omega)$ *telles que la suite* $(d_{p,\Omega}(f, \mathscr{P}_k))$ *soit dans* Exp_r.

On démontre, en utilisant le théorème de plongement de Sobolev, que l'espace $\mathscr{A}_{r,p}(\overline{\Omega})$ est indépendant de p pour $1 \leq p \leq \infty$. La proposition 3 dit en particulier que l'on a $\mathscr{A}(\overline{\Omega}) = \mathscr{A}_{1,p}(\overline{\Omega})$.

Considérons maintenant le cas $p = 2$; nous choisissons une base orthonormale dans $L^2(\Omega)$ constituée de polynômes que nous ordonnons par degré non décroissant; notons J le développement en série de Fourier sur cette base, nous avons facilement:

Proposition 4. *L'opérateur J est un isomorphisme*
de $C^\infty(\Omega)$ sur s,
de $\mathscr{A}(\overline{\Omega})$ sur Exp_n,
de $\mathscr{A}_{r,2}(\overline{\Omega})$ sur Exp_{rn} pour tout $r \geq 1$.

En utilisant maintenant le foncteur d'interpolation $\tilde{\tilde{\Phi}}_r$ et (1), on en déduit:

Proposition 5. *Pour tout $r \geq 1$, l'espace $\mathscr{A}_{r,p}(\overline{\Omega})$ est un espace d'interpolation entre les espaces $C^\infty(\overline{\Omega})$ et $\mathscr{A}(\overline{\Omega})$ et on a, plus précisément:*

$$\tilde{\tilde{\Phi}}_r[C^\infty(\overline{\Omega}), \mathscr{A}(\overline{\Omega})] = \mathscr{A}_{r,p}(\overline{\Omega}).$$

De même, on montrerait que $\mathscr{A}_{r,p}(\overline{\Omega})$ est un espace d'interpolation entre $C^\infty(\overline{\Omega})$ et $\mathscr{A}_{r_0,p}(\overline{\Omega})$ pour $r \geq r_0$ et aussi entre $\mathscr{A}_{r_1,p}(\overline{\Omega})$ et $\mathscr{A}_{r_0,p}(\overline{\Omega})$ pour $r_1 \geq r \geq r_0$.

3. Description des espaces d'interpolation. Les espaces d'interpolation $\mathscr{A}_{r,p}(\overline{\Omega})$ obtenus par approximation polynomiale peuvent être caractérisés par des propriétés différentielles, quand on fait des hypothèses de régularité suffisantes sur la frontière de Ω.

Rappelons d'abord que l'on appelle espace de Gevrey d'ordre r $(r \geq 1)$ sur l'ouvert $\Omega \subset R^n$, l'espace $G_r(\Omega) = \{f \in C^\infty(\Omega);\ \forall K \subset \Omega,\ \exists L > 0$ tel que l'on ait

$$\|D^\alpha f\|_{L^2(K)} \leq L^{|\alpha|+1} (|\alpha|!)^r \quad \text{pour tout} \quad \alpha \in N^n\}.$$

On en déduit comme d'habitude la définition de $G_r(K)$, où K est un compact de R^n, et de $G_r(\Gamma)$, où Γ est une "bonne" variété.

Pour fixer les idées, nous considérons une situation particulière (utile par exemple dans l'étude des problèmes aux limites, cf. [1], [6]):

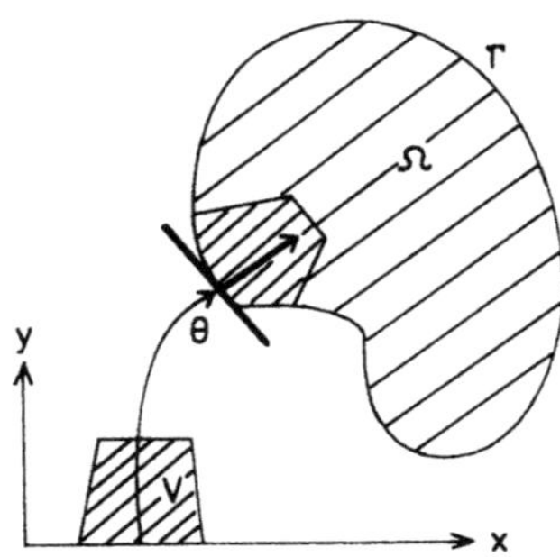

Soit $\bar{\Omega}$ une variété compacte à bord Γ analytique.

Sur Γ, on peut considérer un opérateur auto-adjoint dans $L^2(\Gamma)$, par exemple l'opérateur de Laplace—Beltrami Δ et, en utilisant le développement de Fourier $\mathfrak{F}$ sur les fonctions propres de cet opérateur et le théorème de Kotake et Narasimhan [5] sur une variété sans bord, compacte, analytique, on obtient:

(2)
$$\begin{cases} \text{L'opérateur } \mathfrak{F} \text{ est un isomorphisme} \\ \text{de } C^\infty(\Gamma) \text{ sur } s, \\ \text{de } \mathscr{A}(\Gamma) \text{ sur } \mathrm{Exp}_{n-1}, \\ \text{de } G_r(\Gamma) \text{ sur } \mathrm{Exp}_{r\,(n-1)}. \end{cases}$$

On en déduit que les espaces de Gevrey $G_r(\Gamma)$ sont d'interpolation entre $C^\infty(\Gamma)$ et $\mathscr{A}(\Gamma)$, et on a même plus précisément (d'après (1) et (2)):

(3)
$$\overset{\approx}{\Phi}_r[C^\infty(\Gamma), \mathscr{A}(\Gamma)] = G_r(\Gamma).$$

Par ailleurs, on peut montrer (cf. [4]) que les espaces $G_r(\bar{\Omega})$ pour $r>1$ ne sont pas des espaces d'interpolation entre $C^\infty(\bar{\Omega})$ et $\mathscr{A}(\bar{\Omega})$. On est donc amené à comparer les espaces de Gevrey et les espaces $\mathscr{A}_{r,p}(\bar{\Omega})$, qui sont meilleurs du point de vue de l'interpolation et de l'approximation polynomiale; on a le résultat:

PROPOSITION 6. *Soit* $r \geqq 1$. *L'espace* $\mathscr{A}_{r,p}(\bar{\Omega})$ *est constitué par les fonctions* $f \in C^\infty(\bar{\Omega})$ *telles que:*

(i) $f \in G_r(\Omega)$.

(ii) *Pour toute carte locale* (V, θ) *d'un point de* Γ, *il existe* $L > 0$ *vérifiant pour tous* $k \in N$ *et* $a \in N^{n-1}$

$$\|(D_y y D_y)^k D_x^\alpha \tilde{f}\|_{L^2(V)} \leq L^{|\alpha|+k+1}((|\alpha|+2k)!)^r$$

où x *et* y *désignent respectivement les variables tangentielles et normale et* $\tilde{f} = f \circ \theta$.

Pour $\Omega =]-1, +1[$ on peut obtenir $\mathscr{A}_{r,p}(\bar{\Omega})$ comme suit: soit S la sphère unité de R^3; les fonctions de $G_r(S)$ invariantes par rotation autour d'un diamètre sont en bijection, par projection sur ce diamètre avec les fonctions de $\mathscr{A}_{r,p}([-1, +1])$; dans le cas général, les fonctions de $\mathscr{A}_{r,p}(\bar{\Omega})$ sont au voisinage du bord dans "l'espace $\mathscr{A}_{r,p}$ en variable normale à valeurs G_r en variables tangentielles".

La démonstration de la proposition 6 est assez longue; nous en donnons le schéma (cf. [1], [2] pour les démonstrations complètes):

1°) On construit un opérateur différentiel dégénéré au bord de Ω, elliptique d'ordre 2 à l'intérieur de Ω, et qui constitue une généralisation convenable de l'opérateur de Legendre $-\dfrac{d}{dx}(1-x^2)\dfrac{d}{dx}+1$ sur $]-1, +1[$; on note $(\mathfrak{A}, D(\mathfrak{A}))$ une réalisation autoadjointe de cet opérateur dans $L^2(\Omega)$ et une étude de la régularité de $\mathfrak{A}$ montre que:

(i) L'injection de $D(\mathfrak{A})$ dans $L^2(\Omega)$ est compacte.

(ii) $C^\infty(\bar{\Omega}) = \bigcap_{k \in N} D(\mathfrak{A}^k)$.

(iii) $\mathscr{A}(\bar{\Omega}) = \left\{ f \in C^\infty(\bar{\Omega}); \ \exists L > 0 \ \text{avec} \ \sup_{k \geq 0} \dfrac{\|\mathfrak{A}^k f\|_{L^2(\Omega)}}{L^k(2k)!} < \infty \right\}$

et les mêmes calculs montrent aussi que l'espace des fonctions $f \in C^\infty(\bar{\Omega})$ vérifiant les propriétés (i) (ii) de la proposition 6 est également:

(iv) $\mathscr{A}_r(\bar{\Omega}) = \left\{ f \in C^\infty(\bar{\Omega}); \ \exists L > 0 \ \text{avec} \ \sup_{k \geq 0} \dfrac{\|\mathfrak{A}^k f\|_{L^2(\Omega)}}{L^k(2k!)^r} < \infty \right\}$.

2°) On étudie ensuite la croissance des valeurs propres de l'opérateur $(\mathfrak{A}, D(\mathfrak{A}))$ et, en utilisant (ii) (iii) (iv), on montre que le développement de Fourier $\mathfrak{F}$ sur une base orthonormale de fonctions propres de l'opérateur $(\mathfrak{A}, D(\mathfrak{A}))$ réalise un isomorphisme

de $C^\infty(\bar{\Omega})$ sur s,

de $\mathscr{A}(\bar{\Omega})$ sur Exp_n,

de $\mathscr{A}_r(\bar{\Omega})$ sur Exp_{rn} pour tout $r \geq 1$.

3°) Il en résulte que l'on a $\mathscr{A}_r(\bar{\Omega}) = \mathscr{A}_{r,p}(\bar{\Omega})$, puisque les deux espaces sont des espaces d'interpolation entre $C^\infty(\bar{\Omega})$ et $\mathscr{A}(\bar{\Omega})$ associés au même foncteur d'interpolation.

Ce qui démontre la proposition 6. Remarquons que nous avons besoin de l'interpolation pour montrer l'équivalence des propriétés différentielles ((i) (ii) de la proposition 6 ou bien $\mathscr{A}_r(\bar{\Omega})$) et de la propriété d'approximation polynomiale (définition de $\mathscr{A}_{r,p}(\bar{\Omega})$). Il serait intéressant d'avoir une démonstration plus directe.

BIBLIOGRAPHIE

[1] M. S. Baouendi et C. Goulaouic, *Régularité analytique et itérés d'opérateurs elliptiques dégénérés; applications.* J. Functional Analysis (à paraître).

[2] M. S. Baouendi et C. Goulaouic, *Approximation polynomiale de fonctions C^∞ et analytiques* (à paraître aux Annales de l'Institut Fourier, Grenoble).

[3] S. Bernstein, *Oeuvres complètes.*

[4] C. Goulaouic, *Prolongement de foncteurs d'interpolation et applications.* Ann. Inst. Fourier, Grenoble **18**, 1 (1968), 1—98 et **19**, 2 (1970), 269—278.

[5] T. Kotake et N. S. Narasimhan, *Regularity theorems for fractional powers of a linear elliptic operator.* Bull. Soc. Math. France **90** (1962), 449—471.

[6] J. L. Lions et E. Magenes, *Problèmes aux limites non homogènes.* tome **3**, Dunod, Paris 1970,

[7] M. Zerner, *Développement en série de polynômes orthogonaux des fonctions indéfiniment différentiables.* C. R. Acad. Sci. Paris Sér. A—B **268** (1969), A218—A220.

Nonlinear Transformations with the Conservation of Differential Properties of Functions

By

S. M. NIKOLSKIĬ

STEKLOV MATHEMATICAL INSTITUTE
MOSCOW

Now I am writing a chapter of the book which we three colleagues — O. Besov, V. Il'in and myself — are preparing for print. One can consider this book as a continuation of my recently published book on "Approximation of functions of many variables and imbedding theorems" (in Russian), Moscow 1969.

The chapter mentioned is devoted to the problem of traces of functions of many variables on smooth manifolds. As background material of that chapter I have taken my paper [1]. This problem was solved there for the classes H_p^r which we define afterwards. But to correspond to the general spirit of the book I had to extend these results to the more general Besov-classes $B_{p\theta}^r$. In particular I had to extend my lemma about nonlinear transformations with the conservation of differential properties of functions to Besov's class. The proof of the lemma in the general case demanded a change. I was also successful in deriving more exact results (even for the class H_p^r) concerning the order of differentiability of the transformation. In the proof I apply the method of approximation by the functions of exponential type.

To formulate the Lemma I begin with some definitions. Let $G \subset R_n$ be an open set of the n-dimensional space R_n of points $\mathbf{x} = (x_1, \ldots, x_n)$, G_δ, $\delta > 0$, the set of points $\mathbf{x} \in G$ whose distance from the boundary Γ of G is greater than δ. And as usual let

$$\|F\|_{L_p(G)} = \left(\int_G |f(\mathbf{x})|^p \, d\mathbf{x} \right)^{1/p} \qquad (1 \leq p \leq \infty).$$

We define the value

$$\Omega^k(f^\varrho, t) = \sup_{|h| < t} \|\Delta_h^k f_h^\varrho\|_{L_p(G_k|h|)},$$

where the supremum is taken over all vectors

$$\mathbf{h} = (h_1, \ldots, h_n)$$

with length

$$|\mathbf{h}| = \left(\sum_1^n h_j^2 \right)^{1/2} < t.$$

Here $f_{\mathbf{h}}^{\varrho}$ is the derivative of f of order ϱ in direction of $\mathbf{h}$ and $\varDelta_{\mathbf{h}}^k$ is the symbol for the difference of order k with step $\mathbf{h}$. Also let

$$r>0, \quad r = \varrho+\alpha, \quad \varrho \text{ integer and } \quad 0<\alpha\leqq 1.$$

By definition a function $f\in H_p^r(G)$ if $f\in L_p(G)$ and $\Omega^k(f^\varrho, t)\leqq Mt^\alpha$, where M does not depend on t. Here and further

$$k = \begin{cases} 2 \ (r \text{ integral}) \\ 1 \ (r \text{ non-integral}). \end{cases}$$

By definition a function

$$f\in B_{p\theta}^r(G) \qquad (1\leqq\theta<\infty, \ 1\leqq p\leqq\infty)$$

if the following norm is finite:

$$\|f\|_{B_{p\theta}^r(G)} = \|f\|_{L_p(G)} + \|f\|_{b_{p\theta}^r(G)},$$

with

$$\|f\|_{b_{p\theta}^r(G)}^\theta = \int\limits_0^1 t^{-1-\alpha\theta}\Omega^k(f^\varrho, t)^\theta \, dt.$$

We will also consider a transformation A

$$\mathbf{u}=A\mathbf{x}, \qquad \mathbf{x}\in G, \ \mathbf{u}=(u_1, \ldots, u_n)\in G_*$$

of the type

$$u_s=\varphi_s(x) \qquad (s=1, \ldots, n).$$

The operator A transforms an open boundary set G one-to-one on a set G_*.

By definition $A\in C^r$, $r>0$, if the derivatives φ_s^k of the functions φ_s of order $k\leqq r$ are continuous uniformly on G in case r integral, or if $A\in C^\varrho$, $\varrho=[r]$, and the derivatives of order ϱ satisfy a Lipschitz inequality of degree α ($r = \varrho+\alpha$) on G in case r is non-integral, respectively.

Now we can formulate our lemma.

LEMMA. *Let* $f(u)\in H_p^r(R_n^*)$, $A\in C^{l_r}$,

(I) $$l_r = \begin{cases} r+\varepsilon \ (\varepsilon>0) \ (r \ \textit{integral}) \\ \max(1, r) \ (r \ \textit{non-integral}). \end{cases}$$

Then

$$\varphi(\mathbf{x}) = f(A\mathbf{x})\in H_p^r(G).$$

In this assertion it is possible to substitute H_p^r by $B_{p\theta}^r$, if one substitutes r by $r+\varepsilon$ in equality (I). I note that this lemma may be used to prove the following theorem.

THEOREM. *Let $f(\mathbf{x}) \in B^r_{p\theta}(R_n)$, $S \subset R_n$ be an m-dimensional manifold having differentiability of order l_r, and*

$$\varrho = r - \frac{n-m}{p} > 0 \qquad (1 \leqq m < n).$$

Then there is the trace $f|_S \in B^{\varrho}_{p\theta}(S)$.

Conversely, if a function φ defined on S belongs to the class $B^{\varrho}_{p\theta}(S)$, then there is a function $f \in B^r_{p\theta}(R_n)$ so that

$$f|_S = \varphi.$$

This theorem is known for stronger differentiability conditions upon S. For H^r_p see my paper mentioned in Математический Сборник 1953.

We will discuss the method of proof.

It is known that the function $f(\mathbf{u}) \in B^r_{p\theta}(R_n)$, $(B^r_{p\infty} = H^r_p)$, can be expanded in the series

$$f(\mathbf{u}) = \sum_{s=0}^{\infty} Q_s(\mathbf{u}) \qquad (\mathbf{u} \in R_n),$$

where Q_s are functions of exponential type 2^s and the following series is finite

$$\left(\sum_{s=0}^{\infty} 2^{\theta sr} \|Q_s\|^{\theta}_{L_p(R_n)} \right)^{1/\theta} < \infty.$$

The derivative of $\varphi(\mathbf{x}) = f(A\mathbf{x})$ of order ϱ with respect to $\mathbf{x}$ may be written as a sum

$$\varphi^{(\varrho)}(\mathbf{x}) = \sum_{|\mathbf{k}| \leqq \varrho} \lambda(\mathbf{x}) f^{(\mathbf{k})}(A\mathbf{x}),$$

where $f^{(\mathbf{k})}$ are derivatives of f with respect to $\mathbf{u}$ and the function $\lambda(\mathbf{x})$ does not depend on f.

We then investigate each member of the series

$$\lambda(\mathbf{x}) f^{(\mathbf{k})}(A\mathbf{x}) = \sum_j \lambda(\mathbf{x}) Q_j^{(\mathbf{k})}(A\mathbf{x}),$$

particularly using the Bernstein inequality

$$\|Q_j^{(\mathbf{k})}(\mathbf{u})\|_{L_p(R_n^*)} \leqq 2^{|\mathbf{k}|} \|Q_j(\mathbf{u})\|_{L_p(R_n^*)}.$$

I know of some cases when the numbers l_r are exact. For example, the case H^r_∞ for r non-integral and H^r_p for $1/p < r < 1$.

LITERATURE

[1] С. М. Никольский, *Свойства некоторых классов функций многих переменных на дифференцируемых многообразиях.* Математический Сборник **33** (75):2 1953, 261—326.

[2] О. Бесов, *Исследование одного семейства функциональных пространств в связи с теоремами вложения и продолжения.* Труды МИАН СССР **60** (1961), 42—81.

Positive and Monotone Approximation

By

G. G. LORENTZ

DEPT. OF MATH.
UNIVERSITY OF TEXAS
AUSTIN

This paper consists of two parts, only loosely connected.*)

I. FIXED POINTS OF MONOTONE OPERATORS

1. Introduction

Let T be a linear bounded operator that maps $C[a, b]$ into $C^{(k)}[a, b]$ (with the natural topology of $C^{(k)}$). By $C_+^{(k)}$ we denote the set of all functions $f \in C^{(k)}$ for which $f^{(k)}(x) \geqq 0$, $a \leqq x \leqq b$. We shall say that T is *monotone of order k* (or *k-monotone*), where $k \geqq 0$, if $f \in C_+^{(k)}$ implies $Tf \in C_+^{(k)}$. For example, a polynomial operator of order n, that maps C into the space of all polynomials of degree n, has this property for $k > n$. It is easy to see, on the other hand, that the linear space of all polynomials of degree $k-1$ is preserved by a k-monotone operator T.

Here are examples of monotone operators: 1) The Bernstein operators B_n are k-monotone for each k; 2) The Taylor operator $T_n f(x) = T_{n, x_0}(f, x) = f(x_0) + (x-x_0) f'(x_0) + \cdots + \dfrac{1}{n!}(x-x_0)^n f^{(n)}(x_0)$, $a \leqq x_0 \leqq b$, is monotone of order $k \geqq n$. 3) The operator $S_{n, c}(f)$ defined by $S_{n, c}(f, x) = f(x)$ for $x \geqq c$, by $S_{n, c}(f, x) = T_{n, c}(f, x)$ $x \leqq c$, where $a < c < b$ is fixed, is k-monotone. (The domain of the definition of the operators 2), 3) is $C^{(k)}$).

We shall say that T is *regular*, if the set

$$N(f) = [x : (Tf)^{(k)}(x) = 0]$$

is either $[a, b]$ or is of measure zero for each $f \in C$.

An operator T will be called *k-completely monotone*, where $k \geqq 0$, if T is p-monotone for each $p \geqq k$.

---

*) This work has been supported, in part by Grant no. GP—23 566 of the National Science Foundation.

It is known that a Bernstein operator B_n, $n \geqq 1$, which is 0-completely monotone, has as its only fixed points linear functions (see KELISKY and RIVLIN [3], SIKKEMA [6]). We shall extend this result to fairly arbitrary monotone and completely monotone operators.

2. Monotone operators

LEMMA 1. *Let $f \in C^{(k)}[a, b]$, and let A be a closed set in $[a, b]$, such that $f^{(k)} = 0$ a.e. on A. Then for each $\varepsilon > 0$ there is a neighbourhood U of A and a function $F \in C^{(k)}[a, b]$ with the properties*

$$(1) \qquad\qquad \|f - F\| < \varepsilon$$

$$(2) \qquad\qquad F^{(k)} = 0 \quad on \quad U.$$

PROOF. Let $M = \|f^{(k)}\|$, let $\varepsilon_1 > 0$, $\delta > 0$, let U be the $\frac{1}{2} \delta$-neighbourhood of A, and let U_1 be the δ-neighbourhood of A. Then U_1 contains the closure of U, and for sufficiently small δ, $m(U_1 \setminus U) < \varepsilon_1$.

Let g be the continuous function, equal to $f^{(k)}$ on $[a, b] \setminus U_1$, equal to zero on $\overline{U}$, and obtained by linear interpolation elsewhere. Obviously $\|g\| \leqq M$. Let F be the k-th integral of g, which satisfies $F^{(i)}(x_0) = f^{(i)}(x_0)$ at some $x_0 \in [a, b] \setminus U_1$. Then

$$\|F^{(k-1)} - f^{(k-1)}\| \leqq \int_a^b |g - f^{(k)}| \, dx = \int_{U_1 \setminus (A \cap U)} |g - f^{(k)}| \, dx \leqq 2 M \varepsilon_1 .$$

Integrating further, we obtain $\|F^{(i)} - f^{(i)}\| \leqq 2 M \varepsilon_1 (b-a)^{k-i-1}$, $i = 0, \ldots, k-1$, and therefore

$$\|F - f\| < \varepsilon,$$

if ε_1 is sufficiently small.

LEMMA 2. *Let $f_0 \in C_+^{(k)}$, $k \geqq 1$. If T is a k-monotone operator and $(Tf_0)^{(k)}(x_0) = 0$, then also $(Tf)^{(k)}(x_0) = 0$ for each function $f \in C^{(k)}$, that has the property that $f^{(k)}(x) = 0$ a.e. on $N_0 = N(f_0)$.*

PROOF. Assume that $(Tf)^{(k)}(x_0) \neq 0$. We put

$$(3) \qquad\qquad L(f) = (Tf)^{(k)}(x_0).$$

This is a continuous linear operator on $C[a, b]$, which is positive in the sense that $f \in C_+^{(k)}$ implies $L(f) \geqq 0$. Also, $L(f_0) = 0$, $L(f) \neq 0$. Let F be a function of Lemma 1, corresponding to $A = N_0$ and the function f. Then, for sufficiently small ε, $L(F) \neq 0$. Consider the functions $g = M f_0 \pm F$, $M > 0$. On the set $[a, b] \setminus U$, $f_0^{(k)}(x) \geqq m$ for some

$m > 0$, hence for all large M, $g^{(k)}(x) \geqq 0$ for $x \in [a, b] \backslash U$. On U, $F^{(k)}(x) = 0$, $f_0^{(k)}(x) > 0$, hence again $g^{(k)}(x) \geqq 0$. Thus $g \in C_+^{(k)}$, hence $L(g) \geqq 0$. Therefore, for large M,

$$L(g) = ML(f_0) \pm L(F) = \pm L(F) \geq 0,$$

and $L(F) = 0$, a contradiction.

Interesting corollaries are obtained if f_0 is a fixed point of T.

THEOREM 1. *Let $f_0 \in C_+^{(k)}$ $(k \geqq 1)$ and f be fixed points of a k-monotone regular operator T, $f_0^{(k)} \not\equiv 0$, let $N_0 = N(f_0)$. Then $f^{(k)}$ vanishes on N_0.*

PROOF. We have $mN_0 = 0$, since T is regular. On N_0, $(Tf_0)^{(k)}(x) = f_0^{(k)}(x) = 0$, hence by Lemma 2, $f^{(k)}(x) = (Tf)^{(k)}(x) = 0$, as required.

EXAMPLE. A k-monotone regular operator T on $[0, 1]$ cannot reproduce both functions x^k and x^{k+1}.

We next prove that the set $N_0 = N(f_0)$ has strong localization properties, even if T is not regular. For any function $f \in C^{(k)}$, the values of $(Tf)^{(k)}(x)$, $x \in N_0$ depend only upon the values of $f^{(k)}(x)$, $x \in N_0$. This can be expressed in the following way.

THEOREM 2. *Let $f_0 \in C_+^{(k)}$ $(k \geqq 1)$ be a fixed point of a k-monotone operator T. If $f \in C^{(k)}$ and $f^{(k)}(x) = 0$ on the set $N_0 = N(f_0)$, then also $(Tf)^{(k)}(x) = 0$, $x \in N_0$.*

PROOF. On the set N_0, $(Tf_0)^{(k)}(x) = f_0^{(k)}(x) = 0$, hence $(Tf)^{(k)}(x) = 0$ by Lemma 2.

3. Completely monotone operators

A k-completely monotone operator T will be called *regular* if each set $[x : (Tf)^{(p)}(x) = 0]$, $p \geqq k$ is either of measure 0 or is identical with $[a, b]$, and if $(Tf)^{(p)}$ can have zeros of infinite order only by being identically zero.

Here we have an improvement of Theorem 1.

LEMMA 3. *If for each $p \geqq k$, $(k \geqq 1)$, T is a p-monotone regular operator, and if $f_0 \in C_+^{(p)}$, $p \geqq k$ is a fixed point of T that does not reduce to a polynomial, and if f is another fixed point of T, then each zero of $f_0^{(p)}$ is also a zero of $f^{(p)}$, of at least the same multiplicity.*

This follows immediately from Theorem 1.

THEOREM 3. *Let T be a k-completely monotone $(k \geqq 1)$ regular operator, $f_0 \in C_+^{(p)}$, $p \geqq k$ a fixed point of T, that is not a polynomial of degree $k - 1$. Then all fixed points of T are contained among the functions of the form $Cf_0 + P$, where P is a polynomial of degree $k + r - 1$, r being the multiplicity of a as a zero of $f_0^{(k)}$.*

PROOF. Clearly, $f_0^{(k)}$ can have a zero only at $x=a$. Let r, $0 \leqq r < +\infty$ be the multiplicity of this zero. By Lemma 3, a is a zero of $f^{(k)}$ of multiplicity at least r. Then

$$(4) \qquad \lim_{x \to a} \frac{f^{(k)}(x)}{f_0^{(k)}(x)} = \lim_{x \to a} \frac{f^{(k+r-1)}(x)}{f_0^{(k+r-1)}(x)} = \frac{f^{(k+r)}(a)}{f_0^{(k+r)}(a)} = m$$

exists, and changing the sign of f, we may assume that $m \geqq 0$.

Since $f_0^{(k+r)}(x) \neq 0$, $a < x < b$, there exists

$$(5) \qquad \max_{a \leqq x \leqq b} \frac{f^{(k+r)}(x)}{f_0^{(k+r)}(x)} = A \geqq 0.$$

Let this maximum be attained at x_0, $a \leqq x_0 \leqq b$. Then $g = f - A f_0 \in C_+^{(k+r)}$, and $g^{(k+r)}(x_0) = 0$, $f_0^{(k+r)}(x_0) \neq 0$. If $g^{(k+r)}$ is not identically zero, the pair of functions g, f_0 contradicts Theorem 1. It follows that g is a polynomial of degree $k+r-1$.

EXAMPLE. For each n, the Bernstein operator B_n is regular and 1-completely monotone on $[0, 1]$. It has the fixed point $f_0 = x$, and $x = 0$ is not a zero of f_0'. It follows that the only fixed points of B_n are the linear functions.

Results above are similar to some known theorems about eigenvalues of compact operators. See KREĬN—RUTMAN [7, Theorems 6. 1—6. 3]. However, our conditions and proofs are different.

QUESTION. It is known that the maximal order of convergence of a sequence of positive polynomial operators P_n is at most n^{-2}. Bernstein operators are completely monotone and have maximal order of convergence n^{-1}. Do there exist completely monotone polynomial operators with a better order of convergence?

II. KOROVKIN SYSTEMS

4. Existence of strong Korovkin systems

Let Q be a compact Hausdorff topological space. A system of real continuous functions on Q

$$(6) \qquad F: f_0, f_1, \ldots, f_m$$

is called a *Korovkin system of order* m (compare [4]) if for each sequence of positive linear operators L_n, mapping $C(Q)$ into itself, relations $L_n f_i \to f_i$, $i = 0, \ldots, m$, imply $L_n f \to f$ for all $f \in C(Q)$. Of special interest are systems of the form

$$(7) \qquad G: 1, g_1, \ldots, g_m.$$

Several authors (see ŠAŠKIN [5], FREUD [2]) have proved that F is a Korovkin system if for each $q_0 \in Q$ there exists a polynomial $P_{q_0}(q) = \sum_{k=0}^{m} a_k f_k(q)$ that satisfies

$P_{q_0}(q_0)=0$, $P_{q_0}(q)>0$, $q\neq q_0$; the converse is not true. Systems F which have this property we shall call *strong Korovkin systems*; we would like to prove their existence.

We begin with some remarks. For two compact sets Q, Q', let h be a homeomorphism of Q onto Q'. This defines a linear isometry $Hf=f'$ by means of the formula: $f'(q')=f(h^{-1}q)$, between the spaces $C(Q)$ and $C(Q')$. Then, F is a Korovkin system (strong Korovkin system) on Q if and only if the system F': $Hf_0, \ldots, Hf_m$ is a Korovkin (strong Korovkin) system on Q'. This follows at once by considering the sequence of operators $HL_n H^{-1}$ on $C(Q')$.

ŠAŠKIN [5, Theorem 2″] has shown that a system G is a Korovkin system if and only if the mapping Ψ of Q into R^m given by

$$(8) \qquad\qquad \Psi(q)=\{g_1(q), \ldots, g_m(q)\}$$

is $1-1$, and if no point $p \in N=\Psi(Q)$ is a non-trivial convex combination of other points $p_1, \ldots, p_k$ of N. Other equivalent forms of this requirement are that the set N should be precisely the set of all extreme points of the convex hull K of N, and that the set $1, x_1, \ldots, x_m$ (where the x_i are the coordinates of the point $p \in R^m$) should be a Korovkin system on N.

Of some interest is the question whether it is possible to restrict the number k in the above statement. From the theorem of CARATHÉODORY [1, p. 9] it follows that one can assume $k \leq m+1$; if Q is connected, then an appeal to a theorem of FENCHEL [1] allows the reduction $k\leq m$. For further reduction see § 5.

THEOREM 4. *If Q has a Korovkin system F of order m, then also a strong Korovkin system of type G, of order $\leq m$.*

PROOF. The reduction from the case F to G is due to ŠAŠKIN [5, Theorem 3], so that we can assume that a system G exists on Q, and we can take functions g_i to be linearily independent.

Let $K=\operatorname{co} N$, $N=\Psi(Q)$. If K has no interior points, we consider the linear hull L of N. If $L=R^m$, then a neighbourhood of 0 is contained in the set of all $p = \sum_1^m a_i e_i$, $|a_i|<\varepsilon$, where $e_1, \ldots, e_m$ is a base of L in N. Then also a neighbourhood of the point $m^{-1} \sum_1^m e_i \in K$ belongs to K, a contradiction. If L has no interior points, we can change the coordinates so that L is contained in the subspace $x'_1=$const. This however means that the functions of G are linearily dependent.

Hence K has an interior point p_0, with the ball $\|p-p_0\| \leq r$ contained in K. A ray from p_0 intersects the boundary S of K in exactly one point. For if $p_1 \in S$, and $p = tp_0+(1-t)p_1$, $0<t\leq 1$, then K contains the ball $\|p'-p\| \leq tr$. Thus the projection Π of S onto the unit sphere S_0: $\|p-p_0\| = 1$ is a homeomorphism. Let $\Pi(N)=N_0$. Then no point of N_0 is a convex combination of other points

of N_0, since N_0 is contained in the strictly convex surface S_0, and N_0 is homeomorphic to Q. By Šaškin's theorem, the system $G : 1, \bar{g}_i(q)$, $i=1, \ldots, m$, where $\bar{g}_i$ is the i-th coordinate of the point $\Pi(\Psi(q))$ of N_0, is a Korovkin system for Q. If $q_0 \in Q$, and $a + \sum_{i=1}^{m} a_i x_i = 0$ is the equation of the strictly supporting hyperplane for N_0 at the point $\Pi(\Psi(q_0))$ of N_0, then $P_{q_0} = a + \sum_{1}^{m} a_i \bar{g}_i$ is the required polynomial. This completes the proof.

The projection Π employed above, and the projection used by Šaškin to reduce F to G, are both infinitely differentiable on N. For example, for Π we have $\Pi p = p_0 + (p - p_0)/\|p - p_0\|$. If Q is a manifold, one can define classes of differentiable functions $C^{(p)}(Q)$ for each $p = 1, 2, \ldots$. Hence we obtain: *If a manifold Q carries a Korovkin system F of class $C^{(p)}(Q)$ and order m, then Q carries also a strong Korovkin system G of the same class and of order $\leq m$.*

For a compact space Q that carries a finite Korovkin system, we define $m(Q)$ as *the smallest order of all Korovkin systems carried by Q*. Then we have:

THEOREM 5. *If Q possesses finite Korovkin systems, then*

$$(9) \qquad\qquad m(Q) = d(Q) + 1,$$

where $d(Q)$ is the smallest dimension of a sphere into which Q can be topologically embedded.

PROOF. In the proof of Theorem 4 we have seen that $m(Q) \geq d(Q) + 1$. The inverse inequality follows from the fact that the d-dimensional sphere $\sum_{k=1}^{d+1} x_k^2 = 1$ carries the Korovkin system $1, x_1, \ldots, x_{d+1}$ of order $d+1$, and the fact that if F is a Korovkin system for Q, then also for any compact subset $Q_1 \subset Q$. This last fact can be derived from [5, Theorem 2″].

The number $d(Q)$ is known, for example, for many m-dimensional closed manifolds.

5. A counterexample

The interesting paper of ŠAŠKIN [5] contains the following assertion (Theorem 9): If Q is the closure of a bounded region in R^{m-1}, then a system F given by (6) is a Korovkin system on Q if and only if it is a Chebyshev system of rank $m-2$.

This property of F has a simple geometric interpretation, which we state only for systems of type G. It means that the mapping Ψ defined by (8) is a homeomorphism and that no point $p \in N = \Psi(Q)$ is a convex combination of two other points of N. This in turn is equivalent to the 2-regularity of N, that is, to the property that no three distinct points of N lie on a line.

The proof given by Šaškin is not convincing, in the part dealing with the boundary of N (p. 140, lines 1—8, esp. when $k < m$ for k used there). As we shall see, the theorem itself is incorrect. The proof seems to work if $m = 3$ (when the only possible value of k is 3), also probably if Q has no boundary, that is, if it is a closed $(m-1)$ dimensional manifold. In this case, however, N is also an $(m-1)$ manifold in R^m, that has a strictly supporting plane in each point (see Šaškin's proof of Theorem 9), hence is an $(m-1)$-sphere. Then Q is also a sphere. Therefore, Šaškin's argument proves his assertion for probably only $m-1$ dimensional topological spheres, and for systems of type G. If in addition one recalls [5, Theorem 3], one sees that *no other closed $(m-1)$ manifold can carry a Korovkin system of $m+1$ functions.*

In what follows, we show that Šaškin's theorem is incorrect for an $(m-1)$-dimensional cell Q, $m \geq 4$, even for systems of type G. Our proof will proceed geometrically. We shall construct in R^m a set N homeomorphic to Q; the coordinates of the points of N together with the function 1 will provide the necessary system G. We restrict ourselves to the case $m = 4$.

THEOREM 5. *There exists in R^4 a set N with the following properties. N is homeomorphic to a 3-cell; N is 2-regular, and yet a point of N is a convex combination of three other points of N.*

PROOF. In R^4, let S be the sphere $\sum_{i=1}^{4} x_i^2 = 1$, S^- its lower cup $x_4 \leq 0$. Since S is strictly convex, each subset of S is 2-regular. Let I be the vertical interval $(0, 0, 0, x_4)$ $0 < x_4 \leq 1$, I_i, $i = 1, 2$, 3-intervals parallel to I with bases at the points $(0, \pm 1, 0, 0)$, $(-1, 0, 0, 0)$. Let J be a strictly convex arc in the $x_1 x_4$-plane close to I, with the endpoints $(1, 0, 0, 0)$ and $(1-a, 0, 0, 1)$ with small $a > 0$. If we rotate J around the x_4-axis, the obtained surface of rotation together with S^- can be completed to a strictly convex surface Σ, which contains S inside. The circular arc $C: x_1^2 + x_4^2 = 1$, $x_1 \geq 0$, $x_4 > 0$, lies in the plane $x_2 = x_3 = 0$; a similar arc C' lies in the plane E' with the equations $x_2 = 0$, $x_3 = \alpha x_1$ with small $\alpha > 0$. The set N will consist of S^- plus modified sets I_i, C'.

If p is a point outside of S, there is a critical 3-sphere $S(p)$ around p, which intersects S exactly where the tangents from p touch S. A straight line through p can cut the part of S outside of $S(p)$ only once. In particular, if p is on I_1(or I_2), the radius of $S(p)$ is ≤ 1 (equal to the distance from p to the point $(0, 1, 0, 0)$), while the distance from p to C is > 1. Also the cup S^- lies outside of $S(p)$. For I_3 the situation is similar, but simpler. By compactness arguments, we derive from this:

(A) *There exist convex neighbourhoods I_i^0 of the I_i, $i = 1, 2, 3$ in the half space $x_4 > 0$, and a neighbourhood C^0 of C on $S \backslash S^-$, such that a straight line through a point of I_i^0 can intersect $S^- \cup C^0$ only once.*

We shall assume that $C' \subset C^0$. We take $x > 0$ so small, that the two intervals $I'_i = (x, \pm\sqrt{1-x^2}, 0, x_4)$, $0 < x_4 \leq 1$, $i = 1, 2$ are contained in the corresponding I^0_i, and we put $I'_3 = I_3$. For a pair I'_i, I'_j, $i \neq j$, the two dimensional plane joining them has the equations $x_3 = 0$, $\alpha_1 x_1 + \alpha_2 x_2 = 1$, and does not intersect the plane E'. Hence the curve C' has a non-zero distance to each of the first named planes. It is possible to replace the sets I^0_i, C^0 by smaller sets, so that they remain convex neighbourhoods of I'_i, and a neighbourhood of C', respectively, and that in addition to (A) they satisfy:

(B) *Each straight line connecting two points of two different sets I^0_i does not intersect C^0.*

It is now possible to select the arc J in such a way that its rotations to the points $(x, \pm\sqrt{1-x^2}, 0, 0)$, $(-1, 0, 0, 0)$, which we denote by J_i, $i = 1, 2, 3$ are contained in the corresponding I^0_i. We can now embed J_i, C' into three dimensional strips N_i, N', lying on the surfaces Σ, S respectively, so that they are contained in the constructed neighbourhoods and that the set $S^- \cup \bigcup N_i \cup N' = N$ is a 3-cell.

The point $(0, 0, 0, 1)$ belongs to the convex hull of the three endpoints of the arcs J_i. It remains to show that N is 2-regular.

Let p_1, p_2, p_3 be three different points of N. If none of them belongs to $\bigcup_1^3 N_i$, or to N', then all three are on the strictly convex surface Σ, or S, and cannot be collinear. Therefore, we can assume that $p_1 \in \bigcup N_i$ and $p_2 \in N'$. If $p_3 \in S^- \cup N'$, then the line connecting p_1 and p_2 cannot contain p_3, by (A). If $p_3 \in N_j$ and $p_1 \in N_i$, $i \neq j$, then the line $p_1 p_3$ cannot contain p_2 by (B). Finally, if $p_1, p_3 \in N_i$, then the segment $p_1 p_3$ is inside I^0_i and does not intersect $S^- \cup N'$. And the remaining part of the line $p_1 p_3$ is outside Σ, and hence does not intersect S. This completes the proof.

BIBLIOGRAPHY

[1] T. Bonnesen and W. Fenchel, *Theorie der konvexen Körper*. Ergebnisse der Mathem., Bd. **3**, no. 1, Springer, Berlin 1934.

[2] G. Freud, *Über positive lineare Approximationsfolgen von stetigen reellen Funktionen auf kompakten Mengen.* On Approximation Theory, Proc. of a Conference, ISNM 5, Birkhäuser, Basel 1964, 233—238.

[3] R. P. Kelisky and T. J. Rivlin, *Iterates of Bernstein polynomials.* Pacific J. Math. **21** (1967), 511—520.

[4] P. P. Korovkin, *Linear Operators and Approximation Theory.* Hindustan Publ. Corp., Delhi 1960.

[5] Ju. A. Šaškin, *Korovkin systems in spaces of contionuous functions.* Amer. Math. Soc. Transl. (2) **54** (1966), 125—144 (= Izv. Akad. Nauk SSSR, Ser. Mat. **26** (1962), 495—512).

[6] P. C. Sikkema, *Über Potenzen von verallgemeinerten Bernstein-Operatoren.* Mathematica (Cluj), **8** (31) (1966), 173—180.

[7] M. G. Kreĭn and M. A. Rutman, *Linear operators leaving invariant a cone in a Banach space.* Uspehi Mat. Nauk 3, no. 1 (23) (1948), 3—95.

Sätze vom Bohman—Korowkin-Typ für Banachsche Funktionenräume

Von

M. W. MÜLLER

MATHEM. INSTITUT A
UNIVERSITÄT STUTTGART

Herrn Professor Dr. Werner Meyer-König zum 60. Geburtstag am 26. Mai 1972 gewidmet

1. Einleitung

Als Sätze vom Bohman—Korowkin-Typ bezeichnet man in der Approximations-theorie i.a. hinreichende Bedingungen dafür, daß eine Folge linearer positiver Opera-toren, die einen normierten linearen Raum stetiger Funktionen in sich abbilden, in der zugehörigen Operatornorm gegen die Identität konvergiert. Allen diesen Sätzen ist gemeinsam, daß man — grob gesprochen — nur eine punktweise Kon-vergenz der Operatorfolge auf einer relativ „kleinen" Menge nachzuweisen hat.

Um das Adjektiv „klein" näher präzisieren zu können, führte G. FREUD [4] im Jahre 1963 den Begriff *Probemenge* ein. W. SCHEMPP [11] griff dieses Konzept erst kürzlich wieder auf (unter dem Namen Korowkin—Test-Familie), um mit seiner Hilfe einen sehr allgemeinen Bohman—Korowkin-Satz anzugeben für Folgen positiver Endomorphismen eines linearen Raumes beschränkter reellwertiger und gleichmäßig stetiger Funktionen über einem uniformen Raum, wobei letzterer nicht notwendig in einen endlichdimensionalen linearen Raum eingebettet zu sein braucht.

Das Ziel der vorliegenden Arbeit sind Bohman—Korowkin-Sätze für gewisse Räume meßbarer Funktionen (sogenannter Banachscher Funktionenräume), wobei im 2. Teil insbesondere Funktionen mit einem starken Wachstumsverhalten berück-sichtigt werden sollen.

Einige im weiteren Verlauf der Arbeit nötige Hilfsmittel, so u. a. der Begriff einer allgemeinen Testmenge auf Ω bez. K (eine Verallgemeinerung der Freudschen Probemenge) werden in der nächsten Nummer bereitgestellt.

2. Hilfsmittel

Im folgenden seien stets Ω eine nichtleere (beschränkte oder unbeschränkte) Teilmenge der reellen Zahlengeraden R und K eine feste kompakte Menge mit $\emptyset \neq K \subseteq \Omega$. $C(\Omega)$ bezeichnet den linearen Raum der auf Ω definierten reellwertigen stetigen Funktionen.

Definition 2. 1. Eine Menge $\{f_1, \dots, f_r\}$ aus $C(\Omega)$ wird eine *allgemeine Test-menge* — kurz ATM — *auf Ω bez. K* genannt, wenn gilt:

(T1) $\exists \{a_1, \dots, a_r\} \subset C(\Omega)$ derart, daß die durch

$$P(t, x) := \sum_{i=1}^{r} a_i(x) f_i(t), \qquad (t, x) \in \Omega \times \Omega$$

erklärte Funktion P nichtnegativ ist und Null genau für $t=x$,

(T2) für alle $\delta > 0$ ist

$$\inf \{P(t, x) | t \in \Omega, \quad x \in K \quad \text{mit} \quad |t-x| \geqq \delta\} > 0,$$

(T3) $$\exists Q \in \{P(\cdot, x_1) + P(\cdot, x_2) | x_1, x_2 \in \Omega, \quad x_1 \neq x_2\}$$

mit $q := \inf_{t \in \Omega} Q(t) > 0$.

Ist Ω kompakt, so sind (T2) und (T3) automatisch erfüllt, desgleichen im Falle $K = \Omega$ (wir sprechen dann auch kurz von einer ATM auf K). Wie in [9, Nr. 1. 2] gezeigt wurde, ist es nicht schwer, allgemeine Testmengen auf Ω bez. K zu kon-struieren. Als sehr zweckmäßig erweist sich das von P. W. ERMAKOW [3] eingeführte Tripel $\{p_0, p_1, p_2\}$, wo die p_i für $t \in \Omega$ erklärt sind durch $p_0(t) := 1$, $p_1(t) := t$, $p_2(t) := |t|^s$, $s = 2, 3, \dots$. Speziell für $s = 2$ ergibt sich die bekannte Korowkinsche Testmenge.

Im Falle $p_0 \in \{f_1, \dots, f_r\}$ kann die Bedingung (T3) weggelassen werden.

Zu der vorgegebenen kompakten Menge K seien $\mathfrak{M}(K)$ der lineare Raum der auf K definierten reellwertigen (Lebesgue-) meßbaren Funktionen und $M(K)$ der lineare Quotientenraum von $\mathfrak{M}(K)$ modulo des zugehörigen Nullraumes. Wie üblich werden wir die Elemente des Quotientenraumes der Einfachheit halber Funktionen nennen und mit $f, g, \dots$ bezeichnen.

Definition 2. 2. Ein Banach-Raum $(B(K), \|\cdot\|_B)$ bestehend aus Elementen von $M(K)$ heißt genau dann ein *Banachscher Funktionenraum*, wenn seine Norm den folgenden Bedingungen genügt:

(N1) $$[g \in M(K) \wedge f \in B(K) \wedge |g| \leqq |f|] \Rightarrow [g \in B(K) \wedge \|g\|_B \leqq \|f\|_B],$$

(N2) $$[\{f_n\}_{n=1}^{\infty} \subset B(K) \wedge 0 \leqq f_n \nearrow f \quad \text{mit} \quad f \in B(K)] \Rightarrow [\|f_n\|_B \to \|f\|_B],$$

(N3) $\|f\|_B$ ist umordnungsinvariant für alle $f \in B(K)$.

Die B-Norm besitzt bekanntlich eine Darstellung der Form

(2. 1) $$\|f\|_B = \sup_{c \in C} \int_K c(t) |f(t)| \, dt \qquad (f \in B(K)),$$

wo C eine Klasse positiver meßbarer Funktionen ist, die mit jedem c alle seine Umordnungen enthält (vgl. G. G. LORENTZ—T. SHIMOGAKI [8]).

Als wichtigste Beispiele Banachscher Funktionenräume seien genannt die Lebesgueschen Räume $L^p(K)$ mit $1 \leqq p < \infty$, ferner die Orlicz-Räume $L_M^*(K)$, falls die sogenannte N-Funktion $M(u)$ die Youngsche Δ_2-Bedingung erfüllt (vgl. M. A. KRASNOSEL'SKII und YA. RUTICKII [5, pp. 60—82]), sowie die Lorentz-Räume $\Lambda(\Phi, r)$ mit $1 \leqq r < \infty$ (vgl. G. G. LORENTZ [6]).

Im weiteren Verlauf der Arbeit stützen wir uns gelegentlich auf den folgenden

HILFSSATZ. *Der Raum $C(K)$ ist dicht in $B(K)$ bezüglich der B-Norm.*

Da wir für diesen Hilfssatz in der Literatur keinen allgemeingültigen Beweis finden konnten, skizzieren wir ganz kurz einen Dichtebeweis. Geht man von einer beliebigen Funktion $f \in B(K)$ aus, so liefert der übliche Stutzungsprozess bei Beachtung von (N1) eine nichtfallende Folge $\{f_n\}_{n=1}^\infty$ beschränkter Elemente aus $B(K)$ mit $\lim_{n \to \infty} f_n = f$ (punktweise). Wegen $c|f - f_n| \leqq c|f|$ für alle n und alle $c \in C$ folgt dann auf Grund von (2. 1) und des allgemeinen Lebesgueschen Konvergenztheorems die Aussage $\|f - f_n\|_B \to 0$ $(n \to \infty)$. f liegt somit im $\|\cdot\|_B$-Abschluß der beschränkten Elemente aus $B(K)$. Ohne Einschränkung der Allgemeinheit sei deshalb $f \in B(K)$ mit $\mathop{\mathrm{Sup}}_{x \in K} |f(x)| := M < \infty$. Nach dem Satz von Lusin existiert eine Folge $\{F_n\}_{n=1}^\infty$ aus $C(K)$ derart, daß für alle $n \in N$ gilt $\mathop{\mathrm{Sup}}_{x \in K} |F_n(x)| \leqq M$, wobei das (Lebesguesche) Maß der Menge $K_n := \{x \in K \mid f(x) \neq F_n(x)\}$ kleiner als n^{-1} ist. Für ein beliebiges $c \in C$ folgt dann

$$\int_K c(x) |f(x) - F_n(x)| \, dx \leqq 2M \int_{K_n} c(x) \, dx \to 0 \qquad (n \to \infty)$$

und folglich $\lim_{n \to \infty} \|f - F_n\|_B = 0$, da $c \in C$ beliebig gewählt war.

3. Approximation von Funktionen aus $B(K)$

Der Satz vom Bohman—Korowkin-Typ für Folgen positiver Endomorphismen des Raumes $B(K)$ lautet wie folgt.

SATZ 3. 1. *Es sei $\{L_n\}_{n=1}^\infty$ eine Folge beschränkter linearer positiver Operatoren, die einen linearen Teilraum D von $B(K)$ in $B(K)$ abbilden. D enthalte den Raum $C(K)$, und es gelte*

(a) $\overline{\lim_{n \to \infty}} \|L_n\|_B < \infty,$

(b) $\lim_{n \to \infty} \|L_n f_i - f_i\|_B = 0$ *für die Elemente $\{f_1, \ldots, f_r\}$ einer allgemeinen Testmenge auf K.*

Dann folgt

(c) $\lim\limits_{n \to \infty} \|L_n f - f\|_B = 0$ *für alle* $f \in D$.

Im Falle $D = B(K)$ *sind* (a) *und* (b) *auch notwendig für* (c).

BEWEISSKIZZE. 1. Die Notwendigkeit von (a) im Falle $D = B(K)$ folgt direkt aus dem Satz von Banach—Steinhaus (vgl. [2, p. 60]), während (b) in trivialer Weise erfüllt ist. 2. Der Beweis der Hinlänglichkeit gliedert sich in zwei Schritte. Im 1. Schritt wird mit Hilfe von (b) die Konvergenzaussage (c) zunächst für alle Elemente des in $B(K)$ — und damit auch in D — dicht gelegenen Raumes $C(K)$ nachgewiesen. Unter Beachtung der monotonieerhaltenden Eigenschaft (N1) der B-Norm überträgt man hierbei sinngemäß den entsprechenden Teil eines Beweises, der in [9, Satz 2. 1] u. a. für den Fall einer L^p-Norm geführt wurde. (Für den Fall der Sup-Norm über K findet sich die Grundidee des letztgenannten Beweises im Lehrbuch von G. G. LORENTZ [7, pp. 7—8] bzw. in einer Arbeit von G. FREUD [4].) Im 2. Schritt läßt sich dann die Aussage (c) mit Hilfe eines Standardschlusses leicht von $C(K)$ auf D ausdehnen. Man beachte hierbei nur die Voraussetzung (a) der — grob gesprochen — gleichmäßigen Beschränktheit der Folge der Operatornormen, die Dichtheit von $C(K)$ in D und die auf Grund des 1. Schrittes bereits gesicherte Konvergenz $L_n h \xrightarrow{\|\cdot\|_B} h$ $(n \to \infty)$ für die Elemente $h \in C(K)$.

BEMERKUNG. Im Falle positiver Endomorphismen des mit der Sup-Norm $\|\cdot\|$ versehenen Raumes $C(K)$ (klassischer Satz von Bohman—Korowkin bzw. die im Lehrbuch von G. G. LORENTZ [7, p. 7] angegebene Fassung mit Hilfe allgemeiner Testmengen auf K) sind zwei der in Satz 3. 1 unvermeidlichen Voraussetzungen automatisch erfüllt, nämlich einerseits die Beschränktheit der einzelnen Operatoren L_n und andererseits die Voraussetzung (a), welche wegen $\|L_n\| = \|L_n p_0\|$ eine direkte Folge der r Testbedingungen in (b) ist.

4. Über Anwendungen auf singuläre Integrale

Die im Rahmen einer Theorie meßbarer Funktionen natürlichsten linearen Approximationsverfahren lassen sich durch Folgen singulärer Integrale beschreiben. Es zeigt sich, daß bei Approximation in Banachschen Funktionenräumen, die für die Anwendbarkeit von Satz 3. 1 grundlegende Voraussetzung (a) bereits unter sehr milden Voraussetzungen an die Folge der Kerne erfüllt ist.

Sei $K_n(\cdot, \cdot)$, $n \in N$ ein auf $K \times K$ definierter meßbarer reellwertiger nichtnegativer Kern mit den Eigenschaften

(I) $\qquad \displaystyle\int_K K_n(t, x)\, dx \le M \quad$ für alle $\quad t \in K \quad$ und alle $\quad n \in N$,

(II) $\qquad \displaystyle\int_K K_n(t, x)\, dt \le M \quad$ für alle $\quad x \in K \quad$ und alle $\quad n \in N$, wo M eine von n unabhängige endliche positive Konstante ist. Nach einem bekannten Satz von W. Orlicz ([10]) stellt dann das durch

$$(T_n f)(x) := \int\limits_K K_n(t, x) f(t)\, dt \qquad (f \in L^p(K),\ 1 \leqq p \leqq \infty,\ n \in \mathbf{N})$$

definierte singuläre Integral T_n einen beschränkten linearen positiven Operator von $L^p(K)$ in sich dar mit $\|T_n\|_p \leqq M$. T_n bildet insbesondere also $L^1(K)$ in sich und $L^\infty(K)$ in sich ab mit der Norm $\|T_n\|_1 \leqq M$ bzw. $\|T_n\|_\infty \leqq M$.

Auf Grund eines Interpolationssatzes für lineare Operatoren von A. P. Calderón ([1]) besitzen die drei Paare $L^\infty(K)$, $L^\infty(K)$; $L^1(K)$, $L^1(K)$; $B(K)$, $B(K)$ die *Interpolationseigenschaft*, d.h. der Operator T_n kann ausgedehnt werden zu einer beschränkten Abbildung von $B(K)$ in sich mit der Norm $\|T_n\|_B \leqq M$. Da letztere nicht vom Folgenindex n abhängt, ist für die Folge $\{T_n\}_{n=1}^\infty$ die Voraussetzung (a) im Satz 3. 1 erfüllt.

BEISPIEL. Die Folge der singulären Integrale von Landau—Stieltjes bildet den Raum $B(K)$ in sich ab und besitzt gleichmäßig beschränkte Normen.

5. Approximation von Funktionen aus $B(\Omega, K)$

Satz 3. 1 behandelte die Approximation von Funktionen, für die auf ihrem vollen Definitionsbereich eine B-Norm erklärt ist. Es sollen nun meßbare Funktionen approximiert werden, die ein gewisses Wachstumsverhalten haben, welches zur Folge hat, daß eine B-Norm nur noch für die Restriktion der Funktion auf einen kompakten Teil ihres Definitionsbereiches existiert. Genauer seien nun Ω eine nichtleere (beschränkte oder unbeschränkte) meßbare Teilmenge von $\mathbf{R}$ und K eine feste nichtleere echt in Ω enthaltene kompakte Menge. Mit $B(\Omega, K)$ bezeichnen wir den linearen Raum aller $f \in M(\Omega)$ mit $\mathrm{Rest}_K f \in B(K)$. Durch die Abbildung

$$\| \cdot \|_{B,K} : f \in B(\Omega, K) \longmapsto \|\mathop{\mathrm{Rest}}\limits_K f\|_B$$

wird auf $B(\Omega, K)$ eine Seminorm erklärt.

Wir werden hinreichende Bedingungen dafür angeben, daß die Folge $\{L_n\}_{n=1}^\infty$ linearer positiver Operatoren aus $B(\Omega, K)$ in $B(K)$ Elemente aus $B(\Omega, K)$ auf kompakten Teilmengen von K durch Elemente aus $B(K)$ in der B-Norm approximiert. Das Verhalten einer zu approximierenden Funktion f auf $\Omega - K$ (also außerhalb des Bereichs über dem eine B-Norm existiert) wirkt sich hier darin aus, daß die Abbildungsvorschrift L_n das Verhalten von f in ganz Ω berücksichtigt. Laut Definition wird der Funktion $f \in B(\Omega, K)$ auf $\Omega - K$ keinerlei Wachstumsbeschränkung auferlegt. Wir tun dies im ersten der beiden folgenden Sätze indirekt, indem wir — ähnlich wie H. Walk ([12]) bei der Approximation unbeschränkter stetiger

Funktionen — fordern, daß die Bildfolge von $h(|f|)$ für eine Funktion $h: \mathbf{R}^+ \to \mathbf{R}^+$ mit $h(s)/s \to \infty$ $(s \to \infty)$ existiert und im Sinne der Norm im Bildraum gleichmäßig beschränkt ist.

Da man für die einzelnen Operatoren L_n keine Operatornorm im üblichen Sinne erklären kann, führen wir zunächst einen zur Operatornorm adäquaten Begriff ein. Es sei D ein linearer Teilraum des seminormierten linearen Raumes $B(\Omega, K)$ und $[D, B(K)]$ die Menge der linearen Operatoren A von D in $B(K)$ mit

$$(5.1) \qquad \|A\|_{B,K} := \mathrm{Sup}\,\{\|Af\|_B \mid f \in D,\ \|f\|_{B,K} \leqq 1\} < \infty.$$

Die Menge $[D, B(K)]$ läßt sich in natürlicher Weise zu einem linearen Raum machen, und die durch (5. 1) definierte Abbildung in $\mathbf{R}^+$ besitzt alle Eigenschaften einer Seminorm. Wir nennen einen Operator $A \in [D, B(K)]$ *K-beschränkt* und die nichtnegative Zahl $\|A\|_{B,K}$ die *K-Seminorm von A*.

Noch eine letzte Bezeichnung: $C_0(\Omega, K)$ sei der lineare Raum aller $f: \Omega \to \mathbf{R}$ mit $\mathrm{Rest}_K f \in C(K)$ und $f = 0$ für $t \in \Omega - K$.

Es folgt nun der erste der beiden für diese Nummer angekündigten Sätze vom Bohman—Korowkin-Typ: der zu approximierenden Funktion wird zunächst eine *indirekte* Wachstumsbeschränkung auferlegt.

SATZ 5. 1. *Es sei $\{L_n\}_{n=1}^{\infty}$ eine Folge K-beschränkter linearer positiver Operatoren, die einen linearen Teilraum D von $B(\Omega, K)$ in $B(K)$ abbilden. D enthalte p_0, $C_0(\Omega, K)$ sowie die Elemente einer ATM $\{f_1, \ldots, f_r\}$ auf Ω bez. K. Es gelte*

(a)
$$\varlimsup_{n \to \infty} \|L_n\|_{B,K} < \infty,$$

(b)
$$\lim_{n \to \infty} \|L_n f_i - f_i\|_B = 0 \qquad (i = 1, 2, \ldots, r).$$

Existiert zu $f \in D$ eine Funktion $h: \mathbf{R}^+ \to \mathbf{R}^+$ mit $h(s)/s \to \infty$ $(s \to \infty)$ und $h(|f|) \in D$, gilt ferner

(c)
$$\varlimsup_{n \to \infty} \|L_n h(|f|)\|_B < \infty$$

und $X_K f \in D$, wo X_K die charakteristische Funktion von K in Ω ist, so folgt

$$\lim_{n \to \infty} \|X_{K_u}(L_n f - f)\|_B = 0$$

für jede kompakte Teilmenge K_u von K mit $\mathrm{dist}\,(\Omega - K, K_u) > 0$.

Die letzte Bedingung besagt, daß zwischen den beiden kompakten Mengen K_u und K noch eine Art von Sicherheitszone liegen muß. Über die Approximation auf dieser Zone läßt sich im allgemeinen nichts aussagen.

Daß die Forderung $h(s)/s \to \infty$ $(s \to \infty)$ in (c) nicht zu $h(s) = O(s)$ $(s \to \infty)$ abgeschwächt werden kann, zeigt ein einfaches Gegenbeispiel (vgl. etwa [9, S. 65—66]).

Für global beschränkte Funktionen f ist (c) automatisch erfüllt, falls h in beschränkten Teilmengen von R^+ beschränkt ist.

Eine ausführliche Darstellung des Beweises von Satz 5. 1 folgt in einer späteren Arbeit. Im wesentlichen beruht er auf einer Übertragung von Beweisen in [9, Kapitel 3], in denen speziell die Funktion $h(s)=s^r$, $s\in R^+$, $r>1$ verwendet wird. Bei der hier vorliegenden allgemeinen Funktion h ist zu beachten, daß zu beliebigem $\varepsilon>0$ eine Zahl

$$K(\varepsilon) := \operatorname{Sup}\{s-\varepsilon h(s)\,|\,s\in R^+\} < \infty$$

existiert und dann offensichtlich gilt

$$s \leqq K(\varepsilon)+\varepsilon h(s) \qquad (\forall s\in R^+)$$

(vgl. auch H. WALK [12, S. 17]). Dies führt dann zu einer Aufspaltung von Termen, die in [9, Kapitel 3] mit Hilfe einer Hölderschen Ungleichung für lineare positive Operatoren abgeschätzt werden. Auch hier erweist sich wieder ein zuvor bewiesener Lokalisationssatz als sehr nützlich. Dieser besagt grob gesprochen folgendes: Ersetzt man die Werte einer Funktion $f\in B(\Omega, K)$, die der Bedingung (c) genügt, außerhalb einer geeigneten Obermenge V von K durch Null, d.h. ersetzt man f durch $X_V f$, so unterscheiden sich für hinreichend großes n die L_n-Bilder von f und $X_V f$ im Sinne der B-Norm beliebig wenig voneinander.

Im Gegensatz zu den Voraussetzungen des vorangehenden Satzes wird jetzt der zu approximierenden Funktion f eine *direkte* Wachstumsbeschränkung auferlegt.

SATZ 5. 2 (Dominierte Konvergenz). *Die Aussage von Satz 5. 1 bleibt gültig, wenn man anstelle der Existenz einer Funktion* $h\colon R^+\to R^+$ *mit* $h(s)/s\to\infty$ $(s\to\infty)$ *und* $h(|f|)\in D$, *sowie der Gültigkeit von* (c) *fordert:*

(c_1) *Zu* $f\in D$ *mit* $X_K f\in D$ *existiert eine positive Funktion* $g\in D$ *mit* $X_K g\in D$ *und* $|f(t)|\leqq g(t)$ *für* $t\in\Omega-K$,

und es gelte

(c_2)
$$\lim_{n\to\infty}\|X_{K_u}(L_n g-g)\|_B = 0$$

für jede kompakte Teilmenge K_u *von* K *mit* $\operatorname{dist}(\Omega-K, K_u) > 0$.

Auch der Beweis dieses Satzes wird in einer späteren Arbeit ausführlich dargestellt werden (vgl. hierzu [9, Nr. 3. 5]).

LITERATUR

[1] A. P. Calderón, *Spaces between L^1 and L^∞ and the theorem of Marcinkiewicz*. Studia Math. **26** (1966), 273—299.

[2] N. Dunford and J. T. Schwartz, *Linear Operators*. Vol. I: *General Theory*. Interscience, New York 1958.

[3] P. W. Ermakow, *Über die Bedingungen der Konvergenz linearer positiver Operatoren auf unbeschränkten Mengen* [Russisch]. Studien zeitgenössischer Probleme der Konstr. Funktionentheorie (Fortschr. 2. All-Union Konferenz Baku 1962), Baku 1965, 146—151.

[4] G. Freud, *Über positive lineare Approximationsfolgen von stetigen reellen Funktionen auf kompakten Mengen*. (Proceedings of the Conference in Oberwolfach, 1963.) In: On Approximation Theory, edited by P. L. Butzer and J. Korevaar ISNM 5, 233—238, Birkhäuser, Basel 1964.

[5] M. A. Krasnosel'skii and Ya. B. Rutickii, *Convex Functions and Orlicz Spaces*. Nordhoff Ltd., Groningen 1961.

[6] G. G. Lorentz, *Some new functional spaces*. Ann. of Math. (2) **51** (1950), 37—55.

[7] G. G. Lorentz, *Approximation of Functions*. Holt, Rinehart and Winston, New York 1966.

[8] G. G. Lorentz and T. Shimogaki, *Interpolation theorems for operators in function spaces*. J. Functional Analysis 2 (1968), 31—51.

[9] M. W. Müller, *Approximation durch lineare positive Operatoren bei gemischter Norm*. Habilitationsschrift Stuttgart 1970. 110 S.

[10] W. Orlicz, *Ein Satz über die Erweiterung von linearen Operationen*. Studia Math. 5 (1935), 127—140.

[11] W. Schempp, *A note on Korovkin test families*. (to appear).

[12] W. Walk, *Approximation unbeschränkter Funktionen durch lineare positive Operatoren*. Habilitationsschrift Stuttgart 1970. 110 S.

IV.
Harmonic Analysis and Approximation

Projection métrique de $L^1(T)$ sur des sous-espaces fermés invariants par translation

Par

J.-P. KAHANE

UNIVERSITÉ PARIS-SUD
CENTRE D'ORSAY

$T = R/2\pi Z$ est le cercle, Z est le groupe des entiers, Λ et Λ' sont deux parties complémentaires non vides de Z, et L^1_Λ est le sous-espace fermé de $L^1(T)$ engendré par les exponentielles $\{e^{i\lambda t}\}_{\lambda \in \Lambda}$; tout sous-espace fermé de $L^1(T)$ invariant par translation, propre et $\neq \{0\}$, est un L^1_Λ. On note $\|\ \|$ la norme dans $L^1(T)$. Etant donné $f \in L^1(T)$ et $h \in L^1_\Lambda$, on dit que h est une projection métrique de f sur L^1_Λ si

$$\|f - h\| = \min_{l \in L^1_\Lambda} \|f - l\| = d(f, L^1_\Lambda).$$

L'ensemble des projections métriques de f sur L^1_Λ est un convexe fermé, éventuellement vide, noté $\mathscr{P}_\Lambda f$; si cet ensemble est réduit à un seul élément, on note cet élément $P_\Lambda f$.

On peut se poser les problèmes suivants:

a) *"existence": pour quels Λ a-t-on $\mathscr{P}_\Lambda f \neq \emptyset$ pour tout $f \in L^1(T)$?*

b) *"unicité": la condition d'existence étant satisfaite, pour quels Λ l'ensemble $\mathscr{P}_\Lambda f$ est-il réduit à un seul élément pour tout f?*

c) *"continuité": la condition d'unicité étant satisfaite, l'application P_Λ est-elle continue?*

d) *"continuité uniforme": P_Λ est-elle alors uniformément continue sur les bornés?*

La question a) semble la plus difficile. Nous répondons ci-dessous complètement aux questions b), c), d). En cours de route, nous poserons d'autres problèmes.

Dans le cas $\Lambda = Z^+ = \{0, 1, 2, ...\}$, L^1_Λ est identifiable à la classe de Hardy H^1, et la question d) a été posée récemment par G. HENKIN (Studia Mathematica 38, 1970, problème 53, p. 480); Henkin attribue à V. PTAK et S. HAVINSON la preuve de l'existence et de l'unicité de la projection métrique L^1 sur H^1. H. S. SHAPIRO m'a signalé durant la conférence la priorité de DOOB [1] à ce sujet, ainsi qu'un article de D. J. NEWMAN [4] d'où résulte aisément la continuité de cette projection métrique.

Signalons comme références générales sur les projections métriques [6], [7], [8]. D'autre part la traduction par transformation de Fourier des problèmes d'existence et d'unicité de P_Λ ne sont rien d'autre que les problèmes du prolongement minimal au sens de Beurling, c'est-à-dire du prolongement d'une fonction de l'algèbre $A(\Lambda')$ (constituée par les restrictions à Λ' des fonctions de $A(Z)=\mathfrak{F}L^1(T)$, plus communément désignées comme "suites de coefficients de Fourier—Lebesgue") en une fonction de $A(Z)$ de norme minimum. Le prolongement minimal est souvent considéré dans $A(R)$ plutôt que dans $A(Z)$; pour un aperçu des résultats et des références, voir [7] chap. 7. Signalons enfin, sur la projection métrique de L^1 sur H^1, l'intéressant résultat de H. S. SHAPIRO: si f est analytique sur T, sa projection métrique est analytique sur T. D'autres résultats du même type peuvent être obtenus par nos méthodes.

* * *

Voici quelques remarques sur le problème d'existence.

1. Etant donné f et Λ, il existe une suite $h_n \in L^1_\Lambda$ telle que

$$\lim_{n \to \infty} \|f - h_n\| = d(f, L^1_\Lambda).$$

Quitte à restreindre la suite h_n, on peut supposer qu'elle converge faiblement vers une mesure. Cette mesure, μ, a son spectre dans Λ. De plus, en désignant par $M(T)$ l'espace des mesures complexes sur T, dont $L^1(T)$ est un sous-espace, et par K_N le noyau de Fejér d'ordre N, on a

$$\|f - \mu\|_{M(T)} \leqq \varliminf_{N \to \infty} \|(f - \mu) * K_N\| = \varliminf_{N \to \infty} \lim_{n \to \infty} \|(f - h_n) * K_N\| \leqq d(f, L_\Lambda).$$

Si $\mu \in L^1(T)$ (donc $\mu \in L^1_\Lambda$), c'est donc une projection métrique de f sur L^1_Λ.

Le problème d'existence a donc une solution positive dès que Λ a la propriété suivante: (R) toute mesure sur T à spectre dans Λ est absolument continue. Des exemples classiques de telles suites Λ sont: a) les suites lacunaires à la Hadamard (Sidon), b) $\Lambda = Z^+$ (F. et M. Riesz); à cause de cet important exemple, les Λ satisfaisant à la condition (R) ont été appelés "ensembles de Riesz". Pour d'autres exemples, voir [3] et [5].

2. La condition (R) n'est pas nécessaire pour l'existence. Par exemple, si $\Lambda' = \{0\}$, le prolongement minimal dans $A(Z)$ d'un élément de $A(\Lambda')$ est toujours possible, et en général d'une infinité de façons. De même si Λ' est un sous-groupe, ou un translaté de sous-groupe.

3. Remarquons encore que si Λ' est fini, l'existence a lieu si et seulement si Λ' est réduit à un point; dans le cas contraire en effet, il n'existe pas de prolongement minimal dans $A(Z)$ de la fonction $1|_{\Lambda'}$.

4. Remarquons enfin que, si Λ est un sous-groupe (soit $\Lambda = NZ$), le problème se ramène à définir, à partir de f sommable une fonction mesurable g telle que, pour presque tout t,

$$\sum_{n=1}^{N} \left| g(t) - f\left(t + \frac{2\pi n}{N}\right) \right| = \inf_{\alpha \in C} \sum_{n=1}^{N} \left| \alpha - f\left(t + \frac{2\pi n}{N}\right) \right|.$$

Le problème d'existence a donc toujours une solution positive — de même si Λ est un translaté de sous-groupe —. Nous verrons plus tard que l'unicité dépend de la parité de N.

* * *

Avant de résoudre les problèmes b), c), d), il est utile de caractériser l'ensemble Γ_Λ des $f \in L^1(T)$ qui se projettent en 0:

$$\Gamma_\Lambda = \{f \mid d(f, L_\Lambda^1) = \|f\|\}.$$

D'après un critère classique, simple application du théorème de Hahn—Banach (cf. p. ex. [7], p. 56), on a $f \in \Gamma_\Lambda$ si et seulement s'il existe $u \in L^\infty(T)$ tel que

$$(1) \qquad \|u\|_\infty \leq 1, \quad \int uf = \int |f|, \quad \forall l \in L_\Lambda^1 \ \int ul = 0.$$

Désignons par $\tilde{\Lambda}'$ le symétrique de Λ' par rapport à 0 ($\lambda \in \tilde{\Lambda}' \leftrightarrow -\lambda \in \Lambda'$) et par $L_{\tilde{\Lambda}'}^\infty$ le sous espace de $L^\infty(T)$ constitué par les éléments à spectre dans $\tilde{\Lambda}'$. On peut écrire (1) sous la forme

$$(2) \qquad u \in L_{\tilde{\Lambda}'}^\infty, \quad \|u\|_\infty \leq 1. \quad uf = |f| \quad \text{p. p.}.$$

Supposons — ce qu'on peut toujours réaliser par une translation Λ — que $0 \notin \Lambda$. Alors $0 \in \tilde{\Lambda}'$, et Γ_Λ contient toutes les fonctions positives sommables, ainsi que leurs produits par les $e^{i\lambda t}$ ($\lambda \in \tilde{\Lambda}'$).

Une étude plus poussée des Γ_Λ mène aux questions suivantes, que nous n'aborderons pas ici:

— *pour quels Λ est-il vrai que les seules fonctions $u \in L_{\tilde{\Lambda}'}^\infty$ telles que $|u|=1$ soient les $e^{i\lambda t}$ ($\lambda \in \tilde{\Lambda}'$),*

— *même question, en remplaçant "$|u|=1$" par "$\|u\| \leq 1$ et $|u|=1$ sur un ensemble de mesure >0".*

* * *

Dire que Λ ne satisfait pas à la condition d'unicité (problème b)), c'est dire qu'il existe $f \in \Gamma_\Lambda$ et $h \in L_\Lambda^1$, $h \neq 0$, telles que

$$\|f - h\| = \|f\| \ (= d(f, L_\Lambda^1)).$$

Soit $f \in \Gamma_\Lambda$, $h \in L_\Lambda^1$ et u satisfaisant à (2); alors $uf = |f|$ p. p. et $\int uh = 0$, donc

$$(3) \quad \int |f| = \int (|f| - \Re uh) \leq \int ||f| - \Re uh| \leq \int ||f| - uh| = \int |u| |f - h| \leq \int |f - h|.$$

Remarquons que la différence des deux dernières intégrales est $\int (1 - |u|)|h|$. Pour avoir $\|f\| = \|f - h\|$, il faut et il suffit que

$$(4) \qquad \begin{aligned} &\text{(i)} \quad |f| \geq \Re uh \quad \text{p. p.} \\ &\text{(ii)} \quad \Im uh = 0 \quad \text{p. p.} \\ &\text{(iii)} \quad (1 - |u|)h = 0 \quad \text{p. p..} \end{aligned}$$

Une condition nécessaire et suffisante pour l'unicité est donc que (2) et (4) (où $f \in L^1(T)$ et $h \in L^1_\Lambda$) entrainent $h=0$ p. p..

Supposons $\Lambda = \mathbf{Z}^{++} = \{1, 2, 3, \ldots\}$ (soit $L^1_\Lambda = H^1_0$, sous-espace de H^1 constitué par les fonctions de valeur moyenne nulle); alors $L^\infty_{\Lambda'}$ est la classe de Hardy H^∞. Comme $uh \in H^1_0$, les conditions (2) et (4 ii) entrainent $u=0$ ou $h=0$; si $u=0$, (4 iii) donne $h=0$; ainsi (2) et (4) entrainent $h=0$, donc Λ satisfait à la condition d'unicité (théorème de Doob).

Il en est de même si Λ est n'importe quelle demi-droite d'entiers ($\{a, a+1, a+2, \ldots\}$ ou $\{a, a-1, a-2, \ldots\}$).

Si au contraire il existe un $\lambda' \in \Lambda'$ et un $n \neq 0$ tels que $\lambda'-n$ et $\lambda'+n$ appartiennent à Λ, on peut choisir $u=e^{-i\lambda't}$, $uh=\cos nt$, $uf=|f| \geqq uh$, donc Λ ne satisfait pas à la condition d'unicité. Une condition nécessaire pour l'unicité est donc

$$\forall \lambda_1 \in \Lambda \quad \forall \lambda_2 \in \Lambda \quad \frac{\lambda_1+\lambda_2}{2} \notin \Lambda'.$$

Cela signifie que Λ est une progression arithmétique de raison impaire ϱ.

Supposons maintenant Λ finie et (quitte toujours à translater Λ) $0 \in \Lambda$. Si $\Lambda \subset]-N, N[$, on peut choisir $u=f=\operatorname{sign} \sin Nt$, et $h=1$. De nouveau, il n'y a pas unicité. Pour l'unicité, il est donc nécessaire que Λ soit une progression arithmétique infinie de raison impaire.

Pour voir si c'est suffisant, il suffit d'examiner les deux cas $\Lambda = (2p+1)\mathbf{Z}$ et $\Lambda = (2p+1)\mathbf{Z}^{++}$ (p entier >0). Dans ces deux cas, nous poserons, pour toute function $l \in L^1$,

$$l_j(t) = l\left(t+\frac{2j\pi}{2p+1}\right) \qquad (j=0, 1, \ldots, 2p).$$

Supposons $\Lambda = (2p+1)\mathbf{Z}$. Alors

$$l \in L^1_\Lambda \Leftrightarrow l \in L^1 \quad \text{et} \quad l_1 = \cdots = l_{2p} \quad \text{p. p.}$$

$$u \in L^\infty_{\Lambda'} \Leftrightarrow u \in L^\infty \quad \text{et} \quad \sum_{j=0}^{2p} u_j = 0 \quad \text{p. p..}$$

Soit $h \in L^1_\Lambda$, vérifiant (4 ii) et (4 iii), u vérifiant (2). Supposons $h \neq 0$ sur un ensemble E de mesure >0. D'après (4 ii), on a

(5) $$\arg u_0(t) = \arg u_1(t) = \cdots = \arg u_{2p}(t) = -\arg h(t) \quad (\operatorname{mod} \pi)$$

p. p. sur E (on convient que arg 0 est indéterminé). Supposons (5); comme la somme $\sum_{j=0}^{2p} u_j(t)$ qui est nulle, contient un nombre *impair* de termes de même argument (mod π) et de module $\leqq 1$, l'un des $u_j(t)$ au moins est de module strictement inférieur à 1. Quitte à remplacer E par un sous-ensemble de mesure >0 et à translater

u, on peut supposer $|u(t)| < 1$ sur E. Cela contredit (4 iii). Donc $h=0$ p. p., et on a donc unicité.

Supposons enfin $\Lambda = (2p+1)\mathbf{Z}^{++}$. Posons

$$(6) \qquad\qquad w = \frac{1}{2p+1} \sum_0^{2p} u_j.$$

Supposons toujours (2) et (4). Alors $h \in H_0^1$ et $w \in H^\infty$. Si l'on n'a pas $h=0$ p. p., on a $h \neq 0$ p. p. Faisons cette hypothèse. Il s'ensuit, par (4 iii), $|u|=1$ p. p., donc comme plus haut (5) a lieu p. p. La somme qui définit w contient un nombre impair de termes, tous de module 1 et de même argument (mod π), donc $w \neq 0$ p. p. Or (utilisant de nouveau 4 ii) $\mathfrak{Im}\, wh=0$; c'est impossible puisque $wh \in H_0^1$ et $wh \neq 0$ p. p. La contradiction établit que (2) et (4) entrainent $h=0$ p. p., d'où l'unicité.

En résumé, *les Λ qui répondent à la condition d'unicité b) sont les progressions arithmétiques infinies de raison impaire.*

* * *

On va donc considérer le problème de la continuité c) lorsque Λ est une progression arithmétique infinie de raison impaire. On se ramène immédiatement aux cas suivants:

$$\Lambda = \mathbf{Z}^{++}, \quad \Lambda = (2p+1)\mathbf{Z} \quad \text{et} \quad \Lambda = (2p+1)\mathbf{Z}^{++} \qquad (p \text{ entier } > 0).$$

Soit $f \in \Gamma_\Lambda$, $g \in L^1$ (proche de f), et $h = P_\Lambda g$. On a

$$\|f-h\| \leq \|f-g\| + \|g-h\| \leq \|f-g\| + \|g\| \leq 2\|f-g\| + \|f\|.$$

Il s'agit de montrer que $\|h\|$ est petit. On est donc amené au problème suivant.

Etant donné $f \in \Gamma_\Lambda$, existe-t-il une fonction $\omega_f(\delta) \searrow 0$ ($\delta \searrow 0$) telle que, si $h \in L_\Lambda^1$ et $\|f-h\| \leq \|f\| + \delta$, on ait $\|h\| \leq \omega_f(\delta)$?

On va s'appuyer sur l'inégalité

$$\|\varphi + i\psi\|^2 \geq \|\varphi\|^2 + \|\psi\|^2$$

(φ réelle, ψ réelle) dont la démonstration est laissée au lecteur; on l'écrit

$$(7) \qquad\qquad \|\psi\| \leq \big((\|\varphi + i\psi\| - \|\varphi\|)(\|\varphi + i\psi\| + \|\varphi\|)\big)^{\frac{1}{2}}.$$

L'hypothèse $\|f-h\| \leq \|f\| + \delta$ entraine que deux intégrales consécutives dans (3) sont distantes de moins de δ, donc (en utilisant (7) pour la seconde inégalité et en remarquant pour la troisième que $(1-|u|)f = 0$)

$$(8) \qquad
\begin{aligned}
&\text{(i)} \quad 2 \int (|f| - \mathfrak{Re}\, uh)^- \leq \delta \qquad (x^- = \sup(0, -x)) \\[4pt]
&\text{(ii)} \quad \int |\mathfrak{Im}\, uh| \leq \sqrt{\delta(2\|f\| + \delta)} = \alpha \\[4pt]
&\text{(iii)} \quad \int (1 - |u|)|h| \leq \delta.
\end{aligned}$$

1er cas: $\Lambda = \mathbf{Z}^{++}$. Alors $uh \in H_0^1$. Soit $\beta > 0$, et $F = \{t \mid |\Re\, uh| > \beta\}$. La transformation de Hilbert, qui permet de passer de $\Im\, uh$ à $\Re\, uh$, étant de type L^1 faible (cf. p. ex. [2] p. 66), on a d'après (8 ii) mes $F \leq \alpha/\beta$ (on a normalisé la mesure sur T). On a

$$\int_{CF} |\Re\, uh| \leq \beta$$

et, d'après (8 i),

$$\int_F (\Re\, uh)^+ \leq \int_F |f| + \delta/2.$$

Comme $\int \Re\, uh = 0$, on a

$$\left| \int_F (\Re\, uh)^+ - (\Re\, uh)^- \right| = \left| \int_F \Re\, uh \right| = \left| \int_{CF} \Re\, uh \right| \leq \beta$$

donc

$$\int_F (\Re\, uh)^- \leq \int_F (\Re\, uh)^+ + \beta$$

et

$$\int_F |\Re\, uh| \leq 2 \int_F |f| + \delta + \beta.$$

Posons

$$\overline{\omega}_f(x) = \sup_{\text{mes } X = x} \int_X |f|.$$

On obtient

$$\int |uh| \leq 2\overline{\omega}_f(\alpha/\beta) + \beta + \delta + \alpha.$$

Soit maintenant $G = \{t \mid |u(t)| \leq \frac{1}{2}\}$; on a $f = 0$ sur G, donc, d'après (8 iii),
$\int_G |h| \leq 2\delta$. D'autre part $\int_{CG} |h| \leq 2 \int |uh|$. On obtient $\|h\| \leq \omega_f^*(\delta)$, avec

$$(9) \qquad \omega_f^*(\delta) = 2\big(\sup_{\beta > 0} (2\overline{\omega}_f(\alpha/\beta) + \beta) + 2\delta + \alpha\big)$$

α étant défini par (8 ii).

2ème cas: $\Lambda = (2p+1)\mathbf{Z}$ (p entier > 0). Par l'absurde, on établit facilement l'existence d'un $c > 0$, ne dépendant que de p, tel qu'on ait l'implication

$$(10) \qquad \left. \begin{cases} 1 - c \leq |z_j| \leq 1 \quad (j = 0, 1, \ldots, 2p) \\ \left| \sum_0^{2p} z_j \right| \leq c \end{cases} \right\} \Rightarrow \sum_0^{2p} |\Im\, z_j| \geq c.$$

Soit

$$(11) \qquad E = \{t \mid 1 - c \leq |u_j(t)| \leq 1 \quad (j = 0, 1, \ldots, 2p)\}.$$

Comme $h_0 = h_1 = \cdots = h_{2p} = h$, on a d'après (10)

$$\sum_0^{2p} |\Im\, h_j u_j| \geq c|h| \quad \text{sur} \quad E$$

donc, en tenant compte de (8 ii),

$$\int_E |h| \leq (2p+1)\alpha/c.$$

Sur CE, on a par (8 iii)

$$(12) \qquad \int_{CE} |h| \leq \delta/c.$$

En définitive, on obtient $\|h\| \leq \omega_f^{**}(\delta)$ avec

$$(13) \qquad \omega_f^{**}(\delta) = (2p+1)\alpha/c + \delta/c.$$

Remarquons que la fonction ω_f^{**} ne dépend que de p et de $\|f\|$.

3ème cas: $\Lambda = (2p+1)\mathbf{Z}^{++}$ (p entier >0). On combine les deux raisonnements précédents. Soit w la moyenne des u_j, comme en (6), et soit

$$g = \frac{1}{2p+1} \sum_0^{2p} |f_j|.$$

De (8 i) résulte, compte tenu de $h_0 = h_1 = \cdots = h_{2p} = h$,

$$\int (|f_j| - \mathfrak{Re}\, u_j h)^- \leq \delta$$

d'où, par la convexité de la fonction $x \to x^-$,

$$\int (g - \mathfrak{Re}\, wh)^- \leq \delta.$$

De même, il résulte de (8 ii) que

$$\int |\mathfrak{Im}\, wh| \leq \alpha.$$

D'après l'étude du premier cas, où l'on remplace f par g, u par 1, et h par wh, on a

$$\int |wh| \leq \omega_g^*(\delta)$$

où $\omega_g^*(\delta)$ est défini par (9).

Soit $G = \{t \mid |w(t)| < \gamma\}$, où $0 < \gamma < c/(2p+1)$ (c est le nombre qui figure en (10)). Alors $\int_{CG} |h| \leq \dfrac{1}{\gamma} \omega_g^*(\delta)$, et il ne reste plus qu'à majorer l'intégrale de $|h|$ sur G.

Pour cela, considérons de nouveau l'ensemble E de (11). Sur $E \cap G$, (10) est satisfaite avec $z_j = h_j(t)u_j(t)$, donc $\int_{E \cap G} |h| \leq (2p+1)\alpha/c$. Sur CE, on a (12). Finalement $\|h\| = \omega_f^{***}(\delta)$ avec

$$(14) \qquad \omega_f^{***}(\delta) = \frac{1}{\gamma} \omega_g^*(\delta) + \omega_f^*(\delta).$$

Les formules (9), (13), (14) répondent au problème posé. Ainsi *quand la condition d'unicité est satisfaite, l'application P_Λ est toujours continue.*

Considérons enfin le problème d) (continuité uniforme sur les bornés de L^1), dans chacun des trois cas qui se présentent.

1er cas: $\Lambda = \mathbf{Z}^{++}$. Pour tout $\delta > 0$, il existe une fonction $h \in H_0^1$ telle que $\|h_1\| = 2$ et $\|h_2\| = \delta$. Choisissons $f = h_1^+ = \sup(h_1, 0)$ et $g = f + ih_2$. Alors $P_\Lambda f = 0$ et $P_\Lambda g = P_\Lambda(h + (f - h_1)) = h$. On a donc $\|f\| = 1$, $\|g - f\| < \delta$, $\|P_\Lambda g - P_\Lambda f\| > 2 - \delta$. *L'application P_Λ n'est donc pas uniformément continue sur les bornés de L^1.* C'est la réponse négative au problème de G. Henkin signalé en introduction.

2ème cas: $\Lambda = (2p+1)\mathbf{Z}$ (p entier >0). Il résulte de la remarque faite après (13) que P_Λ *est uniformément continue sur les bornés de L^1.*

3ème cas: $\Lambda = (2p+1)\mathbf{Z}^{++}$ (p entier >0). En remplaçant les fonctions $f(t)$, $g(t)$, $h(t)$ introduites dans le 1er cas par $f((2p+1)t)$, $g((2p+1)t)$, $h((2p+1)t)$ on voit que l'application P_Λ *n'est pas uniformément continue sur les bornés de L^1.*

Cela achève l'étude des problèmes b), c), d).

RÉFÉRENCES

[1] J. L. Doob, *A minimum problem in the theory of analytic functions.* Duke Math. J. **8** (1941), 413—424.

[2] Y. Katznelson, *An introduction to harmonic analysis.* Wiley, New York 1968.

[3] Y. Meyer, *Spectre des mesures et mesures absolument continues.* Studia Math. **30** (1968), 87—99.

[4] D. J. Newman, *Pseudo uniform convexity in H^1.* Proc. Amer. Math. Soc. **14** (1963), 676—679.

[5] W. Rudin, *Trigonometric series with gaps.* J. Math. Mech. **9** (1960), 203—227.

[6] A. Schönhage, *Approximationstheorie.* De Gruyter, Berlin 1971.

[7] H. S. Shapiro, *Topics in approximation theory.* Lecture Notes **187**, Springer, Berlin 1971.

[8] I. Singer, *Best approximation in normed linear spaces by elements of linear subspaces.* Grundlehren **171**, Springer, Berlin 1970.

Bases dans H et bases de sous espaces de dimension finie dans A

Par

P. BILLARD

UNIVERSITÉ DE PROVENCE
MARSEILLE

Introduction: Nous développons ici une méthode de construction d'une base de Schauder dans H^1 et d'une base de Schauder de sous espaces de dimension finie dans A (cf. [2]) car [1] contient une erreur à la page 292 (cf. [4] pour l'historique des problèmes et [3] pour les définitions et propriétés générales).

CHAPITRE 1

Sur les bases de Schauder de sous espaces de dimension finie

THÉORÈME 1-1. *Soit X un espace de Banach de dimension infinie et soit $(P_n)_{n=1,2,\dots}$ une suite de projections continues de X sur des sous espaces de dimension finie telle que*

$$1\text{-}(1) \qquad\qquad x \in X \Rightarrow P_n(x) \xrightarrow[n \to \infty]{} x$$

$$1\text{-}(2) \qquad\qquad n < n' \Rightarrow P_n P_{n'} = P_n,$$

alors X possède une base de Schauder de sous espaces de dimension finie.

La démonstration simple suivante du théorème 1-1 nous a été communiquée par A. PELCZYNSKI (Cf. aussi [6], th 1.3).

LEMME 1-1. *Soit X un espace de Banach de dimension infinie et soit $(T_k)_{k=1,2,\dots}$ une suite de projections continues de X sur des sous espaces de dimension finie telle que*

$$1\text{-}(3) \qquad\qquad x \in X \Rightarrow T_k(x) \xrightarrow[k \to \infty]{} x$$

$$1\text{-}(4) \qquad\qquad \forall k \ \forall k', \quad T_k T_{k'} = T_{\inf(k,k')},$$

alors X possède une base de Schauder de sous espaces de dimension finie.

DÉMONSTRATION: En vertu de $1-(4)$ nous avons $k<k' \Rightarrow T_k(X) \subseteq T_{k'}(X)$ et quitte à remplacer $(T_k)_{k=1,2,\dots}$ par l'une de ses sous-suites, nous pouvons supposer que nous avons

$$1\text{-}(5) \qquad\qquad T_1(X) \neq \{0\} \quad \text{et} \quad k<k' \Rightarrow T_k(X) \neq T_{k'}(X).$$

Posons $T_0 = 0$ et $H_k = (T_k - T_{k-1})(X)$ $(k = 1, 2, \ldots)$ de sorte qu'avec 1-(3) nous avons, pour $x \in X$, la représentation

$$1\text{-}(6) \qquad\qquad x = \sum_{k=1}^{\infty} x_k \qquad (x_k \in H_k, \; k = 1, 2, \ldots)$$

en posant $x_k = (T_k - T_{k-1})(x)$ $(k = 1, 2, \ldots)$; mais inversement si pour $x \in X$ nous avons 1-(6), alors de 1-(4) nous déduisons $x_k = (T_k - T_{k-1})(x)$ $(k = 1, 2, \ldots)$ ce qui, joint à 1-(5), achève la démonstration du lemme 1-1.

Sous les hypothèses du théorème 1-1 nous pouvons former une sous suite $(P_{n_k})_{k=1, 2, \ldots}$ de $(P_n)_{n=1, 2, \ldots}$ telle que

$$\| (P_{n_{k'+1}} P_{n_{k'}} \cdots P_{n_k}) - (P_{n_{k'}} \cdots P_{n_k}) \| \leqq 2^{-k'} \qquad (k \leqq k', \; k = 1, 2, \ldots)$$

ce qui nous permet de poser, pour la topologie de la norme,

$$T_k = \lim_{k' \to \infty} P_{n_{k'}} P_{n_{k'-1}} \cdots P_{n_k} \qquad (k = 1, 2, \ldots);$$

cette suite $(T_k)_{k=1, 2, \ldots}$ vérifie visiblement les hypothèses du lemme 1-1 ce qui achève la démonstration du théorème 1-1.

CHAPITRE 2

Sur l'existence d'une base de Schauder dans H^1

(α1) *Notations et rappels de résultats classiques.* Lorsque nous parlerons du système de Haar, h_0 et $(h_{j,v})$ $(1 \leqq j \leqq 2^v, \; v = 0, 1, 2, \ldots)$ sur le tore T, celui-ci sera souvent assimilé, en dédoublant chaque point de $D = \{2\pi p/2^q \in T, \; p \text{ et } q \text{ entiers} \geqq 0 \text{ et } 0 < 2\pi p/2^q < 2\pi\}$, au groupe $[0, 2\pi]^*$ produit dénombrable de groupes discrets à deux éléments, la mesure de Haar de $[0, 2\pi]^*$ étant normalisée de façon à donner la masse 2π à $[0, 2\pi]^*$ et le systéme $(h_{j,v})$ étant normalisé dans $L^2([0, 2\pi]^*)$; d'ailleurs pour $1 \leqq p \leqq \infty$, les espaces de Banach réels $L^p(T)$ et $L^p([0, 2\pi]^*)$ qui s'identifient canoniquement seront notés L^p et la norme correspondante $\| \; \|_p$. Nous poserons $E = \{x \in T, \; x \notin D \cup \{0\}\}$ et E pourra être assimilé à un sous ensemble de $[0, 2\pi]^*$.

Si v est entier $\geqq 0$, $(\omega_{j,v})_{1 \leqq j \leqq 2^v}$ sera la partition usuelle de $[0, 2\pi]^*$ en 2^v intervalles égaux successifs; si $f \in L^1$, les sommes particulières

$$s_{2^v}(f) = a_0(f) h_0 + \sum_{\substack{0 \leqq v' < v \\ 1 \leqq j' \leqq 2^{v'}}} a_{j'v'}(f) h_{j',v'} \qquad (v = 0, 1, 2, \ldots)$$

de la série de Fourier Haar de f vérifient

$$2\text{-}(1) \qquad s_{2^\nu}(f)(x) = \frac{1}{2\pi} \int\limits_{[0,\,2\pi]^*} f(y)\,W_{2^\nu}(x \overset{*}{-} y)\,dy \qquad (\nu = 0, 1, 2, \ldots)$$

et $\qquad s_{2^\nu}(f) \xrightarrow[\nu \to \infty]{} f$ (dans L^1)

où $x \in [0, 2\pi]^*$, où le signe $*$ rappelle qu'il s'agit de la loi additive de $[0, 2\pi]^*$ et où $W_{2^\nu}(t) = 2^\nu x_{\omega_{1,\nu}}(t)$ ($x_{\omega_{1,\nu}}$ désignant la fonction caractéristique de $\omega_{1,\nu}$).

Considérons le sous ensemble H_r^1 de H^1 (resp. le sous ensemble A_r de A) constitué des $F \in H^1$ (resp. des $F \in A$) dont le développement en série entière $F(z) = \sum\limits_{n=0}^{\infty} c_n z^n$ a son premier coefficient c_0 réel; lorsque H^1 (resp. A) est considéré comme un espace de Banach réel, H_r^1 (resp. A_r) est un sous-espace fermé de H^1 (resp. de A).

D'après [5], chap 7, n° 4, th (4-4), H_r^1 (resp. A_r) est l'espace vectoriel réel des couples $(f, \tilde{f})$ où $f \in L^1$ est telle que la conjuguée $\tilde{f}$ de f (cf. [5], chap 4, n° 3) vérifie $\tilde{f} \in L^1$ (resp. des couples $(f, \tilde{f})$ où f est un élément de l'espace $C = C(T)$ des fonctions réelles continues sur T tel que $\tilde{f} \in C(T)$) et la norme

$$2\text{-}(2) \qquad \|(f, \tilde{f})\| = \sup(\|f\|_1, \|\tilde{f}\|_1) \quad (\text{resp. } \|(f, \tilde{f})\| = \sup(\|f\|_\infty \|\tilde{f}\|_\infty))$$

sur H_r^1 (resp. sur A_r) est équivalente à la norme usuelle. Dans toute la suite de ce chapitre, $M_1, M_1', \ldots; M_2, M_2', \ldots; \ldots$ désigneront des constantes absolues > 0.

(α2) *Une décomposition de* $\overset{\frown}{s_{2^\nu}(f)}$. Dans tout ce paragraphe ainsi que dans les deux suivants, nous fixons arbitrairement $f \in L^1$ telle que $\tilde{f} \in L^1$ (resp. $f \in C$ telle que $\tilde{f} \in C$) ν est entier, $\nu \geqq 3$. Un calcul facile donne

$$2\text{-}(3) \qquad \overset{\frown}{s_{2^\nu}(f)}(x) = \sigma_\nu(f)(x) + \tau_\nu(f)(x)$$

avec

$2\text{-}(4)$

$$\begin{cases} \sigma_\nu(f)(x) = \\ \quad = \dfrac{1}{2\pi} \int\limits_{[0,\,2\pi]^*} f(y) \left(-\dfrac{1}{\pi} \int\limits_0^\pi \left(W_{2^\nu}((x+t) \overset{*}{-} y) - W_{2^\nu}((x-t) \overset{*}{-} y) \right) \left(\dfrac{1}{2\,\mathrm{tg}\,(t/2)} - \dfrac{1}{t} \right) dt \right) dy \\ \tau_\nu(f)(x) = \\ \quad = \dfrac{1}{2\pi} \int\limits_{[0,\,2\pi]^*} f(y) \left(-\dfrac{1}{\pi} \int\limits_0^\pi \dfrac{W_{2^\nu}((x+t) \overset{*}{-} y) - W_{2^\nu}((x-t) \overset{*}{-} y)}{t}\,dt \right) dy \end{cases}$$

et avec

$$2\text{-}(5) \qquad \|\sigma_\nu(f)\|_1 \leqq \frac{M_1}{2\pi} \|f\|_1 \quad (\text{resp. } \|\sigma_\nu(f)\|_\infty \leqq M_1 \|f\|_\infty).$$

Pour $x \in E$ et pour $y \in [0, 2\pi]^*$ posons

$$2\text{-}(6) \qquad I_v(x, y) = -\frac{1}{\pi} \int_0^\pi \frac{W_{2^v}((x+t) \overset{*}{\_} y) - W_{2^v}((x-t) \overset{*}{\_} y)}{t}\, dt$$

et pour $\omega_{j,v}$ pris sous la condition

$$2\text{-}(7) \qquad\qquad\qquad\qquad x + \pi \notin \omega_{j,v}$$

soit $d = d_{j,v}(x)$ la distance (au sens de T) de x à l'extrémité de $\omega_{j,v}$ la plus proche de x. Nous distinguerons les quatre cas suivants:

$$2\text{-}(8) \qquad \begin{cases} \text{cas 1} & x \notin \omega_{j,v}, \text{ et il existe } t, 0 < t < \pi \text{ et } x+t \in \omega_{j,v} \\ \text{cas 2} & x \notin \omega_{j,v}, \text{ et il existe } t, -\pi < t < 0 \text{ et } x+t \in \omega_{j,v} \\ \text{cas 3} & x \in \omega_{j,v} \text{ et } x-d \text{ est extrémité de } \omega_{j,v} \\ \text{cas 4} & x \in \omega_{j,v} \text{ et } x+d \text{ est extrémité de } \omega_{j,v}. \end{cases}$$

Si dans les quatre cas précédents ε_1 prend successivement les valeurs $-1, 1, -1, 1$ et ε_2 les valeurs $1, 1, -1, -1$ nous avons

$$2\text{-}(9) \qquad x \in \omega_{j,v} \Rightarrow I_v(x, y) = \varepsilon_1 \frac{2^v}{\pi} \log\left(\varepsilon_2 + \frac{2\pi\, 2^{-v}}{d}\right).$$

La notion d'intervalle étant prise au sens de T, posons

$$\Delta_v(x) = \text{intervalle de centre } x \text{ et de longueur } 4(2\pi 2^{-v})$$

et écrivons

$$2\text{-}(10) \quad \tau_v(f)(x) = \left[\frac{1}{2\pi} \int_{[0, 2\pi]^* - \Delta_v(x)} f(y) I_v(x, y)\, dy\right] + \left[\frac{1}{2\pi} \int_{\Delta_v(x)} f(y) I_v(x, y)\, dy\right] =$$

$$= \tau_{v,1}(f)(x) + \tau_{v,2}(f)(x).$$

(α3) *Etude de* $\tau_{v,2}(f)$. Pour chaque $\omega_{j,v}$ et chaque $x \in \omega_{j,v} \cap E$, les intervalles $\omega_{j'',v}$ pour lesquels $\omega_{j'',v} \cap \Delta_v(x)$ a des points intérieurs sont parmi les cinq intervalles $\omega_{j_k,v}$ $(k=1, \ldots, 5)$ successif (au sens de T) tels que $\omega_{j,v} = \omega_{j_3,v}$. Puisque $v \geqq 3$, les $\omega_{j_k,v}$ $(k=1, \ldots, 5)$ satisfont $2\text{-}(7)$ pour $x \in \omega_{j,v} \cap E$. Ecrivons $\omega_{j,v} = \omega_{j',v+1} \cup \cup \omega_{j'+1,v+1}$ et, pour $x \in \omega_{j',v+1} \cap E$, en déterminant le cas de $2\text{-}(8)$ dans lequel nous sommes lorsque nous remplaçons, dans $2\text{-}(8)$, $\omega_{j,v}$ par $\omega_{j_k,v}$ $(k=1, \ldots, 5)$ nous obtenons

$$|\tau_{v,2}(f)|(x) \leqq \sum_{1 \leqq k \leqq 5,\, k \neq 3} \left(\frac{1}{2\pi} \int_{\omega_{j_k,v}} |f(y)|\, dy\right) \left(\frac{2^v}{\pi} \left|\log\left(1 + \frac{2\pi 2^{-v}}{d_{j_k,v}(x)}\right)\right|\right) +$$

$$+ \left(\frac{1}{2\pi} \int_{\omega_{j_3,v}} |f(y)|\, dy\right) \left(\frac{2^v}{\pi} \left|\log\left(\frac{2\pi 2^{-v}}{d_{j_3,v}(x)} - 1\right)\right|\right)$$

·d'où, avec une définition des K_k $(k=1, \ldots, 5)$ obtenue en intégrant sur $\omega_{j',\nu+1}$,

$$\int\limits_{\omega_{j',\nu+1}} |\tau_{\nu,2}(f)(x)|\, dx \leq \sum_{k=1}^{5} \left(\frac{1}{2\pi} \int\limits_{\omega_{j_k,\nu}} |f(y)|\, dy \right) K_k.$$

Or les K_k $(k=1, \ldots, 5)$ sont majorés par une constante absolue comme nous le voyons avec des changements de variables immédiats dans les intégrales qui les définissent et, puisque nous avons un résultat analogue sur $\omega_{j'+1,\nu+1}$, résumons ce paragraphe par

2-(11)
$$\begin{cases}
\|\tau_{\nu,2}(f)\|_1 \leq \dfrac{5M_2}{2\pi}\, \|f\|_1 \quad (\text{resp. } \omega_{j,\nu} = \omega_{j',\nu+1} \cup \omega_{j'+1,\nu+1} \text{ et} \\[2mm]
x \in \omega_{j',\nu+1} \cap E \Rightarrow |\tau_{\nu,2}(f)(x)| \leq \dfrac{\|f\|_\infty}{\pi} \sum_{1 \leq k \leq 5,\, k \neq 3} \left| \log\left(1 + \dfrac{2\pi\, 2^{-\nu}}{d_{j_k,\nu}(x)}\right) \right| + \\[2mm]
+ \dfrac{\|f\|_\infty}{\pi} \left| \log\left(\dfrac{2\pi\, 2^{-\nu}}{d_{j_3,\nu}(x)} - 1 \right) \right| \text{ et avec une inégalité analogue} \\[2mm]
\text{pour } x \in \omega_{j'+1,\nu+1} \cap E).
\end{cases}$$

(α4) *Etude de* $\tau_{\nu,1}(f)$. Pour $x \in E$ nous définissons l'entier j_x par $x + \pi \in \omega_{j_x,\nu}$, le nombre réel ξ_x par $0 < \xi_x < \pi$ et $x + \xi_x$ est extrémité de $\omega_{j_x,\nu}$, le nombre réel y_x par $-\pi < y_x < 0$ et $x + y_x$ est extrémité de $\omega_{j_x,\nu}$, l'ensemble $J'_\nu(x)$ comme l'ensemble des indices j pour lesquels

$$\delta'_{j,\nu} = \{x + t,\ 2(2\pi\, 2^{-\nu}) < t < \xi_x \text{ et } x+t \text{ est intérieur à } \omega_{j,\nu}\} \neq \emptyset$$

et l'ensemble $J''_\nu(x)$ comme l'ensemble des indices j pour lesquels

$$\delta''_{j,\nu} = \{x + t,\ -2(2\pi\, 2^{-\nu}) > t > y_x \text{ et } x+t \text{ est intérieur à } \omega_{j,\nu}\} \neq \emptyset.$$

Convenons de prendre la détermination de $x-y$ qui vérifie $|x-y| < \pi$, un calcul facile donne

2-(12)
$$y \in \delta'_{j,\nu} \cup \delta''_{j,\nu} \Rightarrow \left| I_\nu(x,y) - \left(\frac{2}{x-y} \right) \right| \leq M_3 \frac{2^{-\nu}}{|x-y|^2}.$$

Posons

2-(13)
$$f_{1,\nu}(x) = \frac{1}{2\pi} \int\limits_{T - \Delta_\nu(x) - \omega_{j_x,\nu}} f(y)\, I_\nu(x,y)\, dy,$$

$$f_{2,\nu}(x) = \frac{1}{2\pi} \int\limits_{T - \Delta_\nu(x) - \omega_{j_x,\nu}} f(y)\, \frac{2}{x-y}\, dy$$

de sorte qu'avec 2-(12) nous obtenons

2-(14) $$|f_{1,v}(x) - f_{2,v}(x)| \leqq M_4 \int_T |f(y)|\, \varphi(x, y)\, dy$$

où

2-(15) $\quad \varphi(x, y) = \dfrac{1}{|x-y|^2}$ ou 0 selon que $|x-y| \geqq 2(2\pi\, 2^{-v})$ ou non.

De 2-(14) et 2-(15) nous déduisons

2-(16) $\quad \|f_{1,v} - f_{2,v}\|_1 \leqq M_5 \|f\|_1 \quad$ (resp. $|(f_{1,v} - f_{2,v})(x)| \leqq M_5 \|f\|_\infty$).

Posons

2-(17) $$f(v, x) = -\frac{1}{\pi} \int_{2(2\pi\, 2^{-v})}^{\pi} \frac{f(x+t) - f(x-t)}{2\,\mathrm{tg}\,(t/2)}\, dt\,;$$

remarquons que

$$\left| \frac{1}{2\,\mathrm{tg}\,(t/2)} - \frac{1}{t} \right| \leqq M_5'\,,$$

$$\frac{1}{2\pi} \int_{T - \Delta_v(x)} f(y)\, \frac{2}{x-y}\, dy = -\frac{1}{\pi} \int_{2(2\pi\, 2^{-v})}^{\pi} \frac{f(x+t) - f(x-t)}{t}\, dt$$

et que

$$\int_{\omega_{j_x},\,v} |f(y)| \left(|I_v(x, y)| + \frac{2}{|x-y|} \right) dy \leqq M_6 \int_{\omega_{j_x},\,v} |f(y)|\, dy$$

de sorte qu'avec 2-(16) nous obtenons finalement

2-(18) $\|\tau_{v,1}(f) - f(v, \cdot)\|_1 \leqq M_7 \|f\|_1 \quad$ (resp. $|\tau_{v,1}(f)(x) - f(v, x)| \leqq M_7 \|f\|_\infty$).

(α5) *Convergence de* $\left(s_{2^v}(f), \widetilde{s_{2^v}(f)}\right)$ *vers* $(f, \tilde{f})$ *dans* H_r^1. Si $(f, \tilde{f}) \in H_r^1$, 2-(3), 2-(5), 2-(10), 2-(11), 2-(18) puis 2-(1) et $\|\tilde{f} - f(v, \cdot)\|_1 \xrightarrow[v \to \infty]{} 0$ (cf. [5], chap 4, exer 5) montrent que la suite $\left(s_{2^v}(f), \widetilde{s_{2^v}(f)}\right)_{v=3,4,\ldots}$ est bornée dans H_r^1; si l'entier $v \geqq 3$ est fixé, $(f, \tilde{f}) \in H_r^1 \to \left(s_{2^v}(f), \widetilde{s_{2^v}(f)}\right) \in H_r^1$ est visiblement une application linéaire continue de H_r^1 dans H_r^1; remarquons que $\left(s_{2^v}(f), \widetilde{s_{2^v}(f)}\right) \xrightarrow[v \to \infty]{} (f, \tilde{f})$ dans H_r^1 si $f \in L^2$ donc si f est un polynôme trigonométrique, or le sous espace de H_r^1 constitué des $(f, \tilde{f})$ où f est un polynôme trigonométrique est dense dans H_r^1 puisque pour un élément quelconque $(f, \tilde{f})$ de H_r^1 nous avons $\widetilde{F_{e_n}(f)} = F_{e_n}(\tilde{f})$ si $F_{e_n}(f)$ est la moyenne de Fejér de rang n de f (cf [5], chap 7, n° 4, ch 4-4) donc nous avons $\left(F_{e_n}(f), \widetilde{F_{e_n}(f)}\right) =$

$=(F_{e_n}(f), F_{e_n}(\tilde{f})) \xrightarrow[n\to\infty]{} (f,\tilde{f})$ dans H_r^1 (cf. [5], chap 4, n° 5, th 5-5). Finalement le théorème de Banach Steinhaus nous donne

$$2\text{-}(19) \qquad (f,\tilde{f}) \in H_r^1 \Rightarrow (s_{2^\nu}(f), \widetilde{s_{2^\nu}(f)}) \xrightarrow[\nu\to\infty]{} (f,\tilde{f}) \quad (\text{dans } H_r^1).$$

(α6) *Etude de* $\widetilde{s_{2^\nu,n}(f)}$ $\left(s_{2^\nu,n}(f) = \sum_{j=1}^{n-2^\nu} a_{j,\nu}(f) h_{j,\nu}, \; 2^\nu < n < 2^{\nu+1}, \; (f,\tilde{f}) \in H_r^1 \right)$.
Dans tout ce paragraphe nous fixons arbitrairement $(f,\tilde{f}) \in H_r^1$. La disjonction des supports des $h_{j,\nu}$ $(1 \leqq j \leqq 2^\nu)$ donne immédiatement

$$2\text{-}(20) \qquad \sup_{n,\,2^\nu < n < 2^{\nu+1}} \|s_{2^\nu,n}(f)\|_1 \leqq \|s_{2^{\nu+1}}(f) - s_{2^\nu}(f)\|_1 \xrightarrow[\nu\to\infty]{} 0.$$

Si $I_{j,\nu}$ est l'intervalle (au sens de T) qui a la même centre que $\omega_{j,\nu}$ et dont la longueur est $3(2\pi 2^{-\nu})$ un calcul facile donne

$$2\text{-}(21) \qquad \int_{I_{j,\nu}} |\hat{h}_{j,\nu}(x)|\, dx \leqq M_8 2^{-\nu/2} \quad \text{où} \quad \hat{h}_{j,\nu}(x) = -\frac{1}{\pi} \int_{-\pi}^{\pi} \frac{h_{j,\nu}(x+y)}{y}\, dy.$$

Pour obtenir l'estimation

$$2\text{-}(22) \qquad \int_{T-I_{j,\nu}} |\hat{h}_{j,\nu}(x)|\, dx \leqq M_9 2^{-\nu/2}$$

observons d'abord que si $\omega_{j'',\nu}$ est l'intervalle $\omega_{j,\nu}+\pi$ nous avons $\int_{\omega_{j'',\nu}} |\hat{h}_{j,\nu}(x)|\, dx \leqq$
$\leqq M_9' 2^{-\nu/2}$ de sorte que nous pouvons nous restreindre à établir 2-(22) avec $T-I_{j,\nu}-$
$-\omega_{j'',\nu}$ à la place de $T-I_{j,\nu}$; cela étant désignons par α, $\alpha+3(2\pi 2^{-\nu})$ les extrémités de $I_{j,\nu}$ et par β l'extrémité de $\omega_{j'',\nu}$ qui vérifie $\beta+\lambda = \alpha$ avec $0<\lambda<\pi$; l'intervalle $I = \{x+t, \; 0<t<\lambda\}$ est l'un des deux intervalles qui constituent $T-I_{j,\nu}-\omega_{j'',\nu}$; posons $\omega_{j,\nu} = \omega_{j',\nu+1} \cup \omega_{j'+1,\nu+1}$ et soient γ, $\gamma+2\pi 2^{-\nu-1}$ les extrémités de $\omega_{j',\nu+1}$ de sorte que nous avons

$$x \in I \Rightarrow \hat{h}_{j,\nu}(x) = \frac{-2^{\nu/2}}{\pi(2\pi)^{1/2}} \left[\log\left(1 + \frac{2\pi 2^{-\nu-1}}{\gamma-x}\right) - \log\left(1 + \frac{2\pi 2^{-\nu-1}}{\gamma+2\pi 2^{-\nu-1}-x}\right) \right]$$

d'où, après un calcul facile

$$\int_I |\hat{h}_{j,\nu}(x)|\, dx \leqq M_9'' 2^{-\nu/2}$$

d'où 2-(22) en vertu d'un résultat analogue sur l'autre des deux intervalles qui constituent $T-I_{j,\nu}-\omega_{j'',\nu}$. De 2-(21), 2-(22) et $\left| \dfrac{1}{2\,\mathrm{tg}\,(t/2)} - \dfrac{1}{t} \right| \leqq M_5'$, nous tirons

$$\int_T |\tilde{h}_{j,\nu}(x)|\, dx \leqq M_{10} 2^{-\nu/2} \leqq M_{11} \int_T |h_{j,\nu}(x)|\, dx$$

ce qui, joint à 2-(20), nous conduit immédiatement à

$$2\text{-}(23) \qquad \sup_{n,\ 2^\nu < n < 2^{\nu+1}} \|\widetilde{s_{2^\nu,n}(f)}\|_1 \xrightarrow[\nu\to\infty]{} 0.$$

(α7) *Le théorème principal.* Si $f \in L^1$ vérifie $\tilde{f} \in L^1$, nous désignerons par f° l'élément $(f,\tilde{f})$ de H_r^1. Pour chaque entier $n \geqq 2$ nous avons d'une façon et d'une seule $n = 2^{\nu(n)} + j(n)$ où les entiers $\nu(n)$, $j(n)$ vérifient $\nu(n) \geqq 0$ et $1 \leqq j(n) \leqq 2^{\nu(n)}$. D'aprés 2-(19), 2-(20) et 2-(23), nous avons, pour $f^\circ \in H_r^1$,

$$2\text{-}(24) \qquad a_0(f)h_0^0 + a_{1,0}(f)h_{1,0}^0 + \sum_{n=2}^{m} a_{j(n),\nu(n)} h_{j(n),\nu(n)}^0 \xrightarrow[m\to\infty]{} f^\circ \quad \text{(dans } H_r^1)$$

et puisque la suite

$$2\text{-}(25) \qquad h_0^0,\, h_{1,0}^0,\, h_{j(2),\nu(2)}^0,\, \ldots,\, h_{j(n),\nu(n)}^0,\, \ldots$$

est une suite de vecteurs topologiquement indépendants dans H_r^1 du fait que le système de Haar a cette propriété dans L^1, nous obtenons le

LEMME 2-1. *La suite* 2-(25) *est une base de Schauder de* H_r^1.

Si f est une fonction définie sur T à valeurs dans un ensemble quelconque, la fonction $\check{f}$ est définie sur T par

$$2\text{-}(26) \qquad \forall x,\, x \in T \Rightarrow \check{f}(x) = f(-x).$$

Clairement nous avons

$$2\text{-}(27) \qquad \nu \geqq 1 \quad \text{et} \quad 1 \leqq j \leqq 2^{\nu-1} \Rightarrow \check{h}_{j,\nu} = -h_{2^\nu - j + 1,\nu}.$$

Posons $\check{H}_r^1 = \{(f,\tilde{f}) \in H_r^1,\, f = \check{f}\}$ et fixons arbitrairement $(f,\tilde{f}) \in \check{H}_r^1$; nous obtenons, par un calcul facile,

$$2\text{-}(28) \quad a_{1,0}(f) = 0 \quad \text{avec} \quad \nu \geqq 1 \quad \text{et} \quad 1 \leqq j \leqq 2^\nu \Rightarrow a_{j,\nu}(f) = -a_{2^\nu - j + 1,\nu}(f).$$

De 2-(27), 2-(28) et du lemme 2-1 nous tirons immédiatement

$$2\text{-}(29) \qquad a_0(f)(h_0)^0 + \sum_{\nu=1}^{\infty} \left[\sum_{j=1}^{2^{\nu-1}} a_{j,\nu}(f)(h_{j,\nu} + \check{h}_{j,\nu})^0 \right] = f^\circ \quad \text{(dans } \check{H}_r^1)$$

où la série, dans 2-(29), est réorganisée en une série simple à la façon de 2-(24). De 2-(29) et de la topologique indépendance de la suite

$$2\text{-}(30) \qquad (h_0)^0;\ (h_{1,1} + \check{h}_{1,1})^0;\ \ldots;\ (h_{1,\nu} + \check{h}_{1,\nu})^0,\ \ldots,\ (h_{2^{\nu-1},\nu} + \check{h}_{2^{\nu-1},\nu})^0;\ \ldots$$

dans $\check{H}_r^1$ (laquelle découle immédiatement de ce que la suite des fonctions entre parenthèses dans 2-(30) est une suite de vecteurs $\neq 0$ de L^∞ qui est orthogonale dans L^2) nous déduisons le

LEMME 2-2. *La suite* 2-(30) *est une base de Schauder de* $\check{H}_r^1$.

L'espace $\check{H}_r^1$ réinterprété comme espace de fonctions holomorphes dans le disque ouvert $|z| < 1$ est le sous espace fermé de H_r^1 constitué des $F \in H_r^1$ $\left(F(z) = \sum_{n=0}^\infty c_n z^n \right)$ pour lesquelles la suite $(c_n)_{n=0,1,\dots}$ est réelle; d'après le lemme 2-2 désignons par $(F_n)_{n=1,2,\dots}$ une base de Schauder de H_r^1; chaque $G \in H^1$ s'écrit d'une façon et d'une seule sous la forme

$$G = G_1 + iG_2 \qquad (G_1 \in \check{H}_r^1 \text{ et } G_2 \in \check{H}_r^1)$$

donc, d'après le théorème de Banach Schauder, nous avons au sens des espaces vectoriels topologiques réels, la décomposition vectorielle topologique

$$H^1 = \check{H}_r^1 \oplus i\check{H}_r^1$$

ce qui montre immédiatement que G ne s'écrit que d'une seule façon sous la forme

$$G = \sum_{n=1}^\infty \lambda_n F_n \qquad (\lambda_n = \lambda_n' + i\lambda_n'' \text{ où } \lambda_n', \lambda_n'' \text{ sont réels } n = 1, 2, \dots)$$

d'où le

THÉORÈME 2-1. *L'espace* H^1 *possède une base de Schauder.*

CHAPITRE 3

Sur l'existence d'une base de Schauder de sous espace de dimension finie dans A.

(α1) *Quelques nouvelles notations et quelques résultats élémentaires.* Si $f, g \in L^1(T)$, l'expression

$$3\text{-}(1) \qquad (f*g)(x) = \frac{1}{\pi} \int_T f(t) g(x-t) \, dt$$

a un sens pour presque tout $x \in T$ et définit la convoluée $f*g$ de f et de g qui vérifie $f*g \in L^1(T)$; nous utiliserons sans référence les propriétés usuelles de la convolution.

L'expression classique de la série de Fourier usuelle $\mathfrak{F}(f*g)$ de $f*g$ (cf. [5], chap 2, n° 1, 1-(9)) donne immédiatement, en supposant $\tilde{f}, \widetilde{f*g} \in L^1(T)$,

$$3\text{-}(2) \qquad \widetilde{f*g} = \tilde{f}*g.$$

Si f est une fonction réelle définie sur T, la translatée $\gamma_x(f)$ de f par $x \in T$ est définie, pour $t \in T$, par $\gamma_x(f)(t) = f(t-x)$ et $\check{f}$ comme au paragraphe 7 du chapitre 2. Si $f, \check{f}, g \in L^1(T)$ il est immédiat de vérifier, pour $x \in T$,

$$3\text{-}(3) \qquad \widetilde{\gamma_x(f)} = \gamma_x(\check{f}) \quad \text{et} \quad \gamma_x(f) * g = f * \gamma_x(g) = \gamma_x(f * g).$$

Si f est une fonction réelle définie sur T, nous dirons que f est lipschitzienne (ce que nous noterons $f \in \Lambda_1$) s'il existe $\lambda \geqq 0$ tel que

$$3\text{-}(4) \qquad x_1, x_2 \in T \Rightarrow |f(x_1) - f(x_2)| \leqq \lambda |x_1 - x_2|$$

et dans ce cas λ_f désignera la borne inférieure des $\lambda \geqq 0$ vérifiant 3-(4) (visiblement λ_f vérifie aussi 3-(4)).

Si $f \in \Lambda_1$, alors f est absolument continue, donc si

$$\mathfrak{F}(f) \sim \alpha_0/2 + \sum_{n=1}^{\infty} (\alpha_n \cos nx + \beta_n \sin nx)$$

nous avons

$$\mathfrak{F}(f') \sim \sum_{n=1}^{\infty} (-n\alpha_n \sin nx + n\beta_n \cos nx)$$

et de là, en écrivant

$$\|\check{f}\|_\infty \leqq \sum_{n=1}^{\infty} (|\alpha_n| + |\beta_n|) = \sum_{n=1}^{\infty} n^{-1}(n|\alpha_n| + n|\beta_n|)$$

nous obtenons immédiatement

$$3\text{-}(5) \qquad f \in \Lambda_1 \Rightarrow \check{f} \in C \quad \text{et} \quad \|\check{f}\|_\infty \leqq 2\left(\sum_{n=1}^{\infty} n^{-2}\right)^{1/2} \lambda_f.$$

Si f est une fonction réelle définie sur T nous dirons que f est polygonale (ce que nous noterons $f \in \Pi$) si f est continue sur T et s'il existe un entier $v > 0$ tel que f soit affine sur chaque $\omega_{j,v}$ $(1 \leqq j \leqq 2^v)$; clairement $\Pi \subseteq \Lambda_1$. Posons $\varphi_v(t) = \frac{1}{2} W_{2^v}(t + 2\pi 2^{-v-1})$ d'où $\|\varphi_v\|_1 = \pi$ $(v = 0, 1, \ldots)$; nous avons immédiatement le

LEMME 3-1. *Si f est une fonction réelle définie sur T telle que, pour un entier $v > 0$, f soit constante à l'intérieur de chaque $\omega_{j,v}$ $(1 \leqq j \leqq 2^v)$ et si, pour chaque entier j vérifiant $1 \leqq j \leqq 2^v$ nous désignons par j_1 l'entier tel que $\omega_{j_1,v}$ succède à $\omega_{j,v}$ au sens de T, alors $f * \varphi_v$ est l'élément de Π qui, pour chaque entier j vérifiant $1 \leqq j \leqq 2^v$ prend la valeur $f(\xi_{j,v})$ au centre $\xi_{j,v}$ de $\omega_{j,v}$ et qui, pour chaque entier j vérifiant $1 \leqq j \leqq 2^v$, est affine dans l'intervalle dont les extrémités sont les points $\xi_{j,v}$, $\xi_{j_1,v}$ et dont le centre est l'extrémité commune aux intervalles $\omega_{j,v}$, $\omega_{j_1,v}$.*

Considérons le sous espace fermé $\check{A}_r = \{f^\circ \in A_r, f = \check{f}\}$ de A_r et les sous espaces $\check{\Lambda}_1 = \{f \in \Lambda_1, f = \check{f}\}$, $\check{\Pi} = \{f \in \Pi, f = \check{f}\}$ de Λ_1 et Π; un calcul facile s'appuyant sur $\left(s_{2^v}(f)\right) \xrightarrow[v \to \infty]{} f^\circ$ (dans H_r^1) et sur les propriétés usuelles de la convolution donne le

LEMME 3-2. *$\{f^\circ, f \in \Pi\}$ (resp. $\{f^\circ, f \in \check{\Pi}\}$) est un sous espace dense de A_r (resp. $\check{A}_r$).*

(α2) *Etude de $(s_{2^\nu} * \varphi_\nu)^\circ$ et de $(s_{2^\nu} * \varphi_{\nu+1})^\circ$.* Fixons arbitrairement $f^\circ \in A_r$ et écrivons s_{2^ν}, σ_ν, $\tau_{\nu,1}$, $\tau_{\nu,2}$ au lieu de $s_{2^\nu}(f)$, $\sigma_\nu(f)$, $\tau_{\nu,1}(f)$, $\tau_{\nu,2}(f)$. Pour $\nu = 3, 4, \ldots$ et pour presque tout $x \in T$ nous avons (cf. 2-(3) et 2-(10))

$$3\text{-}(6) \quad \begin{cases} |(\widetilde{s_{2^\nu}} * \varphi_\nu)(x)| \le |(\sigma_\nu * \varphi_\nu)(x)| + |(\tau_{\nu,1} * \varphi_\nu)(x)| + |(\tau_{\nu,2} * \varphi_\nu)(x)| \\[2mm] |(\widetilde{s_{2^\nu}} * \varphi_{\nu+1})(x)| \le |(\sigma_\nu * \varphi_{\nu+1})(x)| + |(\tau_{\nu,1} * \varphi_{\nu+1})(x)| + |(\tau_{\nu,2} * \varphi_{\nu+1})(x)|. \end{cases}$$

Puisque $\|f(\nu, \cdot) - \tilde{f}\|_\infty \xrightarrow[\nu \to \infty]{} 0$ (cf. [5], chap 4, exer 5), 2-(5) et 2-(18) donnent

$$3\text{-}(7) \qquad \sup_{\nu = 3, 4, \ldots} [\sup(\|\sigma_\nu * \varphi_\nu\|_\infty, \|\sigma_\nu * \varphi_{\nu+1}\|_\infty, \|\tau_{\nu,1} * \varphi_\nu\|_\infty, \|\tau_{\nu,2} * \varphi_{\nu+1}\|_\infty)] < \infty.$$

Pour un entier $\nu \ge 3$ arbitrairement fixé et pour $x \in E$, définissons, comme au début du α3 du chap 2 les entiers $j_k = j_k(x)$ ($k = 1, \ldots, 5$) par le fait que les $\omega_{j_k, \nu}$ ($k = 1, \ldots, 5$) se succèdent au sens de T et que $x \in \omega_{j_3, \nu}$; désignant toujours par $d_{j_k, \nu}(x)$ la distance, au sens de T, de x à l'extrémité de $\omega_{j_k, \nu}$ la plus proche de x, nous avons, en posant

$$\Psi_{k, \nu}(x) = \left| \log\left(1 + 2\pi 2^{-\nu} d^{-1}_{j_k, \nu}(x)\right)\right| \quad \text{pour} \quad k = 1, \ldots, 5 \text{ avec } k \ne 3$$

et

$$\Psi_{3, \nu}(x) = \left| \log\left(2\pi 2^{-\nu} d^{-1}_{j_3, \nu}(x) - 1\right)\right|,$$

$$3\text{-}(8) \qquad\qquad |\tau_{\nu,2}(x)| \le \pi^{-1} \|f\|_\infty \sum_{k=1}^{5} \Psi_{k, \nu}(x).$$

Les valeurs absolues des expressions $(\Psi_{k, \nu} * \varphi_\nu)(x)$; $(\Psi_{k, \nu} * \varphi_{\nu+1})(x)$ ($\nu = 3, 4, \ldots$, $k = 1, \ldots, 5$) étant majorées par une constante absolue comme nous le voyons avec des changements de variables immédiats dans les intégrales qui les définissent, 3-(6), 3-(7) et 3-(8) nous conduisent au

LEMME 3-3. *Pour chaque f° fixé dans A_r les suites $((s_{2^\nu} * \varphi_\nu)^\circ)_{\nu = 0, 1, \ldots}$, $((s_{2^\nu} * \varphi_{\nu+1})^\circ)_{\nu = 0, 1, \ldots}$ sont bornées dans A_r et ces suites sont des suites de $\check{A}_r$ si $f^\circ \in \check{A}_r$.*

Fixons arbitrairement $f \in \Pi$. Soit ν un entier > 0 tel que f soit affine sur chaque $\omega_{k, \nu}$ ($1 \le k \le 2^\nu$) et soit ν' un entier vérifiant $\nu' > \nu$; $\xi_{j, \nu'}$ étant le centre de l'intervalle $\omega_{j, \nu'}$ ($1 \le j \le 2^{\nu'}$), la fonction $s_{2^{\nu'}}$ est la fonction qui est constante à l'intérieur de chaque intervalle $\omega_{j, \nu'}$ et égale à $f(\xi_{j, \nu'})$ ($1 \le j \le 2^{\nu'}$); par ailleurs, si $I_{j, \nu'}$ désigne l'intervalle concentrique à $\omega_{j, \nu'}$ et de longueur $2\pi 2^{-\nu'-1}$, la fonction $s_{2^{\nu'}} * \varphi_{\nu'+1}$ est la fonction réelle continue sur T qui coïncide avec $s_{2^{\nu'}}$ sur chacun des intervalles $I_{j, \nu'}$ ($1 \le j \le 2_{\nu'}$) et qui est affine sur chacun des intervalles qui constituent $T - \left(\bigcup_{j=1}^{2^{\nu'}} I_{j, \nu'} \right)$.

Cette construction jointe à la construction de $s_{2^{\nu'}} * \varphi_{\nu'}$ (cf. lemme 3-1) nous conduit à

$$3\text{-}(9) \qquad \sup\left(\|f-(s_{2^{\nu'}} * \varphi_{\nu'+1})\|_\infty,\ \|f-(s_{2^{\nu'}} * \varphi_{\nu'})\|_\infty\right) \xrightarrow[\nu'\to\infty]{} 0$$

et à

$$3\text{-}(10) \qquad \sup_{\nu'>\nu}\left(\sup\left(\lambda_{f-(s_{2^{\nu'}} * \varphi_{\nu'+1})},\ \lambda_{f-(s_{2^{\nu'}} * \varphi_{\nu'})}\right)\right) < \infty.$$

Fixons arbitrairement $\varepsilon>0$. Nous pouvons fixer un entier $N_\varepsilon>0$ assez grand pour que

$$3\text{-}(11) \qquad 2\left(\sum_{n=N_\varepsilon+1}^\infty n^{-2}\right)^{1/2}\left(\sup_{\nu'>\nu}\lambda_{f-(s_{2^{\nu'}} * \varphi_{\nu'+1})}\right) \leqq \varepsilon/2.$$

Si $\mathfrak{F}\left(f-(s_{2^{\nu'}} * \varphi_{\nu'+1})\right) \sim \alpha_0/2 + \sum_{n=1}^\infty (\alpha_n \cos nx + \beta_n \sin nx)$, alors d'après 3-(9)

$$1\leqq j \leqq N_\varepsilon \Rightarrow |\alpha_j|+|\beta_j| \xrightarrow[\nu'\to\infty]{} 0$$

donc nous pouvons fixer un entier $\nu'_\varepsilon>\nu$ assez grand pour que

$$3\text{-}(12) \qquad \nu' \geqq \nu'_\varepsilon \Rightarrow \sum_{j=1}^{N_\varepsilon} (|\alpha_j|+|\beta_j|) \leqq \varepsilon/2.$$

De 3-(5), 3-(11) et 3-(12) nous déduisons

$$\nu' \geqq \nu'_\varepsilon \Rightarrow \overline{\|f-(s_{2^{\nu'}} * \varphi_{\nu'+1})\|}_\infty \leqq \sum_{j=1}^{N_\varepsilon} (|\alpha_j|+|\beta_j|) + \sum_{j=N_\varepsilon+1}^\infty j^{-1}(j|\alpha_j|+j|\beta_j|) \leqq$$

$$\leqq \frac{\varepsilon}{2}+\frac{\varepsilon}{2} = \varepsilon$$

d'où

$$\overline{\|f-(s_{2^{\nu'}} * \varphi_{\nu'+1})\|}_\infty \xrightarrow[\nu'\to\infty]{} 0;$$

par la même méthode

$$\overline{\|f-(s_{2^{\nu'}} * \varphi_{\nu'})\|}_\infty \xrightarrow[\nu'\to\infty]{} 0$$

d'où le

LEMME 3-4. *Pour chaque f fixée dans Π, les suites $\left((s_{2^\nu} * \varphi_\nu)^\circ\right)_{\nu=0,1,\dots}$, $\left((s_{2^\nu} * \varphi_{\nu+1})^\circ\right)_{\nu=0,1,\dots}$ convergent dans A_r vers $f^\circ \in A_r$ et ces suites sont des suites de $\check{A}_r$ si $f\in\check{\Pi}$.*

Puisque pour chaque entier $\nu\geqq 0$, les applications linéaires

$$3\text{-}(13) \qquad \begin{cases} S_{\nu,\nu}: f^\circ \in A_r \to \left(s_{2^\nu}(f) * \varphi_\nu\right)^\circ \in A_r \\ S_{\nu,\nu+1}: f^\circ \in A_r \to \left(s_{2^\nu}(f) * \varphi_{\nu+1}\right)^\circ \in A_r \\ \Gamma_{2\pi 2^{-\nu-2}}: f^\circ \in A_r \to \left(\gamma_{2\pi 2^{-\nu-2}}(f)\right)^\circ \in A_r \\ \Gamma_{-2\pi 2^{-\nu-2}}: f^\circ \in A_r \to \left(\gamma_{-2\pi 2^{-\nu-2}}(f)\right)^\circ \in A_r \end{cases}$$

de A_r dans A_r sont continues, les lemmes 3-3 et 3-4 nous conduisent, avec le théorème de Banach Steinhaus, au

LEMME 3-5. *Pour chaque f^0 fixée dans A_r, les quatre suites*

$$\left(S_{v,v}(f^0)\right)_{v=0,1,\dots}, \quad \left(S_{v,v+1}(f^0)\right)_{v=0,1,\dots},$$

$$\left(\Gamma_{2\pi 2^{-v-2}}(f^0)\right)_{v=0,1,\dots}, \quad \left(\Gamma_{-2\pi 2^{-v-2}}(f^0)\right)_{v=0,1,\dots}$$

de A_r convergent dans A_r vers f^0 et les deux premières de ces suites sont des suites de $\check{A}_r$, si $f^0 \in \check{A}_r$.

(α3) *Le théorème principal.* Fixons $f^0 \in A_r$. La fonction $s_{2^v} = s_{2^v}(f)$ est, à l'intérieur de chaque $\omega_{k,v}$, constante et égale à la valeur moyenne $M_{\omega_{k,v}}(f)$ de f sur $\omega_{k,v}$ ($1 \leq k \leq 2^v$); la fonction $s_{2^v} * \varphi_v$ n'a plus, sur chaque $\omega_{k,v}$, la même valeur moyenne que s_{2^v}; en fait, en désignant par k_1 l'entier tel que l'intervalle $\omega_{k_1,v}$ succède à $\omega_{k,v}$ au sens de T et en posant $\omega_{k,v} = \omega_{j_k-1,v+1} \cup \omega_{j_k,v+1}$, $\omega_{k_1,v} = \omega_{j_{k_1},v+1} \cup \omega_{j_{k_1}+1,v+1}$ nous avons, si λ et λ_1 sont les valeurs prises par s_{2^v} à l'intérieur de $\omega_{k,v}$ et $\omega_{k_1,v}$,

$$3\text{-}(14) \quad M_{\omega_{j_k,v+1}}(s_{2^v} * \varphi_v) = \lambda - \frac{1}{4}(\lambda - \lambda_1), \quad M_{\omega_{j_{k_1},v+1}}(s_{2^v} * \varphi_v) = \lambda_1 + \frac{1}{4}(\lambda - \lambda_1).$$

De même en posant

$$\omega_{j_k,v+1} = \omega_{j'_k-1,v+2} \cup \omega_{j'_k,v+2}, \; \omega_{j_{k_1},v+1} = \omega_{j'_{k_1},v+2} \cup \omega_{j'_{k_1}+1,v+2}$$

nous avons

$$3\text{-}(15) \qquad\qquad M_{\omega_{j'_k,v+2}}(s_{2^v} * \varphi_{v+1}) = \lambda - \frac{1}{4}(\lambda - \lambda_1),$$

$$M_{\omega_{j'_{k_1},v+2}}(s_{2^v} * \varphi_{v+1}) = \lambda_1 + \frac{1}{4}(\lambda - \lambda_1).$$

Puisque

$$3\text{-}(16) \quad \begin{cases} M_{\omega_{j_k,v+1}}\left((\gamma_{2\pi 2^{-v-2}} h_{j_k,v+1}) * \varphi_{v+2}\right) = \\[2mm] = \dfrac{3}{8} M_{\omega_{j'_k,v+2}}(\gamma_{2\pi 2^{-v-2}} h_{j_k,v+1}) = \dfrac{3}{8}\sqrt{2^{v+1}}/\sqrt{2\pi} \\[3mm] M_{\omega_{j_{k_1},v+1}}\left((\gamma_{2\pi 2^{-v-2}} h_{j_k,v+1}) * \varphi_{v+2}\right) = -\dfrac{3}{8}\sqrt{2^{v+1}}/\sqrt{2\pi} \end{cases}$$

nous déduisons de 3-(15) que le coefficient de Fourier Haar $a_{j_k,v+1}$ de

$$\gamma_{-2\pi 2^{-v-2}}(s_{2^v} * \varphi_{v+1}) \text{ est donné par } a_{j_k,v+1} = \frac{\sqrt{\pi}}{2\sqrt{2}\sqrt{2^{v+1}}}(\lambda - \lambda_1)$$

ce qui donne, avec 3-(16)

$$3\text{-}(17)\quad\begin{cases} M_{\omega_{j_k,\,v+1}}\big(a_{j_k,v+1}(\gamma_{2\pi\,2^{-v-2}}\,h_{j_k,v+1})*\varphi_{v+2}\big) = \\[2mm] = a_{j_k,v+1}\left(\dfrac{3}{8}\,\sqrt{2^{v+1}}/\sqrt{2\pi}\right) = \dfrac{3}{32}(\lambda-\lambda_1) \\[4mm] M_{\omega_{j_{k_1},\,v+1}}\big(a_{j_k,v+1}(\gamma_{2\pi\,2^{-v-2}}\,h_{j_k,v+1})*\varphi_{v+2}\big) = -\dfrac{3}{32}(\lambda-\lambda_1). \end{cases}$$

En vertu de 3-(14) et 3-(17), la fonction $\Omega_v(f)$ définie par

$$\begin{cases} \Omega_v(f) = \{s_{2^v}(f)*\varphi_v\} + \dfrac{8}{3}\,\gamma_{2\pi\,2^{-v-2}}\big\{\big(s_{2^{v+2}}[\gamma_{-2\pi\,2^{-v-2}}(s_{2^v}(f)*\varphi_{v+1})]*\varphi_{v+2}\big) - \\[2mm] -\big(s_{2^{v+1}}[\gamma_{-2\pi\,2^{-v-2}}(s_{2^v}(f)*\varphi_{v+1})]*\varphi_{v+2}\big)\big\} \end{cases}$$

(qui vérifie visiblement $\overbrace{\Omega_v(f)}=\Omega_v(f)$ si $\check{f}=f$) a sur chaque $\omega_{k,v}$ la même valeur moyenne que $f\,(1\le k\le 2^v)$ et puisque la construction de $\Omega_v(f)$ à partir de f n'utilise que les valeurs moyennes de f sur les $\omega_{k,v}$ $(1\le k\le 2^v)$ nous avons $P_v^2=P_v$ si P_v est l'application linéaire de A_r dans A_r définie pour chaque $f^\circ\in A_r$ par $P_v(f^\circ)= =(\Omega_v(f))^\circ$; autrement dit P_v est une projection de A_r qui, lorsqu'elle est restreinte à $\check{A}_r$, est une projection de $\check{A}_r$.

D'après 3-(13) et la définition de P_v nous avons

$$P_v = S_{v,v} + \frac{8}{3}\,\Gamma_{2\pi\,2^{-v-2}}[(S_{v+2,v+2}\,\Gamma_{-2\pi\,2^{-v-2}}\,S_{v,v+1}) - (S_{v+1,v+2}\,\Gamma_{-2\pi\,2^{-v-2}}\,S_{v,v+1})]$$

d'où la continuité de P_v et d'où, avec le lemme 3-5,

$$3\text{-}(18)\qquad\qquad f^\circ\in\check{A}_r \Rightarrow P_v(f^\circ)\underset{v\to\infty}{\to}f^\circ \quad (\text{dans } \check{A}_r).$$

Si l'entier v' vérifie $v'>v$, la fonction $\Omega_{v'}(f)$ ayant les mêmes valeurs moyennes que f dans les $\omega_{k,v'}$ $(1\le k\le 2^{v'})$, il s'ensuit que $\Omega_{v'}(f)$ a les mêmes valeurs moyennes que f dans les $\omega_{k,v}$ $(1\le k\le 2^v)$ et puisque la construction de $\Omega_v(f)$ à partir de f n'utilise que les valeurs moyennes de f sur les $\omega_{k,v}$ $(1\le k\le 2^v)$ nous avons

$$3\text{-}(19)\qquad\qquad v<v' \Rightarrow P_v P_{v'} = P_v.$$

Maintenant 3-(18) et 3-(19) nous permettent d'appliquer le théorème 1-1 à la suite $(P_v)_{v=1,2,\dots}$ de projections continues de $\check{A}_r$ sur des sous espaces de dimension finie de $\check{A}_r$, d'où le

LEMME 3-6. *L'espace de Banach réel $\check{A}_r$ possède une base de Schauder $(B_n)_{n=1,2,\dots}$ de sous espaces de dimension finie.*

Dans ces conditions $C_n = B_n + iB_n$ est un sous espace de dimension finie de l'espace de Banach complexe A ($n = 1, 2, \ldots$). L'espace de Banach réel $\check{A}_r$ réinterprété comme espace de fonctions holomorphes dans le disque ouvert $|z| < 1$ est le sous espace fermé de l'espace de Banach réel A_r constitué des $F \in A_r$ $\left(F(z) = \sum_{n=0}^{\infty} c_n z^n \right)$ pour lesquelles la suite $(c_n)_{n=0,1,\ldots}$ est réelle. Chaque élément $G \in A$ s'écrit d'une façon et d'une seule sous la forme $G = G_1 + iG_2$ ($G_1, G_2 \in \check{A}_r$) donc, d'après le théorème de Banach Schauder, nous avons au sens des espaces vectoriels topologiques réel, la décomposition vectorielle topologique $A_r = \check{A}_r \oplus i\check{A}_r$ ce qui montre immédiatement que G ne s'écrit que d'une seule façon sous la forme

$$G = \sum_{n=1}^{\infty} F_n \quad \left(\text{au sens de } A, F_n \in C_n \ (n = 1, 2, \ldots) \right)$$

d'où le

THÉORÈME 3-1. *L'espace A possède une base de Schauder de sous espaces de dimension finie.*

REFERENCES

[1] E. J. Akutowicz, *Construction of a Schauder basis in some spaces of holomorphic functions in the unit disc.* Colloq. Math. **15** (1966), 287—296.

[2] P. Billard, *Sur les bases de Schauder dans les espaces de Banach H^1 et A.* C. R. Acad. Sci. Paris Sér. A **271** (1970), 36—38.

[3] C. W. Mc Arthur, *The weak basis theorem.* Colloq. Math. **18** (1967), 71—76.

[4] I. Singer, *Some remarks and problems on bases in Banach spaces.* I S N M — Vol. **10** Birkhäuser, Basel 1969, 130—139.

[5] A. Zygmund, *Trigonometric series.* Second edition — Cambridge 1959.

[6] W. B. Johnson, H. P. Rosenthal and M. Zippin, *On bases, finite dimensidnal decompositions,* Israel. J. Math. **9** (1971), 488—506.

Logarithmic and Exponential Variants of Bernstein's Inequality and Generalized Derivatives

By

E. GÖRLICH

LEHRSTUHL A FÜR MATHEMATIK
TECHNISCHE HOCHSCHULE AACHEN

Dedicated to Professor Fritz Reutter on the occasion of his 60 th birthday on August 26, 1971

Introduction

The first two sections of this paper deal with two parallelisms to the classical Bernstein and Jackson theory of best approximation by trigonometric polynomials. Starting from a generalization of the Bernstein inequality, the classes of functions with a logarithmic or exponential decay of the error of their best approximation are characterized by smoothness properties in terms of the moduli of continuity of suitable semi-group operators. The results may be understood as limiting cases of the classical situation where the best approximation error decays with some power of the parameter.

Correspondingly, there is a "logarithmic" and an "exponential" type of derivative, one being weaker and the other stronger than the ordinary (or fractional) derivative of any order. In Sec. 3 the functions which are differentiable in this sense are characterized by their structural properties such as the existence of a limit of an improper integral similar to one known from the theory of fractional integration or by analyticity properties.

This note contains the main results of the author's paper [20] which also includes more details on some of the proofs given. Several extensions of the present results will be considered in a more systematic treatment to appear later: see [21]. The author gratefully acknowledges helpful discussions with Professor P. L. Butzer, Drs. W. Trebels, K. Scherer, and E. L. Stark; he is also indebted to Professor J.-P. Kahane for his critical remarks concerning the original version of Thm. 6 at the Oberwolfach conference.

1. Inequalities of Bernstein type

Let X denote one of the spaces $C_{2\pi}$ or $L_{2\pi}^p$, $1 \leq p < \infty$ of 2π-periodic functions of a real variable for which the norms

$$\|f\|_C = \max_{-\pi \leq x < \pi} |f(x)|, \quad \|f\|_p = \left\{ \int_{-\pi}^{\pi} |f(x)|^p \, dx \right\}^{1/p}$$

respectively, are finite. Denoting by $\mathbf{Z}$ the set of all integers, the Fourier coefficients of an $f \in X$ are given by

$$f^{\wedge}(k) = \frac{1}{2\pi} \int_{-\pi}^{\pi} f(x) e^{-ikx} dx \qquad (k \in \mathbf{Z}).$$

The classical Bernstein inequality[1]) for the ordinary derivative of a trigonometric polynomial of degree $\leq n$

$$(1.1) \qquad\qquad t_n(x) = \sum_{k=-n}^{n} c_k e^{ikx}$$

with complex coefficients c_k may be written as

$$(1.2) \qquad \|t_n'(x)\|_X \equiv \Big\| \sum_{k=-n}^{n} ik c_k e^{ikx} \Big\|_X \leq n \|t_n(x)\|_X \qquad (n \in \mathbf{N})$$

where $\mathbf{N}$ is the set of non-negative integers. One way of generalizing this inequality is to replace the sequence $\{ik\}$ by other sequences. This problem has already been posed by G. Szegő [32] in 1928. One such generalization is the known fractional Bernstein type inequality[2])

$$(1.3) \qquad \Big\| \sum_{k=-n}^{n} |k|^{\gamma} c_k e^{ikx} \Big\|_X \leq 2n^{\gamma} \|t_n(x)\|_X \qquad (\gamma > 0; \ n \in \mathbf{N}).$$

Here we are interested in the two limiting cases of (1.2), one with a more slowly increasing sequence of factors

$$(1.4) \quad \Big\| \sum_{k=-n}^{n} \log[(1+k^2)^{1/2}] c_k e^{ikx} \Big\|_X \leq \frac{5}{2} \log[(1+n^2)^{1/2}] \|t_n(x)\|_X \qquad (n \in \mathbf{N}),$$

and the other with a more rapidly increasing sequence

$$(1.5) \quad \Big\| \sum_{k=-n}^{n} (e^{\beta|k|} - 1) c_k e^{ikx} \Big\|_X \leq 2(e^{\beta n} - 1) \|t_n(x)\|_X \qquad (\beta > 0; \ n \in \mathbf{N}).$$

The latter exponential Bernstein-type inequality was essentially known to G. Szegő [32]. The proof of (1.4), (1.5) is easy. Indeed, a straightforward modification of the classical F. Riesz proof [30] of (1.2) (see also Ja. L. Geronimus [17]) gives the following

---

[1]) This was proved in $C_{2\pi}$ for even and odd trigonometric polynomials in 1912 by S. Bernstein [4] who attributed it to E. Landau in [6]. According to M. Fekete [16] it is due to L. Fejér.

[2]) Inequalities of this type seem to originate in the paper of P. Civin [14], see also I. I. Ogiewetzki [28], C. Watari [34], and P. L. Butzer—E. Görlich [8]. In these papers (1.3) is proved with constant 2 replaced by some constant depending on γ and tending to ∞ for $\gamma \to 0+$. But also 2 is not yet the sharp constant.

LEMMA 1. *Let $\varphi(x) > 0$ be convex (or concave and monotonely increasing) for $x > 0$ with $\varphi(0) = 0$. Then each $t_n(x)$ of the form* (1. 1) *satisfies*

$$\left\| \sum_{k=-n}^{n} \varphi(|k|) c_k e^{ikx} \right\|_X \leq 2\varphi(n) \|t_n(x)\|_X \qquad (n \in \mathbf{N}).$$

Obviously the constant 2 is not best possible, but this is not important for our applications. The lemma immediately implies (1. 5) and, using a slightly refined argument, it also implies (1. 4)[3]).

2. Theorems of Bernstein and Jackson type

Let us start with the most natural generalization of (1. 2) to the fractional case, namely

$$(2. 1) \qquad \left\| \sum_{k=-n}^{n} (ik)^{\gamma} c_k e^{ikx} \right\|_X \leq C_{\gamma} n^{\gamma} \|t_n(x)\|_X \qquad (\gamma > 0),$$

where C_γ is some constant not depending on n. Here the left hand side may be interpreted as the Liouville derivative $t_n^{(\gamma)}$ of order γ of t_n, where $f^{(\gamma)}(x) \sim \sum_{k=-\infty}^{\infty} (ik)^{\gamma} f^{\wedge}(k) e^{ikx}$ for general $f \in X$. Similarly the inequality (1. 3) is connected with the fractional Riesz derivative $(D^{[\gamma]} f)(x) \sim \sum_{k=-\infty}^{\infty} |k|^{\gamma} f^{\wedge}(k) e^{ikx}$; the latter will be preferable for comparing it with the results to follow because the factor $|k|^{\gamma}$ is even just as are our factors $\varphi(|k|)$. More precisely, a function $f \in X$ is said to have a Riesz derivative of order $\gamma > 0$ in X if $|k|^{\gamma} f^{\wedge}(k)$, $k \in \mathbf{Z}$, are the Fourier coefficients of some function $g \in X$, and in this case one defines $D^{[\gamma]} f = g$. This leads to the following[4])

DEFINITION 1. *A function $f \in X$ is said to have a logarithmic derivative $D_L f$ in X if $\log [(1 + k^2)^{1/2}] f^{\wedge}(k)$, $k \in \mathbf{Z}$, are the Fourier coefficients of some function $g \in X$, and $D_L f$ is defined by $[D_L f]^{\wedge}(k) = g^{\wedge}(k)$, $k \in \mathbf{Z}$. The exponential derivative $D_\beta f$ of order $\beta > 0$ in X is defined by $(e^{\beta |k|} - 1) f^{\wedge}(k) = [D_\beta f]^{\wedge}(k)$, $k \in \mathbf{Z}$, if these are the Fourier coefficients of a function in X.*

Denoting the domain of the operator D_L by $\mathscr{D}(D_L)$ it is easily seen that $\mathscr{D}(D_L)$ is a Banach space under the norm $\|f\|_{\mathscr{D}(D_L)} = \|f\|_X + \|D_L f\|_X$, and that D_L is a closed operator; similarly for D_β, $\beta > 0$.

---

[3]) It may be noted that (1. 3) is also an immediate consequence of Lemma 1. There are many other applications of Lemma 1, e.g. $\varphi(x) = \log (1 + x)$. Moreover, the lemma is easily modified such as to contain also the case $\varphi(x) = e^{\beta x^{\gamma}} - 1$ for $\beta > 0$, $\gamma > 0$.

[4]) Note that this "logarithmic derivative" has nothing in common with the familiar logarithmic derivative f'/f.

Now let $E_n[f] = \inf_{t_n \in \mathscr{T}_n} \| f - t_n \|_X$ be the best approximation of $f \in X$ by elements of the set $\mathscr{T}_n$ of trigonometric polynomials of degree $\leq n$. In the proof of the classical Bernstein theorem, i.e. the result that $E_n[f] = O(n^{-\alpha})$, $n \to \infty$, $0 < \alpha < 1$, implies $\| f(x+t) - f(x) \|_X = O(|t|^\alpha)$, $t \to 0$, the essential idea is the following. Instead of the difference $f(x+t) - f(x)$ one considers an infinite sum of such differences for certain trigonometric polynomials t_n, and then the representation $t_n(x+u) - t_n(x) =$

$$= \int_0^u t_n'(x+v)\, dv$$

allows one to apply inequality (1. 2). Hence if we wish to establish a logarithmic or exponential analog of Bernstein's theorem it is necessary not only to replace (1. 2) by (1. 4) or (1. 5), respectively, but also to replace the difference $f(x+t) - f(x)$ by other differences which are related to the generalized derivatives $D_L f$ or $D_\beta f$ in the same way as the ordinary difference is related to f'.

To start with the logarithmic case, the substitute for $f(x+t)$ should thus be $(T_L(t)f)(x)$, where $\{T_L(t);\ t \geq 0\}$ is a semi-group of bounded linear operators on X into itself with infinitesimal generator $-D_L$. Such a semi-group, which even belongs to the class (C_0) of strongly continuous semi-groups, exists indeed by the Hille—Yosida theorem (see P. L. BUTZER—H. BERENS [7], p. 34) since, apart from the fact that D_L is closed, $\mathscr{D}(D_L)$ is dense in X and the operator norm of the resolvent satisfies $\| R(\lambda; -D_L) \|_{[X, X]} \leq \lambda^{-1}$ for every $\lambda > 0$, as can be shown. Moreover, a corollary to the Hille—Yosida theorem ([7], p. 37) gives that there is only one such semi-group $\{T_L(t);\ t \geq 0\}$ in (C_0) having $-D_L$ as its infinitesimal generator. One readily obtains the explicit representation

$$(2.\,2) \qquad (T_L(t)f)(x) = \frac{1}{2\pi} \int_{-\pi}^{\pi} f(x-u)\chi_{L,t}(u)\, du \qquad (f \in X;\ t > 0)$$

with $T_L(0)f = f$ and kernel[5])

$$\chi_{L,t}(u) = \sum_{k=-\infty}^{\infty} (1+k^2)^{-t/2} e^{iku}.$$

Using this semi-group operator the logarithmic counterpart of the Bernstein theorem can be formulated as follows:

THEOREM 1. *Let* $f \in X$ *and* $0 < \alpha < 1$. *Then*

$$(2.\,3) \qquad E_n[f] = O\big(\{\log[(1+n^2)^{1/2}]\}^{-\alpha}\big) \qquad (n \to \infty)$$

implies

$$\| T_L(t)f - f \|_X = O(t^\alpha) \qquad (t \to 0+).$$

---

[5]) In view of the fact that the kernel is the periodic analog of the well-known Bessel potential (see e.g. the literature cited in [19]), it follows that the kernel is non-negative and hence $T_L(t)$ has operator norm 1 for all $t \geq 0$.

PROOF. Let $\{t_n^*\}$ be a sequence of polynomials such that $\|f-t_n^*\|_X = O(\{\log[(1+n^2)^{1/2}]\}^{-\alpha})$ for $n\to\infty$. Setting $V_0(x)=t_1^*(x)$ and $V_n(x) = t_{a_n}^*(x) - t_{a_{n-1}}^*(x)$ with[6]

$$(2.4) \qquad\qquad a_n = [\![(2^{2^n}-1)^{1/2}]\!]$$

for $n\geqq 1$, then $\lim\limits_{n\to\infty} \left\|f(x)-\sum\limits_{k=0}^{n} V_k(x)\right\|_X = 0$. Thus

$$\|T_L(t)f-f\| = \left\|\sum_{n=0}^{\infty}[T_L(t)V_n - V_n]\right\| \leqq \left\|\sum_{n=0}^{m-1}[T_L(t)V_n-V_n]\right\| + \left\|\sum_{n=m}^{\infty}[T_L(t)V_n-V_n]\right\|$$
$$(t>0).$$

Here m is an arbitrary natural number to be specified later. Using the representation

$$(2.5) \qquad\qquad T_L(t)f-f = -\int_0^t T_L(\tau)D_Lf\,d\tau \qquad (f\in\mathscr{D}(D_L))$$

which is valid for any semi-group of class (C_0) (cf. [7], p. 10), it follows by the logarithmic Bernstein-type inequality (1.4) and by the uniform boundedness of the operators $T_L(t)$ with respect to t (cf. footnote 5) that

$$\|T_L(t)f-f\| \leqq t\sum_{n=0}^{m-1}\|D_L V_n\| + 2\sum_{n=m}^{\infty}\|V_n\| \leqq (3/4)t(\log 2)\sum_{n=0}^{m-1}2^n\|V_n\| + 2\sum_{n=m}^{\infty}\|V_n\|.$$

Now the hypothesis yields

$$\|V_n\| = \|t_{a_n}^* - t_{a_{n-1}}^*\| \leqq \|t_{a_n}^*-f\| + \|t_{a_{n-1}}^*-f\| = O(2^{-\alpha n}) \qquad (n\to\infty),$$

and hence

$$\|T_L(t)f-f\| = O\left(t\sum_{n=0}^{m-1}2^{(1-\alpha)n} + \sum_{n=m}^{\infty}2^{-\alpha n}\right) =$$

$$= O\left(t\frac{2^{(1-\alpha)m}-1}{2^{1-\alpha}-1} + \frac{2^{-\alpha m}}{2^\alpha-1}\right) = O(t2^{(1-\alpha)m}+2^{-\alpha m}) \qquad (t\to 0+)$$

with O independent of m. Choosing now $m=[\![(\log 1/t)(\log 2)^{-1}]\!]$ then $2^{(1-\alpha)m} \leqq t^{\alpha-1}$ and $2^{-\alpha m}\leqq(2t)^\alpha$. This gives

$$\|T_L(t)f-f\| = O(t^\alpha) \qquad (t\to 0+)$$

and the proof is complete.

---

[6] Here $[\![x]\!]$ denotes the greatest integer $\leqq x$.

The case $\alpha \geqq 1$ may be treated similarly. E.g. the hypothesis of Thm. 1 for $\alpha = 1$ implies

$$\|T_L(2t)f - 2T_L(t)f + f\|_X = O(t) \qquad (t \to 0+)$$

which is the analog of the classical Zygmund theorem.

Let us remark that in some sense the particular semi-group operator $T_L(t)$ may be considered as a generalized translation operator since it has the following properties in common with the right translation: (i) For each $t \geqq 0$ $T_L(t)$ is a bounded linear operator on X into itself with norm 1, (ii) for each $f \in X$ one has $\lim_{t \to 0+} \|T_L(t)f - f\|_X = 0$, (iii) $T_L(t)$ is a positive operator, (iv) $T_L(t)T_L(\tau) = T_L(t + \tau)$ for all $t, \tau \geqq 0$, (v) one has the representation (2. 5) as an analog of the fundamental theorem of differential and integral calculus, (vi) the corresponding differential operator D_L is closed. But $T_L(t)$ is not a generalized translation operator in the sense of B. M. Levitan [23], p. 6, since this would require $T_L(t)$ to be at least 2π-periodic with respect to t.

The converse of Thm. 1 is also true, i.e. the following theorem of Jackson type:

Theorem 2. *Let $f \in X$ and $0 < \alpha < 1$. Then $\|T_L(t)f - f\|_X = O(t^\alpha)$, $t \to 0+$ implies* (2. 3).

The proof of Thm. 2 is carried out as in the classical case by using a particular polynomial approximation process. Such a process may be constructed similarly to the typical means[7] by

$$(2. 6) \qquad \mathfrak{L}_n(f; x) = \sum_{k=-n+1}^{n-1}{}' \left\{1 - \frac{\log(1+k^2)^{1/2}}{\log(1+n^2)^{1/2}}\right\} f^\wedge(k) e^{ikx}$$

for $n = 1, 2, \ldots$. It can be shown that

$$(2. 7) \qquad \|\mathfrak{L}_n(f; x)\|_X \leqq (3/2)\|f(x)\|_X \qquad (f \in X)$$

and that the Jackson-type inequality

$$(2. 8) \qquad \|f(x) - \mathfrak{L}_n(f; x)\|_X \leqq M\{\log[(1+n^2)^{1/2}]\}^{-1}\|D_L f\|_X$$

holds for each $f \in \mathscr{D}(D_L)$, where M is a constant independent of n and f. Once (2. 7) and an inequality of type (2. 8) are known the proof of Thm. 2 follows by means of intermediate space theory (see P. L. Butzer—H. Berens [7] and P. L. Butzer—K. Scherer [11]) which allows one to interpolate between the spaces X and $\mathscr{D}(D_L)$. For the present particular instance this interpolation may also be performed by

---

[7]) For $f \in X$, $\varkappa > 0$, $n = 1, 2, \ldots$ the typical means of f are defined by $\mathfrak{T}_n^\varkappa(f; x) = \sum_{k=-n+1}^{n-1}{}' \left\{1 - \left(\frac{|k|}{n}\right)^\varkappa\right\} f^\wedge(k) e^{ikx}$.

using an elementary comparison theorem of N. P. KUPTSOV [22], p. 123. We omit further details.

It should be mentioned that the characterization of (2. 3) given by Theorems 1 and 2 adds a second characterization to a known one given by N. K. BARI and S. B. STEČKIN [3] (see G. G. LORENTZ [24], p. 62) implying that (2. 3) is equivalent to

$$\|f(x+t)-f(x)\|_X = O(\{\log[(1+1/t^2)^{1/2}]\}^{-\alpha}) \qquad (t\to 0).$$

In the exponential case, however, the Bari—Stečkin result does not apply. Here we restrict ourselves to the space $X=L^p_{2\pi}$, $1<p<\infty$, since the proofs of the following two theorems break down for $X=C_{2\pi}$ or $L^1_{2\pi}$.

Again D_β is a closed operator from the Banach space $\mathscr{D}(D_\beta)$ into X for each $\beta>0$, and $\mathscr{D}(D_\beta)$ is dense in X. Moreover, each $\lambda>0$ belongs to the resolvent set of $-D_\beta$ and the operator norm of the resolvent satisfies

$$(2.\,9) \qquad \|R(\lambda;-D_\beta)\|_{[p,p]} \leqq A_p\lambda^{-1} \qquad (\lambda>0,\ \beta>0,\ 1<p<\infty),$$

where A_p is a constant depending on p only. Indeed, defining for each $\lambda>0$ a bounded operator B_λ on $L^p_{2\pi}$ by $[B_\lambda f]\hat{}\,(k)=b^\lambda_k f\hat{}\,(k)$, $k\in\mathbf{Z}$ with $b^\lambda_k = (\lambda+e^{\beta|k|}-1)^{-1}$, it follows by partial summation that

$$\sum_{k=-n}^{n} b^\lambda_k f\hat{}\,(k)e^{ikx} = \sum_{k=0}^{n-1} (b^\lambda_k - b^\lambda_{k+1})S_k(f;x) + b^\lambda_n S_n(f;x),$$

where $S_k(f;x)$ denotes the kth partial sum of the Fourier series of f. Since $b^\lambda_k - b^\lambda_{k+1} \geqq 0$ and ([10], p. 350)

$$(2.\,10) \qquad \|S_k(f;x)\|_p \leqq A_p\|f(x)\|_p \qquad (1<p<\infty;\ k=0,1,\ldots),$$

it follows by a standard argument ([10], p. 233) that $B_\lambda f\in L^p_{2\pi}$ for each $\lambda>0$ and thus

$$\|B_\lambda f\|_p = \lim_{n\to\infty} \Big\|\sum_{k=-n}^{n} b^\lambda_k f\hat{}\,(k)e^{ikx}\Big\|_p \leqq A_p b_0\|f\|_p = A_p\lambda^{-1}\|f\|_p.$$

Now $B_\lambda(\lambda f+D_L f) = f$ for each f in the dense set $\mathscr{D}(D_L)$, thus $B_\lambda=R(\lambda;-D_L)$, proving (2. 9).

Hence the Hille—Yosida theorem and its corollary can be applied again giving that there exists a unique semi-group $\{T_\beta(t);\ t\geqq 0\}$ with infinitesimal generator $-D_\beta$ for each $\beta>0$, namely

$$(2.\,11) \quad (T_\beta(t)f)(x) = \frac{1}{2\pi} \int_{-\pi}^{\pi} f(x-u)\chi_{\beta,t}(u)\,du \qquad (f\in L^p_{2\pi},\ 1<p<\infty,\ t>0)$$

with $T_\beta(0)f=f$ and kernel

$$\chi_{\beta,t}(u) = \sum_{k=-\infty}^{\infty} e^{-t(e^{\beta|k|}-1)}e^{iku}.$$

By the same reasoning as in the proof of Theorem 1, using (1. 5) and replacing (2. 4) by $a_n'=n$ and m by $m'=\llbracket \beta^{-1} \log t^{-1} \rrbracket$, one obtains

THEOREM 3. *Let* $f \in L_{2\pi}^p$, $1<p<\infty$, $\beta>0$, *and* $0<\alpha<1$. *Then* $E_n[f]=O(e^{-\alpha\beta n})$ $n \to \infty$, *implies* $\|T_\beta(t)f-f\|_p = O(t^\alpha)$, $t \to 0+$.

Again, corresponding results for $\alpha \geqq 1$ with higher differences in $T_\beta(t)f$ are also valid.

To prove the converse of Thm. 3 we choose as a testing process

$$(2. 12) \qquad \mathfrak{U}_{\beta,n}(f; x) = \sum_{k=-n+1}^{n-1} \left\{1 - \frac{e^{\beta|k|}-1}{e^{\beta n}-1}\right\} f^\wedge(k) e^{ikx} \qquad (\beta>0).$$

Using (2. 10) and (1. 5) one has immediately part a) of the following

LEMMA 2. a) *For each* $f \in L_{2\pi}^p$, $1<p<\infty$, $\beta>0$ *there holds*

$$\|\mathfrak{U}_{\beta,n}(f; x)\|_p \leqq M_p \|f\|_p,$$

where M_p *is a constant independent on n and f, and*
b) *for each* $f \in \mathcal{D}(D_\beta)$

$$\|f(x) - \mathfrak{U}_{\beta,n}(f; x)\|_p \leqq A_p \|D_\beta f\|_p (e^{-\beta n}-1)^{-1} \qquad (n=1, 2, \ldots),$$

where A_p *is the same constant as in* (2. 10).

To prove b) we put

$$\mu_k(n) = \begin{cases} 1, & 0 \leqq k \leqq n \\ (e^{\beta n}-1)/(e^{\beta|k|}-1), & k \geqq n \end{cases}$$

and obtain as above by partial summation for each $f \in \mathcal{D}(D_\beta)$

$$\|f(x) - \mathfrak{U}_{\beta,n}(f; x)\|_p = (e^{\beta n}-1)^{-1} \left\| \sum_{k=-\infty}^{\infty} (D_\beta f)^\wedge(k) \mu_k(n) e^{ikx} \right\|_p =$$

$$= \lim_{m \to \infty} (e^{\beta n}-1)^{-1} \left\| \sum_{k=0}^{m-1} (\mu_k(n) - \mu_{k+1}(n)) S_k(D_\beta f; x) + \mu_m S_m(D_\beta f; x) \right\|_p \leqq$$

$$\leqq \lim_{m \to \infty} (e^{\beta n}-1)^{-1} \left\{ \sum_{k=0}^{m-1} (\mu_k(n) - \mu_{k+1}(n)) + \mu_m \right\} A_p \|D_\beta f\|_p =$$

$$= \mu_n(n) A_p \|D_\beta f\|_p = A_p \|D_\beta f\|_p.$$

Here we used (2. 10) and the fact that $\mu_k(n) - \mu_{k+1}(n) > 0$ for $k \geqq n$, $\mu_k(n) - \mu_{k+1}(n) = 0$ for $0 \leqq k < n$, and $\lim_{m \to \infty} \mu_m(n) = 0$. This proves Lemma 2.

From this result the desired Jackson-type theorem follows by an interpolation argument as indicated after Theorem 2 above, namely

THEOREM 4. *If $f \in L_{2\pi}^p$, $1 < p < \infty$, $\beta > 0$, $0 < \alpha < 1$, then $\|T_\beta(t)f - f\|_p = O(t^\alpha)$, $t \to 0+$, implies $E_n[f] = O(e^{-\alpha\beta n})$, $n \to \infty$.*

The characterization of the property $E_n[f] = O(e^{-\alpha\beta n})$, $n \to \infty$, given by Thms. 3 and 4 is a counterpart to the classical result of S. N. Bernstein [5] which states that for a function $f \in C_{2\pi}$ and $\beta > 0$ one has $\limsup\limits_{n \to \infty} (E_n[f])^{1/n} = e^{-\beta}$ if and only if $f(x) = F(x)$, where $F(x+iy)$ is analytic in the strip $|y| < \beta$. This result has been considerably extended by S. M. NIKOLSKIĬ [25], S. M. NIKOLSKIĬ—M. K. POTAPOV [26], and M. K. POTAPOV [29]. Correspondingly, our class $\mathscr{D}(D_\beta)$ will be characterized by analyticity properties in Thm. 6 below.

We add two remarks concerning the exponential approximation process (2. 12).
1. The process $\mathfrak{U}_{\beta,n}(f; x)$ is actually known: it is easy to show that it coincides with the truncated Poisson sum $(r = e^\beta > 1)$

$$\sum_{v=0}^{n-1} r^v S_v(f; x) \left\{ \sum_{v=0}^{n-1} r^v \right\}^{-1}$$

which was introduced by M. GHERMANESCO [18]. This was pointed out to the author by E. L. Stark.

The fact that $\mathfrak{U}_{\beta,n}(f; x)$ does not define an approximation process on $C_{2\pi}$ may be derived from the representation of its kernel $(r = e^\beta)$

$$(2.13) \qquad \sum_{k=-n+1}^{n-1} \left\{ 1 - \frac{e^{\beta|k|} - 1}{e^{\beta n} - 1} \right\} e^{ikx} =$$

$$= \frac{r-1}{r^n-1} \frac{r^{n+1} \sin(n-1/2)x - r^n \sin(n+1/2)x + (r+1)\sin x/2}{(\sin x/2)(r^2 - 2r\cos x + 1)}$$

whose $L_{2\pi}^1$-norm is $O(\log n)$ as $n \to \infty$. We omit the elementary proof of this fact.
2. It may be noticed that e.g. in case $f \in L_{2\pi}^2$ the following representation of $\mathfrak{U}_{\beta,n}(f; x)$ as an infinite linear combination of typical means (see footnote 7) is immediate:

$$(2.14) \qquad \mathfrak{U}_{\beta,n}(f; x) = (e^{\beta n} - 1)^{-1} \sum_{v=1}^{\infty} \frac{(\beta n)^v}{v!} \mathfrak{T}_n^v(f; x).$$

3. Characterization of logarithmic and exponential differentiability

The existence of a logarithmic derivative is a less restrictive condition than e.g. the existence of a fractional Riesz derivative $D^{[\gamma]}f$ for any $\gamma > 0$. On the other hand, the existence of the exponential derivative $D_\beta f$ in X for any $\beta > 0$ is stronger than ordinary or Riesz differentiability of arbitrary order. More precisely, as an obvious consequence of the theory of Fourier multipliers one has the following

LEMMA 3. *Let* $f \in X$.
a) *If* $D_\beta f \in X$ *for some* $\beta > 0$, *then* $f^{(n)} \in X$ *for any* $n = 1, 2, \ldots$ *as well as* $D^{[\gamma]}f \in X$ *for any* $\gamma > 0$.
b) $D^{[\gamma]}f \in X$ *for some* $\gamma > 0$ *implies* $D_L f \in X$.

An example of a function $f_0 \in L_{2\pi}^1$ with $f_0^{(n)} \in L_{2\pi}^1$ for $n = 1, 2, \ldots$ which does not possess an exponential derivative $D_\beta f_0$ for any $\beta > 0$ is given by $f_0(x) = \sum_{k=-\infty}^{\infty} \exp(-|k|^{1/2}) e^{ikx}$. Similarly the function $f_1 \in L_{2\pi}^1$ with Fourier coefficients $f_1^{\wedge}(k) = [\log(2+|k|)]^{-2}$, $k \in \mathbf{Z}$, shows that also the converse of part b) does not hold.

In order to characterize the condition $f \in \mathscr{D}(D_L)$ by means of smoothness properties upon f, let us first consider the corresponding characterization of the condition $f \in \mathscr{D}(D^{[\gamma]})$. If $f \in X$ and $0 < \gamma < 1$, the existence of $D^{[\gamma]}f \in X$ is equivalent to the existence of the strong limit[8])

$$(3.1) \quad \frac{1}{\pi} \sin \frac{\pi\gamma}{2} \, \Gamma(\gamma+1) \, \operatorname*{s-lim}_{\varepsilon \to 0+} \int_\varepsilon^\infty \frac{f(x+t) - 2f(x) + f(x-t)}{t^{1+\gamma}} \, dt = (D^{[\gamma]}f)(x).$$

In view of Lemma 3, b) it can be expected that for the logarithmic derivative a similar formula holds with $t^{1+\gamma}$ replaced by a function that behaves like t near the origin. Indeed we have

THEOREM 5. *A function* $f \in X$ *has a logarithmic derivative* $D_L f \in X$ *if and only if in case* $X = C_{2\pi}$ *or* $L_{2\pi}^1$

$$\operatorname*{s-lim}_{\varepsilon \to 0+} \frac{1}{2} \int_\varepsilon^\infty \frac{f(x+t) - 2f(x) + f(x-t)}{t e^t} \, dt \in X$$

---

[8]) For references and further results cf. e.g. P. L. BUTZER—R. J. NESSEL [10], p. 428 f, 416 f, P. L. BUTZER—E. GÖRLICH [9], P. L. BUTZER— W. TREBELS [12], [13], E. GÖRLICH [19], R. L. WHEEDEN [35].

and, in case $X=L_{2\pi}^p$, $1<p<\infty$

$$\left\|\int_\varepsilon^\infty \frac{f(x+t)-2f(x)+f(x-t)}{te^t}\,dt\right\|_p = O(1) \qquad (\varepsilon\to 0+).$$

If one of these properties is satisfied it follows that

$$(D_L f)(x) = \underset{\varepsilon\to 0+}{\text{s-lim}}\ \frac{1}{2}\int_\varepsilon^\infty \frac{f(x+t)-2f(x)+f(x-t)}{te^t}\,dt.$$

The proof of Thm. 5 is similar to the somewhat involved proof of (3. 3), using arguments of R. SALEM—A. ZYGMUND [31] (cf. also [13], § 6 and [9]) and some properties of the Abel—Poisson integral of f. Here the formula[9])

$$\int_0^\infty \frac{1-\cos kt}{te^t}\,dt = \log\,[(1+k^2)^{1/2}] \qquad (k\in\mathbf{Z})$$

is essential. We omit further details.

This result has some connection with the general theory of the logarithm of a closed operator as treated e.g. by V. NOLLAU [27], but his representations are valid under more restrictive assumptions and do not imply Thm. 5.

In order to characterize the condition $f\in\mathcal{D}(D_\beta)$ we use a result of N. I. ACHIESER [1], [2] (see A. F. TIMAN [33], p. 141). It states that a function $f\in L_{2\pi}^p$, $1<p<\infty$, is equal a.e. to a function $F(x)$ having an analytic continuation $F(x+iy) = u(x,y)+$ $+iv(x,y)$ in the strip $|y|<\beta$ and satisfying $\|u(\cdot,y)\|_p=O(1)$ for $y\to\pm\beta$ if and only if there exists a $g\in L_{2\pi}^p$ such that $f^\wedge(k)=e^{-\beta|k|}(1+e^{-2\beta|k|})g^\wedge(k)$, $k\in\mathbf{Z}$. The latter property is easily seen to be equivalent to $f\in\mathcal{D}(D_\beta)$, so that we obtain

THEOREM 6. *A function $f\in L_{2\pi}^p$, $1<p<\infty$, has an exponential derivative $D_\beta f$ of order $\beta>0$ in $L_{2\pi}^p$ if and only if $f(x)=F(x)$ a.e., where $F(x+iy)$ is analytic in the strip $|y|<\beta$ and $\|\mathrm{Re}\,F(\cdot+iy)\|_p=O(1)$ for $y\to\pm\beta$.*

Finally, we remark that, implicitely, the present results also contain the saturation theorems for the approximation processes (2. 6), (2. 12) as well as for the semigroup operators (2. 2), (2. 11) in the spaces under consideration. For example, if $f\in L_{2\pi}^1$ then $\|T_L(t)f-f\|_1 = o(t)$, $t\to 0+$, implies $f=$const. a.e., and $\|T_L(t)f-f\|_1=$ $=O(t)$, $t\to 0+$ is equivalent to $\left\|\int_\varepsilon^\infty [f(x+t)-2f(x)+f(x-t)](te^t)^{-1}\,dt\right\|_1 = O(1)$

---

[9]) A. ERDÉLYI [15], p. 157, (59).

$\varepsilon \to 0+$. This may be interpreted in the following sense. If $T_L(t)f$ is understood as a fractional-type integral of f of order $t > 0$, namely the inverse of the Bessel derivative $f^{\{t\}}(x) \sim \sum_{k=-\infty}^{\infty} (1+k^2)^{t/2} f^{\wedge}(k) e^{ikx}$, then the saturation class of this fractional integral, considered as an approximation process for $t \to 0+$, is given by the relative completion of $\mathscr{D}(D_L)$ with respect to $L_{2\pi}^1$.

REFERENCES

[1] N. I. Achieser, *On best approximation of some classes of continuous periodic functions* (Russian). Dokl. Akad. Nauk SSSR **17** (1937), 451—454.

[2] N. I. Achieser, *On best approximation of analytic functions* (Russian). Dokl. Akad. Nauk SSSR **18** (1938), 241—244.

[3] N. K. Bari and S. B. Stečkin, *Best approximation and differential properties of two conjugate functions* (Russian). Trudy Mosk. Mat. Obšč. **5** (1956), 483—522.

[4] S. N. Bernstein, *Sur l'ordre de la meilleure approximation des fonctions continues par des polynomes de degré donné*. Mémoires Académie Royale de Belgique, Classe des Sci. (2) **4** (1912), 1—103.

[5] S. N. Bernstein, *Sur la valeur asymptotique de la meilleure approximation des fonctions analytiques, admettant des singularités données*. Bull. Acad. de Belgique **2** (1913), 76—90.

[6] S. N. Bernstein, *Leçons sur les propriétés extrémales et la meilleure approximation des fonctions analytiques d'une variable réelle*. Gauthier-Villars, Paris 1926.

[7] P. L. Butzer and H. Berens, *Semi-groups of operators and approximation*. (Grundlehren der Math. Wiss., Vol. **145**). Springer, Berlin 1967.

[8] P. L. Butzer und E. Görlich, *Saturationsklassen und asymptotische Eigenschaften trigonometrischer singulärer Integrale*. Festschrift z. Gedächtnisfeier f. Karl Weierstrass, 339—392. Westdeutscher Verlag, Köln 1965.

[9]. P. L. Butzer und E. Görlich, *Zur Charakterisierung von Saturationsklassen in der Theorie der Fourierreihen*. Tôhoku Math. J. **17** (1965), 29—54.

[10] P. L. Butzer and R. J. Nessel, *Fourier analysis and approximation*. Vol. I. Birkhäuser, Basel — Academic Press, New York 1971.

[11] P. L. Butzer und K. Scherer, *Approximationsprozesse und Interpolationsmethoden*. (Hochschulskripten 826/826a). Bibliographisches Institut, Mannheim 1968.

[12] P. L. Butzer and W. Trebels, *Hilbert transforms, fractional integration and differentiation*. Bull. Amer. Math. Soc. **74** (1968), 106—110.

[13] P. L. Butzer und W. Trebels, *Hilberttransformation, gebrochene Integration und Differentiation*. (Forschungsberichte des Landes Nordrhein-Westfalen, Nr. 1889). Westdeutscher Verlag, Köln 1968.

[14] P. Civin, *Inequalities for trigonometric integrals*. Duke Math. J. **8** (1941), 656—665.

[15] A. Erdélyi, (editor) *Tables of integral transforms* Vol. I. McGraw-Hill, New York 1954.

[16] M. Fekete, *Über einen Satz des Herrn Serge Bernstein*. J. Reine Angew. Math. **146** (1916), 88—94.

[17] Ja. L. Geronimus, *Some remarks on the method of coefficients*. Dokl. Akad. Nauk SSSR **166** (1966), 1274—1276. English transl. Soviet Math. **7** (1966), 263—265.

[18] M. Ghermanesco, *Sur l'intégrale de Poisson I, II*. Bulletin Scientifique de l'École Polytechnique de Timişoara **4** (1932), fasc. 3—4, p. 159—184; **5** (1933), fasc. 1—2. p. 41—74.

[19] E. Görlich, *Distributional methods in saturation theory.* J. Approximation Theory 1 (1968), 111—136.

[20] E. Görlich, *Logarithmische und exponentielle Ungleichungen vom Bernstein-Typ und verallgemeinerte Ableitungen.* Habilitationsschrift, Technische Hochschule Aachen, 1971. 68 pp.

[21] E. Görlich, R. J. Nessel, and W. Trebel, *Bernstein-type inequalities for families of multiplier operators in Banach spaces with Cesàro decompositions* (to appear).

[22] N. P. Kuptsov, *Direct and converse theorems of approximation theory and semigroups of operators.* Uspehi Mat. Nauk 23 (1968), no. 4 (142), 117—178. English transl. Russian Math. Surveys 23 (1968), 115—177.

[23] B. M. Levitan, *Generalized translation operators and some of their applications.* Israel Program of Scientific Translations, Jerusalem 1964.

[24] G. G. Lorentz, *Approximation of functions.* Holt, Rinehart and Winston, New York 1966.

[25] S. M. Nikolskiĭ, *Uniform differential properties of an analytic function in a strip* (Russian). Mathematica (Cluj) 2 (25) (1960), 149—157.

[26] S. M. Nikolskiĭ and M. K. Potapov, *The boundary properties of functions analytic in a strip* (Russian). Mathematica (Cluj) 4 (27) (1962), 123—130.

[27] V. Nollau, *Über den Logarithmus abgeschlossener Operatoren in Banachschen Räumen.* Acta Sci. Math. (Szeged) 30 (1969), 161—174.

[28] I. I. Ogiewetzki, *Generalization of the inequality of P. Civin for the fractional derivative of a trigonometric polynomial to L_p space.* Dopovidi AN URSR 1958, No. 5, 486—488. English transl. Acta Math. Acad. Sci. Hung. 9 (1958), 133—135.

[29] M. K. Potapov, *On the problem of boundary properties of functions analytic in a strip* (Russian). Mathematica (Cluj) 7 (30) (1965), 341—356.

[30] F. Riesz, *Sur les polynomes trigonométriques.* C. R. Akad. Sci. Paris 158 (1914), 1657—1661.

[31] R. Salem and A. Zygmund, *Capacity of sets and Fourier series.* Trans. Amer. Math. Soc. 59 (1946), 23—41.

[32] G. Szegő, *Über einen Satz des Herrn S. Bernstein.* Schriften der Königsberger Gelehrten Gesellschaft 5 (1928), 59—70.

[33] A. F. Timan, *Theory of approximation of functions of a real variable.* Pergamon Press, Oxford 1963.

[34] C. Watari, *A note on saturation and best approximation.* Tôhoku Math. J. 15 (1963), 273—276.

[35] R. L. Wheeden, *On hypersingular integrals and Lebesgue spaces of differentiable functions.* Trans. Amer. Math. Soc. 134 (1968), 421—435.

Fourier Multipliers whose Multiplier Norm is an Attained Value

By

HAROLD S. SHAPIRO

DEPT. OF MATH.
UNIVERSITY OF MICHIGAN
ANN ARBOR

1. Introduction

Let $1 \leqq p \leqq \infty$, and denote by $J_p = J_p(R)$ the set of tempered distributions u on the real line R such that the transformation $\varphi \to u * \varphi$, defined initially for infinitely differentiable φ with compact support, extends to a bounded linear operator (which we shall denote by $[u]$) from $L^p(R)$ to $L^p(R)$ (this for $p < \infty$; for $p = \infty$, the closure of the set of test functions φ is of course the set of *continuous* functions *) in $L^\infty(R)$.) If we introduce a norm in J_p by defining $\|u\|_{J_p}$ to be the operator norm of $[u]$, J_p becomes a Banach space. We denote by $M_p = M_p(\hat{R})$ the set of Fourier transforms of elements of J_p, defining $\|\hat{u}\|_{M_p} = \|u\|_{J_p}$, so that M_p is made into a Banach space isometrically isomorphic to J_p. As is well known (see [2, 3]) the elements of $M_p(\hat{R})$ are bounded measurable functions on $\hat{R}$, called the *Fourier multipliers* (of $L^p(R)^\frown$); moreover, $M_p(\hat{R})$ is a Banach algebra with respect to pointwise multiplication. We take for granted the following basic propositions, proved in [2]:

(a) *$J_1(R)$ is precisely the set of bounded measures on R; so that $M_1(\hat{R})$ is the space (usually called $B(\hat{R})$ in harmonic analysis) of Fourier transforms of bounded measures.*

(b) *$M_2(\hat{R}) = L^\infty(\hat{R})$ (isometrically).*

(c) *For $1 \leqq p_1 < p_2 \leqq 2$, $M_{p_1}(\hat{R}) \subset M_{p_2}(\hat{R})$; the M_{p_2} norm does not exceed the M_{p_1} norm, and in particular*

$$\|\hat{u}\|_{M_p} \leqq \|\hat{u}\|_\infty \qquad (1 \leqq p \leqq 2).$$

(d) *$M_p = M_q$ (isometrically), if $p^{-1} + q^{-1} = 1$.*

(e) *Each affine map $x \to ax + b$ of $\hat{R}$ onto $\hat{R}$ induces an isometry of $M_p(\hat{R})$ on itself.*

The above discussion is valid, *mutatis mutandis*, with R, $\hat{R}$ replaced by any locally compact Abelian group G and its dual $\hat{G}$. Of interest in this paper will be $J_p(T)$ and $M_p(Z)$, where T denotes the "circle" (real numbers (mod 2π)) and Z the integers. A basic proposition connecting $M_p(\hat{R})$ with $M_p(Z)$ see [3, p. 374]) is:

---

*) which moreover vanish at infinity.

(f) *Let U denote a continuous complex-valued function on $\hat{R}$ which belongs to $M_p(\hat{R})$. Then, the sequence $\{U(n)\}_{n=-\infty}^{\infty}$ belongs to $M_p(Z)$, and its norm does not exceed the $M_p(\hat{R})$ norm of U.*

Actually, for (f) to hold, it is not essential that U be continuous, but merely *regulated* (a concept due to DE LEEUW), meaning that $\lim\limits_{a \to 0} (1/2a) \int_{-a}^{a} U(x+\xi)d\xi = U(x)$ for all $x \in \hat{R}$.

The present paper is concerned with the Fourier multipliers U on $\hat{R}$ or Z such that the inequality $\|U\|_{M_p} \leqq \|U\|_{\infty}$ becomes equality, and moreover $\|U\|_{M_p}$ is an attained value of U. Thus, for U defined on Z, we are led (making trivial normalisations) to consider the class

$$W_p = \{u \in J_p(T) : \|u\|_{J_p} = 1 \text{ and } \hat{u}(0) = 1\}.$$

For $p=2$ this class presents no interest; the Fourier transforms of elements of W_p are merely sequences bounded by one, and containing a "1" in the central position. Therefore, *we shall henceforth assume $p \neq 2$.* Consider now $p=1$: a moment's thought shows that W_1 *is the set of positive measures on T having total mass one.* Suppose now $u \in W_1$, and $\hat{u}(-1)=1$. Then, writing $u=d\varrho$ we have $\int_{-\pi}^{\pi} e^{it} d\varrho(t) = \int_{-\pi}^{\pi} d\varrho$, whence it follows that ϱ is a unit point mass concentrated at $t=0$. The main result of this paper is that the analogous proposition holds in W_p, $p \neq 2$:

THEOREM 1. *Let u be a distribution on the circle satisfying $\|u*f\|_p \leqq \|u\|_p$ for trigonometric polynomials f, and $\hat{u}(-1)=\hat{u}(0)=1$. We suppose $1 \leqq p \leqq \infty$, $p \neq 2$. Then $\hat{u}(n)=1$ for all n, or what is the same thing, $u=\delta$ (unit point mass at $t=0$).*

Another way of stating theorem 1 is that a sequence containing two ones "back to back", but not consisting identically of ones, has norm (in $M_p(Z)$) strictly larger than 1. The method of proof of theorem 1 actually furnishes lower bounds for this norm (for details see Section 2. 1 below).

Naturally, Theorem 1 would be of little interest unless W_p contains "genuine" distributions, i.e. distributions which are not measures, for $1<p<\infty$. In view of the following result, this is certainly the case for p in the interval [4/3, 4] and one can also show that it is true for $1<p<\infty$.

THEOREM 2. *Let $u \in J_4(T)$ have norm not exceeding $1/3$, and $\hat{u}(0)=0$. Then $u_0 + u$ has norm 1, where u_0 denotes the function identically equal to 1 on T; in other words, the sequence*

$$... \hat{u}(-2), \hat{u}(-1), 1, \hat{u}(1), \hat{u}(2), ...$$

has $M_4(Z)$ norm equal to 1. (The constant $1/3$ is not sharp).

Since $J_4(T)$ is known to contain distributions which are not measures, we can, letting v denote one of these, choose $u = \varepsilon(v - \hat{v}(0))$ with sufficiently small ε, in order to get the desired example.

Theorem 2 is due to Charles FEFFERMAN (unpublished), who generously consented to my inclusion of the result, and its proof, in the present paper. (Strictly speaking, Fefferman proved his result for convolution operators on *real-valued* $L^4(T)$; the extension to complex-valued $L^4(T)$ involves a slight technical complication).

From theorem 1, and (f) above, we can deduce some analogous results for $M_p(\hat{R})$:

THEOREM 3. *Let U denote a continuous function on $\hat{R}$, not identically equal to 1. Suppose for every $\varepsilon > 0$ there exists a pair of distinct points x_1, x_2 such that $|x_1 - x_2| \leq \varepsilon$, $U(x_1) = U(x_2) = 1$. Then, for $1 \leq p \leq \infty$, $p \neq 2$ the $M_p(\hat{R})$ norm of U exceeds 1.*

COROLLARY. *If $U(x)$ equals 1 throughout an interval, but not identically, its $M_p(\hat{R})$ norm $(p \neq 2)$ exceeds 1.*

Actually, as Nestor RIVIÈRE has pointed out to me, we can by suitably adapting the method used to prove Theorem 1, obtain a result much stronger than the above Corollary, namely:

(∗) *If $U(x) = 1 + O(|x - x_0|^c)$ in a neighbourhood of $x = x_0$, where $c > 4$, and $U(x)$ is not identically 1, its $M_p(\hat{R})$ norm $(p \neq 2)$ exceeds 1.*

Here it is an open question whether 4 is the best constant, at least for certain p. Observe that *for $p = 1$, the much weaker hypothesis*

$$(1) \qquad U(x) = 1 + o(|x - x_0|^2)$$

implies that the $M_1(\hat{R})$ norm of U exceeds 1. Indeed, assuming the norm of U were 1, and supposing $x_0 = 0$, we should necessarily have $U = \hat{\varrho}$ where ϱ is a positive measure on R, implying that U is of positive type ("positive definite") on $\hat{R}$; but, for such U, $U(x) - U(0) = o(x^2)$ implies U is constant, as is well known (see [4], p. 59).

It seems very unlikely that the constant 4 in (∗) is sharp. Quite possibly the hypothesis on U in (∗) could even be replaced by (1), which would then give a definitive result (in view, e.g., of $U(x) = (1 + x^2)^{-1}$ whose $M_1(\hat{R})$ norm is 1). In view of these uncertainties, we shall not include the proof of (∗) in the present paper, but hope to return to the topic in a later work. This topic has an interesting connection with approximation theory, which was in fact the initial stimulus to my studying the circle of questions in the present paper. Therefore, it seems fitting to discuss briefly this connection here, even though a detailed discussion will have to be postponed to some later occasion.

For some purposes it is of interest to consider, instead of the usual "approximate identity"

$$(2) \qquad k_{(a)}(t) = (1/a)k(t/a) \qquad (k \in L^1(R), \ \int k \, dt = 1)$$

(conceived as a convolution operator on $L^p(R)$, i.e. an element of $J_p(R)$) the analogously defined operators with k no longer assumed to be an integrable function, but merely an element of J_p satisfying $\hat{k}(0)=1$. The case when the J_p norm of k equals 1 is then analogous to the case of a "positive kernel", i.e. $k(t) \geqq 0$ in (2). As is well known, the positive kernels (2) are saturated with order a^2, and $(*)$ can be shown to imply that the analogously defined operators just described have a saturation order which in any case cannot be $O(a^c)$ for any $c > 4$. This is always assuming, of course, $p \neq 2$: for $p = 2$, we can nontrivially have $\hat{k}(x) = 1$ throughout a neighbourhood of $x = 0$, and hence no saturation phenomenon.

As a final remark, we wish to point out that our method of proof of Theorem 1 seems susceptible of generalization, and could possibly be made to yield an analogous result for some other discrete groups.

2. Proof of Theorem 1

It is convenient to assemble some (mostly well known) preliminary material. Let us first recall the definition that an element f of a Banach space B is *orthogonal* to the subspace S of B if

$$(1) \qquad \| f - g \| \geqq \| f \|, \quad \text{all} \quad g \in S$$

i.e. if 0 is a closest element to f from S. (For general discussion of this concept see [5].)

LEMMA 1. *Let φ denote a non-null bounded linear functional on B which vanishes on the subspace S. Then, any element $f \neq 0$ of B such that $|\langle \varphi, f \rangle| = \| \varphi \|_{B^*} \| f \|$ is orthogonal to S.*

PROOF. For $g \in S$ we have

$$\| \varphi \|_{B^*} \| f \| = |\langle \varphi, f \rangle| = |\langle \varphi, f - g \rangle| \leqq \| \varphi \|_{B^*} \| f - g \|$$

which implies (1).

REMARK. The converse is also true: *if $f \neq 0$ is an element of B orthogonal to S, there exists a non-null bounded linear functional φ vanishing on S for which $\langle \varphi, f \rangle = \| \varphi \|_{B^*} \| f \|$*. However, we shall not require this fact (whose proof is a simple application of the Hahn—Banach theorem).

LEMMA 2. *Let S denote a closed subspace of $L^p(\varrho)$, $1 \leq p < \infty$ where ϱ is a positive measure on some space, and φ an element of $L^q(\varrho)$ (where $q = p/(p-1)$) satisfying*

$$(2) \qquad \int \varphi h \, d\varrho = 0, \quad \text{for all} \quad h \in S.$$

Then the function f, where

$$f(x) = \begin{cases} |\varphi(x)|^q/\varphi(x), & \varphi(x) \neq 0 \\ 0, & \varphi(x) = 0 \end{cases}$$

belongs to $L^p(\varrho)$ and is orthogonal to S.

PROOF. That $f \in L^p(\varrho)$ is clear, since $|f(x)|^p = |\varphi(x)|^{(q-1)p} = |\varphi(x)|^q$. Now, using the bilinear form

$$\langle \psi, g \rangle = \int \psi g \, d\varrho \qquad (\psi \in L^q(\varrho), \; g \in L^p(\varrho))$$

to identify $L^q(\varrho)$ as the dual of $L^p(\varrho)$, (2) says that φ vanishes on S, and since

$$\langle \varphi, f \rangle = \int |\varphi|^q \, d\varrho = \left(\int |\varphi|^q \, d\varrho \right)^{1/q} \left(\int |\varphi|^q \, d\varrho \right)^{1/p} = \|\varphi\|_q \|f\|_p$$

the desired conclusion follows from Lemma 1.

DEFINITION. *For $0 \leq \lambda < 1$, $1 < p < \infty$*

$$f_{p,\lambda}(t) = |1 + \lambda e^{it}|^q/(1 + \lambda e^{it}).$$

Here, as always henceforth in this paper, $q = p/(p-1)$.

LEMMA 3. *$f_{p,\lambda}$ is orthogonal to the subspace R of $L^p(T)$ defined by:*

$$R = \text{closed subspace spanned by } \{e^{int}\}: \quad n \in Z \setminus \{-1, 0\}.$$

PROOF. Clearly $\varphi(t) = 1 + \lambda e^{it}$ satisfies

$$\int_T \varphi(t) e^{int} \, dt = 0 \qquad (n \in Z \setminus \{-1, 0\}).$$

Thus $\int_T \varphi h \, dt = 0$, $h \in R$ and Lemma 2 implies the desired conclusion.

LEMMA 4. *The Fourier coefficient of rank $n \geq 0$ of $f_{p,\lambda}$ is*

$$(3) \qquad \hat{f}_{p,\lambda}(n) = \lambda^n \sum_{k=0}^{\infty} \binom{q/2}{k} \binom{(q/2)-1}{k+n} \lambda^{2k}.$$

If we are given p, $2 < p < \infty$ and $n \geq 0$ we can find $\lambda = \lambda(p, n)$ so that $\hat{f}_{p,\lambda}(n) \neq 0$.

PROOF. Writing $z = \lambda e^{it}$, we have

$$f_{p,\lambda}(t) = |1 + z|^q/(1 + z) = (1 + z)^{(q/2)-1}(1 + \bar{z})^{q/2},$$

and replacing each factor on the right by its (absolutely convergent) expansion in a binomial series, multiplying out, and collecting terms that contain e^{int}, we get (3). For $2 < p < \infty$, neither $q/2$ nor $(q/2) - 1$ is a non-negative integer, consequently the series in (3) is non-terminating and for suitably chosen $\lambda > 0$ the right side of (3) is not equal to zero.

PROOF OF THEOREM 1. In view of the identity of J_p and J_q, *there is no loss of generality in supposing $p > 2$.* Let now $u \in J_p(T)$ have norm 1, $\hat{u}(-1) = \hat{u}(0) = 1$. We require the following inequality of Clarkson, valid for a pair of elements x, y in any $L^p(\varrho)$ with $2 \leq p < \infty$ (cf. [1, p. 400]):

$$(4) \qquad \|(1/2)(x+y)\|^p + \|(1/2)(x-y)\|^p \leq (1/2)(\|x\|^p + \|y\|^p).$$

Now, fix an integer $n \geq 1$ and choose λ, $0 < \lambda < 1$ so that $\hat{f}_{p,\lambda}(n) \neq 0$, which is possible by virtue of Lemma 4. For notational convenience, we denote $f_{p,\lambda}$ (for this choice of λ) simply by f. We propose to insert f for x, and $u*f$ for y, in (4). Now, recall (Lemma 3) that f is orthogonal to R, and this implies that any element $F \in L^p(T)$ whose Fourier coefficients of rank 0 and -1 coincide with those of f (so that $f - F \in R$) satisfies $\|F\| \geq \|f\|$. In particular, because $\hat{u}(-1) = \hat{u}(0) = 1$, the element $F = (1/2)(f + (u*f))$ has this property and (4) yields

$$\|f\|^p + \left\|(1/2)(f - (u*f))\right\|^p \leq (1/2)(\|f\|^p + \|u*f\|^p) \leq \|f\|^p.$$

Thus, $f - (u*f) = 0$, whence $\hat{f}(n)(1 - \hat{u}(n)) = 0$ implying that $\hat{u}(n) = 1$ for every $n \geq 1$. To conclude the proof of Theorem 1 we have only to observe that the mapping $v \to -v$ induces an isometry of $M_p(Z)$ on itself. (Or alternatively, we could proceed as above, obtaining an explicit formula similar to (3) for the negatively indexed Fourier coefficients of $f_{p,\lambda}$.)

DEDUCTION OF THEOREM 3. Assume U satisfies the hypotheses of Theorem 3, but has norm 1. We deduce that (contrary to hypothesis) $U(x)$ is identically 1. Indeed: given $\varepsilon > 0$, let $x_1 < x_2 \leq x_1 + \varepsilon$ and $U(x_1) = U(x_2) = 1$. Then

$$V(n) = U(x_1 - n(x_2 - x_1)) \qquad (n \in Z)$$

is an element of $M_p(Z)$ of norm 1 with $V(-1) = V(0) = 1$, hence $V(n) = 1$ for all n, i.e. $U(x) = 1$ on a lattice of points with spacing $\leq \varepsilon$. Since ε can be chosen arbitrarily small, we have (by continuity) $U(x) = 1$, all x.

REMARK. The continuity hypothesis in Theorem 3 can be dropped if we make a stronger hypothesis about the set where $U(x) = 1$. Thus, a more recondite argument shows: *if U is any measurable function, equal to 1 on a set of positive measure but not identically, then for $p \neq 2$ the $M_p(\hat{R})$ norm of U exceeds 1.*

2.1 *Quantitative refinement of Theorem 1.* Suppose we have $u \in J_p(T)$ with $\hat{u}(-1) = \hat{u}(0) = 1$; denoting by m the norm of u, we have, by the reasoning used in proving Theorem 1, (again writing f for $f_{p,\lambda}$):

$$\|f\|^p + 2^{-p}\|f - (u * f)\|^p \leq (1/2)(\|f\|^p + m^p\|f\|^p)$$

whence

$$(5) \qquad \|f - (u * f)\| \leq 2^{1/q}(m^p - 1)^{1/p}\|f\|.$$

Now, to get the most out of this inequality, we wish to choose λ so that $|\hat{f}_{p,\lambda}(n)|$ is as large as possible. It is easy to see that (3) remains valid also for $\lambda = 1$; in other words: *the Fourier coefficient of rank $n \geq 1$ of*

$$f(t) = |1 + e^{it}|^q/(1 + e^{it}).$$

is

$$\hat{f}(n) = \sum_{k=0}^{\infty} \binom{q/2}{k}\binom{(q/2)-1}{k+n}.$$

Let us obtain a lower bound for $|\hat{f}(n)|$. Setting $c = 1 - (q/2)$, so that $0 < c < 1/2$,

$$\hat{f}(n) = \sum_{k=0}^{\infty} \binom{-c}{k+n}\binom{1-c}{k} =$$

$$= \binom{-c}{n}\left[1 - \frac{(c+n)(1-c)}{n+1} - \frac{(c+n)(c+n+1)}{(n+1)(n+2)} \cdot \frac{(1-c)c}{2!} - \cdots\right] =$$

$$= \binom{-c}{n}[1 - a_1 - a_2 - \cdots], \text{ say}$$

where $a_k > 0$. Now,

$$a_k \leq (1-c)c(c+1)\cdots(c+k-2)/k! \qquad (k \geq 2).$$

Hence

$$\sum_{k=2}^{\infty} a_k \leq (1-c)\left[\frac{c}{2!} + \frac{c(c+1)}{3!} + \cdots\right] = (1-c)[c/(1-c)] = c.$$

Hence

$$1 - a_1 - a_2 - \cdots \geq 1 - a_1 - c = (1-c)^2/(n+1) \geq 1/4(n+1).$$

Thus, finally, $|\hat{f}(n)| \geq (1/4)\alpha_p(n)$, where

$$(6) \qquad \alpha_p(n) = \left|\binom{(q/2)-1}{n}\right|(n+1)^{-1}.$$

(Observe that $\alpha_p(n)$ is asymptotic, for large n, to $C_p n^{-1-(q/2)}$.) Using this in (5), and observing that $\|g\|$ majorizes $|\hat{g}(n)|$ (where $g = f - (u * f)$) we get

$$(1/4)\alpha_p(n)|1 - \hat{u}(n)| \leq 2^{1/q}(m^p - 1)^{1/p}\|f\|.$$

Observing that, trivially, $\|f\| \leqq 2^{q-1}$, we obtain after some manipulation

Theorem 1'. *If* $u \in J_p(T)$, $2 < p < \infty$ *and* $\hat{u}(-1) = \hat{u}(0) = 1$,

$$\|u\|_{J_p}^p \geqq 1 + [(1/12) \sup_{n \geqq 1} \alpha_p(n) |1 - \hat{u}(n)|]^p$$

where $\alpha_p(n)$ *is given by* (6).

Clearly other variants of the theorem could be obtained using the same idea, e.g. minorize the term $\| f - (u*f)\|$ in (5) by $\| f-(u*f)\|_2 = (\sum_{-\infty}^{\infty} |\hat{f}(n)|^2 |1 - \hat{u}(n)|^2)^{1/2}$ instead of $|\hat{f}(n)| \cdot |1 - \hat{u}(n)|$; or, assume that $\hat{u}(-1)$ equals not 1 but a slightly smaller number (this variant is needed in order to prove (∗), in the Introduction). Since the method of proof should be clear, we leave the elaboration of such variants to the interested reader.

Of greater interest seems the extension of Theorem 1 to discrete groups other than Z. Here one encounters the apparently new problem of finding a function $h \in L^q(G)$ (G: compact group) whose Fourier transform is supported in a certain subset (corresponding to $\{-1, 0\}$ in the above case $\hat{G} = Z$) of the dual group $\hat{G}$, and such that the Fourier transform of $|h|^q/h$ is different from zero at a prescribed point of $\hat{G}$.

Also, L. CARLESON has called our attention to the problem of identifying all elements $u \in W_p$ such that $\hat{u}(k) = 1$ for some $k > 1$. If, for instance, $\hat{u}(2) = 1$ the method of proof of Theorem 1 shows that $\hat{u}(n) = 1$ for all even n, but we do not gain any information concerning $u(n)$ for odd n.

3. Proof of Theorem 2

Lemma 5. *For any complex number* z,

$$(1) \qquad |1+z|^4 \geqq 1 + 2(z+\bar{z}) + |z|^2 + (1/5)|z|^4.$$

Proof. The left side of (1) minus the right side equals (letting $z = re^{it}$)

$$2r^2 \cos 2t + 3r^2 + 4r^3 \cos t + (4/5)r^4$$

and we have to show this is $\geqq 0$, i.e. that

$$(4/5)r^2 + (4 \cos t)r + (3 + 2 \cos 2t) \geqq 0$$

for all t and all $r \geqq 0$. In fact, it is non-negative for all real r, since its discriminant is

$$16 \cos^2 t - (16/5)(3 + 2 \cos 2t) = -(8/5)(1 - \cos 2t) \leqq 0.$$

Lemma 6. *For any complex number* z,

$$(2) \qquad |1+z|^4 \leqq 1 + 2(z+\bar{z}) + 8|z|^2 + 3|z|^4.$$

PROOF. By straightforward expansion, and trivial estimates

$$|1+z|^4 \leq 1+2(z+\bar{z})+6|z|^2+4|z|^3+|z|^4$$

and since $|z|^3 \leq (1/2)(|z|^2+|z|^4)$ we get (2).

LEMMA 7. *If ϱ is a positive measure of total mass one on some space, and φ a complex-valued function satisfying $\int \varphi\, d\varrho=0$, then*

$$(3) \qquad \int |1+\varphi|^4\, d\varrho \geq 1+\int |\varphi|^2\, d\varrho +(1/5)\int |\varphi|^4\, d\varrho$$

$$(4) \qquad \int |1+\varphi|^4\, d\varrho \leq 1+8\int |\varphi|^2\, d\varrho +3\int |\varphi|^4\, d\varrho.$$

PROOF. Immediate from Lemmas 5 and 6.

It is now easy to prove Theorem 2. Let $f\in L^4(T)$. We must show, under the hypotheses of the theorem,

$$\int |(u_0+u)*f|^4\, d\varrho \leq \int |f|^4\, d\varrho$$

where ϱ denotes normalized Haar measure on T. Since this is obvious in case $u_0*f = \hat{f}(0)$ vanishes, we may assume without loss of generality that $\hat{f}(0)=1$, i.e. $f = 1+g$ where $\int g\, d\varrho=0$, and the proposition to be proved is

$$(5) \qquad \int |1+(u*g)|^4\, d\varrho \leq \int |1+g|^4\, d\varrho.$$

Now, substituting $u*g$ for φ in (4) (observe that $\int (u*g)\, d\varrho = 0$) gives

$$(6) \qquad \int |1+(u*g)|^4\, d\varrho \leq 1+8\int |u*g|^2\, d\varrho +3\int |u*g|^4\, d\varrho$$

and since $\int |u*g|^4\, d\varrho \leq 3^{-4}\int |g|^4$ by hypothesis, and $\int |u*g|^2\, d\varrho \leq 3^{-2}\int |g|^2\, d\varrho$ (since the $J_2(T)$ norm of u cannot exceed its $J_4(T)$ norm, which is $\leq 1/3$) the right side of (6) is bounded by

$$1+(8/9)\int |g|^2\, d\varrho +(1/27)\int |g|^4\, d\varrho$$

and comparison with (3) shows that this does not exceed $\int |1+g|^4\, d\varrho$. Thus (5) holds, and the theorem is proved.

REMARKS. To prove an analogue of Theorem 2 for $p=2n$, where n is a positive integer (this restriction is not really essential) does not present great difficulties; what is involved is proving an inequality analogous to (1), i.e.

$$(7) \qquad |1+z|^{2n} \geq 1+n(z+\bar{z})+a_n|z|^2+b_n|z|^{2n}$$

for suitable positive constants a_n, b_n (the analog of (2), which is also needed, is completely straightforward). The inequality (7) is a generalization of Bernoulli's inequality (namely, $(1+x)^{2n} \geq 1+2nx$)) in two ways: passing from the real variable

x to the complex variable z, and (what is more interesting) the additional terms on the right side. Our proof of (7), which shall not be presented here, does not seem to yield optimal values for a_n and b_n, and further pursuit of this matter may have a certain technical interest.

We also point out that Theorem 2 enables us to construct elements $u \in W_4$ (see the Introduction) which are not measures and such that $\hat{u}(-1)$ is as close to 1 as we please. Indeed, if $v \in M_4$ is not a measure, $\hat{v}(-1) = \hat{v}(0) = 0$, and the norm of v does not exceed $1/3$, then $u = (1-\varepsilon)\delta + \varepsilon(u_0 + v)$ has norm 1 in M_4, and $\hat{u}(0) = 1$, $\hat{u}(-1) = 1 - \varepsilon$.

Finally, observe that Theorem 2 is valid (with the identical proof) with T replaced by any compact Abelian group.

Since the completion of this paper, the following developments have taken place:

a) It has been observed, first by Leon Brown and Gregory Bachelis (jointly), and later, independently, by Bert Schreiber and myself (jointly) that the analog of Theorem 1 when the circle is replaced by an arbitrary compact Abelian group is valid. The proof is essentially a duplication, *mutatis mutandis*, of that given in the present paper. Thus, the question raised in the concluding paragraph of Section 2 is settled in the affirmative.

b) In the proposition ($*$) following Theorem 3, the hypothesis $c > 4$ can be replaced by $c > 3$; this I have shown in a paper "Regularity properties of the element of closest approximation", to appear in the *Transactions A.M.S.* There now seem to be good grounds for conjecturing that one can further weaken the hypothesis in question to $c > 2$.

c) Professor L. Leindler has given the optimal values for a_n and b_n in (7), in a paper "A generalization of Bernoulli's inequality", to be published in *Acta Sci. Math.*, Szeged.

d) The detailed proof of the generalization of Theorem 2 to $J_p(T)$, $1 < p < \infty$, is contained in the forthcoming note "A planar face on the unit sphere of the multiplier space M_p, $1 < p < \infty$", by Charles Feffermann and the author to appear in *Proceedings A. M. S.*

BIBLIOGRAPHY

[1] J. A. Clarkson, *Uniformly Convex Space*. Trans. Amer. Math. Soc. **40** (1936), 396—414.

[2] L. Hörmander, *Estimates for translation invariant operators in L^p spaces*. Acta Math. **104** (1960), 93—140.

[3] K. de Leeuw, *On L^p multipliers*. Ann. of Math. **81** (1965), 364—379.

[4] E. Lukács, *Characteristic Functions*. Griffin, London 1960.

[5] H. S. Shapiro, *Topics in Approximation Theory*. Lecture Notes in Math. **187** Springer, Berlin 1971.

Nikolskiĭ Constants for Positive Singular Integrals of Perturbed Fejér-Type

By

EBERHARD L. STARK

LEHRSTUHL A FÜR MATHEMATIK
TECHNISCHE HOCHSCHULE AACHEN

Concerning the approximation of periodic functions by means of positive singular integrals, the problem of determining the best asymptotic or Nikolskiĭ constant for the measure of approximation with respect to Lipschitz classes was solved in general provided the corresponding kernel is of Fejér's type. But there is still a number of important examples to which this theory does not apply. It is shown that for kernels of perturbed Fejér-type the associate Nikolskiĭ constant may be evaluated by a simple comparison theorem.

1. Introduction

For the class $C_{2\pi}$ of 2π-periodic functions $f(x)$ continuous on the whole real axis $(-\infty, \infty)$ a singular convolution integral with parameter ϱ is defined by

$$(1) \qquad I_\varrho(\chi; f; x) = \frac{1}{\pi} \int_{-\pi}^{\pi} f(x-t)\chi_\varrho(t)\, dt \qquad (\varrho > 0,\ \varrho \to \infty);$$

the function $\chi_\varrho(x)$ is assumed to be a positive kernel as $\varrho \to \infty$, i.e., for all $\varrho > 0$ the function is real-valued*) with $\chi_\varrho(x) \geqq 0$ (a.e.), $\chi_\varrho \in L_{2\pi}^1$, and normalized by $\int_{-\pi}^{\pi} \chi_\varrho(t)\, dt = \pi$. However, concerning the examples given below the results will be applied to continuous kernels which may be represented as a Fourier cosine series

$$(2) \qquad \chi_\varrho(x) = \frac{1}{2} + \sum_{k=1}^{\infty} \lambda_{k,\varrho}(\chi)\cos kx \geqq 0$$

with convergence factors given by the real Fourier coefficients

$$(3) \qquad \lambda_{k,\varrho}(\chi) = \frac{2}{\pi} \int_{0}^{\pi} \chi_\varrho(t)\cos kt\, dt \qquad (k = 0, 1, 2, \ldots).$$

---

*) $L_{2\pi}^1$ is the set of all 2π-periodic functions, Lebesgue integrable over $[-\pi, \pi]$; $L^1(-\infty, \infty)$ is defined analogously.

The positive linear operator (1) is a strong approximation process, thus

$$\lim_{\varrho \to \infty} A_\varrho(\chi;f) = 0, \quad A_\varrho(\chi;f) = \|I_\varrho(\chi;f;x) - f(x)\|$$

(with the usual maximum norm) if and only if $\lim_{\varrho \to \infty} \lambda_{1,\varrho}(\chi)=1$. For the Lipschitz classes (normalized by 1 and 2, respectively)

$$\mathrm{Lip}_1\, \alpha = \{f \in C_{2\pi};\ |f(x+h)-f(x)| \leq |h|^\alpha\} \qquad (0<\alpha\leq 1),$$

$$\mathrm{Lip}_2^*\, \alpha = \{f \in C_{2\pi};\ |f(x+h)-2f(x)+f(x-h)| \leq 2|h|^\alpha\} \qquad (0<\alpha\leq 2)$$

the measure of approximation is defined by

$$(4) \qquad (\mathrm{i})\quad \varDelta_\varrho(\chi;\alpha) = \sup_{f \in \mathrm{Lip}_1\alpha} A_\varrho(\chi;f), \qquad (\mathrm{ii})\quad \varDelta_\varrho^*(\chi;\alpha) = \sup_{f \in \mathrm{Lip}_2^*\alpha} A_\varrho(\chi;f),$$

respectively. If for some $\alpha>0$

$$(5) \qquad \lim_{\varrho \to \infty} \varrho^\alpha \varDelta_\varrho^*(\chi;\alpha) = N^*(\chi;\alpha) > 0$$

then $N^*(\chi,\alpha)$ is called the Nikolskiĭ constant for the measure of approximation (4, ii). $N(\chi,\alpha)$ for (4, i) is defined in the same way. To operate with (4) the αth algebraic moment of the kernel is needed, namely

$$M_\varrho(\chi;\alpha) \equiv \frac{2}{\pi} \int_0^\pi t^\alpha \chi_\varrho(t)\,dt \qquad (\alpha>0).$$

Then a basic result of S. M. NIKOLSKIĬ [23] states that

$$(6) \qquad \varDelta_\varrho(\chi;\alpha) = M_\varrho(\chi;\alpha)$$

in case $0<\alpha\leq 1$; for $0<\alpha\leq 2$ one only has (cf. [21] and the literature cited there)

$$(7) \qquad M_\varrho(\chi;\alpha) - R_\varrho(\chi;\alpha) \leq \varDelta_\varrho^*(\chi;\alpha) \leq M_\varrho(\chi;\alpha),$$

$$(8) \qquad R_\varrho(\chi;\alpha) \equiv \frac{2}{\pi} \int_0^\pi [t^\alpha - \sin^\alpha t]\chi_\varrho(t)\,dt.$$

Furthermore it is known that

$$(9) \qquad \varDelta_\varrho^*(\chi;\alpha) \equiv \varDelta_\varrho(\chi;\alpha) \qquad (0<\alpha\leq 1).$$

Finally, let us note that after obvious modifications all assertions concerning the measure of approximation remain valid if $C_{2\pi}$ is replaced by $L_{2\pi}^1$ (cf. e.g. [21, p. 28], [14, p. 25] and the references given there). For these general preliminaires one may consult [7].

2. Kernels of perturbed Fejér-type

In order to define kernels of perturbed Fejér-type we first of all need the definition of a kernel of Fejér's type [7], [2]: *Let $\Phi \in L^1(-\infty, \infty)$ with $\Phi(x) \geqq 0$ (a.e.) be normalized by $\int_{-\infty}^{\infty} \Phi(t)\, dt = \pi$. Then*

$$(10) \qquad F_\varrho(x) = \sum_{k=-\infty}^{\infty} \varrho \Phi(\varrho[x+2k\pi]) \qquad (\varrho > 0)$$

is said to be a kernel of Fejér's type. (For the fact that (10) is indeed a kernel see e.g. [21].) If the generating $\Phi(x)$ is even then (10) is even. The corresponding approximation process (1) is called a singular integral of Fejér's type; for periodic functions it may be written as

$$(11) \qquad I_\varrho(\mathscr{F};f;x) = \frac{1}{\pi} \int_{-\infty}^{\infty} f\left(x - \frac{t}{\varrho}\right) \Phi(t)\, dt \qquad (f \in C_{2\pi}).$$

Roughly speaking the main advantage of these integrals is that they allow to solve "periodic" problems by unrolling them onto the real axis. Now for singular integrals of Fejér's type the Nikolskiĭ constants are determined for all α with $0 < \alpha \leqq 2$ by the following theorem due to R. J. NESSEL [21; 22]:

Let $F_\varrho(x)$ be a kernel of Fejér's type with an even generating function $\Phi(\mathscr{F}; x)$. If

$$(12) \qquad m(\Phi(\mathscr{F}); \alpha) = \frac{2}{\pi} \int_0^{\infty} t^\alpha \Phi(\mathscr{F}; t)\, dt < \infty \qquad (0 < \alpha \leqq 2),$$

then for the measure of approximation

$$(13) \qquad \Delta_\varrho^*(\mathscr{F}; \alpha) = m(\Phi(\mathscr{F}); \alpha)\varrho^{-\alpha} + o(\varrho^{-\alpha}) \qquad (\varrho \to \infty),$$

$$(14) \qquad N^*(\mathscr{F}; \alpha) = m(\Phi(\mathscr{F}); \alpha) \qquad (0 < \alpha \leqq 2).$$

In what follows we consider kernels which belong to the class

$$(15) \qquad \mathfrak{R}^2 = \left\{ \chi_\varrho(x) \geqq 0;\ \lim_{\varrho \to \infty} \varrho^2(1 - \lambda_{k,\varrho}(\chi)) = c_s(\chi)k^2,\quad k = 1, 2, 3, \ldots \right\},$$

thus positive singular integrals (1) with given saturation order $O(\varrho^{-2})$, $\varrho \to \infty$, saturation class Lip* 2, and saturation constant $c_s(\chi)$. Then the starting point of our considerations is a simple relation between Nikolskiĭ constants and saturation constants, respectively ([14; 13], [19], [7, p. 83]): *If $\chi_\varrho \in \mathfrak{R}^2$, then with $\alpha = 2$ in (5) one has*

$$(16) \qquad N^*(\chi; 2) = 2c_s(\chi).$$

Hence, in the particular (saturation) case $\alpha = 2$ the Nikolskiĭ constants are easily evaluated for kernels of $\mathfrak{R}^2$. The decisive fact now is that there exist *pairs* of kernels $\chi_\varrho^{(1)}$ and $\chi_\varrho^{(2)}$ of class $\mathfrak{R}^2$, in short $[\chi_\varrho^{(1)}; \chi_\varrho^{(2)}] \in \mathfrak{R}^2$, such that these have equal saturation constants, thus *equal* Nikolskiĭ constants for $\alpha = 2$ by (16), i.e.,

$$(17) \qquad\qquad N^*(\chi^{(1)}; 2) = N^*(\chi^{(2)}; 2),$$

and secondly that *one* kernel of this pair is of Fejér's type, thus with Nikolskiĭ constants known by (14) for all α with $0 < \alpha \leq 2$.

This paper is concerned with the following problem: Given a pair of (periodic) kernels $[F_\varrho; \chi_\varrho] \in \mathfrak{R}^2$ with $F_\varrho(x)$ of Fejér's type and $N^*(\mathscr{F}; 2) = N^*(\chi; 2)$. Will the Nikolskiĭ constants also coincide for the remaining fractional values of α, i.e., $N^*(\chi; \alpha) = N^*(\mathscr{F}; \alpha)$ for $0 < \alpha < 2$? The answer will be seen to be positive if certain restrictions are posed upon the comparison kernel $\chi_\varrho(x)$. This leads to the definition of kernels of perturbed Fejér-type as they may be called for convenience.

Let $F_\varrho(x) \in \mathfrak{R}^2$ be a kernel of Fejér's type (10) with associated generating function $\Phi(\mathscr{F}; x)$ which allows the series representation

$$F_\varrho(x) = \frac{1}{2} + \sum_{k=1}^{\infty} \lambda_{k,\varrho}(\mathscr{F}) \cos kx.$$

With an arbitrary kernel $\chi_\varrho(x) \in \mathfrak{R}^2$ given by (2) the pair of kernels $[F_\varrho; \chi_\varrho] \in \mathfrak{R}^2$ is called a *comparison pair* if the additional assumption

$$\lim_{\varrho \to \infty} \frac{1 - \lambda_{1,\varrho}(\mathscr{F})}{1 - \lambda_{1\cdot\varrho}(\chi)} = 1$$

is fulfilled, or equivalently, if the basic relation (17) is satisfied with $N^*(\mathscr{F}; 2) = N^*(\chi; 2)$. Moreover, the *comparison kernel* $\chi_\varrho(x)$ is assumed to be decomposable in the form

$$(18) \qquad\qquad F_\varrho(x) = \chi_\varrho(x)\{1 + \varepsilon_\varrho(x)\} \qquad (\varrho > 0)$$

with *perturbation term* $\varepsilon_\varrho(x)$ (a decomposition which will be seen to be most suitable in the examples below). The comparison kernel is said to be a *kernel of perturbed Fejér-type* if the perturbation term satisfies

$$(19) \qquad r_\varrho(\chi; \alpha) \equiv \frac{2}{\pi} \int_0^\pi t^\alpha |\varepsilon_\varrho(t)| \chi_\varrho(t)\, dt = o(\varrho^{-2}) \qquad (0 < \alpha \leq 2; \ \varrho \to \infty).$$

THEOREM. *Let* $[F_\varrho; \chi_\varrho] \in \mathfrak{R}^2$ *be a comparison pair,* $\chi_\varrho(x)$ *being a kernel of perturbed Fejér-type. Then*

$$(20) \qquad\qquad N^*(\chi; \alpha) = N^*(\mathscr{F}; \alpha) \qquad (0 < \alpha \leq 2).$$

Let us remark that in view of (15) the 2nd moment (12) of the Fejér-type kernel

F_ϱ exists (see [7, p. 461 (Probl. 12. 3. 3)]) and so does the αth moment for all α with $0 < \alpha \leq 2$.

PROOF. In view of (18) and (19) it follows that

$$\varrho^\alpha M_\varrho(\mathscr{F}; \alpha) = \varrho^\alpha M_\varrho(\chi; \alpha) + o(\varrho^{\alpha-2}) \qquad (0 < \alpha \leq 2; \ \varrho \to \infty).$$

Hence in any case

(21)
$$\lim_{\varrho \to \infty} \varrho^\alpha M_\varrho(\mathscr{F}; \alpha) = \lim_{\varrho \to \infty} \varrho^\alpha M_\varrho(\chi; \alpha) \qquad (0 < \alpha \leq 2).$$

In case $0 < \alpha \leq 1$ it is easily seen by means of (6), (9), and (13) that

(22)
$$\lim_{\varrho \to \infty} \varrho^\alpha M_\varrho(\mathscr{F}; \alpha) = \lim_{\varrho \to \infty} \varrho^\alpha \Delta_\varrho(\mathscr{F}; \alpha) = N^*(\mathscr{F}; \alpha) = m(\Phi(\mathscr{F}); \alpha) \qquad (0 < \alpha \leq 1);$$

thus by (21)

(23)
$$N^*(\chi; \alpha) = N^*(\mathscr{F}; \alpha) \qquad (0 < \alpha \leq 1).$$

In case $1 < \alpha \leq 2$, concerning the remainder term $R_\varrho(\chi; \alpha)$ in (7) it has to be shown that*)

(24)
$$\lim_{\varrho \to \infty} \varrho^\alpha R_\varrho(\chi; \alpha) = 0 \qquad (1 < \alpha \leq 2).$$

But in view of (8) and (18) the following relation holds

$$\varrho^\alpha R_\varrho(\chi; \alpha) = -\varrho^\alpha \frac{2}{\pi} \int_0^\pi t^\alpha \left[1 - \left(\frac{\sin t}{t}\right)^\alpha\right] \varepsilon_\varrho(t) \chi_\varrho(t)\, dt + \varrho^\alpha \frac{2}{\pi} \int_0^\pi [t^\alpha - \sin^\alpha t] F_\varrho(t)\, dt;$$

since the second term on the right-hand side is $o(1)$, $\varrho \to \infty$, by the method of proof given in [21] one has

$$|\varrho^\alpha R_\varrho(\chi; \alpha)| \leq \varrho^\alpha \frac{2}{\pi} \int_0^\pi t^\alpha \left[1 - \left(\frac{\sin t}{t}\right)^\alpha\right] |\varepsilon_\varrho(t)| \chi_\varrho(t)\, dt + o(1) \leq \varrho^\alpha r_\varrho(\chi; \alpha) + o(1)$$

$$(\varrho \to \infty)$$

in view of $0 \leq 1 - (\sin t/t)^\alpha \leq 1$ for $0 \leq t \leq \pi$, $\alpha > 0$. Hence (24) is true by hypothesis. Finally (20) follows for all values of α with $0 < \alpha \leq 2$ from (7) by using (24) and (21).

Instead of the indirect condition (19) upon $\varepsilon_\varrho(x)$ it is suitable to have a simple direct criterion for the perturbation term.

---

*) Let us mention that in [9] there is to appear to following result: *If $\chi_\varrho \in \mathfrak{R}^2$, $1 < \alpha < 2$, then* $\Delta_\varrho^*(\chi; \alpha) = M_\varrho(\chi; \alpha) + o(M_\varrho(\chi; \alpha))$, $\varrho \to \infty$.

COROLLARY. *For the comparison pair* $[F_\varrho; \chi_\varrho] \in \Re^2$ *the estimate*)*

$$(25) \qquad |\varepsilon_\varrho(x)| \leqq C\{x^2 + \varrho^{-2}\} \qquad (\varrho > 0;\ \varrho \to \infty)$$

is sufficient for (19) *to be valid.*

To prove this fact a lemma is needed.

LEMMA. *If* $\chi_\varrho \in \Re^2$, *then for* $\varrho \to \infty$

$$(26) \qquad \int_0^\pi t^\alpha \chi_\varrho(t)\,dt = O(\varrho^{-\alpha}) \qquad (0 < \alpha \leqq 2),$$

$$(27) \qquad \int_0^\pi t^{2+\alpha} \chi_\varrho(t)\,dt = o(\varrho^{-2}) \qquad (\alpha > 0).$$

PROOF. For arbitrary positive kernels $\chi_\varrho(x)$ (not necessarily of class $\Re^2$) the following estimate holds [34, p. 664]

$$\int_0^\pi t^\alpha \chi_\varrho(t)\,dt \leqq C[1 + \lambda_{1,\varrho}(\chi)]^{\alpha/2} \qquad (0 < \alpha \leqq 2).$$

Hence (26) is a consequence of (15). — If for $f \in C_{2\pi}$ there exists the second Riemann derivative $D^2 f(x_0)$, then in view of (15) it is known (cf. [7, p. 75], [14, p. 30]) that for singular integrals (1) an expansion of Voronovskaja-type holds:

$$I_\varrho(\chi; f; x_0) = f(x_0) + (1 - \lambda_{1,\varrho}(\chi)) D^2 f(x_0) + o(1 - \lambda_{1,\varrho}(\chi)) \qquad (\varrho \to \infty),$$

thus in case $f(-x) = f(x)$ and $x_0 = 0$

$$\frac{2}{\pi} \int_0^\pi f(t) \chi_\varrho(t)\,dt = f(0) + (1 - \lambda_{1,\varrho}(\chi)) D^2 f(0) + o(1 - \lambda_{1,\varrho}(\chi)) \qquad (\varrho \to \infty).$$

Now, for the particular instance $f(x) = |x|^{2+\alpha} \in C_{2\pi}$ with $\alpha > 0$ one has $f(0) = D^2 f(0) = 0$ and therefore

$$\int_0^\pi t^{2+\alpha} \chi_\varrho(t)\,dt = o(1 - \lambda_{1,\varrho}(\chi)) = o(\varrho^{-2}) \qquad (\varrho \to \infty).$$

This proves (27).

Now the *proof of the Corollary* is an immediate consequence of the Lemma. For (19) it follows by means of (25) that

$$|r_\varrho(\chi; \alpha)| \leqq \frac{2C}{\pi} \int_0^\pi t^\alpha \{t^2 + \varrho^{-2}\} \chi_\varrho(t)\,dt = o(\varrho^{-2}) \qquad (\alpha > 0;\ \varrho \to \infty).$$

---

*) Throughout the paper C will denote absolute positive constants which are in general distinct.

3. Applications

1. A first example, thus the comparison pair $[Z_{n-1}; K_{n-2}]$, is built up from the Fejér-type kernel $Z_{n-1}(x)$ of BOHMAN—ZHENG WEI-XING ([5, p. 141], [37, p. 361]) and the well-known kernel $K_{n-2}(x)$ of FEJÉR—KOROVKIN (see e.g. [7]) (for detailed references to these two kernels compare [14]).

The generating function of the first-named kernel is given by

$$(28) \qquad \Phi(\mathscr{Z}; x) = 4\pi^2 \frac{\cos^2(x/2)}{(\pi^2 - x^2)^2} \geqq 0$$

(for the normalization see e.g. [37, p. 354]). Setting $\varrho = n = 1, 2, 3, \dots$ in (10) and using well-known expansions in the theory of meromorphic functions, thus

$$(29) \quad 8\pi^2 n \sum_{k=-\infty}^{\infty} \frac{1}{[\pi^2 - n^2(x + 2k\pi)^2]^2} = \frac{1}{\pi}\left[\cot \frac{1}{2}\left(x + \frac{\pi}{n}\right) - \cot \frac{1}{2}\left(x - \frac{\pi}{n}\right)\right] +$$

$$+ \frac{1}{2n}\left[\frac{1}{\sin^2 \frac{1}{2}\left(x + \frac{\pi}{n}\right)} + \frac{1}{\sin^2 \frac{1}{2}\left(x - \frac{\pi}{n}\right)}\right],$$

the calculations finally result in the closed representation of the kernel of Bohman—Zheng Wei-xing, namely (cf. also [37, p. 361] for another formulation)

$$(30) \qquad Z_{n-1}(x) = \frac{1}{n}\cos^2 \frac{nx}{2}\left\{\frac{\frac{n}{\pi}\sin \frac{\pi}{n}}{\cos x - \cos \frac{\pi}{n}} - \frac{1 - \cos \frac{\pi}{n}\cos x}{\left(\cos x - \cos \frac{\pi}{n}\right)^2}\right\} \geqq 0$$

$$(n = 1, 2, 3, \dots),$$

the positivity of this rather unwieldy expression is guaranteed by (28). On the other hand, since (30) is an even trigonometric polynomial of degree $d(\mathscr{Z}; n) = n - 1$ (this fact follows immediately from (33) below) one has the cosine polynomial representation

$$(31) \qquad Z_{n-1}(x) = \frac{1}{2} + \sum_{k=1}^{n-1} \lambda_{k,n-1}(\mathscr{Z})\cos kx$$

with convergence factors

$$(32) \qquad \lambda_{k,n-1}(\mathscr{Z}) = \left(1 - \frac{k}{n}\right)\cos \frac{k\pi}{n} + \frac{1}{\pi}\sin \frac{k\pi}{n} \qquad (0 \leqq k \leqq n - 1).$$

Concerning (32) it may be noted that for the function

$$(33) \qquad \psi(v) = \begin{cases} (1 - |v|)\cos \pi v + \dfrac{1}{\pi}\sin |\pi v| & (|v| < 1), \\ 0 & (|v| \geqq 1) \end{cases}$$

the Fourier transform of (33) delivers (28), namely [5, p. 112 f]

$$\psi^{\wedge}(x) = \frac{1}{2} \int_{-\infty}^{\infty} \psi(v) e^{-ivx}\, dv = \Phi(\mathscr{Z}; x);$$

this is of importance since by means of the Poisson summation formula (see [7, p. 202 f]) the convergence factors (32) immediately follow from (33). This has the advantage of avoiding an application of (29).

The kernel of Fejér—Korovkin, written as a cosine polynomial of degree $d(\mathscr{K}; n) = n-2$, is given by

$$(34) \qquad K_{n-2}(x) = \frac{1}{2} + \sum_{k=1}^{n-2} \lambda_{k,n-2}(\mathscr{K}) \cos kx \qquad (n = 2, 3, 4 \ldots),$$

$$(35) \qquad \lambda_{k,n-2}(\mathscr{K}) = \left(1 - \frac{k}{n}\right) \cos \frac{k\pi}{n} + \frac{1}{n} \cot \frac{\pi}{n} \sin \frac{k\pi}{n} \qquad (0 \leq k \leq n-2)$$

so that its convergence factors are somewhat more complicated than (32). Nevertheless, an obvious connection between (35) and (32) is given by the fact that $1/\pi$ is the first term in the series expansion of $\frac{1}{n} \cot \frac{\pi}{n}$ so that the Fourier coefficients (32) are a truncated version of (35), cf. [14, p. 36]. On the other hand the closed representation of (34), namely

$$(36) \qquad K_{n-2}(x) = \frac{1}{n} \cos^2 \frac{nx}{2} \frac{\sin^2(\pi/n)}{(\cos x - \cos(\pi/n))^2} \geq 0 \qquad (n = 2, 3, 4, \ldots)$$

is quite transparent in contrast to (30).

Both kernels belong to the class $\mathfrak{K}^2$; this follows immediately from the leading term of the expansion of (32), namely

$$(37) \qquad 1 - \lambda_{k,n-1}(\mathscr{Z}) = \frac{\pi^2}{2} \frac{k^2}{n^2} + O(n^{-3}) \qquad (n \to \infty)$$

as well as from the completely analogous first-term-expansion of (35). The decomposition (18) for this comparison pair is readily established as

$$(38) \qquad Z_{n-1}(x) = K_{n-2}(x)\{1 + \varepsilon(n; x)\}$$

with perturbation term

$$(39) \qquad \varepsilon(n; x) = (\cos x - \cos(\pi/n))\delta(n),$$

$$\delta(n) = \frac{(n/\pi) - \cot(\pi/n)}{\sin(\pi/n)} \qquad (n = 2, 3, 4, \ldots).$$

(Of course, $\varepsilon(n; x)$ is a first degree cosine polynomial, but the simple structure of

(38) and (39) justifies the fact that we investigate a comparison pair with kernels of different degree; however, since $d(\mathscr{L}; n)/d(\mathscr{K}; n) \to 1$ for $n \to \infty$, this obviously has no influence upon the general theorem.) In order to apply the Corollary a simple estimate yields

$$|\varepsilon(n, x)| \leq C\{x^2 + n^{-2}\} \qquad (0 \leq x \leq \pi; \quad n = 2, 3, 4, \ldots)$$

since $1/3 \leq \delta(n) \leq 2/\pi$. Finally, the Nikolskiĭ constants are calculated via (28) from (20) and (14), thus

$$(40) \qquad N^*(\mathscr{K}; \alpha) = N^*(\mathscr{L}; \alpha) = 8\pi \int_0^\pi t^\alpha \frac{\cos^2(t/2)}{(\pi^2 - t^2)^2} \, dt \qquad (0 < \alpha \leq 2).$$

Only in case $\alpha = 1$ and $\alpha = 2$ the explicit value of the latter integral seems to be known

$$(41) \qquad N^*(\mathscr{K}; 1) = \frac{2}{\pi}(\pi \, \mathrm{Si}[\pi] - 2) \qquad \left(\mathrm{Si}[\pi] = \int_0^\pi \frac{\sin x}{x} \, dx\right),$$

$$(42) \qquad N^*(\mathscr{K}; 2) = \pi^2$$

(a direct integration of (40) for $\alpha = 1$ is carried out in [37, p. 354]; (42) is obvious from (37) and (16)).

The Nikolskiĭ constants for the kernel of Fejér—Korovkin have already been evaluated several times over using very individual properties of the kernel: $N^*(\mathscr{K}; 2)$ is part of the investigations in [16], [19], [10], [31], [14, 13], [12]. — The constant $N(\mathscr{K}; 1)$ goes back to [24] (cf. [17]); in [31] it was moreover shown that

$$(43) \qquad \Delta_n^*(\mathscr{K}; 1) = \frac{2}{\pi}(\pi \, \mathrm{Si}[\pi] - 2)\frac{1}{n} + \frac{4}{\pi}(\pi \, \mathrm{Si}[\pi] - 2)\frac{1}{n^2} + o(n^{-2}) \qquad (n \to \infty)$$

starting off the investigations of higher order Nikolskiĭ constants of this kernel. — The general result (40) was established for the first time in [25] (yet the proof given there for $1 < \alpha \leq 2$ was incorrect; see [14, p. 25], [9]).

Fortunately, (36) is one of the two important "natural" kernels of class $\mathfrak{K}^2$ arising in approximation theory (the second one is the original kernel of Jackson; see Example 2) for which there exists a so-called "generating φ-function" in the sense of P. P. KOROVKIN [18] (cf. [14, p. 46 ff], [7, p. 85 f]). A series of general theorems on Nikolskiĭ constants may be expressed in terms of such generating functions. For the sake of completeness let us collect these results for

$$(44) \qquad \varphi(\mathscr{K}; x) = \sin \pi x \qquad (0 \leq x \leq 1)$$

of the Fejér—Korovkin kernel. In [34, p. 665 f] the following more refined expansions may be found

$$\Delta_n^*(\mathscr{K}; 1) = \frac{2}{\pi}(\pi \, \mathrm{Si}[\pi] - 2)\frac{1}{n} + O(n^{-2}),$$

$$\Delta^*(\mathscr{K}; 2) = \frac{\pi^2}{n^2} + O(n^{-3}) \qquad (n \to \infty),$$

cf. (43); another general expansion is [35, p. 771]

$$\Delta_n^*(\mathcal{K};\alpha) = \frac{M(\alpha)\Gamma(\alpha)\sin(\alpha\pi/2)}{n^\alpha} + \begin{cases} O(n^{-\alpha-1}) & (0<\alpha\le 1), \\ O(n^{-2}\log n) & (1<\alpha<2), \end{cases}$$

with

$$M(\alpha) = \frac{1}{\pi}\int\limits_0^\pi \left[\left(\frac{\pi}{t}\right)^\alpha - \left(\frac{\pi}{t}\right)^{\alpha-1}\right]\sin t\, dt.$$

By a formula occuring in [26; p. 151] the particular constant (41) was obtained once more. Further representations of the general constant (40) are contained in [3, p. 151 f; 4, p. 202 ff]. Finally, the most elegant formulation [2, p. 357 ff] is given by

$$N^*(\mathcal{K};\alpha) = \frac{1}{\pi\int\limits_0^1 \varphi^2(\mathcal{K};t)\,dt}\int\limits_0^\infty u^\alpha \left|\int\limits_0^1 \varphi(\mathcal{K};t)e^{itu}\,dt\right|^2 du \qquad (0<\alpha\le 2)$$

which gives (40) after a few steps.

Thus the (first) Nikolskiĭ constants for the particular kernels of this example are known, to each of them at least one general method applies (of course, this is not true for all comparison pairs; see Example 3). However, the method described in this paper delivers the background why the constants are the same for both kernels. Furthermore, in answer to an unsolved question, this comparison method reveals that (28) may be regarded as the real line version (concerning the approximation of nonperiodic functions, see [7]) of the periodic Fejér—Korovkin kernel; compare also [37].

2. The singular integral of (Jackson-) DE LA VALLÉE POUSSIN [36, p. 45] is of type (11), thus

$$I_\varrho(\mathscr{P};f;x) = \frac{3}{2\pi}\int\limits_{-\infty}^\infty f\left(x-\frac{2t}{\varrho}\right)\left(\frac{\sin t}{t}\right)^4 dt \qquad (\varrho>0),$$

with generating function

$$(45) \qquad \Phi(\mathscr{P};x) = \frac{3}{4}\left(\frac{\sin(x/2)}{x/2}\right)^4 \ge 0.$$

An application of (10) again with $\varrho=n=1, 2, 3, \ldots$ (cf. [6, p. 378], [7, p. 130], [21, p. 21]) delivers the periodic (version of this) kernel as a polynomial of degree $d(\mathscr{P};n) = = 2n-1$, namely

$$(46) \qquad P_{2n-1}(x) = \frac{2+\cos x}{4n^2}\left(\frac{\sin(nx/2)}{\sin(x/2)}\right)^4$$

with convergence factors (see e.g. [7, p. 131, 517])

$$(47) \qquad \lambda_{k,2n-1}(\mathscr{P}) = \begin{cases} 1 - \dfrac{3}{2}\left(\dfrac{k}{n}\right)^2 + \dfrac{3}{4}\left(\dfrac{k}{n}\right)^3 & (1 \le k \le n), \\[2mm] \dfrac{1}{4}\left(2 - \dfrac{k}{n}\right)^3 & (n \le k \le 2n-1), \\[2mm] 0 & (k \ge 2n). \end{cases}$$

(Here a similar remark as for (33) holds; see [7, p. 205].)

The comparison kernel is the original kernel of JACKSON (see e.g. [7, p. 517])

$$(48) \qquad J_{2n-2}(x) = \frac{3}{2n(2n^2+1)}\left(\frac{\sin(nx/2)}{\sin x/2}\right)^4 \ge 0,$$

thus a polynomial of degree $d(\mathscr{J}; n) = 2n-2$. The saturation limit in (15) for the (more complicated) convergence factors of (48) is known to be [7, p. 450]

$$(49) \qquad \lim_{n \to \infty} n^2\left(1 - \lambda_{k,2n-2}(\mathscr{J})\right) = \frac{3}{2}k^2.$$

Since this is the same as that for (47) the comparison pair $[P_{2n-1}; J_{2n-2}]$ belongs to $\mathfrak{R}^2$. Finally, the decomposition (18) follows as

$$P_{2n-1}(x) = J_{2n-2}(x)\left\{1 + \varepsilon(n; x)\right\},$$

$$\varepsilon(n; x) = \frac{2+\cos x}{6n^2} - \frac{1-\cos x}{3}$$

so that (25) holds.

In view of (20), (14), (12), and (45) it follows that

$$(50) \qquad N^*(\mathscr{J}; \alpha) = N^*(\mathscr{P}; \alpha) = \frac{3 \cdot 2^\alpha}{\pi} \int_0^\infty \frac{\sin^4 t}{t^{4-\alpha}}\,dt \qquad (0 < \alpha \le 2)$$

and after integrating (cf. [15, p. 13], [25, p. 190])

$$(51) \qquad N^*(\mathscr{J}; \alpha) = \frac{6(2^{1-\alpha}-1)}{\Gamma(4-\alpha)\cos(\alpha\pi/2)} \qquad (0 < \alpha \le 2;\ \alpha \ne 1)$$

$$(52) \qquad = \frac{2(2^{1-\alpha}-1)\Gamma(\alpha)\sin(\alpha\pi/2)}{\pi(3-\alpha)(2-\alpha)(1-\alpha)} \qquad (0 < \alpha \le 2;\ \alpha \ne 1, 2).$$

For the case of nonfractional α de L'Hospital's rule may be applied to these formulae giving

$$(53) \qquad \text{(i)} \quad N^*(\mathscr{J}; 1) = \frac{6}{\pi}\log 2, \qquad \text{(ii)} \quad N^*(\mathscr{J}; 2) = 3.$$

In case $0 < \alpha \leqq 1$ the constant $N(\mathscr{J}; \alpha)$ of (50) is exactly that of [20, p. 276], (52) was given in [25, p. 191] (cf. [3, p. 53 f]); both results have been obtained via individual properties of the kernel under consideration. Furthermore, $N^*(\mathscr{J}; \alpha)$ is evaluated among others in [9] via the mentioned general theorem; it may also be deduced from general theorems in [2], [4] (cf. [21, p. 22 f]) using the generating φ-function of (48), namely $\varphi(\mathscr{J}; x) = 1 - 2|x - 1/2|$. For the Nikolskiĭ constants of (46) in the form (51) see [21, p. 21]. Concerning the integer values of α there is a result of another type for the measure of approximation of the singular integrals with kernels (46) and (48) in [8, p. 232 ff] (cf. [32]) in deriving so-called essential asymptotic expansions, for example

$$\Delta_n^*(\mathscr{J}; 1) = \frac{6}{\pi} \log 2 \cdot \frac{1}{n} + \frac{3}{2\pi} \frac{\log n}{n^3} + \frac{1}{2\pi}\left(3\gamma - 4\log 2 + \frac{5}{2}\right)\frac{1}{n^3} - \frac{3}{4\pi}\frac{\log n}{n^5} + O(n^{-5}) \qquad (n \to \infty),$$

$$\Delta_n^*(\mathscr{P}; 1) = \frac{6}{\pi} \log 2 \frac{1}{n} - \frac{1}{4\pi}\frac{1}{n^3} + O(n^{-5}) \qquad (n \to \infty)$$

(with γ being Euler's constant). Only the first term of these expansion coincides. On the other hand, it is this fact which uncovers at the same time that a comparison of these kernels is limited to the *first order* Nikolskiĭ constants as given by the Theorem. — The isolated constant $N^*(\mathscr{J}; 1)$ was given for the first time in [24], for $N^*(\mathscr{P}; 1)$ compare [38]. $N^*(\mathscr{J}; 2)$ is contained in [19, p. 14], [8, p. 232], [32, p. 157], [14, p. 40], $N^*(\mathscr{P}; 2)$ may be found in [8, p. 234], [4, p. 45].

Whereas for (48) there exists the generating function $\varphi(\mathscr{J}; x)$, this is not true (cf. [14, p. 48]) for the many known generalizations of this kernel as e.g. higher degree Jackson and Jackson—Matsuoka kernels. Thus no general results via φ-functions are available. But here the Theorem applies if one deals with the corresponding generalizations of (46) as comparison kernels. Finally, there are some results on Nikolskiĭ constants of these generalized Jackson—de La Vallée Poussin kernels: for fractional α we refer to [25], [9], for $\alpha = 1, 2$ see [27—30], [32], [14].

3. As a final example we consider a comparison pair of singular integrals with kernels which are generalizations of the Abel—Poisson kernel, thus of

$$p_r(x) = \frac{1}{2}\frac{1 - r^2}{1 - 2r\cos x + r^2} = \frac{1}{2} + \sum_{k=1}^{\infty} r^k \cos kx \geqq 0 \qquad (r \to 1-)$$

as treated in [33]. For these nonpolynomial kernels there exists no generating φ-function. The Fejér-type singular integral is that of ANGHELUTZA [1, p. 140], namely

$$I_r(\mathscr{A}; f; x) = \frac{2}{\pi} \int_{-\infty}^{\infty} f(x + t\log r) \frac{dt}{(1 + t^2)^2} \qquad \left(\frac{1}{\varrho} = -\log r;\ r \to 1-\right)$$

with

$$\Phi(\mathscr{A}; x) = \frac{2}{(1 + x^2)^2} \geqq 0.$$

The seemingly somewhat intricate closed representation of the corresponding periodic
kernel is given as

$$A_r(x) = p_r^2(x) \frac{4r}{(1-r^2)^2} \left[(1+r^2) \log \frac{1}{r} - (1-r)^2 \right] \cdot$$

$$\cdot \left[\cos x - \frac{1}{2r} \frac{4r^2 \log \frac{1}{r} - (1-r)^4}{(1+r^2) \log \frac{1}{r} - (1-r^2)} \right] \geqq 0,$$

whereas the series representation is

$$A_r(x) = \frac{1}{2} + \sum_{k=1}^{\infty} r^k \left\{ 1 + k \log \frac{1}{r} \right\} \cos kx \geqq 0.$$

The comparison kernel is that of GHERMANESCO [11, p. 76 ff] with the very
simple closed representation

$$(54) \quad G_r(x) = p_r^2(x) \frac{2(1-r)}{1+r} = \frac{1}{2} + \sum_{k=1}^{\infty} r^k \left\{ 1 + k \frac{1-r^2}{1+r^2} \right\} \cos kx \geqq 0 \qquad (r \to 1-).$$

Concerning (15) one has (cf. [33])

$$(55) \qquad \lim_{r \to 1-} \frac{1 - \lambda_{k,r}(\mathscr{A})}{(1-r)^2} = \lim_{r \to 1-} \frac{1 - \lambda_{k,r}(\mathscr{G})}{(1-r)^2} = \frac{1}{2} k^2 \qquad (k = 1, 2, 3, \ldots)$$

(thus with saturation order $O([1-r]^2), r \to 1-$) so that $[A_r; G_r] \in \mathfrak{R}^2$. The decomposition (18) becomes

$$A_r(x) = G_r(x) \{ 1 + \varepsilon(r; x) \},$$

$$\varepsilon(r; x) = \frac{2r}{(1+r)^3} \frac{(1+r)^2 \log r - (1-r)^2}{(1-r)^3} \{ (1+r^2)(1 - \cos x) - (1-r)^2 \}$$

from which one derives

$$|\varepsilon(r; x)| \leqq C \{ x^2 + (1-r)^2 \} \qquad (r \to 1-).$$

Thus (25) is verified. This, in turn, delivers the Nikolskiǐ constants

$$N^*(\mathscr{G}; \alpha) = N^*(\mathscr{A}; \alpha) = \frac{2}{\pi} B\left(\frac{1+\alpha}{2}; \frac{3-\alpha}{2} \right) = \frac{1-\alpha}{\cos(\alpha\pi/2)} \qquad (0 < \alpha \leqq 2, \ \alpha \neq 1)$$

as well as $N^*(\mathscr{G}; 1) = 2/\pi$ (e.g. again by de L'Hospital's rule). For $N^*(\mathscr{G}; 2) = 1$

compare (55) and (16). — Further comparison kernels of type (54), (55) may be taken from [33].

The author wishes to take this opportunity to express his thanks to Professors P. L. BUTZER and R. J. NESSEL for their valuable advice and attention given to this paper.

REFERENCES*)

[1] T. Anghelutza, *Une remarque sur l'intégrale de Poisson.* Bull. Sci. Math. Bibliothèque École Hautes Études = Darboux Bull. (2) **48** (1924), 138—140; FdM **50** (1924), 330.

[2] V. A. Baskakov, *The degree of approximation of differentiable functions by certain positive operators* (Russ.). Mat. Sb. (N. S.) **76 (118)** (1968), 344—361; MR 33#4488.

[3] L. I. Bausov, *The order of approximation of functions of class Z_α by positive linear polynomial operators* (Russ.). Uspehi Mat. Nauk 7, no. **1 (103)** (1962), 149—155; MR 27#1756.

[4] L. I. Bausov, *On the order of approximation of functions of class Z_α by linear positive operators* (Russ.). Mat. Zametki 4 (1968), 201—210 = Transl. Math. Notes 4 (1968), 612—617.

[5] H. Bohman, *Approximate Fourier analysis of distribution functions.* Ark. Mat. 4 (1960), 99—157; MR 23#A 3963.

[6] P. L. Butzer und E. Görlich, *Saturationsklassen und asymptotische Eigenschaften trigonometrischer singulärer Integrale.* In: Festschrift zur Gedächtnisfeier für Karl Weierstrass 1815—1965 (Ed. H. Behnke—K. Kopfermann, Wiss. Abh. Arbeitsgemeinschaft für Forschung des Landes Nordrhein—Westfalen 33) Westdeutsch. Verl., Köln—Oplande 1966, 612 pp.; 339—392; MR 33#4555.

[7] P. L. Butzer und R. J. Nessel, *Fourier Analysis and Approximation.* Vol. 1. *One-Dimensional Theory.* Birkhäuser Verl., Basel—Stuttgart and Academic Press, New York—London 1971, xiv+553 pp.

[8] P. L. Butzer und E. L. Stark, *Wesentliche asymptotische Entwicklungen für Approximationsmaße trigonometrischer singulärer Integrale.* Math. Nachr. 39 (1969), 223—237; MR 40#3139.

[9] R. DeVore, *Approximation of Continuous Functions by Positive Linear Operators.* Preprint (1971).

[10] G. A. Fomin, *On the best approximation of functions of classes Z_2 and Z_1 by certain linear operators* (Russ.). In: Studies of Contemporary Problems of Constructive Theory of Functions (Russ.). (Proc. Second All-Union Conf., Baku, 1962, Ed. I. I. Ibragimov) Izdat. Akad. Nauk Azerbaĭdžan. SSR, Baku 1965, 638 pp., 207—211; MR 33#6243.

[11] M. Ghermanesco, *Sur l'intégrale de Poisson. Sur l'intégrale de Poisson (suite).* Bull. Sci. École Polytechnique de Timişoara 4,fasc. 3—4 (1932), 159—184, 5, fasc. 1—2 (1933), 41—74; FdM **58** (1932), 1068.

[12] E. Görlich, *Über optimale Approximationsprozesse.* In: Constructive Function Theory (Proc. Int. Conf., Golden Sands, Varna, 1970, Ed. B. Penkov—D. Vačov) Izdat. Bolg. Akad. Nauk 1972, 363 pp., 187—191.

[13] E. Görlich and E. L. Stark, *A unified approach to three problems on approximation by positive linear operators.* In: Proceedings of the Conference on Constructive Theory of Functions

---

*) If available, the respective review of the *Mathematical Reviews* (MR) or of the *Fortschritte der Mathematik* (FdM) is added.

(Approximation Theory) (Budapest, 1969, Ed. G. Alexits—S. B. Stechkin) Akadémiai Kiadó, Budapest 1972, 538 pp; 201—208.

[14] E. Görlich und E. L. Stark, *Über beste Konstanten und asymptotische Entwicklungen positiver Faltungsintegrale und deren Zusammenhang mit dem Saturationsproblem.* Jber. Deutsch. Math.-Verein. **72** (1970), 18—61.

[15] B. L. Golinskiĭ, *Approximation on the entire number axis of two functions which are conjugate in the sense of Riesz by integral operators of singular type* (Russ.). Mat. Sb. (N. S.) **66 (108)** (1965), 3—34; MR **30** #2280.

[16] P. P. Korovkin, *An asymptotic property of positive methods of summation of Fourier series and best approximation of functions of class* Z_2 *by linear positive polynomial operators* (Russ.). Uspehi Mat. Nauk **13**, no. **6 (84)** (1958), 99—103; MR **21** #253.

[17] P. P. Korovkin, *Linear Operators and Approximation* (Russ.). Gos. Izdat. Fiz.-Mat. Lit., Moscow 1959, 211 pp. (=Transl. Hindustan Publ. Corp., Delhi 1960, vii+222 pp.); MR **27** #561.

[18] P. P. Korovkin, *Asymptotic properties of positive methods of summation of Fourier series* (Russ.). Uspehi Mat. Nauk **5**, no. **1 (91)** (1960), 207—217; MR **22** #6975.

[19] Y. Matsuoka, *On the degree of approximation of functions by some positive linear operators.* Sci. Rep. Kagoshima Univ. **9** (1960), 11—16; MR **23** #A 1189.

[20] I. P. Natanson, *On the accuracy of representation of continuous periodic functions by singular integrals* (Russ.). Dokl. Akad. Nauk SSSR **73** (1950), 273—276; MR **2**, 94.

[21] R. J. Nessel, *Über Nikolskiĭ-Konstanten von positiven Approximationsverfahren bezüglich Lipschitz-Klassen.* Jber. Deutsch. Math.-Verein. **73** (1971), 6—47.

[22] R. J. Nessel, *Nikolskiĭ constants of positive operators for Lipschitz classes.* In: Constructive Function Theory (Proc. Int. Conf., Golden Sands/Varna, 1970, Ed. B. Penkov—D. Vačov) Izdat. Bolg. Akad. Nauk 1972, 363 pp., 239—244.

[23] S. M. Nikolskiĭ, *Sur l'allure asymptotique du reste dans l'approximation au moyen des sommes de Fejér des fonctions vérifiant la condition de Lipschitz* (Russ.; French sum.) Izv. Akad. Nauk SSSR Ser. Mat. **4** (1940), 501—508; MR **2**, 279.

[24] I. M. Petrov, *Order of approximation of functions belonging to the class* Z *by some polynomial operators* (Russ.). Uspehi Mat. Nauk **3**, no. **6 (84)**, (1958), 127—131; MR **21** #732.

[25] I. M. Petrov, *The order of approximation of functions of class* Z_α *by certain polynomial operators* (Russ.). Izv. Vysš. Učebn. Zaved. Matematika **1960**, no. **1 (14)** (1960), 188—193; MR **24** #A 967.

[26] I. M. Petrov, *On the order of approximation of class* Z_1 *by positive linear polynomials* (Russ.). Uspehi Mat. Nauk **19**, no. **2 (116)** (1964), 151—154; MR **29** #413.

[27] F. Schurer, *Some remarks on the approximation of functions by some positive linear operators.* Monatsh. Math. **67** (1963), 353—358; MR **28** #411.

[28] F. Schurer, *On linear positive operators in approximation theory* (Dutch sum.). Thesis, Technische Hogeschool te Delft, 1965; Uitgeverij Waltman, Delft 1965, iv+79 pp.; MR **34** #6389.

[29] F. Schurer and F. W. Steutel, *On linear positive operators of the Jackson type.* Math. Communication, Technological Univ. Twente **1** (1966) 45 pp.

[30] F. Schurer and F. W. Steutel, *On linear positive operators of the Jackson type.* Mathematica (Cluj) **9 (32)** (1967), 155—184; MR **36** #583.

[31] E. L. Stark, *Über einige Konstanten der singulären Integrale von Dirichlet, Rogosinski, Fejér und Fejér—Korovkin.* Diplomarbeit, Rheinisch-Westfälische Technische Hochschule Aachen, 1966, viii+105 pp.

[32] E. L. Stark, *Über die Approximationsmaße spezieller singulärer Integrale* (Engl. sum.). Computing **4** (1969), 153—159; MR **39** #7336.

[33] E. L. Stark, *On a generalization of Abel—Poisson's singular integral having kernels of finite oscillation*. Studia Sci. Math. Hungar. (to appear).

[34] R. Taberski, *Some properties of (K, φ)-summability*. Bull. Acad. Polon. Sci. Sér. Sci. Math. Astronom. Phys. **9** (1961), 659—666; MR **24** #3446.

[35] R. Taberski, *More about (K, φ)-summability*. Bull. Acad. Polon. Sci. Sér. Sci. Math. Astronom. Phys. **9** (1961), 769—774; MR **24** #3447.

[36] C. de La Vallée Poussin, *Leçons sur L'Approximation des Fonctions d'une Variable Réelle*. Gauthier-Villars Ed., Paris 1919, 1952 II, vi+151 pp.

[37] Zheng Wei-xing, *On the extreme property of the operator $\mathscr{B}_\sigma(f, x)$* (Chin.). Acta Math. Sinica **5** (1965), 54—62 =Transl. Chinese Math. **8** (1965), 353—362; MR **33** #473.

[38] Review in Referativnyĭ Žurnal, Matematika **1960**, no. 266: Cheng Wee-shing, *Asymptotic formula for the approximation of class Z^* by polynomials* (Chin., Eng. Sum.). Nanjing Daxue Xuebao, Ziran Kexue=Acta Univ. Nankin. Sci. Nat. **1959**, no. 3 (1959), 1—6.

A Pointwise "*o*" Saturation Theorem
for Positive Convolution Operators

By

RONALD A. DEVORE

DEPT. OF MATH.
OAKLAND UNIVERSITY
ROCHESTER, MICHIGAN

1. Introduction

We wish to consider the saturation of positive convolution operators in the space $C^*[-\pi, \pi]$ of 2π-periodic and continuous functions. For this purpose, let (L_n) be a sequence of operators given by the convolution formulae

$$(1) \qquad L_n(f, x) = (f * d\mu_n)(x) = \frac{1}{\pi} \int_{-\pi}^{\pi} f(x+t)\, d\mu_n(t)$$

where each $d\mu_n$ is a non-negative, even, Borel measure on $[-\pi, \pi]$ with $\dfrac{1}{\pi} \int_{-\pi}^{\pi} d\mu_n(t) = 1$.

We also suppose that the Fourier—Stieltjes coefficients $\varrho_{k,n} = \dfrac{1}{\pi} \int_{-\pi}^{\pi} \cos kt\, d\mu_n(t)$ satisfy

$$(2) \qquad \lim_{n \to \infty} \frac{1 - \varrho_{k,n}}{1 - \varrho_{1,n}} = \psi_k \neq 0. \qquad k = 1, 2, \ldots .$$

The requirement (2) is a standard assumption for saturation theorems. In parrticular, under these assumptions, we have that for $f \in C^*$

$$(3) \qquad \| f - L_n(f) \| = o(1 - \varrho_{1,n}), \quad \text{if and only if } f \text{ is constant,}$$

where $\| - \|$ denotes the supremum norm. This is the "*o*" part of the general saturation theorem of SUNOUCHI—WATARI [8] and is easily proved using transforms. Namely, if $\| f - L_n(f) \| = o(1 - \varrho_{1,n})$ then it follows by taking transforms that $\hat{f}(k) - \hat{f}(k)\varrho_{k,n} = o(1 - \varrho_{1,n})$. Because of (2) we have $\hat{f}(k) = 0$, $k \neq 0$.

The situation becomes more difficult if we seek a characterization of the functions f which satisfy a pointwise "*o*" condition

$$(4) \qquad f(x) - L_n(f, x) = o_x(1 - \varrho_{1,n}) \quad \text{for each} \quad x \in [-\pi, \pi].$$

Here, the simplest case occurs when $\psi_k = k^2$. This is equivalent [7] to having for each $S_\varepsilon = [-\pi, \pi] \setminus (-\varepsilon, \varepsilon)$, $\varepsilon > 0$

$$(5) \qquad \int_{S_\varepsilon} d\mu_n(t) = o(1 - \varrho_{1,n}).$$

Still another equivalent formulation is that the asymptotic formula

$$(6) \qquad \lim_{n\to\infty} (1-\varrho_{1,n})^{-1}(L_n(f, x)-f(x)) = f''(x)$$

should hold for all functions f with two continuous derivatives. Because of (5), it is possible to use the parabola technique of BAJSANSKI and BOJANIC [2] to obtain a pointwise "o" theorem.

A sketch of the argument is as follows. If f is a continuous function on $[-\pi, \pi]$ which satisfies (4) then subtracting a constant if necessary, we can assume $f(-\pi)= =f(\pi)=0$. We also suppose that $f(x_0)>0$ for some $x_0 \in (-\pi, \pi)$. So that from Lemma 1 of [2], it follows that there is a point $y \in (-\pi, \pi)$ and a parabola $Q(x) = \alpha(x-y)^2+ +\beta(x-y)+f(y)$, with $\alpha<0$, such that $Q(x)\geq f(x)$ for $x\in[-\pi, \pi]$. Therefore,

$$(7) \qquad L_n(f, y)-f(y) = \frac{1}{\pi} \int_{-\pi}^{\pi} [f(x)-f(y)]\,d\mu_n(x-y) \leq$$

$$\leq \frac{1}{\pi} \int_{-\pi}^{\pi} (Q(x)-Q(y))\,d\mu_n(x-y) =$$

$$= \frac{\alpha}{\pi} \int_{-\pi}^{\pi} (x-y)^2\,d\mu_n(x-y)+\frac{\beta}{\pi} \int_{-\pi}^{\pi} (x-y)\,d\mu_n(x-y).$$

It is easy to show that the second term on the right hand side of (7) is $o(1-\varrho_{1,n})$ by using (5) and the fact that $d\mu_n$ is even. Therefore,

$$L_n(f, y)-f(y) \leq \frac{\alpha}{\pi} \int_{-\pi}^{\pi} (x-y)^2\,d\mu_n(x-y)+o(1-\varrho_{1,n}) \leq$$

$$\leq \alpha\pi \int_{-\pi}^{\pi} \sin^2\left(\frac{x-y}{2}\right) d\mu_n(x-y)+o(1-\varrho_{1,n}) = \alpha\pi^2(1-\varrho_{1,n})+o(1-\varrho_{1,n}),$$

where we have used the inequality $\sin\dfrac{t}{2} \geq \dfrac{t}{\pi}$ for $0\leq t\leq \pi$.

Since $\alpha<0$, this contradicts (4) for the point y. Thus $f(x)\leq 0$ for all $x \in [-\pi, \pi]$. Replacing f by $-f$ in the above argument we conclude that $f(x)\geq 0$ on $[-\pi, \pi]$ and thus $f\equiv 0$ on $[-\pi, \pi]$ as desired.

The parabola technique cannot be applied directly when (5) does not hold since in this case the terms $\int_{-\pi}^{\pi} (x-y)\,d\mu_n(x-y)$ are not negligible. The object of this paper is to prove a general pointwise "o" theorem with no restrictions on (ψ_k).

2. Main results

THEOREM: *Let (L_n) be a sequence of positive convolution operators of the form
(1) where the Fourier coefficients of $d\mu_n$ satisfy (2). If $f \in C^*$ then*

$$(8) \qquad f(x) - L_n(f, x) = o_x(1 - \varrho_{1,n}) \quad \text{for each} \quad x \in [-\pi, \pi]$$

if and only if f is constant on $[-\pi, \pi]$.

PROOF: The "if" part of the theorem is obvious. The proof of the "only if"
part is based on a trigonometric analogue of the parabola technique of BAJSANSKI
BOJANIC. However, we must first prove two lemmas which give some properties of
functions which satisfy (8). Of course, ultimately we wish to show that such functions
are constant.

If x is a point in $[-\pi, \pi]$, such that, for each neighbourhood I of x we have
$\int_I d\mu_n(t) \neq o(1 - \varrho_{1,n})$, then we shall say x is an essential point. Otherwise, we say
x is a negligible point. Let f be a function which satisfies (8). We set $M = \max\limits_{-\pi \leq t \leq \pi} f(t)$
and when $x_0 \in [-\pi, \pi]$, with $f(x_0) = M$, we let $\mathfrak{M}(x_0) = \{t:\ t$ is an essential point
and $f(x_0 + t) = M\}$. Also, let $\mathfrak{M} = \bigcap\limits_{x_0} \mathfrak{M}(x_0)$. We consider all points modulo 2π.

LEMMA 1: *If $x \notin \mathfrak{M}$, then x is a negligible point.*

PROOF: Suppose $x \notin \mathfrak{M}(x_0)$ for some x_0 with $f(x_0) = M$. Then either x is negligible
or $f(x_0 + x) < M$. In the latter case, let $I = \{y: f(x_0 + y) < \frac{1}{2}[M + f(x_0 + x)]\}$. I is
a neighbourhood of x and

$$\tfrac{1}{2}[M - f(x_0 + x)] \int_I d\mu_n(t) \leq \int_I [f(x_0) - f(x_0 + t)]\, d\mu_n(t).$$

However

$$\frac{1}{\pi} \int_I [f(x_0) - f(x_0 + t)]\, d\mu_n(t) \leq \frac{1}{\pi} \int_{-\pi}^{\pi} [f(x_0) - f(x_0 + t)]\, d\mu_n(t) =$$

$$= f(x_0) - L_n(f, x_0) = o(1 - \varrho_{1,n}).$$

This shows that $\int_I d\mu_n(t) = o(1 - \varrho_{1,n})$ and thus x is negligible.

LEMMA 2: *If f is not constant, then $\mathfrak{M}$ has only a finite number of points. Also,
if x is any point in $\mathfrak{M}$ then $x = 2\pi\alpha$ where α is rational.*

PROOF: Suppose first that (x_n) is a sequence of distinct points each of which is
in $\mathfrak{M}$. Choosing a subsequence if necessary we can assume $x_n \to x$ where $0 \leq x < 2\pi$.
We write $x_n = 2\pi\alpha_n$ and $x = 2\pi\alpha$.

Let x_0 be any point in $[-\pi, \pi]$, where $f(x_0) = M$. Then, $f(x_0 + x_n) = M$ and
$x_n \in \mathfrak{M}(x_0 + x_n)$, so that $f(x_0 + 2x_n) = M$. More generally, for each positive integer

k, we have $f(x_0+kx_n) = M$. By continuity, $f(x_0+kx) = M$ for each positive integer k. If α is irrational then the points $k\alpha$ taken modulo 1 are dense in $[0, 1]$ so that the points kx taken modulo 2π are dense in $[0, 2\pi]$. Thus, in this case, $f=M$ on a set of points which are dense in $[x_0, x_0+2\pi]$ and therefore $f=M$ on $[x_0, x_0+2\pi]$. From periodicity, we conclude that f is constant.

If α is rational then the points $k\alpha_n = k(\alpha+\delta_n)$ where $0\neq\delta_n\to 0$, taken modulo 1, are dense in $[0, 1]$. Therefore, we again have that f is constant. This shows that $\mathfrak{M}$ has no limit point in $[0, 2\pi]$ and hence must consist of only a finite number of points.

Finally, if a point of the form $x=2\pi\alpha$ with α irrational were in $\mathfrak{M}$ then, as we have mentioned before, $f(x_0+kx) = M$ for each positive integer k so that $f=M$ on a set of points which is dense in $[x_0, x_0+2\pi]$. This again gives that f is constant.

PROOF OF THE THEOREM: Let f be a function which satisfies (8) and suppose f is not constant. By subtracting a constant, if necessary, we can suppose that $f(-\pi)==f(\pi)=0$. Also suppose $M>0$. Then, it follows from Lemma 2 that there is a positive integer m such that $\mathfrak{M} \subseteq \left\{\dfrac{k\pi}{m}: k=0, \pm 1, \pm 2, \ldots, \pm m\right\}$. Let

$$I = \bigcup_{k=-m}^{m} \left[\frac{k\pi}{m} - \frac{\pi}{8m}, \frac{k\pi}{m} + \frac{\pi}{8m}\right] \cap [-\pi, \pi].$$

Since each point $y\notin\mathfrak{M}$ is negligible we could use a compactness argument to show that for $S=[-\pi, \pi]\setminus I$

$$\tag{9} \int_S d\mu_n(t) = o(1-\varrho_{1,n}).$$

The function $h(x) = -M \sin^2 mx + 2M$ is $\geq f(x)$ on $[-\pi, \pi]$. Let $c = \min_{t\in I}(h(t)-f(t))$. Then, $h(x)-c \geq f(x)$ on I and for some $y\in I$

$$h(y)-c = f(y).$$

Therefore,

$$\tag{10} \int_I [h(x)-h(y)]\,d\mu_n(x-y) \geq \int_I [f(x)-f(y)]\,d\mu_n(x-y).$$

But $h(x)-h(y) = -M \cos(2my) \sin^2 m(x-y) - \dfrac{M}{2} \sin(2my) \sin 2m(x-y)$ and thus

$$\frac{1}{\pi} \int_{-\pi}^{\pi} [h(x)-h(y)]\,d\mu_n(x-y) = -\frac{M}{\pi} \cos(2my) \int_{-\pi}^{\pi} \sin^2 m(x-y)\,d\mu_n(x-y) =$$

$$= -\frac{M}{2} \cos 2my(1 - \varrho_{2m,n}).$$

Here, we have used the fact that $d\mu_n$ is even to drop the term involving $\sin 2m(x-y)$ which would normally appear. Because of (9), we have

$$\frac{1}{\pi} \int\limits_{I} [h(x)-h(y)]\,d\mu_n\,(x-y) =$$

$$= \frac{1}{\pi} \int\limits_{-\pi}^{\pi} [h(x)-h(y)]\,d\mu_n\,(x-y) - \frac{1}{\pi} \int\limits_{S} [h(x)-h(y)]\,d\mu_n\,(x-y) =$$

$$= -\frac{M}{2}\cos 2my(1-\varrho_{2m,n}) + o(1-\varrho_{1,n}).$$

Finally, from (10)

$$L_n(f, y) - f(y) = \frac{1}{\pi} \int\limits_{I} [f(x)-f(y)]\,d\mu_n\,(x-y) + o(1-\varrho_{1,n}) \leqq$$

$$\leqq \frac{1}{\pi} \int\limits_{I} [h(x)-h(y)]\,d\mu_n\,(x-y) + o(1-\varrho_{1,n}) =$$

$$= -\frac{M}{2}\cos 2my(1-\varrho_{2m,n}) + o(1-\varrho_{1,n}) = -\frac{M}{2}\cos 2my(1-\varrho_{1,n})\psi_{2m} + o(1-\varrho_{1,n})$$

where for the last equality we have used (2).

Since $\cos 2my > 0$ and $\psi_{2m} > 0$, we must have $f(y) - L_n(f, y) \neq o(1-\varrho_{1,n})$ which is the desired contradiction and therefore $M=0$. This shows $f(x) \leqq 0$ on $[-\pi, \pi]$. To see that $f(x) \geqq 0$ on $[-\pi, \pi]$, we merely work with the function $-f$ in place of f in the above argument. This completes the proof of the theorem.

3. Remarks. The first pointwise saturation theorem appears to be the "o" theorem for the Fejér operators (σ_n) which was given by ANDRIENKO [1]. In this case, it is possible to weaken (4) by discarding certain small sets (e.g. countable) and still conclude that f is constant.

This cannot be done in general. For example, if (L_n) is any sequence satisfying (2) with $\psi_k = k^2$ then the function $f(t) = |t|$ satisfies

$$f(x) - L_n(f, x) = o_x(1-\varrho_{1,n}) \qquad (-\pi < x \leqq \pi, \quad x \neq 0).$$

ANDRIENKO has also given local "o" theorems for the Fejér operators. Namely, if

$$f(x) - \sigma_n(f, x) = o_x(n^{-1}) \qquad (a \leqq x \leqq b)$$

then f is constant on $[a, b]$. To see that this cannot be done in general, let

$$\frac{1}{\pi}\,d\mu_n = \frac{1}{2}\left(1 - \frac{1}{2}\,n^{-2}\right)(d\varrho_{n-1} + d\varrho_{-n-1}) + \frac{n^{-2}}{4}\,(d\varrho_{\pi/2} + d\varrho_{-\pi/2})$$

where $d\varrho_{x_0}$ denotes the Dirac measure at x_0. Then

$$1-\varrho_{1,n} = n^{-2}+o(n^{-2})$$

and

$$\lim_{n\to\infty} \frac{(1-\varrho_{k,n})}{1-\varrho_{1,n}} = \frac{k^2}{2}+\sin^2\frac{k\pi}{4}.$$

If f is any function in C^* which is twice continuously differentiable on $\left(-\dfrac{\pi}{8},\dfrac{\pi}{8}\right)$ then

$$\lim_{n\to\infty} 2n^2(L_n(f,x)-f(x)) = f''(x)+\frac{1}{2}\left[f\left(x+\frac{\pi}{2}\right)+f\left(x-\frac{\pi}{2}\right)-2f(x)\right]$$

for $x\in\left(-\dfrac{\pi}{8},\dfrac{\pi}{8}\right)$. Thus, we can take any such f and define it outside $\left(-\dfrac{\pi}{8},\dfrac{\pi}{8}\right)$ in such a way that

$$f\left(x+\frac{\pi}{2}\right)+f\left(x-\frac{\pi}{2}\right) = 2f(x)-2f''(x) \qquad \left(-\frac{\pi}{8}<x<\frac{\pi}{8}\right),$$

and then f will satisfy

$$f(x)-L_n(f,x) = o_x(1-\varrho_{1,n}) \qquad \left(-\frac{\pi}{8}<x<\frac{\pi}{8}\right).$$

In an unpublished paper, H. BERENS has shown the pointwise "o" theorem for the case when there is an $0<\alpha\leqq2$ such that $\psi_k=k^\alpha$, $k=1, 2, \ldots$. Here, the proof is based on knowing that the function $h_\alpha(t)$ whose Fourier coefficients are $k^{-\alpha}$ has the property that

$$\int_a^b h_\alpha(t)\,dt > 0$$

for each $-\pi\leqq a<b\leqq\pi$.

In this same vein, if we let $d\lambda_n(t) = 2(1-\varrho_{1,n})^{-1}\sin^2\dfrac{t}{2}\,d\mu_n(t)$, then $\dfrac{1}{\pi}\int |d\lambda_n(t)|=1$. Therefore, there is a subsequence (n_k) and a measure $d\lambda$ such that $d\lambda_{n_j}\to d\lambda$, weak $*$. Our essential points are just the points in the support of $d\lambda$. If the support of $d\lambda$ is all of $[-\pi,\pi]$, the proof of the theorem can be given as we argued in Section 1.

For the Cesàro means [3] and more generally the typical means [4] of the Fourier series, H. Berens has given both pointwise "o" and "O" theorems. When $\psi_k=k^2$, $k=1, 2, \ldots$, then there is a companion pointwise "O" theorem which also is due to BERENS [5]. However, there is no general pointwise "O" theorem which is companion to our "o" theorem. For such a "O" theorem, it will be necessary to assume the multiplier condition

$$\left(\frac{1-\varrho_{k,n}}{\psi_k(1-\varrho_{1,n})}\right)_{k=0}^{\infty} \in (L_\infty, L_\infty),$$

since this is needed even in the norm case. (See for example DeVore [6].)

Regarding the form such a "O" would most likely take, we note that when the multiplier condition holds then there is an asymptotic formula for (L_n). If D is the distribution with $\hat{D}(k)=\psi_k$, then there is a subsequence (n_j) such that for each $f \in C^*$ with $f*D$ continuous we have

$$\lim_{n\to\infty} (1-\varrho_{1,n_j})^{-1}(L_{n_j}(f, x) - f(x)) = f*D.$$

This asymptotic condition indicates that the "O" theorem should read: If $g \in L_1$ and

$$\varliminf_{n\to\infty} (1-\varrho_{1,n})^{-1}(L_n(f, x) - f(x)) \leq g(x) \leq \varlimsup_{n\to\infty} (1-\varrho_{1,n})^{-1}(L_n(f, x) - f(x)),$$

then $f*D \in L_1$ and $f*D = g$ a.e.

4. Acknowledgement

We would like to thank Professor H. Berens for many helpful comments.

REFERENCES

[1] V. A. Andrienko, *Approximation of functions by Fejér means.* Siberian Math. J. 9 (1968), 1—8.

[2] B. Bajsanski and R. Bojanic, *A note on approximation by Bernstein polynomials.* Bull. Amer. Math. Soc. 70 (1964), 675—677.

[3] H. Berens, *On the saturation theorem for the Cesàro means of the Fourier series.* (to appear).

[4] H. Berens, *On pointwise approximation of Fourier series by typical means.* Tôhoku Math. J. 23 (1971), 147—154.

[5] H. Berens, *Pointwise saturation of positive operators.* (to appear).

[6] R. DeVore, *On the direct theorem of saturation.* Tôhoku Math. J. 23. (1971), 363—370.

[7] E. Görlich and E. Stark, *Über beste Konstanten und asymptotische Entwicklungen positiver Faltungsintegrale und deren Zusammenhang mit dem Saturationsproblem.* Jber. Deutsch. Math.-Verein. 72 (1970), 18—61.

[8] G. Sunouchi and C. Watari, *On determination of the class of saturation in the theory of approximation of functions.* Proc. Japan Acad. 34 (1958), 477—481.

Convolution Operators for Fourier—Jacobi Expansions

By

HERMAN BAVINCK

MATHEMATISCH CENTRUM
AMSTERDAM

1. Introduction

1. 1. In some recent work [3], [4], the author has used the convolution structure for Jacobi series, introduced by Askey and Wainger [2], in order to study the summation of Jacobi series by classical summability methods. Many of these summability methods, in fact, can be interpreted as convolution operators and it is possible to investigate the order of approximation of these operators by the same techniques as are used for trigonometric convolution operators. In this paper some new summability kernels are introduced, which can be written in a simple closed form by means of Jacobi polynomials. Even in the case of Fourier series ($\alpha = \beta = -\frac{1}{2}$) these kernels induce new approximation processes. The saturation order and the saturation class of these processes are obtained.

1. 2. By X we denote one of the following function spaces on $[-1, 1]$: the space C of continuous functions with the norm ($x = \cos \theta$)

$$\|f\|_C = \sup_{0 \leqq \theta \leqq \pi} |f(\cos \theta)|$$

or the L^p spaces ($1 \leqq p < \infty$) with respect to the weight function

$$(1.1) \qquad \varrho^{(\alpha,\beta)}(\theta) = \left(\sin \frac{\theta}{2}\right)^{2\alpha+1} \left(\cos \frac{\theta}{2}\right)^{2\beta+1} \qquad (\alpha \geqq \beta \geqq -\tfrac{1}{2})$$

endowed with the norm

$$\|f\|_p = \left[\int_0^\pi |f(\cos \theta)|^p \, \varrho^{(\alpha,\beta)}(\theta) \, d\theta\right]^{1/p}.$$

Functions belonging to X can be expanded in terms of Jacobi polynomials $P_n^{(\alpha,\beta)}(\cos \theta)$, the polynomials which are orthogonal with respect to (1. 1). If we take

$$R_n^{(\alpha,\beta)}(\cos \theta) = \frac{P_n^{(\alpha,\beta)}(\cos \theta)}{P_n^{(\alpha,\beta)}(1)},$$

then with $f \in X$ we associate

$$(1.2) \qquad f(\cos \theta) \sim \sum_{n=0}^\infty f^\wedge(n) \, \omega_n^{(\alpha,\beta)} \, R_n^{(\alpha,\beta)}(\cos \theta),$$

where

$$(1.3) \qquad f^\wedge(n) = \int_0^\pi f(\cos\theta) R_n^{(\alpha,\beta)}(\cos\theta) \varrho^{(\alpha,\beta)}(\theta)\, d\theta \qquad (n=0,1,\dots),$$

and

$$(1.4) \qquad \omega_n^{(\alpha,\beta)} = \left[\int_0^\pi \{R_n^{(\alpha,\beta)}(\cos\theta)\}^2 \varrho^{(\alpha,\beta)}(\theta)\, d\theta\right]^{-1} =$$

$$= \frac{(2n+\alpha+\beta+1)\Gamma(n+\alpha+\beta+1)\Gamma(n+\alpha+1)}{\Gamma(n+\beta+1)\Gamma(n+1)\Gamma(\alpha+1)\Gamma(\alpha+1)}.$$

ASKEY and WAINGER [2] have introduced a generalized translation and a convolution structure for Jacobi series. The translation $T_\Phi f$ maps a function $f \in X$ with the expansion (1.2) into

$$(1.5) \qquad T_\Phi f(\cos\theta) \sim \sum_{n=0}^\infty f^\wedge(n)\, \omega_n^{(\alpha,\beta)} R_n^{(\alpha,\beta)}(\cos\theta) R_n^{(\alpha,\beta)}(\cos\Phi).$$

GASPER [7] has shown that T_Φ is a positive operator and consequently has operator norm 1. For $f_1, f_2 \in L^1$ the convolution $f_1 * f_2$ is defined by

$$(1.6) \qquad (f_1 * f_2)(\cos\theta) = \int_0^\pi T_\Phi f_1(\cos\theta) f_2(\cos\Phi) \varrho^{(\alpha,\beta)}(\Phi)\, d\Phi$$

and has the following properties.

1.3. For $f_1, f_2, f_3 \in L^1$ and $g \in X$

$$(1.7) \qquad \begin{cases} \text{i)} & f_1 * f_2 = f_2 * f_1, \\ \text{ii)} & f_1 * (f_2 * f_3) = (f_1 * f_2) * f_3, \\ \text{iii)} & \|f_1 * g\|_X \leq \|f_1\|_1 \|g\|_X, \\ \text{iv)} & (f_1 * f_2)^\wedge(n) = f_1^\wedge(n) f_2^\wedge(n). \end{cases}$$

In [3] we have defined a summability kernel.

1.4. DEFINITION. Let $K_\lambda \in L^1$, $\lambda > 0$, satisfying the conditions

$$\text{(a)} \qquad \int_0^\pi K_\lambda(\cos\theta) \varrho^{(\alpha,\beta)}(\theta)\, d\theta = 1,$$

$$\text{(b)} \qquad K_\lambda(\cos\theta) \geq 0, \qquad \lambda > 0,\ 0 \leq \theta \leq \pi,$$

$$\text{(c)} \qquad \lim_{\lambda\to\infty} K_\lambda^\wedge(n) = 1, \qquad n = 0,1,\dots.$$

Then we call K_λ a positive summability kernel. If instead of b we merely have

$$\text{(b')} \qquad \int_0^\pi |K_\lambda(\cos\theta)| \varrho^{(\alpha,\beta)}(\theta)\, d\theta \leq N, \quad \text{uniformly in } \lambda,$$

with $N \geq 1$, we call K_λ a quasi-positive kernel.

Condition c is often replaced by

(c')$$\lim_{\lambda \to \infty} \int_h^\pi |K_\lambda(\cos \theta)|\, \varrho^{(\alpha,\beta)}(\theta)\, d\theta = 0, \quad \text{for each } h, \quad 0 < h \leqq \pi.$$

We have ([3], theorems 3. 3 and 3. 4)

1. 5. THEOREM. *If $f \in X$ and $K_\lambda \in L^1$, $\lambda > 0$, satisfies the conditions a, b (or b') and c (or c') of definition 1. 4, then*

$$\|K_\lambda * f\|_X \leqq N \|f\|_X \quad \text{uniformly in } \lambda$$

and

$$\lim_{\lambda \to \infty} \|K_\lambda * f - f\|_X = 0.$$

The convolution operators $\{K_\lambda, \lambda > 0\}$ are said to be saturated if there exists a positive non-increasing function $\Phi(\lambda)$ on $0 < \lambda < \infty$ with $\lim_{\lambda \to \infty} \Phi(\lambda) = 0$ such that

i)$$\|K_\lambda * f - f\|_X = o\big(\Phi(\lambda)\big) \qquad (\lambda \to \infty)$$

if and only if f belongs to some 'trivial' subspace of X;

ii) there exists a 'non-trivial' element $f_0 \in X$ satisfying

$$\|K_\lambda * f_0 - f_0\|_X = O\big(\Phi(\lambda)\big) \qquad (\lambda \to \infty).$$

The function $\Phi(\lambda)$ is then called the saturation order and the set $F(X, K_\lambda)$, which consists of all the elements of X which satisfy ii, is called the saturation class or Farvard class of $\{K_\lambda\}$ (compare [6, p. 434]).

The Lipschitz classes with respect to the generalized translation are defined by

(1. 8)$$\text{Lip}\,(\gamma, X) = \{f \in X : \exists c > 0, \ \sup_{0 < \Psi < \Phi} \|T_\Psi f - f\|_X \leqq c\Phi^\gamma\} \qquad (0 < \gamma \leqq 2).$$

If for $f \in X$ with the expansion (1. 2) there exists an element $Af \in X$ such that

(1. 9)$$Af \sim \sum_{n=0}^\infty n(n+\alpha+\beta+1) f^\wedge(n)\, \omega_n^{(\alpha,\beta)}\, R_n^{(\alpha,\beta)}(\cos \theta),$$

then we say that $f \in D(A)$ and we call A the operator which maps $D(A)$ into X by $f \to Af$. The operator A is the realization in A of the differential operator

$$-\frac{1}{\varrho^{(\alpha,\beta)}(\theta)} \frac{d}{d\theta} \left\{ \varrho^{(\alpha,\beta)}(\theta) \frac{d}{d\theta} \right\}$$

374 H. BAVINCK

with boundary conditions $\dfrac{d}{d\theta} = 0$ at $\theta = 0$ and $\theta = \pi$ in view of the differential equation for Jacobi polynomials

$$-\frac{1}{\varrho^{(\alpha,\beta)}(\theta)}\frac{d}{d\theta}\left\{\varrho^{(\alpha,\beta)}(\theta)\frac{d}{d\theta}R_n^{(\alpha,\beta)}(\cos\theta)\right\} = n(n+\alpha+\beta+1)R_n^{(\alpha,\beta)}(\cos\theta).$$

There is a close connection between the generalized translation operator and the operator A, as LÖFSTRÖM and PEETRE showed in [8]. For $f \in D(A)$, in fact, we have

$$\lim_{\Phi\to 0+}\left\|\frac{f-T_\Phi f}{C_1(\Phi)}-Af\right\|_X = 0,$$

where

$$C_1(\Phi) = \int_0^\Phi \frac{1}{\varrho^{(\alpha,\beta)}(\theta)}\left(\int_0^\theta \varrho^{(\alpha,\beta)}(\tau)\,d\tau\right)d\theta = O(\Phi^2)\qquad (\Phi\to 0^+).$$

Moreover, if the K function norm, introduced by PEETRE [9] is given by

$$K(\Phi,f;X,D(A)) = \inf_{\substack{f=f_0+f_1\\ f_0\in X\\ f_1\in D(A)}}(\|f_0\|_X+\Phi\|f_1\|_{D(A)}),$$

where

$$\|f\|_{D(A)} = \|f\|_X+\|Af\|_X,$$

then it can be shown (LÖFSTRÖM—PEETRE [8], BAVINCK [3]) that the spaces

$$(X,D(A))_{\theta,\infty;K} = \left\{f\in X : \sup_{\Phi>0}\Phi^{-\theta}K(\Phi,f;X,D(A)) < \infty\right\}\qquad (0<\theta\leq 1)$$

coincide with the spaces Lip $(2\theta, x)$, defined by (1.8).

NOTATION. We shall use the notation $a_n \approx b_n$ $(n\to\infty)$ if there are positive numbers c_1 and c_2 such that $c_1 a_n \leq b_n \leq c_2 a_n$.

2. An oscillating kernel

The following formula is due to SZEGŐ ([11], section 9. 4)

$$(2.1)\qquad R_n^{(\alpha+k+1,\beta)}(\cos\theta) = \frac{\Gamma(n+\beta+1)\Gamma(n+1)\Gamma(\alpha+k+2)\Gamma(\alpha+1)}{\Gamma(n+\alpha+\beta+k+2)\Gamma(n+\alpha+k+2)\Gamma(k+1)}\times$$

$$\times\sum_{v=0}^n \frac{\Gamma(n+v+\alpha+\beta+k+2)\Gamma(n-v+k+1)}{\Gamma(n+v+\alpha+\beta+2)\Gamma(n-v+1)}\,\omega_v^{(\alpha,\beta)}R_v^{(\alpha,\beta)}(\cos\theta)\qquad (k>0).$$

We shall study here the polynomial kernel $J_{n,1}^{(\alpha+k+1,\beta)}(\cos\theta)$, $n\geq 0$, defined by

(2.2)

$$J_{n,1}^{(\alpha+k+1,\beta)}(\cos\theta) = \frac{\Gamma(n+\alpha+\beta+2)\,\Gamma(n+\alpha+k+2)\,\Gamma(k+1)}{\Gamma(n+\beta+1)\,\Gamma(n+k+1)\,\Gamma(\alpha+k+2)\,\Gamma(\alpha+1)}\, R_n^{(\alpha+k+1,\beta)}(\cos\theta) =$$

$$= O(n^{2\alpha+2})\, R_n^{(\alpha+k+1,\beta)}(\cos\theta).$$

We first show that $J_{n,1}^{(\alpha+k+1,\beta)}(\cos\theta)$ satisfies the conditions a, b′ and c′ in definition 1.4. Taking the term $\nu=0$ in (2.1) we see that condition a is satisfied. For the proof of condition b′ we use the well-known estimates (SZEGŐ [11] (7.32.5))

(2.3)
$$R_n^{(\alpha,\beta)}(\cos\theta) = \begin{cases} \theta^{-\alpha-\frac{1}{2}}\,O(n^{-\frac{1}{2}-\alpha}) & \text{if } \quad cn^{-1}\leq\theta\leq\dfrac{\pi}{2} \\[2mm] O(1) & \text{if } \quad 0\leq\theta\leq cn^{-1}. \end{cases}$$

Then

$$\int_0^\pi |R_n^{(\alpha+k+1,\beta)}(\cos\theta)|\,\varrho^{(\alpha,\beta)}(\theta) = \int_0^{1/n} + \int_{1/n}^{\pi/2} + \int_{\pi/2}^{\pi} = I_1 + I_2 + I_3\,.$$

$$I_1 = O(1)\int_0^{1/n} \theta^{2\alpha+1}\,d\theta = O(n^{-2\alpha-2}),$$

$$I_2 = O(n^{-\frac{1}{2}-\alpha-k-1})\int_{1/n}^{\pi/2} \theta^{\alpha-k-\frac{1}{2}}\,d\theta = O(n^{-2\alpha-2}),\quad \text{if}\quad k>\alpha+\tfrac{1}{2},$$

$$I_3 = O(n^{\beta-\alpha-k-1})\int_0^{\pi/2} |R_n^{(\beta,\alpha+k+1)}(\cos\theta)|\,\varrho^{(\alpha,\beta)}(\theta)\,d\theta = O(n^{-\alpha-k-\frac{1}{2}}).$$

Hence

$$\int_0^\pi |J_{n,1}^{(\alpha+k+1,\beta)}(\cos\theta)|\,\varrho^{(\alpha,\beta)}(\theta)\,d\theta \leq M, \quad \text{if}\quad k>\alpha+\tfrac{1}{2}.$$

For the proof of c′ we take $n^{1/2} > \dfrac{1}{h}$. Then for $k > \alpha+\dfrac{1}{2}$

$$O(n^{2\alpha+2})\int_h^\pi |R_n^{(\alpha+k+1,\beta)}(\cos\theta)|\,\varrho^{(\alpha,\beta)}(\theta)\,d\theta \leq O(n^{2\alpha+2})\left(\int_{n^{-1/2}}^{\pi/2} + \int_{\pi/2}^{\pi} \right) =$$

$$= O(n^{\frac{1}{2}(\alpha+\frac{1}{2}-k)}) + O(n^{(\alpha+\frac{1}{2}-k)}) = o(1).$$

When we use the notation $\lambda_n = n(n+\alpha+\beta+1)$, the kernel $J_{n,1}^{(\alpha+k+1,\beta)}(\cos\theta)$ has the following representation in the case k is an integer

(2.4)
$$J_{n,1}^{(\alpha+k+1,\beta)}(\cos\theta) = \sum_{\nu=0}^{n} \prod_{j=0}^{k-1}\left(1-\frac{\lambda_\nu}{\lambda_{n+j+1}}\right) \omega_\nu^{(\alpha,\beta)}\, R_\nu^{(\alpha,\beta)}(\cos\theta).$$

This representation shows that this kernel is essentially a generalization of the typical means (see BUTZER—NESSEL [6], p. 262). Also, as is shown by SZEGŐ [11], section 9. 41, this kernel is closely related to the Cesàro means. In the case of Fourier series $(\alpha = \beta = -\tfrac{1}{2})$ we can choose $k=1$ and we get

$$(2.\,5) \qquad J_{n,1}^{(3/2,\,-1/2)}(\cos\theta) = \frac{1}{\pi}\left\{1 + 2\sum_{v=1}^{n}\left(1 - \left(\frac{v}{n+1}\right)^2\right)\cos v\theta\right\},$$

which are typical means. From the relation

$$R_n^{(\alpha+1,\beta)}(\cos\theta) = \frac{(\alpha+1)}{2n+\alpha+\beta+2}\,\frac{R_n^{(\alpha,\beta)}(\cos\theta) - R_{n+1}^{(\alpha,\beta)}(\cos\theta)}{\sin^2(\theta/2)}$$

we derive

$$J_{n,1}^{(3/2,\,-1/2)}(\cos\theta) = \frac{1}{4\pi(n+1)^2}\left[\frac{(2n+3)\sin(2n+1)(\theta/2) - (2n+1)\sin(2n+3)(\theta/2)}{(\sin(\theta/2))^3}\right].$$

For the other values of $k>0$, the kernel $J_{n,1}^{(\frac{1}{2}+k,\,-\frac{1}{2})}(\cos\theta)$ differs slightly from the Riesz means.

For the convolution of a function $f\in X$ with the kernel (2. 2) the following relations are valid.

$$(2.\,6) \qquad J_{n,1}^{(\alpha+k+2,\beta)}*f - J_{n-1,1}^{(\alpha+k+2,\beta)}*f =$$

$$= \frac{(k+1)(2n+k+\alpha+\beta+2)}{n(n+\alpha+\beta+1)(n+k+1)(n+\alpha+\beta+k+2)}\,J_{n,1}^{(\alpha+k+1,\beta)}*Af,$$

where A is the operator defined by (1. 9). In the case $k=$integer, formula (2. 6) is easy to check by using the representation (2. 4). Furthermore, we have

$$(2.\,7) \quad J_{n,1}^{(\alpha+k+2,\beta)}*f - J_{n,1}^{(\alpha+k+1,\beta)})*f = -\frac{1}{(n+k+1)(n+\alpha+\beta+k+2)}\,J_{n,1}^{(\alpha+k+1,\beta)}*Af.$$

By theorem 1. 5 and by repeated application of (2. 6) we obtain

$$f - J_{n,1}^{(\alpha+k+2,\beta)}*f =$$

$$= (k+1)\sum_{l=n+1}^{\infty}\frac{(2l+k+\alpha+\beta+2)}{l(l+\alpha+\beta+1)(l+k+1)(l+\alpha+\beta+k+2)}\,J_{l,1}^{(\alpha+k+1,\beta)}*Af.$$

Then, by (2. 7)

$$f - J_{n,1}^{(\alpha+k+1,\beta)}*f = -\frac{1}{(n+k+1)(n+\alpha+\beta+k+2)}\,J_{n,1}^{(\alpha+k+1,\beta)}*Af +$$

$$+ (k+1)\sum_{l=n+1}^{\infty}\frac{(2l+k+\alpha+\beta+2)}{l(l+\alpha+\beta+1)(l+k+1)(l+\alpha+\beta+k+2)}\,J_{l,1}^{(\alpha+k+1,\beta)}*Af.$$

If we put

$$c_n = k \sum_{l=n+1}^{\infty} \frac{(2l+k+\alpha+\beta+2)}{l(l+\alpha+\beta+1)(l+k+1)(l+\alpha+\beta+k+2)} = O(n^{-2}) \qquad (n \to \infty),$$

then

$$\|c_n^{-1}(f-J_{n,1}^{(\alpha+k+1,\beta)}*f)-Af\|_X \leq \frac{(k+1)}{k} \sup_{l \geq n+1} \|J_{l,1}^{(\alpha+k+1,\beta)}*Af-Af\|_X +$$

$$+ \frac{c_n^{-1}}{(n+k+1)(n+\alpha+\beta+k+2)} \|J_{n,1}^{(\alpha+k+1,\beta)}*Af-Af\|_X +$$

$$+ \|Af\|_X \left| \frac{1}{k} - \frac{c_n^{-1}}{(n+k+1)(n+\alpha+\beta+k+2)} \right| = o(1) \qquad (n \to \infty).$$

Hence,

$$\lim_{n\to\infty} [c_n^{-1}\{f-J_{n,1}^{(\alpha+k+1,\beta)}*f\} - Af] = 0 \quad \text{in} \quad X.$$

Since for the operator A an inequality of the Bernstein type is valid (see STEIN [10]), there is a constant M such that

$$\|A(J_{n,1}^{(\alpha+k+1,\beta)}*f)\|_X \leq Mn^2 \|f\|_X.$$

Therefore, the approximation process $J_{n,1}^{(\alpha+k+1,\beta)}*f$ satisfies all the conditions of the general theorems on approximation processes in Banach spaces treated in BERENS [5] (see also BUTZER—NESSEL* [6], 13. 4. 1). It follows that the process $J_{n,1}^{(\alpha+k+1,\beta)}*f$ is saturated with order n and the saturation class is $\text{Lip}(2, X)=(X, D(A))_{1,\infty;K}$. The spaces of non-optimal approximation can also be characterized in terms of the intermediate spaces $(X, D(A))_{\theta,q;K}$.

3. A positive kernel

We now consider the positive polynomial kernel $J_{n,2}^{(\alpha+k+1,\beta)}(\cos\theta)$, which is given by

$$(3.1) \qquad J_{n,2}^{(\alpha+k+1,\beta)}(\cos\theta) = c_n[R_n^{(\alpha+k+1,\beta)}(\cos\theta)]^2,$$

where

$$(3.2) \qquad c_n^{-1} = \int_0^{\pi} [R_n^{(\alpha+k+1,\beta)}(\cos\theta)]^2 \varrho^{(\alpha,\beta)}(\theta) \, d\theta.$$

The following useful estimate can be derived by means of (2. 3) and some asymptotic

*) Compare also P. L. Butzer—K. Scherer: *Abstract Spaces and Approximation*. Proceedings of the Oberwolfach Conference. July 13—27 (1968). ISNM **10** Birkhäuser, Basel 1969.

formulas for Jacobi polynomials (see SzEGŐ [11], (7. 34. 1) for similar estimates).
For $\alpha+k+1$, $\alpha+l$, β each greater than $-\frac{1}{2}$, we have for $n \to \infty$

$$(3.\,3) \quad \int_0^\pi [R_n^{(\alpha+k+1,\beta)}(\cos\theta)]^2 \varrho^{(\alpha+l,\beta)}(\theta)\, d\theta \approx \begin{cases} n^{-2\alpha-2l-2}, & 2l < 2k+1, \\ n^{-2\alpha-2k-3}\log n, & 2l = 2k+1, \\ n^{-2\alpha-2k-3}, & 2l > 2k+1. \end{cases}$$

From (3. 3) it follows that $c_n \approx n^{2\alpha+2}$, if $k > -\frac{1}{2}$. For the kernel (3. 1) the conditions
a and b of definition 1. 4 are trivially satisfied. We verify condition c', choosing
$n > h^{-2}$. By (2. 3)

$$O(n^{2\alpha+2}) \int_h^\pi [R_n^{(\alpha+k+1,\beta)}(\cos\theta)]^2 \varrho^{(\alpha,\beta)}(\theta)\, d\theta = O(n^{2\alpha+2}) \left[\int_{n^{-1/2}}^{\pi/2} + \int_{\pi/2}^\pi \right] =$$

$$= O(n^{-2k-1}) \left[\int_{n^{-1/2}}^{\pi/2} \theta^{-2k-2}\, d\theta + \int_0^{\pi/2} d\theta \right] = o(1) \qquad (n \to \infty,\ k > -\tfrac{1}{2}).$$

Hence for $k > -\frac{1}{2}$ the convolution of a function $f \in X$ with $J_{n,2}^{(\alpha+k+1,\beta)}$ $(k > -\frac{1}{2})$
approximates the function f in the X norm as $n \to \infty$. The trigonometric moments of
order σ for a kernel $K_\lambda (\cos\theta)$ are defined by

$$T(K_\lambda; \sigma) = \int_0^\pi \left(\sin\frac{\theta}{2}\right)^\sigma K_\lambda(\cos\theta)\varrho^{(\alpha,\beta)}(\theta)\, d\theta = \int_0^\pi K_\lambda(\cos\theta)\varrho^{(\alpha+\sigma/2,\beta)}(\theta)\, d\theta.$$

Thus, formula (3. 3) enables us to survey the asymptotic behavior of the trigonometric
moments of the kernel $J_{n,2}^{(\alpha+k+1,\beta)}$. If $k > \frac{1}{2}$, then it follows that

$$T(J_{n,2}^{(\alpha+k+1,\beta)}; 4) = o\big(T(J_{n,2}^{(\alpha+k+1,\beta)}; 2)\big).$$

We may conclude by BAVINCK [4], theorems 1. 5 and 2. 3, that the process $J_{n,2}^{(\alpha+k+1,\beta)}$,
$k > \frac{1}{2}$, is saturated with order n^{-2} and that the saturation class is Lip $(2, X)$. The
kernel $J_{n,2}^{(\alpha+k+1,\beta)}$, with k sufficiently large, shows the same behavior as the higher
order Jackson kernel. We have, in fact,

$$T(J_{n,2}^{(\alpha+k+1,\beta)}; \sigma) \approx n^{-\sigma} \qquad (0 \leqq \sigma < 2k+1).$$

For the Jackson kernel of order r

$$L_{n,r}(\theta) = \lambda_{n,r}^{-1} \left(\frac{\sin(n\theta/2)}{\sin(\theta/2)}\right)^{2r}$$

the same relation is valid (see BAVINCK [3], (4. 19))

$$T(L_{n,r}; \sigma) \approx n^{-\sigma} \qquad (2r > 2\alpha+\sigma+2),$$

but for larger values of α a high order of the Jackson kernel is necessary to compete
with the kernel $J_{n,2}^{(\alpha+k+1,\beta)}$. In the case of Fourier series $(\alpha = \beta = -\frac{1}{2})$ the kernel

$J_{n,2}^{(1/2,\,-1/2)}$ (the case $k=0$) coincides with the Fejér kernel, but the relatively simple kernel, $J_{n,2}^{(3/2,\,-1/2)}$, which has the same optimal properties as the Jackson kernel has never been considered, as far as the author knows.

The case $k=1$ will be studied in some more detail. In this case the constant c_n (see (3. 2)) and the trigonometric moments can be computed explicitly by means of Parseval's formula. Substituting $k=1$ in formula (2. 4), we have

$$\int_0^\pi [J_{n,1}^{(\alpha+2,\beta)}(\cos\theta)]^2 \varrho^{(\alpha,\beta)}(\theta)\,d\theta = \sum_{v=0}^n \left(1-\frac{\lambda_v}{\lambda_{n+1}}\right)^2 \omega_v^{(\alpha,\beta)}.$$

The sum at the right-hand side can be evaluated, if one uses (2. 4) for different values of k at the point $\theta=0$. After some calculation one obtains

$$c_n^{-1} = \frac{\Gamma(n+\beta+1)\Gamma(n+1)\Gamma(\alpha+3)\Gamma(\alpha+1)}{(\alpha+3)\Gamma(n+\alpha+\beta+3)\Gamma(n+\alpha+3)}(2n^2+2n(\alpha+\beta+3)+(\alpha+3)(\alpha+\beta+2)).$$

The second and the fourth trigonometric moments are easily computed.

$$T(J_{n,2}^{(\alpha+2,\beta)};2) = c_n\frac{(2n+\alpha+\beta+3)}{(\alpha+2)\omega_n^{(\alpha+2,\beta)}} = \frac{(\alpha+3)(\alpha+1)}{p_{2n}}$$

and

$$T(J_{n,2}^{(\alpha+2,\beta)};4) = c_n\frac{1}{\omega_n^{(\alpha+2,\beta)}} = \frac{(\alpha+3)(\alpha+2)(\alpha+1)}{(2n+\alpha+\beta+3)p_{2n}},$$

where

$$p_{2n} = 2n^2+2n(\alpha+\beta+3)+(\alpha+3)(\alpha+\beta+2).$$

The author has not succeeded in calculating the Fourier—Jacobi coefficients of the kernel $J_{n,2}^{(\alpha+2,\beta)}$, except the first few, which follow from the trigonometric moments. In the case $\alpha = \beta = -\frac{1}{2}$, $k=1$, we have

$$J_{n,2}^{(3/2,\,-1/2)} = \frac{15}{\pi(n+1)(2n+1)(2n+3)(4n^2+8n+5)}\times$$

$$\times\left[\frac{(2n+3)\sin(2n+1)(\theta/2)-(2n+1)\sin(2n+3)(\theta/2)}{4(\sin(\theta/2))^3}\right]^2$$

and the Fourier coefficients can be calculated by squaring the series (2. 5), since we have

$$\cos n\theta\cos m\theta = \tfrac{1}{2}(\cos(n+m)\theta+\cos(n-m)\theta).$$

A simple representation of the product $R_n^{(\alpha,\beta)}(\cos\theta)R_m^{(\alpha,\beta)}(\cos\theta)$ in the general case is not known. However, in some important special cases (α,α), $(\alpha,-\frac{1}{2})$ and $(\alpha+1,\alpha)$ the coefficients $a(k,m,n)$ in the representation

$$R_n^{(\alpha,\beta)}(x)R_m^{(\alpha,\beta)}(x) = \sum_{k=|n-m|}^{n+m}{}' a(k,m,n)R_k^{(\alpha,\beta)}(x)$$

are available (see ASKEY [1], p. 11).

The Fourier coefficients of the kernel $J_{n,2}^{(3/2,\,-1/2)}$ can also be computed by means of contour integration. In fact,

$$(J_{n,2}^{(3/2,\,-1/2)})^{\smallfrown}(m) = \frac{15}{\pi(n+1)(2n+1)(2n+3)(4n^2+8n+5)}$$

$$\int_0^\pi \left[\frac{(2n+3)\sin(2n+1)(\theta/2) - (2n+1)\sin(2n+3)(\theta/2)}{4(\sin\theta/2)^3}\right]^2 \cos m\theta\, d\theta =$$

$$= \frac{15}{4\pi i(n+1)(2n+1)(2n+3)(4n^2+8n+5)} \cdot$$

$$\oint_{|z|=1} \left(\sum_{k=-n}^{n}(n+k+1)(n-k+1)z^{n+k}\right)^2 \frac{(z^{2m}+1)}{z^{2n+m+1}}\, dz.$$

By using the residue theorem, one obtains after some elementary calculations:

$$(J_{n,2}^{(3/2,\,-1/2)})^{\smallfrown}(2q+1) =$$

$$= \frac{(n-q+1)(2n-2q+1)2(n-q)(4n^2-3nq+6q^2+14n+26q+12)}{(n+1)(2n+1)(2n+3)(4n^2+8n+5)}, \qquad 0 \leq q < n,$$

$$(J_{n,2}^{(3/2,\,-1/2)})^{\smallfrown}(2q) =$$

$$= \frac{(n-q+1)(2n-2q+1)(2n-2q+3)(4n^2+12nq+4q^2+8n+12q+5)}{(n+1)(2n+1)(2n+3)(4n^2+8n+5)}, \qquad 0 \leq q \leq n.$$

REFERENCES

[1] R. Askey, *Eight lectures on orthogonal polynomials.* Mathematisch Centrum TC **51**/70, 1970.

[2] R. Askey and S. Wainger, *A convolution structure for Jacobi series.* Amer. J. Math. **91** (1969), 463—485.

[3] H. Bavinck, *Approximation processes for Fourier—Jacobi expansions.* Math. Centrum, Amsterdam, report TW **126** (1971).

[4] H. Bavinck, *On positive convolution operators for Jacobi series.* Tôhoku Math. J. **24** (1972), 55—69.

[5] H. Berens, *Interpolationsmethoden zur Behandlung von Approximationsprozessen auf Banachräumen.* Lecture notes in Math. **64**, Springer, Berlin 1968.

[6] P. L. Butzer and R. J. Nessel, *Fourier analysis and approximation.* Vol. I. Birkhäuser, Basel—Stuttgart 1971.

[7] G. Gasper, *Positivity and the convolution structure for Jacobi series.* Ann. of Math. **93** (1971), 112—118.

[8] J. Löfström and J. Peetre, *Approximation theorems connected with generalized translations.* Math. Ann. **181** (1969), 255—268.

[9] J. Peetre, *A theory of interpolation of normed spaces.* Notes Universidade de Brasilia, 1963.

[10] E. M. Stein, *Interpolation in polynomial classes and Markoff's inequality.* Duke Math. J. **24** (1957), 467—476.

[11] G. Szegő, *Orthogonal polynomials.* Amer. Math. Soc. Coll. Publ. **23** (1967), third edition.

V.
Spline- and Algebraic Approximation

Cardinal Interpolation and Spline Functions IV.
The Exponential Euler Splines*)

By

I. J. SCHOENBERG

MATH. RESEARCH CENTER
UNIVERSITY OF WISCONSIN
MADISON

For George Pólya on his 85th birthday, December 13, 1972

Contents

Introduction

As background for our discussion we recall a result from [10]. First a few definitions. Let $\mathscr{S}_n$ denote the class of cardinal spline functions $S(x)$ of degree n ($n \geqq 1$) having their knots at the integer points of the real axis. This means that $S(x) \in \mathscr{S}_n$, provided that the restriction of $S(x)$ to every unit interval $(v, v+1)$ is a polynomial of degree n at most, and that

$$(1) \qquad S(x) \in C^{n-1}(-\infty, \infty).$$

---

*) Sponsored by the U. S. Army under Contract No. DA—31—124—ARO—D—462.

Furthermore, let

(2) $$\mathscr{S}_n^* = \{S(x); S(x+\tfrac{1}{2})\in\mathscr{S}_n\}.$$

The elements of $\mathscr{S}_n^*$ are again cardinal spline functions of degree n, but having their knots at $v+\tfrac{1}{2}$ half way between consecutive integers.

Let $y=(y_v)$ $(-\infty<v<\infty)$ be an infinite sequence of real or complex numbers. We call cardinal spline interpolation the problem of finding $S(x)$ in $\mathscr{S}_n$, or perhaps in $\mathscr{S}_n^*$, such that

(3) $$S(v)=y_v \quad \text{for all integers } v.$$

This problem has a unique solution subject to certain conditions to be described presently.

Let

(4) $$\tilde{\mathscr{Y}} = \{y=(y_v); y_v=O(|v|^s) \text{ as } v\to\pm\infty, \text{ for some appropriate } s\geqq 0\},$$

(5) $$\tilde{\mathscr{S}}_n = \{S(x); S(x)\in\mathscr{S}_n, S(x)=O(|x|^s) \text{ as } x\to\pm\infty, \text{ for some appropriate } s\geqq 0\}.$$

We define similarly the class $\tilde{\mathscr{S}}_n^*$ if we replace in (5) the condition $S(x)\in\mathscr{S}_n$ by $S(x)\in\mathscr{S}_n^*$. We may briefly describe the sequences y and functions $S(x)$ of these classes as being of power growth.

The main result of [10] is the following

THEOREM 1. 1. *Let n be odd. If*

(6) $$y = (y_v)\in\tilde{\mathscr{Y}},$$

then the interpolation problem (3) has a unique solution $S(x)$ such that

(7) $$S(x)\in\tilde{\mathscr{S}}_n.$$

2. *Let n be even. If (6) holds, then the problem (3) has a unique solution $S(x)$ such that*

(8) $$S(x)\in\tilde{\mathscr{S}}_n^*.$$

Only if $n=1$ is this result trivial, the graph of the solution $S(x)$ being the continuous polygonal line obtained by linear interpolation between consecutive data.

The purpose of the present paper is to study the possibility of spline interpolation of the geometric progression

(9) $$y_v = t^v \quad (-\infty<v<\infty; \ t \text{ fixed}, \ t\neq 0).$$

If t is such that $|t|=1$, then $y=(t^v)$ is a bounded sequence so that Theorem 1 applies.

If $|t| \neq 1$, then (t^ν) grows exponentially and a different special approach is needed to solve this problem. We shall see that already Euler has provided the essential tools for its solution.

I acknowledge with pleasure the helpful advice of my colleagues T. N. E. GRE-VILLE and A. SHARMA.

I. The cardinal spline interpolation of a geometric progression

1. *The exponential splines.* If we seek to find within the class $\mathscr{S}_n$ a not identically vanishing element having smallest support, then we find the so-called *B*-splines, up to a constant factor. The forward *B*-spline is the function

$$(1.1) \quad Q(x) = Q_{n+1}(x) = \frac{1}{n!} \left\{ x_+^n - \binom{n+1}{1}(x-1)_+^n + \cdots + (-1)^{n+1}(x-n-1)_+^n \right\},$$

where $x_+ = \max(0, x)$. This is an element of $\mathscr{S}_n$ having several properties, the most obvious being that

$$(1.2) \qquad\qquad Q(x) > 0 \quad \text{if} \quad 0 < x < n+1, \quad Q(x) = 0 \quad \text{elsewhere},$$

and its symmetry about the point $x = (n+1)/2$ expressed by the identity

$$(1.3) \qquad\qquad\qquad Q(x) = Q(n+1-x).$$

Its most important property we state as

LEMMA 1. *Every* $S(x) \in \mathscr{S}_n$ *allows a unique representation of the form*

$$(1.4) \qquad\qquad\qquad S(x) = \sum_{-\infty}^{\infty} c_j Q(x-j),$$

with constant coefficients c_j*. Conversely, for every sequence* (c_j) *(1.4) defines an element of* $\mathscr{S}_n$ *(see* [3] *for a more general Theorem 4 on page 80).*

We turn now to the problem of interpolating the sequence (9) by an element of $\mathscr{S}_n$. Equivalently, we are to interpolate at the integers the exponential function $f(x) = t^x$ by an element of $\mathscr{S}_n$. Because this exponential satisfies the functional equation

$$(1.5) \qquad\qquad f(x+1) = tf(x) \quad \text{for all real } x,$$

it seems natural to try to interpolate the sequence (9) by an element of $\mathscr{S}_n$ satisfying the same functional equation (1.5).

Such elements are described by

LEMMA 2. *The most general element $S(x)$ of $\mathscr{S}_n$ satisfying the functional equation*

$$(1.6) \qquad S(x+1) = tS(x)$$

is of the form

$$(1.7) \qquad S(x) = c_0 \sum_{-\infty}^{\infty} t^j Q(x-j) \qquad (c_0 \text{ constant}).$$

PROOF. Starting from the expansion (1.4) we find that

$$(1.8) \qquad S(x+1) = \sum_j c_j Q(x+1-j) = \sum c_{j+1} Q(x-j),$$

and

$$(1.9) \qquad tS(x) = \sum_j tc_j Q(x-j).$$

Since the right sides of (1.8) and (1.9) are to be identical by the requirement (1.6), the unicity in Lemma 1 shows that we must have $c_{j+1}=tc_j$ for all j, or $c_j=c_0 t^j$ for all j, and the representation (1.4) reduces to (1.7). Conversely, the same argument shows that the function (1.7) satisfies (1.6).

DEFINITION 1. *The elements of $\mathscr{S}_n$ of the form* (1.7) *are called* exponential splines. For convenience we write

$$(1.10) \qquad \Phi_n(x; t) = \sum_{-\infty}^{\infty} t^j Q_{n+1}(x-j).$$

Similar arguments show that

$$(1.11) \qquad S^*(x) = c_0 \Phi_n(x + \tfrac{1}{2}; t)$$

is the most general element of $\mathscr{S}_n^*$ satisfying (1.6).

2. *A preliminary solution of the interpolation problem.* A solution of the problem of interpolating the geometric progression (9) becomes now obvious: It all depends on whether

$$(2.1) \qquad \Phi_n(0; t) \neq 0,$$

or

$$(2.2) \qquad \Phi_n(0; t) = 0,$$

and we may state the following: *If* (2.1) *holds, then*

$$(2.3) \qquad S_n(x; t) = \Phi_n(x; t)/\Phi_n(0; t)$$

is a solution of the problem. If (2.2) *holds, then the attempt at interpolation fails.*

PROOF. If (2.1) holds, then $S_n(0; t)=1$ and the relation (1.6) shows that $S_n(v; t)=t^v$ for all integers v.

25 Linear Operators and Approximation

Let us express the critical quantity $\Phi_n(0; t)$ in terms of the B-spline. Using (1. 10) and (1. 3) we find that

$$\Phi_n(0; t) = \sum t^j Q(-j) = \sum t^j Q(n+1+j) = \sum t^{j-n} Q(j+1)$$

and in view of (1. 2), we obtain

$$(2. 4) \qquad \Phi_n(0; t) = t^{-n} \sum_0^{n-1} Q(j+1) t^j.$$

LEMMA 3. *The reciprocal algebraic equation of degree* $n-1$

$$(2. 5) \qquad \sum_0^{n-1} Q(j+1) t^j = 0$$

has only simple and negative roots λ_v, *which we label so as to satisfy the inequalities*

$$(2. 6) \qquad \lambda_{n-1} < \lambda_{n-2} < \cdots < \lambda_2 < \lambda_1 < 0.$$

This was established in [9, Lemma 8 on page 182] for odd values of n only. In Lemma 7 below we derive this for all n and point out that it was known to FROBENIUS [5] in 1910.

Similarly, an attempt of interpolation of (9) by an exponential spline (1. 11), of $\mathscr{S}_n^*$, depends on the condition

$$(2. 7) \qquad \Phi_n(\tfrac{1}{2}; t) \neq 0,$$

in which case

$$(2. 8) \qquad S_n(x; t) = \Phi_n(x+\tfrac{1}{2}; t)/\Phi_n(\tfrac{1}{2}; t)$$

is a solution in $\mathscr{S}_n^*$. From

$$\Phi_n(\tfrac{1}{2}; t) = \sum t^j Q(\tfrac{1}{2}-j) = \sum t^j Q(n+\tfrac{1}{2}+j) = \sum t^{j-n} Q(j+\tfrac{1}{2})$$

we obtain

$$(2. 9) \qquad \Phi_n(\tfrac{1}{2}; t) = t^{-n} \sum_0^{n} Q(j+\tfrac{1}{2}) t^j.$$

The relevant result here is

LEMMA 3*. *The reciprocal algebraic equation of degree* n

$$(2. 10) \qquad \sum_0^{n} Q(j+\tfrac{1}{2}) t^j = 0$$

has only simple and negative roots μ_v *satisfying*

$$(2. 11) \qquad \mu_n < \mu_{n-1} < \cdots < \mu_2 < \mu_1 < 0.$$

This was established in [9, Lemma 8, page 182] for even values of n only. For a general proof see Lemma 7* below (§ 5).

We may summarize our results as follows.

THEOREM 2. 1. *If*

$$(2.\ 12) \qquad t \neq \lambda_\nu \qquad (\nu = 1, \ldots, n-1),$$

then

$$(2.\ 13) \qquad S_n(x;\ t) = \Phi_n(x;\ t)/\Phi_n(0;\ t)$$

is an element of $\mathscr{S}_n$ interpolating the sequence (9).

2. If

$$(2.\ 14) \qquad t \neq \mu_\nu \qquad (\nu = 1, \ldots, n),$$

then

$$(2.\ 15) \qquad S_n^*(x;\ t) = \Phi_n(x + \tfrac{1}{2};\ t)/\Phi_n(\tfrac{1}{2};\ t)$$

is an element of $\mathscr{S}_n^*$ interpolating the sequence (9).

DEFINITION 2. *We call* (2. 13) *the* exponential Euler spline of degree n, *while* (2. 15) *is the* midpoint exponential Euler spline. *Note the conditions* (2. 12) *and* (2. 14), *respectively, under which these functions exist.*

Here are the reasons for these terms:

1. Assume that n is odd. Since $n-1$ is even and because the λ_ν are reciprocal in pairs, it follows from (2. 6) that $-1 \neq \lambda_\nu$ for $\nu = 1, \ldots, n-1$. By Theorem 2 we may therefore consider the function $S_n(x;\ -1)$ and find it to be identical with the odd degree *Euler spline* interpolating the sequence $y_\nu = (-1)^\nu$ (see [10, § 1B]).

2. Let n be even. From (2. 11), and the fact that the μ_ν are reciprocal in pairs, we conclude that $-1 \neq \mu_\nu$ for $\nu = 1, \ldots, n$. By Theorem 2 we may therefore consider the function $S_n^*(x;\ -1)$ which is identical with the even degree Euler spline interpolating the sequence $y_\nu = (-1)^\nu$ (see [10, § 1 B]).

Further reasons for the term "exponential Euler splines" will become obvious in Part II of the present paper.

We add that the exponential splines

$$(2.\ 16) \quad \Phi_n(x;\ \lambda_\nu) \qquad (\nu = 1, \ldots, n-1) \quad \text{and} \quad \Phi_n(x + \tfrac{1}{2};\ \mu_\nu) \qquad (\nu = 1, \ldots, n)$$

are the eigensplines of the classes $\mathscr{S}_n$ and $\mathscr{S}_n^*$, respectively, and that they play a fundamental role in the proof of Theorem 1 (see [10]).

Theorem 2 furnishes an explicit spline interpolant for the geometric progression (9) within $\mathscr{S}_n$, as well as $\mathscr{S}_n^*$. However, we are not yet satisfied with this solution, which we call "preliminary", and this for several reasons. Firstly, the evaluation of the exponential Euler splines (2. 13) and (2. 15), is laborious. Secondly, it seems difficult to decide from these expressions how the spline functions $S_n(x; t)$ and $S_n^*(x; t)$ behave as we let $n \to \infty$. Indeed, the feeling seems inescapable that the relations

$$\lim_{n \to \infty} S_n(x; t) = \lim_{n \to \infty} S_n^*(x; t) = t^x$$

should hold, at least if $t > 0$, when the conditions (2. 12) and (2. 14) are automatically fulfilled. For this reason we present below a computationally sounder approach to the exponential Euler splines. It will also show clearly the role of Euler's work in this matter.

II. The exponential Euler polynomials

3. *The exponential Euler polynomial $A_n(x; t)$.* We assume throughout that $t \neq 1$. Since n and t are kept fixed we shall sometimes omit them and write $\Phi_n(x; t) = \Phi_n(x) = \Phi(x)$.

LEMMA 4. *The polynomial $P(x)$ representing the spline function $\Phi_n(x; t)$ in the interval $[0, 1]$, is of exact degree n.*

PROOF. Let us assume that

(3. 1) $$P(x) \in \pi_{n-1}.$$

From the functional equation

(3. 2) $$\Phi_n(x+1; t) = t\Phi_n(x; t),$$

we obtain $\Phi_n(x+v) = t^v \Phi(x)$ and therefore also

$$\Phi(x) = t^v \Phi(x-v) = t^v P(x-v) \quad \text{if} \quad v < x < v+1.$$

From (3. 1) we conclude that all polynomial components of $\Phi(x)$ are of degree $n-1$. But then the basic requirement (1) implies that the entire function $\Phi_n(x; t)$ reduces to a single polynomial of degree $n-1$, which is absurd in view of (3. 2) and the assumption that $t \neq 1$.

DEFINITION 3. *If*

$$\Phi_n(x; t) = P(x) = C_0(t)x^n + \text{lower degree terms}, \quad \text{in} \quad 0 \leq x \leq 1,$$

we define

$$A_n(x; t) = \frac{1}{C_0(t)} P(x) = x^n + \text{lower degree terms},$$

and call it the exponential Euler polynomial *of degree n.*

By differentiation of (3. 2) and setting $x = 0$ we obtain

(3. 3) $$\Phi^{(v)}(1) = t\Phi^{(v)}(0) \qquad (v = 0, 1, \ldots, n-1).$$

From the Definition 3 we conclude that $A_n(x; t)$ has the following properties:

$$(3.4) \qquad A_n(x; t) = x^n + \text{lower degree terms,}$$

$$(3.5) \qquad A_n^{(v)}(1; t) = t A_n^{(v)}(0; t) \qquad (v = 0, 1, \ldots, n-1).$$

LEMMA 5. *The exponential Euler polynomial $A_n(x; t)$ is uniquely defined by the properties* (3.4) *and* (3.5).

PROOF. Indeed, suppose we had two distinct polynomials $\tilde{A}(x)$ and $\tilde{\tilde{A}}(x)$, both having the properties (3.4) and (3.5). By means of the definition

$$(3.6) \qquad \tilde{S}(x) = t^v \tilde{A}(x-v) \quad \text{if} \quad v \leq x < v+1, \quad \text{and all integer} \quad v,$$

we obtain a piecewise polynomial function $\tilde{S}(x)$ defined for all real x. The properties $\tilde{A}^{(v)}(1) = t\tilde{A}^{(v)}(0)$, $(v = 0, \ldots, n-1)$, and (3.6), will easily show that $\tilde{S}(x)$ is an element of $\mathscr{S}_n$ satisfying the relation $\tilde{S}(x+1) = t\tilde{S}(x)$. Similarly we obtain $\tilde{\tilde{S}}(x) \in \mathscr{S}_n$, such that $\tilde{\tilde{S}}(x) = \tilde{\tilde{A}}(x)$ in $[0, 1]$ and satisfies $\tilde{\tilde{S}}(x+1) = t\tilde{\tilde{S}}(x)$. It follows that $s(x) = \tilde{S}(x) - \tilde{\tilde{S}}(x)$ is also an element of $\mathscr{S}_n$ satisfying $s(x+1) = ts(x)$. Moreover, $s(x) \not\equiv 0$. By Lemma 2 we conclude that

$$s(x) = c_0 \Phi_n(x; t) \qquad (c_0 \neq 0).$$

However, now we reach a contradiction with Lemma 4, because

$$S(x) = \tilde{A}(x) - \tilde{\tilde{A}}(x) \in \pi_{n-1} \quad \text{in} \quad [0, 1].$$

Let us determine the generating function of the $A_n(x; t)$. In order to use more effectively the relations (3.5) for this purpose, we write

$$(3.7) \qquad A_n(x; t) = x_n + \binom{n}{1} a_1 x^{n-1} + \binom{n}{2} a_2 x^{n-2} + \cdots + a_n,$$

where the coefficients $a_v = a_v(t)$ depend on t and presumably also on n. Observe, that the relations (3.5) are equivalent to the relations

$$(3.8) \qquad 1 + \binom{v}{1} a_1 + \binom{v}{2} a_2 + \cdots + a_v = t a_v \qquad (v = 1, 2, \ldots, n).$$

Indeed, the relation (3.5) is equivalent to the relation (3.8) with v replaced by $n - v$.

The relations (3.8) show a surprising fact: Since $t \neq 1$, the coefficients $a_v = a_v(t)$ do not depend on n at all, but rather the coefficients $a_0 = 1, a_1, \ldots, a_n$ form the section of an infinite sequence $a_0 = 1, a_1, a_2, \ldots$ defined by the infinite system of relations

$$(3.9) \qquad 1 + \binom{v}{1} a_1 + \cdots + a_v = t a_v \qquad (v = 1, 2, 3, \ldots).$$

Let us determine this infinite sequence. The structure of the relations (3. 9) suggest the use of the generating function

$$(3.10) \qquad g(z) = \sum_{0}^{\infty} \frac{a_\nu}{\nu!} z^\nu.$$

The relations (3. 9) are equivalent to the identity

$$t - 1 = \sum_{0}^{\infty} \frac{t a_\nu}{\nu!} z^\nu - \sum_{0}^{\infty} \frac{1 + \binom{\nu}{1} a_1 + \cdots + a_\nu}{\nu!} z^\nu,$$

or

$$t - 1 = t g(z) - e^z g(z),$$

whence

$$(3.11) \qquad g(z) = \frac{t-1}{t-e^z} = \sum_{0}^{\infty} \frac{a_\nu(t)}{\nu!} z^\nu.$$

Combining (3. 7) and (3. 11) we obtain

LEMMA 6. *The exponential Euler polynomials*

$$(3.12) \qquad A_n(x; t) = x_n + \binom{n}{1} a_1(t) x^{n-1} + \cdots + a_n(t)$$

form an Appell sequence of polynomials generated by the expansion

$$(3.13) \qquad \frac{t-1}{t-e^z} e^{xz} = \sum_{0}^{\infty} \frac{A_n(x; t)}{n!} z^n.$$

L. CARLITZ [2] uses the notation $A_n(x; t) = H_n(x|t)$ and calls them Eulerian polynomials. If $t = -1$, then (3. 13) becomes

$$\frac{2}{e^z + 1} e^{xz} = \sum_{0}^{\infty} \frac{A_n(x; -1)}{n!} z^n$$

showing that $A_n(x; -1) = E_n(x)$ are the standard Euler polynomials (see NÖRLUND [7, 35]).

We introduce now a new set of polynomials $\Pi_n(z)$ by their generating function

$$(3.14) \qquad \frac{t-1}{t-e^z} = \sum_{0}^{\infty} \frac{\Pi_n(t)}{(t-1)^n} \frac{z^n}{n!} \qquad (\Pi_0(t) = 1).$$

In view of (3. 11), we see that

$$(3.15) \qquad a_n(t) = \frac{\Pi_n(t)}{(t-1)^n}.$$

We shall see that $\Pi_n(t)$ $(n \geqq 1)$ is a reciprocal monic polynomial of degree $n-1$ having integer coefficients. These are remarkable polynomials indeed. They were introduced by EULER [4, Chap. VII, § 178] and were studied by several mathematicians prior to FROBENIUS [5] who showed that $\Pi_n(t)$ has simple negative zeros. For this reason we call them *Euler—Frobenius polynomials*. They were rediscovered by a number of mathematicians, in particular in connection with spline interpolation (see [8], [9], [1]). Also by E. HILLE [6, 47]. We retained the notation $\Pi_n(t)$ used by QUADE and COLLATZ in their outstanding paper [8].

LEMMA 7. 1. *The Euler—Frobenius polynomial $\Pi_n(t)$ defined by the generating function* (3. 14) *is a monic reciprocal polynomial of degree $n-1$ having integer coefficients. It is related to the B-spline* (1. 1) *by the relation*

$$(3.16) \qquad \Pi_n(t) = n! \sum_0^{n-1} Q_{n+1}(v+1) t^v.$$

2. $\Pi_n(t)$ *can be independently defined by the expansion*

$$(3.17) \qquad \frac{\Pi_n(t)}{(1-t)^{n+1}} = \sum_0^{\infty} (v+1)^n t^v.$$

3. *It satisfies the recurrence relation*

$$(3.18) \qquad \Pi_{n+1}(t) = (1+nt)\Pi_n(t) + t(1-t)\Pi_n'(t) \qquad (\Pi_1(t)=1).$$

4. *The zeros λ_v of $\Pi_n(t)$ are simple and negative. We label them so that*

$$(3.19) \qquad \lambda_{n-1} < \lambda_{n-2} < \cdots < \lambda_2 < \lambda_1 < 0.$$

PROOF OF 2: Assuming $|t| < 1$ and $|z|$ sufficiently small, the following expansions and interchanges of summations are valid

$$\frac{t-1}{t-e^z} = e^{-z}(1-t)/(1-te^{-z}) = e^{-z}(1-t) \sum_0^{\infty} t^v e^{-vz} = (1-t) \sum_0^{\infty} t^v e^{-(v+1)z} =$$

$$= (1-t) \sum_0^{\infty} t^v \sum_{n=0}^{\infty} \frac{(-1)^n (v+1)^n}{n!} z^n,$$

$$\frac{t-1}{t-e^z} = (1-t) \sum_{n=0}^{\infty} (-1)^n \frac{z^n}{n!} \sum_{v=0}^{\infty} (v+1)^n t^v.$$

On comparing with (3. 14) we find that (3. 17) holds.

PROOF OF 1. In (3. 17) we may replace $(v+1)^n$ by $(v+1)_+^n$ and then sum over all integer values of v. Thus (3. 17) yields

$$\Pi_n(t) = \sum_{i=0}^{n+1} (-1)^i \binom{n+1}{i} t^i \sum_{v=-\infty}^{\infty} (v+1)_+^n t^v.$$

Setting $i+v = j$, whence $v = j-i$, and summing first by i we obtain

$$\Pi_n(t) = \sum_{j=0}^{\infty} t^j \sum_{i=0}^{n+1} (-1)^i \binom{n+1}{i} (j+1-i)_+^n .$$

In terms of the B-spline

$$Q_{n+1}(t) = \frac{1}{n!} \sum_{i=0}^{n+1} (-1)^i \binom{n+1}{i} (t-i)_+^n$$

we obtain

$$\Pi_n(t) = n! \sum_{j=0}^{\infty} t^j Q_{n+1}(j+1)$$

which establishes (3. 16) in view of (1. 2).

PROOF OF 3. Multiplying (3. 17) by t and differentiating we easily obtain (3. 18).

PROOF OF 4. We establish this by induction in n, as first done by FROBENIUS in [5]. Assuming that the zeros of $\Pi_n(t)$ satisfy (3. 19), we find from (3. 18) that

$$\Pi_{n+1}(\lambda_v) = \lambda_v(1-\lambda_v)\Pi_n'(\lambda_v) \qquad (v=1, \ldots, n-1).$$

Here $(-1)^{v-1}\Pi_n'(\lambda_v)>0$ for all v, hence

$$(-1)^v \Pi_{n+1}(\lambda_v) > 0 \qquad (v=1, \ldots, n-1),$$

and it follows $\Pi_{n+1}(t)$ has a zero in each of the n intervals

$$(-\infty, \lambda_{n-1}), (\lambda_{n-1}, \lambda_{n-2}), \ldots, (\lambda_1, 0).$$

Using (3. 18) we easily find that

$$\Pi_0(t) = \Pi_1(t) = 1, \quad \Pi_4(t) = t^3 + 11t^2 + 11t + 1,$$

$$\Pi_2(t) = t+1, \qquad \Pi_5(t) = t^4 + 26t^3 + 66t^2 + 26t + 1,$$

$$\Pi_3(t) = t^2 + 4t + 1, \quad \Pi_6(t) = t^5 + 57t^4 + 302t^3 + 302t^2 + 57t + 1.$$

4. *Solution of the interpolation problem in $\mathscr{S}_n$.* From its Definition 3 we know that the exponential Euler polynomial

(4. 1)
$$A_n(x; t) = x_n + \binom{n}{1} a_1(t)x^{n-1} + \cdots + a_n(t)$$

is, up to a non-vanishing factor, identical with the polynomial component of $\Phi_n(x, t)$ in $[0, 1]$. Moreover, by

$$A_n(0; t) = a_n(t) = (t-1)^{-n}\Pi_n(t)$$

and the identity (3. 16), we know that the λ_v in (2. 12) are identical with those in (3. 19). We may therefore state the

COROLLARY 1. *If* $t \neq 1$ *and*

$$(4.2) \qquad t \neq \lambda_\nu \qquad (\nu = 1, \ldots, n-1),$$

then the exponential Euler spline (2.13) *may also be defined by*

$$(4.3) \qquad S_n(x; t) = A_n(x; t)/A_n(0; t) \quad \text{in the interval } [0, 1].$$

The extension of $S_n(x; t)$ *to a real* x *may be accomplished by means of the functional equation*

$$(4.4) \qquad S_n(x+1; t) = t S_n(x; t).$$

In § 11 below we give numerical results for the particularly simple case when $t = 2$.

5. *The case of the class* $\mathscr{S}_n^*$. The case of $\mathscr{S}_n^*$ does not require an essentially new discussion, in view of the definition (2) of $\mathscr{S}_n^*$. A shift of origin to $x = \frac{1}{2}$ is all that is required. Writing

$$(5.1) \qquad B_n(x; t) = A_n(x + \tfrac{1}{2}; t) = x_n + \binom{n}{1} b_1(t) x^{n-1} + \cdots + b_n(t),$$

we obtain from (3.13) the generating function

$$(5.2) \qquad \frac{t-1}{t-e^z} e^{(x+\frac{1}{2})z} = \sum_0^\infty \frac{B_n(x; t)}{n!} z^n,$$

whence

$$(5.3) \qquad \frac{t-1}{t-e^z} e^{z/2} = \sum_0^\infty \frac{b_n(t)}{n!} z^n.$$

We now define a new sequence of polynomials $P_n(t)$ by setting

$$(5.4) \qquad b_n(t) = \frac{P_n(t)}{2^n (t-1)^n}.$$

As an analogue of Lemma 7 we have here

LEMMA 7*. 1. *The* $P_n(t)$ *defined by the generating function*

$$(5.5) \qquad \frac{t-1}{t-e^z} e^{z/2} = \sum_0^\infty \frac{P_n(t)}{2^n (t-1)^n} \frac{z^n}{n!}$$

are monic reciprocal polynomials of degree n having integer coefficients. In terms of the B-spline (1.1) *we may write*

$$(5.6) \qquad P_n(t) = 2^n n! \sum_{\nu=0}^n Q_{n+1}(\nu + \tfrac{1}{2}) t^\nu.$$

2. $P_n(t)$ *can be independently defined by the expansion*

$$(5.7) \qquad \frac{P_n(t)}{(1-t)^{n+1}} = \sum_{v=0}^{\infty} (2v+1)^n t^v.$$

3. $P_n(t)$ *satisfies the recurrence relation*

$$(5.8) \qquad P_{n+1}(t) = (1+(2n+1)t)P_n(t) + 2t(1-t)P_n'(t) \qquad (P_0(t)=1).$$

4. *The zeros* μ_v *of* $P_n(t)$ *are simple and negative and we may assume that*

$$(5.9) \qquad \mu_n < \mu_{n-1} < \cdots < \mu_2 < \mu_1 < 0.$$

As in the case of Lemma 7, these statements are established in the order 2., 1., 3., and 4. As the arguments are entirely analogous, we may omit details. (We begin the proof of 3. by replacing t by t^2 in (5.7)).

The polynomials $P_n(t)$ may be called *midpoint Euler—Frobenius polynomials*. We find by means of (5.8)

$$P_0(t) = 1, \qquad\qquad P_3(t) = t^3 + 23t^2 + 23t + 1,$$

$$P_1(t) = t+1, \qquad\qquad P_4(t) = t^4 + 76t^3 + 230t^2 + 76t + 1.$$

$$P_2(t) = t^2 + 6t + 1,$$

In view of the identity (5.6) it is clear that the μ_v in (2.14) are identical with those in (5.9). We may therefore state

COROLLARY 1*. *If* $t \neq 1$ *and*

$$(5.10) \qquad t \neq \mu_v \qquad (v=1, \ldots, n),$$

then the exponential Euler spline (2.15) *may be written as*

$$(5.11) \qquad S_n^*(x; t) = B_n(x; t)/B_n(0; t) \quad \text{in the interval} \quad -\tfrac{1}{2} \leq x \leq \tfrac{1}{2}.$$

The extension of $S_n^*(x; t)$ *to all real* x *is made by the functional equation*

$$(5.12) \qquad S_n^*(x+1; t) = tS_n^*(x; t).$$

6. *A characteristic extremum property of the polynomial* $A_n(x; t)/A_n(0; t)$ *if* $t > 1$. We consider the polynomial

$$(6.1) \qquad \varphi_n(x) = \varphi_n(x; t) = A_n(x; t)/A_n(0; t),$$

which is the polynomial representing the exponential Euler spline (4.3) in the interval [0, 1]. Here we assume that

$$(6.2) \qquad t > 1.$$

Similar results hold if $0 < t < 1$, but we omit their discussion.

From (3. 5) we see that $\varphi_n(x)$ has the following properties:

(6. 3)
$$\varphi_n(0) = 1,$$

(6. 4)
$$\varphi_n^{(v)}(1) = t\varphi_n^{(v)}(0) \qquad (v=0, 1, \ldots, n-1).$$

We define the class F_n of functions $f(x)$ satisfying the following three conditions:

(6. 5) $\quad f(x) \in C^{n-1}[0, 1]$ and $f^{(n-1)}(x)$ satisfies a Lipschitz condition,

(6. 6)
$$f(0) \geqq 1,$$

(6. 7)
$$f^{(v)}(1) \geqq tf^{(v)}(0) \qquad (v=0, 1, \ldots, n-1).$$

THEOREM 3. *The polynomial $\varphi_n(x)$ is the unique element of the class F_n that minimizes the norm*

(6. 8)
$$\|f^{(n)}\| = \operatorname*{essential\ sup}_{0 \leqq x \leqq 1} |f^{(n)}(x)| \quad \text{if} \quad f(x) \in F_n,$$

giving it its least value

(6. 9)
$$\|\varphi_n^{(n)}\| = \frac{n!}{a_n(t)} = n!(t-1)^n/\Pi_n(t).$$

PROOF. We define the function $K(x)$ as a polynomial of degree $n-1$ by setting

(6. 10)
$$K(x) = \varphi_{n-1}(x; t^{-1}) = A_{n-1}(x; t^{-1})/A_{n-1}(0; t^{-1}).$$

We know that the $A_n(x; t)$ form an Appell sequence. This means that $A_n'(x; t) = nA_{n-1}(x; t)$. Repeated differentiation of (6. 10) therefore gives

$$K^{(v)}(0) = n(n-1) \cdots (n-v+1) A_{n-1-v}(0; t^{-1})/a_{n-1}(t^{-1}) =$$

$$= v!\binom{n}{v} a_{n-1-v}(t^{-1})/a_{n-1}(t^{-1}) = v!\binom{n}{v}(t^{-1}-1)^v \Pi_{n-1-v}(t^{-1})/\Pi_{n-1}(t^{-1})$$

in view of (3. 15). Since $\Pi_k(x) > 0$ if $x > 0$, and $t^{-1}-1 < 0$, by (6. 2), we conclude that

(6. 11)
$$(-1)^v K^{(v)}(0) > 0 \qquad (v=0, 1, \ldots, n-1).$$

Moreover, $K(x)$ satisfies, by (6. 10), (6. 3), and (6. 4), the relations

(6. 12)
$$K(0) = 1,$$

(6. 13)
$$K^{(v)}(1) = t^{-1} K^{(v)}(0) \qquad (v=0, 1, \ldots, n-2).$$

Evidently, we are to establish the following: *If*

(6. 14)
$$f(x) \in F_n \quad \text{and} \quad \|f^{(n)}\| \leqq \|\varphi_n^{(n)}\|,$$

then

(6. 15)
$$f(x) = \varphi_n(x) \quad \text{in} \quad [0, 1].$$

We first derive a certain identity by integrations by parts:

$$\int_0^1 Kf^{(n)}\,dx = \int_0^1 K\,df^{(n-1)} = K(1)f^{(n-1)}(1) - K(0)f^{(n-1)}(0) - \int_0^1 K'f^{(n-1)}\,dx =$$

$$= t^{-1}K(0)\left(f^{(n-1)}(1) - tf^{(n-1)}(0)\right) - \int_0^1 K'\,df^{(n-2)} =$$

$$= t^{-1}K(0)\left(f^{(n-1)}(1) - tf^{(n-1)}(0)\right) - t^{-1}K'(0)\left(f^{(n-2)}(1) - tf^{(n-2)}(0)\right) + \int_0^1 K''\,df^{(n-3)}$$

and we continue in like manner until we reach the relation

$$\int_0^1 K(x)f^{(n)}(x)\,dx =$$

$$= \sum_{v=0}^{n-2} t^{-1}(-1)^v K^{(v)}(0)\left(f^{(n-v-1)}(1) - tf^{(n-v-1)}(0)\right) + (-1)^{n-1}\int_0^1 K^{(n-1)}(x)f'(x)\,dx.$$

But $K^{(n-1)}(x) = K^{(n-1)}$ is a constant and therefore

$$(6.16)\quad \int_0^1 K(x)f^{(n)}(x)\,dx = \sum_{v=0}^{n-2} t^{-1}(-1)^v K^{(v)}(0)\left(f^{(n-v-1)}(1) - tf^{(n-v-1)}(0)\right) +$$

$$+ (-1)^{n-1}K^{(n-1)}(f(1) - f(0)).$$

This is the identity that we need. We apply it first to $f(x) = \varphi_n(x)$. From (6. 3) and (6. 4) we obtain that

$$(6.17)\quad \int_0^1 K(x)\varphi_n^{(n)}(x)\,dx = (-1)^{n-1}K^{(n-1)}(\varphi_n(1) - \varphi_n(0)) = |K^{(n-1)}|(t-1),$$

by (6. 11). Observe next that by (6. 11) and (6. 7) all terms of the sum in (6. 16) are ≥ 0. Moreover, by (6. 6) and (6. 7)

$$(6.18)\qquad\qquad f(1) - f(0) \geq tf(0) - f(0) = (t-1)f(0) \geq t-1$$

and therefore

$$(6.19)\qquad\qquad f(1) - f(0) \geq t-1.$$

Using these facts as well as the assumptions (6. 14) we get the following string of inequalities and relations

$$(6.20)\quad |K^{(n-1)}|(t-1) \leq \sum_0^{n-2} t^{-1}(-1)^v K^{(v)}(0)\left(f^{(n-v-1)}(1) - tf^{(n-v-1)}(0)\right) +$$

$$+ (-1)^{n-1}K^{(n-1)}(f(1) - f(0)) =$$

$$= \int_0^1 K(x)f^{(n)}(x)\,dx \leq \int_0^1 K(x)\varphi_n^{(n)}(x)\,dx = |K^{(n-1)}|(t-1)$$

by (6. 19), (6. 16), (6. 14) and (6. 17). Since the extreme terms of (6. 20) are equal, we must have everywhere equalities, also in (6. 19). Therefore, by $f(1)-f(0) = t-1$ and (6. 18), we get

$$(6. 21) \qquad\qquad f(0)=1,$$

$$(6. 22) \qquad\qquad f^{(v)}(1) = tf^{(v)}(0) \qquad (v=0, ..., n-1).$$

Moreover, the relation

$$\int_0^1 K(x)\,(\varphi_n^{(n)}(x)-f^{(n)}(x))\,dx = 0,$$

in view of $K(x)>0$ in $[0, 1]$, shows that $f^{(n)}(x)=\varphi_n^{(n)}(x)$ almost everywhere. We conclude that $f(x)$ is a polynomial of degree n having the same highest degree term at $\varphi_n(x)$. By a reasoning already used in the proofs of Lemmas 4 and 5, we conclude that $f(x)=\varphi_n(x)$ in $[0, 1]$. Thus (6. 15) is established and therefore also Theorem 3.

III. The limit of the exponential Euler spline $S_n(x; t)$ as $n \to \infty$

7. *The Fourier expansion of* $S_n(x; t)t^{-x}$. Let

$$(7. 1) \qquad t=\tau e^{i\alpha}, \ \tau>0, \ -\pi<\alpha\leqq\pi, \quad \text{hence} \quad \tau=|t|,$$

and let t^x be the exponential function defined by

$$(7. 2) \qquad t^x=\tau^x e^{i\alpha x} \quad \text{for all real} \quad x.$$

LEMMA 8. *The function* $\Omega(x)$ *defined by the relation*

$$(7. 3) \qquad S_n(x; t)=t^x\,\Omega(x) \quad \text{for all real } x,$$

is a periodic function of period 1.

PROOF. This fact is evident, as both $S_n(x; t)$ and t^x satisfy the same functional equation (1. 5). Explicitly

$$\Omega(x+1) = S_n(x+1; t)t^{-x-1} = tS_n(x; t)t^{-1}t^{-x} = \Omega(x).$$

Let us find the Fourier series expansion of $\Omega(x)$.
For this purpose we assume that

$$(7. 4) \qquad\qquad 0\leqq x\leqq 1,$$

so that

$$(7. 5) \qquad\qquad S_n(x; t)=A_n(x; t)/A_n(0; t).$$

In writing this relation we assume that

(7. 6)
$$A_n(0; t) = a_n(t) = \Pi_n(t)(t-1)^{-n} \neq 0,$$

hence

(7. 7)
$$t \neq 1, \quad \text{and} \quad t \neq \lambda_v \quad (v=1, \ldots, n-1).$$

where the λ_v are the zeros of the Euler—Frobenius polynomial $\Pi_n(t)$.

Our main tool is Euler's generating function (3. 13). Dividing both sides by z^{n+1} we obtain

(7. 8)
$$F(z) = \frac{t-1}{t-e^z} \frac{e^{xz}}{z^{n+1}} = \sum_{v=0}^{\infty} \frac{A_v(x; t)}{v!} \frac{z^v}{z^{n+1}}.$$

The left side is a meromorphic function having a pole of order $n+1$ at the origin and simple poles at the zeros of e^z-t. By (7. 1), these poles are at the points

(7. 9)
$$t_k = \log \tau + i\alpha + 2k\pi i \quad (k=0, \pm1, \pm2, \ldots).$$

Notice that $t_k \neq 0$, for all k, for if $\log \tau = 0$ and $\alpha = 0$, then $t=1$, a case excluded by (7. 7).

Let us evaluate the residues of $F(z)$. From (7. 8) we find that

(7. 10)
$$\text{Residue of } F(z) \text{ at } z = 0 = \frac{A_n(x; t)}{n!},$$

(7. 11)
$$\text{Residue at } t_k = -\frac{t-1}{e^{t_k}} \frac{e^{xt_k}}{t_k^{n+1}} = -\frac{t-1}{t} \frac{e^{xt_k}}{t_k^{n+1}}.$$

LEMMA 9. *The sum of all the residues of $F(z)$ in the entire finite plane vanishes*

PROOF. Let $z = u+iv$, u and v real. Let L and l be large and positive, l being an integer, and let $R_{L,l}$ denote a rectangle bounded by the lines $u = \pm L$, $v = \alpha \pm \pm(2l+1)\pi$. By Cauchy's residue theorem we obtain

(7. 12)
$$A_n(x; t)/n! - \sum_{k=-l}^{l} (t-1)t^{-1} e^{xt_k} t_k^{-n-1} = \frac{1}{2\pi i} \int_{R_{L,l}} F(z)\, dz.$$

Our assumption (7. 4) implies that the integral on the right side $\to 0$ as L and $l \to +\infty$. We may omit the simple details. Moreover, the resulting series on the left side converges absolutely, if $n \geq 1$, and we obtain the expansion

(7. 13)
$$A_n(x; t)/n! = \sum_{-\infty}^{\infty} (t-1)t^{-1} e^{xt_k} t_k^{-n-1}.$$

Here

$$e^{xt_k} = e^{x(\log \tau + i\alpha + 2k\pi i)} = t^x e^{2k\pi ix}$$

and (7. 13) becomes

$$(7.14) \qquad A_n(x; t)/n! = (t-1)t^{-1}t^x \sum_{-\infty}^{\infty} e^{2k\pi i x}/t_k^{n+1} \qquad (0 \leq x \leq 1).$$

Using (7. 14), (7. 5), and (7. 3) we obtain

THEOREM 4. *The periodic function* $\Omega(x)$ *of Lemma 8 admits the Fourier series expansion*

$$(7.15) \qquad \Omega(x) = \sum_{-\infty}^{\infty} \frac{1}{t_k^{n+1}} e^{2\pi i k x} \Big/ \sum_{-\infty}^{\infty} \frac{1}{t_k^{n+1}} \quad \textit{for all real} \quad x.$$

This is obtained by setting $x=0$ in (7. 14) and using (7. 5) and (7. 3).

8. *The case when* $t>0$, $t \neq 1$. In this case $t=\tau$, $\alpha=0$, and (7. 9) becomes

$$(8.1) \qquad t_k = \log t + 2\pi i k,$$

and from (7. 15) we obtain

$$(8.2) \qquad \Omega_n(x) = \frac{1 + 2 \sum\limits_{k=1}^{\infty} \mathrm{Re}\left\{\left(\dfrac{\log t}{\log t + 2\pi i k}\right)^{n+1} e^{2\pi i k x}\right\}}{1 + 2 \sum\limits_{k=1}^{\infty} \mathrm{Re}\left(\dfrac{\log t}{\log t + 2\pi i k}\right)^{n+1}}.$$

To obtain more explicit results we assume that

$$(8.3) \qquad\qquad t > 1,$$

and write

$$(8.4) \qquad t_k = \log t + 2\pi i k = \varrho_k e^{i\gamma_k} \qquad (-\pi < \gamma_k < \pi).$$

Thus

$$(\log t + 2\pi i k)^{-n-1} e^{2\pi i k x} = \varrho_k^{-n-1} e^{(2\pi k x - (n+1)\gamma_k)i},$$

and (8. 2) may be written as

$$(8.5) \qquad \Omega_n(x) = \frac{1 + 2 \sum\limits_{k=1}^{\infty} (\varrho_0/\varrho_k)^{n+1} \cos\left(2\pi k x - (n+1)\gamma_k\right)}{1 + 2 \sum\limits_{k=1}^{\infty} (\varrho_0/\varrho_k)^{n+1} \cos\left(n+1\right)\gamma_k}.$$

Notice that the first ratio

$$(8.6) \qquad \varrho_0/\varrho_1 = \left(1 + \frac{4\pi^2}{(\log t)^2}\right)^{-\frac{1}{2}} < 1$$

is the largest of the ratios ϱ_0/ϱ_k.

Observe that the relative error in the approximation of t^x by $S_n(x;t)$ is by (7. 3) equal to

$$(8.\ 7) \qquad \frac{t^x - S_n(x;t)}{t^x} = 1 - \Omega_n(x).$$

From (8. 5) we immediately obtain the

COROLLARY 2. *The relative error* (8. 7) *is estimated by*

$$(8.\ 8) \qquad 1 - \Omega_n(x) = O(\varrho_0/\varrho_1)^n \quad as \quad n \to \infty,$$

uniformly for all real x.

EXAMPLE. For $t=2$ we find

$$(8.\ 9) \qquad \varrho_0/\varrho_1 = \left(1 + \frac{4\pi^2}{(\log 2)^2}\right)^{-\frac{1}{2}} = 0.1096.$$

This shows that $S_n(x;2)$ will approximate 2^x to about n significant figures. Again from (8. 5) we obtain the following more precise result.

COROLLARY 3.

$$\Omega_n(x) = \frac{1 + 2(\varrho_0/\varrho_1)^{n+1}\cos\left(2\pi x - (n+1)\gamma_1\right)}{1 + 2(\varrho_0/\varrho_1)^{n+1}\cos(n+1)\gamma_1} + O(\varrho_0/\varrho_2)^{n+1} \quad as \quad n \to \infty.$$

9. *The convergence theorem for the case when t is not negative.* We return to the case of a general t assuming, however, that t is not negative and (7. 1) becomes

$$(9.\ 1) \qquad t = |t|e^{i\alpha} \qquad (-\pi < \alpha < \pi).$$

If we mark the poles (7. 9) on the vertical line $\operatorname{Re} z = \log|t|$ we see that

$$(9.\ 2) \qquad |t_0| < |t_k| \quad \text{if} \quad k \neq 0.$$

This shows that in the two sums of (7. 15) the terms corresponding to $k=0$ dominate all others. In fact

$$(9.\ 3) \qquad \Omega_n(x) = \{1 + \sum_{k \neq 0} (t_0/t_k)^{n+1} e^{2\pi i k x}\}/\{1 + \sum_{k \neq 0} (t_0/t_k)^{n+1}\} = 1 + O(\gamma^n)$$

where

$$(9.\ 4) \qquad \gamma = \max\left(|t_0/t_1|, |t_0/t_{-1}|\right) < 1.$$

This proves

THEOREM 5. *If t is not negative and* (9. 1) *holds, then*

$$(9.\ 5) \qquad \lim_{n \to \infty} S_n(x;t) = |t|^x e^{i\alpha x}.$$

From (9. 3) and (9. 4) it is clear that the rate of convergence deteriorates as t approaches the negative side of the real axis.

10. *The case when t is real and negative.* Now it becomes especially important to assume that t is such that

(10. 1)
$$\Pi_n(t) \neq 0,$$

and Theorem 4 is valid. In (7. 9) we now have that $\alpha = \pi$, hence

(10. 2)
$$t_k = \log|t| + (2k+1)\pi i$$

and

(10. 3)
$$\bar{t}_k = t_{-k-1}.$$

We also write

(10. 4)
$$t_k = \varrho_k e^{i\gamma_k}, \quad \varrho_n > 0 \qquad (-\pi < \gamma_k < \pi).$$

Now (7. 13) becomes

$$\frac{|t|}{1+|t|} \frac{A_n(x; t)}{n!} = 2 \operatorname{Re} \frac{e^{xt_0}}{t_0^{n+1}} + 2 \operatorname{Re} \frac{e^{xt_1}}{t_1^{n+1}} + \cdots$$

and using (10. 4) we find that

(10. 5) $S_n(x; t) = |t|^x \dfrac{\cos\left(\pi x - (n+1)\gamma_0\right) + (\varrho_0/\varrho_1)^{n+1} \cos\left(3\pi x - (n+1)\gamma_1\right) + \cdots}{\cos(n+1)\gamma_0 + (\varrho_0/\varrho_1)^{n+1} \cos(n+1)\gamma_1 + \cdots}.$

Notice that if $t = -1$, hence by (10. 1) that n is odd, or $n = 2m-1$, then $\log|t| = 0$ and (10. 2) shows that $\gamma_0 = \gamma_1 = \gamma_2 = \cdots = \dfrac{\pi}{2}$. Then (10. 5) becomes

$$S_n(x; -1) = \frac{\cos(\pi x - m\pi) + (\varrho_0/\varrho_1)^{2m} \cos(3\pi x - m\pi) + \cdots}{\cos m\pi + (\varrho_0/\varrho_1)^{2m} \cos m\pi + \cdots}.$$

This is seen to converge to $\cos \pi x$ as $n = 2m-1 \to \infty$. Thus

(10. 6)
$$\lim S_n(x; -1) = \cos \pi x \quad \text{as} \quad n = 2m-1 \to \infty.$$

If t is negative, $t \neq -1$, then the question as to when we have convergence becomes much more delicate. Let us at least discuss a single example.

THEOREM 6. *If*

(10. 7)
$$t = -e,$$

then the

(10. 8)
$$\lim_{n \to \infty} S_n(x; -e) \ does \ not \ exist.$$

PROOF. Here $\log |t| = \log e = 1$, while (10. 2) and (10. 4) show that

(10. 9) $\gamma_0 = \arctan \pi$.

To begin with we observe that $t = -e$ is a transcendental number and therefore

(10. 10) $\Pi_n(-e) \neq 0$ for all n.

Moreover, we claim that

(10. 11) γ_0 *is not a rational multiple of* π.

Indeed, suppose that $\gamma_0 = r\pi$, where r is rational. It follows that $e^{\gamma_0 i} = e^{r\pi i}$ is a root of unity, hence that $e^{\gamma_0 i}$ is an algebraic number. By (10. 9) we conclude that

$$\pi = \tan \gamma_0 = \frac{1}{i} \frac{e^{\gamma_0 i} - e^{-\gamma_0 i}}{e^{\gamma_0 i} + e^{-\gamma_0 i}}$$

is an algebraic number, which it is not. This contradiction establishes (10. 11).

Now we return to (10. 5) and may argue as follows: If we choose the angle ψ such that

(10. 12) $\psi \not\equiv \frac{\pi}{2} \;(\mathrm{mod}\ \pi)$

hence

(10. 13) $\cos \psi \neq 0,$

then we can let n grow to infinity through an infinite sequence N such that

(10. 14) $\lim_{\substack{n \to \infty \\ n \in N}} (n + 1)\gamma_0 \equiv \psi \;(\mathrm{mod}\ 2\pi)$

or

(10. 15) $\lim_{\substack{n \to \infty \\ n \in N}} \cos (n+1)\gamma_0 = \cos \psi \neq 0.$

This is the case because the numbers $(n+1)\gamma_0$ are everywhere dense, mod 2π, in fact they are equidistributed, mod 2π. Now (10. 5) and (10. 15) show that

(10. 16) $\lim_{\substack{n \to \infty \\ n \in N}} S_n(x; -e) = e^x \frac{\cos (\pi x - \psi)}{\cos \psi} = e^x(\cos \pi x + \sin \pi x \tan \psi).$

Since ψ was arbitrary, subject to (10. 13), the theorem is established.

Notice, however, that the limits on the right side of (10. 16) satisfy the functional equation

$$f(x+1) = -ef(x).$$

11. *Approximating the exponential 2^x by exponential Euler splines.* In the present last section we wish to apply Corollary 1 of § 4 to the special case when $t=2$. The corresponding exponential 2^x is remarkable in several ways. It is the finite difference analogue of e^x because $\Delta 2^x = 2^x$. We also mention the beautiful theorem of Polya and Hardy: "2^x *is the transcendental entire function of least growth which assumes integer values at $x=0, 1, 2, \dots$*." (See [11, 55].)

From (3.15) we see that if $t=2$ then $a_n(2)=\Pi_n(2)$ are all integers. They are readily computed from the recurrence relations (3.9) which become

$$(11.1) \qquad a_n = 1 + \binom{n}{1}a_1 + \cdots + \binom{n}{n-1}a_{n-1} \qquad (n=1, 2, \dots, a_1=1).$$

We find the values

n	0	1	2	3	4	5	6	7	8
a_n	1	1	3	13	75	541	4683	47293	545835

From (4.1) we see that the corresponding exponential Euler polynomials $A_n(x; 2)$ are monic polynomials with integer coefficients. In the interval [0, 1] the relation (4.3) is valid. Using the above values for the a_n, we find in [0, 1]

$$S_3(x; 2) = (x^3 + 3x^2 + 9x + 13)/13,$$

$$S_5(x; 2) = (x^5 + 5x^4 + 30x^3 + 130x^2 + 375x + 541)/541,$$

$$S_7(x; 2) = (x^7 + 7x^6 + 63x^5 + 455x^4 + 2625x^3 + 11{,}361x^2 + 32{,}781x + 47{,}293)/47{,}293.$$

For the other intervals $(\nu, \nu+1)$ (ν integer $\neq 0$), the spline function $S_n(x; 2)$ is to be computed from the relation

$$(11.2) \qquad\qquad S_n(x; 2) = 2^\nu S_n(x-\nu; 2).$$

REMARKS 1. From (8.8) and (8.9) we see that $S_n(x; 2)$ approximates 2^x in [0, 1] to almost n accurate decimal places. Using (11.2) we get an approximation to almost n significant figures for all real x.

2. A much better approximation is obtained if we replace in [0, 1] $S_n(x; 2)$ by other appropriate polynomials of degree n, for instance the polynomial of best approximation to 2^x in [0, 1] afterwards extending the approximation to all real x by means of (11.2). However, the global approximating function so obtained is discontinuous for all integer values of x, while our exponential Euler spline satisfies $S_n(x; 2) \in C^{n-1}(-\infty, \infty)$.

3. Similar results are obtained for the case when $t = -1$. In this case we obtain the approximation of $\cos \pi x$ for all real x by the classical Euler splines.

26*

REFERENCES

[1] J. H. Ahlberg and E. N. Nilson, *Polynomial splines on the real line.* J. Approximation Theory **3** (1970), 398—409.

[2] L. Carlitz, *Eulerian numbers and polynomials.* Math. Mag. **32** (1959), 247—260.

[3] H. B. Curry and I. J. Schoenberg, *On Pólya frequency functions IV. The fundamental spline functions and their limits.* J. Analyse Math. **17** (1966), 71—107.

[4] L. Euler, *Institutionis Calculi Differentialis*, vol. II. Petersburg 1755.

[5] G. Frobenius, *Über die Bernoullischen Zahlen und die Eulerschen Polynome.* Sitzungsberichte der Preussischen Akad. der Wiss. Phys.-Math. Kl. (1910), 809—847. Also in volume 3 of Collected Works, 440—478.

[6] E. Hille, *Analytic function theory*, vol. II. Ginn and Company, Boston 1962.

[7] N. E. Nörlund, *Vorlesungen über Differenzenrechnung.* Springer, Berlin 1924.

[8] W. Quade and L. Collatz, *Zur Interpolationstheorie der reellen periodischen Funktionen.* Sitzungsber. der Preuss. Akad. der Wiss., Phys.-Math. Kl, (1938), 383—429.

[9] I. J. Schoenberg, *Cardinal interpolation and spline functions.* J. Approximation Theory **2** (1969), 167—206.

[10] I. J. Schoenberg, *Cardinal interpolation and spline functions II. Interpolation of data of power growth.* MRC-Tech. Sum. Rep. **1104**, Sept. 1970. To appear in J. Approximation Theory.

[11] J. M. Whittaker, *Interpolatory functions theory.* Cambridge Tract **33**, University Press. Cambridge 1935.

Применения сплайнов в теории приближений

Ю. Н. СУББОТИН

СВЕРДЛОВСКОЕ ОТДЕЛЕНИЕ
МАТЕМАТИЧЕСКОГО ИНСТИТУТА
ИМ. В. А. СТЕКЛОВА АН ССР

1. Об одной экстремальной задаче функциональной интерполяции

Пусть r-фиксированное натуральное число и в точках $x_k = kh$, $k = 0, \pm 1$, $\pm 2, \ldots$, заданы значения $y_k = f(kh)$ такие, что

$$(1) \qquad |\Delta^r y_k| = |\Delta^r_h f(kh)| = \left\| \sum_{m=0}^{r} (-1)^{r-m} C_r^m y_{m+k} \right\| \leqq M < \infty$$

$$\left(k = 0, \pm 1, \pm 2, \ldots, \quad C_r^m = \frac{r!}{m!(r-m)!} \right).$$

Для заданного выпуклого модуля непрерывности $\omega(\delta)$ $(0 \leqq \delta < \infty)$ ищется функция $f(x)$, проходящая через точки (kh, y_k) для модуля непрерывности r-ой производной которой справедливо неравенство

$$\omega(\delta, f^{(r)}) \leqq C_r \omega(\delta) \qquad (0 < \delta < \infty)$$

с возможно меньшей константой C_r для всего класса последовательностей $\{y_k\}$, удовлетворяющих неравенству (1).

Пусть норма

$$\|\Delta^r y_k\| = \sup_{-\infty < k < \infty} |\Delta^r y_k|,$$

$\mathcal{Y}$ — последовательность $\{y_k\}$,

Δ^r — класс последовательностей $\mathcal{Y}$, удовлетворяющих неравенству

$$\|\Delta^r y_k\| \leqq M,$$

$F(\mathcal{Y})$ — класс функций $f(x)$, определенных на всей числовой прямой, проходящих через точки (kh, y_k) $(k = \pm 1, \pm 2, \ldots)$.

В [1] доказана следующая теорема.

Теорема 1. *Если $\omega(\delta)$ $(0 \leqq \delta < \infty)$ выпуклый модуль непрерывности, то справедливо следующее равенство*

$$(2) \qquad B_n(\omega) = \sup_{Y \in \Delta^r} \inf_{f \in F(Y)} \sup_{\delta > 0} \frac{\omega(\delta, f^{(r)})}{\omega(\delta)} = A_r(\omega) M,$$

где

$$(3) \qquad A_r(\omega) = \dfrac{\pi^r}{(2h)^r \left| \displaystyle\int_0^\pi \omega\left(\dfrac{xh}{\pi}\right) \sum_{m=0}^\infty (-1)^{m(r+1)} \dfrac{\sin\dfrac{2m+1}{2}x}{(2m+1)^r}\, dx \right|}.$$

При доказательстве зтой теоремы применяется следующий результат автора о нулях алгебраического полинома [2].

Теорема А. *Если полином $P_{2n}(x)$ представим в одной из следующих двух форм*

$$\dfrac{P_{2n}(x)}{(1-x)^{2n+1}} = \int_0^1 f(t) \sum_{l=0}^\infty (l+t)^{2n} x^l\, dt$$

или

$$\dfrac{P_{2n}(x)}{(1-x)^{2n}} = \int_0^{\frac{1}{2}} f(t) \sum_{l=0}^\infty (l+t)^{2n-1} x^l\, dt + \int_{\frac{1}{2}}^1 f(t) \sum_{l=0}^\infty (l+t)^{2n-1} x^l\, dt,$$

где функция $f(t) \geqq 0$ интегрируема, отлична от нуля на множестве положительной меры и удовлетворяет условию $f(t)=f(1-t)$, $0 \leqq t \leqq \dfrac{1}{2}$, то полином $P_{2n}(x) \equiv$

$\equiv x^{2n} P_{2n}\left(\dfrac{1}{x}\right)$ и все его нули отрицательны и различны, $x = -1$ не является нулем

полинома $P_{2n}(x)$.

При выводе оценки $B_n(\omega)$ снизу применяется формула (24) из работы автора [2]. Пусть задана последовательность $\{y_m\}$, $m=0, \pm 1, \pm 2, \ldots$ и функция $f(x)$, $f(m)=y_m$ имеет $n-1$-ю абсолютно непрерывную производную. Тогда имеет место следующее равенство

$$(4) \qquad \sum_{m=0}^N (-1)^m \Delta^n y_m = \dfrac{1}{(n-1)!} \int_0^1 \sum_{p=0}^{n-1} (-1)^p f^{(n)}(t+p-1) \sum_{l=0}^p (-1)^l a_l(t)\, dt +$$

$$+ \dfrac{1}{(n-1)!} \int_0^1 \sum_{p=n}^N (-1)^p f^{(n)}(t+p-1) \sum_{l=0}^n (-1)^l a_l(t)\, dt +$$

$$+ \dfrac{1}{(n-1)!} \int_0^1 \sum_{p=N+1}^{N+n} (-1)^p f^{(n)}(t+p-1) \sum_{l=p-N}^n (-1)^l a_l(t)\, dt,$$

где

$$a_l(t) = \sum_{k=0}^{n-l} (-1)^{n-k-l} C_n^{k+l} (k+1-t)^{n-1}$$

и

$$(5) \qquad Q_n^*(t) = \sum_{l=0}^{n+1} (-1)^l a_l(t), \quad Q_n(x) = Q_n^*\left(\frac{1+x}{2}\right).$$

Доказательство теоремы 1 будет состоять в нахождении совпадающих оценок сверху и снизу для величины

$$B_n(\omega) = \sup_{Y \in \Delta^n} \inf_{f \in F(Y)} \sup_{\delta>0} \frac{\omega(\delta, f^{(n)})}{\omega(\delta)}.$$

Выведем оценку для $B_n(\omega)$. Рассмотрим вспомогательную функцию

$$(6) \qquad \varphi(t) = \begin{cases} \frac{1}{2}\,\omega(2th), & 0 \leq t \leq \frac{1}{2} \\ \frac{1}{2}\,\omega(2h-2th), & \frac{1}{2} \leq t \leq 1. \end{cases}$$

Рассмотрим сначала случай $n = 2s+1$. Положим

$$(7) \qquad f^{(2s+1)}(mh+th) = Z_m^{(2s+1)} \varphi(t), \qquad (0 \leq t \leq 1),$$

$m = 0,\ \pm 1,\ \pm 2, \ldots$ Имеем очевидное неравенство

$$(8) \qquad \omega(\delta, f^{(2s+1)}) \leq \|Z_m^{(2s+1)}\|\,\omega(\delta).$$

Будем искать функцию $f(x)$ в виде формулы Тейлора с остаточным членом в интегральной форм Коши, тогда получим

$$\int_0^1 \sum_{l=0}^{2s} Z_{m+l}^{(2s+1)} \varphi(t) \sum_{k=l+1}^{2s+1} (-1)^{2s+1-k} C_{2s+1}^k (k-l-t)^{2s}\, dt = \frac{\Delta^{2s+1} y_m}{h^{2s+1}} (2s)!$$

или

$$(9) \qquad \int_0^1 \sum_{l=0}^{2s} Z_{m+l}^{(2s+1)} \varphi(t) \sum_{k=0}^{l} (-1)^k C_{2s+1}^k (k-l-t)^{2s}\, dt = \frac{\Delta^{2s+1} y_m}{h^{2s+1}} (2s)!$$

$$(m = 0,\ \pm 1,\ \pm 2, \ldots).$$

Характеристический полином $P_{2s}(x)$ разностного уравнения (9) удовлетворяет условиям теоремы А.

В силу теоремы 1 из [2] для ограниченного решения разностного уравнения (9) справедлива оценка

$$\|Z_m^{(2s+1)}\| = \frac{\|\Delta^{2s+1} y_m\|}{h^{2s+1}} (2s)! \frac{1}{|P_{2s}(-1)|} =$$

$$= \frac{(2s)!\,\|\Delta^{2s+1} y_m\|}{h^{2s+1}} \cdot \frac{1}{\left| \int_0^1 \varphi(t) \sum_{l=0}^{2s} (-1)^l \sum_{k=0}^{l} (-1)^k C_{2s+1}^k (l-k+t)^{2s}\, dt \right|}.$$

Преобразуем правую часть последнего неравенства. Имеем

$$\sum_{l=0}^{2s}(-1)^l\sum_{k=0}^{l}(-1)^k C_{2s+1}^k(l-k+t)^{2s}=$$

$$=-\sum_{l=0}^{2s+1}(-1)^l\sum_{k=0}^{2s+1-l}(-1)^{2s+1-k-l}C_{2s+1}^{k+l}(k+1-t)^{2s}=-Q_{2s}^*(t).$$

Положив $t=(1+x)/2$ и воспользовавшись равенством (6) и леммой 4 из [2], окончательно получим

$$(10)\qquad \|Z_m^{(2s+1)}\|\leqq\frac{2M(2s)!}{h^{2s+1}\int\limits_0^1\omega[(1-x)h]|Q_{2s}(x)|\,dx}.$$

Потребовав, чтобы $f(kh)=y_k$ $(k=0,1,\dots,2s)$, получим функцию $f^*(x)$, $f^*(kh)=y_k$ $(k=0,\pm1,\pm2,\dots)$, у которой модуль непрерывности $2s+1$-ой производной удовлетворяет неравенству (8) с константой, не превосходящей правой части неравенства (10).

Рассмотрим случай $n=2s$. Положим

$$(11)\qquad f^{(2s)}\big(mh+th-(h/2)\big)=Z_m^{(2s)}\varphi(t)\qquad (0\leqq t\leqq1,\ m=0,\pm1,\pm2,\dots).$$

Для модуля непрерывности функции $f^{(2s)}(x)$ имеем неравенство

$$(12)\qquad \omega(\delta,f^{(2s)})\leqq\|Z_m^{(2s)}\|\,\omega(\delta).$$

Поступая как в предыдущем случае, получим

$$(13)\qquad \frac{(2s-1)!\,\Delta^{2s}y_m}{h^{2s}}=\sum_{l=0}^{2s}a_{s,l}Z_{m+l}^{(2s)}\qquad (m=0,\pm1,\pm2,\dots),$$

где

$$a_{s,l}=\int\limits_{1/2}^{1}\varphi(t)\sum_{k=l}^{2s}(-1)^k C_{2s}^k(k-l+1-t)^{2s-1}\,dt+$$

$$+\int\limits_{0}^{1/2}\varphi(t)\sum_{k=l+1}^{2s}(-1)^k C_{2s}^k(k-l-t)^{2s-1}\,dt=$$

$$=\int\limits_{1/2}^{1}\varphi(t)\sum_{k=0}^{l-1}(-1)^k C_{2s}^k(l+t-k-1)^{2s-1}\,dt+$$

$$+\int\limits_{0}^{1/2}\varphi(t)\sum_{k=0}^{l}(-1)^k C_{2s}^k(l-k+t)^{2s-1}\,dt,$$

при $l=0$ считаем равным нулю первое слагаемое, при $l=2s$ — второе.

Пусть $P_{2s}^*(x)$ — характеристический полином разностного уравнения (13). Тогда

$$(1-x)^{2s}\left\{\int_0^{1/2}\varphi(t)\sum_{l=0}^\infty(l+t)^{2s-1}x^l\,dt+x\int_{1/2}^1\varphi(t)\sum_{l=0}^\infty(l+t)^{2s-1}x^l\,dt\right\}=$$

$$=\int_0^{1/2}\varphi(t)\sum_{l=0}^{2s-1}x^l\sum_{k=0}^l(-1)^kC_{2s}^k(l-k+t)^{2s-1}\,dt+$$

$$+\int_{1/2}^1\varphi(t)\sum_{l=0}^{2s-1}x^{l+1}\sum_{k=0}^l(-1)^kC_{2s}^k(l-k+t)^{2s-1}\,dt=$$

$$=\sum_{l=0}^{2s}x^l\left[\int_0^{1/2}\varphi(t)\sum_{k=0}^l(-1)^kC_{2s}^k(l-k+t)^{2s-1}\,dt+\right.$$

$$\left.+\int_{1/2}^1\varphi(t)\sum_{k=0}^{l-1}(-1)^kC_{2s}^k(l-k-1+t)^{2s-1}\,dt\right]=\sum_{l=0}^{2s}a_{s,l}x^l=P_{2s}^*(x).$$

Из теоремы 1 из [2] и теоремы А выводим, что разностное уравнение (13) имеет единственное ограниченное решение и для нормы зтого решения справедлива оценка

$$\|Z_m^{(2s)}\|\leqq\frac{\|\Delta^{2s}y_m\|(2s-1)!}{h^{2s}|P_{2s}^*(-1)}.$$

Вычислив значение характеристического полинома $P_{2s}^*(x)$ в точке $x=-1$, окончательно имеем

$$(14)\qquad\|Z_m^{(2s)}\|\leqq\frac{(2s-1)!2M}{h^{2s}\int_0^1\omega(xh)|Q_{2s-1}(x)|\,dx}.$$

Потребовав, чтобы $f(kh)=y_k$ ($k=0,1,\ldots.2s-1$), получим функцию $f^*(x)$, $f^*(kh)=y_k$ ($k=0,\pm1,\pm2,\ldots$), модуль непрерывности $2s$-ой производной которой удовлетворяет неравенству (12), с константой, не превосходящей правой части неравенства (14).

Принимая во внимание неравенства (8), (10), (12), и (14) выводим, что какова бы ни была последовательность $У=\{y_m\}$, $У\in\Delta^n$, существует функция $f(x)\in F(У)$, для модуля неперерывности n-ой производной которой справедливо неравенство

$$\omega(\delta,f^{(n)})\leqq A_n(\omega)M\omega(\delta),$$

то есть

$$(15)\qquad B_n(\omega)\leqq MA_n(\omega),$$

410   Ю. Н. Субботин

где

$$
(16) \qquad A_n(\omega) =
\begin{cases}
\dfrac{2(2s-1)!}{h^{2s}\displaystyle\int_0^1 \omega(xh)\,|Q_{2s-1}(x)|\,dx}, & n = 2s, \\[4ex]
\dfrac{2(2s)!}{h^{2s+1}\displaystyle\int_0^1 \omega[(1-x)h]\,|Q_{2s}(x)|\,dx}, & n = 2s+1.
\end{cases}
$$

Разложим полином $Q_n(x)$ в ряд Фурье [2] и воспользуемся леммой 4 из работы [2], тогда получим

$$
(17) \qquad A_n(\omega) = \frac{\pi^n}{2^n h^n \left| \displaystyle\int_0^1 \omega(yh) \sum_{m=0}^{\infty} (-1)^m \frac{\cos\left(\dfrac{2m+1}{2}\pi x - \dfrac{n-1}{2}\pi \right)}{(2m+1)^n}\, dx \right|}
$$

где $y=x$ при $n=2s$ и $y=1-x$ при $n=2s+1$. Заменив в первом случае x на t/π, во втором $1-x$ на t/π выводим

$$
A_n(\omega) = \frac{\pi^n}{2^n h^n \left| \displaystyle\int_0^\pi \omega\left(\frac{xh}{\pi}\right) \sum_{m=0}^{\infty} (-1)^{m(n+1)} \frac{\sin\dfrac{2m+1}{2}x}{(2m+1)^n}\, dx \right|}.
$$

Покажем, что найденные оценки сверху неулучшаемы. Пусть последовательность $\tilde{Y}=\{y_m\}$ такова, что $\tilde{y}_m = (-1)^m M/2^n$, тогда $\|\Delta^n \tilde{y}_m\| = M$. При этом справедливо неравенство

$$
(18) \qquad B_n(\omega) \geqq \inf_{f \in F(\tilde{Y})} \sup_{\delta > 0} \frac{\omega(\delta, f^{(n)})}{\omega(\delta)}.
$$

Мы можем брать нижнюю грань по функциям $f(x) \in F(\tilde{Y})$, у которых n-я производная ограничена одним и тем же числом $M_1 \leqq M A_n(\omega)\omega[(n+2)h]$. Предположим противное. Пусть существует точка x_1, что $|f^{(n)}(x_1)| > M A_n(\omega)\omega[(n+2)h]$. Так как $f(x) \in F(\tilde{Y})$, то между точками mh и $(m+1)h$, $m=0,\ \pm 1,\ \pm 2,\ \dots$, функция $f(x)$ обращается в нуль. Отсюда по теореме Ролля на каждом отрезке длины $(n+2)h$ функция $f^{(n)}(x)$ обращается в нуль. Следовательно, для точки x_1 найдется точка x_0, что $f^{(n)}(x_0)=0$ и $|x_1-x_0| \leqq (n+2)h$. Отсюда

$$
\omega[(n+2)h, f^{(n)}] \geqq |f^{(n)}(x_1)| > M A_n(\omega)\omega[(n+2)h],
$$

что противоречит оценке сверху. Итак, мы можем считать, что существует такое число $M_1 > 0$, что нижнюю грань в (18) можно брать по функциям, у которых $\|f^{(n)}(x)\| \leqq M_1$.

Справедливо тождество

$$\sum_{m=0}^{N} (-1)^m \Delta^n y_m = \frac{h^n}{(n-1)!} \int_0^1 \sum_{p=0}^{n-1} (-1)^p f^{(n)}(th+ph) \sum_{l=0}^{p} (-1)^l a_l(t)\, dt +$$

$$+ \frac{h^n}{(n-1)!} \int_0^1 \sum_{p=n}^{N} (-1)^p f^{(n)}(th+ph) Q_{n-1}^*(t)\, dt +$$

$$+ \frac{h^n}{(n-1)!} \int_0^1 \sum_{p=N+1}^{N+n} (-1)^p f^{(n)}(th+ph) \sum_{l=p-N}^{n} (-1)^l a_l(t)\, dt,$$

где $Q_{n-1}^*(t)$ определяется равенством (4).

Рассмотрим случай $n=2s$. Принимая во внимание связь между $Q_n^*(t)$ и $Q_n(x)$ и лемму 4 из [2] имеем равенство

$$Q_{2s-1}^*(\tfrac{1}{2}-t) = -Q_{2s-1}^*(\tfrac{1}{2}+t).$$

Для последовательности $\{\tilde{y}_m\}$ справедливы оценки

$$M(N+1) \leqq C(n) M_1 h^n + \frac{h^n}{(n-1)!} \left| \int_0^{1/2} \sum_{p=n}^{N} (-1)^p f^{(n)}(th+ph) Q_{n-1}^*(t)\, dt + \right.$$

$$\left. + \int_{1/2}^{1} \sum_{p=n}^{N} (-1)^p f^{(n)}(th+ph) Q_{n-1}^*(t)\, dt \right| = C(n) M_1 h^n +$$

$$+ \frac{h^n}{(n-1)!} \left| \int_0^{1/2} \sum_{p=n}^{N} (-1)^p f^{(n)}\left(ph + \frac{h}{2} - th \right) Q_{n-1}^*\left(\frac{1}{2} - t \right) dt + \right.$$

$$\left. + \int_0^{1/2} \sum_{p=n}^{N} (-1)^p f^{(n)}\left(ph + \frac{h}{2} + th \right) Q_{n-1}^*\left(\frac{1}{2} + t \right) dt \right| = C(n) M_1 h^n +$$

$$+ \frac{h^n}{(n-1)!} \left| \int_0^{1/2} \sum_{p=n}^{N} (-1)^p \left[f^{(n)}\left(ph + \frac{h}{2} + th \right) - f^{(n)}\left(ph + \frac{h}{2} - th \right) \right] Q_{n-1}^*\left(\frac{1}{2} + t \right) dt \right|.$$

$$n = 2s.$$

Пусть

$$\left| f^{(2s)}\left(ph + \frac{h}{2} + th \right) - f^{(2s)}\left(ph + \frac{h}{2} - th \right) \right| \leqq C_p^{(2s)} \omega(2th),$$

тогда

$$M(N+1) \leqq C(2s)M_1 h^{2s} + \frac{h^{2s}\|C_p^{(2s)}\|(N+1-2s)}{(2s-1)!} \int\limits_0^{1/2} \omega(2th)\left|Q_{2s-1}^*\left(\frac{1}{2}+t\right)\right| dt.$$

Отсюда

$$\|C_p^{(2s)}\| \geqq \frac{M(N+1)-C(2s)M_1 h^{2s}}{(N+1-2s)\dfrac{h^{2s}}{(2s-1)!}\int\limits_0^{1/2}\omega(2th)\left|Q_{2s-1}^*\left(\frac{1}{2}+t\right)\right| dt}.$$

Перейдя к пределу при $N \to \infty$ и положив $t=x/2$ находим

$$(19) \qquad \|C_p^{(2s)}\| \geqq \frac{2M(2s-1)!}{h^{2s}\int\limits_0^1 \omega(xh)|Q_{2s-1}(x)|\,dx}.$$

Рассмотрим случай нечетного $n=2s+1$. Имеем

$$M(N+1) \leqq C_1(n)M_1 h^n + \frac{h^n}{(n-1)!}\left| -\int\limits_0^{1/2}\sum_{p=n}^{N-1}(-1)^p f^{(n)}(ph+h+th)Q_{n-1}^*(t)\,dt + \right.$$

$$\left. + \int\limits_{1/2}^{1}\sum_{p=n}^{N-1}(-1)^p f^{(n)}(ph+th)Q_{n-1}^*(t)\,dt \right| =$$

$$= C_1(n)M_1 h^n + \frac{h^n}{(n-1)!}\left|\int\limits_0^{1/2}\sum_{p=n+1}^{N}(-1)^p[f^{(n)}(ph+ht)-f^{(n)}(ph-th)]Q_{n-1}^*(1-t)\,dt\right|,$$

$n=2s+1$. Мы заменили во втором интеграле t на $1-t$ и воспользовались равенством

$$Q_{2s}^*(1-t) = Q_{2s}^*(t), \qquad 0 \leqq t \leqq \tfrac{1}{2}.$$

Пусть

$$|f^{(2s+1)}(ph+th)-f^{(2s+1)}(ph-th)| \leqq C_p^{(2s+1)}\omega(2th),$$

тогда, полагая $t=(1-x)/2$, имеем

$$M(N+1) \leqq C_1(2s+1)M_1 h^{2s+1} - \frac{h^{2s+1}}{(2s)!}\|C_p^{(2s+1)}\|(N-2s-1)\times$$

$$\times \int\limits_0^{1/2}\omega(2th)|Q_{2s}^*(1-t)|\,dt = C_1(2s+1)M_1 h^{2s+1} +$$

$$+ \frac{h^{2s+1}}{2(2s)!}(N-2s-1)\|C_p^{(2s+1)}\|\int\limits_0^1\omega[(1-x)h]|Q_{2s}(x)|\,dx.$$

Таким образом для любой функции $f(x) \in F(\tilde{Y})$ выполняется неравенство

$$(20) \qquad \|C_p^{(2s+1)}\| \geqq \frac{2(2s)!\,M}{h^{2s+1} \int_0^1 \omega[(1-x)h]\,|Q_{2s}(x)|\,dx}.$$

Из неравенств (19) и (20) выводим

$$(21) \qquad B_n(\omega) \geqq \inf_{f \in F(\tilde{Y})} \sup_{0 < \delta < h} \frac{\omega(\delta, f^{(r)})}{\omega(\delta)} \geqq \frac{2M(n-1)!}{h^n \int_0^1 \omega(yh)\,|Q_{n-1}(x)|\,dx}$$

где $y = x\,(n = 2s)$ и $y = 1 - x$ $(n = 2s + 1)$. Из неравенств (15) и (21) и равенства (17) выводим утверждение теоремы 1.

2. О поперечнике в $L(0, 2\pi)$ класса V^r

Пусть $L = L(0, 2\pi)$ — пространство измеримых 2π-периодических, действительных суммируемых функций $f(x)$ с нормой

$$\|f\|_L = \|f\|_{L(0, 2\pi)} = \int_0^{2\pi} |f(x)|\,dx,$$

$H_n - n$-мерное подпространство из L, V^r — класс функций $f(x) \in L(0, 2\pi)$ и имеющих $(r-1)$ абсолютно-непрерывную производную и r-ю производную, ограниченной вариации

$$(22) \qquad \bigvee_0^{2\pi} f^{(r)}(t) = \int_0^{2\pi} |df^{(r-1)}(t)| \leqq 1.$$

Здесь будет найден $2n - 1$-мерный поперечник в пространстве L для класса функций, удовлетворяющих неравенству (22). Из результатов С. М. Никольского [4] следует, что

$$d_{2n-1}(W^r, L) \leqq K_r n^{-r}.$$

Здесь

$$d_n(V^r, L) = \inf_{H_n \subset L} \sup_{f(t) \in V^r} \inf_{\varphi(t) \in H_n} \|f(t) - \varphi(t)\|,$$

где $H_n - n$-мерное подпространство порстранства $L(0, 2\pi)$.

Пусть $S_r(x, \pi/n)$ — семейство 2π-периодических сплайн-функций степени r с узлами $\{\pi_i/n\}$, $i = 0, 1, \ldots, n$. В. М. Тихомиров [5] показал, что

$$(23) \qquad \|S_r^{(r)}(x, \pi/n)\|_{C(0, 2\pi)} \leqq K_r^{-1} n^r \|S_r(x, \pi/n)\|_{C(0, 2\pi)},$$

где K_r — константа Фавара:

$$K_r = \frac{4}{\pi} \sum_{m=0}^{\infty} \frac{(-1)^{m(r+1)}}{(2m+1)^{r+1}} \qquad (r = 1, 2, \ldots).$$

Для нахождения оценки $d_{2n-1}(W^r, L)$ снизу мы методом Штейна [6] доказываем аналог неравенства (23) в метрике $L(0,2\pi)$ для 2π-периодических „сплайн"-функций $S_r(x, \pi/n)$, образующих $2n$-мерное подпространство. Имеем [7],

$$(24) \qquad \bigvee_0^{2\pi} S_r^{(r)}(x, \pi/n) \leqq K_r^{-1} n^r \|S_r(x, \pi/n)\|_L,$$

Отсюда с помощью метода В. М. Тихомирова [5] выводим, что

$$d_{2n-1}(V^r, L) \geqq K_r n^{-r},$$

т.е.

$$d_{2n-1}(V^r, L) = K_r n^{-r}.$$

Это равенство вытекает также из общих торем двойственности, а также доказано в [8].

Отметим, что неравенство (24) является наилучшим в смысле Л. В. Тайкова [9] неравенством Бернштейна т. к. предположение о существовании $2n$-мерного подпространства функций из $L(0,2\pi)$, для которых аналог неравенства (24) справедлив с константой, меньшей чем $K_r^{-1} n^r$, приводим к оценке поперечника снизу, превышающей оценку сверху.

3. О базисе в $C(0, 2\pi)$.

Как обычно $C(0,2\pi)$ — обозначает пространство непрерывных 2π-периодических функций. Справедлива следующая теорема.

Теорема 2. *Для любого натурального $k \geqq 1$ существует такой базис $\{\varphi_s(t)\}$ $s = 0, 1, \ldots$ пространства $C(0,2\pi)$, что для любой функции $f(t) \in C(0,2\pi)$*

$$(25) \qquad \|f(t) - S_n(f, t)\|_{C(0,2\pi)} \leqq A_k \omega_{k+1}(f, 1/n),$$

где $S_n(f, t)$ — частичная сумма разложения функции $f(t)$ по элементам базиса, A_k — постоянная, зависящая только от k и $\omega_k(f, \delta)$ — k-ый модуль непрерывности функции $f(t)$

$$\omega_k(f, \delta) = \sup_{|h| \leqq \delta} \sup_{0 \leqq x \leqq 2\pi} \left| \sum_{s=0}^{k} (-1)^{k-s} C_k^s f(x + sh) \right|,$$

$C_k^s = \dfrac{k!}{s!(k-s)!}$. *Если, кроме того, функция $f(t)$ имеет непрерывную r-ю производную $(r \leq k+1)$, то*

$$(26) \qquad \|f^{(i)}(t) - S_n^{(i)}(f, t)\|_{C(0,2\pi)} \leq A_{k,i} h^{r-i} \omega_{k-r+1}(f^{(r)}, 1/n) \qquad (i = 0, 1, \ldots, r),$$

где $A_{k,i}$ — константа, зависящая только от k и i.

Отметим, что элементы базиса состоят из 2π — периодических полиномиальных сплайн-функций степени k с равноотстоящими узнами. Аналогичные оценки справедливы для интерполяционных 2π — периодических сплайн-функций, при этом случай $r=k$ хорошо изучен, см., например, [10].

Ради упрощений проведем доказательство для нечетного k.

Положим $\varphi_0(t) = 1$. Разделим отрезок $[0,2\pi]$ на две равные части и построим 2π — периодический сплайн $\varphi_1(t)$ степени $2m-1$ с узлами в точках $0, \pi$ и 2π и такой, что $s(0) = s(2\pi) = 0$, $s(\pi) = 1$ и положим $\varphi_1(t) = s(t)$. Пусть отрезок $[0,2\pi]$ разбит на 2^{n+1} частей $(n \geq 0)$, тогда обозначим через $\varphi_{2^n + i}(t)$ сплайн степени $2m+1$ с узлами $t_s = \dfrac{\pi}{2^n} s$ $(s = 0, 1, \ldots, 2^{n+1})$ и такой, что $\varphi_{2^n + i}(t_s) = \delta_{2i-1, s}$ где $\delta_{i,s}$ — символ Кронекера, $(i = 1, \ldots, 2^n)$. Таким образом система функций $\{\varphi_s(t)\}$ — построена. Покажем, что она обладает нужными свойствами. Докажем вначале единственность разложения любой функции из $C(0,2\pi)$ по злементам $\{\varphi_s(t)\}$. Пусть существует функция, для которой

$$(27) \qquad f(t) \equiv \sum_{s=0}^{\infty} \alpha_s \varphi_s(t) \equiv \sum_{s=0}^{\infty} \beta_s \varphi_s(t),$$

причем ряды равномерно сходятся и не все $\alpha_s \neq \beta_s$. Пусть s_0 первый номер, когда $\alpha_{s_0} \neq \beta_{s_0}$. Если $s_0 = 0$, то, полагая $t = 0$, имеем по определению $\varphi_s(0) = 0$, $s \geq 1$, а тогда из (27) следует, что $\alpha_0 = \beta_0$, и мы приходим к противоречию. При $s_0 = 1$, приходим к противоречию, положив $t = 1$. Пусть $s_0 = 2^{n_0} + m_0$, $1 \leq m_0 \leq 2^{n_0}$, $n_0 = 1, 2, \ldots$, тогда, положив t_0 равным $\dfrac{\pi(2m_0 - 1)}{2^{n_0}}$, вновь приходим к противоречию с предположением, что $\alpha s_0 \neq \beta s_0$, так как по предположению все предыдущие слагаемые совпадают, а по построению $\{\varphi_k(t)\}$, $\varphi_{s_0}(t_0) = 1$ и $\varphi_s(t_0) = 0$, $s > s_0$.

Поставим функции $f(t)$ в соответствие ряд

$$(28) \qquad \sum_{s=0}^{\infty} \alpha_s(f) \varphi_s(t),$$

где $\alpha_0 = \varphi_0(0)$, $\alpha_1 = f(\pi) - \alpha_0 \varphi_0(\pi)$, ...

$$(29) \qquad \alpha_{2^{n_0}+m_0} = f\left[\frac{\pi(2m_0-1)}{2^{n_0}}\right] - \sum_{s=0}^{2^{n_0}+m_0-1} \alpha_s \varphi\left[\frac{\pi(2m_0-1)}{2^{n_0}}\right],$$

$$m_0 = 1, 2, ..., 2^{n_0}, \quad n_0 = 1, 2, ...$$

Докажем, что ряд (28) с коэффициентами $\alpha_s = \alpha_s(f)$ определенными равенствами (29), равномерно сходится к $f(t)$ и для n-ой частичной суммы этого ряда справедлива оценка (25).

Пусть

$$S_n(f, t) = \sum_{s=0}^{n} \alpha_s(f)\varphi_s(t) = S_{2^{n_0}}(f, t) + [S_n(f, t) - S_{2^{n_0}}(f, t)],$$

где $n = 2^{n_0}+m_0$, $1 \leq m_0 \leq 2^{n_0}$, $n_0 = 1, 2, ...$

Тогда в силу определения коэффициентов α_s и функций $\varphi_s(t)$ выражение, стоящее в квадратных скобках, является сплайном степени $(2m-1)$ с равноотстоящими узлами $x_i = \pi_i/2^{n_0}$, $i = 0, 1, ..., 2^{n_0+1}$, и этот сплайн принимает в узлах значения либо равные нулю, либо $f(x_i) - S_{2^{n_0}}(f, x_i)$. Поэтому в силу следствия 1 из [7]

$$(30) \qquad \|f(t) - S_n(f, t)\|_C \leq \bar{A}_k \|f(t) - S_{2^{n_0}}(f, t)\|_C,$$

$2^{n_0} \leq n < 2^{n_0+1}$, где $\bar{A}_k$ — зависит только от k. Далее $S_{2^{n_0}}(f, t)$ — сплайн степени $2m-1$ с равноотстоящими узлами $t_i = \pi_i/2^{n_0-1}$, и в силу определения коэффициентов α_s он интерполирует функцию $f(t)$ в узлах $\{t_i\}$. Для $2m-1$-ой производной такого сплайна в [7], см. также [2] дано явное выражение

$$(31) \qquad S_{2^{n_0}}^{(2m-1)}(t) = \sum_{k=-\infty}^{\infty} \alpha_k \frac{\Delta_h^{2m-1} f[t_i + (k-m)h]}{h^{2m-1}},$$

где $t_i < t \leq t_{i+1}$, $h = t_{i+1} - t_i$, $i = 0, 1, ..., 2^{n_0}$ и $|\alpha_k| \to 0$ при $|k| \to \infty$ со скоростью геометрической прогрессии и не зависят от f и интерполяционный сплайн $S_{2^{n_0}}(f, t)$ формулой (31) определяется с точностью до полинома $P_{2m-2}(t)$ степени не выше $2m-2$, т. е.

$$(32) \qquad S_{2^{n_0}}^{(2m-1)}(f, t) = P_{2m-2}(t) + \frac{1}{(2m-2)!} \int_0^t (t-u)^{2m-2} S_{2^{n_0}}^{(2m-1)}(f, u)\, du.$$

При выводе неравенства (25), не нарушая общности, можно считать

$$0 \leq t \leq \pi/2^{n_0-1} = h.$$

Для определения полинома $P_{2m}(t)$ воспользуемся тем, что $S_{2^{n_0}}(f, t)$ интерполирует функцию $f(t)$ в узлах $\{t_i\}$, $t_i = \pi_i/2^{n_0-1}$,

$$(33) \qquad f(t_i) = S_{2^{n_0}}(f, t_i).$$

В дальнейшем нас будут интересовать лишь $i=0, 1, \ldots, 2m-1$. Из (32) и (33) с помощью интерполяционной формулы Лагранжа находим

$$P_{2m-2}(t) = \sum_{i=0}^{2m-1} \frac{\omega(t) f(t_i)}{\omega'(t_i)(t-t_i)} - \frac{1}{(2m-2)!} \int\limits_0^{\bar{t}} \sum_{i=0}^{2m-1} \frac{\omega(t)(t_i-u)_+^{2m-2}}{\omega'(t_i)(t-t_i)} S_{2^{n_0}}^{(2m-1)}(f, u)\, du,$$

где $\bar{t} = t_{2m-1}$, $\omega(t) = (t-t_0)(t-t_1), \ldots, (t-t_{2m-1})$,

и

$$(x-u)_+^m = \begin{cases} (x-u)^m, & x \geqq u \\ 0, & x < u \end{cases} \qquad (m \geqq 0).$$

Таким образом

$$(34) \qquad |f(t) - S_{2^{n_0}}(t)| \leqq \left| \sum_{i=0}^{2m-1} \frac{\omega(t)[f(t)-f(t_i)]}{\omega'(t_i)(t-t_i)} \right| +$$

$$+ \left| \int\limits_0^{\bar{t}} \left\{ (t-u)_+^{2m-2} - \sum_{i=0}^{2m-1} \frac{\omega(t)(t_i-u)_+^{2m-2}}{\omega'(t_i)(t-t_i)} \right\} S_{2^{n_0}}^{(2m-1)}(f, u)\, du \right|.$$

В силу неравенства Лебега первое слагаемое не превосходит $L_{2m-1} \cdot E_{2m-1}(f; 0, \bar{t})$ где L_m — константа Лебега и $E_n(f; a, b]$ — величина наилучшего приближения функции f на отрезке $[a, b]$. В силу неравенства Уитни $E_n(f; 0, \bar{t}) \leqq C_n \omega_{n+1}(f, \bar{t}/n)$, следовательно, первое слагаемое в правой части неравенства (34) не превосходит

$$A'_m \omega_{2m}\left(f, \frac{2m-1}{2^{n_0}} \pi\right) \leqq A''_m \omega_{2m}(f, 1/2^{n_0}),$$

где A''_m зависит только от m. Для оценки второго слагаемого заметим, что интеграл от 0 до $\bar{t}$ от выражения

$$\varphi(t, u) = (t-u)_+^{2m-2} - \sum_{i=0}^{2m-1} \frac{\omega(t)(t_i-u)_+^{2m-2}}{\omega'(t_i)(t-t_i)}$$

равен нулю. Поэтому

$$\left| \int\limits_0^{\bar{t}} \varphi(t, u) S_{2^{n_0}}^{(2m-1)}(f, u)\, du \right| = \left| \int\limits_0^{\bar{t}} \varphi(t, u)[S_{2^{n_0}}^{(2m-1)}(f, u) - S_{2^{n_0}}^{(2m-1)}(f, u_0)]\, du \right| \leqq$$

$$\leqq \bar{t} \max_{0 \leqq t,\, u \leqq \bar{t}} |\varphi(t, u)| \max_{0 \leqq u \leqq \bar{t}} |S_{2^{n_0}}^{(2m-1)}(f, u) - S_{2^{n_0}}^{(2m-1)}(f, u_0)|,$$

где u_0 — фиксированное число, $0 \leqq u_0 \leqq \bar{t}$. Так как

$$\max_{0 \leqq t,\, u \leqq \bar{t}} |\varphi(t, u)| \leqq A'''_m h^{2m-2}, \quad \bar{t} \leqq (2m-1)h,$$

и в силу (31)

$$\max_{0 \le u \le i} \left| S_{2^{n_0}}^{(2m-1)}(f, u) - S_{2^{n_0}}^{(2m-1)}(f, u_0) \right| \le$$

$$\le (2m-1)\max_i \sum_{k=-\infty}^{\infty} |\alpha_k| \frac{\Delta_n^{2m} f[t_i + (k-m)h]}{h^{2m-1}} \le B_m \omega_{2m}(f, h) h^{-2m+1}$$

$$(h = t_i - t_{i-1} = \pi/2^{n_0 - 1})$$

Таким образом и второе слагаемое в (33) не превосходит $B_m' \omega_{2m}(f, 1/2^{n_0})$, где B_m' — зависит только от m, так как n_0 произвольное число, то неравенство (25) доказано. Для доказательства неравенства (26) следует i — раз применить теорему Ролля, а затем провести аналогичные рассуждения, тогда мы прийдем к неравенству

$$\|f^{(i)}(t) - S_n^{(i)}(f, t)\|_{C(0,2\pi)} \le A_{k,i} \omega_{2m-i}(f^{(i)}, 1/n),$$

применение неравенств $\omega_{k+l}(f, \delta) \le \delta^l \omega_k(f^{(l)}, \delta)$ завершает доказательство теоремы.

ЛИТЕРАТУРА

[1] Ю. Н. Субботин, *Приближение ,,сплайн''-финкциями и оценки поперелников.* Труды Матем. ин-та АН СССР, **CIX** (1971), 35—60.

[2] Ю. Н. Субботин, *Функциональная интерполяция в среднем с наименьшей п-ой производной.* Труды Матем. ин-та АН СССР, **LXXXYШ** (1967), 30—60.

[3] Ю. Н. Субботин, *О связи между конечными разностями и соответствующими производными.* Труды Матем. ин-та АН СССР, **LXXYШ** (1965), 24—42.

[4] С. М. Никольский, *Приближение функций тригонометрическими полиномами в среднем,* Изв. АН СССР. сер. матем., **10** (1946), 207—256.

[5] В. М. Тихомиров, *Поперечники мнпжеств в функциональных пространствах и теория наилучших приближений,* УМН, ХУ, в. 3 (93), **(1960)**, 81—120.

[6] E. M. Stein, *Functions of exponential type.* Ann. of Math. (2) **65** N 3 (1957), 582—592.

[7] Ю. Н. Субботин, *Поперечник класса $W^r L$ в $L(0,2\pi)$ и приближение сплайн-функциями,* Матем. заметки, т. 7, № 1 **(1970)**, 43—52.

[8] Ю. И. Маказов, *Поперечники некоторых функциональных классов в пространстве L.* Вести АН Белорусской ССР сер. физ.-матем. наук, № 4 **(1969)**, 19—28.

[9] Л. В. Тайков, *О наилучшем приближении в среднем некоторых классов аналитических функций,* Матем. заметки, 1, № 2 **(1967)**, 155—162.

[10] J. H. Ahlberg, E. N. Nilson and I. L. Walsh, *The theory of splines and their applications.* Academic Press, New York—London 1967.

The Fundamental Theorem of Algebra for Monosplines with Multiplicities

By

CHARLES MICCHELLI

MATHEMATICAL SCIENCES DEPT.
IBM RESEARCH CENTER
YORKTOWN HEIGHTS, NEW YORK

Introduction

This paper is devoted to the study of monosplines which have a maximum number of zeros.*) Our results extend those of KARLIN and SCHUMAKER [6] and SCHOENBERG [10]. These authors deal with monosplines which have only simple knots. We relax the continuity restriction at the knots and allow monosplines with multiplicities.

An extended spline of degree $n-1$ $(n \geqq 1)$ is a function $s(x)$ which is a polynomial of degree $n-1$ in each of the intervals

$$(-\infty, \xi_1), [\xi_1, \xi_2), \ldots, [\xi_r, \infty).$$

ξ_i is said to be a knot of multiplicity m_i, $1 \leqq m_i \leqq n$, provided $s \in C^{n-m_i-1}$ in a neighbourhood of ξ_i but not in C^{n-m_i}. Every extended spline of degree $n-1$ with knots $\{\xi_i\}_1^r$ and corresponding multiplicities $\{m_i\}_1^r$ can be represented as

$$(1) \qquad s(x) = \sum_{i=0}^{n-1} \lambda_i x^i + \sum_{i=1}^{r} \sum_{j=1}^{m_i} c_{ij}(x - \xi_i)_+^{n-j}$$

where $c_{im_i} \neq 0$, $i = 1, \ldots, r$ (see [4]).

A monospline of degree n is a function of the form

$$(2) \qquad M(x) = \frac{x^n}{n!} + s(x).$$

We denote the class of monosplines of degree n with knots $\{\xi_i\}_1^r$ of multiplicity $m = (m_1, \ldots, m_r)$, respectively, as $M_{n,m}(\xi_1, \ldots, \xi_r)$. Let us also set $\sigma_i = 1$, if m_i is odd, and zero otherwise.

Our main result is:

THEOREM 1: *Every monospline M in $M_{n,m}(\xi_1, \ldots, \xi_r)$ has no more than $n +$ $+ \sum_{i=1}^{r} (m_i + \sigma_i)$ zeros counting multiplicities. Conversely, given $N = n + \sum_{i=1}^{r} (m_i + \sigma_i)$*

---

*) We thank Professor KARLIN for bringing the problem studied here to our attention.

27*

distinct points $-\infty < t_1 < \cdots < t_N < +\infty$, there exists an M in $M_{n,m}(\xi_1, \ldots, \xi_r)$ which vanishes at t_i, $i=1, \ldots, N$. M is uniquely determined by its zeros if and only if the m_i's are all odd.

Theorem 1 is referred to as the fundamental theorem of algebra for monosplines. When all the knots are simple, $m_i = 1$, $i = 1, \ldots, r$, Theorem 1 was announced in [10] and subsequently proven in [6].

The proof we give of Theorem 1 adopts the approach in [6] but at the same time simplifies their analysis. Our proof also encompasses the case of monosplines generated by extended complete Tchebycheff systems ([6]) but we restrict our presentation to polynomial splines.

Theorem 1 ($m_i \equiv 1$) has been applied to characterizing the monospline which has minimum Tchebycheff norm with variable knots, see JOHNSON [2] and SCHUMAKER [12]. Its relationship to best approximation in the L^2-norm by splines with variable knots is described in KARLIN [3] (see also POWELL [9]) and to Gaussian quadrature formulae in [10].

Theorem 1 in the general case is related to quadrature formulae with multiple Gaussian nodes which were first studied by TURÁN [13] and later by POPOVICIU [8]. Our result however does not include the work of these authors as we must restrict the zeros of the monospline to be distinct in the converse statement of Theorem 1. We conjecture that Theorem 1 remains true when zeros of multiplicity $\leq n+1$ are allowed, as in [6] and [10]. Some of our results do hold when there are multiple zeros and we will indicate when this is the case.

As a byproduct of the proof of Theorem 1 we answer a question left unsettled in [13] (Theorem 3).

The proof of Theorem 1 requires several preliminary results about zeros of monosplines. We begin by defining a zero of order α of M where α is allowed to be as large as $n+1$.

First of all, if x is not a knot of M then the ordinary definition of a zero of order α is adopted

$$(3) \qquad\qquad M(x) = \cdots = M^{(\alpha-1)}(x) = 0, \quad M^{(\alpha)}(x) \neq 0.$$

Let $M_+(x)(M_-)$ be the monospline which agrees with $M(x)$ to the right (left) of ξ_i and has no knots in $(-\infty, \xi_i]$ $([\xi_i, \infty))$. Suppose M_+ has a zero of order α at ξ_i (convention (3) applies) and M_- has a zero of order β at ξ_i. Then M has a zero of order max (α, β) provided the sign of $M_-^{(n-m_i)}(x)$ to the right of ξ_i is the same as the sign of $M_+^{(n-m_i)}(x)$ to the left of ξ_i. Otherwise, M has a zero of order max $(\alpha, \beta)+1$.

Denote the number of zeros of M in (a, b), counting multiplicities, by $Z(M; (a, b))$.

Let $S^+(x_1, \ldots, x_n)$ stand for the maximum number of sign changes in the sequence $x_1, \ldots, x_n$ where a zero entry is allowed to be $+1$ or -1.

PROPOSITION 1: *If M is in $M_{n,m}(\xi_1, \ldots, \xi_r)$ with $a < \xi_1 < \cdots < \xi_r < b$ then*

$$(4) \qquad Z(M; (a, b)) \leq$$

$$\leq n + \sum_{i=1}^{r} (m_i + \sigma_i) - S^+(M(a), \ldots, (-1)^n M^{(n)}(a)) - S^+(M(b), \ldots, M^{(n)}(b)).$$

The proof of this proposition depends on the following generalization of the classical result of Fourier—Budan.

LEMMA 1: *If p is a polynomial of exact degree n then*

$$(5) \quad Z(p; (a, b)) \leq n - S^+(p(a), -p'(a), \ldots, (-1)^n p^{(n)}(a)) - S^+(p(b), \ldots, p^{(n)}(b)).$$

A proof of this lemma can be found in [4]. We also note that Proposition 1 generalizes a result in [5].

PROOF OF PROPOSITION 1: Set $\xi_0 = a$ and $\xi_{r+1} = b$, then applying Lemma 1 on the interval (ξ_i, ξ_{i+1}) we obtain

$$(6) \qquad Z(M; (\xi_i, \xi_{i+1})) \leq$$

$$\leq n - S^+(M(\xi_i^+), \ldots, (-1)^n M^{(n)}(\xi_i^+)) - S^+(M(\xi_{i+1}^-), \ldots, M^{(n)}(\xi_{i+1}^-)) \quad (i = 1, \ldots, r).$$

Let α_i denote the multiplicity of the zero of M at ξ_i. Then summing (6) over i yields

$$Z(M; (a, b)) \leq$$

$$\leq n + \sum_{i=1}^{r} (m_i + \sigma_i) - S^+(M(a), \ldots, (-1)^n M^{(n)}(a)) - S^+(M(b), \ldots, M^{(n)}(b)) -$$

$$- \sum_{i=1}^{r} (T_n(M, \xi_i) - 1 + \sigma_i)$$

where

$$T_n(M, \xi_i) \equiv$$

$$\equiv S^+(M(\xi_i), \ldots, M^{(n)}(\xi_i^-)) + S^+(M(\xi_i^+), \ldots, (-1)^n M^{(n)}(\xi_i^+)) + m_i + 1 - n - \alpha_i.$$

Thus the proof is complete provided that

$$(7) \qquad T_n(M, \xi_i) \geq 1 - \sigma_i \qquad (i = 1, \ldots, r).$$

To prove (7), we first require a simple fact about $S^+(x_0, \ldots, x_n)$.

LEMMA 2:

$$S^+(x_0, \ldots, x_n) + S^+(x_0, \ldots, (-1)^n x_n) \geq n.$$

Equality holds if and only if $x_j \neq 0$, $j = 0, 1, \ldots, n$.

The proof of (7) requires considering several cases. We restrict ourselves to the case when $0 \leq \alpha_i \leq n - m_i - 1$. The other cases follow by a careful application of the zero convention which we adopted.

Since $M \in C^{n-m_i-1}$ in a neighbourhood of ξ_i and M has a zero of order α_i at ξ_i we have

(8)
$$T_n(M, \xi_i) = m_i + 1 - n + \alpha_i + S^+\left(M^{(\alpha_i)}(\xi_i^-), \ldots, M^{(n)}(\xi_i^-)\right) +$$
$$+ S^+\left(M^{(\alpha_i)}(\xi_i^+), \ldots, (-1)^{n-\alpha_i} M^{(n)}(\xi_i^+)\right) \geqq$$
$$\geqq m_i + 1 - n + \alpha_i + S^+\left(M^{(\alpha_i)}(\xi_i), \ldots, M^{(n-m_i-1)}(\xi_i)\right) +$$
$$+ S^+\left(M^{(\alpha_i)}(\xi_i), \ldots, (-1)^{n-m_i-1} M^{(n-m_i-1)}(\xi_i)\right) \geqq$$
$$\geqq m_i + 1 - n + \alpha_i + n - m_i - 1 - \alpha_i = 0.$$

In accordance with Lemma 2, $T_n(M, \xi_i)=0$ if and only if

(9)
$$M^{(j)}(\xi_i^-) > 0 \qquad (j = n - m_i - 1, \ldots, n),$$

(10)
$$(-1)^{n+j} M(\xi_i^+) > 0 \qquad (j = n - m_i - 1, \ldots, n),$$

(11)
$$M^{(j)}(\xi_i) \neq 0 \qquad (j = \alpha_i, \ldots, n - m_i - 1).$$

Thus, if $T_n(M, \xi_i)=0$ then $M^{(n-m_i-1)}(\xi_i)>0$ and $(-1)^{m_i} M^{(n-m_i-1)}(\xi_i)<0$. Therefore, in this case m_i must be odd.

COROLLARY 1: *Let M be a member of $M_{n,m}(\xi_1, \ldots, \xi_r)$ having the form*

$$M(x) = \frac{x^n}{n!} + \sum_{i=0}^{n-1} \lambda_i x^i + \sum_{i=1}^{r} \sum_{j=1}^{m_i} c_{ij}(x - \xi_i)_+^{n-j}$$

then

1) $\quad Z(M; (-\infty, +\infty)) \leqq n + \sum_{i=1}^{r} (m_i + \sigma_i).$

Moreover, when equality holds in 1), *the next two statements are valid.*
2) *If m_i is odd then $c_{ij}<0$, $j=1, 3, \ldots, m_i$.*
3) *If x_1 and x_N are the smallest and largest zero of M, respectively, then*

$$M^{(i)}(x)>0 \qquad (i=0, 1, \ldots, n, \quad x>x_N)$$
$$(-1)^{n-i} M^{(i)}(x)>0 \qquad (i=0, 1, \ldots, n, \quad x>x_1).$$

PROOF: 1) and 3) are an immediate consequence of Proposition 1. To prove 2) we note that the hypothesis implies

(12)
$$T_n(M, \xi_i) = 1 - \sigma_i \qquad (i=1, \ldots, r),$$

and so from (9) and (10), it follows that

$$M^{(n-j)}(\xi_i^-) > 0, \quad M^{(n-j)}(\xi_i^+) < 0 \qquad (j=1, 3, \ldots, m_i).$$

COROLLARY 2: *Let M be a monospline in $M_{n,m}(\xi_1, \ldots, \xi_r)$ which vanishes at* $t_1 < \cdots < t_N$, $N = n + \sum_{i=1}^{r} (m_i + \sigma_i)$. *If $m_i < n$ then*

(13) $$t_{k(i)} < \xi_i < t_{n+k(i-1)+1},$$

where $k(i) = \sum_{j=1}^{i} (m_j + \sigma_j)$.

In the case that $m_i = n$ and n is odd, then

(14) $$\xi_i = t_{k(i)}.$$

PROOF: Suppose $m_i < n$ and $\xi_i \leqq t_{k(i)}$. Since M is continuous at ξ_i then the monospline M_+ which agrees with M on $[\xi_i, \infty)$ has at least $n + \sum_{j=i+1}^{r} (m_j + \sigma_j) + 1$ zeros. But M_+ has only $r - i$ knots. This contradicts Proposition 1 and so the first inequality of (13) must hold. The remaining assertions follow in a similar manner.

REMARK. Corollary 2 generalizes to include monosplines which have multiple zeros. Specifically, if we repeat each as many times as its multiplicity then

(15) $$t_{k(i)} \leqq \xi_i \leqq t_{n+k(i-1)+1}.$$

Strict inequality prevails in (15) whenever the multiplicity of the zero at ξ_i does not exceed $n - m_i$.

We will now use Corollaries 1 and 2 to show that it is sufficient to prove Theorem 1 under the assumption that all the m_i's are odd. To do this we first must recall certain properties of the kernel $\Phi_n(x, \xi) = (x - \xi)_+^{n-1}$. Define

(16) $$\Phi_n \begin{pmatrix} t_1, \ldots, t_s \\ \xi_1, \ldots, \xi_s \end{pmatrix} = \begin{vmatrix} \Phi_n(t_1, \xi_1) \ldots \Phi_n(t_1, \xi_s) \\ \Phi_n(t_2, \xi_1) \ldots \Phi_n(t_2, \xi_s) \\ \vdots \qquad\qquad \vdots \\ \Phi_n(t_s, \xi_1) \ldots \Phi_n(t_s, \xi_s) \end{vmatrix}.$$

In the event that there are repeated values of x's or ξ's, we modify (16) to include derivatives of $\Phi_n(x, \xi)$. For instance, if $t_1 = t_2 < t_3 < \cdots < t_s$ then we replace the second row with

$$\frac{\partial}{\partial t} \Phi_n(t_1, \xi_1), \frac{\partial}{\partial t} \Phi_n(t_1, \xi_2), \ldots, \frac{\partial}{\partial t} \Phi_n(t_1, \xi_s).$$

Similarly for coincidences in the ξ's.

PROPOSITION 2: *Let t_i, ξ_i, $(i = 1, \ldots, s)$ satisfy the following conditions:*
(a) $\xi_i \leqq \xi_2 \leqq \cdots \leqq \xi_s$, $t_1 \leqq t_2 \leqq \cdots \leqq t_s$.
(b) *Whenever l of the t_i's and m of the ξ's coincide, so as to be equal to x, then*
   $m + l \leqq n + 1$.
(c) *No more than n consecutive ξ's or t's coincide.*

Then

(17)
$$\Phi_n\begin{pmatrix} t_1, & \dots, & t_s \\ \xi_1, & \dots, & \xi_s \end{pmatrix} \geqq 0$$

with strict inequality prevailing in (17) *if and only if*

(18)
$$\xi_{i-n} < t_i < \xi_i, \qquad i = 1, \dots, r.$$

For $i \leqq n$, the first inequality of (18) *is omitted.*

A proof of this proposition can be found in [4].

LEMMA 3: *Suppose Theorem 1 is true whenever all the knots have odd multiplicities. Then it is also true for arbitrary multipicities satisfying $m_i \leqq n-2$, $i = 1, \dots, r$.*

PROOF: Suppose $m_1, \dots, m_r$ and $\{t_i\}_{i=1}^N$ are given where $N = n + \sum_{i=1}^{r} (m_i + \sigma_i)$. Define $I = \{i : m_i$ is even$\}$ and $J = \{i : m_i$ is odd$\}$. We need only consider the case when $I \neq \emptyset$.

Given $t_1 < \cdots < t_N$, the hypothesis assures the existence of a unique monospline $\overline{M}(x)$

$$\overline{M}(x) = \frac{x^n}{n!} + \sum_{i=0}^{n-1} \overline{\lambda}_i x^i + \sum_{i \in J} \sum_{j=1}^{m_i} \overline{c}_{ij}(x - \overline{\xi}_i)_+^{n-j} + \sum_{i \in I} \sum_{j=1}^{m_i-1} \overline{c}_{ij}(x - \overline{\xi}_i)_+^{n-j}$$

which has a zero at t_i,

(19)
$$\overline{M}(t_i) = 0 \qquad (i = 1, \dots, N).$$

By a change of origin we can assume without loss of generality that $t_1 > 0$.

Set $p = n + \sum_{i=1}^{r} (m_i + 1)(> N)$ and define a mapping ψ from R^p into R^N as follows:

$$\psi : (\lambda_0, \dots, \lambda_{n-1}, c_{11}, \dots, c_{1m_i}, \xi_1, c_{21}, \dots, c_{2m_2}, \xi_2, \dots, c_{r1}, \dots, c_{rm_r}, \xi_r)$$

$$\rightarrow (M(t_1), M(t_2), \dots, M(t_N))$$

where

(20)
$$M(x) = \frac{x^n}{n!} + \sum_{i=0}^{n-1} \lambda_i x^i + \sum_{i=1}^{r} \sum_{j=1}^{m_i} c_{ij}(x - \xi_i)_+^{n-j}.$$

Equation (19) means ψ vanishes at

$$\overline{x} = (\overline{\lambda}_0, \dots, \overline{\lambda}_{n-1}, \overline{c}_{11}, \dots, \overline{c}_{1m_1}, \overline{\xi}_1, \dots, \overline{c}_{r1}, \dots, \overline{c}_{rm_r}, \overline{\xi}_r)$$

where we define $\bar{c}_{im_i}=0$ when $i\in I$. The Jacobian of ψ with respect to all of its variables except the $p-N$ knots in $\{\xi_i : i\in I\}$ is

$$J(x) = d \prod_{i\in J} c_{im_i} \Phi_n \begin{pmatrix} t_1, & & \cdots\cdots, & & t_N \\ \underbrace{0, \ldots, 0}_{n}, \underbrace{\xi_1, \ldots, \xi_1}_{m_1+\sigma_1}, \underbrace{\xi_2, \ldots, \xi_2}_{m_2+\sigma_2}, \ldots, \underbrace{\xi_r, \ldots, \xi_r}_{m_r+\sigma_r} \end{pmatrix}$$

where d is a nonzero constant. Our hypothesis insures us that J is a continuous function of all its arguments. Moreover, since M was constructed to have a maximum number of zeros we can invoke Corollaries 1 and 2 with Proposition 2 to conclude that $J(\bar{x})\neq 0$. Hence, by the implicit function theorem, there exists a neighbourhood about each knot in $\{\bar{\xi}_i ; i\in I\}$ such that the monospline (20) satisfies (19).

We must still deal with the possibility that $m_i \geq n-1$ for some i. This is the subject of the next lemma.

LEMMA 4: *Suppose Theorem 1 is true whenever $m_i \leq n-2$, $i=1, \ldots, r$. Then it is also true for arbitrary multiplicities.*

PROOF: Suppose there is just one index i such that $m_i \geq n-1$. First consider the case $m_i = n-1$ and n even. Set $k = \sum_{j=1}^{i} (m_j+\sigma_j) = n+\sum_{j=1}^{i-1} (m_j+\sigma_j)$. Corollary 2 implies that the monospline we are looking for is constructible by piecing together two monosplines whose existence is assured by our hypothesis. Specifically, let M_1 be a monospline with $i-1$ knots satisfying

$$M_1(t_i)=0 \qquad (i=1, \ldots, k)$$

and M_2 a monospline with $r-i$ knots satisfying

$$M_2(t_i)=0 \qquad (i = k+1, \ldots, N).$$

Corollary 1 implies $M_1(t)>0$ for $t>t_k$ and $M_2(t)>0$, $t<t_{k+1}$. Thus there exists a unique ξ_i in (t_k, t_{k+1}) such that $M_1(\xi_i)=M_2(\xi_i)$. Hence

$$M(t) = \begin{cases} M_1(t), & t\leq\xi_i \\ M_2(t), & t\geq\xi_i \end{cases}$$

is the required monospline.

If n were odd then we cannot expect a uniquely constructed M. In fact, in this case $k = n-1+\sum_{j=1}^{i-1} (m_j+\sigma_j)$ and so for any $x\in(t_{k+1}, t_{k+2})$ we can construct M_3 vanishing at $t_1, \ldots, t_k, x$. Again there exists a $\xi_i\in(t_{k+1}, t_{k+2})$ such that $M_3(\xi_i)= = M_2(\xi_i)$. Thus

$$M(t) = \begin{cases} M_3(t), & t\leq\xi_i \\ M_2(t), & t\geq\xi_i \end{cases}$$

is the required monospline.

If $m_i = n$ then similar constructions will yield the desired conclusion. Finally, if there is more than one index satisfying $m_i \geq n-1$ then the previous analysis just serves as an induction step on the number of such indices. Q.e.d.

We need one final property of monosplines to complete the proof of Theorem 1. This fact states if a monospline has a maximum number of zeros on some bounded interval and the m_i's are all odd then M has bounded coefficients.

PROPOSITION 3: *Given any $K > 0$ there exists a $\lambda > 0$ such that whenever* $M \in M_{n,m}(\xi_1, \ldots, \xi_r)$

$$M(x) = \frac{x^n}{n!} + \sum_{i=0}^{n-1} \lambda_i x^i + \sum_{i=1}^{r} \sum_{j=1}^{m_i} c_{ij}(x - \xi_i)_+^{n-j}$$

and M has $n + \sum_{i=1}^{r} (m_i + 1)$ distinct zeros in $(-K, +K)$ then

$$|\lambda_i| \leq \lambda \qquad (i = 0, 1, \ldots, n-1)$$

$$|c_{ij}| \leq \lambda \qquad (j = 1, \ldots, m_i; \quad i = 1, \ldots, r).$$

REMARK: It is essential in this proposition that all the m_i's are odd as can be seen from easily constructed counter-examples.

PROOF: The proof proceeds by simultaneous induction on n and r. The case $r = 0$, $n \geq 1$ is obvious. If $n = 1$ and $r \geq 1$ then $m_i \equiv 1$, $i = 1, \ldots, r$. This case is handled in [6].

Now suppose the proposition is true for all monosplines of degree $n-1$ with r knots and all monosplines of degree n with $r-1$ knots. Let M be a monospline of degree n with r knots. Let us first consider the case when $m_i < n$, $i = 1, \ldots, r$. Define $D_+ M(x) = \lim_{h \to 0+} \dfrac{M(x+h) - M(x)}{h}$ then $D_+ M \in M_{n-1, m}(\xi_1, \ldots, \xi_r)$. By Rolle's theorem and Corollary 1, $D_+ M$ has $n - 1 + \sum_{i=1}^{r} (m_i + \sigma_i)$ distinct zeros. Therefore the induction hypothesis implies that all the coefficients of $D_+ M$ are bounded. Hence the same is true for M except possibly for the constant term λ_0. Since M certainly has at least one zero and all of its knots are in $(-K, +K)$, we see that λ_0 is also bounded.

In the case that $m_i = n$ for some i we can appeal to Corollary 2 to conclude that the two monosplines, M_+ and M_- (for their definition, see the description following (3)), both have a maximum number of zeros in $(-K, +K)$. Applying the induction hypothesis to M_+ and M_- we again conclude that M has bounded coefficients.

We have now completed the essential preliminaires and are now ready to prove Theorem 1.

PROOF OF THEOREM 1: Theorem 1 is obviously true when $r=0$. We will employ an induction on r to prove the general case. Recall that Lemmas 3 and 4 enable us to assume that all the m_i's are odd and $m_i \leqq n-2$, $i=1, 2, \ldots, r$. Moreover, by a change of origin we may assume $0 < t_1 < t_2 < \cdots < t_N$, $N = n + \sum_{i=1}^{r} (m_i + 1)$.

Assume the theorem is true for monosplines with $r-1$ knots. We will advance the induction by proving it valid for monosplines with r knots.

Hence, by the induction hypothesis, there exists a unique $\underline{M} \in M_{n,m}(\underline{\xi}_1, \ldots, \underline{\xi}_{r-1})$

$$\underline{M}(x) = \frac{x^n}{n!} + \sum_{i=0}^{n-1} \lambda_i x^i + \sum_{i=1}^{r-1} \sum_{j=1}^{m_i} \underline{c}_{ij}(x - \underline{\xi}_i)_+^{n-j}$$

which satisfies $\underline{M}(t_i)=0$, $i = 1, \ldots, p = n + \sum_{i=1}^{r-1} (m_i + 1)$. We add an additional knot to $\underline{M}$ so that it will also vanish at $t_{p+1}, \ldots, t_{N-1}$. More generally, for any $M \in M_{n,m}(\xi_1, \ldots, \xi_{r-1})$, we define

$$(21) \qquad M(x, \xi) = M(x) + \sum_{j=1}^{m_r} c_{jm_r}(x - \xi)_+^{n-j} \qquad (\xi = \xi_r).$$

(21) is a family of monosplines in $M_{n,m}(\xi_1, \ldots, \xi_{r-1}, \xi)$. Choosing $M = \underline{M}$ and $\xi \geqq t_p$ then $M(x, \xi)$ will also vanish at $t_1, \ldots, t_p$. Under the further restriction

$$(22) \qquad t_p \leqq \xi < t_{p+1}$$

we can determine $c_{1m_1}, \ldots, c_{1m_r}$ uniquely so that

$$(23) \qquad M(t_i, \xi)=0 \qquad (i=1, \ldots, N-1).$$

In fact, if we let $\{l_i(x)\}_{i=1}^{m_r}$ be the Lagrange polynomials of degree m_r-1 which satisfy

$$l_i(t_{p+j}) = \delta_{ij} \qquad (i, j = 1, \ldots, m_r)$$

then

$$(24) \qquad M(x, \xi) = \underline{M}(x) - (x - \xi)_+^{n-m_r} \sum_{j=1}^{m_r} \frac{\underline{M}(t_{p+j})}{(t_{p+j} - \xi)^{n-m_r}} l_j(x)$$

is the unique monospline which satisfies (23) under the restriction (22). Moreover, since $l_i(t_N)>0$, $i=1, \ldots, m_r$ and $\underline{M}(t_{p+j})>0$, $j=1, \ldots, m_r$, we see that $\lim_{\xi \to t_{p+1}} M(t_N, \xi) = = -\infty$. Hence we cannot extend the definition of $M(x, \xi)$ continuously to include $\xi = t_{p+1}$, if we desire to preserve (23). The remainder of the proof consists in verifying that we can extend the definition of $M(x, \xi)$ to the left in ξ until we reach the unique monospline $\overline{M} \in M_{n,m}(\xi_1, \ldots, \xi_r)$ which satisfies $\overline{M}(t_i)=0$, $i=1, \ldots, N$. Henceforth, we will let the coefficients of M in the representation (21) also depend on ξ. In other words, in the proof of the theorem, all the $N-1$ coefficients of $M(x, \xi)$

will be constructed as functions of ξ which is the largest knot of $M(x, \xi)$. The construction is uniquely determined by (22) and the additional requirement

$$(25) \qquad\qquad M(t_N, \xi) < 0.$$

We begin by choosing a $\xi^* \in (t_p, t_{p+1})$ so that (25) is satisfied. Define $\bar{\xi}_r$ to be the infimum of all τ such that $M(t, \xi)$ is defined as a C^1 function on $(-\infty, +\infty) \times [\tau, \xi^*]$, which satisfies (23) and (25) for $\xi \in [\tau, \xi^*]$. Thus $M(t, \xi)$ is C^1 on $(-\infty, +\infty) \times I$, $I = (\bar{\xi}_r, \xi^*)$. (25) implies there is a unique $t_N(\xi)$, $t_N(\xi) > t_N$ such that

$$M\big(t_N(\xi), \xi\big) = 0 \qquad (\xi \in I).$$

We claim that $t_N(\xi)$ is a nondecreasing function. To prove this it is sufficient to show that $\dfrac{\partial}{\partial \xi} M(t, \xi) < 0$ for $\xi \in I$, $t \in (t_N, \infty)$. If we differentiate (20) we obtain

$$(26)$$
$$\frac{\partial}{\partial \xi} M(t, \xi) = \sum_{i=0}^{n-1} \lambda_i'(\xi) t^i + \sum_{i=1}^{r} \sum_{j=1}^{m_i} d_{ij}(\xi)(t - \xi_i)_+^{n-j} + \sum_{i=1}^{r} c_{im_i} \xi_i'(\xi) \frac{\partial}{\partial \xi_i}(t - \xi_i)_+^{n-m_i}$$

where $d_{ij}(\xi)$ are some functions whose explicit form is of no importance. Substituting $t = t_1, t_2, \ldots, t_{N-1}$ in (26) and using (23) we can solve the linear equations and obtain

$$(27) \quad c_{im_i} \xi_i' = \frac{(-1)^{m_i} \dfrac{\partial}{\partial \xi} M(t, \xi)\, \Phi_n \begin{pmatrix} t_1, & t_2, & & \cdots & , & & & t_{N-1} \\ 0, \ldots, 0, & \underbrace{\xi_1, \ldots, \xi_1}, & \cdots, & \underbrace{\xi_i, \ldots, \xi_i}, & \underbrace{\xi_r, \ldots, \xi_r} \end{pmatrix}}{\Phi_n \begin{pmatrix} t_1, & & \cdots, & & t_{N-1}, & t \\ 0, \ldots, 0, & \underbrace{\xi_1, \ldots, \xi_1}, & \cdots, & \underbrace{\xi_r, \ldots, \xi_r} \end{pmatrix}}.$$

Since $M(t, \xi)$ has a maximum number of zeros, we know from Corollaries 1 and 2 that $c_{im_i} < 0$, $i = 1, \ldots, r$ and

$$(28) \qquad\qquad t_{k(i)} < \xi_i < t_{n+k(i-1)+1}.$$

These inequalities are precisely the conditions needed to insure that the denominator in (27) is positive, as can be seen from Proposition 2. Choosing $i = r$ in (27), we obtain

$$\frac{\partial}{\partial \xi} M(t, \xi) > 0 \qquad (\xi \in I,\ t \in (t_N, \infty)).$$

We also have the additional result $\xi_i'(\xi) \geq 0$, $i = 1, \ldots, r-1$.

We now know that $t_N(\xi)$ is a nondecreasing function and so all the zeros of $M(t, \xi)$, $\xi \in I$ are bounded. By Proposition 3, we can select a sequence $\{\xi_n\}$, such that $\xi_n \to \bar{\xi}_r^+$ and $\lim\limits_{n \to \infty} M(t, \xi_n) = \overline{M}(t)$. $\overline{M}$ is necessarily a monospline satisfying (23) and also $\overline{M}(t_N) \leq 0$. Thus $\overline{M}$ must have r distinct knots of multiplicity m_i, $i = 1, \ldots, r$

otherwise we would contradict Corollary 1. We will now show that $\overline{M}(t_N)=0$. To do this, we define a mapping $\Phi: R^N \to R^{N-1}$ as

$$\Phi: (\lambda_0, \ldots, \lambda_{n-1}, c_{11}, \ldots, c_{1m}, \xi_1, c_{21}, \ldots, c_{rm_r} \xi) \to (M(t_1), \ldots, M(t_{N-1}))$$

$$M(x) = \frac{x^n}{n!} + \sum_{i=0}^{n-1} \lambda_i x^i + \sum_{i=1}^{r} \sum_{j=1}^{m_i} c_{ij}(t-\xi_i)_+^{n-j}.$$

Suppose to the contrary that $\overline{M}(t_N)<0$. Then Corollaries 1, 2 and Proposition 2 imply the Jacobian of Φ with respect to its first $N-1$ variables is nonzero when evaluated at the coefficients of $\overline{M}$. By the implicit function theorem, we can extend the definition of $M(t, \xi)$ to the left of $\bar{\xi}_r$ preserving (23) and (25). This contradicts the definition of $\bar{\xi}_r$ and so $\overline{M}(t_N)=0$.

To complete the proof we need to show that $\overline{M}$ is unique. Let M^* be any other monospline in $M_{n,m}(\xi_1^*, \ldots, \xi_r^*)$ such that $M^*(t_i)=0$, $i=1, \ldots, N$. Using analysis identical to that previously employed we can deform $M^*(t)$ continuously, preserving (23) and (25) until its largest knot is $\geq t_p$. Since uniqueness was assumed valid for monosplines with $r-1$ knots, the deformation must be identical with (24). However, (24) was deformed uniquely into $\overline{M}$ by the implicit function theorem. Hence $\overline{M}=M^*$.

We end the paper with two applications of Theorem 1. The first concerns the existence of an equioscillating monospline.

THEOREM 2: *Suppose all the m_i's are odd; then there exists an $M \in M_{n,m}(\xi_1, \ldots, \xi_r)$ and $N = n + \sum_{i=1}^{r} (m_i+1)+1$ points, $-1=t_0<t_1<\cdots<t_{N-1}<t_N=+1$, such that* $M(t_i)=(-1)^i \max_{-1 \leq t \leq +1} |M(t)|$, $i=0, 1, \ldots, N$.

The proof of Theorem 2 is a straightforward application of Theorem 1 and the techniques developed in [1]. The monospline whose existence is asserted here is conjectured to be the unique monospline of smallest sup norm with r variable knots of fixed multiplicities $\{m_i\}_{i=1}^{r}$. In [11], the question was raised as to whether the results in [2] generalize to monosplines with multiplicities. Theorem 2 is a start in answering this question affirmatively.

Our last theorem is about quadrature formulae.

THEOREM 3: *Suppose all the m_i's are odd; then there exists a unique quadrature formula of the form*

(29)
$$\int_{-1}^{+1} f(x)\,dx = \sum_{i=1}^{r} \sum_{j=0}^{m_i-1} \lambda_{ij} f^{(j)}(\xi_i)$$

which is exact for polynomials of degree $n-1$, $n = \sum_{i=1}^{r} (m_i+1)$. *Moreover,*

(30)
$$\lambda_{ij}>0 \qquad (j=0, 2, \ldots, m_i-1, \quad i=1, \ldots, r).$$

PROOF: The existence and uniqueness of (29) are proven in [8]. (30) is proven by observing that the hypothesis implies the monospline

$$M(x) = \frac{(x+1)^n}{n} - \sum_{i=1}^{r} \sum_{j=0}^{m_i} \lambda_{ij} \frac{n!}{j!} (-1)^j (x - \xi_i)_+^{n-j-1}$$

has a zero of order n at $+1$ and -1. Therefore (30) follows from Corollary 1.

When $m_i = 3$, $i = 1, 2, \ldots, r$ it was proven in [13] that $\lambda_{i,2} > 0$, $i = 1, \ldots, r$, while the corresponding question for $\lambda_{i,0}$, $\lambda_{i,1}$ was left unsettled. (30) implies that $\lambda_{i,0}$ is also positive for $i = 1, \ldots, r$.

In [7] the quadrature formulae in Theorem 3 were explicitly constructed for the weight function $1/\sqrt{1-x^2}$ and $m_i = 2k+1$, $i = 1, \ldots, r$. For $k = 2$ it was found that $\lambda_{i,1}$, $i = 1, \ldots, r$ are not of one sign. Since Theorem 3 generalizes to include any weight function, it seems that (30) is the strongest statement which is valid for all weight functions.

BIBLIOGRAPHY

[1] C. Fitzgerald and L. L. Schumaker, *A Differential Equation Approach to Interpolation at Extremal Points*. J. Analyse Math. **22** (1969), 117—134.

[2] R. S. Johnson, *On Monosplines of Least Deviation*. Trans. Amer. Math. Soc. **96** (1966), 458—477.

[3] S. Karlin, *The Fundamental Theorem of Algebra for Monosplines Satisfying Certain Boundary Conditions and Applications to Optimal Quadrature Formulas*. In *Approximations with Special Emphasis on Spline Functions*, I. J. Schoenberg, Editor, Academic Press, New York 1969.

[4] S. Karlin, *Total Positivity*. Standford University Press, Stanford, California 1968.

[5] S. Karlin and C. Micchelli, *The Fundamental Theorem of Algebra for Monosplines Satisfying Boundary Conditions*. (to appear).

[6] S. Karlin and Schumaker, *The Fundamental Theorem of Algebra for Tchebycheffian Monosplines*. J. Analyse Math. **20** (1967), 233—270.

[7] C. Micchelli and T. Rivlin, *The Turán Formulae and Highest Precision Quadrature Rules for Chebyshev Coefficients*. to appear in IBM J. Res. Develop.

[8] T. Popoviciu, *Sur une généralisation de la formule d'intégration numérique de Gauss*. Acad. R. P. Romîne Fil. Iaşi. Stud. Cerc. Şti. **6** (1955), 29—57.

[9] M. J. D. Powell, *On Best L_2 Spline Approximation*. Numerische Mathematik Differentialgleichungen, Approximationtheorie, Sonderdruck aus ISNM, Vol. **9**, Birkhäuser, Basel 1968, 317—339.

[10] I. J. Schoenberg, *Spline Functions, Convex Curves and Mechanical Quadrature*. Bull. Amer. Math. Soc. **64** (1958), 352—357.

[11] I. J. Schoenberg and Zvi. Ziegler, *On Cardinal Monosplines of Least L_∞-Norm on the Real Axis*. J. Analyse Math. **XXIII** (1970), 409—436.

[12] L. Schumaker, *Uniform Approximation by Chebyshev Spline Functions*. II *Free Knots*. SIAM J. Numer. Anal. **5** (1968), 647—656.

[13] P. Turán, *On the Theory of the Mechanical Quadrature*. Acta Sci. Math. (Szeged) **12** Par. A (1950), 30—37.

A Contribution to the Problem of Weighted Polynomial Approximation

By

GÉZA FREUD

MAGYAR TUDOMÁNYOS AKADÉMIA
MATEMATIKAI KUTATÓ INTÉZETE
BUDAPEST

1. Introduction

Let $W(x)$ $(-\infty < x < \infty)$ be a nonnegative continuous function satisfying

$$(1.1) \qquad \lim_{n \to \infty} |x|^n W(x) = 0 \qquad (n = 0, 1, \ldots),$$

and let L_p $(1 \leqq p \leqq \infty)$ be the Banach-space of integrable functions with norm

$$(1.2) \qquad \|f\|_p = \left\{ \int_{-\infty}^{\infty} |f(x)|^p \, dx \right\}^{1/p} \qquad (1 \leqq p < \infty)$$

resp. for $p = \infty$ the space of bounded measurable functions with norm

$$(1.3) \qquad \|f\|_\infty = \operatorname{vrai\,max} |f(x)|.$$

We denote by $\mathscr{P}_n$ the set of polynomials (in a single variable) of degree n at most. By (1.1) $qW \in L_p$ for every $q \in \mathscr{P}_n$ $(1 \leqq p \leqq \infty;\ n = 0, 1, \ldots)$. For $Wf \in L_p$ let

$$(1.4) \qquad \varepsilon_n^{(p)}(W; f) = \inf_{q \in \mathscr{P}_n} \|W(f - q)\|_p.$$

In the present note we discuss the rate of decrease of $\varepsilon_n^{(p)}(W_\beta; f)$ for

$$(1.5) \qquad W_\beta(x) = (1 + x^2)^{\beta/2} e^{-x^2/2} \qquad (\beta \geqq 0).$$

The weights w for which $\varepsilon_n^{(p)}(w; f) \to 0$ for every f satisfying $wf \in L_p$ were determined for $p = 2$ by M. Riesz [29] (see also our book [13], § II. 4) and for $p = 1$ by M. A. Naimark [25].

For $p = \infty$ the additional condition

$$(1.6) \qquad \lim_{|x| \to \infty} x^n f(x) = 0 \qquad (n = 0, 1, \ldots)$$

is required. The weights for which $\varepsilon_n^\infty(w; f) \to 0$ for every $wf \in L_\infty$ satisfying (1.6) were characterized in the papers of H. Pollard [27], [28], N. I. Achieser [1] and S. N. Mergelian [24]. For the weights W_β $(\beta \geqq 0)$ we obtain $\varepsilon_n^{(p)}(W_\beta; f) \to 0$ for every f satisfying $W_\beta f \in L_p$ if $1 \leqq p < \infty$ resp. for every f satisfying $W_\beta f \in L_\infty$ and (1.6), as a corollary of our main result. Concerning other investigations (which apply to much more general w, but only for more restricted classes of functions f) we refer to M. M. Dzrbasian [7], [8], A. S. Dzafarov [6] and G. Freud [14].

Let

$$(1.7) \qquad \tau(x) = \begin{cases} |x| & (|x| \leqq 1) \\ 1 & (|x| > 1). \end{cases}$$

We introduce the generalized continuity modulus[1])

$$(1.8) \quad \omega(p; W_\beta; f, h) = \sup_{|t| \leqq h} \|W_\beta(x+t)f(x+t) - W_\beta(x)f(x)\|_p + \|\tau(hx)W_\beta(x)f(x)\|_p$$

and the generalized Zygmund modulus

$$(1.9) \quad \omega_2(p; W_\beta; f, h) = \sup_{|t| \leqq h} \|W_\beta(x+t)f(x+t) + W_\beta(x-t)f(x-t) - W_\beta(x)f(x)\|_p +$$

$$+ \sup_{|t| \leqq h} \|\tau(hx)[W_\beta(x+t)f(x+t) - W_\beta(x-t)f(x-t)]\|_p +$$

$$+ \|\tau^2(hx)W_\beta(x)f(x)\|_p + h^2\|W_\beta(x)f(x)\|_p.$$

Our main result is that the order of magnitude of $\varepsilon_n^{(p)}(W_\beta; f)$ is determined by these generalized moduli.

In particular, *if $1 \leqq p \leqq \infty$, r is a nonnegative integer and $0 < \alpha < 1$, a necessary and sufficient condition that for a fixed $\beta \geqq 0$ $\varepsilon_n^{(p)}(W_\beta; f) = O(n^{-(r+\alpha)/2})$ should hold is that f is the r-times iterated integral of a function $f^{(r)}$ which satisfies $W_\beta f^{(r)} \in L_p$ and $\omega(p; W_\beta; f, h) = O(h^\alpha)^2$.*

For $\beta = 0$ this was proved in the papers [15], [16], [18] of the author. The extension to $\beta > 0$ is new.

Further, $\varepsilon_n^{(p)}(W_\beta; f) = O(n^{-r/2})$ holds for a natural r if and only if f is the $r-1$ times iterated integral[2] of a function $f^{(r-1)}$ satisfying $\omega_2(p; W_\beta; f^{(r-1)}, h) = O(h)$.

In our present lecture we give the complete proof of the direct ("Jackson type") part of our theorem (see Theorem 3.9). The converse ("Bernstein type") part was proved for the special case $\beta = 0$, i.e. $W(x) = e^{-x^2/2}$ in our papers [15], [18]. We intend to publish soon our Bernstein-type result for $\beta > 0$ somewhere else. Our present analysis is based on properties of certain orthogonal polynomials.*) These properties are treated in part 2. Our Jackson-type theorem is proved in part 3. In this last part of our lecture some arguments concerning linear operators in Banach spaces are essential.

For fixing our notations, we denote by "c" positive numbers depending at most on $\beta \geqq 0$, but not necessary the same even if they figure several times in the same formula. By $c(\cdot)$ we denote positive numbers depending on the parameters indicated in brackets.

---

[1]) The norms $\|\cdot\|$ refer to the expressions $(\cdot)$ as functions of x.

[2]) For $r = 0$ we set $f^{(0)} = f$.

*) It was published in *Mathem. Zeitschrift* **126** (1972), 123—134. (Note added in proof.)

By $A \sim B$ we denote that A/B is between two positive bounds which depend at most on β. We say that "f is w-orthogonal to g" if $\int\limits_{-\infty}^{\infty} f(t)g(t)W(t)\,dt = 0$.

Note added in proof (30. August 1972). We proved after our lecture the following more general theorem: Let $\alpha > 0$, k a natural integer,

$$W_{\alpha,k}(x) = (1 + x^{2k})^{\alpha} \exp\{-\tfrac{1}{2} x^{2k}\}$$

and

$$\omega(L_p; W_{\alpha,k}; f, h) \sup_{|t| \leq h} \|W_{\alpha,k}(x+t)f(x+t) - W_{\alpha,k}(x)f(x)\|_p + \|\tau(h^{\frac{1}{2k-1}}x)W_{\alpha,\kappa}(x)f(x)\|_p.$$

Then we have

$$\varepsilon_n^{(p)}(W_{\alpha,k}; f) \leq d_1 e^{d_2 r} n^{-r\left(1 - \frac{1}{2k}\right)}.$$

Also a corresponding Bernstein type converse of this theorem is valid. We are returning to these results elsewhere.

2. Lemmata on orthogonal polynomials

We denote by $p_n(w; x)$ the n-th orthogonal polynomial with respect to the weight $w(x)$. Let $v_\beta(x) = |x|^{2\beta} e^{-x^2}$ so that

$$(2.1) \qquad W_\beta^2(x) \sim v_0(x) + v_\beta(x) \qquad (\beta \geq 0).$$

Let

$$u_\alpha(x) = x^\alpha e^{-x} \qquad (x \geq 0), \quad u_\alpha(x) = 0 \qquad (x < 0).$$

We have

$$(2.2) \qquad p_n(u_\alpha; x) = (-1)^n \left[\Gamma(\alpha+1)\binom{n+\alpha}{n}\right]^{-1/2} L_n^\alpha(x) \qquad (\alpha > -1),$$

where L_n^α denotes the n-th Laguerre polynomial (see G. Szegő [30]), further

$$(2.3) \qquad p_{2n}(v_\beta; x) = p_n(u_{\beta-1/2}; x^2), \quad p_{2n+1}(v_\beta; x) = x p_n(u_{\beta+1/2}; x^2)$$

(compare e.g. [13] problems 1. 13 and 1. 14).

LEMMA 2. 1. *The leading coefficients* $\gamma_n(W_\beta^2)$ *of* $p_n(W_\beta^2; x)$ *satisfy*

$$(2.4) \qquad 0 < \frac{\gamma_{n-1}(W_\beta^2)}{\gamma_n(W_\beta^2)} < cn^{1/2} \qquad (n = 1, 2, \ldots).$$

PROOF. We have

$$(2.5) \qquad p_{2n+1}(W_\beta^2; x) = x p_n(U_\beta; x^2)$$

where

$$U_\beta(x) = x^{1/2}(1+x)^\beta e^{-x} \qquad (x \geq 0), \quad U_\beta(x) = 0 \qquad (x < 0).$$

The quotient $u_{\beta+1/2}(x)/U_\beta(x) = (x/(1+x))^\beta$ is a nondecreasing function of x for $\beta \geqq 0$, so that by a theorem of A. A. MARKOV [23] (see also G. SZEGŐ [30], Theorem 6. 12. 2) the greatest zeros of the corresponding orthogonal polynomials $p_n(U_\beta)$ resp. $p_n(u_{\beta+1/2})$ satisfy

$$(2.6) \qquad x_{1n}(u_\beta) \leqq x_{1n}(u_{\beta+1/2}) < 4n + c$$

(for the second part of the inequality see e.g. G. SZEGŐ, [30] Theorem 6. 31. 2). From (2. 5) and (2. 6) we see that

$$(2.7) \qquad x_{1n}(W_\beta^2) < \sqrt{2n + c}$$

is valid for odd n. Since $\{x_{1n}(W_\beta^2)\}$ is an increasing sequence, (2. 7) is valid also for even n. Since U_β^2 is an even function, we have for the smallest zero of $p_n(U_\beta^2)$

$$(2.8) \qquad x_{nn}(W_\beta^2) = -x_{1n}(W_\beta^2).$$

Now (2. 4) is obtained from (2. 7), (2. 8) and the well known inequality

$$0 < \frac{\gamma_{n-1}(w)}{\gamma_n(w)} \leqq \frac{1}{2}[x_{1n}(w) - x_{nn}(w)]$$

(see [13], Problem 1. 10).

We turn now to estimates for the Christoffel functions

$$\lambda_n(W_\beta^2; \xi) = \left\{ \sum_{k=0}^{n-1} {}' \, p_k^2(W_\beta^2; \xi) \right\}^{-1}$$

(see [13]).

LEMMA 2. 2. *We have for* $\beta \geqq 0$, $n = 1, 2, \ldots$

$$(2.9) \qquad \lambda_n(W_\beta^2; \xi) \leqq cn^{-1/2} W_\beta^2(\xi) \qquad (n^{-1/2} \leqq |\xi| \leqq n^{1/2}).$$

PROOF. By the minimum property of Christoffel functions

$$(2.10) \qquad \lambda_n(w; \xi) = \min_{\substack{q \in \mathscr{P}_{n-1} \\ q(\xi)=1}} \int_{-\infty}^{\infty} q^2(x) w(x)\, dx$$

and applying the transformation $t = x^2$ we get

$$(2.11) \qquad \lambda_{2n}(W_\beta^2; \xi) \leqq \min_{\substack{q \in \mathscr{P}_{n-1} \\ \xi q(\xi^2)=1}} \int_{-\infty}^{\infty} [xq(x^2)]^2 W_\beta^2(x)\, dx =$$

$$= \min_{\substack{q \in \mathscr{P}_{n-1} \\ \xi q(\xi^2)=1}} \int_{0}^{\infty} q^2(t) U_\beta(t)\, dt = \xi^{-2} \lambda_n(U_\beta; \xi^2).$$

Now by Theorem 4. 2 of our paper [17][3])

$$(2.12) \qquad \lambda_n(U_\beta; \eta) \leqq cU_\beta(\eta)\sqrt{\frac{\eta}{n}} \qquad \left(\frac{1}{3n} \leqq \eta \leqq 3n\right).$$

By (2. 11) and (2. 12) we see that (2. 9) holds for even n and $(\frac{3}{2}n)^{-1/2} \leqq \xi \leqq \leqq (\frac{3}{2}n)^{1/2}$. To end the proof of Lemma 2. 2 it is enough to observe that

$$\lambda_{2n+1}(W_\beta^2; \xi) \leqq \lambda_{2n}(W_\beta^2; \xi).$$

LEMMA 2. 3. *We have for every natural n and* $-\infty < x < \infty$

$$(2.13) \qquad \sum_{v=0}^{n-1} p_v^2(u_\alpha; x) \leqq \frac{c(\alpha)n^{\alpha+1}}{1+(nx)^{\alpha+1/2}} e^{x-x/4n}.$$

PROOF. The Mehler generator series of $L_n^\alpha(x)L_n^\alpha(y)$ (see formula 5. 1. 15 of G. SZEGŐ [30]) furnishes by (2. 2)

$$(2.14) \quad \sum_{v=0}^{\infty} p_v^2(u_\alpha; x)z^v = (1-z)^{-1}\exp\left\{-\frac{2xz}{1-z}\right\}(-x^2z)^{-\alpha/2}J_\alpha\left\{\frac{2ix\sqrt{z}}{1-z}\right\} \qquad (|z|<1).$$

Inserting $z = 1-(1/n)$, $t_n = 2n\sqrt{1-(1/n)}\,x$ in (2. 14) we obtain

$$(2.15) \qquad \sum_{v=0}^{n-1} p_v^2(u_\alpha; x) < \left(1-\frac{1}{n}\right)^{-n} \sum_{v=0}^{\infty} p_v^2(u_\alpha; x)\left(1-\frac{1}{n}\right)^v <$$

$$< c(\alpha)n^{\alpha+1}e^{-2(n-1)x}(it_n)^{-\alpha}J_\alpha(it_n).$$

The Bessel function $J_\alpha(it)$ is asymptotically $c(\alpha)(it)^\alpha$ for $t \to 0$ and asymptotically $c(\alpha)t^{-1/2}e^t$ for $t \to \infty$, consequently

$$(2.16) \qquad (it_n)^{-\alpha}J_\alpha(it_n) < \frac{c(\alpha)}{1+t_n^{\alpha+1/2}}e^{t_n} < \frac{c(\alpha)}{1+(nx)^{\alpha+1/2}}e^{t_n}.$$

We arrive at (2. 13) by combining (2. 15), (2. 16) and the elementary inequality

$$-2(n-1)x+t_n = \left(-2n+2+2n\sqrt{1-\frac{1}{n}}\right)x < \left(1-\frac{1}{4n}\right)x.$$

LEMMA 2. 4. *We have for $\beta \geqq 0$*

$$(2.17) \qquad \lambda_n^{-1}(W_\beta^2; \xi) = \sum_{k=0}^{n-1} p_k^2(W_\beta^2; \xi) \leqq cn^{1/2}W_\beta^{-2}(\xi)e^{-\xi^2/4n}.$$

---

[3]) The conditions of this theorem are satisfied since $e^x U_\beta(x)$ is increasing and $x^{-1-\beta}e^x U_\beta(x)$ is decreasing.

PROOF. By (2. 1) and (2. 10) we have (see Theorem I. 4. 2 in [13])

$$(2.18) \qquad \lambda_n(W_\beta^2; \xi) \geqq \max\,[\lambda_n(v_0; \xi),\, \lambda_n(v_\beta; \xi)].$$

Combining (2. 3) and (2. 13) we get

$$(2.19) \qquad \lambda_n^{-1}(v_\beta; \xi) = \sum_{k=0}^{n-1} p_k^2(v_\beta; \xi) \leqq$$

$$\leqq \sum_{v=0}^{n-1} p_v^2(u_{\beta-1/2}; \xi^2) + \xi^2 \sum_{v=0}^{n-1} p_v^2(u_{\beta+1/2}; \xi^2) \leqq c n^{1/2} |\xi|^{-2\beta} e^{\xi^2 - \xi^2/4n}.$$

From (2. 18) and (2. 19) (applied also to $\beta=0$!) we arrive at (2. 17).

Let $W_\beta f \in L_\infty$. Since $W_\beta p_v(W_\beta^2) \in L_1$ $(v=0, 1, \ldots)$ hence the coefficients a_v of the following orthogonal expansion exist:

$$(2.20) \qquad f(x) \approx \sum a_v(W_\beta^2; f) p_v(W_\beta^2; x).$$

The partial sum of degree v of (2. 20) is denoted by $s_v(W_\beta^2; f; x)$.

We are going to give an estimation of the strong $(C, 1)$ sums of the orthogonal series (2. 20). This estimate is based on an idea of T. CARLEMAN [5], who developed it for the trigonometric Fourier series. It was first applied to orthogonal polynomial series by K. TANDORI [32], [33]. Tandori's analysis was modified by the author [11], [12] (see also § IV. 3 and § IV. 4 in [13]). In the proof of the next theorem we apply the idea in this modified form:

THEOREM 2. 5. *We have*

$$(2.21) \qquad \left\| W_\beta(x) \frac{1}{n} \sum_{v=0}^{n-1} |s_v(W_\beta^2; f; x)| \right\|_\infty \leqq c e^{-x^2/8n} \|W_\beta f\|_\infty.$$

Apart from the exponential factor, the special case $\beta=0$ of (2. 21) was given by G. FREUD—S. KNAPOVSKI [19].

PROOF. We have by the Christoffel—Darboux formula

$$(2.22) \qquad s_v(W_\beta^2; f; x) = \int_{-\infty}^{\infty} K_{v+1}(W_\beta^2; x; t) f(t) W_\beta^2(t)\, dt$$

$$(2.23) \qquad K_{v+1}(W_\beta^2; x; t) = \sum_{k=0}^{v} p_k(W_\beta^2; x) p_k(W_\beta^2; t) =$$

$$= \frac{\gamma_v(W_\beta^2)}{\gamma_{v+1}(W_\beta^2)} \, \frac{p_v(W_\beta^2; t) p_{v+1}(W_\beta^2; x) - p_{v+1}(W_\beta^2; t) p_v(W_\beta^2; x)}{x-t}.$$

Let $I_n(x) = [x - n^{-1/2}, x + n^{-1/2}]$, $J_n(x) = (-\infty, \infty) \setminus I_n(x)$, $f_n(t) = f(t)$ for $t \in I_n(x)$ and $f_n(t) = 0$ for $t \in J_n(x)$ resp. $f_n^*(t) = 0$ for $t \in I_n(x)$ and $f_n^*(t) = f(t)$ for $t \in J_n(x)$. Clearly $f = f_n + f_n^*$ and

$$(2.24) \qquad s_n(W_\beta^2; f; x) = s_n(W_\beta^2; f_n; x) + s_n(W_\beta^2; f_n^*; x).$$

We have for $v \leq n - 1$

$$|s_v(W_\beta^2; f_n; x)|^2 \leq \int_{-\infty}^{\infty} f_n^2(t) W_\beta^2(t)\, dt \cdot \int_{-\infty}^{\infty} K_{v+1}^2(W_\beta^2; x, t) W_\beta^2(t)\, dt =$$

$$= \int_{x-n^{-1/2}}^{x+n^{-1/2}} [f(t) W_\beta(t)]^2\, dt \cdot \sum_{k=0}^{v} p_k^2(W_\beta^2; x) \leq 2n^{-1/2} \|W_\beta f\|_\infty^2 \sum_{k=0}^{n-1} p_k^2(W_\beta^2; x),$$

and by Lemma 2.4

$$(2.25) \qquad |s_v(W_\beta^2; f_n; x)| \leq c \|W_\beta f\|_\infty W_\beta^{-1}(x) e^{-x^2/8n}.$$

Let $F_n(t) = (x - t)^{-1} f_n^*(t)$. By (2.22) and (2.23) we have

$$(2.26)$$

$$s_v(W_\beta^2; f_n^*; x) = \frac{\gamma_v(W_\beta^2)}{\gamma_{v+1}(W_\beta^2)} [p_{v+1}(W_\beta^2; x) a_v(W_\beta^2; F_n) - p_v(W_\beta^2; x) a_{v+1}(W_\beta^2; F_n)].$$

The Fourier coefficients $a_v(W_\beta^2; F_n)$ of the orthogonal expansion

$$F_n(x) \approx \sum a_v(W_\beta^2; F_n) p_v(W_\beta^2; x)$$

satisfy Bessel's inequality, so that

$$(2.27) \quad \sum_{v=0}^{n} a_v^2(W_\beta^2; F_n) \leq \int_{-\infty}^{\infty} F_n^2(t) W_\beta^2(t)\, dt = \int_{J_n(x)} \frac{[f(t) W_\beta(t)]^2}{(x-t)^2}\, dt \leq 2n^{1/2} \|W_\beta f\|_\infty^2.$$

By (2.26) and (2.27) we obtain, applying Lemma 2.1 and Lemma 2.4

$$(2.28) \qquad \left\{ \sum_{v=0}^{n-1} |s_v(W_\beta^2; f_n^*; x)| \right\}^2 \leq cn \sum_{v=0}^{n} p_v^2(W_\beta^2; x) \sum_{v=0}^{n} a_v^2(W_\beta^2; F_n) \leq$$

$$\leq cn^2 \|W_\beta f\|_\infty^2 W_\beta^{-2}(x) e^{-x^2/4n}.$$

By (2.25) and (2.28), (2.21) is proved.

Let us build the de La Vallée—Poussin type means

$$(2.29) \qquad v_n(W_\beta^2; f; x) = \frac{1}{n - [n/2]} \sum_{v=[n/2]}^{n-1} s_v(W_\beta^2; f; x).$$

LEMMA 2. 6. *The de La Vallée—Poussin means enjoy the following properties:*

a) $v_n(W_\beta^2; q) = q$ *for every* $q \in \mathscr{P}_{[n/2]}$;
b) $f - v_n(W_\beta^2; f)$ *is* W_β^2-*orthogonal to* $\mathscr{P}_{[n/2]}$ *for every* f *with* $W_\beta f \in L_\infty$;
c) *if* $W_\beta f \in L_\infty$, *then*

$$(2.30) \qquad \|W_\beta v_n(W_\beta^2; f)\|_\infty \leq c \|W_\beta f\|_\infty;$$

d) *if* $W_\beta f \in L_1$, *then*

$$(2.31) \qquad \|W_\beta v_n(W_\beta^2; f)\|_1 \leq c \|W_\beta f\|_1.$$

PROOF. Property a) is a consequence of the fact that $s_\nu(W_\beta^2; q) = q$ for every $q \in \mathscr{P}_\nu$, and property b) of the fact that $f - s_\nu(W_\beta^2; f)$ is W_β^2-orthogonal to $\mathscr{P}_\nu$. Property c) follows from Theorem 2. 5. It remained to prove property d). We refer to the relation, valid under the condition $Wf \in L_1$, $Wg \in L_\infty$,

$$(2.32) \qquad \int_{-\infty}^{\infty} g(t) s_\nu(W^2; f; t) W^2(t)\, dt = \sum_{k=0}^{\nu} a_k(W^2; f) a_k(W^2; g) =$$

$$= \int_{-\infty}^{\infty} f(t) s_\nu(W^2; g; t) W^2(t)\, dt,$$

and consequently

$$\int_{-\infty}^{\infty} g(t) v_n(W_\beta^2; f; t) W_\beta^2(t)\, dt = \int_{-\infty}^{\infty} f(t) v_n(W_\beta^2; g; t) W_\beta^2(t)\, dt.$$

Making use of property c) we obtain

$$\|W_\beta v_n(W_\beta^2; f)\|_1 = \sup_{\|W_\beta g\|_\infty \leq 1} \int_{-\infty}^{\infty} g(t) v_n(W_\beta^2; f; t) W_\beta^2(t)\, dt =$$

$$= \sup_{\|W_\beta g\|_\infty \leq 1} \int_{-\infty}^{\infty} f(t) v_n(W_\beta^2; g; t) W_\beta^2(t)\, dt \leq c \int_{-\infty}^{\infty} |f(t)| W_\beta(t)\, dt = c \|W_\beta f\|_1.$$

All the four stated properties of $v_n(W_\beta^2; f)$ are proved.

3. The "direct" part of the approximation theorems

Let $-\infty < \xi < \infty$ and

$$(3.1) \qquad \Gamma_\xi(t) = \begin{cases} 0 & (t < \xi) \\ 1 & (t \geq \xi). \end{cases}$$

LEMMA 3. 1. *We have for every real* ξ *and* $\beta \geq 0$

$$(3.2) \qquad \varepsilon_n^{(1)}(W_\beta; \Gamma_\xi) \leq c n^{-1/2} W_\beta(\xi) \qquad (n = 1, 2, \ldots).$$

PROOF. Since $W_\beta(\sqrt{2}\,x) \sim W_{\beta/2}^2(x)$ get by the transformation $t = \sqrt{2}\,x$

$$\varepsilon_n^{(1)}(W_{2\beta}; \Gamma_\xi) \sim \varepsilon_n^{(1)}(W_\beta^2; \Gamma_{\xi/\sqrt{2}}).$$

Consequently (3. 2) (for β replaced by 2β, ξ replaced by $\sqrt{2}\xi$) is equivalent to

$$(3.\ 3) \qquad \varepsilon_n^{(1)}(W_\beta^2; \Gamma_\xi) < cn^{-1/2} W_\beta^2(\xi).$$

It is enough to prove (3. 3) for even n.

For $\xi > \sqrt{n}$ we approximate Γ_ξ by $q_0(x) \equiv 0$ and for $\xi < -\sqrt{n}$ by $q_1(x) \equiv 1$. In both cases we get

$$\int_{-\infty}^{\infty} |\Gamma_\xi(x) - q_i(x)| W_\beta^2(x)\, dx \leqq \int_{|\xi|}^{\infty} W_\beta^2(x)\, dx \leqq c|\xi|^{-1/2} W_\beta^2(\xi) \leqq$$

$$\leqq cn^{-1/2} W_\beta^2(\xi) \qquad (i=0, 1).$$

For $n^{-1/2} \leqq |\xi| \leqq n^{1/2}$ we approximate $\Gamma_\xi(x)$ by one of the Markov—Stieltjes polynomials $\varphi_n(W_\beta^2; x; \xi)$ or $\Phi_n(W_\beta^2; x; \xi)$, (see [13] § I. 5), which satisfy $\varphi_n \in \mathscr{P}_{2n-2}$, $\Phi_n \in \mathscr{P}_{2n-2}$

$$\varphi_n(W_\beta^2; x; \xi) \leqq \Gamma_\xi(x) \leqq \Phi_n(W_\beta^2; x; \xi) \qquad (-\infty < x < \infty)$$

and

$$\int_{-\infty}^{\infty} [\Phi_n(W_\beta^2; x; \xi) - \varphi_n(W_\beta^2; x; \xi)] W_\beta^2(x)\, dx = \lambda_n(W_\beta^2; \xi).$$

For $n^{-1/2} \leqq |\xi| \leqq n^{1/2}$ we have by Lemma 2. 2

$$\varepsilon_{2n}^{(1)}(W_\beta; \Gamma_\xi) \leqq \int_{-\infty}^{\infty} [\Gamma_\xi(x) - \varphi_n(W_\beta^2; x; \xi)] W_\beta^2(x)\, dx \leqq$$

$$\leqq \int_{-\infty}^{\infty} [\Phi_n(W_\beta^2; x; \xi) - \varphi_n(W_\beta^2; x; \xi)] W_\beta^2(x)\, dx = \lambda_n(W_\beta^2; \xi) < cn^{-1/2} W_\beta^2(\xi).$$

Finally take $|\xi| \leqq n^{-1/2}$, then

$$\varphi_n(W_\beta^2; x; n^{-1/2}) \leqq \Gamma_{n^{-1/2}}(x) \leqq \Gamma_\xi(x) \leqq \Gamma_{-n^{-1/2}}(x),$$

and again by Lemma 2. 2

$$\varepsilon_{2n}^{(1)}(W_\beta^2; \Gamma_\xi) \leqq \int_{-\infty}^{\infty} [\Gamma_\xi(x) - \varphi_n(W_\beta^2; x; n^{-1/2})] W_\beta^2(x)\, dx \leqq$$

$$\leqq \int_{-\infty}^{\infty} [\Gamma_\xi(x) - \Gamma_{n^{-1/2}}(x)] W_\beta^2(x)\, dx + \int_{-\infty}^{\infty} [\Gamma_{n^{-1/2}}(x) - \varphi_n(W_\beta^2; x, n^{-1/2})] W_\beta^2(x)\, dx \leqq$$

$$\leqq \int_{-n^{-1/2}}^{n^{-1/2}} W_\beta^2(x)\, dx + \lambda_n(W_\beta^2; n^{-1/2}) < cn^{-1/2} < cn^{-1/2} W_\beta^2(\xi) \qquad (|\xi| \leqq n^{-1/2});$$

this completes our proof.

LEMMA 3. 2. (S. M. Nikol'skiĭ [26].) *We have for every nonnegative weight $W \in L$ satisfying* (1. 1)

$$(3. 4) \qquad \varepsilon_n^{(1)}(W; f) = \sup_{\substack{\|Wh\|_\infty \leq 1 \\ h \in R_n}} \int_{-\infty}^{\infty} h(t)f(t)W^2(t)\, dt$$

where R_n is composed of the functions h with $Wh \in L_\infty$, which are W^2-orthogonal to $\mathscr{P}_n$.

PROOF.[4]) We denote the "sup" expression in (3. 4) by $\delta_n(W; f)$. We have for every $h \in R_n$ and $q \in \mathscr{P}_n$

$$\int_{-\infty}^{\infty} h(t)f(t)W^2(t)\, dt = \int_{-\infty}^{\infty} h(t)[f(t) - q(t)]W^2(t)\, dt \leq$$

$$\leq \|Wh\|_\infty \int_{-\infty}^{\infty} |f(t) - q(t)|\, W(t)\, dt$$

so that

$$(3. 5) \qquad \delta_n(W; f) \leq \varepsilon_n^{(1)}(W; f).$$

By $L_p(W)$ we denote the Banach space of functions with norm $\|Wf\|_p$.

$\delta_n(W; f)$ is the norm of the functional

$$A(g) = \int_{-\infty}^{\infty} g(t)f(t)W^2(t)\, dt$$

as defined for $g \in R_n \subset L_\infty(W)$. By the Hahn—Banach theorem $A(g)$ has an extension $\bar{A}(g)$ to the whole of $L_\infty(W)$ with $\|\bar{A}\| = \delta_n(W; f)$. Since by (1. 1) $\mathscr{P}_n \subset L_\infty(W)$, for every $g \in L_\infty(W)$ we have $g - s_n(W^2; g) \in R_n$.

$$\bar{A}(g) = \bar{A}[g - s_n(W^2; g)] + \bar{A}[s_n(W^2; g)] =$$

$$= A[g - s_n(W^2; g)] + \sum_{v=0}^{n} \bar{A}[p_v(W^2)]a_v(W^2; g) =$$

$$= \int_{-\infty}^{\infty} [g(t) - s_n(W^2; g; t)]f(t)W^2(t)\, dt + \int_{-\infty}^{\infty} g(t)\sum_{v=0}^{n} \bar{A}[p_v(W^2)]p_v(W^2; t)W^2(t)\, dt.$$

---

[4]) For an alternative proof see R. C. BUCK [4].

Considering (2. 32) we see that for a certain $P \in \mathscr{P}_n$ we have

$$\bar{A}(g) = \int_{-\infty}^{\infty} g(t)[f(t) - P(t)]W^2(t)\, dt$$

i.e.

$$(3.\,6) \qquad \delta_n(W; f) = \|\bar{A}\| = \int_{-\infty}^{\infty} |f(t) - P(t)| W(t)\, dt.$$

Now (3. 4) is obtained from (3. 5) and (3. 6).

THEOREM 3. 3. *Let* $n \geqq 1$, $g \in L$, $|g(x)| \leqq W_\beta(x)$ $(-\infty < x < \infty)$ *and*

$$(3.\,7) \qquad \int_{-\infty}^{\infty} g(t)P(t)\, dt = 0 \quad \text{for every} \quad P \in \mathscr{P}_n;$$

then we have

$$(3.\,8) \qquad \left| \int_{x}^{\infty} g(t)\, dt \right| = \left| \int_{-\infty}^{x} g(t)\, dt \right| \leqq cn^{-1/2} W_\beta(x).$$

This Theorem is an analogue of H. Bohr's well known inequality; see H. BOHR [3], J. FAVARD [10], B. SZ.-NAGY—A. STRAUSZ [31], L. HÖRMANDER [21], C. G. ESSEEN [9]. The author recently proposed a systematic study of polynomial, weighted Bohr-type inequalities. A first result (corresponding to $\beta = 0$ of Theorem 3. 3) was given by the author in [15]. For finite interval of integration the problem was investigated recently by B. AURELL [2], G. FREUD—J. SZABADOS [20] and L. LANDBERG [22].

PROOF. Since $\int_{-\infty}^{\infty} g(t)\, dt = 0$ by assumption, the two integrals in (3. 7) have equal moduli. We have by (3. 7) for every $P \in \mathscr{P}_n$

$$\left| \int_{x}^{\infty} g(t)\, dt \right| = \left| \int_{-\infty}^{\infty} \Gamma_x(t)g(t)\, dt \right| =$$

$$= \left| \int_{-\infty}^{\infty} [\Gamma_x(t) - P(t)]g(t)\, dt \right| \leqq \int_{-\infty}^{\infty} |\Gamma_x(t) - P(t)| W_\beta(t)\, dt,$$

and by Lemma 3. 1

$$\left| \int_{x}^{\infty} g(t)\, dt \right| \leqq \varepsilon_n^{(1)}(W_\beta; \Gamma_x) \leqq cn^{-1/2} W_\beta(x).$$

THEOREM 3. 4. *Let* $F(t)$ *be of bounded variation over every compact interval, then for every* $\beta \geqq 0$ *for which the integral on the right of* (3. 9) *is finite we have*

$$(3.\,9) \qquad \varepsilon_n^{(1)}(W_\beta; F) \leqq cn^{-1/2} \int_{-\infty}^{\infty} W_\beta(t)|dF(t)|.$$

PROOF. By Lemma 3. 2 we need an estimate independent of g of

$$(3.10) \qquad B_n(g) = \int\limits_{-\infty}^{\infty} g(t) F(t)\, dt$$

where $g(t)=h(t)W_\beta^2(t)$ satisfies the conditions a) $|g(x)|\leq W_\beta(x)$ b) for every $P\in\mathscr{P}_n$

$$\int\limits_{-\infty}^{\infty} g(t) P(t)\, dt = 0.$$

Let $G(x) = \int\limits_{x}^{\infty} g(t)\, dt$, then by Theorem 3. 3

$$(3.11) \qquad |G(x)| \leq c n^{-1/2} W_\beta(x).$$

From the finiteness of the integral in (3. 9) we infer $\lim\limits_{|x|\to\infty} W_\beta(x)F(x)=0$, consequently the partial integration

$$(3.12) \qquad B_n(g) = \int\limits_{-\infty}^{\infty} g(t) F(t)\, dt = \int\limits_{-\infty}^{\infty} G(t)\, dF(t)$$

is legitimate. By (3. 11) and (3. 12) we have

$$B_n(g) \leq c n^{-1/2} \int\limits_{-\infty}^{\infty} W_\beta(t)\, |dF(t)|.$$

By Lemma 3. 2 $\sup\limits_{|g(x)|\leq W_\beta(x)} B_n(g)=\varepsilon_n^{(1)}(W_\beta; F)$, this proves (3. 9).

LEMMA 3. 5. *For a $\beta\geq 0$ let $\|W_\beta f\|_\infty<\infty$; then $F(x) = \int\limits_{0}^{x} f(t)\, dt$ satisfies*

$\lim\limits_{|x|\to\infty} W_\beta(x)f(x)=0$ *and*

$$(3.13) \qquad \varepsilon_n^{\infty}(W_\beta; F) \leq c\|W_\beta f\|_\infty n^{-1/2} \qquad (n=1, 2, \ldots).$$

PROOF. Clearly

$$(3.14) \qquad |F(x)| \leq \|W_\beta f\|_\infty \int\limits_{0}^{x} W_\beta^{-1}(t)\, dt < c\|W_\beta f\|_\infty |x|^{-1} W_\beta^{-1}(x).$$

We consider the de La Vallée—Poussin means $v_n(W_\beta^2; f; x)$. By (2. 30)

$$(3.15) \qquad \|W_\beta[f - v_n(W_\beta^2; f)]\|_\infty \leq c\|W_\beta f\|_\infty.$$

Introducing the auxiliary function[5]

$$\Psi_x(t) = \begin{cases} W_\beta^{-2}(t)\, \operatorname{sign} t & (t\in[0, x]) \\ 0 & (t\notin[0, x]) \end{cases}$$

---

[5]) For $x<0$ we mean by $t\in[0, x]$ that $t\in[x, 0]$.

we have by Lemma 2. 6, property b) for every $P \in \mathscr{P}_{[n/2]}$

$$(3.16) \quad F(x) - \int_0^x v_n(W_\beta^2; f; t)\, dt = \int_{-\infty}^\infty [f(t) - v_n(W_\beta^2; f; t)]\, \Psi_x(t) W_\beta^2(t)\, dt =$$

$$= \int_{-\infty}^\infty [f(t) - v_n(W_\beta^2; f; t)]\, [\Psi_x(t) - P(t)] W_\beta^2(t)\, dt.$$

According to (3. 15) and (3. 16)

$$(3.17) \qquad\qquad \left| F(x) - \int_0^x v_n(W_\beta^2; f; t)\, dt \right| \leq$$

$$\leq c \|W_\beta f\|_\infty \inf_{P \in \mathscr{P}_{[n/2]}} \int_{-\infty}^\infty |\Psi_x(t) - P(t)|\, W_\beta(t)\, dt = c \|W_\beta f\|_\infty \varepsilon_n^{(1)}(W_\beta; \Psi_x).$$

Applying Theorem 3. 4

$$W_\beta(x) \left| F(x) - \int_0^x v_n(W_\beta^2; f; t)\, dt \right| \leq$$

$$\leq c \|W_\beta f\|_\infty n^{-1/2} W_\beta(x) \int_{-\infty}^\infty W_\beta(t)\, |d\Psi_x(t)| \leq c \|W_\beta f\|_\infty n^{-1/2}.$$

Since $\int_0^x v_n(W_\beta^2; f; t)\, dt \in \mathscr{P}_n$, (3. 13) is valid.

LEMMA 3. 6. *If $\beta \geq 0$, $W_\beta f \in L_p$ and $F(x) = \int_0^x f(t)\, dt$ then*

$$(3.18) \qquad\qquad \varepsilon_n^{(p)}(W_\beta; F) \leq c \|W_\beta f\|_p n^{-1/2}.$$

PROOF. We proved (3. 18) for $p=1$ in Theorem 3. 4 and for $p=\infty$ in Lemma 3. 5. Having this in mind, we get (see Lemma 2. 6 a) and c))

$$(3.19) \quad \|W_\beta[F - v_n(W_\beta^2; F)]\|_\infty = \inf_{P \in \mathscr{P}_{[n/2]}} \{\|W_\beta(F - P)\|_\infty + \|W_\beta v_n(W_\beta^2; F - P)\|_\infty\} \leq$$

$$\leq c \inf_{P \in \mathscr{P}_{[n/2]}} \|W_\beta(F - P)\|_\infty = c \varepsilon_{[n/2]}^{(\infty)}(W_\beta; F) \leq c \|W_\beta f\|_\infty n^{-1/2},$$

and by parts a) and d) of Lemma 2. 6

$$(3.20) \qquad\qquad \|W_\beta[F - v_n(W_\beta^2; F)]\|_1 \leq c \|W_\beta f\|_1 n^{-1/2}.$$

We consider the linear operator $V_n(\varphi)$ transforming the function $\varphi = W_\beta f$ to

$$(3.21) \qquad\qquad V_n(\varphi) = W_\beta[F - v_n(W_\beta^2; F)].$$

By (3. 19) and (3. 20) V_n has an (L_∞, L_∞)-norm as well as an (L_1, L_1)-norm bounded by $cn^{-1/2}$. By the Riesz—Thorin interpolation theorem there exists a bounded extension of this linear operator mapping L_p to L_p $(1 < p < \infty)$. The right side of 3. 21

is an L_p-continuous function of φ. Since (3. 21) holds for $\varphi \in L_\infty \cap L_p$ and $L_\infty \cap L_p$ is dense in L_p, this extension is represented by (3. 21) for every $1 < p < \infty$.

By the Riesz—Thorin interpolation theorem the (L_p, L_p)-norm of V_n is less than $cn^{-1/2}$, i.e.

$$(3.22) \qquad \|W_\beta[F - v_n(W_\beta^2; F)]\|_p \leqq c \|W_\beta f\|_p n^{-1/2}.$$

Since $v_n(W_\beta^2; F) \in \mathscr{P}_n$, (3. 18) is proved.

THEOREM 3. 7. *Let $f(x)$ be for a fixed $\beta \geqq 0$ and $1 \leqq p \leqq \infty$ the r-times iterated integral of an $f^{(r)}$ with $W_\beta f^{(r)} \in L_p$, then*

$$(3.23) \qquad \varepsilon_{n+r}^{(p)}(W_\beta; f) \leqq e^{cr} n^{-r/2} \varepsilon_n^{(p)}(W_\beta; f^{(r)}) \qquad (n = 1, 2, \ldots).$$

PROOF. We select a $P \in \mathscr{P}_{n+1}$ for which $\|W_\beta(f' - P')\| \leqq 2\varepsilon_n^{(p)}(W_\beta; f')$. By Lemma 3. 6

$$\varepsilon_{n+1}^{(p)}(W_\beta; f) = \varepsilon_{n+1}^{(p)}[W_\beta; f - P - f(0) + P(0)] \leqq c \|W_\beta(f' - P')\|_p n^{-1/2} \leqq$$

$$\leqq e^c n^{-1/2} \varepsilon_n^{(p)}(W_\beta; f')$$

i.e. (3. 23) is valid for $r = 1$. The case $r > 1$ is reduced by induction to the case $r = 1$.

Let us fix $\beta \geqq 0$ and $1 \leqq p \leqq \infty$, and let $W_\beta f \in L_p$. Let n be a natural integer, $h = n^{-1/2}$

$$(3.24) \qquad g_n(x) = \begin{cases} W_\beta(x)f(x) & (|x| \leqq n^{1/2}) \\ 0 & (|x| > n^{1/2}) \end{cases}$$

$$(3.25) \qquad \varphi_n(x) = W_\beta^{-1}(x) \frac{1}{h} \int_x^{x+h} g_n(t)\, dt$$

$$(3.26) \qquad \Phi_n(x) = W_\beta^{-1}(x) \frac{1}{h^2} \int_{x-h}^{x+h} (h - |t - x|) g_n(t)\, dt.$$

LEMMA 3. 8. *We have*

$$(3.27) \qquad \|W_\beta(f - \varphi_n)\|_p \leqq \omega(p; W_\beta; f; n^{-1/2})$$

$$(3.28) \qquad \|W_\beta \varphi_n'\|_p \leqq c\omega(p; W_\beta; f; n^{-1/2}) n^{1/2}$$

$$(3.29) \qquad \|W_\beta(f - \Phi_n)\|_p \leqq c\omega_2(p; W_\beta; f; n^{-1/2})$$

$$(3.30) \qquad \|W_\beta \Phi_n''\|_p \leqq c\omega_2(p; W_\beta; f; n^{-1/2}) n.$$

(See (1. 7), (1. 8), (1. 9).)

PROOF. We use repeatedly Minkovski's inequality.

$$W_\beta(x)[f(x) - \varphi_n(x)] =$$

$$-\frac{1}{h} \int_x^{x+h} [W_\beta(x+t)f(x+t) - W_\beta(x)f(x)]\, dt + \frac{1}{h} \int_x^{x+h} [W_\beta(x+t)f(x+t) - g_n(x+t)]\, dt.$$

Taking (3. 24) and (1. 7) in consideration, we have by (1. 8) and $h=n^{-1/2}$

$$\|W_\beta(f-\varphi_n)\|_p \leq \sup_{0\leq t\leq h} \|W_\beta(x+t)f(x+t) - W_\beta(x)f(x)\|_p +$$

$$+ \sup_{0\leq t\leq h} \|\tau(hx)W_\beta(x)f(x)\|_p \leq \omega(p;W_\beta;f;n^{-1/2}).$$

Hence (3. 27) is proved. From (3. 25) and (1. 5) we obtain by differentiation

$$(3.31)\quad W_\beta(x)\varphi_n'(x) = h^{-1}[g_n(x+h)-g_n(x)] + \left(1+\frac{\beta}{1+x^2}\right)xh^{-1}\int_x^{x+h} g_n(t)\,dt =$$

$$= h^{-1}[g_n(x+h)-g_n(x)] + \left(1+\frac{\beta}{1+x^2}\right)x\left\{h^{-1}\int_0^h [g_n(x+t)-g_n(x)]\,dt + g_n(x)\right\}.$$

Since $g_n(x)=0$ for $|x|>n^{1/2}=h^{-1}$, we get by (3. 24)

$$(3.32)\qquad\qquad \sup_{0\leq t\leq h} \|g_n(x+t)-g_n(x)\|_p \leq \omega(p;W_\beta;f;h)$$

and from (3. 31)

$$h\|W_\beta\varphi_n'\|_p \leq$$

$$\leq \|g_n(x+h)-g_n(x)\|_p + (\beta+1)\|\tau(hx)W_\beta(x)f(x)\|_p + \sup_{0\leq t\leq h} \|g_n(x+t)-g_n(x)\|_p \leq$$

$$\leq (3+\beta)\omega(p;W_\beta;f,h),$$

which proves (3. 28). The proof of (3. 29) and (3. 30) is similar.

THEOREM 3. 9. *Let f be the r-times iterated integral of a function $f^{(r)}$ satisfying $W_\beta f^{(r)}\in L_p$ for a $\beta\geq 0$ and a $1\leq p\leq\infty$, then*

$$(3.33)\qquad\qquad \varepsilon_{n+r}^{(p)}(W_\beta;f) \leq e^{c(r+1)}n^{-r/2}\omega(p;W_\beta;f^{(r)},n^{-1/2})$$

and

$$(3.34)\qquad\qquad \varepsilon_{n+r}^{(p)}(W_\beta:f) \leq e^{c(r+1)}n^{-r/2}\omega_2(p;W_\beta;f^{(r)},n^{-1/2}).$$

PROOF. By Theorem 3. 7 it is sufficient to consider the case $r=0$. We have by (3. 27)

$$\varepsilon_n^{(p)}(W_\beta;f) \leq \varepsilon_n^{(p)}(W_\beta;\varphi_n) + \omega(p;W_\beta;f,n^{-1/2}).$$

We infer from (3. 28) and Lemma 3. 6

$$\varepsilon_n^{(p)}(W_\beta;\varphi_n) \leq cn^{-1/2}\|W_\beta;\varphi_n'\|_p \leq c\omega(p;W_\beta;f,n^{-1/2}),$$

and this proves (3. 33). (3. 34) is obtained in a similar way from (3. 29) and (3. 30).

REFERENCES

[1] S. N. Achiezer, *On weighted approximation of continuous functions by polynomials on the whole real axis.* Uspehi Mat. Nauk **11** (1956), 3—43.

[2] B. Aurell, *A variant of Bohr's inequality.* Master's thesis, University of Gothenburgh 1970.

[3] Bohr, *Ein allgemeiner Satz über die Integration eines trigonometrischen Polynoms.* Collected Works, vol. II, Kobenhavn 1952, 273—288.

[4] R. C. Buck, *Applications of duality in approximation theory.* Proc. of the symposium on *Approximation of Functions*, edited by H. L. Garabedian, Elsevier Publ. Co. Amsterdam—London—New York 1965, 27—42.

[5] T. Carleman, *A theorem concerning Fourier series.* Proc. London Math. Soc. **21** (1923), 483—492.

[6] A. S. Dzafarov, *On the weighted-bestpossible approximation of functions of several variables by polynomials* (Russian). Trudy Inst. Phys. Mathem. Akad. Nauk. Auserbeidsan SSR ser. matem. **8** (1959), 117—133.

[7] M. M. Dzrbasian, *On weighted best polynomial approximation on the real axis* (Russian). Dokl. Akad. Nauk SSSR **84** (1952), 1123—1126.

[8] M. M. Dzrbasian, *Questions of the theory of weighted polynomial approximation in complex regions.* Mat. Sb., **36** (78) (1955), 353—440.

[9] C.-G. Esseen, *Fourier analysis of distribution functions.* Acta Math. **77** (1945), 1—125.

[10] J. Favard, *Application de la formule sommatoire d'Euler à la demonstration de quelques propriétés des integrales des fonctions périodiques et presque-périodiques.* Matem. Tidsskrift B. 1936, 81—94.

[11] G. Freud, *Über die starke (C, 1) Summierbarkeit von orthogonalen Polynomreihen.* Acta Math. Acad. Sci. Hungar. **3** (1952), 83—88.

[12] G. Freud, *Über die (C, 1)-Summen der Entwicklungen nach orthogonalen Polynomen.* Acta Math. Acad. Sci. Hungar. **14** (1963), 197—208.

[13] G. Freud, *Orthogonale Polynome.* Birkhäuser, Basel 1969. English translation by I. Földes: *Orthogonal Polynomials.* Pergamon Press, London—Toronto—New York 1971.

[14] G. Freud, *On weighted polynomial approximation on the whole real axis.* Acta Math. Acad. Sci. Hungar. **20** (1969), 223—225.

[15] G. Freud, *On weighted approximation by polynomials on the real line* (Russian). Dokl. Akad. Nauk SSSR **191** (1970), 293—294.

[16] G. Freud, *On a Markov-type inequality* (Russian). Dokl. Akad. Nauk SSSR **197** (1971), 790—793.

[17] G. Freud, *On a class of orthogonal polynomials* (Russian). Mat. Zametki **9** (1971), 511—520.

[18] G. Freud, *On polynomial approximation with the weight $e^{-x^2/2}$* (Russian). Dokl. Akad. Nauk SSSR (in print).

[19] G. Freud and S. Knapovski, *On linear processes of approximation*, III. Studia Math. **25** (1965), 373—383.

[20] G. Freud and J. Szabados, *Remark concerning a theorem of Bohr* (Hungarian). Mat. Lapok **201** (1971), 1292—1294.

[21] L. Hörmander, *A new proof and a generalization of an inequality of Bohr.* Math. Scand. **2** (1954), 33—45.

[22] L. Landberg, *Some variants of Bohr's inequality.* Report No 1971—17, Chalmers Inst. of Techn. and The University of Gothenburgh 1971.

[23] A. A. Markov, *Sur les racines de certaines équations*, II. Math. Ann. **27** (1886), 177—182.

[24] S. N. Mergelian, *Weighted approximation with polynomials* (Russian). Uspehi Mat. Nauk **11**, 5 (1956), 107—152.

[25] M. A. Naimark, *Extremal spectral functions of symmetric operators* (Russian). Isv. Akad. Nauk SSSR **11** (1947).

[26] S. M. Nikol'skiĭ, *Approximation of functions in the mean by trigonometric polynomials* (Russian). Isv. Akad. Nauk SSSR ser. mat. **10** (1946), 207—256.

[27] H. Pollard, *Solution of Bernstein's approximation problem*. Proc. Amer. Math. Soc. **4** (1953), 869—875.

[28] H. Pollard, *The Bernstein approximation problem*. Proc. Amer. Math. Soc. **6** (1955), 402—411.

[29] M. Riesz, *Sur le problème et le théorème de Parseval correspondant*. Acta Sci. math. Szeged **1** (1922—-23), 209—225.

[30] G. Szegő, *Orthogonal Polynomials*. 2nd ed. Amer. Math. Soc., New York 1959.

[31] B. Sz.-Nagy and A. Strausz, *On a theorem of Bohr* (Hungarian). MTA Matem. és Term. Tud. Értesítője **57** (1938), 121—135.

[32] K. Tandori, *Über die Cesàrosche Summierbarkeit der orthogonalen Polynomreihen*, I. Acta Math. Acad. Sci. Hungar. **3** (1952), 73—82.

[33] K. Tandori, *Über die Cesàrosche Summierbarkeit der orthogonalen Polynomreihen*, II. Acta Math. Acad. Sci. Hungar. **5** (1954), 236—253.

Zur Konvergenz der Stufenpolynome
über den Nullstellen der Legendre-Polynome

Von

A. SCHÖNHAGE

FACHBEREICH MATHEMATIK
UNIVERSITÄT TÜBINGEN

Vorbemerkung: Im folgenden wird ein von G. Freud während der Tagung gestelltes Problem behandelt, das an eine Arbeit von FEJÉR [1] aus dem Jahre 1916 anschließt.

Für $f \in C[-1, 1]$ und $n \geq 1$ bezeichne $S_n f$ das über den Nullstellen $x_1 < x_2 < \cdots$ $\cdots < x_n$ des Legendre-Polynoms P_n durch die Bedingungen

$$(1) \qquad (S_n f)(x_k) = f(x_k), \quad (S_n f)'(x_k) = 0 \qquad (1 \leq k \leq n)$$

bestimmte „Stufenpolynom" vom Grade $\leq 2n - 1$.

Die Frage nach dem Konvergenzbereich des durch die Folge der Polynomoperatoren $S_1, S_2, \ldots$ definierten Interpolationsprozesses beantwortet folgender

SATZ: *Notwendig und hinreichend für die Konvergenz $S_n f \to f$ in $C[-1, 1]$ ist die* **Bedingung**

$$(2) \qquad f(-1) = f(+1) = \tfrac{1}{2} \int_{-1}^{1} f(t)\, dt.$$

Zum Beweis benötigen wir außer den mit

$$(3) \qquad l_k(x) = \frac{P_n(x)}{P_n'(x)\,(x - x_k)}$$

gegebenen Elementen der Lagrange-Interpolation die zugehörigen elementaren Stufenpolynome $h_k = S_n l_k$ mit $h_k(x_j) = \delta_{kj}$ und $h_k'(x_j) = 0$. Damit hat man allgemein

$$(4) \qquad (S_n f)(x) = \sum_{k=1}^{n} f(x_k) h_k(x),$$

was durch die explizite Darstellung (vgl. [2], pp. 401, 402)

$$(5) \qquad h_k(x) = \left(1 - \frac{2x_k}{1 - x_k^2}(x - x_k)\right) l_k^2(x)$$

ergänzt wird. Für konstantes f_0 gilt $S_n f_0 = f_0$, also

$$(6) \qquad \sum_{k=1}^{n} h_k(x) = \sum_{k=1}^{n} \left[1 - \frac{2x_k}{1-x_k^2}(x-x_k) \right] l_k^2(x) = 1.$$

Die Umformung

$$1 - \frac{2x_k}{1-x_k^2}(x-x_k) = \frac{1-x^2+(x-x_k)^2}{1-x_k^2}$$

zeigt $h_k(x) \geqq 0$ für $|x| \leqq 1$, woraus sich

$$\sum_{k=1}^{n} |h_k(x)| = 1 \quad \text{für} \quad |x| \leqq 1$$

und damit die Abbildungsnorm $|S_n| = 1$ ergibt. Außerdem gewinnt man mittels der genaueren Abschätzung

$$1 - \frac{2u}{1-u^2}(x-u) \geqq \sqrt{1-x^2} \quad \text{für} \quad |x| \leqq 1, \ |u| < 1$$

aus (6) auch die später noch anzuwendende Ungleichung

$$(7) \qquad (1-x^2) \sum_{k=1}^{n} l_k^2(x) \leqq \sqrt{1-x^2} \quad \text{für} \quad |x| \leqq 1.$$

Zum Nachweis der Notwendigkeit von (2) benutzen wir die explizite Darstellung der Gauß-Quadratur

$$(8) \qquad \gamma_n(f) = \sum_{k=1}^{n} \frac{2}{1-x_k^2} \cdot \frac{1}{(P_n'(x_k))^2} f(x_k) \quad \text{(vgl. [2], p. 443)}.$$

Speziell für $x = \pm 1$ berechnet man nach (4), (5) und (3)

$$(S_n f)(\pm 1) = \sum_{k=1}^{n} \frac{(\pm 1 - x_k)^2}{1-x_k^2} \left(\frac{P_n(\pm 1)}{P_n'(x_k)(\pm 1 - x_k)} \right)^2 f(x_k) = \frac{1}{2}\gamma_n(f).$$

Weil für jedes $f \in C[-1, 1]$ und $n \to \infty$ bekanntlich $\gamma_n(f) \to \int_{-1}^{1} f(t)\,dt$ gilt, impliziert $S_n f \to f$ demnach die Gültigkeit von (2).

Beim Schluß von (2) auf die Konvergenz $S_n f \to f$ kann normierend

$$(9) \qquad f(-1) = f(+1) = 0 \quad \text{und} \quad \int_{-1}^{1} f(t)\,dt = 0$$

vorausgesetzt werden, weil die S_n konstante Funktionen identisch abbilden. Außerdem genügt es wegen der gleichmäßig beschränkten Abbildungsnormen $|S_n| = 1$, den

Konvergenzbeweis für die in dem mit (9) beschriebenen Unterraum von $C[-1, 1]$ dicht liegenden Polynome zu führen.

Sei also f ein Polynom der Form

$$(10) \qquad f(x) = (1 - x^2)\varphi(x) \quad \text{mit} \quad \int_{-1}^{1} (1 - x^2)\varphi(x)\, dx = 0.$$

Für $2n > \operatorname{grad} f$ (was im weiteren stets vorausgesetzt wird) gilt dann auch $\gamma_n(f) = 0$, mittels (8) also

$$\sum_{k=1}^{n} \frac{\varphi(x_k)}{(P'_n(x_k))^2} = 0.$$

Dies zeigt in Verbindung mit (3), daß in

$$(S_n f)(x) = \sum_{k=1}^{n} (1 - x_k^2)\varphi(x_k)\left(1 - \frac{2x_k}{1 - x_k^2}(x - x_k)\right) l_k^2(x) =$$

$$= (1 - x^2) \sum_{k=1}^{n} \varphi(x_k) l_k^2(x) + \sum_{k=1}^{n} (x - x_k)^2 \varphi(x_k) l_k^2(x)$$

die zweite Summe den Wert 0 hat; so folgt

$$(11) \qquad (S_n f)(x) = (1 - x^2) \sum_{k=1}^{n} \varphi(x_k) l_k^2(x).$$

Nahe den Randpunkten ± 1 verwenden wir die Abschätzungen (vgl. (10), (7) und (11))

$$|f(x)| \leqq |\varphi|(1 - x^2) \quad \text{und} \quad |(S_n f)(x)| \leqq |\varphi|\sqrt{1 - x^2}.$$

Danach gibt es zu vorgegebenem $\varepsilon > 0$ ein $\eta > 0$ ($\eta < 1$) mit

$$(12) \qquad |(S_n f)(x) - f(x)| < \varepsilon \quad \text{für} \quad 1 - \eta < |x| \leqq 1.$$

Für die Diskussion des Bereichs $-(1 - \eta) \leqq x \leqq 1 - \eta$ zeigen wir vorbereitend die Ungleichungen

$$(13) \qquad (1 - x^2) \sum_{|x - x_k| > \delta} l_k^2(x) \leqq \frac{1}{\delta^2 n}$$

$$(14) \qquad (1 - x^2)\left| \sum_{k=1}^{n} l_k^2(x) - 1 \right| \leqq \frac{2}{\delta n} + \frac{2\delta}{\eta - \delta} \qquad \text{für} \quad 0 < \delta < \eta,\ |x| \leqq 1 - \eta.$$

Sie beruhen auf der Ungleichung

$$(1 - x^2)(P_n(x))^2 \leqq 1/n$$

und Formel (8) für konstantes f, nämlich

$$\sum_{k=1}^{n} \frac{1}{1-x_k^2} \cdot \frac{1}{(P_n'(x_k))^2} = 1.$$

So folgt (13) aus

$$\sum_{|x-x_k|>\delta} l_k^2(x) \leqq \frac{1}{\delta^2} \sum_{k=1}^{n} (x-x_k)^2 l_k^2(x) \leqq \frac{1}{\delta^2} \sum_{k=1}^{n} \frac{(P_n(x))^2}{(1-x_k^2)(P_n'(x_k))},$$

und für (14) erhält man mittels (6) und (7)

$$\left| \sum_{k=1}^{n} l_k^2(x) - 1 \right| = \left| \sum_{k=1}^{n} \frac{2x_k(x-x_k)}{1-x_k^2} l_k^2(x) \right| \leqq$$

$$\leqq \sum_{|x-x_k|>\delta} \frac{2(P_n(x))^2}{(1-x_k^2)(P_n'(x_k))^2 |x-x_k|} + \sum_{|x-x_k|\leqq\delta} \frac{2|x-x_k|}{1-x_k^2} l_k^2(x) \leqq$$

$$\leqq \frac{2}{\delta n} \cdot \frac{1}{1-x^2} + \frac{2\delta}{\eta-\delta} \cdot \frac{1}{\sqrt{1-x^2}}.$$

Mit (10), (11), (7), (13) und (14) folgt nunmehr für $|x| \leqq 1-\eta$

$$|(S_n f)(x) - f(x)| = (1-x^2) \left| \sum_{k=1}^{n} \varphi(x_k) l_k^2(x) - \varphi(x) \right| \leqq$$

$$\leqq (1-x^2) \left(\sum_{|x-x_k|\leqq\delta} |\varphi(x_k) - \varphi(x)| l_k^2(x) + \sum_{|x-x_k|>\delta} |\varphi(x_k) - \varphi(x)| l_k^2(x) + \right.$$

$$\left. + |\varphi(x)| \left| \sum_{k=1}^{n} l_k^2(x) - 1 \right| \right) \leqq \omega(\varphi, \delta) \sqrt{1-x^2} + 2|\varphi| \cdot \frac{1}{\delta^2 n} + |\varphi| \left(\frac{2}{\delta n} + \frac{2\delta}{\eta-\delta} \right),$$

und dieser Ausdruck wird $< \varepsilon$, wenn zuerst $\delta > 0$ hinreichend klein und dann n hinreichend groß gewählt wird. Zusammen mit (12) beweist dies die gleichmäßige Konvergenz $|S_n f - f| \to 0$ für $n \to \infty$.

LITERATUR

[1] L. Fejér, *Über Interpolation.* Göttinger Nachrichten (**1916**), 66—91, bzw. Werke Bd. II, 25—48 (1970).
[2] I. P. Natanson, *Konstruktive Funkti onentheorie.* Akademie-Verlag, Berlin 1955.

29*

Chebyshev Semi-Discrete Approximations
for Linear Parabolic Problems*)**)

By

RICHARD S. VARGA

DEPT. OF MATH.
KENT STATE UNIVERSITY
KENT, OHIO

1. Introduction

Consider the solution $u(x, t)$ of the heat equation

(1.1)
$$\begin{cases} u_t(x, t) = u_{xx}(x, t) + r(x), & 0 < x < 1, \ t > 0, \\ u(x, 0) = \tilde{u}(x), & 0 \leq x \leq 1, \\ u(0, t) = u(1, t) = 0, & t > 0. \end{cases}$$

Leaving time continuous, consider the particular spatial discretization of (1.1) brought about by the usual three-point difference approximation to u_{xx}, i.e.,

$$u_{xx}(ih, t) \doteq \frac{u((i+1)h, t) - 2u(ih, t) + u((i-1)h, t)}{h^2} \qquad ((N+1)h = 1).$$

The resulting approximation $w(ih, t)$ to the solution $u(x, t)$ of (1.1), called the *semi-discrete* approximation of $u(x, t)$, satisfies

(1.2)
$$\begin{cases} \dfrac{dw(ih, t)}{dt} = \dfrac{w((i+1)h, t) - 2w(ih, t) + w((i-1)h, t)}{h^2} + r(ih), & 1 \leq i \leq N, \ t > 0, \\ w(ih, 0) = \tilde{u}(ih), & 0 \leq i \leq N+1, \\ w(0, t) = w((N+1)h, t) = 0, & t > 0. \end{cases}$$

Written equivalently in matrix notation, this becomes

(1.3)
$$\begin{cases} \dfrac{d\mathbf{w}(t)}{dt} = -A\mathbf{w}(t) + \mathbf{r}, & t > 0, \\ \mathbf{w}(0) = \tilde{\mathbf{u}}, \end{cases}$$

where $\mathbf{w}(t)$, $\mathbf{r}$, and $\tilde{\mathbf{u}}$ are column vectors with N components, with $\mathbf{w}(t) = (w_1(t), \ldots, w_N(t))^T$ where $w_i(t) \equiv w(ih, t)$. Note that $\mathbf{r}$ and $\tilde{\mathbf{u}}$ are determined from

---

*) Research supported in part by AEC Grant AT (11-1)-2075.
**) The contents of this paper also can be found in [12, Ch. 9].

given quantities, and A is the familar tridiagonal Hermitian and positive definite $N \times N$ matrix, given by

$$(1.4) \qquad A = \frac{1}{h^2} \begin{bmatrix} 2 & -1 & & & & 0 \\ -1 & 2 & -1 & & & \\ & & \cdot & \cdot & \cdot & \\ & & & \cdot & \cdot & \cdot \\ & & & & \cdot & \cdot & -1 \\ 0 & & & & -1 & 2 \end{bmatrix}$$

In what is to follow, only the Hermitian positive definite character of the $N \times N$ matrix A is essential, *and we henceforth assume that our semi-discretization results in* (1.3) *with A Hermitian and positive definite*. In particular, this assumption is valid for linear parabolic problems in n spatial variables of the form

$$(1.5) \qquad \begin{cases} u_t(x, t) = \sum_{i=1}^{n} (K_i(x) u_{x_i}(x, t))_{x_i} - \sigma(x) u(x, t) + r(x), & \text{for} \quad t > 0, \ x \in \Omega, \\ u(x, 0) = \tilde{u}(x), & x \in \Omega, \\ u(x, t) = g(x), & x \in \partial\Omega, \ t > 0, \end{cases}$$

where Ω is a bounded region in R^n, and the quantities $K_i(x)$, $\sigma(x)$, are positive in $\bar{\Omega}$, provided that a suitable $(2n+1)$-point difference approximation of (1.5) is used (cf. [10, p. 253]).

Returning to (1.3), the solution $\mathbf{w}(t)$ can obviously be expressed as

$$(1.6) \qquad \mathbf{w}(t) = A^{-1}\mathbf{r} + \exp(-tA)\{\tilde{\mathbf{u}} - A^{-1}\mathbf{r}\} \qquad (t \geq 0),$$

where as usual, $\exp(-tA) \equiv \sum_{k=0}^{\infty} (-tA)^k/k!$. The solution of (1.6) is commonly approximated by means of matrix Padé rational approximations of $\exp(-tA)$, and these give, as special cases, the well-known forward difference, backward difference, and Crank—Nicolson methods for such parabolic problems (cf. [10, § 8. 3]). Our interest in the next section will be on *Chebyshev*, rather than *Padé*, rational approximations of $\exp(-tA)$. This is because Padé rational approximations of e^{-x}, being defined as *local* approximations of e^{-x} at $x=0$, are generally poor approximations of e^{-x} for large x, and this leads to restrictions (for reasons of stability and/or accuracy) on the time step that can be taken. Chebyshev rational approximations of e^{-x}, in contrast, are defined *globally* with respect to the interval $[0, +\infty)$, and do not have such time step restrictions, as we shall see.

2. Chebyshev semi-discrete approximations

To define the Chebyshev semi-discrete approximations of (1.6), we consider the following approximation problem. If π_m denotes all real polynomials $p(x)$ of degree at most m, and $\pi_{m,n}$ analogously denotes all real rational functions $r_{m,n}(x) = p(x)/q(x)$ with $p \in \pi_m$, $q \in \pi_n$, then let

$$(2.1) \qquad \lambda_{m,n} \equiv \inf_{\pi_{m,n}} \|e^{-x} - r_{m,n}(x)\|_{L_\infty[0,\infty]} = \inf_{\pi_{m,n}} \{\sup_{x \geq 0} |e^{-x} - r_{m,n}(x)|\}.$$

These constants $\lambda_{m,n}$ are called the *Chebyshev constants for e^{-x}* with respect to the interval $[0, +\infty)$. It is obvious that $\lambda_{m,n}$ is finite if and only if $0 \leq m \leq n$, and moreover, given any pair (m, n) of nonnegative integers with $0 \leq m \leq n$, it is known (cf. ACHIESER [1, p. 55]) that, after dividing out possible common factors, there exists a unique $\hat{r}_{m,n} \in \pi_{m,n}$ with

$$(2.2) \qquad \hat{r}_{m,n}(x) = \hat{p}_{m,n}(x)/\hat{q}_{m,n}(x)$$

and with $\hat{q}_{m,n}(x) > 0$ on $[0, \infty)$, such that

$$(2.3) \qquad \lambda_{m,n} = \|e^{-x} - \hat{r}_{m,n}(x)\|_{L_\infty[0,\infty]}.$$

Since $\hat{q}_{m,n}(tA) = \sum_{j=0}^{n} c_j(tA)^j$ is a real polynomial in the $N \times N$ matrix A, it is evident from the fact that $\hat{q}_{m,n}(x)$ is positive on $[0, +\infty)$ that $\hat{q}_{m,n}(tA)$ is a Hermitian and positive definite $N \times N$ matrix for each $t \geq 0$. Thus, in analogy with (1.6), we define the (m, n)-*th Chebyshev semi-discrete approximation* $\mathbf{w}_{m,n}(t)$ of the solution $\mathbf{w}(t)$ of (1.3) as

$$(2.4) \qquad \mathbf{w}_{m,n}(t) = A^{-1}\mathbf{r} + \left(\hat{q}_{m,n}(tA)\right)^{-1}\left(\hat{p}_{m,n}(tA)\right)\{\tilde{\mathbf{u}} - A^{-1}\mathbf{r}\} \qquad (t \geq 0).$$

For the practical computation of $\mathbf{w}_{m,n}(t)$ for a fixed finite $t \geq 0$, assume first that the steady-state solution $\hat{\mathbf{w}} \equiv A^{-1}\mathbf{r}$ of (1.3) has been determined, which amounts to solving the matrix equation $A\hat{\mathbf{w}} = \mathbf{r}$. Then, we write (2.4) equivalently as

$$(2.5) \qquad \hat{q}_{m,n}(tA)\mathbf{w}_{m,n}(t) = \mathbf{v}_0; \quad \mathbf{v}_0 \equiv \hat{q}_{m,n}(tA)\hat{\mathbf{w}} + \hat{p}_{m,n}(tA)\{\tilde{\mathbf{u}} - \hat{\mathbf{w}}\},$$

where $\mathbf{v}_0$ is determined from the known initial vector $\hat{\mathbf{u}}$ (cf. (1.3)), and the known steady-state vector $\hat{\mathbf{w}} = A^{-1}\mathbf{r}$. Since $\hat{q}_{m,n} \in \pi_n$ is positive on $[0, +\infty)$, $\hat{q}_{m,n}$ can be factored into real linear and quadratic factors:

$$(2.6) \qquad \hat{q}_{m,n}(x) = \prod_{i=1}^{s_1} l_i(x) \cdot \prod_{j=1}^{s_2} m_j(x), \quad s_1 + 2s_2 = n,$$

where $l_i \in \pi_1$, $m_j \in \pi_2$, and where the l_i and m_j are also positive on $[0, +\infty)$. Thus, the matrices $l_i(tA)$ and $m_j(tA)$ are again Hermitian and positive definite for each

$t \geq 0$, and the solution $\mathbf{w}_{m,n}(t)$ of (2.5) can be obtained by solving recursively the matrix problems

$$(2.7) \qquad \begin{cases} m_j(tA)\mathbf{v}_j = \mathbf{v}_{j-1}, & 1 \leq j \leq s_2, \\ l_i(tA)\mathbf{v}_{s_2+i} = \mathbf{v}_{s_2+i-1}, & 1 \leq i \leq s_1, \end{cases}$$

and then defining $\mathbf{w}_{m,n}(t) \equiv \mathbf{v}_{s_2+s_1}$. In particular, when A is tridiagonal as in (1.4), the matrices of (2.7) are either tridiagonal or five-diagonal positive definite matrices. As such, the solution of (2.7) by means of Gaussian elimination with no pivoting is both computationally fast and numerically accurate.

For computational efficiency, one should always choose $m=n$ in (2.4) for applications of the Chebyshev semi-discrete method to actual problems. The reason for this is quite clear: the bulk of the work in finding the solution $\mathbf{w}_{m,n}(t)$ of (2.5) comes from the inversion of the polynomial $\hat{q}_{m,n}(tA)$ of degree n in the matrix A, and the work involved in this inversion in practice is virtually independent of the choice of m. For further discussion of such computational aspects of the Chebyshev semi-discrete method, see [11].

To estimate the error in $\mathbf{w}(t) - \mathbf{w}_{m,n}(t)$ we use vector l_2-norms, i.e., if $\mathbf{v} = (v_1, \ldots, v_N)^T$, then $\|\mathbf{v}\|_2^2 \equiv \sum_{i=1}^{N} |v_i|^2$. If, for any $N \times N$ matrix C, $\|C\|_2$ denotes the induced operator norm (or spectral norm) of C, i.e.,

$$(2.8) \qquad \|C\|_2 \equiv \sup_{\mathbf{v} \neq 0} \left\{ \frac{\|C\mathbf{v}\|_2}{\|\mathbf{v}\|_2} \right\},$$

it is well known (cf. [10, p. 11]) when C is Hermitian with (real) eigenvalues μ_i, $1 \leq i \leq N$, that $\|C\|_2$ can be expressed as

$$(2.9) \qquad \|C\|_2 = \max_{1 \leq i \leq N} |\mu_i|.$$

Consequently, if $\{\lambda_i\}_{i=1}^{N}$ denotes the (positive) eigenvalues of A, the assumed Hermitian character of A allows us to conclude from (2.9) that

$$(2.10) \qquad \|\exp(-tA) - \hat{r}_{m,n}(tA)\|_2 = \max_{1 \leq i \leq N} |e^{-t\lambda_i} - \hat{r}_{m,n}(t\lambda_i)|, \quad \text{for all} \quad t \geq 0.$$

But as $t\lambda_i \geq 0$ for all $1 \leq i \leq N$ and for all $t \geq 0$, it follows from (2.3) that

$$\|\exp(-tA) - \hat{r}_{m,n}(tA)\|_2 \leq \lambda_{m,n}, \quad \text{for all} \quad t \geq 0.$$

Consequently, from (1.6) and (2.4),

$$(2.11) \qquad \|\mathbf{w}(t) - \mathbf{w}_{m,n}(t)\|_2 \leq \|\exp(-tA) - \hat{r}_{m,n}(tA)\|_2 \cdot \|\tilde{\mathbf{u}} - A^{-1}r\|_2 \leq$$

$$\leq \lambda_{m,n} \|\tilde{\mathbf{u}} - A^{-1}r\|_2, \quad \text{for all} \quad t \geq 0.$$

Note that since the right-hand side of (2.11) is *independent* of t, we have an error bound for $\mathbf{w}(t) - \mathbf{w}_{m,n}(t)$ for *all* $t \geq 0$. In contrast with the familiar Padé methods which restrict the size of t for reasons of accuracy and/or stability, the Chebyshev semi-discrete method can be used for very large values of t. The difference, of course, comes from the fact that Padé rational approximations of e^{-x} are designed to approximate e^{-x} well in a neighbourhood of $x=0$, whereas Chebyshev rational approximations of e^{-x} are designed to approximate e^{-x} over $[0, +\infty)$.

In general, the error of the spatial discretization leading to (1. 3) must be bounded to give the total error (i.e., space and time) of these Chebyshev semi-discrete approximations. Such spatial discretization errors are discussed in [12], for example.

3. The Chebyshev constants for e^{-x}

The utility of the Chebyshev semi-discrete approximations depends, from (2. 11), on the behavior of the Chebyshev constants $\lambda_{m,n}$ of (2. 1), as $n \to \infty$. From (2. 1), it is clear that

$$(3. 1) \qquad 0 < \lambda_{n,n} \leqq \lambda_{n-1,n} \leqq \cdots \leqq \lambda_{0,n} \qquad (n \geqq 0).$$

Based on elementary arguments, the following result was proved in CODY, MEINARDUS, and VARGA [4].

THEOREM 1. *Let* $\{m(n)\}_{n=0}^{\infty}$ *be any sequence of nonnegative integers with* $0 \leqq$ $\leqq m(n) \leqq n$ *for each* $n \geqq 0$. *Then*,

$$(3. 2) \qquad \varlimsup_{n \to \infty} (\lambda_{m(n),n})^{1/n} \leqq \frac{e^{-\alpha}}{2} < \frac{1}{2},$$

where $\alpha=0.13923\ldots$ *is the real solution of* $2\alpha e^{2\alpha+1}=1$. *Moreover*,

$$(3. 3) \qquad \varlimsup_{n \to \infty} (\lambda_{0,n})^{1/n} \geqq \frac{1}{6}.$$

The results of (3. 2) and (3. 3) establish the *geometric convergence to zero* of the Chebyshev constants $\lambda_{m,n}$ for e^{-x} in $[0, \infty)$. In particular, if $m(n)=n$, then the Chebyshev constants $\lambda_{n,n}$ for e^{-x} in $[0, +\infty)$ are from [4]:

n	$\lambda_{n,n}$	n	$\lambda_{n,n}$	n	$\lambda_{n,n}$
0	5.00(−01)	5	9.35(−06)	10	1.36(−10)
1	6.69(−02)	6	1.01(−06)	11	1.47(−11)
2	7.36(−03)	7	1.09(−07)	12	1.58(−12)
3	7.99(−04)	8	1.17(−08)	13	1.70(−13)
4	8.65(−05)	9	1.26(−09)	14	1.83(−14)

where $\alpha(-\beta)$ denotes $\alpha \cdot 10^{-\beta}$ in the table above. Thus, the rate of convergence to zero of the $\lambda_{n,n}$ appears to be much better than that given by the upper bound of (3. 2). Also, the quantities $\lambda_{0,n}$, $0 \leq n \leq 9$, as tabulated in [4], would lead one to conjecture that $\lim_{n \to \infty} (\lambda_{0,n})^{1/n}$ exists, and that

$$(3.\ 4) \qquad \lim_{n \to \infty} (\lambda_{0,n})^{1/n} = \frac{1}{3} .$$

This in fact has been recently shown by SCHÖNHAGE [8].

4. Chebyshev constants for other entire functions

The preceeding results on the geometric convergence to zero of the Chebyshev constants $\lambda_{m,n}$ for $1/e^x$ in (3. 2) and (3. 3) hold for a wider class of entire functions than just $f(z) = e^z$. A generalization of the results of Theorem 1 has been recently given in MEINARDUS and VARGA [7], and can be described as follows.

Let $f(z) = \sum_{k=0}^{\infty} a_k z^k$ be an entire function (i.e., analytic for every finite z) with $M_f(r) \equiv \sup_{|z|=r} |f(z)|$ its maximum modulus function. Then, f is of *perfectly regular growth* (ϱ, β) (cf. BOAS [2, p. 8] and VALIRON [9, p. 45]) if there exist two (finite) positive numbers ϱ (the order) and B (the type) such that

$$(4.\ 1) \qquad \lim_{r \to \infty} \frac{\ln M_f(r)}{r^\varrho} = B.$$

We then have (cf. [7])

THEOREM 2. *Let* $f(z) = \sum_{k=0}^{\infty} a_k z^k$ *be an entire function of perfectly regular growth* (ϱ, B) *with* $a_k \geq 0$ *for all* $k \geq 0$, *and for any pair* (m, n) *of nonnegative integers with* $0 \leq m \leq n$, *let*

$$(4\ .2) \qquad \lambda_{m,n} \equiv \inf_{\pi_{m,n}} \left\| \frac{1}{f(x)} - r_{m,n}(x) \right\|_{L_\infty[0, \infty]}$$

be its associated Chebyshev constants. Then, for any sequence $\{m(n)\}_{n=0}^{\infty}$ *of nonnegative integers with* $0 \leq m(n) \leq n$ *for each* $n \geq 0$,

$$(4.\ 3) \qquad \varlimsup_{n \to \infty} (\lambda_{m(n),n})^{1/n} \leq 2^{-1/\varrho} < 1.$$

Moreover,

$$(4.\ 4) \qquad \varlimsup_{n \to \infty} (\lambda_{0,n})^{1/n} \geq 2^{-2-1/\varrho}.$$

As special cases of Theorem 2, we have of course $f(z) = e^z$, $f(z) = \sinh(z^p)$ and $f(z) = J_p(iz)$ for p a nonnegative integer, where J_p denotes the Bessel function of the first kind. For $f(z)=e^z$, for which $\varrho=B=1$ in (4. 1), the results of (4. 3) and (4. 4) are slightly weaker than those of (3. 2) and (3. 3) of Theorem 1.

The proofs of Theorems 1 and 2 depend upon estimating

$$\frac{1}{s_n(x)}-\frac{1}{f(x)}$$

where $s_n(z) = \sum_{k=0}^{n} a_k z^k$ is the n-th partial sum of $f(z)$. It is shown in [7] that, under the hypotheses of Theorem 2,

$$\lim_{n\to\infty}\left(\left\|\frac{1}{s_n}-\frac{1}{f}\right\|_{L_\infty[0,\infty]}\right)^{1/n} = 2^{-1/\varrho},$$

so that the upper bound of (4. 3) cannot be improved using this specific technique.

Upon examining Theorem 2, we see that the bounds of (4. 3) and (4. 4) depend upon ϱ, but not on B, and this suggests the possibility of extensions of Theorem 2 to entire functions which are of finite order, but not of perfectly regular growth. Such extensions have been considered in MEINARDUS, REDDY, TAYLOR, and VARGA [6], and we state a representative result which generalizes Theorem 2. For notation, let $\varepsilon(r, s)$, for given $r>0$ and $s>1$, denote the unique open ellipse in the complex plane with foci at $x=0$ and $x=r$ and semi-major and semi-minor axes a and b such that $b/a = (s^2-1)/(s^2+1)$. If $f(z)$ is any entire function, we set

$$(4. 5) \qquad \tilde{M}_f(r, s)=\sup\ \{|f(z)|:z\in\varepsilon(r, s)\}.$$

THEOREM 3. *Let $f(z) = \sum_{k=0}^{\infty} a_k z^k$ be an entire function with nonnegative Taylor coefficients and $a_0>0$. If there exist real numbers $s>1$, $A>0$, $\theta>0$ and $r_0>0$ such that*

$$(4. 6) \qquad \tilde{M}_f(r, s) \leqq A(\|f\|_{L_\infty[0,r]})^\theta \quad \text{for all}\quad r\geqq r_0,$$

then there exist a real number $q\geqq s^{1/(1+\theta)}>1$ and a sequence of real polynomials $\{p_n(x)\}_{n=0}^{\infty}$ with $p_n\in\pi_n$ for each $n\geqq0$ such that

$$(4. 7) \qquad \varlimsup_{n\to\infty}\left\{\left\|\frac{1}{f(x)}-\frac{1}{p_n(x)}\right\|_{L_\infty[0,\infty]}\right\}^{1/n} = \frac{1}{q} < 1.$$

Note that (4. 7) implies the geometric convergence to zero of the Chebyshev constants $\{\lambda_{m(n),n}\}_{n=0}^{\infty}$ of $1/f$ when $0\leqq m(n)\leqq n$.

To motivate the next result, it is convenient to recall some classical results of Bernstein for polynomial approximation on *finite* intervals. Given a real-valued function $f\in C^0[-1, +1]$, let

$$(4. 8) \qquad E_n(f) \equiv \inf_{\pi_n} \|f-p_n\|_{L_\infty[-1,+1]}.$$

If f is the restriction to $[-1, +1]$ of a function analytic in an ellipse in the complex plane with foci -1 and $+1$, then Bernstein proved (cf. MEINARDUS [5, p. 91]) that there exists a real number $q > 1$ such that

$$(4.9) \qquad \overline{\lim_{n \to \infty}} \, E_n^{1/n}(f) = \frac{1}{q} < 1.$$

Conversely, if (4.9) holds, Bernstein proved the *inverse* result (cf. MEINARDUS [5, p. 92]) that f is necessarily the restriction to $[-1, +1]$ of a function analytic in an ellipse in the complex plane with foci at -1 and $+1$. Consider then the results of of Theorems 2 and 3. These give *sufficient* conditions on the entire function $f(z)$ so that the Chebyshev constants $\lambda_{m,n}$ of $1/f$, for $0 \leq m \leq n$, converge geometrically to zero as $n \to \infty$. In the spirit of Bernstein's classical inverse theorems, the following result of [6] gives *necessary* conditions for this geometric convergence.

THEOREM 4. *Let $f(x) > 0$ be a real continuous function on $[0, \infty)$, such that there exist a sequence of real polynomials $\{p_n(x)\}_{n=0}^{\infty}$ with $p_n \in \pi_n$ for all $n \geq 0$, and a real number $q > 1$ such that*

$$(4.10) \qquad \overline{\lim_{n \to \infty}} \left(\left\| \frac{1}{p_n} - \frac{1}{f} \right\|_{L_\infty[0, \infty]} \right)^{1/n} = \frac{1}{q} < 1.$$

Then, there exists an entire function $F(z)$ with $F(x) = f(x)$ for all $x \geq 0$. Moreover, F is of finite order, i.e.,

$$\overline{\lim_{r \to \infty}} \, \frac{\ln \ln M_F(r)}{\ln r} = \varrho < \infty.$$

In addition, for each $s > 1$, there exist real numbers $K = K(q, s) > 0$, $\theta = \theta(q, s) > 1$, and $r_0 = r_0(q, s) > 0$ such that

$$(4.11) \qquad \tilde{M}_F(r, s) \leq K(\|f\|_{L_\infty[0, r]})^\theta \quad \text{for all} \quad r \geq r_0.$$

Finally, to complement the preceding results of this section, it is shown in [6] that there exist entire functions $f(z)$, of finite order which are positive on $[0, +\infty)$, for which the Chebyshev constants $\lambda_{m,n}$ of $1/f$, for $0 \leq m \leq n$, *cannot* converge geometrically to zero as $n \to \infty$.

REFERENCES

[1] N. I. Achieser, *Theory of Approximation.* Frederick Ungar Publishing Co., New York 1956.

[2] R. P. Boas, *Entire Functions.* Academic Press, Ind., New York 1954.

[3] J. C. Cavendish, W. E. Culham, and R. S. Varga, *A comparison of Crank-Nicolson and Chebyshev rational methods for numerically solving linear parabolic equations.* J. Computational Physics (to appear).

[4] W. J. Cody, G. Meinardus, and R. S. Varga, *Chebyshev rational approximation to e^{-x} in $[0, +\infty]$ and applications to heatconduction problems.* J. Approximation Theory 2 (1969), 50—65.

[5] G. Meinardus, *Approximation of Functions: Theory and Numerical Methods.* Springer, New York 1967.

]6] G. Meinardus, A. R. Reddy, G. D. Taylor, and R. S. Varga, *Converse theorems and extensions in Chebyshev rational approximation to certain entire functions in $[0, \infty)$.* Bull. Amer. Math. Soc. 77 (1971), 460—461.

[7] G. Meinardus and R. S. Varga, *Chebyshev rational approximations to certain entire functions in $[0, \infty)$.* J. Approximation Theory 3 (1970), 300—309.

[8] A. Schönhage, *Zur rationalen Approximierbarkeit von e^{-x} über $[0, \infty)$.* J. Approximation Theory (to appear).

[9] G. Valiron, *Lectures on the General Theory of Integral Functions.* Chelsea Publishing Co., New York 1949.

[10] R. S. Varga, *Matrix Iterative Analysis.* Prentice-Hall, Englewood Cliffs, New Yersey 1962.

[11] R. S. Varga, *Some results in approximation theory with applications to numerical analysis,* Numerical Solution of Partial Differential Equations-II (B. E. Hubbard, ed.). Academic Press, Inc., New York pp. 623—649, 1971.

[12] R. S. Varga, *Functional Analysis and Approximation Theory in Numerical Analysis.* Regional Conference Series in Applied Mathematics, #3, Society for Industrial and Applied Mathematics, Philadelphia, Pa. 1971.

On the Convergence Theory of Padé Approximants

By

HANS WALLIN

DEPARTMENT OF MATHEMATICS
UNIVERSITY OF UMEÅ
UMEÅ, SWEDEN

1. Introduction

The purpose of this paper is to inform about the Padé approximation. The problem was studied in detail first by FROBENIUS in 1881 [4] and PADÉ in 1892 [7] but goes back to JACOBI and even to CAUCHY. The Padé approximation is a kind of rational approximation which in general differs from the best uniform rational approximation. It has been used extensively by physicists in recent years (see [1] and [2]).

Early studies of the Padé approximants have to a large extent been devoted to deducing algebraic relations satisfied by the approximants, including connections to continued fractions (compare [8] and [9]). In this paper we give a short survey on the convergence theory of the Padé approximants including, in particular, some unsolved problems and some recent results. We also comment on the definition of the Padé approximants.

2. Different definitions of the Padé approximants

Let

$$(1) \qquad f(z) = \sum_{i=0}^{\infty} a_i z^i$$

be a formal power series with complex coefficients a_i. If n and v are natural numbers the (n, v) *Padé approximant of f* is a rational function P_n/Q_v where

$$P_n(z) = \sum_{0}^{n} \alpha_i z^i \quad \text{and} \quad Q_v(z) = \sum_{0}^{v} \beta_i z^i, \quad Q_v(z) \not\equiv 0,$$

and the coefficients α_i and β_i are determined so that

$$(2) \qquad f(z)Q_v(z) - P_n(z) = Az^{n+v+1} + \text{higher degree terms},$$

where A is a constant (which may be zero). This means that the $n+v+1$ first coefficients in the power series expansion of the left member shall be zero, i.e. that the numbers α_i and β_i shall satisfy the system of equations

$$(3) \qquad \sum_{i=0}^{v} a_{j-i}\beta_i = \begin{cases} \alpha_j, & 0 \leq j \leq n \\ 0, & n+1 \leq j \leq n+v \end{cases}$$

(those a_k that may occur with negative indexes k shall be replaced by 0). Since the last v equations in (3) have $v+1$ unknown β_i it is always possible to choose β_i so that these v equations are satisfied. After that the numbers α_j are determined from the $n+1$ first equations of (3).

In spite of the fact that the numbers β_i and α_i — and hence P_n and Q_v — are clearly not uniquely determined, *the (n, v) Padé approximant P_n/Q_v is unique* which is proved in the following way (FROBENIUS [4, p. 2]): Let P_n^*/Q_v^* be another (n, v) Padé approximant to f and consider the expression

$$(fQ_v - P_n)Q_v^* - (fQ_v^* - P_n^*)Q_v = Q_v P_n^* - Q_v^* P_n.$$

In the left member all the coefficients of terms of degree at most $n+v$ are zero by the definition of the (n, v) Padé approximant. However, the right member is a polynomial of degree at most $n+v$ and consequently the expression vanishes identically which means that $P_n/Q_v = P_n^*/Q_v^*$, i.e. the (n, v) Padé approximant is unique.

REMARK. Observe that it need not be possible to choose the numbers β_i so that $Q_v(0) = \beta_0 \neq 0$. The case when $f(z) = 1+z^2$ and $n = v = 1$ gives an example of this as is easily checked. There seems to have been sometimes some confusion on this point. However, we always have $\beta_0 \neq 0$ if f is such that, for all n and v, the (n, v) Padé approximant P_n/Q_v of f in irreducible form has numerator and denominator of degree exactly n and v, respectively — in that case f is called normal (compare WALL [9, p. 399]). In fact, if $\beta_0 = Q_v(0) = 0$, then $\alpha_0 = P_n(0) = 0$, as is seen from (3), and then P_n and Q_v would have a common factor which cannot occur if f is normal. The concept of normality has played an important role in the theory (see [9]).

WALSH has proved [11] that under suitable conditions the Padé approximant of a function f which is holomorphic in a neighbourhood of the origin is the limit of the corresponding function of best uniform rational approximation of f in the disk $\{z \mid |z| \leq \varepsilon\}$, as $\varepsilon \to 0$.

The convergence problem for the Padé approximation consists of determining when the (n, v) Padé approximant of f converges, and when it converges to f, as one or both of n and v tend to infinity through some subsequence of the positive integers. In particular, the "diagonal" case $n = v$ is of interest.

The Padé approximant can be defined in two other ways which we state in the following two propositions.

PROPOSITION 1. *Let f be the formal power series (1). Let P_n^* and Q_v^*, $Q_v^*(z) \not\equiv 0$, be polynomials of degree at most n and v, respectively, determined such that*

$$f(z)Q_v^*(z) - P_n^*(z) = Az^\mu + higher\ degree\ terms,$$

where A is a constant and the integer μ is as large as possible. (If no largest μ exists, we put $\mu = \infty$.) Then P_n^*/Q_v^* is the (n, v) Padé approximant of f.

PROOF. It follows from the definition of the (n, v) Padé approximant that $\mu \geqq n+v+1$, i.e. P_n^*/Q_v^* is the unique (n, v) Padé approximant.

PROPOSITION 2. *Let the polynomials P_n^* and Q_v^*, $Q_v^*(z) \not\equiv 0$, of degrees at most n and v, be determined so that*

$$(4) \qquad f(z) - \frac{P_n^*(z)}{Q_v^*(z)} = Az^\mu + higher\ degree\ terms,$$

where A is a constant and the integer μ as large as possible (or infinite). Then P_n^*/Q_v^* is the (n, v) Padé approximant of f.

This means that the (n, v) Padé approximant of f is the unique rational function of type (n, v) which has contact with f of highest order at the origin. In this way the Padé approximants are the analogues among rational functions to the Taylor expansions among polynomials.

REMARK. It should be observed that the number μ obtained in Proposition 2 may very well be smaller than the exponent $n+v+1$ occurring in the definition (2) of the (n, v) Padé approximant. Again the case when $f(z) = 1+z^2$ and $n=v=1$ may be used as an example; it gives $n+v+1 = 3$ but the number μ in Proposition 2 is 2.

PROOF OF PROPOSITION 2. (PADÉ [7, p. 12]) Suppose that the polynomials P_n and Q_v are determined so that (2) is satisfied and that Q_v has a zero of order ω at the origin. Then it is easy to see from (2) that P_n has a zero at the origin of order at least ω. Put $R_n(z) = P_n(z)/z^\omega$ and $S_v(z) = Q_v(z)/z^\omega$. Then (1) gives

$$(5) \qquad f(z) - \frac{R_n(z)}{S_v(z)} = Bz^{n+v+1-\omega} + higher\ degree\ terms,$$

where B is a constant. Hence, the integer μ in (4) satisfies $\mu \geqq n+v+1-\omega$. Subtraction of (5) from (4) gives, with some constant C

$$Q_v^*(z)R_n(z) - P_n^*(z)S_v(z) = Cz^{n+v+1-\omega} + higher\ degree\ terms.$$

Since R_n and S_v are polynomials of degrees at most $n-\omega$ and $v-\omega$, the left member is a polynomial of degree at most $n+v-\omega$. However, all terms in the right member of degree at most $n+v-\omega$ vanish and consequently

$$P_n^*/Q_v^* = R_n/S_v = P_n/Q_v$$

is the (n, v) Padé approximant of f.

3. A convergence conjecture

A series of Stieltjes is a series of the form

$$f(z) = \sum_0^\infty a_i(-z)^i \quad \text{where} \quad a_i = \int_0^\infty t^i \, d\Phi(t)$$

and Φ is a bounded, nondecreasing function taking on infinitely many values. If f is such a series, special methods are applicable and there is a rather detailed convergence theory for the Padé approximation, a theory which was initiated by Stieltjes and Wall. We refer to WALL [9] and BAKER ([1] and [2]) for this theory.

If f is not a series of Stieltjes, it has been difficult to obtain satisfactory convergence results. If we consider Padé approximants where the degree of the denominator is bounded by a fixed number the problem is of course easier. In 1902 MONTESSUS DE BALLORE ([5]; see also [8, p. 265]) proved the following theorem:

THEOREM 1 (Montessus de Ballore). *Let the power series $f(z) = \sum a_i z^i$ represent a function f which is holomorphic for $|z| \leq R$ except for m poles in $\{z \mid 0 < |z| < R\}$. Then the (n, m) Padé approximant of f converges uniformly to f, as $n \to \infty$, in the domain obtained by removing from $|z| \leq R$ small disks with centers at the poles of f.*

WILSON [12] extended this result to the case when f has also a non-polar singularity on $|z| = 1$ at which f is continuous. Wilson even gets uniform convergence of the $(n, m+\tau)$ Padé approximant, as $n \to \infty$, for any fixed non-negative integer τ. Montessus de Ballore considered also functions f that are meromorphic in the whole complex plane. In 1970 BAKER [2, p. 8] wanted to generalize the result of Montessus de Ballore for functions that are meromorphic in the whole complex plane and have infinitely many poles to a sequence of (n, v) Padé approximants where both n and v tend to infinity. However, Baker's proof is not complete. On the other hand a consequence of Theorem 1 is the following result which essentially contains Baker's theorem (a special case of Theorem 2 was given by PERRON [8, p. 269—270]).

THEOREM 2. *Let f be holomorphic in the disk $|z| < R$ (where $R \leq \infty$), except for infinitely many poles in $\{z \mid 0 < |z| < R\}$. Let N be the set of poles of f in $|z| < R$. Then there exist sequences n_i and v_i, $i = 1, 2, \ldots$, satisfying*

$$n_i \to \infty \quad \text{and} \quad v_i \to \infty, \quad \text{as} \quad i \to \infty,$$

such that the sequence of (n_i, v_i) *Padé approximants of* f, $i=1, 2, \ldots$, *converges uniformly to* f, *as* $i \to \infty$, *on any compact subset of* $\{z|\ |z| < R\} \backslash N$.

PROOF OF THEOREM 2. We use the notation $S(c, r) = \{z|\ |z - c| \leqq r\}$. Let (ε_i) be a sequence of positive numbers tending monotonically to 0 and (R_i), $R_i < R$, a sequence tending monotonically to R. Put

$$S_i = \bigcup_{c \in N} S(c, \varepsilon_i) \quad \text{and} \quad K_i = S(0, R_i) \backslash S_i.$$

Choose, for $i=1, 2, \ldots$, an integer v_i, where $v_i \to \infty$, as $i \to \infty$, such that f has exactly v_i poles (multiplicity included) in an open disk containing $S(0, R_i)$ and having center at the origin. This is possible since f has infinitely many poles. As a consequence of Theorem 1, the (n, v_i) Padé approximant P_n/Q_{v_i} of f converges uniformly to f in K_i, as $n \to \infty$. Choose n_i satisfying $n_i \to \infty$, as $i \to \infty$, and such that

$$|P_{n_i}/Q_{v_i} - f| < \varepsilon_i \quad \text{on} \quad K_i \qquad (i=1, 2, \ldots).$$

Since $\varepsilon_i \to 0$ and every fixed compact subset of $S(0, R) \backslash N$ is a subset of K_i if i is large enough, Theorem 2 clearly follows.

Since Wilson's theorem contains an arbitrary non-negative integer τ, it is obviously possible to prove a similar corollary to his result by the same method.

A satisfactory investigation about the generalization of the result by Montessus de Ballore to Padé approximants (n, v) where both n and v tend to infinity is still lacking. However, there is a conjecture given by BAKER, GAMMEL and WILLS in the following form (see [1, p. 23—34] for a discussion of the conjecture).

CONJECTURE (Baker, Gammel and Wills). *Let* $f(z) = \sum a_i z^i$ *be a power series representing a function* f *which is holomorphic in* $|z| \leqq 1$, *except for finitely many poles in* $0 < |z| < 1$ *and except for* $z = 1$ *where the function is to be continuous if only points satisfying* $|z| \leqq 1$ *are considered. Then there exists a subsequence of the sequence of* (n, n) *Padé approximants of* f *that converges uniformly to* f *in the domain formed by removing from* $|z| \leqq 1$ *small disks with centers at the abovementioned poles.*

No proof and no counterexamples are known of this conjecture.

4. Convergence in measure

In 1970 NUTTALL [6] proved the following general theorem.

THEOREM 3. (Nuttall) *If* f *is meromorphic in the whole complex plane and does not have a pole at the origin, then the* (n, n) *Padé approximant of* f *converges in measure to* f *in any bounded subset of the complex plane, as* $n \to \infty$.

Nuttall's proof depends on a lemma which states that, for any $H>0$, the absolute value of a polynomial of degree n with leading coefficient one is larger than H^n except on a set having Lebesgue measure not exceeding πH^2. By using this on the denominator Q_n of the (n, n) Padé approximant P_n/Q_n combined with a certain expansion of the meromorphic function f and with the definition of the Padé approximant, it is possible to prove that

$$f - \frac{P_n}{Q_n} = \frac{fQ_n - P_n}{Q_n}$$

is "small" except on a set of "small" Lebesgue measure.

5. Convergence and the size of the power series coefficients

Recently I have investigated the connection between the convergence of the Padé approximants of a power series and the size of the power series coefficients; detailed proofs will appear in [10]. My results imply the following convergence theorem.

THEOREM 4. *Let α be a positive number and n_v, $v = 1, 2, \ldots$, a sequence of positive integers tending to infinity with v. Let*

$$f(z) = \sum_0^\infty a_i z^i$$

be a power series which is convergent in a disk D and satisfies

$$(6) \qquad \sum_{v=1}^\infty (\max_{n_v < i \le 2n_v} |a_i|)^{\alpha/n_v} < \infty.$$

Then the (n_v, n_v) Padé approximant of f converges to f in $D \backslash E$, as $v \to \infty$, where E is a set having α-dimensional Hausdorff measure zero. (This means that E is such that for every $\varepsilon > 0$, E can be covered by means of denumerably many disks with radii r_i such that $\sum_i r_i^\alpha < \varepsilon$.)

REMARK. There is an analogous theorem when the (n_v, n_v) Padé approximants are replaced by (n_v, m_v) Padé approximants. Also, the convergence is uniform on compact subsets of D except on a set having arbitrarily small α dimensional measure.

The proof of Theorem 4 depends on the following lemma by H. CARTAN [3, p. 279] which gives the exceptional set E. The lemma is similar to but gives more detailed information than the above-mentioned lemma used by Nuttall.

LEMMA. (H. Cartan) *For any $H>0$ and $\alpha>0$, the absolute value of a polynomial of degree n with leading coefficient 1 is larger than H^n except on a set which can be covered by at most n disks with radii r_i such that*

$$\sum_i r_i^\alpha < e(2H)^\alpha.$$

The idea of the proof of Theorem 4 is this. Let P_ν/Q_ν be the (n_ν, n_ν) Padé approximant of f and hence — which is easy to check — to

$$g_\nu(z) = \sum_0^{2n_\nu} a_i z^i.$$

If f is holomorphic in D, $g_\nu \to f$ in D, as $\nu \to \infty$. In order to prove Theorem 2 it is therefore enough to prove that

$$(7) \qquad g_\nu - \frac{P_\nu}{Q_\nu} = \frac{g_\nu Q_\nu - P_\nu}{Q_\nu}$$

tends to zero except on a set E having α-dimensional Hausdorff measure zero. By the definition of the (n_ν, n_ν) Padé approximant the numerator in the right member of (7) is a polynomial where all the terms have degree at least $2n_\nu+1$. If, for every ν, Cartan's lemma is used with H equal to a suitably chosen H_ν on the denominator Q_ν in the right member of (7), it turns out that (6) is exactly the right condition to guarantee that (7) tends to zero in $\complement E$, as $\nu \to \infty$.

Theorem 4 has the following corollary which should be compared to the conjecture stated in section 3. It should also be observed that Nuttall's result gives—even for meromorphic functions — this corollary for $\alpha=2$.

COROLLARY. *Let f be an entire function and α a positive number. Then there exists a subsequence of the sequence of (n, n) Padé approximants of f that converges to f everywhere except on a set having α-dimensional Hausdorff measure zero.*

The corollary follows because if f is entire, there clearly exists, for any fixed $\alpha>0$, a sequence n_ν, $\nu=1, 2, 3, \ldots$, of integers tending to infinity such that (6) is satisfied.

Theorem 4 is best possible in the sense that for any $\alpha>0$ there exist integers n_ν and an entire function $f(z)=\sum a_i z^i$ satisfying (6) — and even a condition a little stronger than (6)—such that the sequence of (n_ν, n_ν) Padé approximants is divergent and even unbounded everywhere on a prescribed set E having α-dimensional Hausdorff measure zero and not containing the origin (see [10]). The construction of such a function f is possible by starting with the requirement that the denominator Q_ν of the (n_ν, n_ν) Padé approximant shall have the simple form

$$Q_\nu(z) = (z-b_\nu)^{m_\nu}$$

where b_v and m_v are suitably chosen. A corollary to this result is the following theorem (take $\alpha > 2$; then the set consisting of the whole complex plane has α-dimensional Hausdorff measure zero).

THEOREM 5. *There exists an entire function f such that the sequence of (n, n) Padé approximants of f is divergent — and even unbounded — everywhere except at the origin.*

This theorem should be compared to the corollary to Theorem 4.

6. The poles of the approximants

Parallel to the convergence problem there is a problem concerning the location of the poles of the Padé approximants. Even for entire functions the poles of the (n, n) Padé approximants may cluster everywhere. PERRON [8, p. 270] gave a simple example of an entire function such that the poles of the $(n, 1)$ Padé approximants cluster everywhere in the complex plane.

In [10] I proved the following simple theorem:

Let n_v, $v = 1, 2, 3, \ldots$, be a sequence of positive integers satisfying $n_v > 2n_{v-1}$ for all v. Then there exists an entire function with power series coefficients tending to zero arbitrarily fast such that the (n_v, n_v) Padé approximant of f has a pole at any prescribed point $b_v \neq 0$ of the complex plane, for $v = 1, 2, \ldots$.

A similar result has been proved by H. S. SHAPIRO (personal communication). A general result giving information about the poles of the Padé approximants of functions belonging to general classes of functions is still lacking.

REFERENCES

[1] G. A. Baker, Jr., *Advances in Theoretical Physics.* (K. A. Brueckner, ed.), Vol. 1, Academic Press, New York 1965, 1—58.

[2] G. A. Baker, Jr., *The Padé Approximant in Theoretical Physics.* (G. A. Baker, Jr. and J. L. Gammel, ed.), Academic Press, New York 1970, 1—39.

[3] H. Cartan, *Sur les systèmes de fonctions holomorphes à variétés linéaires et leurs applications.* Ann. Sci. École Norm. Sup. (3) **45** (1928), 255—346.

[4] G. Frobenius, *Über Relationen zwischen den Näherungsbrüchen von Potenzreihen.* J. für reine und angew. Math. **90** (1881), 1—17.

[5] R. de Montessus de Ballore, *Sur les fractions continues algébriques.* Bull. Soc. Math. France **30** (1902), 28—36.

[6] J. Nuttall, *The convergence of Padé approximants of meromorphic functions.* J. Math. Anal. Appl. **31** (1970), 147—153.

[7] H. Padé, *Sur la représentation approchée d'une fonction par des fonctions rationnelles.* Ann. Sci. École Norm. Sup. (3) **9** (1892), supplément, 1—93.

[8] O. Perron, *Die Lehre von den Kettenbrüchen.* Band II, B. G. Teubner, Stuttgart 1957.

[9] H. S. Wall, *Analytic theory of continued fractions.* Van Nostrand, New York 1948.

[10] H. Wallin, *The convergence of Padé approximants and the size of the power series coefficients.* (To appear in Applicable Anal.).

[11] J. L. Walsh, *Padé approximants as limits of rational functions of best approximation.* J. Math. Mech. **13** (1964), 305—312.

[12] R. Wilson, *Divergent continued fractions and non-polar singularities.* Proc. London Math. Soc. **30** (1930), 38—57.

Singular Self-Adjoint Multipoint Boundary Value Problems: Solutions and Approximations

By

JOSEPH W. JEROME

DEPT. OF MATH.
NORTHWESTERN UNIVERSITY
EVANSTON, ILLINOIS

1. Introduction

Let Λ be a linear, self-adjoint, possibly singular differential operator with real-valued coefficients of the form

$$\Lambda f = \sum_{j=0}^{m} (-1)^j D^j (a_j D^j f) \qquad (m \geq 1),$$

where

(a) $\qquad a_m(x) > 0, \qquad x \in (a, b),$

(b) $\qquad \dfrac{1}{a_m} \in L^1(a, b),$

(1. 1)

(c) $\qquad a_m \in C^m(a,)b,$

(d) $\qquad a_j \in C^j[a, b], \qquad 0 \leq j < m.$

The differential operator

$$\Lambda f(x) = (-1)^m D^m [x^\sigma (1-x)^\varrho D^m f(x)]$$

on $[0, 1]$, where $0 \leq \sigma$, $\varrho < 1$ is an example of such an operator.

With Λ we associate the real linear space $H[a, b] = \{f : f \in C^{m-1}[a, b], D^{m-1}f$ absolutely continuous on $[a, b]$, $\sqrt{a_m} D^m f \in L^2(a, b)\}$ which is a real Hilbert space under a suitable inner product to be defined shortly. Now let $B(u, v)$ be the bilinear form on $H[a, b]$ defined by

(1. 2) $$B(u, v) = \sum_{j=0}^{m} (a_j D^j u, D^j v)_{L^2(a,b)}.$$

It will be shown in section 3 that, for all $f \in H[a, b]$,

(1. 3) $$B(f,f) + C_0(f,f)_{L^2(a,b)} \geq \alpha \left[(a_m D^m f, D^m f)_{L^2(a,b)} + \sum_{j=0}^{m-1} (D^j f(a))^2 \right]$$

for some positive constants C_0 and α. In particular, since the right hand side of (1. 3) defines a Hilbert space norm in which convergence implies uniform convergence of derivatives through order $m-1$, then (1. 2) and (1. 3) imply that the left hand side of (1. 3) could be used to define a Hilbert space norm on $H[a, b]$. Since the Euler operator associated with the left hand side of (1. 3) is $\Lambda + C_0 I$, we could then carry out the following analysis for multipoint boundary value problems in which the underlying differential operator is $\Lambda + tI$, $t \geq C_0$, with no assumptions whatsoever on the positivity of the bilinear form, given the Garding-type inequality of (1. 3). This we shall do in section 4. However, in section 3, we prefer to adopt the approach that Λ is the underlying operator and we look for assumptions on $B(f, f)$ which will guarantee the solvability of the multipoint boundary-value problem. In particular we assume explicitly in sections 3 and 5,

$$(1.4) \qquad\qquad B(f, f) \geq 0, \quad f \in H[a, b].$$

With this assumption, there is no loss of generality in assuming that $C_0 = 1$ in (1. 3) and we define

$$(1.5) \qquad\qquad (f, g)_{H[a, b]} = B(f, g) + (f, g)_{L^2(a, b)}.$$

The injection

$$I : H[a, b] \to L^2(a, b)$$

is compact since sets bounded in the norm of the right hand side of (1. 3) have compact closure in $L^2(a, b)$. More generally, such sets have compact closure in the Sobolev spaces $H^j[a, b]$, $0 \leq j < m$, with inner product

$$(1.6) \qquad\qquad (f, g)_{H^j[a, b]} = \sum_{i=0}^{j} (D^i f, D^i g)_{L^2(a, b)}.$$

Documentation of these facts regarding the norm determined by the right hand side of (1. 3) may be found in [6].

Now let $a = x_0 < x_1 < \cdots < x_k < x_{k+1} = b$ describe a mesh of $[a, b]$ where boundary conditions are prescribed. Specifically, let $\mathscr{E}^{(v)}$ be a (non-empty) subset of $\{0, 1, \ldots, m-1\}$ for $0 < v < k+1$, and, for $\mu \in \mathscr{E}^{(v)}$, let

$$(1.7) \qquad\qquad E_\mu^{(v)} f = \sum_{j=0}^{\mu} \alpha_{j\mu}^{(v)} D^j f(x_v) \qquad (f \in H[a, b]),$$

where $\alpha_{j\mu}^{(v)}$ are specified real numbers with $\alpha_{\mu\mu}^{(v)} \neq 0$. Finally, define $\mathscr{E}^{(0)} = \{0, 1, \ldots, m-1\} = \mathscr{E}^{(k+1)}$ with $E_\mu^{(0)} f = D^\mu f(x_0 +)$, $E_\mu^{(k+1)} f = D^\mu f(x_{k+1} -)$, $0 \leq \mu \leq m-1$ and denote by $H_0[a, b]$ the closed linear subspace of $H[a, b]$ annihilated by the continuous boundary operators $\{\lambda_i\}_1^N = \{E_\mu^{(v)} : 0 \leq v \leq k+1, \mu \in \mathscr{E}^{(v)}\}$.

In order to define the multipoint boundary value problem considered in this paper, it is of interest to consider the variational problem

$$(1.8) \qquad\qquad \min \{B(f, f) : \lambda_i f = r_i, \quad 1 \leq i \leq N\}$$

where $B(f, f) \geq 0$ on $H[a, b]$. Indeed the following result is proved in [6].

THEOREM. *If* $B(f,f)\geq 0$ *on* $H[a, b]$ *then the minimization problem* (1.8) *possesses a unique solution* $f_* \in H[a, b]$. f_* *is characterized by the conditions:*

$$\text{(i)} \qquad \Lambda f_*(x) = 0 \qquad x \in \bigcup_{v=0}^{k} (x_v, x_{v+1})$$

$$\text{(ii)} \qquad \lambda_i f_* = r_i \qquad 1 \leq i \leq N$$

(1.9)

$$\text{(iii)} \qquad f_* \in H[a, b] \cap C^{2m}\left(\bigcup_{v=0}^{k} (x_v, x_{v+1})\right)$$

$$\text{(iv)} \qquad [R_\mu^{(v)} f_*]_{x_v} = 0 \quad \text{if} \quad \mu \notin \mathscr{E}^{(v)}, \qquad 1 \leq v \leq k$$

where the operators in (iv) describe certain continuity conditions on the m-th through the $(2m-1)$th derivatives of f_* at the interior mesh points $x_1, \ldots, x_k$. These operators will be described shortly. f_*, then, serves to solve uniquely the multipoint boundary value problem (1.9) with homogeneous equation and inhomogeneous boundary conditions. In [6], f_* is called a (generalized) spline function determined by the singular self-adjoint differential operator Λ. Thus, in the remainder of this paper we will be concerned with the solution of the following multipoint boundary value problem with inhomogeneous equation and homogeneous boundary conditions.

$$\text{(i)} \qquad \Lambda f_* = g_* \qquad g_* \in L^2(a, b)$$

$$\text{(ii)} \qquad \lambda_i f_* = 0 \qquad 1 \leq i \leq N$$

(1.10)

$$\text{(iii)} \qquad f_* \in H[a, b] \cap H_\Lambda\left(\bigcup_{v=0}^{k} (x_v, x_{v+1})\right)$$

$$\text{(iv)} \qquad [R_\mu^{(v)} f_*]_{x_v} = 0 \quad \mu \notin \mathscr{E}^{(v)}, \qquad 1 \leq v \leq k.$$

Here $H_\Lambda\left(\bigcup_{v=0}^{k} (x_v, x_{v+1})\right)$ denotes the class of functions f in $C^{2m-1}\left(\bigcup_{v=0}^{k} (x_v, x_{v+1})\right)$ such that $D^{2m-1}f$ is absolutely continuous on each interval (x_v, x_{v+1}), $1 \leq v \leq k-1$, and such that $D^{2m-1}f$ is absolutely continuous on compact subintervals of $(x_0, x_1]$ and $[x_k, x_{k+1})$, with $\Lambda f \in L^2(a, b)$.

Finally, we define the operators of (1.10 iv). Let $U^{(v)}$, $0 \leq v \leq k+1$, denote the nonsingular lower triangular matrices such that

$$(1.11) \qquad U^{(v)}\begin{bmatrix} D^0 f(x_v) \\ \vdots \\ D^{m-1} f(x_v) \end{bmatrix} = \begin{bmatrix} E_0^{(v)} f \\ \vdots \\ E_{m-1}^{(v)} f \end{bmatrix}.$$

Here, we have set $E_\mu^{(v)} f = D^\mu f(x_v)$ if $\mu \notin \mathscr{E}^{(v)}$. Now define

$$(1.12) \qquad Y^{(v)} = (U^{(v)T})^{-1}$$

and define the operators $\{R_\mu^{(v)}\}_{\mu=0}^{m-1}$, $1 \leq v \leq k$, by

$$(1.13) \qquad Y^{(v)} \begin{bmatrix} P_0 f \\ \vdots \\ P_{m-1} f \end{bmatrix} = \begin{bmatrix} R_0^{(v)} f \\ \vdots \\ R_{m-1}^{(v)} f \end{bmatrix}$$

for all appropriate f on $\bigcup\limits_{v=0}^{k} (x_v, x_{v+1})$ where

$$(1.14) \quad P_\mu f(x) = \sum_{i=0}^{m-\mu-1} (-1)^{i+1} D^i (a_{i+\mu+1} D^{i+\mu+1} f)(x) \qquad \left(x \in \bigcup_{v=0}^{k} (x_v, x_{v+1}) \right).$$

Finally

$$(1.15) \qquad [R_\mu^{(v)} f]_{x_v} = R_\mu^{(v)} f(x_v +) - R_\mu^{(v)} f(x_v -) \qquad (1 \leq v \leq k).$$

We remark that the operators of (1. 15) facilitate the integration by parts formula

$$(1.16) \qquad B(\varphi, f) = \int_{(a,b)} \varphi \Lambda f + \sum_{v=1}^{k} \sum_{\mu=0}^{m-1} E_\mu^{(v)} \varphi [R_\mu^{(v)} f]_{x_v}$$

where $\varphi \in H[a, b]$ has compact support on (a, b) and f satisfies (1. 10 iii).

In section 2, we will derive interpolation and Sobolev-type inequalities, necessary for the sequel, which are of some interest in their own right. In section 3 we completely solve the multipoint boundary value problem (1. 10) for $g_* \in L^2(a, b)$ by constructing an inverse G for the (self-adjoint) operator Λ with domain a suitable linear subspace of the domain defined by (1. 10) ii), (1. 10 iii), (1. 10 iv) and range $L^2(a, b)$.

In section 4 we relax the assumption (1. 4) and replace Λ by $\Lambda + tI$. In this connection we explicitly describe the properties of the (Green's) kernel for G. Finally, in section 5 we discuss optimal approximation schemes to approximate the solution of (1. 10) in terms of the eigenfunctions of Λ, where convergence rates are presented in terms of n-widths.

2. Generalized Sobolev and interpolation inequalities

We will establish the following fundamental inequalities for later use. Theorem 1 and Corollary 2 are interpolation inequalities and Theorem 3 and Corollary 4 are Sobolev-type inequalities.

THEOREM 1. *Suppose* $0 < \varepsilon \leq 1$ *and* $m \geq 2$. *Then, for* $f \in H[a, b]$ *and* $1 \leq j \leq m-1$,

$$(2.1) \qquad \|f\|_{H^j[a,b]}^2 \leq \gamma \left(\zeta(\varepsilon, m-j) \left\| \sqrt{a_m}\, D^m f \right\|_{L^2(a,b)}^2 + \varepsilon^{-j} \|f\|_{L^2(a,b)}^2 \right) \leq$$

$$\leq \gamma \left(\zeta(\varepsilon, m-j) \||f\||_{H[a,b]}^2 + \varepsilon^{-j} \|f\|_{L^2(a,b)}^2 \right),$$

474   J. W. JEROME

where γ depends only on m, and $(b-a)$,

$$|||f|||^2_{H[a,b]} = \left\|\sqrt{a_m}\, D^m f\right\|^2_{L^2(a,b)} + \|f\|^2_{L^2(a,b)},$$

and

$$\zeta(\varepsilon, m-j) = \varepsilon^{m-j-1/2}\,\omega\left(1/a_m, \sqrt{\varepsilon}\,(b-a)\right),$$

where $\omega(1/a_m, h)$ is defined for all $0\leq h\leq b-a$ by,

$$\omega(h) = \omega(1/a_m, h) = \sup\left\{\left|\int_x^y \frac{dt}{a_m(t)}\right| : |y-x| \leq h\right\}.$$

COROLLARY 2. *If $m\geq 1$ and $f\in H[a, b]$ then, for $f\neq 0$ and $1\leq j\leq m$, if τ_f denotes the quotient $\|f\|_{L^2(a,b)}/|||f|||_{H[a,b]}$,*

$$(2.2) \qquad \|f\|_{H^j[a,b]} \leq \gamma\|f\|^{1-j/m}_{L^2(a,b)}\, |||f|||^{j/m}_{H[a,b]} \cdot \left\{1 + \frac{\omega(\tau_f^{1/m}c)}{\tau_f^{1/m}}\right\}^{1/2},$$

where γ depends only on m and $(b-a)$ and $c = b-a$.

THEOREM 3. *Let $r\geq 1$, $m\geq 1$ and $0\leq j\leq m-1$. Then, if $f\in H[a, b]$ and $a\leq x$, $y\leq b$,*

$$(2.3) \qquad |D^j f(x)| \leq \gamma r^{-(m-j-1)}\,\Omega(r)\left[\left\|\sqrt{a_m}\,D^m f\right\|_{L^2(a,b)} + \frac{r^{m-1/2}}{\Omega(r)}\,\|f\|_{L^2(a,b)}\right]$$

where γ depends only on m and $(b-a)$, and where

$$\Omega(r) = \left\{\omega(1/a_m, c(a, b)r^{-1})\right\}^{1/2}, \quad c(a, b) = \min(1, (b-a)/2).$$

COROLLARY 4. *Let $r\geq 1$, $m\geq 1$ and $0\leq j\leq m-1$. Then, we have, if $f\in H[a, b]$, $a\leq x\leq b$ and $\omega(1/a_m, h)$ satisfies*

$$\omega(1/a_m, h) \leq c_2 h^{1-\varrho} \qquad (0<c_2;\ 0\leq\varrho<1),$$

$$(2.4) \qquad |D^j f(x)| \leq \gamma r^{-(m-j-1/2-\varrho/2)}[|||f|||_{H[a,b]} + r^{m-\varrho/2}\|f\|_{L^2(a,b)}],$$

where γ depends only on m and $(b-a)$.

Corollary 2 follows directly from Theorem 1 if, for $f\neq 0$, we set $0<\varepsilon = =(\|f\|_{L^2(a,b)}/|||f|||_{H[a,b]})^{2/m}\leq 1$. Then, by Theorem 1, if $m\geq 2$ and $1\leq j\leq m$,

$$\|f\|^2_{H^j[a,b]} \leq \gamma\left\{\|f\|^{2-(2j+1)/m}_{L^2(a,b)}|||f|||^{2(j+1/2)/m}_{H[a,b]}\,\omega(\tau_f^{1/m}) + \|f\|^{2(1-j/m)}_{L^2(a,b)}|||f|||^{2j/m}_{H[a,b]}\right\}.$$

Thus,

$$\|f\|_{H^j[a,b]} \leq (\gamma)^{1/2}\|f\|^{1-j/m}_{L^2(a,b)}|||f|||^{j/m}_{H[a,b]}\left\{1 + \frac{\omega(\tau_f^{1/m}c)}{\tau_f^{1/m}}\right\}^{1/2}$$

and the corollary follows with constant $(\gamma)^{1/2}$. Corollary 4 follows directly from Theorem 3 and the inequalities

$$\omega\big(1/a_m,\, c(a,b)r^{-1}\big) \le c_2\big(c(a,b)r^{-1}\big)^{1-\varrho},$$

$$\big\|\sqrt{a_m}\,D^m f\big\|_{L^2(a,b)} \le \||f|\|_{H[a,b]},$$

the first two holding by hypotesis.

It remains then to prove Theorem 1 and Theorem 3. For the proof of the former, we first prove the following.

LEMMA 5. *If* $f \in C^1[a,b]$, *with* Df *absolutely continuous on* $[a,b]$ *and* $\sqrt{a_m}\,D^2 f \in L^2(a,b)$, *and* $0 < \varepsilon \le 1$, *then*

$$(2.5)\qquad \int_a^b |Df(x)|^2\,dx \le \gamma\Big[\zeta(\varepsilon,1)\int_a^b a_m(x)\cdot|D^2 f(x)|^2\,dx + \varepsilon^{-1}\int_a^b |f(x)|^2\,dx\Big],$$

$$\gamma = 54\,\max\,[(b-a),\ 4/(b-a)^2].$$

PROOF. We establish first the inequality

$$(2.6)\qquad \int_a^b |Df(x)|^2\,dx \le 54\Big[(b-a)\Big(\int_a^b \frac{dx}{a_m(x)}\Big)\int_a^b a_m(x)|D^2 f(x)|^2\,dx + \frac{1}{(b-a)^2}\int_a^b |f(x)|^2\,dx\Big]$$

which, when $a_m(x) \equiv 1$, reduces to Theorem 3.1 of [1]. As established in the proof of that theorem, if $a=0$, $b=1$ and $0 < \eta < 1/2$, then, for $0 \le x \le 1$,

$$|Df(x)|^2 \le \Big\{\frac{1}{\eta(1-2\eta)}\int_0^\eta |f(\xi)|\,d\xi + \int_{1-\eta}^1 |f(\xi)|\,d\xi + \Big(\int_0^1 |D^2 f(\xi)|\,d\xi\Big)\Big\}^2$$

from which it follows that

$$(2.7)\qquad |Df(x)|^2 \le \frac{4}{\eta}\Big(\frac{1}{1-2\eta}\Big)^2\Big\{\int_0^\eta |f(\xi)|^2\,d\xi + \int_{1-\eta}^1 |f(\xi)|^2\,d\xi\Big\} + 2\Big(\int_0^1 \frac{d\xi}{a_m(\xi)}\Big)\int_0^1 a_m(\xi)|D^2{}_J(\xi)|^2\,d\xi.$$

Integrating (2.7) over $[0,1]$ and noticing, as remarked in [1], that $\dfrac{4}{\eta}\Big(\dfrac{1}{1-2\eta}\Big)^2$ has its minimum on $(0,1/2)$ at $\eta=1/6$, yield (2.6) when $a=0$ and $b=1$. The affine transformation $\xi = x(b-a)+a$ yields (2.6) for general a and b. To obtain (2.5), which generalizes equation (3.3) of [1], let $a=t_0 < < t_1 < \cdots < t_k = b$ be a partition of $[a,b]$ such that $\frac{1}{2}\sqrt{\varepsilon}(b-a) < \max_{0 \le i \le k-1} (t_{i+1}-t_i) < \sqrt{\varepsilon}(b-a)$. Then, by (2.6), for $0 \le i \le k-1$,

$$\int_{t_i}^{t_{i+1}} |Df(x)|^2\,dx \le 54\Big[\sqrt{\varepsilon}\,(b-a)\omega(\sqrt{\varepsilon}\,(b-a))\int_{t_i}^{t_{i+1}} a_m(x)|D^2 f(x)|^2\,dx + \frac{4}{\varepsilon(b-a)^2}\int_{t_i}^{t_{i+1}} |f(x)|^2\,dx\Big] \le$$

$$\le \gamma\Big[\zeta(\varepsilon,1)\int_{t_i}^{t_{i+1}} a_m(x)|D^2 f(x)|^2\,dx + \varepsilon^{-1}\int_{t_i}^{t_{i+1}} |f(x)|^2\,dx\Big]$$

and (2.5) follows by addition.

PROOF OF THEOREM 1. We will show that, if $f \in H[a, b]$,

$$(2.8) \qquad |f|_j^2 \leqq \gamma \left(\zeta(\varepsilon, m-j) \|\sqrt{a_m}\, D^m f\|_{L^2(a,b)}^2 + \varepsilon^{-j} \|f\|_{L^2(a,b)}^2 \right)$$

for $1 \leqq j \leqq m-1$, where $|f|_j^2 = \|D^j f\|_{L^2(a,b)}^2$ and where γ depends only on m, $(b-a)$ and $\displaystyle\int_a^b \frac{dx}{a_m(x)}$. (2.8)

then implies the first inequality of (2.1) with γ replaced by $(j+1)\gamma$ in (2.1). The verification of (2.8) proceeds by induction on $m-j$. First, suppose $m-j = 1$, i.e., $j = m-1$. Then, by Lemma 5,

$$(2.9) \qquad |f|_{m-1}^2 \leqq \gamma \left(\zeta(\varepsilon, 1) \|\sqrt{a_m}\, D^m f\|_{L^2[a,b]}^2 + \varepsilon^{-1} |f|_{m-2}^2 \right)$$

where $\gamma = 54 \max\left[(b-a),\, 4/(b-a)^2 \right]$. Now by Theorem 3.3 of [1], (2.9) implies

$$(2.10) \qquad |f|_{m-1}^2 \leqq \gamma \left(\zeta(\varepsilon, 1) \|\sqrt{a_m}\, D^m f\|_{L^2(a,b)}^2 + \varepsilon^{-1} \gamma^* \{ \varepsilon^* |f|_{m-1}^2 + \varepsilon^{*-m+2} \|f\|_{L^2(a,b)}^2 \} \right)$$

where $\varepsilon^* = \varepsilon\delta$, $\delta = \min(1/2\gamma\gamma^*, 1)$ and $\gamma^* > 0$ depends only on m and $(b-a)$. Thus,

$$|f|_{m-1}^2 \leqq \gamma\zeta(\varepsilon, 1) \|\sqrt{a_m}\, D^m f\|_{L^2(a,b)}^2 + \frac{1}{2}|f|_{m-1}^2 + \frac{1}{2}\varepsilon^{-m+1}\delta^{-m+1}\|f\|_{L^2(a,b)}^2$$

which implies that

$$|f|_{m-1}^2 \leqq 2\gamma\zeta(\varepsilon, 1) \|\sqrt{a_m}\, D^m f\|_{L^2(a,b)}^2 + \varepsilon^{-m+1}\delta^{-m+1}\|f\|_{L^2(a,b)}^2$$

which establishes (2.8) when $j = m-1$ with γ in (2.8) replaced by $\max(2\gamma, \delta^{-m+1})$. Assuming now that (2.8) has been established for all $m-1 > m-j \geqq k \geqq 1$ we show that (2.8) is valid for $m-j = k+1$, i.e., $j = m-k-1$. By Theorem 3.3 of [1],

$$|f|_j^2 = |f|_{m-k-1}^2 \leqq \gamma^* \left(\varepsilon^k |f|_{m-1}^2 + \varepsilon^{-(m-k-1)} \|f\|_{L^2(a,b)}^2 \right)$$

and, applying (2.8) for $j = m-1$, we obtain

$$|f|_j^2 \leqq \gamma^* \left[\varepsilon^k \gamma\zeta(\varepsilon, 1) \|\sqrt{a_m}\, D^m f\|^2 + \varepsilon^k \gamma \varepsilon^{-(m-1)} \|f\|_{L^2(a,b)}^2 + \varepsilon^{-(m-k-1)} \|f\|_{L^2(a,b)}^2 \right] \leqq$$

$$\leqq \gamma(\gamma^* + 1) \left(\zeta(\varepsilon, k+1) \|\sqrt{a_m}\, D^m f\|_{L^2(a,b)}^2 + \varepsilon^{-(m-k-1)} \|f\|_{L^2(a,b)}^2 \right)$$

which establishes (2.8) for $j = m-k-1$. This completes the proof of Theorem 1.

PROOF OF THEOREM 3. The proof is an adaptation of the proof of Theorem 3.9 (Sobolev's inequality) in [1]. Let $f \in H[a, b]$ and $x \in [a, b]$. Then Taylor's theorem with integral remainder gives, for each ξ such that $x + \xi \in [a, b]$,

$$f(x) = \sum_{j=0}^{m-1} \xi^j f_j(x+\xi) + m \int_x^{x+\xi} (t-x)^{m-1} f_m(t)\, dt$$

where $f_j(t) = \dfrac{(-1)^j}{j!} D^j f(t)$. Thus,

$$(2.11) \qquad |f(x)|^2 \leqq 2^m \left\{ \sum_{j=0}^{m-1} \xi^{2j} |f_j(x+\xi)|^2 + m^2 \left| \int_x^{x+\xi} (t-x)^{m-1} f_m(t)\, dt \right|^2 \right\}.$$

Now if $r \geqq 1$ is given, choose h_0 such that

$$(2.12) \qquad \begin{cases} |h_0| = \min(1, (b-a)/2) \\ x + h_0 \in [a, b] \end{cases}$$

and define

$$(2.13) \qquad h = \frac{h_0}{r}.$$

Since

$$\left|\int_x^{x+\xi}(t-x)^{m-1}f_m(t)\,dt\right|^2 \leq \left|\int_x^{x+\xi}\frac{(t-x)^{2m-2}}{a_m(t)}\,dt\right|\cdot\left|\int_x^{x+\xi}a_m(t)\,|f_m(t)|^2\,dt\right| \leq$$

$$\leq \xi^{2m-2}\omega(1/a_m,|\xi|)\left|\int_x^{x+\xi}a_m(t)\,|f_m(t)|^2\,dt\right|$$

we obtain from (2. 11), for ξ between 0 and $0+h$,

$$(2.\,14)\qquad |f(x)|^2 \leq 2^m\left\{\sum_j h^{2j}|f_j(x+\xi)|^2+m^2h^{2m-2}\omega(1/a_m,|h|)\left|\int_x^{x+h}a_m(t)\,|f_m(t)|^2\,dt\right|\right\}$$

and, integrating both sides with respect to ξ as $x+\xi$ varies over the interval I_x with endpoints x and $x+h$, we get

$$|h|\,|f(x)|^2 \leq 2^m\left\{\sum_{j=0}^{m-1} h^{2j}\int_{I_x}|f_j(x+\xi)|^2\,d(x+\xi)+m^2|h|^{2m-1}\omega(1/a_m,|h|)\left|\int_x^{x+h}a_m(t)\,|f_m(t)|^2\,dt\right|\right\}.$$

Thus,

$$(2.\,15)\qquad |f(x)|^2 \leq c\left\{\sum_{j=0}^{m-1}|h|^{2j-1}|f|_j^2+|h|^{2m-2}\omega(1/a_m,|h|)\,\|\sqrt{a_m}\,D^m f\|_{L^2[a,b]}^2\right\}$$

where $1<c=m^2 2^m$. By (2. 8) with $\varepsilon=h^2$, and $0\leq j<m$,

$$(2.\,16)\qquad |f|_j^2 \leq \gamma\big(|h|^{2(m-j)-1}\omega(1/a_m,|h|)\,\|\sqrt{a_m}\,D^m f\|_{L^2(a,b)}^2+h^{-2j}\|f\|_{L^2(a,b)}^2\big)$$

so that (2. 15) and (2. 16) imply

$$|f(x)|^2 \leq m\gamma c\big(|h|^{2m-2}\omega(1/a_m,|h|)\,\|\sqrt{a_m}\,D^m f\|_{L^2(a,b)}^2+|h|^{-1}\|f\|_{L^2(a,b)}^2\big)+$$

$$+ch^{2m-2}\omega(1/a_m,|h|)\,\|\sqrt{a_m}\,D^m f\|_{L^2(a,b)}^2\,.$$

Thus, using the inequality $(|\alpha|^2+|\beta|^2)^{1/2} \leq |\alpha|+|\beta|$, we have

$$|f(x)| \leq \gamma'\left\{|h|^{m-1}(\omega(1/a_m,|h|))^{1/2}\,\|\sqrt{a_m}\,D^m f\|_{L^2(a,b)}+|h|^{-1/2}\|f\|_{L^2(a,b)}\right\}.$$

Using (2. 13) yields (2. 3) with $j=0$. If $0\leq j<m$, a similar analysis gives

$$|h|\,|D^j f(x)|^2 \leq 2^{m-j}\left\{\sum_{\nu=0}^{m-j-1} h^{2\nu}\int_{I_x}|(D^j f)_\nu(x+\xi)|^2\,d(x+\xi)+\right.$$

$$\left.+(m-j)^2|h|^{2m-2j-1}\omega(1/a_m,|h|)\cdot\left|\int_x^{x+h}a_m(t)\,|(D^j f)(t)|_{m-j}^2\,dt\right|\right\},$$

which leads to the inequality

$$(2.\,17)\quad |D^j f(x)|^2 \leq c\left\{\sum_{\nu=0}^{m-j-1}|h|^{2\nu-1}|f|_{\nu+j}^2+h^{2m-2j-2}\omega(1/a_m,|h|)\,\|\sqrt{a_m}\,D^m f\|_{L^2(a,b)}^2\right\}.$$

Using (2. 8), with $\varepsilon=h^2$ and $0 \leq \nu+j < m$, we obtain from (2. 17) the inequality

$$|D^j f(x)|^2 \leq \gamma(m-j)c\{|h|^{2m-2j-2}\omega(1/a_m,|h|)\,\|\sqrt{a_m}\,D^m f\|_{L^2(a,b)}^2+|h|^{-(2j+1)}\|f\|_{L^2(a,b)}^2\}+$$

$$+ch^{2m-2j-2}\omega(1/a_m,|h|)\,\|\sqrt{a_m}\,D^m f\|_{L^2(a,b)}^2$$

so that

$$|D^j f(x)| \leq \gamma'[|h|^{m-j-1}\omega(1/a_m,|h|)^{1/2}\,\|\sqrt{a_m}\,D^m f\|_{L^2(a,b)}+|h|^{-(j+1/2)}\|f\|_{L^2(a,b)}]$$

which implies (2. 3) upon using (2. 13).

3. The boundary value problem

It was established in [6] that there exist positive constants α and C such that

$$(3.1) \quad B(f,f) + C(f,f)_{H^{m-1}[a,b]} \geqq \alpha\left[(a_m D^m f, D^m f)_{L^2(a,b)} + \sum_{j=0}^{m-1}(D^j f(a))^2\right]$$

for all $f \in H[a, b]$ for any bilinear form B satisfying (1.2) and (1.1). (1.3) then follows from (2.1) and (3.1). However, rather than solve (1.10) with Λ replaced by $\Lambda + tI$ for sufficiently large t, we have chosen, in this section, as remarked in the introduction, to assume (1.4), which leads to the inner product (1.5) on the Hilbert space $H[a, b]$.

We now formally define a symmetric operator Λ_0 on $L^2(a, b)$ by defining $\mathscr{D}_{\Lambda_0}$ to be the class of functions with compact support on (a, b) satisfying (1.10 ii, iii, iv) with

$$\Lambda_0 f(x) = \Lambda f(x), \quad x \in \bigcup_{v=0}^{k}(x_v, x_{v+1}), f \in \mathscr{D}_{\Lambda_0}.$$

The integration by parts formula (1.16) implies that Λ_0 is symmetric on $L^2(a, b)$.

THEOREM 6. *The operator Λ_0 from $L^2(a, b)$ to $L^2(a, b)$ with domain $\mathscr{D}_{\Lambda_0}$ has a unique self-adjoint extension Λ_1 with domain a subspace of $H_0[a, b]$. Λ_1 satisfies*

$$(i) \quad \mathscr{D}_{\Lambda_1} \subset \{f : f \text{ satisfies } (1.10 \text{ ii, iii, iv})\}$$

(3.2)

$$(ii) \quad \Lambda_1 f(x) = \Lambda f(x), \quad x \in \bigcup_{v=0}^{k}(x_v, x_{v+1}), f \in \mathscr{D}_{\Lambda_1}.$$

Finally, Λ_1 is invertible with a compact inverse G, defined on all of $L^2(a, b)$.

As an immediate corollary to Theorem 6 we have

COROLLARY 7. *The solution of the multipoint boundary value problem (1.10) is given by $f_* = Gg_*$, where $G = \Lambda_1^{-1}$.*

PROOF OF THEOREM 6. By the Friedrichs extension construction [7, section 124] there exists a self-adjoint operator Λ_0^1, extending the symmetric operator $\Lambda_0 + I$, where

$$(3.3) \quad (f, g)_{H[a,b]} = (\Lambda_0^1 f, g)_{L^2(a,b)} \quad (f \in \mathscr{D}_{\Lambda_0^1}, g \in H_0[a, b])$$

and $\mathscr{D}_{\Lambda_0^1} \subset H_0[a, b]$ is determined by (3.3). We define

$$(3.4) \quad \Lambda_1 = \Lambda_0^1 - I.$$

Λ_1 is clearly a positive self-adjoint extension of Λ_0 satisfying

$$(3.5) \qquad B(f, g) = (\Lambda_1 f, g)_{L^2(a,b)} \qquad (f \in \mathscr{D}_{\Lambda_1}, g \in H_0[a, b])$$

with $\mathscr{D}_{\Lambda_1} \subset H_0[a, b]$ defined by (3. 5).

We will now show that Λ_1 satisfies (3. 2). Suppose then $f \in \mathscr{D}_{\Lambda_1}$. We first show $f \in H_\Lambda(\bigcup_{\nu=0}^{k} (x_\nu, x_{\nu+1}))$. Indeed, if $u = \Lambda_1 f$, then by (3. 5) if $g = \varphi_\nu \in C^\infty(a, b)$ has compact support in one of the intervals $(x_\nu, x_{\nu+1})$, we have

$$(u, \varphi_\nu)_{L^2(a,b)} = B(f, \varphi_\nu) = (f, \Lambda \varphi_\nu)_{L^2(a,b)}$$

i.e., f is a solution, in the sense of distributions, of the equation

$$(3.6) \qquad \Lambda f(x) = u(x) \qquad (x \in (x_\nu, x_{\nu+1})).$$

It follows from [3, section 8] that $f \in H_\Lambda(\bigcup_{\nu=0}^{k} (x_\nu, x_{\nu+1}))$ and that (3. 6) holds classically, i.e., for almost all $x \in (x_\nu, x_{\nu+1})$. Thus (3. 2 ii) holds. We need only show that $[R_\mu^{(\nu)} f]_{x_\nu} = 0$ if $\mu \notin \mathscr{E}^{(\nu)}$, $1 \leq \nu \leq k$. Fix indices $1 \leq \nu_* \leq k$ and $\mu_* \notin \mathscr{E}^{(\nu*)}$ (if they exist). Choose an infinitely differentiable function φ with support on $(x_{\nu_*-1}, x_{\nu_*+1})$ such that $E_\mu^{(\nu*)} \varphi = 0$ if $\mu \neq \mu_*$ and $D^{\mu_*} \varphi(x_{\nu_*}) = E_{\mu_*}^{(\nu*)} \varphi = 1$. Then, by (1. 16), (3. 2 ii) and (3. 5),

$$(\Lambda f, \varphi)_{L^2(a,b)} = (\Lambda_1 f, \varphi)_{L^2(a,b)} = B(f, \varphi) = (\Lambda f, \varphi)_{L^2(a,b)} + [R_{\mu_*}^{(\nu*)} f]_{x_{\nu_*}}$$

so that $[R_{\mu_*}^{(\nu*)} f]_{x_{\nu_*}} = 0$.

Since the injection

$$I: H[a, b] \to L^2(a, b)$$

is compact it follows that the self-adjoint operator Λ_1 has the spectral representation

$$(3.7) \qquad \Lambda_1 = \sum_{n=1}^{\infty} \lambda_n P_n,$$

where $0 \leq \lambda_n \nearrow \infty$ and P_n are finite-dimensional projectors onto the eigenspaces corresponding to λ_n. We claim that $\lambda_1 > 0$. Indeed, if $\Lambda_1 \varphi = 0$ then it follows that $\varphi \in C^{2m}(x_\nu, x_{\nu+1})$ for each $0 \leq \nu \leq k$ and $D^j \varphi(x_\nu) = 0$ for $0 \leq j \leq m-1$ and $0 \leq \nu \leq k+1$ and $\Lambda \varphi(x) = 0$ for $x \in \bigcup_{\nu=0}^{k} (x_\nu, x_{\nu+1})$. Now the dimension of the null space of Λ, restricted to an arbitrary interval $(x_\nu, x_{\nu+1})$, is precisely $2m$. Since $D^j(\cdot)(x_\nu)$, $0 \leq j \leq m-1$ and $D^j(\cdot)(x_{\nu+1})$ constitute $2m$ linearly independent functionals vanishing over this subspace it follows that $\varphi(x) \equiv 0$ in each $(x_\nu, x_{\nu+1})$ and hence $\varphi(x) \equiv 0$ on $[a, b]$ by continuity. Thus, 0 cannot be an eigenvalue of Λ_1. Thus,

$$(3.8) \qquad B(f, f) = \sum_{n=1}^{\infty} \lambda_n \|P_n f\|^2_{L^2(a,b)} \geq \lambda_1 \sum_{n=1}^{\infty} \|P_n f\|^2_{L^2(a,b)} = \lambda_1 (f, f)_{L^2(a,b)}$$

for all $f \in \mathcal{D}_{A_1}$. It follows from this that the range of A_1 is $L^2(a, b)$. In fact, since A_1 is self-adjoint, its Friedrich's extension, guaranteed to be onto by (3. 8), coincides with itself. Thus, A_1 is invertible with a self-adjoint inverse G, bounded as an operator from $L^2(a, b)$ to $\mathcal{D}_{A_1} \subset H_0[a, b]$ and hence compact. We must show, finally, that A_1 is a unique self-adjoint extension with domain a subspace of $H_0[a, b]$. If $\bar{A}_0$ is any self-adjoint extension of A_0 with domain $\mathcal{D}_{\bar{A}_0} \subset H_0[a, b]$ then by (3. 5), for each fixed $f \in \mathcal{D}_{\bar{A}_0}$ and for all $g \in \mathcal{D}_{A_0}$,

$$B(G\bar{A}_0 f, g) = (\bar{A}_0 f, g)_{L^2(a,b)} = (f, \bar{A}_0 g)_{L^2(a,b)} =$$

$$= (f, A_0 g)_{L^2(a,b)} = (A_0 g, f)_{L^2(a,b)} = (A_1 g, f)_{L^2(a,b)} = B(f, g).$$

Thus, $B(G\bar{A}_0 f - f, g) = 0$ for all g in $\mathcal{D}_{A_0}$. But $\mathcal{D}_{A_0}$ is dense in $H_0[a, b]$ when the norm $B(u, u)$ is chosen. Hence, $f = G\bar{A}_0 f$ which demonstrates that $f \in \mathcal{D}_{A_1}$; also, $A_1 f = (A_1 G)\bar{A}_0 f = \bar{A}_0 f$, so that A_1 is a self-adjoint extension of $\bar{A}_0$; thus, $A_1 = \bar{A}_0$. This completes the proof of the theorem.

4. The Green's kernel

In this section we consider the solution of the multipoint boundary-value problem

(4. 1)

(i) $\qquad (A + tI)f_* = g_* \qquad (g_* \in L^2(a, b))$

(ii) $\qquad \lambda_i f_* = 0 \qquad (1 \leq i \leq N)$

(iii) $\qquad f_* \in H[a, b] \cap H_A\left(\bigcup_{\nu=0}^{k} (x_\nu, x_{\nu+1})\right)$

(iv) $\qquad [R_\mu^{(\nu)} f_*]_{x_\nu} = 0 \quad \text{if} \quad \mu \notin \mathscr{E}^{(\nu)} \qquad (1 \leq \nu \leq k),$

where $t \geq C_0$ in (1. 3). Here we do not assume that (1. 4) holds. Also, the operators in (4. 1 iv) are defined relative to $A + tI$; in particular, in (1. 14), a_0 is replaced by $a_0 + t$.

THEOREM 8. *The multipoint boundary value problem* (4. 1) *has a unique solution. Furthermore, there exists a symmetric real-valued function* $G_t(x, y)$ *in* $C([a, b] \times [a, b])$ *satisfying*

(4. 2)

(i) $\quad (A_1 + tI)^{-1} g(x) = (g, G_t(x, \cdot))_{L^2(a,b)}, \, g \in L^2(a, b),$

(ii) $\qquad G_t(x, \cdot) \in H_0[a, b] \quad \text{for all} \quad x \in [a, b],$

(iii) $\qquad |G_t(x, y)| \leq C\Phi(t) \quad (x, y) \in [a, b] \times [a, b],$

where C does not depend on t, x, or y, and $\Phi(t)$ is a strictly decreasing function of t on $[C_0, \infty)$ given by (4. 5) and (4. 6) below

$$\text{(iv)} \quad \frac{\partial^i G_t(x, y)}{\partial x^i} \quad and \quad \frac{\partial^j G_t(x, y)}{\partial y^j} \quad are \ in \ C([a, b] \times [a, b]),$$

$$0 \leq i, j \leq m - 1$$

$$\text{(v)} \quad G_t(x, y) = \sum_{j=1}^{\infty} \frac{\varphi_j(x)\varphi_j(y)}{\lambda_j + t},$$

where the convergence is uniform in (x, y). Here, Λ_1 is a self-adjoint extension of the semi-bounded, symmetric operator Λ_0 defined in section 3, $(\Lambda_1 + tI)^{-1}$ is compact and $\{\lambda_j\}_1^{\infty}$, $\{\varphi_j\}_1^{\infty}$ are the eigenvalues and L^2-normalized eigenfunctions of Λ_1, the latter being complete and orthogonal in both $L^2(a, b)$ and $H_0[a, b]$, with the inner product in this space given by

$$(4. 3) \qquad (f, g)_t = B(f, g) + t(f, g)_{L^2(a,b)}.$$

Finally, the domain of Λ_1 is a subspace of the class satisfying (4. 1 ii, iii, iv).

PROOF. As in the proof of Theorem 6, the Friedrich's extension construction is utilized to obtain a self-adjoint extension $\Lambda_1 + tI$ of $\Lambda_0 + tI$ with domain $\mathscr{D}_{\Lambda_1} = \mathscr{D}_{\Lambda_1 + tI}$ a subspace of the class satisfying (4. 1 ii, iii, iv). $\Lambda_1 + tI$ is a compact, invertible, self-adjoint operator on $L^2(a, b)$. In particular, (4. 1) has a unique solution.

Now since the linear functional $x \to f(x)$ is continuous on $H_0[a, b]$, for each fixed $x \in [a, b]$, it follows from the Riesz representation theorem that there exists $G_{t,x} \in H_0[a, b]$ such that

$$(4. 4) \qquad f(x) = (f, G_{t,x})_t \quad for \ all \ f \in H_0[a, b].$$

Since $\mathscr{R}_{\Lambda_1 + tI} = L^2(a, b)$, (4. 4) implies (4. 2 i) with $g = (\Lambda_1 + tI)f$ and $G_t(x, y) = G_{t,x}(y)$. This also establishes (4. 2 ii). To establish (4. 2 iii), define, for r in $[1, \infty)$ the function

$$(4. 5) \qquad t = \left[\frac{r^{m-1/2}}{\Omega(r)}\right]^2 \Omega^2(1)C_0,$$

and, by (4. 4) and (2. 3) with $j=0$, we conclude

$$\|G_{t,x}\|_t^4 = |G_{t,x}(x)|^2 \leq 2\gamma^2 \Phi(t)\|G_{t,x}\|_t^2 \max(1, 1/C_0\Omega^2(1)),$$

where $\Phi(t)$, $t \geq C_0$, is a strictly decreasing function of t, given by,

$$(4. 6) \qquad \Phi(t) = \frac{r(t)}{t} = \frac{\Omega^2(r(t))}{[r(t)]^{2m-2}}\left(\frac{1}{C_0\Omega^2(1)}\right).$$

In (4. 6), $r(t)$ is the unique value r on $[1, \infty)$ such that (4. 5) holds for t in $[C_0, \infty)$. Thus,

$$(4.\ 7) \qquad\qquad \|G_{t,x}\|_t^2 \leqq C\Phi(t).$$

Now,

$$(4.\ 8) \qquad\qquad |G_t(x, y)| = |(G_{t,x}, G_{t,y})_t| \leqq \|G_{t,x}\|_t \|G_{t,y}\|_t$$

and (4. 2 iii) follows from (4. 7) and (4. 8).

To establish (4. 2 iv), let $D^\alpha G_t(x, t)$ denote a partial derivative of $G_t(x, y)$ with respect to y, of order $0 \leqq \alpha \leqq m-1$. Suppose that $(x_n, y_n) \to (x_0, y_0) \in [a, b] \times [a, b]$. Then, for $n \geqq 1$,

$$|D^\alpha G_t(x_n, y_n) - D^\alpha G_t(x_0, y_0)| \leqq$$

$$\leqq |D^\alpha G_t(x_n, y_n) - D^\alpha G_t(x_0, y_n)| + |D^\alpha G_t(x_0, y_n) - D^\alpha G_t(x_0, y_0)| \leqq$$

$$\leqq \sup_{y \in [a, b]} |D^\alpha G_{t,x_n}(y) - D^\alpha G_{t,x_0}(y)| + w(D^\alpha G_{t,x_0}, |y_n - y_0|),$$

where

$$w(D^\alpha G_{t,x_0}, \delta) = \sup_{\substack{|y'-y''| \leqq \delta \\ y', y'' \in [a, b]}} |D^\alpha G_{t,x_0}(y') - D^\alpha G_{t,x_0}(y'')|.$$

Now the sequence $\{G_{t,x_n}\}_{n \geqq 1}$ is clearly weakly convergent in $L^2(a, b)$ to G_{t,x_0} from the representation

$$(g, G_{t,x_0})_{L^2(a,b)} = f(x_0) \leftarrow f(x_n) = (g, G_{t,x_n}),$$

where g is an arbitrary element of $L^2(a, b)$ and $f = (\Lambda_1 + tI)^{-1}g$. We will now show that $\{G_{t,x_n}\}_n$ is bounded in $H_0[a, b]$. Indeed, for $n \geqq 1$, we have from (4. 2 iii),

$$(G_{t,x_n}, G_{t,x_n})_t = |G_t(x_n, x_n)| \leqq C\Phi(t),$$

where C does not depend on n. Thus, $\{G_{t,x_n}\}$ is bounded in $H_0[a, b]$ and weakly convergent in $L^2(a, b)$. These facts, plus the compactness of the injection of $H_0[a, b]$ into $C^{m-1}[a, b]$ (see [6]) imply the convergence of G_{t,x_n} in $C^{m-1}[a, b]$ to G_{t,x_0}. Indeed, any subsequence of $\{G_{t,x_n}\}_{n \geqq 1}$ has a subsequence convergent in $C^{m-1}[a, b]$ to an element of $C^{m-1}[a, b]$, which must be G_{t,x_0} by the weak convergence property. This establishes that G_{t,x_n} is convergent to G_{t,x_0} in $C^{m-1}[a, b]$ and, in particular,

$$\sup_{y \in [a, b]} |D^\alpha G_{t,x_n}(y) - D^\alpha G_{t,x_0}(y)| \to 0 \quad \text{as} \quad n \to \infty.$$

Moreover, it is obvious that $w(D^\alpha G_{t,x_0}, |y_n - y_0|) \to 0$ as $n \to \infty$ and it follows that, $D^\alpha G_t(x, y)$ is continuous on $[a, b] \times [a, b]$. In particular, $G_t(x, y)$ is continuous and this implies that $G_t^*(x, y) = G_t(y, x)$ is the kernel of the adjoint of $(\Lambda_1 + tI)^{-1}$, i.e.,

$$(4.\ 9) \qquad [(\Lambda_1 + tI)^{-1}]^* g(x) = (g, G_t^*(x, \cdot))_{L^2(a,b)} \quad \text{for all} \quad g \in L^2(a, b).$$

Since $(\Lambda_1 + tI)^{-1}$ is self-adjoint, it follows from (4. 2 i) and (4. 9) that, for all $g \in L^2(a, b)$,

$$(4. 10) \qquad (g, G_t^*(x, \cdot) - G_t(x, \cdot))_{L^2(a,b)} = 0.$$

(4. 10) implies, that, for each fixed $x \in [a, b]$,

$$G_t(x, \cdot) = G_t^*(x, \cdot) \quad \text{as} \quad L^2(a, b) \quad \text{functions.}$$

From the definition of G_t^* and the continuity of $G_t^*(x, \cdot)$ and $G_t(x, \cdot)$, we deduce the symmetry of $G_t(x, y)$. This completes the proof of Theorem 8, since (4. 2 v) follows from Mercer's theorem.

REMARK. Although perhaps not immediately apparent, the problem of section 3 is a special case of that of section 4. Indeed, if (1. 4) holds, then $\lambda_1 > 0$ and

$$B_*(f, f) = B(f, f) - \lambda_1(f, f)_{L^2(a, b)}$$

is non-negative over $H_0[a, b]$. In particular, the inequality

$$B_*(f, f) + t(f, f)_{L^2(a, b)} \geq \alpha_t(a_m D^m f, D^m f)_{L^2(a, b)}$$

holds for all $f \in H_0[a, b]$ and all $t > 0$. Choosing $t = \lambda_1$, we deduce

$$B(f, f) = B_*(f, f) + \lambda_1(f, f)_{L^2(a, b)} \geq \alpha(a_m D^m f, D^m f)_{L^2(a, b)}.$$

Since $\Lambda = \Lambda_* + \lambda_1 I$, then $\Lambda^{-1} = (\Lambda_* + \lambda I)^{-1} = G$ has a Green's function described by Theorem 8.

The final theorem of this section provides additional information about the function $G_t(x, y)$ of Theorem 8.

THEOREM 9. *Suppose $\xi \in [a, b] - \{x_i\}_0^{k+1}$. Then, for $t \geq C_0$,*

$$(4. 11) \quad \begin{aligned} &\text{(i)} \quad (\Lambda_x + tI)G_t(x, \xi) = 0 \qquad (x \in [a, b] - \{x_i\}_0^{k+1} - \{\xi\}) \\ &\text{(ii)} \quad G_t(\cdot, \xi) \in H_0[a, b] \cap C^{2m}([a, b] - \{x_i\}_0^{k+1} - \{\xi\}) \\ &\text{(iii)} \quad [R_\mu^{(v)} G_t(\cdot, \xi)]_{x_v} = 0 \text{ if } \mu \notin \mathscr{E}^{(v)} \qquad (1 \leq v \leq k). \\ &\text{(iv)} \quad G_t(\cdot, \xi) \text{ is } C^{2m-2} \text{ in a neighbourhood of } \xi. \end{aligned}$$

PROOF. For each $\xi \in [a, b] - \{x_i\}_0^{k+1}$ define $\Gamma_t(\cdot, \xi)$ as the unique solution of the minimization problem

$$(4. 12) \qquad \min \{B(f, f) + t(f, f)_{L^2(a,b)} = (f, f)_t : f \in H_0[a, b], f(\xi) = 1\}.$$

By standard arguments (see [6]) one shows that (i), (ii), (iii) and (iv) hold if G_t is replaced by Γ_t, and that the minimum of (4. 12) is positive. We assert that

$$(4. 13) \qquad G_t(x, \xi) = \frac{\Gamma_t(x, \xi)}{(\Gamma_t(\cdot, \xi), \Gamma_t(\cdot, \xi))_t},$$

$$(x, \xi) \in [a, b] \times [a, b] - \{x_i\}_0^{k+1}.$$

Observe first that, since $\Gamma_t(\cdot, \xi)$ is a basis for the one-dimensional subspace of $H_0[a, b]$ consisting of those functions orthogonal to all functions vanishing at ξ it suffices to show that $G(\cdot, \xi)$ is orthogonal to all functions vanishing at ξ and that $H_t(\cdot, \xi)$, defined by

$$H_{t,\xi}(x) = G_t(x, \xi) - \frac{\Gamma_t(x, \xi)}{(\Gamma_t(\cdot, \xi), \Gamma_t(\cdot, \xi))_t},$$

is orthogonal to $\Gamma_t(\cdot, \xi)$. The first of these two facts follows from (4. 4) and the symmetry of $G_t(x, \xi)$. The second follows from the equations

$$(H_{t,\xi}, \Gamma_t(\cdot, \xi))_t = \Gamma_t(\xi, \xi) - \frac{(\Gamma_t(\cdot, \xi), \Gamma_t(\cdot, \xi))_t}{(\Gamma_t(\cdot, \xi), \Gamma_t(\cdot, \xi))_t} = 1 - 1 = 0$$

for all $\xi \in [a, b] - \{x_i\}_0^{k+1}$. These remarks serve to prove Theorem 9.

5. n-widths and approximations

In this section we assume that (1. 4) holds. From Theorem 6, it follows that the eigenvalue problem

$$\text{(5. 1)}$$

$$\text{(i)} \qquad Au(x) = \lambda u(x) \qquad (x_v < x < x_{v+1})$$

$$\text{(ii)} \qquad E_\mu^{(v)} u(x_v) = 0 \ \text{ if } \ \mu \in \mathscr{E}^{(v)} \qquad (0 \leq v \leq k+1)$$

$$\text{(iii)} \qquad u \in H[a, b] \cap H_A\left(\bigcup_{v=0}^k (x_v, x_{v+1})\right)$$

$$\text{(iv)} \qquad [R_\mu^{(v)} u]_{x_v} = 0 \ \text{ if } \ \mu \notin \mathscr{E}^{(v)} \qquad (1 \leq v \leq k)$$

has positive eigenvalues $0 < \lambda_1 < \lambda_2 \leq \cdots$ of finite multiplicity tending to infinity. The associated eigenfunctions $\varphi_1, \varphi_2, \ldots$ form a complete orthonormal system in $L^2(a, b)$ and a complete orthogonal system in $H_0[a, b]$.

Now if f_* is the solution of (1. 10) then we have the estimate

$$B(f_*, f_*) = (g_*, f_*)_{L^2(a,b)} = \int_a^b g_*(x) B(f_*, G_x) \leq$$

$$\leq \left\{\int_a^b g_2^*(x)\right\}^{1/2} B^{1/2}(f_*, f_*) \sup_{x \in [a,b]} \{G_x(x)\}^{1/2}(b-a)^{1/2}$$

which implies that $B(f_*, f_*) \leq \varrho^2$ where $\varrho^2 = (g_*, g_*)_{L^2(a,b)}(b-a) \max_{x \in [a,b]} G(x, x)$. Now, if we define

$$\text{(5. 2)} \qquad \mathscr{R} = \{f : f \text{ satisfies (5. 1 ii, iii, iv)} \quad \text{and} \quad B(f, f) \leq \varrho^2\},$$

$f^* \in \mathscr{R}$.

We now define the n-width, $d_n(\mathcal{R})$, of $\mathcal{R}$ in $L^2(a, b)$ for $n \geq 0$ by

$$(5.\,3) \qquad d_n(\mathcal{R}) = \inf\{E(\mathcal{R}, \mathcal{M}): \dim \mathcal{M} = n, \quad \mathcal{M} \subset L^2(a, b)\}$$

where

$$E(\mathcal{R}, \mathcal{M}) = \inf_{f \in \mathcal{R}} \sup_{g \in \mathcal{M}} \|f - g\|_{L^2(a,b)}$$

$\mathcal{M}$ is called an optimal approximating subspace if dimension $\mathcal{M} \leq n$ and $E(\mathcal{R}, \mathcal{M}) = d_n(\mathcal{R})$. The following result follows immediately from Theorem 3. 3 of [4].

THEOREM 10. *The n-widths of $\mathcal{R}$ are given by*

$$(5.\,4) \qquad d_n(\mathcal{R}) = \varrho^2 \lambda_{n+1}^{-1/2} \qquad (n \geq 0)$$

where $0 < \lambda_1 \leq \lambda_2 \leq \cdots$ are the eigenvalues of (5. 1). Optimal approximating manifolds of dimension n are given by

$$(5.\,5) \qquad \mathcal{M}_n = \operatorname{span}\{\varphi_1, \ldots, \varphi_n),$$

where φ_v is a normalized eigenfunction of (5. 1) corresponding to λ_v, $v \geq 1$.

We now define the approximations

$$(5.\,6) \qquad \Psi_n = \sum_{v=1}^{n} \frac{1}{\lambda_v} \left\{ \int_a^b g_*(\xi)\varphi_v(\xi)\,d\xi \right\} \varphi_v, \qquad (n \geq 1).$$

It is readily seen that Ψ_n are the Fourier expansions of f_* along the set $\{\varphi_v\}_{v \geq 1}$ in the inner product $B(f, g)$. Moreover, it follows immediately from Theorem 10 that the approximations of (5. 6) are optimal in the sense that $\Psi_n \in \mathcal{M}_n$ and

$$\inf_{g \in \mathcal{M}_n} \|f_* - g\|_{L^2(a,b)} = \|f_* - \Psi_n\|_{L^2(a,b)}.$$

As a result of this discussion, Corollary 2 and the equivalence of the norms $\{B(f,f)\}^{1/2}$ and $|||f|||$ we have

COROLLARY 11. *The approximations defined by (5. 6) satisfy*

$$\text{(i)} \qquad |||f_* - \Psi_n|||_{H[a,b]} = o(1)$$

$$(5.\,7) \qquad \text{(ii)} \quad \|f_* - \Psi_n\|_{H^j[a,b]} = o(\lambda_{n+1}^{(j-m)/2m}\{\omega(\lambda_{n+1}^{-1/2m})\lambda_{n+1}^{1/2m}\}^{1/2}) \qquad (0 < j < m)$$

$$\text{(iii)} \qquad \|f_* - \Psi_n\|_{L^2(a,b)} = O(\lambda_{n+1}^{-1/2}) \qquad (n \to \infty).$$

REMARKS. (1) In sections 3 and 5 the hypothesis (1. 4) need only hold for $f \in H_0[a, b]$.

(2) The author conjectures that the estimates (5. 7) imply that the convergence in (iii) is of order $O(n^{-m})$.

(3) In section 3, the hypothesis (1. 4) can be omitted if it is assumed that 0 is not an eigenvalue of Λ_1. Analogous statements, with respect to t not an eigenvalue, apply in section 4.

REFERENCES

[1] S. Agmon, *Lectures on Elliptic Boundary Value Problems.* Van Nostrand, Princeton, N. J. 1965.

[2] M. Golomb and J. Jerome, *Linear ordinary differential equations with boundary conditions on arbitrary point sets.* Trans. Amer. Math. Soc. **153** (1971), 235—264.

[3] I. Halperin, *Introduction to the Theory of Distributions.* University of Toronto Press, Toronto 1952.

[4] J. W. Jerome, *On the $\mathscr{L}_2$ n-width of certain classes of functions of several variables.* J. Math. Anal. Appl. **20** (1967), 110—123.

[5] J. W. Jerome, *Linear, self-adjoint, multipoint boundary value problems and related approximation schemes.* Numer. Math. **15** (1970), 433—449.

[6] J. W. Jerome and J. Pierce, *On spline functions determined by singuiar, self-adjoint, differential operators.* J. Approximation Theory, **5** (1972), 15—40.

[7] F. Riesz and B. Sz.-Nagy, *Functional Analysis.* Ungar, New York 1955.

Über die Konvergenz von natürlichen interpolierenden Splines

Von

KARL SCHERER[1]

LEHRSTUHL A FÜR MATHEMATIK
TECHNISCHE HOCHSCHULE AACHEN

Gegeben sei eine Zerlegung

$$(1) \qquad \Delta : a = x_0 < x_1 < \cdots < x_N = b$$

des Intervalls $[a, b]$ mit $\bar{\Delta} = \max_i (x_{i+1} - x_i)$ und $\underline{\Delta} = \min_i (x_{i+1} - x_i)$. Auf dem Raum $C[a, b]$ der auf $[a, b]$ stetigen Funktionen f mit Norm $\|f\| = \sup_{x \in [a, b]} |f(x)|$ definiert man als *natürliche interpolierende Splinefunktion zu f der Ordnung* $2n-1$ ($n=1, 2, \ldots$) bezüglich Δ diejenige Funktion $S_{2n-1,\Delta}(f; x)$, die folgende Bedingungen erfüllt:

$$(2) \qquad \begin{cases} S_{2n-1,\Delta}(f; x) \in \mathscr{P}_{2n-1}, & x \in (x_i, x_{i+1}), \quad 0 \le i \le N-1 \\ S_{2n-1,\Delta}(f; x) \in C^{(2n-2)}[a, b], \end{cases}$$

$$(3) \qquad S_{2n-1,\Delta}(f; x_i) = f(x_i), \qquad 0 \le i \le N,$$

$$(4) \qquad S^{(k)}_{2n-1,\Delta}(f; a) = S^{(k)}_{2n-1,\Delta}(f; b) = 0, \qquad n \le k \le 2n-2.$$

In (2) bezeichnet $\mathscr{P}_{2n-1}$ die Klasse aller algebraischen Polynome $(2n-1)$-ten Grades und $C^{(2n-2)}[a, b]$ den Raum aller Funktionen mit auf $[a, b]$ stetiger Ableitung $(2n-2)$-ter Ordnung. (2) besagt dann, daß $S_{2n-1,\Delta}(f; x)$ eine Splinefunktion der Ordnung $2n-1$ bezüglich Δ ist. $S_{2n-1,\Delta}(f; x)$ heißt interpolierend zu f wegen (3) und „natürlich" wegen der Bedingung (4), die eine der vielen möglichen Randbedingungen ist, um die Eindeutigkeit von $S_{2n-1,\Delta}(f; x)$ sicherzustellen. Sie ist speziell deswegen von Interesse, weil die Interpolation durch natürliche Splines eine Verallgemeinerung der Lagrange-Interpolation ist. Im Falle einer Zerlegung (1) mit $N = n-1$ fällt nämlich $S_{2n-1,\Delta}(f; x)$ mit dem Lagrangschen Interpolationspolynom $L_N(f; x) \in \mathscr{P}_N$ zusammen.

Da nun der Operator $L_n(f) \equiv L_n(f; x)$ ein schlechtes Konvergenzverhalten aufweist, d.h. nach dem Satz von Faber existiert zu jeder Folge von Zerlegungen (1),

---

[1] Diese Arbeit entstand im Rahmen eines von der Görresgesellschaft (Köln) geförderten Habilitandenstipendiums.

$n \to \infty$, eine Funktion $\tilde{f} \in C[a, b]$ mit

$$\lim_{n \to \infty} \|L_n(\tilde{f}) - \tilde{f}\| \neq 0, \tag{5}$$

stellt sich natürlicherweise die Frage[2]), ob für jedes $f \in C[a, b]$

$$\lim_{N \to \infty} \|S_{2n-1,\Delta}(f) - f\| = 0 \tag{6}$$

für den durch (1)—(4) definierten natürlichen Spline-Interpolationsoperator $S_{2n-1,\Delta}(f) \equiv S_{2n-1,\Delta}(f; x)$ gilt. Der Grenzübergang in (6) ist dabei so zu verstehen, daß der der Polynom-Approximation entsprechende Parameter n fest bleibt und der für die Spline-Approximation charakteristische Parameter N so gegen ∞ strebt, daß $\bar{\Delta} \to 0$ gilt.

Außer im trivialen Fall $n=1$ scheint (6) im kubischen Fall $n=2$ unter der Einschränkung

$$\bar{\Delta}/\underline{\Delta} \leq K < \infty \qquad (\bar{\Delta} \to 0+) \tag{7}$$

bekannt zu sein (vgl. [1, p. 22]), sowie im Falle $n>2$ für äquidistante Zerlegungen der Form

$$\Delta_h : a = x_0 < x_1 < \cdots < x_N = b, \quad x_i = a + ih, \quad 0 \leq i \leq N, \qquad (h \to 0+), \tag{8}$$

nach einem allgemeineren Ergebnis von Swartz—Varga [9].

In dieser Arbeit wird ein zweiter, unabhängiger und einfacherer Beweis dieser Aussage gegeben. Dabei wird ein Vergleichssatz bewiesen, der die Approximationsordnung von interpolierenden Splines, die (2) und (3) erfüllen, unter verschiedenen Randbedingungen vergleicht, und der für sich von Interesse ist.

Es sei $T_{2n-1,\Delta}(f; x)$ eine interpolierende Splinefunktion zu $f \in C[a, b]$ im Sinne von (2) und (3), die die Randbedingungen

$$\begin{cases} T^{(j_i)}_{2n-1,\Delta}(f; a) = \alpha_i \\ T^{(k_i)}_{2n-1,\Delta}(f; b) = \beta_i \end{cases} \quad (1 \leq i \leq n-1, \; n \geq 2) \tag{9}$$

erfüllen, wobei j_i, k_i natürliche Zahlen mit $1 \leq j_1 < j_2 < \cdots < j_{n-1} \leq 2n-2$ und $1 \leq k_1 < k_2 < \cdots < k_{n-1} \leq 2n-2$ sind, sowie α_i, β_i vorgegebene reelle Zahlen. Diese Randbedingungen stellen die Eindeutigkeit von $T_{2n-1,\Delta}(f; x)$ sicher.

SATZ 1: *Es sei $f \in C^m[a, b]$, wobei m eine ganze Zahl mit $0 \leq m \leq 2n$ sei, und $T_{2n-1,\Delta}(f; x)$ die Splinefunktion, die (2), (3) und (9) erfüllt. Ferner sei $R_{2n-1,\Delta}(f; x)$*

---

[2]) Vom Verfasser auf der Oberwolfacher Tagung, August 1971, als ungelöstes Problem vorgelegt.

eine beliebige weitere interpolierende Splinefunktion, die (2) und (3) erfüllt. Dann gilt[3] unter der Voraussetzung (7)

$$(10) \qquad \|T_{2n-1,\Delta}(f) - f\| \leqq$$

$$\leqq C_1 \Big\{ \sup_{0 \leqq l \leqq m} \bar{\Delta}^l \|R_{2n-1,\Delta}^{(l)}(f) - f^{(l)}\| + \sup_{m < l \leqq 2n-2} \bar{\Delta}^l \|R_{2n-1,\Delta}^{(l)}(f)\| \Big\} +$$

$$+ C_1 \Big\{ \sup_{1 \leqq j_i, k_i \leqq m} [\bar{\Delta}^{j_i} |f^{(j_i)}(a) - \alpha_i| + \bar{\Delta}^{k_i} |f^{(k_i)}(b) - \beta_i|] + \sup_{m < j_i, k_i \leqq 2n-2} [\bar{\Delta}^{j_i} |\alpha_i| + \bar{\Delta}^{k_i} |\beta_i|] \Big\}.$$

BEWEIS: Wir untersuchen die Differenz $u(x) = R_{2n-1,\Delta}(f; x) - T_{2n-1,\Delta}(f; x)$, die an den Punkten x_i, $0 \leqq i \leqq N$, verschwindet. Nach dem Satz von Rolle und der Schwarzschen Ungleichung folgt daher in elementarer Weise (vgl. z. B. Jerome—Varga [4, p. 119])

$$\|u(x)\| \leqq C_2 \bar{\Delta}^{n-1/2} \Big\{ \int_a^b [u^n(x)]^2 \, dx \Big\}^{1/2}.$$

Eine $(n-1)$ fache partielle Integration der rechten Seite ergibt

$$\|u\|^2 \leqq C_2^2 \bar{\Delta}^{2n-1} \Big\{ \sum_{i=1}^{n-1} (-1)^{i-1} [u^{(n+i-1)} u^{(n-i)}]_a^b + (-1)^{n-1} \int_a^b u^{(2n-1)}(x) u'(x) \, dx \Big\}.$$

Da $u^{(2n-1)}$ stückweise konstant ist, verschwindet wegen $u(x_i) = 0$, $0 \leqq i \leqq N$, das Integral auf der rechten Seite, so daß

$$\|u\|^2 \leqq C_2^2 \bar{\Delta}^{2n-1} \sum_{i=1}^{n-1} (-1)^{n-i-1} [u^{(2n-1-i)} u^{(i)}]_a^b.$$

Da in dieser Summe die Indizes der Ableitungen jede ganze Zahl aus $[1, 2n-2]$ durchlaufen, kann man weiter abschätzen

$$(11) \qquad \|u\|^2 \leqq C_2^2 \bar{\Delta}^{2n-1} \sum_{i=1}^{n-1} \{|u^{(2n-1-j_i)}(a) u^{(j_i)}(a)| + |u^{(2n-1-k_i)}(b) u^{(k_i)}(b)|\}.$$

Nun beachten wir, daß nach der Markoff-Ungleichung für algebraische Polynome (s. Timan [10, p. 236]) für die stückweise polynomiale Funktion $u(x)$ gilt

$$\|u^{(i)}\| \leqq 2^i (2n-1)^{2i} \underline{\Delta}^{-i} \|u\| \qquad (1 \leqq i \leqq 2n-2).$$

Damit folgt aus (11)

$$\|u\|^2 \leqq C_3 (\bar{\Delta}/\underline{\Delta})^{2n-1} \|u\| \sum_{i=1}^{n-1} \{\underline{\Delta}^{j_i} |u^{(j_i)}(a)| + \underline{\Delta}^{k_i} |u^{(k_i)}(b)|\},$$

also mit (7)

$$(12) \qquad \|u\| \leqq C_4 \sum_{i=1}^{n-1} \{\bar{\Delta}^{j_i} |u^{(j_i)}(a)| + \bar{\Delta}^{k_i} |u^{(k_i)}(b)|\}.$$

---

[3] Hier und im folgenden bezeichnen C_1, C_2, ... Konstanten, die nicht von f und Δ abhängen.

Nach (9) gilt nun für $j_i \leqq m$ die Abschätzung

$$(13) \qquad |u^{(j_i)}(a)| \leqq |R^{(j_i)}_{2n-1,\Delta}(f;a) - f^{(j_i)}(a)| + |f^{(j_i)}(a) - \alpha_i| \leqq$$
$$\leqq \|R^{(j_i)}_{2n-1,\Delta}(f) - f^{(j_i)}\| + |f^{(j_i)}(a) - \alpha_i|,$$

und analog für $k_i \leqq m$

$$(14) \qquad |u^{(k_i)}(a)| \leqq \|R^{(k_i)}_{2n-1,\Delta}(f) - f^{(j_i)}\| + |f^{(k_i)}(b) - \beta_i|.$$

Für j_i, $k_i > m$ verwenden wir die Abschätzungen

$$(15) \qquad |u^{(j_i)}(a)| \leqq \|R^{(j_i)}_{2n-1,\Delta}(f)\| + |\alpha_i|,$$

$$(16) \qquad |u^{(k_i)}(b)| \leqq \|R^{(k_i)}_{2n-1,\Delta}(f)\| + |\beta_i|.$$

Schließlich folgt aus der Dreiecksungleichung $\|T_{2n-1,\Delta}(f) - f\| \leqq \|u\| + \|R_{2n-1,\Delta}(f) - f\|$, woraus sich (10) durch Einsetzen von (12)—(16) ergibt.

Für äquidistante Zerlegungen der Form (8) und den natürlichen Spline-Interpolationsoperator $S_{2n-1,h}(f) \equiv S_{2n-1,\Delta_h}(f)$ gilt speziell die

FOLGERUNG: *Ist $f \in C^{(m)}[a, b]$, $0 \leqq m \leqq 2n$, so gilt*

$$(17) \qquad \|S_{2n-1,h}(f) - f\| \leqq C_5 h^m \|f^{(m)}\| + C_6 \sup_{n \leqq i \leqq \min(m,\, 2n-2)} h^i [|f^{(i)}(a)| + |f^{(i)}(b)|].$$

Dies folgt offensichtlich aus Satz 1, falls $T_{2n-1,\Delta}(f) = S_{2n-1,h}(f)$ angenommen wird (also $j_i = k_i = n-1+i$ und $\alpha_i = \beta_i = 0$, $1 \leqq i \leqq n-1$, nach (4)), und eine interpolierende Vergleichssplinefunktion $R_{2n-1,\Delta_h}(f) \equiv R_{2n-1,h}(f)$ existiert, die

$$(18) \qquad \|R^{(l)}_{2n-1,h}(f) - f^{(l)}\| \leqq C_l h^{m-l} \|f^{(m)}\| \qquad (0 \leqq l \leqq m)$$

und

$$(19) \qquad \|R^{(l)}_{2n-1,h}(f)\| \leqq C_l h^{m-l} \|f^{(m)}\| \qquad (m \leqq l \leqq 2n-2)$$

für $f \in C^{(m)}[a, b]$, $0 \leqq m \leqq 2n$, erfüllt. (Man beachte, daß (18) und (19) für $l=m$ äquivalent sind.) Als einfachste derartige Splinefunktion scheint sich eine solche in Subbotin [7, p. 47] anzubieten, für die (18) bewiesen ist und für die sich (19) mit einer geringen Modifikation des dortigen Beweises ergibt. Weitere Beispiele liefern die in Swartz [8] betrachteten, an äquidistanten Knoten interpolierenden Splines. Es sei ferner bemerkt, daß sowohl obige Folgerung als auch der folgende Satz 2 mit den vorliegenden Methoden allgemeiner nur unter der Einschränkung (7) an die Knoten bewiesen werden kann, falls unter dieser Voraussetzung für eine interpolierende Splinefunktion (18), (19) gezeigt werden kann.

Aus obiger Folgerung, die im Falle $n=2$ Atkinson [2] unter Angabe von genaueren Konstanten gezeigt hat (vgl. auch Kershaw [6]), geht hervor, daß z. B. für $f \in C^{(2n)}[a, b]$ die natürlichen interpolierenden Splines mit der Ordnung $O(h^{2n})$ approximieren, vorausgesetzt daß f die Randbedingungen $f^{(k)}(a) = f^{(k)}(b) = 0$ für $n \leqq k \leqq 2n-2$ erfüllt. Ohne die Voraussetzung besonderer Randbedingungen an f folgt aus (17) als beste Approximationsordnung für glatte Funktionen

$$(20) \qquad \|S_{2n-1,h}(f) - f\| \leq C_7 h^n \|f^{(n)}\| \qquad (f \in C^{(n)}[a, b]),$$

sowie die gleichmäßige Beschränktheit in h, d.h.

$$(21) \qquad \|S_{2n-1,h}(f) - f\| \leq C_5 \|f\| \qquad (f \in C[a, b]).$$

Mit Hilfe eines Argumentes aus der Theorie der intermediären Räume (vgl. [3]) kann man aus diesen Aussagen auch Konvergenzaussagen für weniger glatte Funktionen gewinnen.

SATZ 2. *Es gilt für jedes* $f \in C[a, b]$

$$\|f - S_{2n-1,h}(f)\| \leq C_8 \,\omega_n(f; h),$$

wobei

$$\omega_n(f; h) = \sup_{|\delta| \leq h} \ \sup_{x,\, x+n\delta \in [a,b]} \left| \sum_{i=0}^{n} (-1)^i \binom{n}{i} f(x + i\delta) \right|$$

der n-te Stetigkeitsmodul von f ist.

BEWEIS. Es sei $(f - g) + g = f$ eine beliebige Zerlegung von f, worin $g \in C^{(n)}[a, b]$ ist. Dann gilt nach (20), (21)

$$\|f - S_{2n-1,h}(f)\| \leq C_9 \{\|f - g\| + h^n \|g^{(n)}\|\}.$$

Daraus folgt

$$\|f - S_{2n-1,h}(f)\| \leq C_9 K(h^n, f),$$

wobei das K-Funktional durch

$$K(h^n, f) \equiv K(h^n, f; C[a, b], C^{(n)}[a, b]) = \inf_{g \in C^{(n)}[a,b]} \{\|f - g\| + h^n \|g^{(n)}\|\}$$

definiert ist. Nach einem Argument von H. Johnen [5] gilt aber

$$K(h^n, f) \leq C_{10}\, \omega_n(f, h),$$

woraus die Behauptung des Satzes folgt.

KOROLLAR: *Es gilt die Aussage* (6) *für alle* $f \in C[a, b]$ *und äquidistante Zerlegungen* Δ_h *der Form* (8), $h \to 0+$.

Der Verfasser dankt Herrn P. L. Butzer für die Förderung und Herrn E. Görlich für die kritische Durchsicht dieser Arbeit, sowie den Herren M. Marsden (East Lansing) und Ju. N. Subbotin (Sverdlovsk) für ihre Hinweise bezüglich Literatur.

LITERATUR

[1] J. H. Ahlberg, E. N. Nilson, and J. L. Walsh, *The Theory of Splines and their Applications* Acad. Press, New York 1967.

[2] K. E. Atkinson, *On the order of convergence of natural cubic spline interpolation.* SIAM J. Numer. Anal. **5** (1968), 89—101.

[3] P. L. Butzer and H. Berens, *Semi-Groups of Operators and Approximation.* Springer, Berlin— Heidelberg—New York 1967.

[4] J. W. Jerome and R. S. Varga, *Generalizations of Spline Functions and Applications to Non-linear Boundary Value and Eigenvalue Problems.* In: Theory and Applications of Spline Functions (Proc. Seminar Math. Res. Center, Univ. Wisconsin Oct. 1968. Ed. by T. N. E. Greville), pp. 103—155. Acad. Press. New York 1969.

[5] H. Johnen, *Inequalities connected with the moduli of smoothness* (in print).

[6] D. Kershaw, *A note on the convergence of interpolatory cubic splines.* SIAM J. Numer. Anal. **8** (1971), 67—74.

[7] Ju. N. Subbotin, *Diameter of class $W^r L$ in $L(0, 2\pi)$ and spline function approximation.* Mat. Zametki **7** (1970), 43—52 (Math. Notes **7** (1970), 27—32).

[8] B. K. Swartz, $O(h^{2n+2-l})$-*bounds on some spline interpolation errors.* Los Alamos Scientific Laboratory Report La-3886, 1967.

[9] B. K. Swartz and R. S. Varga, *Error bounds for spline and L-spline interpolation* J. Approximation Theory **6** (1972), 6—49.

[10] A. F. Timan, *Theory of Approximation of Functions of a Real Variable.* Pergamon Press, New York 1963.

New and Unsolved Problems

1. P. L. Butzer and J. R. Dorroh

What is the saturation class for approximation by nonlinear semi-groups?
Let X be a real Banach space and let A be an accretive subset of $X \times X$ such that $R(I+\lambda A) \supset \overline{D(A)}$ for $0 < \lambda \leq \lambda_0$, and A is $\overline{D(A)}$-maximal accretive and closed. Let T be the unique semi-group belonging to $Q(\overline{D(A)})$ such that

$$T(t)x = \lim_{n \to \infty} \left(I + \frac{t}{n}A\right)^{(-n)} x$$

for x in $D(A)$ and $t > 0$. Then

$$\|T(t)x - x\| = O(t)$$

for $x \in D(A)$. If $x \in \overline{D(A)}$ and

$$\|T(t)x - x\| = o(t),$$

then $T(t)x = x$ for all $t \geq 0$. Let K be the saturation class of all $x \in X$ for which the best possible non trivial approximation holds; i.e.,

$$K = \{x \in \overline{D(A)}: \|T(t)x - x\| = O(t)\}.$$

CONJECTURE. K is the set of all x in $\overline{D(A)}$ such that $x = \lim x_n$, where $\{x_n\} \subset$
$\subset D(A)$, and the sequence $\{|Ax_n|\}$ is bounded, where

$$|Ax_n| = \inf \{\|y\| : y \in Ax_n\}.$$

This is correct if T is linear, see [1] or [2], or if X and X^* are uniformly convex,
sec [3].

[1] P. L. Butzer and H. Berens, *Semi-groups and Approximation Theory*. Springer, Berlin 1967.

[2] H. Berens, Lecture Notes in Math. **64**, Springer, Berlin 1968.

[3] M. Crandall and A. Pazy, *Semi-groups and dissipative sets*. J. Functional Analysis **3** (1969).

Added in proof:

This problem was solved by U. Westphal in the note: *Sur la saturation pour des semi-groupes non lineaires.* C. R. Acad. Sc. Paris, t. **274**, p. 1351—1353 (3 mai 1972). M. G. Crandall in his preprint *"A generalized domain for semigroup generators"* independently obtained the main part of this note, but from a different point of view.

2. J. L. B. Cooper

Let $f(x)$ be a real continuous function on a Banach space such that, for each t, $\{x: f(x) \geqq t\} = K_t$ is convex. Find conditions, in terms of the set of directions D_x at each x defined by $D_x = \{d: |d| = 1, f(x+td) > f(x)$ for a $t > 0\}$ in order that K_t have a *unique* support hyperplane at each point of its boundary.

NOTE. The problem on measures I posed in the 1968 meeting (Problem 12, p. 420 in: *Abstract Spaces and Approximation*, ed. P. L. Butzer—B. Sz.-Nagy, ISNM 10, Basel 1969) can be solved by using results in MACKEY's paper on *Induced representations of locally compact groups* I, Ann. Math. **55** (1952), 101—139.

3. F. Deutsch

(This problem was communicated to me orally by B. BROSOWSKI.) Let f be a real valued continuous function on $[0, 1]$. Choose $t_0 \in [0, 1]$ such that $|f(t_0)| = = \|f\| (\equiv \max |f(t)|)$, and let P_0 be the constant polynomial which interpolates to f at t_0, i.e. $p_0(t) \equiv f(t_0)$. At the n-th step $n = 1, 2, \ldots$, we choose a point $t_n \in [0, 1]$ such that $|f(t_n) - p_{n-1}(t_n)| = \|f - p_{n-1}\|$, and then we choose the polynomial p_n of degree at most n which interpolates to f at the points $t_0, t_1, \ldots, t_n$.

There are examples (communicated to me by J.-P. KAHANE) where the sequence of polynomials $\{p_n\}$ thus constructed does not converge to f pointwise.

QUESTION. Can one characterize the classes of functions f for which $\{p_n\}$ does converge to f? At least, can one define large classes of functions for which convergence holds?

REMARK. This problem would seem to have some practical significance.

4. R. A. DeVore

1. If $\mathfrak{A}$ is a subclass of $C_{2\pi}$ and g is a non-negative, even function in $\mathfrak{A}$ with $g(o) = o$ and the property that for each $f \in \mathfrak{A}$

$$|f(x+t) + f(x-t) - 2f(x)| \leqq 2g(t) \qquad (-\pi \leqq x, \quad t \leqq \pi),$$

then we say g is a extreme function for $\mathfrak{A}$. The extreme functions for $\mathrm{Lip}_1 \alpha$ and $\mathrm{Lip}_1^* \alpha$ are $|t|^\alpha$, when $0 < \alpha \leq 1$. The extreme function for $\mathrm{Lip}_2^* 2$ is

$$g_2(t) = \begin{cases} t^2 & 0 \leq |t| \leq \pi/2. \\ \pi^2/2 - (t-\pi)^2 & |t| \geq \pi/2. \end{cases}$$

What are the extreme functions for $\mathrm{Lip}_1^* \alpha$, when $1 < \alpha < 2$.

One important reason for determining extreme functions is its connection with approximation of classes. Let (L_n) be a sequence of positive convolution operators $\left(L_n(f) = f * d\mu_n,\ d\mu_n \geq 0 \text{ and even } \dfrac{1}{\pi} \int_{-\pi}^{\pi} d\mu_n = 1 \right)$. If g is an extreme function for $\mathfrak{A}$ then

$$E(\mathfrak{A}, L_n) \equiv \sup_{f \in \mathfrak{A}} \|f - L_n(f)\| = \frac{1}{\pi} \int_{-\pi}^{\pi} g(t)\, d\mu_n(t).$$

2. *A problem of* G. SUNOUCHI. If $k \in L_1(R)$, $k \geq 0$ and $0 < \alpha < 2$ are such that

$$\lim_{t \to 0} \frac{1 - \hat{k}(t)}{|t|^\alpha} = C \neq 0$$

then is $h(t) = \dfrac{1 - \hat{k}(t)}{|t|^\alpha}$ a Fourier—Stieltjes transform? The condition that h is a Fourier—Stieltjes transform is the usual multiplier condition which must be assumed in order to characterize the saturation class for the family of operators given by condition with the dilates of k.

5. G. Freud

At the Conference in 1963 (see ISNM 5, p. 189) H. S. SHAPIRO communicated the following problem due to D. J. NEWMAN.

Let $\varrho_n(f) = \inf \| f - R \|_{C[-1, 1]} (f \in C[-1, 1])$, where R ranges over ratios of polynomials of degree n at most. The conjecture is that

$$(1) \qquad \varrho_n(f) = o(n^{-1}) \quad \text{for every } f \in \mathrm{Lip}\, 1.$$

We have the following remark concerning (1): Let V' denote the set of absolutely continuous functions $f \in C[-1, 1]$ for which f' has a bounded variation $V(f')$ in $[-1, 1]$ and let

$$\lambda_n = \sup_{f \in V'} \varrho_n(f)/V(f').$$

We proved in [2] that (1) is the consequence of the other unproved hypothesis

$$(2) \qquad \lambda_n = O(n^{-2}).$$

Probably it would be easier to prove (2) than (1). It is known that

$$c_1 n^{-2} < \lambda_n < c_2 \log^2 n \cdot n^{-2}$$

(see [1] and [2]).

[1] G. Freud, Acta Math. Ac. Sci. Hung. **17** (1966), 313—324.

[2] G. Freud, Studia Sci. Math. Hung. **3** (1968), 383—386. (See also correction in Zentralblatt für Math. **165** (1969), 389).

6. K. Gustafson

1. A bounded normal operator A on a separable complex Hilbert space can be written in the form $A = D+K$, D diagonal, K compact, and in the self-adjoint case K can be taken to be Hilbert—Schmidt. What can be said about K in the normal case? Reference: Weyl's theorems, these proceedings.

2. A bounded operator A on a complex Banach space has numerical range $V(A)=\{x'Ax\,|\,x'x=\|x\|=\|x'\|=1,\ x'\in X',\ x\in X\}$. Is $V(A)$ simply connected? Reference: F. F. Bonsall and J. Duncan, *Numerical Ranges of Operators on Normed Spaces and of Elements of Normed Algebras.* Cambridge University Press 1971.

3. An unbounded operator A in a complex Banach space has numerical range $V(A)=\{x'Ax\,|\,x'x=\|x\|=\|x'\|=1,\ x'\in X',\ x\in D(A)\}$. Is $V(A)$ unbounded?

7. G. G. Lorentz

Problem $\mathscr{P}_1$ is the approximation of (increasing) functions f on $[a, b]$ by increasing polynomials on $[a, b]$ of degree n. In the problem $\mathscr{P}_k$, condition $P_n'(x)\geqq 0$ is replaced by $P_n^{(k)}(x)\geqq 0$. The following problems can often be formulated for any k, but we restrict ourselves to the case $k=1$. Let $E_n(f)$ be the degree of approximation of f by arbitrary polynomials, $E^*(f)$ the same for increasing polynomials of degree n.

1. For a fixed n, is

$$\varrho_n = \sup_f \frac{E_n^*(f)}{E_n(f)}$$

finite? If yes, find an upper bound for ϱ_n.

2. For real numbers $p, q > 0$ we say that $\mathscr{P}_1 \in (p, q)$ if each increasing function f of smoothness p (this means that the continuous derivative $f^{(\lfloor p \rfloor)}$ exists and belongs to the class Lip $(p-[p]))$ has $E_n^*(f) \leqq C\dfrac{1}{n^q}$. For which p, q is this true? The following is known:

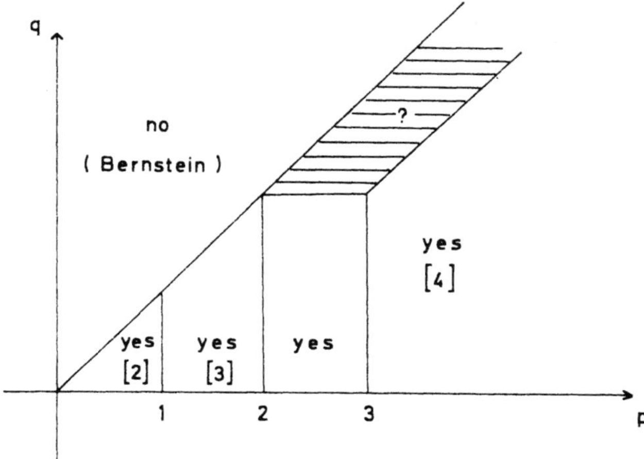

Remove the terra incognita.

3. Let $f = x^{2n+1}$ on $[-1, 1]$. The increasing polynomials of best approximation (their existence follows from [2]) are odd, $P_n = a_0 x^{2n-1} + \cdots + a_{n-1} x$. The polynomials $M_n = f - P_n$ are similar to Chebyshev polynomials but not Chebyshev polynomials for large n. Determine $\|M_n\|$? It is known that $\dfrac{1}{2^{2n}} \leqq \|M_n\| \leqq Cn\dfrac{1}{2^{2n}}$. (LORENTZ [3].)

In what sense are the M_n similar to the Chebyshev polynomials?

4. Let $\Phi_1, \ldots, \Phi_n$ be a given system of functions on $[a, b]$. Under what conditions is it true that each continuous (increasing) function f on $[a, b]$ has a *unique* increasing polynomial $P = a_1 \Phi_1 + \cdots + a_n \Phi_n$ of best approximation? This is true if $\Phi_i = x^{i-1}$, $i = 1, \ldots, n$. (LORENTZ and ZELLER [2].)

5. In connection with problem 4, one can ask when a theorem of the type of ATKINSON and SHARMA [1] about the poisedness of incidence matrices of the Birkhoff interpolation, is true for a given system of functions $\Phi_1, \ldots, \Phi_n$.

6. An operator Pf, applicable to a function f on $[a, b]$, is monotone of order k, if $f^{(k)} \geqq 0$ implies $(Pf)^{(k)} \geqq 0$. One can also define completely monotone operators, for which this is true for each k. There exists an extensive theory of positive operators.

What are results from the theory of monotone operators? (Example: the Bernstein operators $B_n f$ are completely monotone.)

[1] F. V. Atkinson and A. Sharma, SIAM J. Numer. Anal. **6** (1969), 230—236.
[2] G. G. Lorentz and K. Zeller, Trans. Amer. Math. Soc. **149** (1970), 1—18.
[3] G. G. Lorentz, *Monotone Approximation*. Inequalities III, editor O. Shisha, 1971, 201—215 (to appear).
[4] O. Shisha, Pacific J. Math. **15** (1965), 667—671.

8. C. Micchelli and T. Rivlin

Let $\omega(x)$ be a polynomial of degree n which has n distinct zeros in $(-1, 1)$. Suppose also that

$$\left|\omega\left(\cos\frac{\pi k}{n}\right)\right| = 1 \qquad (k = 0, 1, \ldots, n).$$

Is it true that ω must be equal to $\pm T_n$, the n-th Chebychev polynomial?

9. S. M. Nikolskiĭ

1. Let $y = Ax$, $x, y \in E$ be a perfectly continuous linear operator, where E is a separable Banach space.

Is it possible for every $\varepsilon > 0$ to find a finite dimensional operator

$$B(x) = \sum_{j=1}^{N} f_j(x) x_j, \qquad x \in E, \ x_j \in E'$$

(E' — the space of functionals f on E) so that

$$\|A - B\| = \sup_{\|x\| \leqq 1} \|Ax - Bx\| < \varepsilon.$$

2. There is a series of inequalities in the theory of the embedding theorems; see for example the book by S. M. Nikolskiĭ *"Approximation of functions of many variables and imbedding theorems"* (Приближение функций многих переменных и теорема вложения), Moscow 1969.

Those who obtained these inequalities do not take care of exactness of their constants. It would be interesting to obtain these constants, at first for the all n-dimensional space and its coordinate subspace, and then for open sets $G \subset R_n$. In the last case it leads to variational problems with variation on boundaries of G.

10. Ju. A. Rozanov

1. Let H_t, $t \leq t_0$, be a family of subspaces in some Hilbert space:

$$H_s \subseteq H_t \quad \text{for} \quad s \leq t,$$

and P_t, $t \leq t_0$, a family of projection operators: P_t is the self-adjoint projection on H_t. Let N be the multiplicity of the family P_t, $t \leq t_0$, i.e. N is the minimal number of orthogonal cyclic subspaces in H_{t_0}.

Let A be some fixed subspace and P_A the projection on A. Let $\tilde{H}_t = P_A H_t$ and $\tilde{N}$ the multiplicity of a new family $\tilde{P}_t$, $t \leq t_0$, where $\tilde{P}_t$ is projection on H_t.

When do we have $\tilde{N} \leq N$?

Note that there is an example with $N=1$ where for any $\tilde{N}=1, 2, \ldots$ we can find A with this $\tilde{N}$.

2. Let $U_t = \int e^{i\lambda t} \, dE_\lambda$ be the spectral resolution of a unitary group in a Hilbert space. A is a fixed closed subspace, and

$$H(T) = \bigvee_{t \in T} U_t A$$

(closed subspace generated by $\{U_t A\}$), where T is a set of real numbers. P_A is the projection on A and we set

$$F_\lambda = P_A E_\lambda P_A.$$

What properties of $\{F_\lambda\}$ are necessary and sufficient that $\bigcap_n H(T_n) = \{0\}$ for a given sequence of sets $T_1 \supset T_2 \supset T_3 \supset \cdots$ with $\bigcap T_n = \emptyset$?

3. The proximity of $H(T_0)$ and $H(T)$, where T_0 and T are given sets, can be described by $\pi_0 = P_0 P P_0$, $\pi = P P_0 P$, where P_0 and P are the projections on $H(T_0)$ and $H(T)$ respectively. What properties of $\{F_\lambda\}$ are necessary and sufficient for π to be a nuclear operator?

Let $T_1 \supset T_2 \supset \cdots$, $\bigcap_n T_n = \emptyset$ and $\pi_n = P_n P P_n$ where P_n projects on $H(T_n)$. Suppose π_0 is a nuclear operator, so that its trace is finite. Does it follow that $\text{Tr } \pi_n \to 0$ as $n \to \infty$?

11. I. J. Schoenberg

Let $f(z)$ be analytic and regular in an open simply connected domain containing the interval $I=[-1, 1]$. Let $P_m(z)$ denote the polynomial of degree $\leq 2m-1$ satisfying the 2-point conditions

$$P_m^{(v)}(-1) = f^{(v)}(-1), \quad P_m^{(v)}(1) = f^{(v)}(1) \qquad (v=0, 1, \ldots, m-1).$$

It is known that

$$\lim_{m \to \infty} P_m(z) = f(z)$$

will hold in two appropriate neighbourhoods of the points $+1$ and -1.

Here we discuss the extension of this problem from polynomials to spline functions in the following sense. If

$$(1) \qquad\qquad -1 < x_1 < x_2 < \cdots < x_k < +1,$$

then it is known that there exists a uniquely defined $P_{m,k}(x)$, which is piecewise polynomial of degree $\leq 2m-1$, having the knots $x_1, \ldots, x_k$, belonging to $C^{2m-2}(I)$, and satisfying the interpolatory conditions

$$P_{m,k}^{(v)}(-1) = f^{(v)}(-1), \quad P_{m,k}^{(v)}(1) = f^{(v)}(1) \qquad (v = 0, \ldots, m-1),$$

$$P_{m,k}(x_i) = f(x_i) \qquad (i = 1, \ldots, k).$$

PROBLEM. Does there exist a set of knots (1), depending on $f(z)$ but not on m, such that

$$\lim_{m \to \infty} P_{m,k}(x) = f(x) \quad \text{if} \quad -1 \leq x \leq 1?$$

12. I. Segal

1. Special cases of the following problem arise in quantum field theory.

Let $L_p^a(R^m)$ denote the space of a-times differentiable pth-power integrable functions on R^m, as defined by Calderon. For any given measurable function f on R^m, let $M_{a,p;b,q}(f)$ denote the (partially defined) operator of multiplication by f, as an operator in $L_p^a(R^m)$ with values in $L_q^b(R^m)$.

i) For which values of a, p, b, q, and n is $[f \colon \|M_{a,p;b,q}(f)\| < \infty]$ non-trivial, in e.g. the sense of being dense in the space of all measurable functions, in the topology of convergence in measure.

ii) Bound $\|M_{a,p;b,q}(f)\|$ in terms of conventional norms on f.

iii) Do i) with the replacement of the condition that $\|M_{a,p;b,q}(f)\| < \infty$ by the condition that $M_{a,p;b,q}(f)$ be compact.

iv) When $p = q = 2$, the same for the Hilbert—Schmidt, trace, and analogous norms of index r, $1 \leq r < \infty$ (the case $r = \infty$ being i)).

2. In his forthcoming doctoral dissertation at M.I.T. Stephen BERMAN has investigated the one-parameter groups defined on function spaces over R^n, the functions having values in a given finite dimensional space W, under the assumption of invariance under translations in R^n, and natural continuity conditions, together with the assumption that the group is of "finite propagation velocity" in a simple sense. The results are quite precise; the "finite propagation velocity" condition is extremely restrictive; the results suggest (and in some contexts, prove) that the only such groups are the known ones associated with constant coefficient hyperbolic equations.

Another important and physically intuitive property of such a group is that of being "weakly Huyghens": the group T_t is such if for any given compact subsets C and C' of R^n, there exists t_0 such that if u is an initial vector of support C, then $T_t u$ vanishes on C' for $|t| > t$. Intuitively, this is a formulation of the total disappearance of light signals in free space after a finite time; mathematically, the property is important in scattering theory.

QUESTION. Do there exist any continuous translation-invariant weakly Huyghens groups (or semigroups) other than the well known ones (i.e. those defined by hyperbolic constant-coefficient partial differential equations with lacunas).

13. H. S. Shapiro

1. Let S denote a closed, convex translation-invariant subset of $L^p(\mathbf{T})$ $(1 < p < \infty)$, and, for each $f \in L^p(\mathbf{T})$ denote by $P_s f$ the nearest element to f from S.

I can prove: if $f \in \mathrm{Lip}^*(\alpha, p)$, where $2 \leq p < \infty$; $0 < \alpha \leq 2$ then $P_s f \in \mathrm{Lip}^*(\alpha/p, p)$ (and analogous result for $p < 2$).

PROBLEM 1. Is α/p best possible?

REMARK. I have an example of $f \in C^\infty$ (hence, a fortiori, $f \in \mathrm{Lip}^*(2, p)$), S a subspace (of codimension 2) such that $P_s f \notin \mathrm{Lip}^*(\beta, p)$ for any $\beta > p' - 1 + 1/p$. Thus, there is a gap between $p' - 1 + 1/p$ and the (smaller) number $2/p$ to be filled.

PROBLEM 2. Can one find all (or even some) subspaces S such that $f \in C^\infty \Rightarrow \Rightarrow P_s f \in C^\infty$? $S = H^p$ is such a subspace; so is $\{f : \hat{f}(0) = 0\}$.

2. Let $k \in L^1(\mathbf{R}^n)$, $Sk = 1$. Let $k_{(a)}$ denote the usual "approximate identity",

$$k_{(a)}(t) = a^{-n} k(t/a).$$

PROBLEM 3. Find sufficient conditions on k such that

$$(1) \qquad (f * k_{(a)})(t) \to f(t) \qquad (\text{a.e. for every } f \in L^1(\mathbf{R}^n)).$$

REMARKS. a) One sufficient condition, the only general one known to me, is

$$|k(t)| \leq \varphi(|t|)$$

where $\int_{\mathbf{R}^n} \varphi(|t|)\, dt < \infty$, i.e. k has an integrable radial majorant. This is sufficient for some purpose, e.g. to handle the Poisson kernel of a half-space, $k(t) = c_n(1+|t|^2)^{-(n+1)/2}$ but fails to handle e.g. the "product Poisson kernel", $k(t) = \prod_{v=1}^{n} (\pi(1+t_v^2))^{-1}$ as soon as $n \geq 2$. This kernel, crucial for the boundary behaviour of analytic functions in polydiscs, is known to satisfy (1), by virtue of a theorem of MARCINKIEWICZ and ZYGMUND (cf. RUDIN, Function Theory in Polydiscs). It would obviously be desirable to prove (1) for a general class of kernels that includes this one.

b) By virtue of a theorem of E. Stein, (1) is equivalent to a weak-type $(1, 1)$ estimate for the maximal function corresponding to k; i.e.

$$(Mf)(t) = \sup_{a>0} |(f*k_{(a)})(t)|.$$

The problem is to show

$$\text{meas}\,\{t : (Mf)(t) > \lambda\} \leq \frac{C\|f\|_1}{\lambda}$$

for $\lambda > 0$, where $C = C(k)$ does not depend on f.

14. B. Sz.-Nagy

1. Let us suppose that T is a linear, bounded operator on Hilbert space, such that *both* T and T^* (the adjoint of T) have cyclic vectors. Is it then true that the commutant of T (i.e. the set of operators commuting with T) is abelian?

2. Let $\sigma_{\delta n}(f)$ denote the n-th Cesàro-mean of (integral) order δ ($\delta = 1, 2, ...$) of the 2π-periodic continuous function f. Consider the extremum problems

$$P_{\delta n}^{(r)} \dots \sup_{\|f^{(r)}\|_\infty \leq 1} \|f - \sigma_{\delta n}(f)\|_\infty$$

and

$$\tilde{P}_{\delta n}^{(r)} \dots \sup_{\|\tilde{f}^{(r)}\|_\infty \leq 1} \|f - \sigma_{\delta n}(f)\|_\infty,$$

where ~ denotes trigonometric conjugate.

For every problem the extremal functions are the Euler splines defined, respectively, by $f^{(r)}(x) = \pm 1$ and $\tilde{f}^{(r)}(x) = \pm 1$, with constance intervals of length π. Exceptions are the problems

$$\tilde{P}_{1n}^{(1)} \quad \text{and} \quad \tilde{P}_{2n}^{(1)} \qquad (n = 1, 2, ...),$$

where no extremal function is known, cf. B. SZ.-NAGY, Acta Sci. Math., **11** (1946), 71—84.

PROBLEM. What are the extremal functions in these cases (if any).

15. A. C. Zaanen

A Riesz space L is said to have the principal projection property if to every principal band B there belongs another band B_1 such that $B \cap B_1 = \{0\}$ and $B + B_1 = L$. The principal projection property is a sufficient condition for a Riesz space in order that every proper prime ideal shall contain only one minimal prime ideal, but probably (restricting ourselves to archimedean spaces) it is not a necessary condition. It is asked to find a simple necessary and sufficient condition for Archimedean Riesz spaces.

ALPHABETICAL LIST OF PAPERS